Natural and Synthetic Biomedical Polymers

Natural and Synthetic Biomedical Polymers

Edited By

Sangamesh G. Kumbar

Cato T. Laurencin

Meng Deng

ELSEVIER

AMSTERDAM • BOSTON • HEIDELBERG • LONDON • NEW YORK • OXFORD
PARIS • SAN DIEGO • SAN FRANCISCO • SINGAPORE • SYDNEY • TOKYO

Elsevier
30 Corporate Drive, Suite 400, Burlington, MA 01803, USA
525 B Street, Suite 1800, San Diego, CA 92101-4495, USA

First edition **2014**

Library of Congress Cataloging-in-Publication Data
Natural and synthetic biomedical polymers/edited by Sangamesh Kumbar, Cato Laurencin, Meng Deng. — First edition.
pages cm
Summary: "Polymer scientists have made an extensive research for the development of biodegradable polymers which could find enormous applications in the area of medical science. Today, various biopolymers have been prepared and utilized in different biomedical applications. Despite the apparent proliferation of biopolymers in medical science, the Science and Technology of biopolymers is still in its early stages of development. Tremendous opportunities exist and will continue to exist for the penetration of biopolymers in every facet of medical science through intensive Research and Development. Therefore, this chapter addresses different polymerization methods and techniques employed for the preparation of biopolymers. An emphasis is given to cover the general properties of biopolymers, synthetic protocols and their biomedical applications. In order to make the useful biomedical devices from the polymers to meet the demands of medical science, various processing techniques employed for the development of devices have been discussed. Further, perspectives in this field have been highlighted and at the end arrived at the conclusions. The relevant literature was collected from different sources including Google sites, books and reviews"— Provided by publisher.
Includes bibliographical references and index.
ISBN 978-0-12-396983-5 (hardback)
1. Biopolymers. 2. Biodegradable plastics. I. Kumbar, Sangamesh, editor of compilation. II. Laurencin, Cato, editor of compilation. III. Deng, Meng, editor of compilation.
TP248.65.P62N38 2014
610.28—dc23 2014000085

British Library Cataloguing in Publication Data
A catalogue record for this book is available from the British Library

For information on all **Elsevier** publications
visit our web site at store.elsevier.com

Printed and bound by CPI Group (UK) Ltd, Croydon, CR0 4YY

Transferred to digital print 2013

ISBN: 978-0-12-396983-5

Contents

Dedication

Sangamesh G. Kumbar—To my parents (Mr. and Mrs. G. B. Kumbar), wife Swetha, and daughter Gauri.

Cato T. Laurencin—To my wife Cynthia, and my children Ti, Michaela, and Victoria.

Contributors

Aja Aravamudhan *Department of Orthopaedic Surgery, Institute for Regenerative Engineering, Raymond and Beverly Sackler Center for Biomedical, Biological, Physical and Engineering Sciences, The University of Connecticut, Farmington, CT, USA*

Brittany L. Banik *Department of Bioengineering, The Pennsylvania State University, PA, USA*

Mark R. Battig *Department of Bioengineering, College of Engineering, The Pennsylvania State University, PA, USA*

Steve Brocchini *UCL School of Pharmacy, University College London, London, UK*

Justin L. Brown *Department of Bioengineering, The Pennsylvania State University, PA, USA*

Karen Burg *Institute for Biological Interfaces of Engineering, Clemson, USA*

Diane J. Burgess *Department of Pharmaceutical Sciences, School of Pharmacy, University of Connecticut, Storrs, CT, USA*

Sheiliza Carmali *UCL School of Pharmacy, University College London, London, UK*

Tram T. Dang *Center for Biomedical Engineering, Department of Medicine, Brigham and Women's Hospital, Harvard Medical School, Boston, MA, USA*

David H. Koch Institute for Integrative Cancer Research, Massachusetts Institute of Technology, Cambridge, MA, USA

Meng Deng *Department of Orthopaedic Surgery, Institute for Regenerative Engineering, Raymond and Beverly Sackler Center for Biomedical, Biological, Physical and Engineering Sciences, The University of Connecticut, Farmington, CT, USA*

Abraham (Avi) Domb *School of Pharmacy-Faculty of Medicine, The Hebrew University of Jerusalem, Jerusalem, ISR*

Lakshmi Sailaja Duvvuri *Department of Pharmaceutics, National Institute of Pharmaceutical Education and Research, Hyderabad, India*

Muntimadugu Eameema *Department of Pharmaceutics, National Institute of Pharmaceutical Education and Research, Hyderabad, India*

Jennifer Elisseeff *Johns Hopkins School of Medicine, Translational Tissue Engineering Center, Wilmer Eye Institute and Department of Biomedical Engineering, Baltimore, MD, USA*

Sahar E. Fard *Department of Chemistry, Chemical Biology, and Biomedical Engineering, Stevens Institute of Technology, Hoboken, NJ, USA*

Bing Gu *Department of Pharmaceutical Sciences, School of Pharmacy, University of Connecticut, Storrs, CT, USA*

Jinshan Guo *Department of Bioengineering, Materials Research Institute, The Huck Institute of The Life sciences, The Pennsylvania State University, PA, USA*

Umesh Gupta *Department of Pharmaceutical Sciences, College of Pharmacy, South Dakota State University, Brookings, SD, USA*

Matthew D. Harmon *Department of Orthopaedic Surgery, Department of Material Science and Engineering, Institute for Regenerative Engineering, Raymond and Beverly Sackler Center for Biomedical, Biological, Physical and Engineering Sciences, The University of Connecticut, Farmington, CT, USA*

Markus Heiny *Institute for Macromolecular Chemistry, University of Freiburg, Freiburg, Germany*

Anjana Jain *Biomedical Engineering Department, Worcester Polytechnic Institute, Worcester, MA, USA*

Roshan James *Department of Orthopaedic Surgery, Institute for Regenerative Engineering, Raymond and Beverly Sackler Center for Biomedical, Biological, Physical and Engineering Sciences, The University of Connecticut, Farmington, CT, USA*

Tao Jiang *Department of Medicine, Institute for Regenerative Engineering, Raymond and Beverly Sackler Center for Biomedical, Biological, Physical and Engineering Sciences, The University of Connecticut, Farmington, CT, USA*

Ravindra R. Kamble *Department of Studies in Chemistry, Karnatak University, Dharwad, Karnataka, India*

Lohitash Karumbaiah *Department of Biomedical Engineering, Georgia Institute of Technology, Atlanta, Georgia, USA*

Ali Khademhosseini *Center for Biomedical Engineering, Department of Medicine, Brigham and Women's Hospital, Harvard Medical School, Boston, MA, USA*

Harvard-MIT Division of Health Sciences and Technology, Massachusetts Institute of Technology, Cambridge, MA, USA

Wyss Institute for Biologically Inspired Engineering, Harvard University, Boston, MA, USA

Wahid Khan *Department of Pharmaceutics, National Institute of Pharmaceutical Education and Research, Hyderabad, India*

School of Pharmacy-Faculty of Medicine, The Hebrew University of Jerusalem, Jerusalem, ISR

Sangamesh G. Kumbar *Department of Orthopaedic Surgery, Department of Material Science and Engineering, Department of Biomedical Engineering, Institute for Regenerative Engineering, Raymond and Beverly Sackler Center for Biomedical, Biological, Physical and Engineering Sciences, The University of Connecticut, Farmington, CT, USA*

Cato T. Laurencin *University Professor, Albert and Wilda Van Dusen Distinguished Professor of Orthopaedic Surgery, Professor of Chemical, Materials and Biomolecular Engineering; Chief Executive Officer, Connecticut Institute for Clinical and Translational Science; Director, The Raymond and Beverly Sackler Center for Biomedical, Biological, Engineering and Physical Sciences; Director, Institute for Regenerative Engineering, The University of Connecticut, Farmington, CT, USA*

Paul Lee *Department of Chemistry, Chemical Biology, and Biomedical Engineering, Stevens Institute of Technology, Hoboken, NJ, USA*

Adnan Memic *Center for Biomedical Engineering, Department of Medicine, Brigham and Women's Hospital, Harvard Medical School, Boston, MA, USA*

Harvard-MIT Division of Health Sciences and Technology, Massachusetts Institute of Technology, Cambridge, MA, USA

Center of Nanotechnology, King Abdulaziz University, Jeddah, Saudi Arabia

Sara K. Murase *Departament d'Enginyeria Química, Universitat Politècnica de Catalunya, Barcelona, ESP*

Ahmed A. Nada *Department of Orthopaedic Surgery, Institute for Regenerative Engineering, Raymond and Beverly Sackler Center for Biomedical, Biological, Physical and Engineering Sciences, The University of Connecticut, Farmington, CT, USA*

Rajaram K. Nagarale *Department of Chemical Engineering, Indian Institute of Technology Kanpur, Uttar Pradesh, India*

Dianna Y. Nguyen *Department of Bioengineering, Materials Research Institute, The Huck Institute of The Life sciences, The Pennsylvania State University, PA, USA*

Mehdi Nikkhah *Center for Biomedical Engineering, Department of Medicine, Brigham and Women's Hospital, Harvard Medical School, Boston, MA, USA*

Harvard-MIT Division of Health Sciences and Technology, Massachusetts Institute of Technology, Cambridge, MA, USA

Meera Parthasarathy *School of Chemical & Biotechnology, SASTRA University, Centre for Nanotechnology & Advanced Biomaterials, Thanjavur, Tamil nadu, India*

Omathanu Perumal *Department of Pharmaceutical Sciences, College of Pharmacy, South Dakota State University, Brookings, SD, USA*

Jordi Puiggalí *Departament d'Enginyeria Química, Universitat Politècnica de Catalunya, Barcelona, ESP*

Walid P. Qaqish *Department of Biomedical Engineering, The University of Akron, Akron, Ohio, USA*

Daisy M. Ramos *Department of Orthopaedic Surgery, Department of Material Science and Engineering, Institute for Regenerative Engineering, Raymond and Beverly Sackler Center for Biomedical, Biological, Physical and Engineering Sciences, The University of Connecticut, Farmington, CT, USA*

Department of Chemical, Materials and Biomedical Engineering, University of Connecticut, CT, USA

Dina Rassias *Biomedical Engineering Department, Worcester Polytechnic Institute, Worcester, MA, USA*

Tarun Saxena *Department of Biomedical Engineering, Georgia Institute of Technology, Atlanta, Georgia, USA*

Swaminathan Sethuraman *Centre for Nanotechnology & Advanced Biomaterials, School of Chemical & Biotechnology, Sastra University, Thanjavur, India*

Kush N. Shah *Department of Biomedical Engineering, The University of Akron, Akron, Ohio, USA*

Venkatram Prasad Shastri *Hermann Staudinger Haus, University of Freiburg, Freiburg, DEU*

Namdev B. Shelke *Department of Orthopaedic Surgery, Institute for Regenerative Engineering, Raymond and Beverly Sackler Center for Biomedical, Biological, Physical and Engineering Sciences, The University of Connecticut, Farmington, CT, USA*

Anuradha Subramaniam *Centre for Nanotechnology & Advanced Biomaterials, Sastra University, Thanjavur, India*

Xiaoyan Tang *Department of Orthopaedic Surgery, Department of Material Science and Engineering, Institute for Regenerative Engineering, Raymond and Beverly Sackler Center for Biomedical, Biological, Physical and Engineering Sciences, The University of Connecticut, Farmington, CT, USA*

Shalumon Kottappally Thankappan *Department of Orthopaedic Surgery, Institute for Regenerative Engineering, Raymond and Beverly Sackler Center for Biomedical, Biological, Physical and Engineering Sciences, The University of Connecticut, Farmington, CT, USA*

Katelyn Tran *Department of Chemistry, Chemical Biology, and Biomedical Engineering, Stevens Institute of Technology, Hoboken, NJ, USA*

Richard T. Tran *Department of Bioengineering, Materials Research Institute, The Huck Institute of The Life sciences, The Pennsylvania State University, PA, USA*

Chandra M. Valmikinathan *Global Surgery Group, Johnson and Johnson, Somerville, NJ, USA*

Yong Wang *Department of Bioengineering, College of Engineering, The Pennsylvania State University, PA, USA*

Iwen Wu *Department of Biomedical Engineering, Johns Hopkins University; Translational Tissue Engineering Center, Wilmer Eye Institute*

Jonathan Johannes Wurth *Institute for Macromolecular Chemistry, University of Freiburg, Freiburg, Germany; BIOSS – Centre for Biological Signalling Studies, University of Freiburg, Freiburg, Germany*

Zhiwei Xie *Department of Bioengineering, Materials Research Institute, The Huck Institute of The Life sciences, The Pennsylvania State University, PA, USA*

Jian Yang *Department of Bioengineering, Materials Research Institute, The Huck Institute of The Life sciences, The Pennsylvania State University, PA, USA*

Yuan Yin *Biomedical Engineering Department, Worcester Polytechnic Institute, Worcester, MA, USA*

Xiaojun Yu *Department of Chemistry, Chemical Biology, and Biomedical Engineering, Stevens Institute of Technology, Hoboken, NJ, USA*

Yang H. Yun *Dept. of Biomedical Engineering, University of Akron, Akron, OH, USA*

Foreword

I am truly delighted to write the foreword for *Natural and Synthetic Biomedical Polymers* edited by well-established leaders and pioneers in the field, Professors Dr. Kumbar, Dr. Laurencin, and Dr. Deng. This book should prove extremely useful as a reference source for all those working in the fields of polymer chemistry and physics, biomaterial science, tissue engineering, drug delivery, and regenerative medicine. Polymeric materials are routinely used in clinical applications, ranging from surgical sutures to drug-eluting devices to implants. In particular, implants and drug delivery devices fabricated using biodegradable polymers provide the significant advantage of being degraded and/or resorbed after they have served their function. Yet, biomedical polymers must satisfy several design criteria, including physical, chemical, biomechanical, biological, and degradation properties when serving as an active implant material. Several natural and synthetic degradable polymers have been developed and are used clinically today. However, a wide range of new polymers, as well as modifications to existing polymers, are constantly being developed and applied to meet on-going and evolving challenges in biomedical applications. For example, polymeric nanostructures, implants, scaffolds, and drug delivery devices are allowing unprecedented manipulation of cell-biomaterial interactions, promotion of tissue regeneration, targeting of therapies, and combined diagnostic and imaging modalities.

This timely book provides a well-rounded and articulate summary of the present status of natural and synthetic biomedical polymers, their structure and property relationships, and their biomedical applications including regenerative engineering and drug delivery. Polymers that are both synthetic and natural in origin have been widely used as biomaterials for a variety of biomedical applications and greatly impacted the advancement of modern medicine. In this regard, 23 concise and comprehensive chapters are prepared by experts in their fields from different parts of the world. The chapters encompass numerous topics that appear prominently in the modern biomaterials literature and cover a wide range of traditional synthetic, natural, and semi-synthetic polymers and their applications. In my opinion, this book presents an excellent overview of the subject that will appeal to a broad audience and will serve as a valuable resource to those working in the fields of polymer science, tissue engineering, regenerative medicine, or drug delivery. I believe that this textbook will be a welcome addition to personal collections, libraries, and classrooms throughout the world.

Kristi S. Anseth
Professor, Department of Chemical and Biological Engineering, University of Colorado

Chapter 1

Polymer Synthesis and Processing

Mahadevappa Y. Kariduraganavar[*], Arjumand A. Kittur[†], Ravindra R. Kamble[*]
[*]*Department of Studies in Chemistry, Karnatak University, Dharwad, India*
[†]*Department of Chemistry, SDM College of Engineering & Technology, Dharwad, India*

Chapter Outline

1.1 INTRODUCTION

Polymers are the most versatile class of biomaterials, being extensively used in biomedical applications such as contact lenses, pharmaceutical vehicles, implantation, artificial organs, tissue engineering, medical devices, prostheses, and dental materials [1–3]. This is all due to the unique properties of polymers that created an entirely new concept when originally proposed as biomaterials. For the first time, a material performing a structural application was designed to be completely resorbed and become weaker over time. This concept was applied for the first time with catgut sutures successfully and, later, with arguable results, on bone fixation, ligament augmentation, plates, and pins [4,5].

Current research on new and improved biodegradable polymers is focused on more sophisticated biomedical applications to solve the patients' problems with higher efficacy and least possible pains. One example is *tissue engineering*, wherein biodegradable scaffolds seeded with an appropriate cell type provide a substitute for damaged human tissue while the natural process of regeneration is completed [6,7]. Another important application of biodegradable polymer is in the *gene therapy* that provides a safer way of gene delivery than use of viruses as vectors [8,9].

Recently, an implant prepared from biodegradable polymer played a tremendous beneficial role in replacing the stainless steel implant during the surgery [10]. This has not necessitated a second surgical event for the removal. In addition to this, the biodegradation may offer other advantages. For example, a fractured bone, fixated with a rigid, nonbiodegradable stainless steel implant, has a tendency for refracture upon removal of the implant. The bone does not carry sufficient load during the healing process, since the load is mainly carried by the rigid stainless steel. However, an implant prepared from biodegradable polymer can be engineered to degrade at a rate that will slowly transfer load to the healing bone [11]. Another exciting application for which biodegradable polymers offer tremendous applications is the basis for the drug delivery, either as drug delivery system alone or in conjunction with functioning as a medical device. In orthopedic applications, the delivery of a polymer-bound morphogenic protein may be used to speed up the healing process after a fracture or delivery of an antibiotic may help to prevent *osteomyelitis* following surgery [12–14]. Biodegradable polymers also make possible targeting of drugs into sites of inflammation or tumors. Prodrugs with macromolecular carriers have also been used for such purposes. The term *prodrug* has been coined to describe a harmless molecule, which undergoes a reaction inside the body to release the active drug. Polymeric *prodrugs* are obtained by conjugating biocompatible polymeric molecules with appropriate drugs. Such macromolecular conjugate accumulates positively in tumors, since the permeability of cell membranes of tumor cells is higher than that of normal cells [1,15,16].

Polymers used as biomaterials can be naturally occurring and synthetic or combination of both. Natural polymers are abundant, usually biodegradable, and offer good biocompatibility [11,17]. The biocompatibility of a polymer depends on the specific adsorption of protein to the polymer surface and the subsequent cellular interactions. These interactions with the surrounding medium are governed mostly by the distribution of functional groups on the surface of biomaterial. Several useful biocompatible polymers of microbial origin are being produced from natural sources by fermentation processes. They are nontoxic and truly biodegradable [18]. Biodegradation is usually catalyzed by enzymes and may involve both hydrolysis and oxidation. Aliphatic chains are more flexible than aromatic ones and can easily fit into the active sites of enzymes, and hence, they are easier to biodegrade. Crystallinity hinders polymer degradation. Irregularities in chain morphology prevent crystallization and favor degradation [19].

Considering the significance and relevance of biodegradable polymers in the area of medical science, we have made an attempt to discuss the different polymerizations and their techniques employed for the preparation of polymers, synthetic methods of both natural and synthetic polymers including their properties and biomedical applications. At the end of the chapter, the methods of polymer processing for the preparation of films, objects, and fibers have also been discussed.

1.2 TYPES OF POLYMERIZATION

Polymerizations are generally classified according to the types of reactions involved in the synthesis [20,21]. There are mainly three types of polymerizations.

1.2.1 Addition Polymerization

In this polymerization process, the addition polymers are prepared from monomers without the loss of small molecules. Usually, unsaturated monomers such as olefins, acetylenes, aldehydes, or other compounds undergo addition polymerization. It is also called *chain-growth polymerization* since reactions are known to proceed in a stepwise fashion by way of reactive intermediates. The process of polymerization is usually exothermic by 8-20 kcal/mol since a π-bond in the monomer is converted to a sigma bond in the polymer. The reaction quickly leads to a polymer with very high molecular weight. The most common and thermodynamically favored chemical transformations of olefins are the addition reactions. Generally, these polymers can be prepared using bulk, solution, suspension, and emulsion polymerization techniques. Sometimes cross-linking can also be achieved using monomers with two double bonds.

Many well-known thermoplastics are the addition-type polymers. Figure 1.1 illustrates some addition polymerization processes.

The properties and biomedical applications of some of the important addition polymers are given in Table 1.1 [22,23].

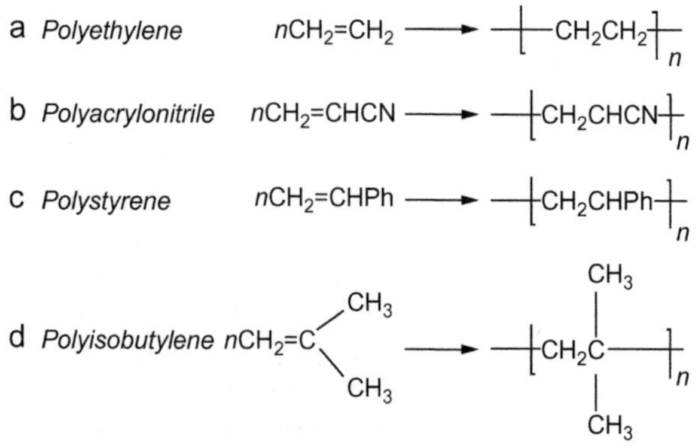

FIGURE 1.1 Common examples of addition polymerization.

TABLE 1.1 Properties and Biomedical Uses of Some Common Addition Polymers

Polymer Name(s)	Properties	Biomedical Uses
Polyethylene low density (LDPE)	Soft, waxy solid	Films, blood bags
Polyethylene high density (HDPE)	Rigid, translucent solid	Hip joints
Polyvinyl chloride (PVC)	Strong rigid solid	Reinforcement of artery
Polytetrafluoroethylene (PTFE, Teflon)	Resistant, smooth solid	Heart pumps, reinforcement of artery and blood vessels
Polymethyl methacrylate (PMMA, Lucite, and Plexiglas)	Hard, transparent solid	Contact lenses, heart pumps

1.2.2 Condensation Polymerization

It is a process in which two different monomers join together by the elimination of small molecules like water, ammonia, methanol, and HCl. It is also known as *step-growth polymerization*. The type of end product resulting from a condensation polymerization is dependent on the number of functional end groups of the monomer that can react. The monomers that are involved in condensation polymerization are not the same as those in addition polymerization. They have two main characteristics: these monomers have functional groups like –OH, $-NH_2$, or –COOH instead of double bonds and each monomer has at least two reactive sites. In this process, high molecular weight can be attained only at high conversions. Most of the reactions have high ΔE_a and hence heating is usually required.

Monomers with only one reactive group terminate a growing chain and thus give end products with a lower molecular weight. Linear polymers are created using monomers with two reactive end groups, and monomers with more than two end groups give three-dimensional polymers that are cross-linked. Dehydration synthesis often involves joining monomers with an –OH group and a freely ionizable –H on either end (such as a hydrogen from the $-NH_2$ in nylon or proteins). Normally, two or more different monomers are used in the reaction. The bonds between the hydroxyl group, the hydrogen atom, and their respective atoms break forming water from the hydroxyl and hydrogen and the polymer.

Polyester is created through ester linkages between monomers, which involve the functional groups like carboxyl and hydroxyl (an organic acid and an alcohol monomer). Nylon is another common condensation polymer, which can be prepared by reacting diamines with carboxyl derivatives. In this example, the derivative is a dicarboxylic acid, but buta-diacyl chlorides are also used. Another approach used is the reaction of difunctional monomers with one amine and one carboxylic acid group on the same molecule. An example of condensation polymerization is given in Figure 1.2.

The carboxylic acids and amines link to form peptide bonds, also known as amide groups. Proteins are the condensation polymers made from amino acid monomers. Carbohydrates are also condensation polymers made from sugar monomers such as glucose and galactose. Condensation polymerization is occasionally used to form simple hydrocarbons. This method, however, is expensive and inefficient, so the addition polymer of ethene, i.e., polyethylene, is generally used. Condensation polymers, unlike addition polymers, may be biodegradable. The peptide or ester bonds between monomers can be hydrolyzed by acid catalysts or bacterial enzymes breaking the polymer chain into smaller pieces. The most commonly known condensation polymers are proteins and fabrics such as nylon, silk, or polyester.

As before, a water molecule is removed, and an amide linkage is formed. An acid group remains on one end of the chain, which can react with another amine monomer. Similarly, an amine group remains on the other end of the chain, which can react with another acid monomer. Thus, monomers can continue to join by amide linkages to form a long chain. Because of the type of bond that links the monomers, this polymer is called a polyamide. The polymer made from these two six-carbon monomers is known as nylon 6,6 (Figure 1.3).

Similarly, a carboxylic acid monomer and an alcohol monomer can join together to form an ester linkage followed by a loss of water molecule. The monoester thus

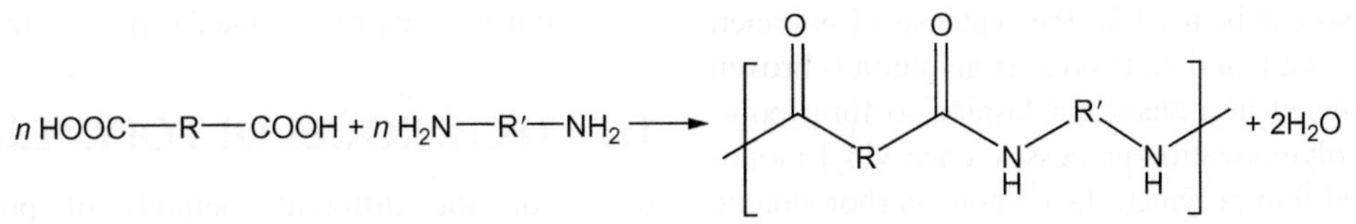

FIGURE 1.2 An example of condensation polymerization.

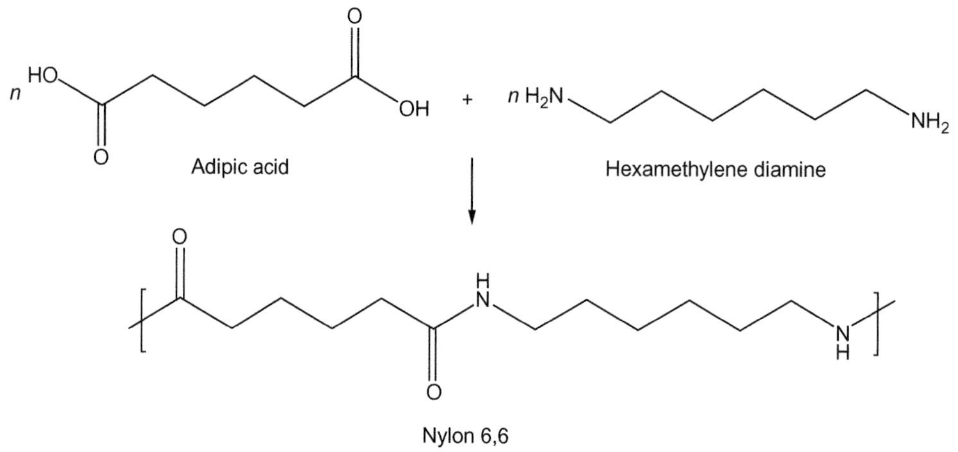

FIGURE 1.3 Preparation of nylon 6,6 as an example of condensation polymerization.

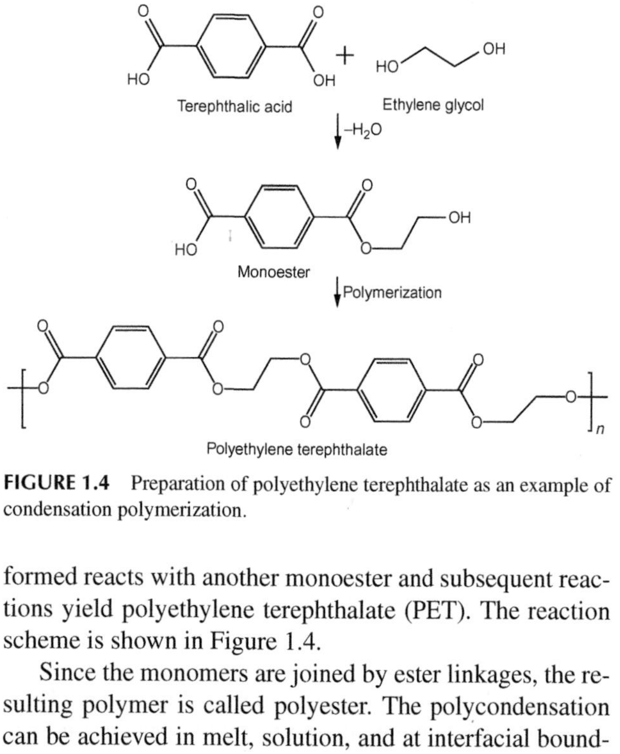

FIGURE 1.4 Preparation of polyethylene terephthalate as an example of condensation polymerization.

formed reacts with another monoester and subsequent reactions yield polyethylene terephthalate (PET). The reaction scheme is shown in Figure 1.4.

Since the monomers are joined by ester linkages, the resulting polymer is called polyester. The polycondensation can be achieved in melt, solution, and at interfacial boundary between two liquids in which the respective monomers are dissolved. It is a slow step addition process and molecular weight is >1,00,000 and highly dependent on monomer stoichiometry. The addition of little amount of tri- or multifunctional monomers develops extensive cross-linking.

1.2.3 Metathesis Polymerization

Olefin metathesis can be used for the synthesis of polymer, wherein carbon−carbon double bond in an olefin is broken and then rearranged in a statistical fashion to form polymer. In other polymerization processes, once vinyl monomer is converted into polymer, the carbon−carbon double bond does not remain in the polymer backbone. However, in metathesis polymerization, the carbon−carbon double bond remains in the polymer backbone chain and such polymers are called polyalkenamers [24]. The mechanism of metathesis polymerization is illustrated in Figure 1.5.

The commonly accepted mechanism for the olefin metathesis reaction was proposed by Chauvin. It involves a [2+2] cycloaddition reaction between transition metal alkylidene complex and the olefin to form an intermediate metallocyclobutane. This metallocycle then breaks up in the opposite fashion to afford a new alkylidene and new olefin. If this process is repeated, eventually, an equilibrium mixture of olefins will be obtained.

The following are the two different types of metathesis polymerization [25]:

(a) Acyclic diene metathesis (ADMET) polymerization
(b) Ring-opening metathesis polymerization (ROMP)

(a) *ADMET polymerization* ADMET starts with an acyclic diene such as 1,5-hexadiene and ends up in a polymer with a double bond in the backbone chain and ethylene as a by-product. The reaction is shown in Figure 1.6.

(b) *ROMP* In this polymerization, a cyclic olefin such as cyclopentene is used to make a polymer that does not have cyclic structures in its backbone and therefore it is called ROMP. Similarly, norbornene is polymerized by ROMP to get polynorbornene (Figure 1.7). Using ROMP, molecules like *endo*-dicyclopentadiene can also be polymerized to get a polymer with a cyclic olefin in a pendant group and the product is called polydicyclopentadiene. This is used to make big things in one piece. This can also undergo vinyl polymerization to give a cross-linked thermoset material.

1.3 TECHNIQUES OF POLYMERIZATION

Based on the different methods of preparation, the polymerization techniques can be classified broadly into

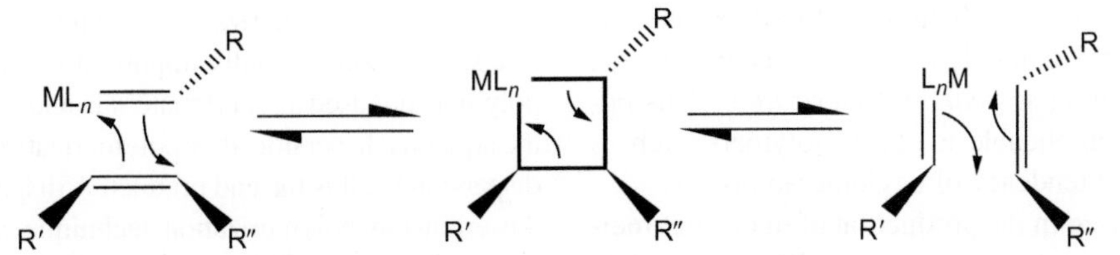

FIGURE 1.5 Mechanism of metathesis polymerization.

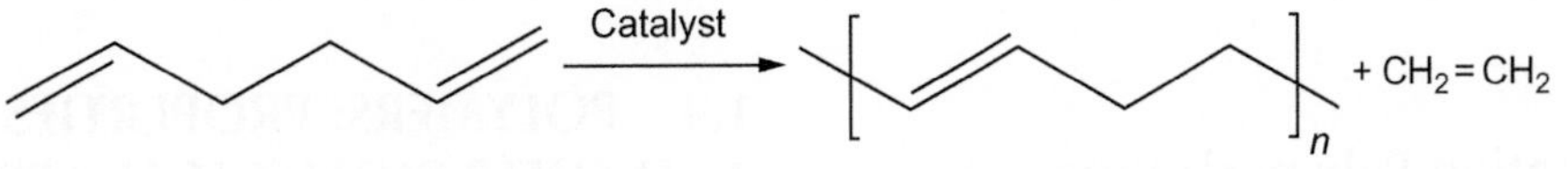

FIGURE 1.6 The reaction of acyclic diene metathesis polymerization.

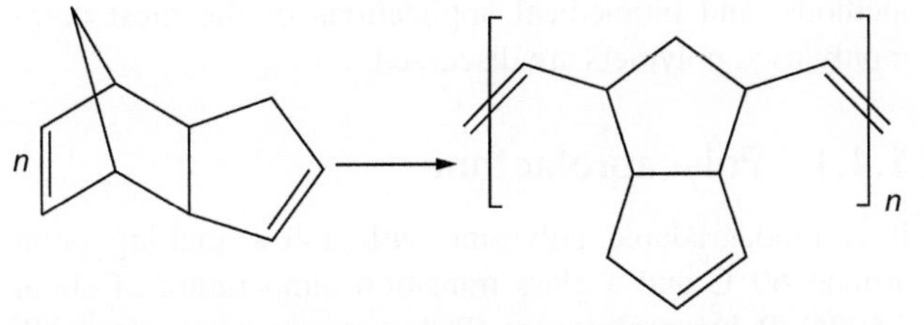

FIGURE 1.7 The reaction of ring-opening metathesis polymerization.

homogeneous and heterogeneous [26–28]. For homogeneous process, the diluted or pure monomers are added directly to one another and the reaction occurs in the media created when mixing the reactants. With heterogeneous process, a phase boundary exists, which acts as an interphase where the reaction occurs.

1.3.1 Solution Polymerization

It is an industrial polymerization technique, wherein a monomer is dissolved in a nonreactive solvent that contains a catalyst. In this method, both the monomer and the resulting polymer are soluble in the solvent. The heat released during the reaction is absorbed by the solvent and thus reduces the reaction rate. Once the maximum or desired conversion is reached, excess solvent is to be removed in order to obtain the pure polymer. The products obtained by this method are relatively low molecular weights because of the possibility of chain transfer. This process is suitable for the production of wet polymers since the removal of excess solvent is difficult and also the solvent is occluded and firmly traps the polymer. Therefore, this polymerization technique is applied when solutions of polymers are required (for ready-made use) for technical applications such as lacquers, adhesives, and surface coatings.

This process is used in the production of sodium polyacrylate, a superabsorbent polymer and neoprene used in disposable diapers and wetsuits, respectively. The polymers produced using this method are generally polyacrylonitrile (PAN), polyacrylic acid, and polytetrafluoroethylene.

1.3.2 Bulk (Mass) Polymerization

Bulk polymerization occurs within the monomer itself. The reaction is catalyzed by additives such as initiator and transfer agents under the influence of heat or light. Since this polymerization process is highly exothermic, it is difficult to control and hence the polymer obtained is generally of nonuniform molecular mass distribution. However, molecular-weight distribution can be easily changed by the use of chain transfer agent. The temperature and pressure can also be varied to control the properties of the final polymer. If the polymer is insoluble in its monomer, it is obtained as a powdery or porous solid. Since the recipe contains primarily the monomers, the polymer formed is usually pure. This is suitable for liquid (or liquefiable) monomers, which can be carried in batch or continuous mode. The product obtained has higher optical clarity, which is suitable for casting especially for clear products (e.g., polymethyl methacrylate (PMMA) films). Low-molecular-weight polymers can also be prepared by this method for adhesives, plasticizers, and lubricants.

1.3.3 Suspension Polymerization

It is a heterogeneous radical polymerization process. Step-growth polymers such as polyesters are manufactured using this technique. In this polymerization, the monomer containing initiator, modifier, etc., is dispersed in a solvent (generally water) by vigorous stirring. The monomer and initiator are insoluble in the liquid phase, so they form beads within the liquid matrix. A suspension agent such as PVA or methyl cellulose is usually added to stabilize the monomer droplets and hinder monomer drops from coming together. The reaction mixture usually has a volume ratio of monomer to liquid phase of 0.10-0.50. A major advantage is that heat transfer is very efficient and the reaction is therefore easily controlled. The reactions are usually carried out in a stirred tank reactor that continuously mixes the solution using turbulent pressure or viscous shear forces. The stirring action helps to keep the monomer droplets separated and creates a more uniform suspension, which leads to a more narrow size distribution of the final polymer beads. The beads look like pearls, hence the name *pearl polymerization*. This polymerization is not applicable to tacky polymers such as elastomers due to the tendency of agglomerations.

This process is used in the production of many commercial resins, including polyvinyl chloride (PVC), a widely used plastic; styrene resins including polystyrene, expanded polystyrene, and high-impact polystyrene; and PAN and PMMA.

1.3.4 Precipitation Polymerization

It is a heterogeneous polymerization process that begins initially as a homogeneous system in the continuous phase where the monomer and the initiator are completely soluble, but upon initiation, the formed polymer is insoluble and thus precipitates. The precipitated polymer can be separated in the form of a gel or powder by centrifugation or simple filtration. The degree of polymerization is high as there is no problem in heat dissipation. Polyvinyl esters and polyacrylic esters are obtained commercially using hydrocarbons as solvents. PAN is prepared using water as solvent.

1.3.5 Emulsion Polymerization

It is a type of radical polymerization in which the liquid monomer is dispersed in an insoluble liquid leading to an emulsion. The most common type of emulsion polymerization is an oil-in-water emulsion, wherein droplets of monomer (the oil) are emulsified (with surfactants) in a continuous phase of water. Water soluble polymers, such as certain polyvinyl alcohols or hydroxyethyl celluloses, can also be used to act as emulsifiers/stabilizers. The polymerization takes place in the latex particles that form spontaneously in the first few minutes of the process. These latex particles are typically 100 nm in size and are made of many individual polymer chains. The particles are stopped from coagulating with each other because each particle is surrounded by the surfactant (soap); the charge on the surfactant repels other particles electrostatically. When water-soluble polymers are used as stabilizers instead of soap, the repulsion between particles arises as these water-soluble polymers form a *hairy layer* around a particle that repels other particles, because pushing particles together would involve in compressing these chains. Since polymer molecules are contained within the particles, the viscosity of the reaction medium remains close to that of water and is not dependent on molecular weight. Emulsion polymerizations are designed to operate at high conversion of monomer to polymer. This can result in significant chain transfer to polymer. For dry (isolated) polymers, water removal is an energy-intensive process.

Emulsion polymerization technique is used to manufacture several commercially important polymers. Many of these polymers are used as solid materials and must be isolated from the aqueous dispersion after polymerization. In other cases, the dispersion itself is the end product. A dispersion resulting from the emulsion polymerization technique is often called latex (especially if derived from a synthetic rubber) or an emulsion (even though emulsion strictly speaking refers to a dispersion of an immiscible liquid in water).

1.4 POLYMERS: PROPERTIES, SYNTHESIS, AND THEIR BIOMEDICAL APPLICATIONS

Under this section, general properties, different synthetic methods, and biomedical applications of the most commonly used polymers are discussed.

1.4.1 Polycaprolactone

It is biodegradable polyester with a low melting point around 60 °C and a glass transition temperature of about −60 °C. Polycaprolactones (PCLs) impart good water, oil, solvent, and chlorine resistance to the polyurethanes (PUs) produced. It is commonly used in the manufacture of specialty PUs. PCL is degraded by hydrolysis of its ester linkages in physiological conditions and has therefore received a great deal of attention for use as an implantable biomaterial. It is especially interesting for the preparation of long-term implantable devices, owing to its degradation, which is even slower than that of polylactide.

This polymer is often used as an additive for resins to improve their processing characteristics and their end-use properties (e.g., impact resistance). Being compatible with a range of other materials, PCL can be mixed with starch to lower its cost and increase biodegradability or it is also added as a polymeric plasticizer to PVC. PCL was approved by the Food and Drug Administration (FDA) in specific applications used in the human body as a drug delivery device and surgical suture (sold under the brand name Monocryl) [29–31]. It is being investigated as a scaffold for tissue repair via tissue engineering and guided bone regeneration membrane. It has been used as the hydrophobic block of amphiphilic synthetic block copolymers used to form the vesicle membrane of polymersomes. In odontology or dentistry (as composite named Resilon), it is used as a component of *night guards* (dental splints) and in root canal filling. It performs like gutta-percha, has the same handling properties, and for retreatment purposes may be softened with heat or dissolved with solvents like chloroform. The major difference between the PCL-based root canal filling material (Resilon and Real Seal) and gutta-percha is that the PCL-based material is biodegradable but the gutta-percha is not.

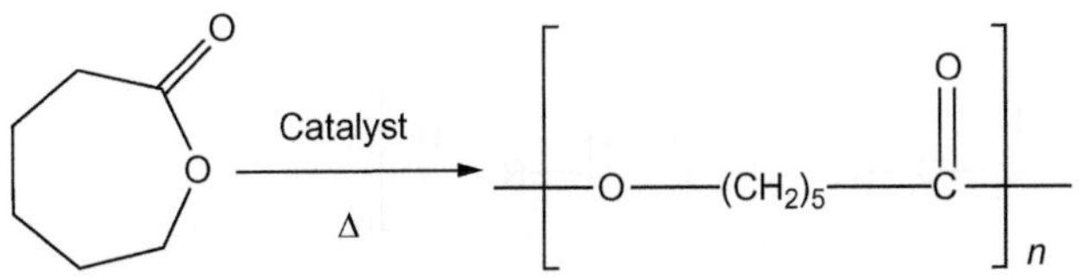

FIGURE 1.8 Synthesis of polycaprolactone.

PCL is prepared by ring-opening polymerization (ROP) of ε-caprolactone using dibutylzinc-triisobutylaluminum systems as a catalyst [32] such as stannous octoate (Figure 1.8). Recently, a wide range of catalysts for the ROP of caprolactone have been reported [33].

1.4.2 Polyethylene Glycol

It is known as polyethylene oxide (PEO) or polyoxyethylene, depending on its molecular weight. Polyethylene glycol (PEG) is the basis of a number of laxatives (e.g., macrogol-containing products, such as Movicol and PEG 3350, Softlax, MiraLAX, or GlycoLax). Whole bowel irrigation with PEG and added electrolytes is used for bowel preparation before surgery or colonoscopy. The preparation is sold under the brand names GoLYTELY, GaviLyte-C, NuLytely, GlycoLax, Fortrans, TriLyte, Colyte, Halflytely, Softlax, Lax-a-Day, ClearLax, and MoviPrep. MiraLAX and Dulcolax Balance are sold without prescription for short-term relief of chronic constipation. MiraLAX is currently FDA-approved for adults for a period of 7 days and is not approved for children [34]. The patients suffering from constipation had a better response to these two medications than to tegaserod [35]. These medications soften the fecal mass by osmotically drawing water into the gastrointestinal tract. It is generally well tolerated; however, side effects include bloating, nausea, gas, and diarrhea with excessive use. When attached to various protein medications, PEG allows a slowed clearance of the carried protein from the blood. This makes for a longer-acting medicinal effect and reduces toxicity and allows longer dosing intervals. Examples include PEG-interferon alpha, which is used to treat hepatitis C, and pegfilgrastim (Neulasta), which is used to treat neutropenia. It has been shown that PEG can improve healing of spinal injuries in dogs [36]. It is also noticed that PEG can aid in nerve repair and is also commonly used to fuse β-cells with myeloma cells in monoclonal antibody production.

PEG is used as an excipient in many pharmaceutical products. Lower-molecular-weight variants are used as solvents in oral liquids and soft capsules, whereas solid variants are used as ointment bases, tablet binders, film coatings, and lubricants [37]. When labeled with PEG a near-infrared fluorophore, it has been used in preclinical work as a vascular, lymphatic, and general tumor-imaging agent by exploiting the enhanced permeability and retention effect of tumors [38]. High-molecular-weight PEG (e.g., PEG 8000) has been shown to be a dietary preventive agent against colorectal cancer in animal models [39]. The chemoprevention database shows PEG is the most effective known agent for the suppression of chemical carcinogenesis in rats. The injection of PEG 2000 into the bloodstream of guinea pigs after spinal cord injury leads to rapid recovery through molecular repair of nerve membranes [40]. It is also reported that using PEG can mask antigens without damaging the function and shape of the cell. PEG is being used in the repair of motor neurons damaged in crush or laceration incidents *in vivo* and *in vitro*. When coupled with melatonin, 75% of damaged sciatic nerves were rendered viable [41].

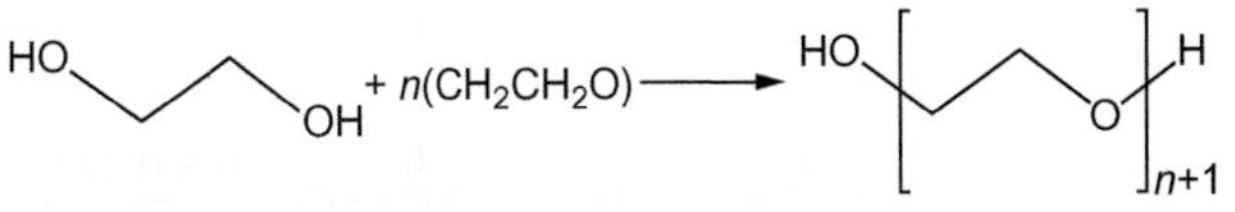

FIGURE 1.9 Synthesis of polyethylene glycol.

PEG is produced by the interaction of ethylene oxide with water, ethylene glycol, or ethylene glycol oligomers [42]. The reaction is catalyzed by acidic or basic catalysts (Figure 1.9). Ethylene glycol and its oligomers are used as a starting material instead of water, as they allow the creation of polymers with a low polydispersity (narrow molecular-weight distribution). Polymer chain length depends on the ratio of reactants.

Depending on the type of catalyst, the mechanism of polymerization can be cationic or anionic. The anionic mechanism is preferable since it allows forming PEG with a low polydispersity. Polymerization of ethylene oxide is an exothermic process. Overheating or contaminating ethylene oxide with catalysts such as alkalis or metal oxides can lead to runaway polymerization, which can end in an explosion after a few hours. PEO, or high-molecular-weight PEG, is synthesized by suspension polymerization technique. It is necessary to hold the growing polymer chain in solution in the course of the polycondensation process. The reaction is catalyzed by magnesium-, aluminum-, or calcium-organo element compounds. To prevent coagulation of polymer chains from solution, chelating additives such as dimethylglyoxime are added. Alkali catalysts such as sodium hydroxide (NaOH), potassium hydroxide (KOH), or sodium carbonate (Na_2CO_3) are used to prepare low-molecular-weight PEG [43].

1.4.3 Polyurethane

This polymer is composed of a chain of organic units joined by urethane links. Most of the PUs are thermosetting polymers, which do not melt upon heating. PUs can outperform many other polymers in flexibility, tear resistance, and abrasion resistance. This is because many devices that are used in these areas can rub against other materials and bend repeatedly. Without PUs, the continued rubbing and bending could result in the device

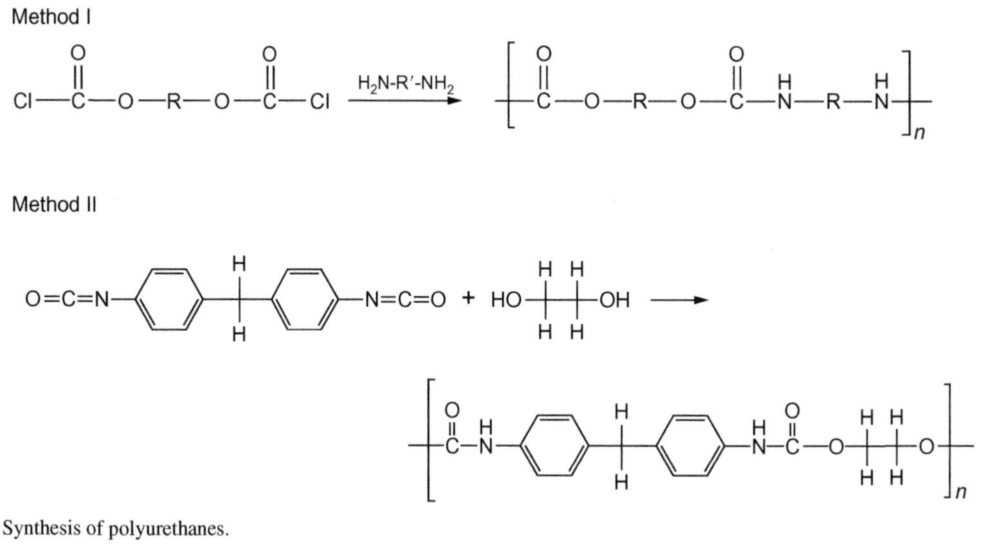

FIGURE 1.10 Synthesis of polyurethanes.

weakening or could cause failure in extreme cases. Today's PUs have been formulated to provide good biocompatibility, flexural endurance, high strength, high abrasion resistance, and processing versatility over a wide range of applications. These attributes are important in supporting new applications continually being found by medical device manufacturers including artificial hearts, catheter tubing, feeding tubes, surgical drapes and drains, intra-aortic balloon pumps, dialysis devices, nonallergenic gloves, medical garments, hospital bedding, wound dressings, and so on [44]. Thermoplastic polyurethanes (TPUs), also called PU elastomers, have molecular structures similar to that of human proteins. Protein absorption, which is the beginning of the blood coagulation cascade, was found to be slower or less than other materials. This makes them ideal candidates for a variety of medical applications requiring adhesive strength and unique biomimetic and antithrombotic properties. For example, TPUs are currently being used as a special sealant to bind bundles of hollow fibers in artificial dialysis cylinders. With the advent of new surgical implants, biomedical PUs can lead the way to eliminate some acute and chronic health challenges. PUs are often used in cardiovascular devices due to their good biocompatibility and their mechanical properties [45]. Patients generally prefer to use PU medical devices compared to other materials due to their comfort. By virtue of their wide range of properties, PUs made significant contributions to the medical industry and will continue to play an important role in the future of science and medicine.

There are two principal methods of forming PUs [46,47], viz., the reaction of bischloroformates prepared from dihydroxy compounds with excess phosgene and with diamines (Method I) and more important from the industrial perspective and the reaction of diisocyanates with dihydroxy compounds (Method II) that has the advantage owing to the nonformation of by-products (Figure 1.10). PU products often are simply called *urethanes*, but should not be confused with ethyl carbamate, which is also called urethane. PUs neither contain nor are produced from ethyl carbamate.

1.4.4 Polydioxanone or Poly-*p*-Dioxanone

It is a colorless, crystalline, biodegradable synthetic polymer. Chemically, polydioxanone is a polymer of multiple repeating ether-ester units. It is characterized by a glass transition temperature in the range of −10 and 0 °C and a crystallinity of about 55%. Polydioxanone is generally extruded into fibers; however, care should be taken to process the polymer to the lowest possible temperature in order to avoid its spontaneous depolymerization back to the monomer. The ether oxygen group in the backbone of the polymer chain is responsible for its flexibility [48,49].

Polydioxanone is used in the preparation of surgical sutures. Other biomedical applications include orthopedics, plastic surgery, drug delivery, cardiovascular devices, and tissue engineering [48,49]. It is degraded by hydrolysis, and the end products are mainly excreted in urine, the remainder being eliminated by digestive or exhaled as CO_2. The biomaterial is completely reabsorbed in 6 months and can be seen only a minimal foreign body reaction tissue in the vicinity of the implant [50].

It is obtained by ROP of the monomer *p*-dioxanone. The process requires heat and an organometallic catalyst like zirconium acetylacetonate or zinc L-lactate. The reaction is shown in Figure 1.11.

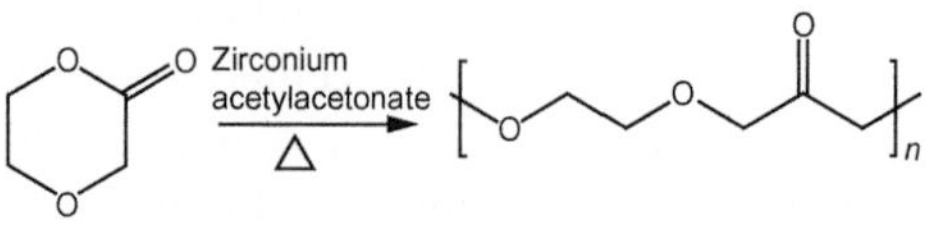

FIGURE 1.11 Synthesis of polydioxanone.

1.4.5 Polymethyl Methacrylate

It is a transparent thermoplastic, often used as a lightweight or shatter-resistant alternative to glass [51]. It has a density of 1.17-1.30 g/cm^3, which is less than half that of glass. It also has good impact strength higher than both glass and polystyrene. However, the impact strength of PMMA is still significantly lower than polycarbonate and some engineered polymers. Although it is not technically a type of glass, the substance has sometimes historically been called acrylic glass. Chemically, it is the synthetic polymer of methyl methacrylate. It has also been sold under many different names, including ACRYLITE®, Lucite, Plexiglass, and Perspex. The glass transition temperature (T_g) of atactic PMMA is 105 °C. The T_g values of commercial grades of PMMA range from 85 to 165 °C, the range being is so wide because of the vast number of commercial compositions that are copolymers with comonomers other than methyl methacrylate. PMMA is thus an organic glass at room temperature; i.e., it is below its T_g [52]. All common molding processes may be used to prepare biomedical devices including injection molding, compression molding, and extrusion. The highest quality PMMA sheets are produced by cell casting, but in this case, the polymerization and molding steps occur concurrently. The strength of the material is higher than molding grades owing to its extremely high molecular mass. Rubber toughening has been used to increase the strength of PMMA due to its brittle behavior in response to applied loads. PMMA swells and dissolves in many organic solvents. It also has poor resistance to many other chemicals on account of its easily hydrolyzed ester groups. Nevertheless, its environmental stability is superior to most other plastics such as polystyrene and polyethylene. PMMA has a maximum water absorption ratio of 0.3-0.4% by weight. Tensile strength decreases with increased water absorption [53]. Its coefficient of thermal expansion is relatively high at 5-10 $\times$ 10^{-5} K^{-1} [54].

PMMA has a good compatibility with human tissue, and it is used in the manufacture of rigid intraocular lenses in the eye when the original lens has been removed in the treatment of cataracts [55]. Historically, hard contact lenses were frequently made of this material. Soft contact lenses are often made of a related polymer, where acrylate monomers containing one or more hydroxyl groups make them hydrophilic. In orthopedic surgery, PMMA bone cement is used to affix implants and to remodel lost bone. PMMA has also been linked to cardiopulmonary events in the operating room due to hypotension [56]. Bone cement acts like a grout and not so much like a glue in arthroplasty. Although sticky, it does not bond to either the bone or the implant; it primarily fills the spaces between the prosthesis and the bone preventing motion. The disadvantage of this bone cement is that it heats up to 82.5 °C while setting that may cause thermal necrosis of neighboring tissue. A major consideration when using PMMA cement is the effect of stress shielding. Since PMMA has a Young's modulus between 18 and 31 GPa [57], which is greater than that of natural bone (around 14 GPa for human cortical bone) [58], the stresses are loaded into the cement, and hence, the bone no longer receives the mechanical signals to continue bone remodeling and so resorption will occur [59].

Dentures are often made of PMMA and can be color-matched to the patient's teeth and gum tissue. PMMA is also used in the production of ocular prostheses, such as the osteo-odonto-keratoprosthesis. In cosmetic surgery, tiny PMMA microspheres suspended in some biological fluid are injected under the skin to reduce wrinkles or scars permanently. A large majority of white dental filling materials (composites) have PMMA as their main organic component. Emerging biotechnology and biomedical research uses PMMA to create microfluidic lab-on-a-chip devices, which require 100-μm-wide geometries for routing liquids. These small geometries are amenable to use PMMA in a biochip fabrication process and offer moderate biocompatibility. Bioprocess chromatography columns use cast acrylic tubes as an alternative to glass and stainless steel. These are pressure-rated and satisfy stringent requirements of materials for biocompatibility, toxicity, and extractable.

PMMA is produced by emulsion polymerization, solution polymerization, and bulk polymerization. Generally, radical initiation is used (including living polymerization methods), but anionic polymerization of PMMA can also be performed. PMMA produced by radical polymerization (all commercial PMMA) is atactic and completely amorphous. Free-radical polymerization is carried out using benzoyl peroxide as an initiator in the presence of *N*,*N*-dimethyl-*p*-toluidine (Figure 1.12) [60].

1.4.6 Polyglycolic Acid or Polyglycolide

It is linear aliphatic polyester and undergoes biodegradable. It has been known since 1954 as a tough fiber-forming thermoplastic polymer. Polyglycolide has a glass transition temperature in the range of 35-40 °C and its melting point is reported to be in the range of 225-230 °C. Polyglycolic Acid (PGA) also exhibits an elevated degree of crystallinity of about 45-55%, thus resulting in insolubility in water [61]. The solubility of this polyester also depends on its molecular weight. The low-molecular-weight oligomers

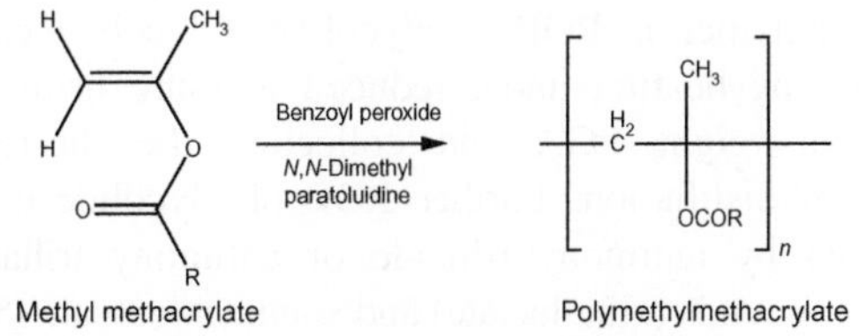

FIGURE 1.12 Synthesis of polymethyl methacrylate.

are sufficiently soluble in organic solvents, whereas high-molecular-weight polymers are insoluble in almost all common organic solvents such as acetone, dichloromethane, chloroform, ethyl acetate, and tetrahydrofuran. However, polyglycolide is soluble in highly fluorinated solvents like hexafluoroisopropanol and hexafluoroacetone sesquihydrate that can be used to prepare solutions of the high-molecular-weight polymer for melt spinning and film preparation. Fibers of PGA exhibit high strength and modulus (7 GPa) and are particularly stiff [62].

Owing to its hydrolytic instability, initially, its use was limited. Currently, polyglycolide and its copolymers poly(lactic-*co*-glycolic acid) with lactic acid, poly(glycolide-*co*-caprolactone) with ε-caprolactone, and poly(glycolide-*co*-trimethylene carbonate) with trimethylene carbonate) are widely used to develop synthetic absorbable sutures that were marketed under the trade name of Dexon and are now sold as Surgicryl [61].

PGA suture is absorbable and braided multifilament. It is coated with *N*-laurin and L-lysine, which render the thread extremely smooth, soft, and safe for knotting. It is also coated with magnesium stearate and finally sterilized with ethylene oxide gas. It is naturally degraded in the body within 60-90 days by hydrolysis. Elderly, anemic, and malnourished patients may absorb the suture more quickly. It is commonly used for subcutaneous sutures, intracutaneous closures, and abdominal and thoracic surgeries. The traditional role of PGA as a biodegradable suture material has led to its evaluation in other biomedical fields. Implantable medical devices have been produced with PGA, including anastomosis rings, pins, rods, plates, and screws [61]. It has also been explored for tissue engineering or controlled drug delivery. Tissue engineering scaffolds made with polyglycolide have been generally obtained in the form of nonwoven meshes [63].

Polyglycolide can be obtained through several different processes starting with different materials: polycondensation of glycolic acid, ROP of glycolide, and solid-state polycondensation of halogenoacetates. Polycondensation of glycolic acid is the simplest process available to prepare PGA, but it is not the most efficient since it yields a low-molecular-weight product. The glycolic acid is heated around 175-185 °C at atmospheric pressure until water ceases to distil. Subsequently, pressure is reduced to 150 mmHg at the same temperature for about 2 h to obtain low-molecular-weight polyglycolide [63,64].

The most common method to prepare a high-molecular-weight polymer is ROP of glycolide. Glycolide can be prepared by heating under reduced pressure using low-molecular-weight PGA and collecting the diester by means of distillation. Further ROP of glycolide can be catalyzed by antimony trioxide or antimony trihalides, zinc compounds (zinc lactate) and stannous octoate (Sn(II) 2-ethylhexanoate), or tin alkoxides. Stannous octoate is the

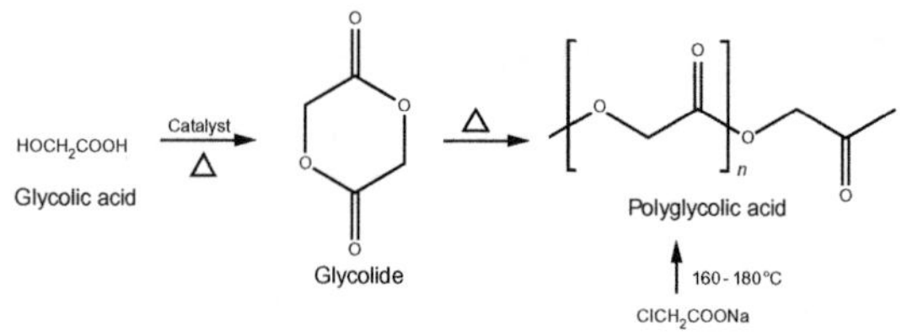

FIGURE 1.13 Synthesis of polyglycolic acid.

most commonly used initiator, as it is approved by the FDA as a food stabilizer. Usage of other catalysts has also been reported. Among these, aluminum isopropoxide, calcium acetylacetonate, and several lanthanide alkoxides (e.g., yttrium isopropoxide) are important. The procedure followed for ROP is briefly outlined: a catalytic amount of initiator is added to glycolide under a nitrogen atmosphere at a temperature of 195 °C. The reaction is allowed to proceed for about 2 h and then temperature is raised to 230 °C for about half an hour. After solidification, the resulting high-molecular-weight polymer is collected [64–66]. The reaction scheme for the preparation of PGA is shown in the Figure 1.13.

In the thermally induced solid-state polycondensation of halogenoacetates ($X\text{-}CH_2COO^-M^+$ where M is a monovalent metal-like sodium and X is a halogen-like chlorine), resulting in the formation of polyglycolide and small crystals of a salt, polycondensation is carried out by heating halogenoacetate, like sodium chloroacetate, at temperature around 160-180 °C while continuously purging nitrogen through the reaction vessel. During the reaction, polyglycolide is formed along with sodium chloride, which precipitates within the polymeric matrix; the salt can be conveniently removed by washing the PGA with water [67].

PGA can also be obtained by reacting carbon monoxide, formaldehyde, or one of its related compounds like paraformaldehyde or trioxane, in the presence of an acidic catalyst. The autoclave is loaded with the catalyst (chlorosulfonic acid), dichloromethane, and trioxane, and then it is charged with carbon monoxide until a specific pressure is reached. The reaction mixture is stirred and allowed to proceed at a temperature of about 180 °C for 2 h. Upon completion, the unreacted carbon monoxide is discharged and a mixture of low- and high-molecular-weight polyglycolide is collected [68].

1.4.7 Polylactic Acid or Polylactide

It is an aliphatic polyester derived from renewable resources, such as corn starch, tapioca roots, chips or starch, or sugarcane. Polylactic acid or polylactide (PLA) can withstand temperatures up to 110 °C [69]. PLA is soluble in chlorinated solvents, hot benzene, tetrahydrofuran, and dioxane [70]. It can be processed like other thermoplastics into fiber (for example, using conventional melt spinning processes) and film. Due to the chiral nature of lactic acid, several distinct forms of polylactide exist:

poly-L-lactide (PLLA) is the product resulting from polymerization of L,L-lactide (also known as L-lactide). PLLA has a crystallinity of around 37%, a glass transition temperature in the range 60-65 °C, a melting temperature around 173-178 °C, and a tensile modulus about 2.7-16 GPa [71,72]. The melting temperature of PLLA can be increased up to 40-50 °C, and its heat deflection temperature can be increased from approximately 60-190 °C by physically blending the polymer with PDLA (poly-D-lactide). PDLA and PLLA form a highly regular stereocomplex with increased crystallinity. The temperature stability is maximized when a 50:50 blend is used, but even at lower concentrations of 3-10% of PDLA, there is still a substantial improvement. In the latter case, PDLA acts as a nucleating agent, thereby increasing the crystallization rate. Biodegradation of PDLA is slower than that of PLA due to higher crystallinity of PDLA.

PLA is prone to degrade into innocuous lactic acid and it is used as medical implants in the form of screws, pins, rods, and as a mesh. Depending on the type of implants, it breaks down in the body within 6 months to 2 years. This gradual degradation is desirable for a support structure, because it gradually transfers the load to the body (e.g., the bone) as that organ heals. Pure PLLA is primarily used for lipoatrophy of cheeks (facial volume enhancer) [73]. The preparation scheme of PLA is given in Figure 1.14.

There are two important methods for PLA synthesis: direct polycondensation of lactic acid and ROP of lactic acid cyclic dimer, known as lactide. In direct condensation, solvent is used and higher reaction times are required. The resulting polymer is a material of low to intermediate molecular weight. ROP of the lactide needs catalyst but results in PLA with controlled molecular weight. Depending on the monomer and reaction conditions, it is possible to control the ratio and sequence of D- and L-lactic acid units in the final polymer [74,75].

1.4.8 Polylactic-*co*-Glycolic Acid

It is a copolymer used in therapeutic devices owing to its biocompatibility and biodegradability. Depending on the ratio of lactide to glycolide used for the polymerization, different forms of polylactic-*co*-glycolic acid (PLGA) can be obtained. Generally, all PLGAs are amorphous rather than crystalline and show a glass transition temperature in the range of 40-60 °C. Unlike the homopolymers of lactic acid (polylactide) and glycolic acid (polyglycolide) that show poor solubility, PLGA can be dissolved in a wide range of common solvents, including chlorinated solvents, tetrahydrofuran, acetone, or ethyl acetate [76].

PLGA degrades by hydrolysis of its ester linkages in the presence of water. It has been shown that the time required for degradation of PLGA is related to the monomers' ratio used in the production. The higher the content of glycolide units, the lower the time required for degradation. An exception to this rule is the copolymer with 50:50 monomers' ratio, which exhibits the faster degradation (about 2 months). In addition, polymers that are end-capped with esters (as opposed to the free carboxylic acid) demonstrate longer degradation half-lives.

PLGA has been successful as a biodegradable polymer as it undergoes hydrolysis in the body to produce the original monomers, lactic acid, and glycolic acid. These two monomers under normal physiological conditions are by-products of various metabolic pathways in the body. Since the body effectively deals with the two monomers, there is minimal systemic toxicity associated with using PLGA for drug delivery or biomaterial applications. Also, the possibility to tailor the polymer degradation time by altering the ratio of the monomers used during synthesis has made PLGA a common choice in the production of a variety of biomedical devices such as grafts, sutures, implants, prosthetic devices, and micro- and nanoparticles and also used in various

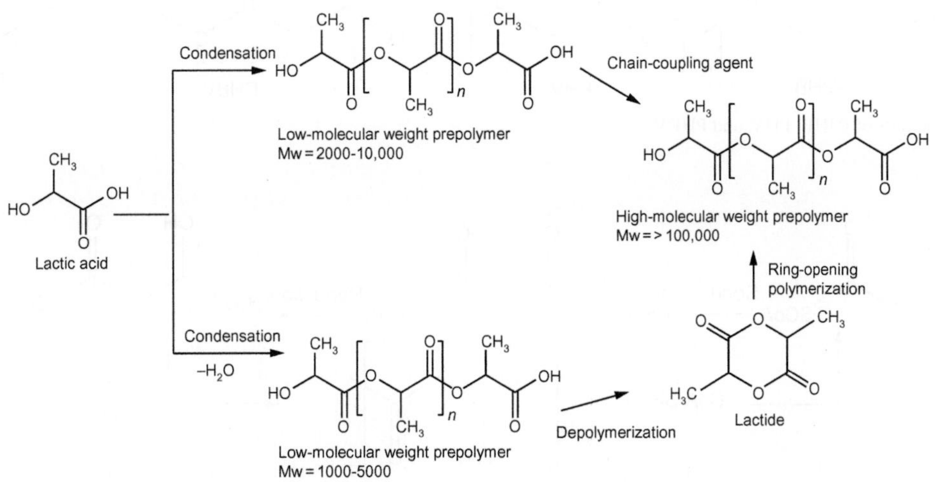

FIGURE 1.14 Synthesis of polylactic acid.

therapeutic aspects. As an example, a commercially available drug delivery device using PLGA is Lupron Depot® for the treatment of advanced prostate cancer [77,78].

PLGA is synthesized by means of random ring-opening copolymerization of two different monomers, the cyclic dimers (1,4-dioxane-2,5-diones) of glycolic acid and lactic acid. Common catalysts used in the preparation of this polymer include tin(II) 2-ethylhexanoate, tin(II) alkoxides, or aluminum isopropoxide. During polymerization, successive monomeric units of glycolic or lactic acid are linked together in PLGA by ester linkages, thus yielding linear aliphatic polyester [48]. Figure 1.15 illustrates the reaction scheme for PLGA preparation.

1.4.9 Polyhydroxybutyrate

It is a polyhydroxyalkanoate, a polymer belonging to the polyesters class that are of interest as bioderived, nontoxic, and biodegradable plastics [79]. It is water-insoluble and relatively resistant to hydrolytic degradation. This differentiates polyhydroxybutyrate (PHB) from most other currently available biodegradable plastics, which are either water-soluble or moisture-sensitive. It exhibits good oxygen permeability and good ultraviolet resistance but poor resistance to acids and bases. It is soluble in chloroform and other chlorinated hydrocarbons. It melts at 175 °C with a glass transition temperature of about 2 °C. The tensile strength is about 40 MPa and close to that of polypropylene. It sinks in water, while polypropylene floats and therefore facilitates anaerobic biodegradation in sediments. The poly-3-hydroxybutyrate (P3HB) form of PHB is probably the most common type of polyhydroxyalkanoate, but other polymers of this class are produced by a variety of organisms. These include poly-4-hydroxybutyrate, polyhydroxyvalerate (PHV), polyhydroxyhexanoate, polyhydroxyoctanoate, and their copolymers. The structures of P3HB, PHV, and PHBV are shown in Figure 1.16.

PHB is produced by microorganisms (such as *Ralstonia eutropha* or *Bacillus megaterium*) apparently in response to conditions of physiological stress. The polymer is primarily a product of carbon assimilation (from glucose or starch) and is employed by microorganisms as a form of energy storage molecule to be metabolized when other common energy sources are not available. Since it is biocompatible, it is suitable for biomedical applications. PHB is sold in the trade name of Biopol. It is currently used in the medical industry for internal suture. It is nontoxic and biodegradable, so it does not have to be removed after recovery [80,81]. Microbial biosynthesis of PHB starts with the condensation of two molecules of acetyl-CoA to give acetoacetyl-CoA, which is subsequently reduced to hydroxybutyryl-CoA. The latter compound is then used as a monomer to polymerize PHB [80,82]. The biosynthesis of PHB is shown in Figure 1.17.

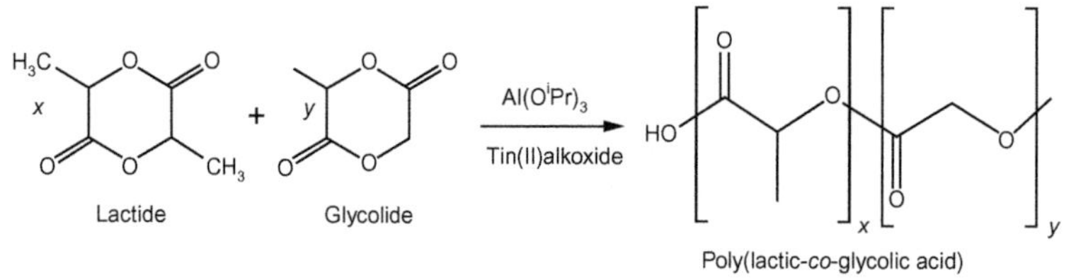

FIGURE 1.15 Synthesis of poly(lactic-*co*-glycolic acid).

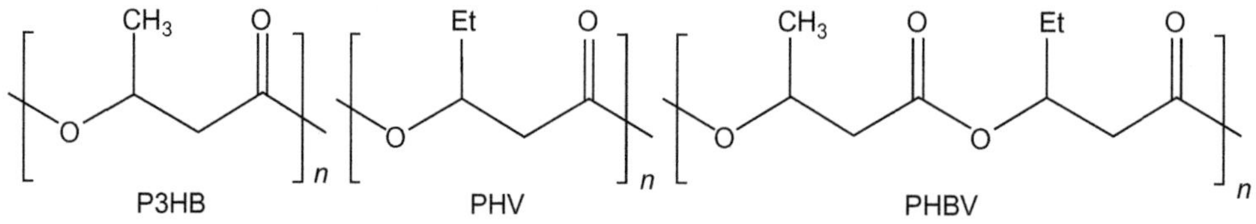

FIGURE 1.16 Structures of P3HB, PHV, and PHBV.

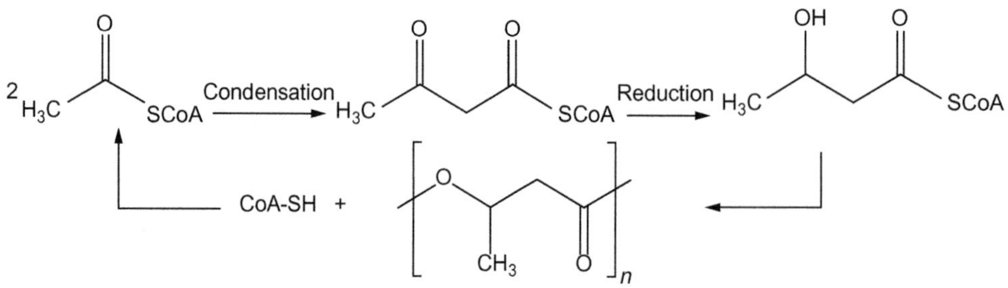

FIGURE 1.17 Biosynthesis of polyhydroxybutyrate.

1.4.10 Polycyanoacrylates

α-Cyanoacrylate is a nontoxic acrylic resin that rapidly polymerizes in the presence of water (hydroxide ion) forming long, straight chains, joining the bonded surfaces together. α-Polycyanoacrylates have the general formula as shown in Figure 1.18, where the group R can be methyl, butyl, hexyl, octyl, and so on [83,84].

The α-cyanoacrylates polymerize by an anionic mechanism in the presence of water (Figure 1.19). Higher alkyl derivatives polymerize more rapidly. The cyanoacrylates when exposed to normal level of humidity in the air cause polymerization rapidly. Because of this property, α-cyanoacrylate is applied thinly to ensure that reaction proceeds rapidly and forms a strong bond within a reasonable time [85].

Usually for medical uses, cyanoacrylates with larger alkyl ester groups such as poly(octyl cyanoacrylate) are preferred since short alkyl groups like methyl groups can irritate tissues. Cyanoacrylate adhesives provide unique benefits for their use in medical device manufacturing processes. The usage of cyanoacrylate medical adhesives as a replacement for the classical suture is in view of its good cosmetic effect, reduced pain and recovery period, and preference by the patients. Hence, these are more efficient, offer low surgery time, and therefore reduced cost. Cyanoacrylates find applications in drug delivery and targeting systems. In addition to their use as skin adhesives, they have been used as adhesives in corneal and retinal surgery and as an adjunct to suturing in internal surgery. They have also shown to be effective in skin burns and bone and cartilage grafts. Dentists use cyanoacrylates in dental cements and fillings [86].

1.4.11 Polyvinylpyrrolidone

Polyvinylpyrrolidone (PVP), commonly called polyvidone or povidone, is a water-soluble polymer made from the monomer *N*-vinylpyrrolidone [87,88]. Dry PVP is a light flaky hygroscopic powder and readily absorbs up to 40%

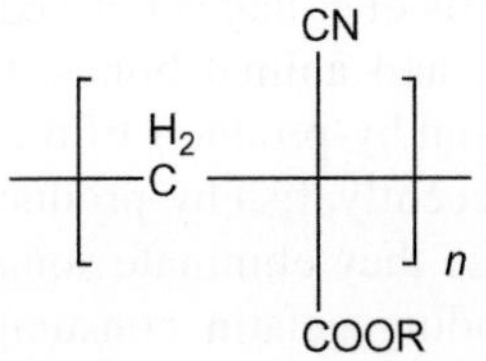

FIGURE 1.18 General formula of polycyanoacrylates.

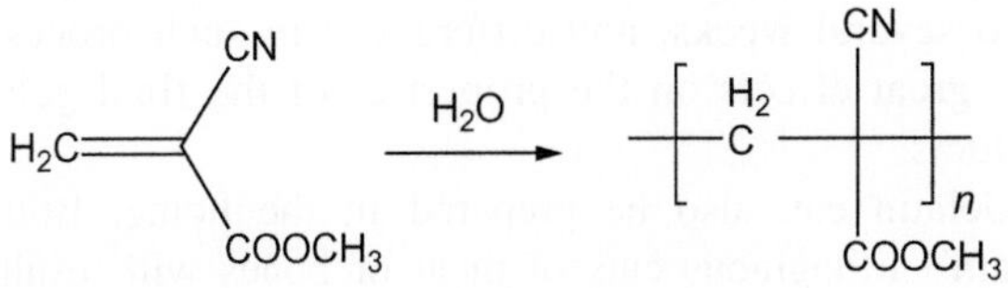

FIGURE 1.19 Polymerization of α-cyanoacrylate.

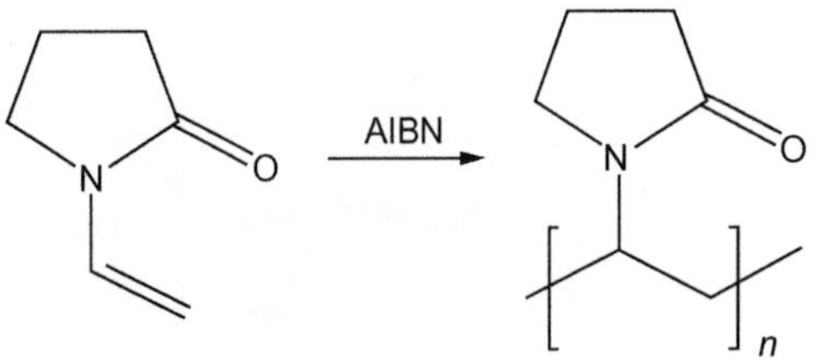

FIGURE 1.20 Synthesis of polyvinylpyrrolidone.

of water by its weight. In solution, it has excellent wetting properties and readily forms films, which makes it good as a coating or an additive to coatings. PVP can be prepared by free-radical polymerization from its monomer *N*-vinylpyrrolidone in the presence of AIBN as an initiator as shown in Figure 1.20 [87,88].

The PVP was used as a blood plasma expander for trauma victims. It is used as a binder in many pharmaceutical tablets and it simply passes through the body when it is administered orally [89]. However, autopsies have found that crospovidone does contribute to pulmonary vascular injury in substance abusers who have injected pharmaceutical tablets intended for oral consumption [90]. PVP added to iodine forms a complex called povidone-iodine that possesses disinfectant properties. This complex is used in various products like solutions, ointment, pessaries, liquid soaps, and surgical scrubs. It is known under the trade name Betadine and Pyodine. It is used in pleurodesis (fusion of the pleura because of incessant pleural effusions). For this purpose, povidone-iodine is equally effective and safe as talc and may be preferred because of easy availability and low cost [91]. It is used as an aid for increasing the solubility of drugs in liquid and semiliquid dosage forms (syrups and soft gelatin capsules) and as an inhibitor of recrystallization.

1.4.12 Chitosan

It is a linear polysaccharide composed of randomly distributed β-(1-4)-linked D-glucosamine (deacetylated unit) and *N*-acetyl-D-glucosamine (acetylated unit). It is made by treating shrimp and other crustacean shells such as crabs and krills with the alkali NaOH. Chitosan is a naturally abundant and renewable polymer and has excellent property such as biodegradability, biocompatibility, nontoxicity, and good adsorption [92]. The structure of chitosan is given in Figure 1.21.

Chitosan has a number of biomedical uses. In medicine, it may be useful in bandages to reduce bleeding and as an antibacterial agent. It can also be used to help deliver drugs through the skin. Properties of chitosan allow it to rapidly clot blood, and it has recently gained approval for use in bandages and other hemostatic agents. Chitosan hemostatic products have been used to quickly stop bleeding and to reduce blood loss and hence result in 100% survival [93]. Chitosan hemostatic

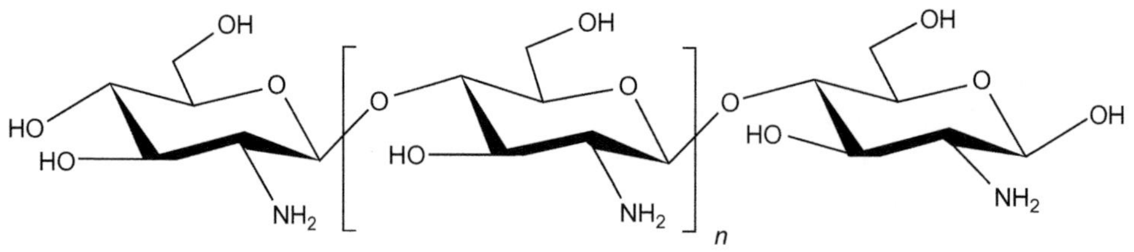

FIGURE 1.21 Structure of chitosan.

products reduce blood loss in comparison to gauze dressings and increase patient survival [94]. Chitosan is hypoallergenic and has natural antibacterial properties, which further support its use in field bandages [95].

Chitosan hemostatic agents are often chitosan salts made from mixing chitosan with an organic acid (such as succinic or lactic acid) [96]. The hemostatic agent works by an interaction between the cell membrane of erythrocytes (negative charge) and the protonated chitosan (positive charge) leading to involvement of platelets and rapid thrombus formation [97]. The chitosan salts can be mixed with other materials to make them more absorbent (such as mixing with alginate) [98] or to vary the rate of solubility and bioabsorbability of the chitosan salt [96]. The chitosan salts are biocompatible and biodegradable, making them useful as absorbable hemostats. The protonated chitosan is broken down by lysozyme in the body to glucosamine [97], and the conjugate bases of the acid (such as lactate or succinate) are substances naturally found in the body. The chitosan salt may be placed on an absorbable backing [99].

Chitosan can be used in transdermal drug delivery. It is mucoadhesive in nature and reactive (so it can be produced in many different forms) and, most importantly, has a positive charge under acidic conditions. This positive charge comes from protonation of its free amino groups. Lack of a positive charge means chitosan is insoluble in neutral and basic environments. However, in acidic environments, protonation of the amino groups leads to an increase in solubility. The implications of this are very important to biomedical applications. This molecule will maintain its structure in a neutral environment but will solubilize and degrade in an acidic environment. It means chitosan can be used to transport a drug to an acidic environment, where the chitosan packaging will then degrade, releasing the drug to the desired environment. One example of this drug delivery has been the transport of insulin [100,101].

Chitosan membranes have been proposed as an artificial kidney membrane because of their suitable permeability and high tensile strength. Chitosan and its same derivatives are used to prepare scaffolds for tissue engineering applications. It can also be used for designing artificial skin, treatment of brain-scalp damage, and in plastic skin surgery. Chitosan has replaced the synthetic polymers in ophthalmological applications due to its characteristic properties such as optical clarity, adequate mechanical stability, sufficient optical correction, wettability, and compatibility.

Crustacean shells consist of 30-40% proteins, 30-50% calcium carbonate, and 20-30% chitin and also contain pigments of a lipidic nature such as carotenoids (astaxanthin, canthaxanthin, lutein, and β-carotene). These proportions vary with species and with season. On the other hand, chitin is associated with a higher protein content but lower carbonate concentration. Chitin is extracted by acid treatment to dissolve the calcium carbonate followed by alkaline extraction to dissolve the proteins and by a depigmentation step to obtain a colorless product mainly by removing the astaxanthin [102].

Chitosan is prepared by hydrolysis of acetamide groups of chitin. This is normally conducted by severe alkaline hydrolysis treatment due to the resistance of such groups imposed by the *trans* arrangement of the C_2-C_3 substituent in the sugar ring [103]. Thermal treatments of chitin under strong aqueous alkali are usually needed to give partially deacetylated chitin (lower than 30%), regarded as chitosan. Usually, NaOH or KOH are used at a concentration of 30-50% (w/v) at high temperature (100 °C). The steps involved in the extraction of chitosan are illustrated in the Figure 1.22.

1.4.13 Gelatin

Gelatin is a translucent, natural, nontoxic, colorless, brittle (when dry), flavorless solid polymer. This is obtained by partial hydrolysis of collagen derived from skin, white connective tissue, and animal bones. On a large scale, gelatin is made from by-products of the meat and leather industry [104]. Recently, fish by-products have also been considered because they eliminate some of the religious obstacles surrounding gelatin consumption [105]. The raw materials are prepared by different curing, acid, and alkali processes, which are employed to extract the dried collagen hydrolysate. These processes [106] may take up to several weeks, and differences in such processes have great effects on the properties of the final gelatin products.

Gelatin can also be prepared in the home. Boiling certain cartilaginous cuts of meat or bones will result in gelatin being dissolved into the water. Depending on the

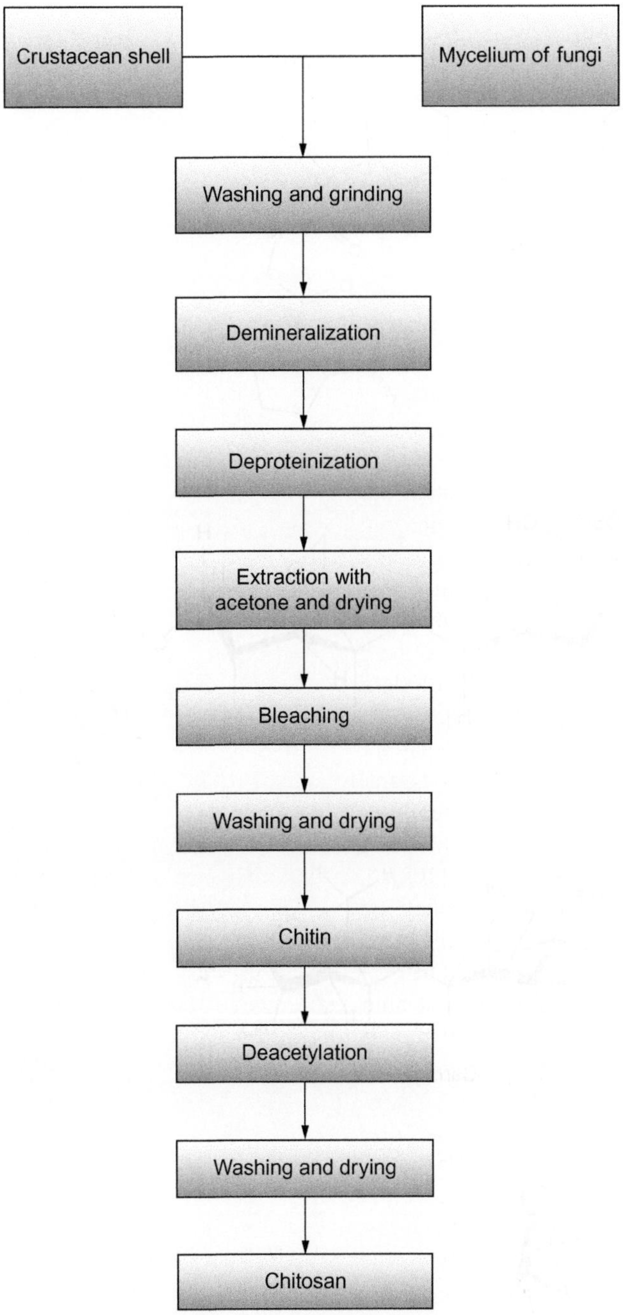

FIGURE 1.22 Extraction process of chitosan. (For color version of this figure, the reader is referred to the online version of this chapter.)

concentration, the resulting stock (when cooled) will naturally form a jelly or gel. This process is used for aspic. While there are many processes whereby collagen can be converted to gelatin, they all have several factors in common. The intermolecular and intramolecular bonds that stabilize insoluble collagen rendering it insoluble must be broken, and the hydrogen bonds that stabilize the collagen helix must also be broken [107]. The manufacturing processes of gelatin consist of three main stages:

a. Pretreatments to make the raw materials ready for the main extraction step and to remove impurities that may have negative effects on physicochemical properties of the final gelatin product
b. The main extraction step, which is usually done with hot water or dilute acid solutions as a multistage extraction to hydrolyze collagen into gelatin
c. The refining and recovering treatments including filtration, clarification, evaporation, sterilization, drying, rutting, grinding, and sifting to remove the water from the gelatin solution, to blend the gelatin extracted, and to obtain dried, blended, and ground final product

Although gelatin is 98-99% protein by dry weight, it has less nutritional value than many other complete protein sources. Gelatin is unusually high in the nonessential amino acids glycine and proline (i.e., those produced by the human body) while lacking certain essential amino acids (i.e., those not produced by the human body). It contains no tryptophan and is deficient in isoleucine, threonine, and methionine.

The approximate amino acid composition of gelatin is as follows: glycine 21%, proline 12%, hydroxyproline 12%, glutamic acid 10%, alanine 9%, arginine 8%, aspartic acid 6%, lysine 4%, serine 4%, leucine 3%, valine 2%, phenylalanine 2%, threonine 2%, isoleucine 1%, hydroxylysine 1%, methionine and histidine <1%, and tyrosine <0.5%. These values vary, especially the minor constituents, depending on the source of the raw material and processing technique [108]. The structure of gelatin unit is given in Figure 1.23.

Gelatin is also a topical hemostatic and is applied on bleeding wound and tied in bandage. This hemostatic action is based on platelet damage at the contact of blood with gelatin, which activates the coagulation cascade. Gelatin also causes a tamponading effect—blood flow stoppage into a blood vessel by a constriction of the vessel by an outside force. Gelatin has also been claimed to promote general joint health and found that it relieved knee joint pain and stiffness in athletes [109]. Oral gelatin consumption has beneficial therapeutic effect on hair loss in both men and women [110,111] and has beneficial effect for some fingernail changes and diseases [112–114].

1.4.14 Carrageenan

Carrageenans or carrageenins are a family of linear sulfated polysaccharides that are extracted from red seaweeds. Carrageenans are large, highly flexible molecules that curl forming helical structures, and therefore, they have an ability to form a variety of different gels at room temperature. All carrageenans are high-molecular-weight polysaccharides made up of repeating galactose units and 3,6-anhydrogalactose, both sulfated and nonsulfated. The units are joined by

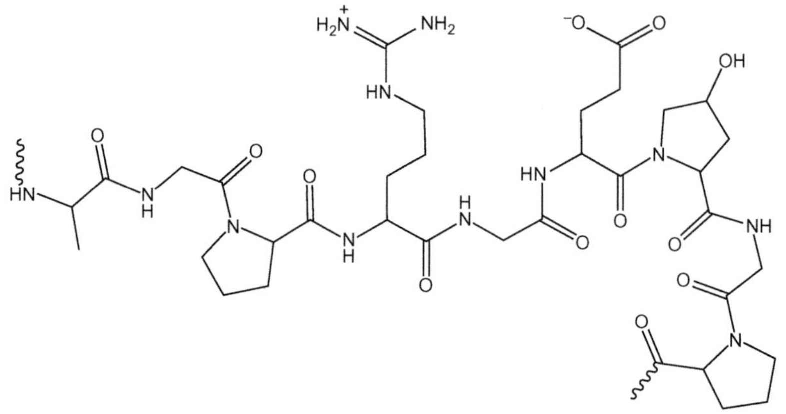

FIGURE 1.23 Structure of gelatin unit.

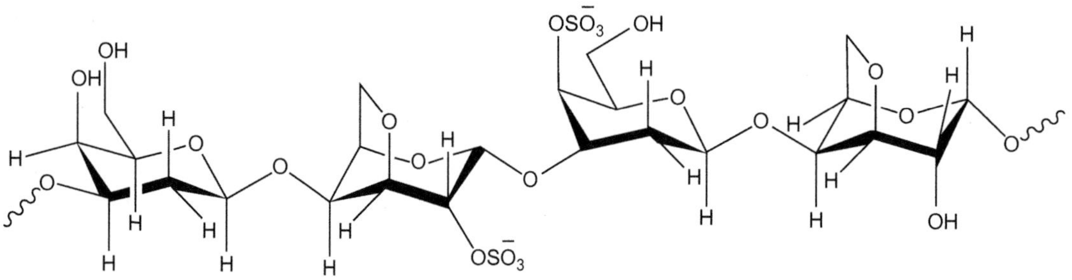

FIGURE 1.24 Structure of carrageenan.

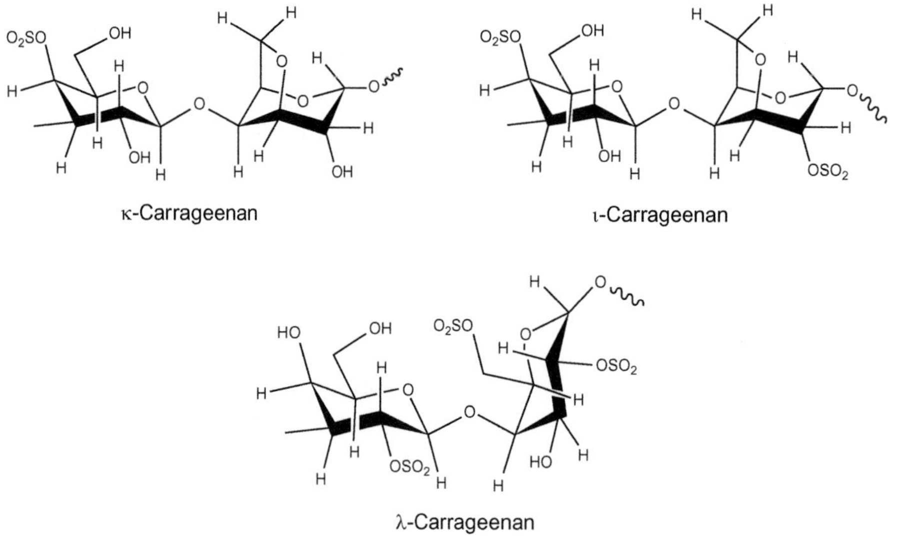

FIGURE 1.25 Structures of kappa, iota, and lambda carrageenans.

alternating alpha 1-3 and beta 1-4 glycosidic linkages [115]. In view of their gelling, thickening, and stabilizing properties, they are widely used in the food industry. Their main application is in dairy and meat products, due to their strong interactions with protein. The structure of carrageenan is given in Figure 1.24.

There are three main commercial classes of carrageenan (Figure 1.25).

a. *Kappa* forms strong, rigid gels in the presence of potassium ions; it reacts with dairy proteins. It is sourced mainly from *Kappaphycus alvarezii* [116].

b. *Iota* forms soft gels in the presence of calcium ions. It is produced mainly from *Eucheuma denticulatum* [116].

c. *Lambda* does not form gel and is used to thicken dairy products. The most common source is *Gigartina* from South America.

The primary differences that influence the properties of *kappa*, *iota*, and *lambda* carrageenans are the number and position of the ester sulfate groups on the repeating galactose units. Many red algal species produce different types of carrageenans during their developmental history. For instance, the genus *Gigartina* produces mainly *kappa* carrageenan during its gametophytic stage and *lambda* carrageenan during its sporophytic stage. All are soluble in hot water, but in cold water, only the lambda forms (and the sodium salts of the other two) are soluble. Carrageenan is mainly composed of dietary fiber, which balances the nutrition better [117]. Technically, carrageenan is considered a dietary fiber [118].

Gelatinous extracts of the *Chondrus crispus* (Irish moss) seaweed have been used as food additives for hundreds of years. Carrageenan is a vegetarian and veg-alternative to gelatin in some applications, although it cannot replace gelatin in confectionery like jelly babies. Scientists have raised serious concerns about the safety of carrageenan in food, based on laboratory animal studies showing gastrointestinal inflammation, ulcerations, and colitis-like disease in animals given food-grade carrageenan in their drinking water or diet [119–121]. Some physicians advise avoiding consumption of foods with carrageenan, especially for people with gastrointestinal symptoms [122].

There are three types of industrial extraction processes and are illustrated in Figure 1.26.

a. *Semirefined*: This is only performed using *Eucheuma cottonii* or *Eucheuma spinosum*. The raw weed is first sorted and crude contaminants are removed by hand. The weed is then washed to remove salt and sand and then cooked in hot alkali to increase the gel strength. The cooked weed is washed, dried, and milled. *E. spinosum* undergoes a much milder cooking cycle as it dissolves quite readily. This product is called semirefined carrageenan [116].
b. *Refined*: In this refining process, the carrageenan is first dissolved and filtered to remove cell wall debris. This is then precipitated from the clear solution either by isopropanol or by potassium chloride.
c. *Mixed processing*: It is a hybrid technology in which seaweed is treated heterogeneously as in the semirefined process, but alcohol or high salt levels are used to inhibit dissolution.

Carrageenan-based gels offer protection against HSV-2 transmission by binding to the receptors on the herpes virus, thus preventing the virus from binding to cells [123]. Carrageenans are extremely potent inhibitors of HPV infection *in vitro* and in animal challenge models [124]. Clinical studies have shown that carrageenans are the first substances found to be active against common cold viruses (not just the symptoms of the cold). Marinomed Biotechnologie, an Austrian company, conducted a study in 2010 showing that their nasal spray, which contains carrageenan, was effective

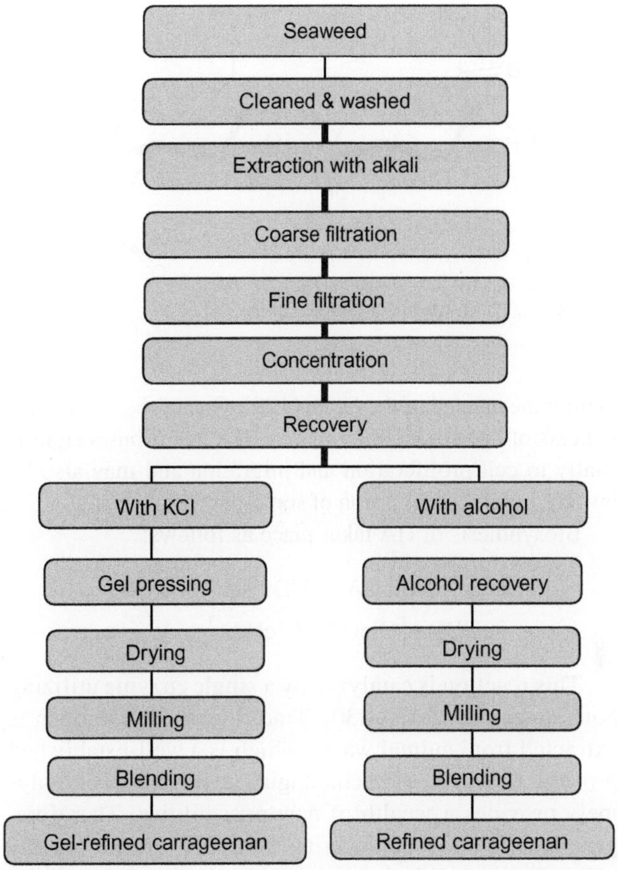

FIGURE 1.26 Steps involved in industrial extraction of carrageenan. (For color version of this figure, the reader is referred to the online version of this chapter.)

as a prophylactic treatment against the viral cause (rather than treating the symptoms) of the common cold [125].

1.4.15 Hyaluronic Acid

Hyaluronan or hyaluronic acid (HA) or hyaluronate is a linear polysaccharide composed of a repeating disaccharide unit of β(1,4)-glucuronic acid (GlcUA)-β(1,3)-*N*-acetylglucosamine. Both individual carbohydrate residues of HA adopt the stable chair conformation, which determines the conformation of the polymer in solution. Chemical structure of HA is given in Figure 1.27. It is a simple linear polymer with high molecular mass and exceptional rheological properties [126] and is the only glycosaminoglycan member that is not sulfated and is not covalently bound to a proteoglycan core protein. The average 70 kg person has roughly 15 g of HA in the body, one-third of which is turned over (degraded and synthesized) every day [127]. HA acid is also a component of the group A streptococcal extracellular capsule [128] and is believed to play a role in virulence [129,130]. It is distributed widely throughout connective, epithelial, and neural tissues. It is unique among glycosaminoglycans since it is nonsulfated and formed in the plasma

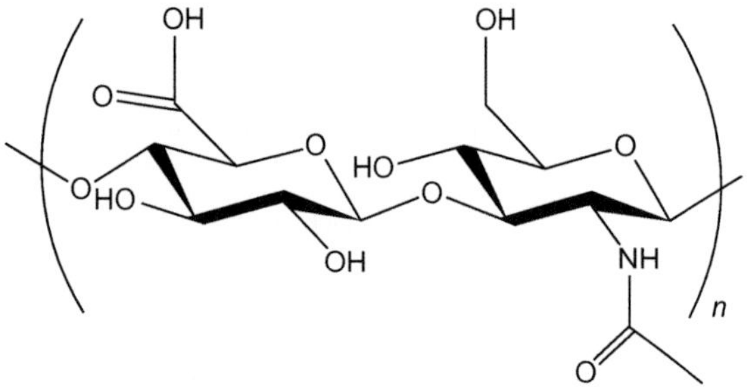

FIGURE 1.27 Structure of hyaluronic acid.

membrane instead of the Golgi [129]. One of the chief components of the extracellular matrix, HA, contributes significantly to cell proliferation and migration and may also be involved in the progression of some malignant tumors.

Biosynthesis of HA takes place as follows:

$$n\text{UDP} - \text{GlcUA} + n\text{UDP} - \text{GlcNAc} \rightarrow$$
$$2n\text{UDP} + \left[\text{GlcUA} + \text{GlcNAc}\right]_n$$

This reaction is catalyzed by a single enzyme utilizing both sugar substrates [130]. Traditionally, hyaluronan is extracted from animal waste, which is a well-established process. However, biotechnological synthesis of biopolymers provides a wealth of new possibilities. Therefore, genetic/metabolic engineering has been applied in the area of tailor-made hyaluronan synthesis. Another approach is the controlled artificial (*in vitro*) synthesis of hyaluronan by enzymes. Advantage of using microbial and enzymatic synthesis for hyaluronan production is the simpler downstream processing and a reduced risk of viral contamination [131].

HA is involved in creating flexible and protective layers in tissues and in many signaling pathways during embryonic development, wound healing, inflammation, and cancer. HA is an important component of active pharmaceutical ingredients for treatment of arthritis and osteoarthritis [132]. Dry and scaly skin (xerosis) caused by atopic dermatitis (eczema) may be treated with a prescription skin lotion containing sodium hyaluronate as its active ingredient. In some cancers, hyaluronan levels correlate well with malignancy and poor prognosis. Hyaluronan is, thus, often used as a tumor marker for prostate and breast cancer [133,134]. Hyaluronan may also be used postoperatively to induce tissue healing, especially after cataract surgery [135]. Current models of wound healing propose the larger polymers of HA appear in the early stages of healing to physically make room for white blood cells, which mediate the immune response.

Hyaluronan has also been used in the synthesis of biological scaffolds for wound-healing applications. These scaffolds typically have proteins such as fibronectin attached to the hyaluronan to facilitate cell migration into the wound. This is particularly important for individuals with diabetes suffering from chronic wounds [136].

1.4.16 Xanthan Gum

It is a polysaccharide secreted by the bacterium *Xanthomonas campestris*. It is the sodium, potassium, or calcium salt of a high-molecular-weight polysaccharide containing D-glucose, D-mannose, and D-glucuronic acid. It also contains not less than 1.5% of pyruvic acid. It is a cream-colored powder soluble in water [137].

Xanthan is prepared by inoculating a sterile aqueous solution of carbohydrates, a source of nitrogen, dipotassium phosphate, and some trace elements. The medium is well aerated and stirred. The polymer is produced extracellularly into the medium. The final concentration of xanthan thus produced will vary greatly depending on the method of production, strain of bacteria, and random variation. After fermentation, that can vary in time from 1 to 4 days and thus the polymer is precipitated from the medium by the addition of isopropyl alcohol. The resulting precipitate is dried and milled to give a powder that is readily soluble in water or brine [138]. The structure of xanthan is shown in Figure 1.28.

The xanthan gum is used as stabilizer, thickener, and emulsifier that are extensively used in pharmaceutical and cosmetic industries [139]. The pseudoplastic properties of this gum enable ointments to hold their shape and to spread readily. It is a suitable suspending vehicle for delivering antispasmodics topically along the length of esophagus in patients with esophageal spa.

1.4.17 Acacia Gum

It is a mixture of polysaccharides and glycoproteins. It is exudated from the stem and branches of *Acacia senegal*

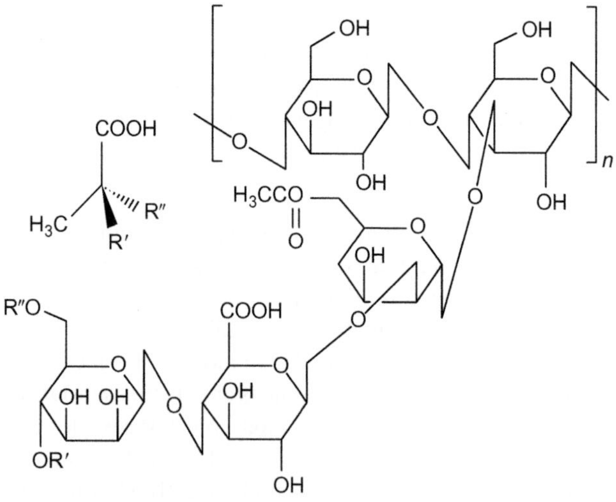

FIGURE 1.28 Structure of xanthan.

wild (*Family Mimosae*). It is also called *senegal gum.* It is a dietary fiber that is soluble in water leaving only a small residue of vegetable articles but insoluble in alcohol and ethers [140,141].

Acacia is used as a pharmaceutical ingredient in medication for throat and stomach inflammation and as a film-forming agent in peel-off skin masks. This gum is often applied to affected skin to treat minor wounds and scrapes. It also acts as a demulcent, which soothes the mucus membranes. Acacia gum is useful in treating diarrhea and other intestinal ailments. Tea made with this gum helps to cure sore throat. This is also useful in reduction of cholesterol level when administered orally. It acts as suspending and emulsifying agent and as a tablet binder. It is used in the pharmaceutical industry as binding agents in the manufacture of cough pastilles and other medical preparations and also as a coating for pills [139].

1.4.18 Alginate

Alginic acid, also called algin or alginate, is an anionic polysaccharide distributed widely in the cell walls of brown algae. It binds with water to form a viscous gum. It absorbs water quickly and is capable of absorbing 200-300 times of water by its weight [142]. It is biocompatible and less toxic with relatively low cost and forms mild gelation by the addition of cations such as Na^+ and Ca^{2+} [143].

Alginic acid is a linear copolymer with homopolymeric blocks of (1-4)-linked β-D-mannuronate (M) and its C-5 epimer α-L-glucuronate (G) residues, respectively, covalently linked together in different sequences or blocks. The different forms of alginic acid are represented in Figure 1.29. The monomers can appear in homopolymeric blocks of consecutive G-residues (G-blocks), consecutive M-residues (M-blocks), or alternating M- and G-residues (MG-blocks).

Commercially available alginate is typically extracted from brown algae (*Phaeophyceae*), including *Laminaria hyperborea*, *Laminaria digitata*, *Laminaria japonica*, *Ascophyllum nodosum*, and *Macrocystis pyrifera* [144], by treatment with aqueous alkali solutions (e.g., NaOH) [145]. The extract is filtered, and either sodium or calcium chloride is added to the filtrate in order to precipitate alginate. The resulting alginate salt is transformed into alginic acid by treatment with dilute HCl. After purification and conversion, water-soluble sodium alginate powder is obtained [146]. On a dry-weight basis, the alginate contents are 22-30% for *A. nodosum* and 25-44% for *L. digitata* [147].

Alginate is widely used in controlled drug and protein delivery [148], wound dressings [149], tissue regeneration [150], etc. It is also used for blood vessel, bone, and cartilage regeneration. Alginate is used in pharmaceutics with several applications. It is compounded into tablets to accelerate disintegration of tablet for faster release of medicinal component. Alginate can form gel in the high acidic stomach thus protects stomach mucosa. It is used in cosmetics with several applications due to its functionality of thickener and moisture retainer.

1.5 PROCESSING OF POLYMERS FOR BIOMEDICAL DEVICES

Polymer materials play an important role irrespective of any field; wherever we see, we find the materials generally made of polymers. Particularly in the recent years, polymeric materials have become boon to the society and replaced many things either wood, glass, or metals. They have become an ultimate substitute for many things in the field of medical, agriculture, aerospace industry, automobile industry, and in the house hold items. It is all due to varied properties of polymer materials, i.e., low cost, chemical inertness, light weight (easy transportation), insulating property (nonconductors), high toughness, good strength, better elegance, easily moldable character, and good mechanical strength. In view of all these versatile properties, the polymer chemists and chemical engineers are gaining a considerable pleasure in converting polymers into useful products. This process is known as polymer fabrication. Therefore, in the subsequent sections, the processing of polymers for various biomedical devices is described.

1.5.1 Fabrication of Polymer Films

Polymer films have unique properties including high aspect ratio, high flexibility, physical adhesion to ubiquitous surfaces, and attractive structural colors. The films can be formed from biodegradable and biocompatible polymer

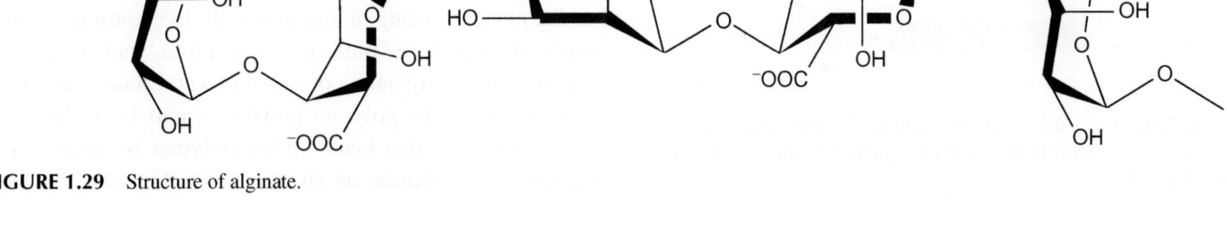

FIGURE 1.29 Structure of alginate.

materials. Such films have been paid a significant attention in the medicine due to potential applications ranging from biocompatible implant coatings to tissue engineering scaffolds.

Polymer films can be prepared by various methods [151,152]. The important methods are discussed in the succeeding text.

1.5.1.1 *Solution Casting*

In this method, polymer is dissolved in a suitable solvent in order to obtain viscous solution, which is then poured onto a flat or nonadhesive surface. The solvent is allowed to evaporate or removed and the dry films can be stripped (peeled) out from the flat surface. In order to avoid the formation of weak conglomerates in the polymer solution, a suitable molecular weight is required (higher molecular weight). To make the solution casting easier, the solvent must be sufficiently volatile so as to evaporate at reasonable rate at room temperature or slightly above. Also, the solvent must not vaporize rapidly; otherwise, it leads to form bubbles or semicrystalline precipitate. The rapid volatilization also causes cooling of the film, which could then cause crazing or condensation of water from the atmosphere. To get good films, usually the solvents must have the boiling points between 60 and 100 °C.

Generally, the polymer solution needs to be filtered before casting and filtration must be done by pressure techniques instead of gravity or vacuum filtration. In a small scale, the polymer film of uniform thickness can be prepared in a laboratory just by spreading the polymer solution over the glass surface and rolled with a glass rod as shown in Figure 1.30.

On a large scale, the process is slightly different. Commercially available polymer films can be prepared by solution casting method. In this process, the polymer solution is fed continuously through slit die, which is then passed in between two oppositely rotating metal drums. During rotation, the solvent gets evaporated and the dry films of polymer are formed (Figure 1.31). In another approach, the polymer solution is fed continuously through slit die onto a moving metal belt. The solvents easily get evaporated during rotation and the dry film of the polymer is obtained. This process is shown in Figure 1.32.

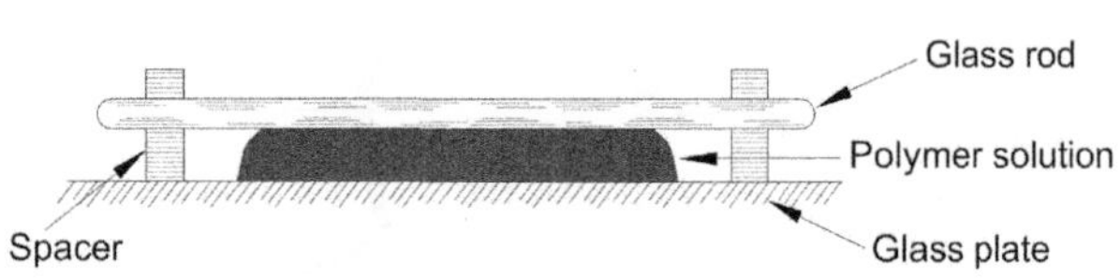

FIGURE 1.30 Laboratory-scale preparation of film by solution casting. (For color version of this figure, the reader is referred to the online version of this chapter.)

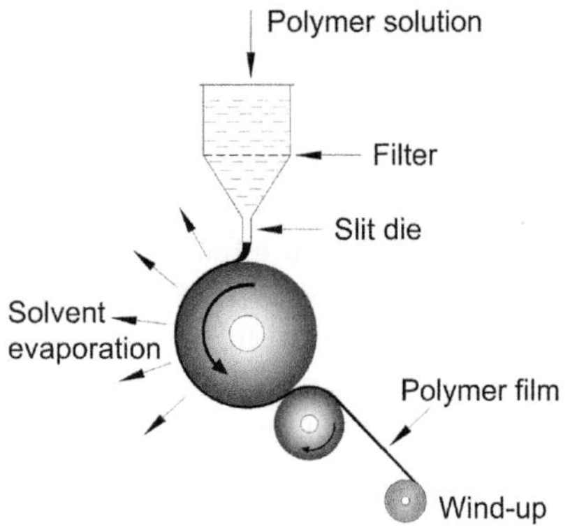

FIGURE 1.31 Large-scale preparation of film by solution casting. (For color version of this figure, the reader is referred to the online version of this chapter.)

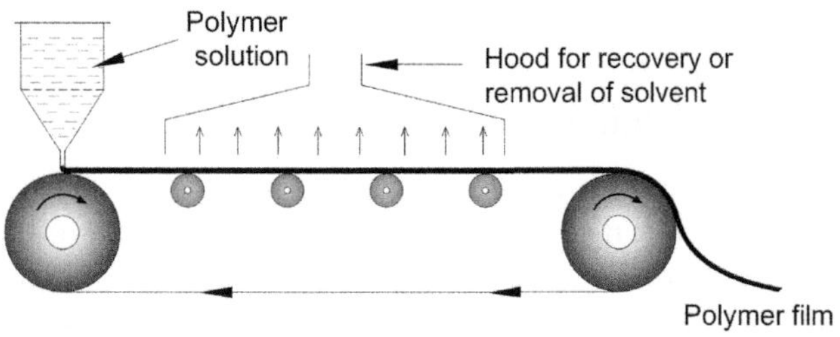

FIGURE 1.32 Preparation of films using moving belt system. (For color version of this figure, the reader is referred to the online version of this chapter.)

1.5.1.2 *Melt Pressing*

This method is suitable for thermally stable polymers. It is more often used in the laboratory than in the manufacturing plant, because large films are difficult to prepare by this method and the process is discontinuous (Figure 1.33). This technique consists of two electrically heated platens. One platen can be forced against the other by means of a hydraulic unit or hand pump. The polymer powder is placed between two sheets of aluminum or copper foil, and this sandwich is placed between the two heated platens. A hydraulic pressure of about 2000 to 5000 psi is applied for about 30 s. The sandwich is cooled and removed, and the film is separated from the foil. In practice, the temperature and pressure must be determined by trial and error. If the temperature is too high, the polymer simply flows out of the sandwich. If it is too low, the film may be opaque or weak because of inadequate fusion.

1.5.1.3 *Melt Extrusion*

For the manufacturing of any materials, the continuous processes are generally preferred. Therefore, melt extrusion method is more advantageous as it is a continuous process. In this method, the polymer powder or pellets are fed into a screw extruder that heats up the polymer by means of a rotating screw spindle as shown in the Figure 1.34. The

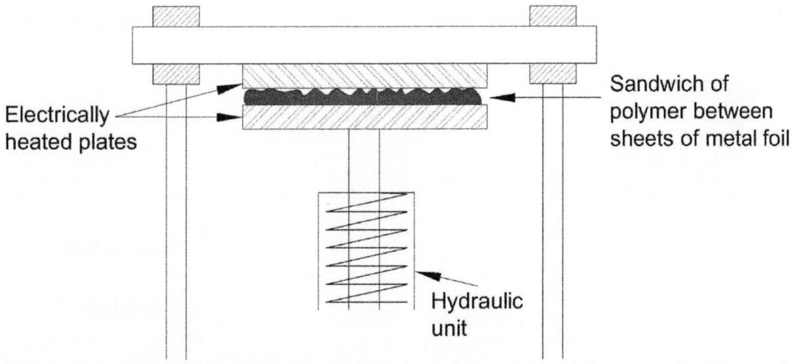

FIGURE 1.33 Preparation of films by melt pressing. (For color version of this figure, the reader is referred to the online version of this chapter.)

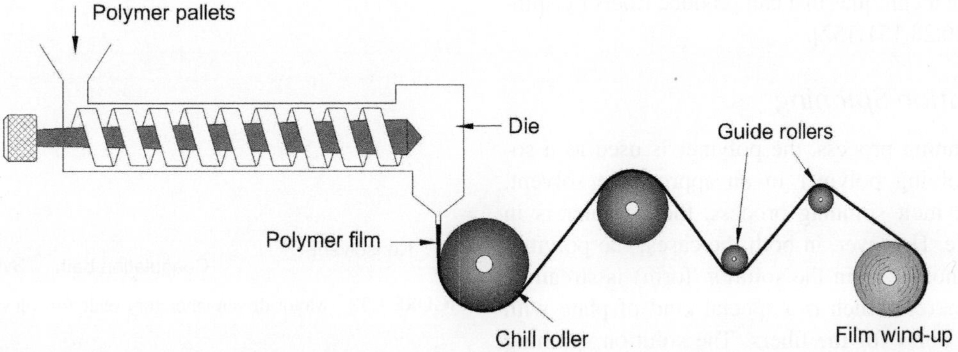

FIGURE 1.34 Preparation of film by melt extrusion. (For color version of this figure, the reader is referred to the online version of this chapter.)

molten polymer is forced into the die and the flat sheet of molten polymer is collected by a rotating drum, which cools the films below its melting point. The subsequent rollers complete the cooling effect and orientation process. The thickness of the film ranging from 0.01 to 0.1 mm can be obtained by this method.

1.5.1.4 Bubble Blown Method

This method can be used as an alternative to melt extrusion method. The polymer powder is fed into a screw extruder, which heats up the polymer. The molten polymer from screw extruder is forced through an annular die. An inert gas is then passed through the tube, so that due to gas pressure, the tube blows into a cylindrical bubble. The bubble is then flattened by a series of rollers to form a continuous film that can be wound up by specially designed rollers. This process is shown schematically in Figure 1.35.

1.5.2 Spinning Industrial Polymers

Spinning process previously was confined to textile industry for the production of fibers. Nowadays, it is extended to production of thin polymeric fibers. These thin fibers have enormous biomedical applications like tissue engineering, drug release, wound dressing, enzyme immobilization,

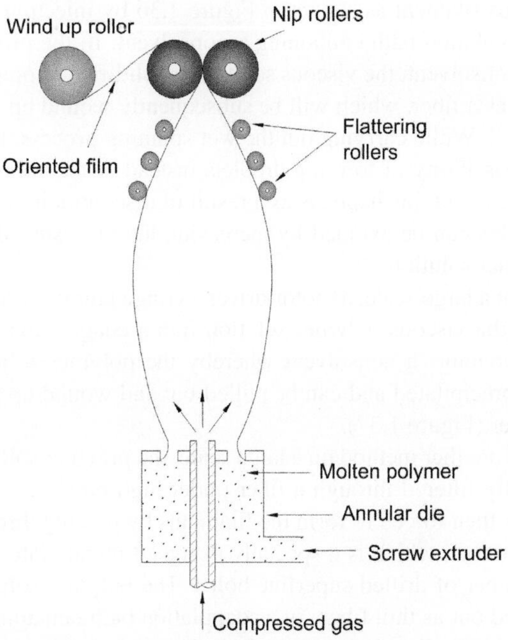

FIGURE 1.35 Bubble blown setup for preparation of film. (For color version of this figure, the reader is referred to the online version of this chapter.)

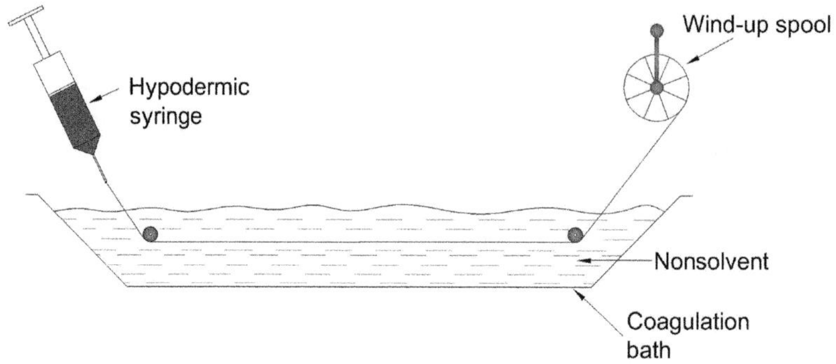

FIGURE 1.36 Laboratory scale for wet spinning of fibers. (For color version of this figure, the reader is referred to the online version of this chapter.)

biosensors, dental restoration, and medical implants. There are mainly three techniques that can produce fibers by spinning process [26,28,151,152].

1.5.2.1 Solution Spinning

In solution spinning process, the polymer is used as a solution by dissolving polymer in an appropriate solvent, whereas in the melt spinning process, the polymer is in the molten state. However, in both the cases, the polymer (either in the molten or in the solution form) is streamed through a spinneret, which is a special kind of plate with extremely fine holes for the fibers. The solution spinning may be of three types.

1.5.2.1.1 Wet Spinning

In this method, a viscous solution of the polymer is taken in a hypodermic syringe and then it is extruded through a continuous filament as shown in Figure 1.36 by injecting into a coagulation bath containing a nonsolvent. In the process of a nonsolvent, the viscous solution is solidified or precipitated as a fiber, which will be subsequently wound up with a spool. While carrying out the wet spinning process, there is a possibility of forming droplets instead of a continuous filament and this happens as a result of discontinuity in the jet. This can be avoided by increasing the viscosity of the polymer solution.

On a large scale, a motor-driven syringe pump is used to push the viscous polymer solution into a coagulation bath that contains a nonsolvent whereby the polymer solution gets precipitated and can be pulled out and wound up with a roller (Figure 1.37).

In another method on a large scale, the polymer solution initially filtered through a filter under high pressure pump and is then forced to form the filaments by passing through a spinneret, which is a specially designed metal plate with a number of drilled superfine holes. The polymer solution ejected out as thin fibers in a coagulation bath containing a nonsolvent. The fibers formed can be washed followed by finishing and finally wound up by means of a roller. This is illustrated in Figure 1.38.

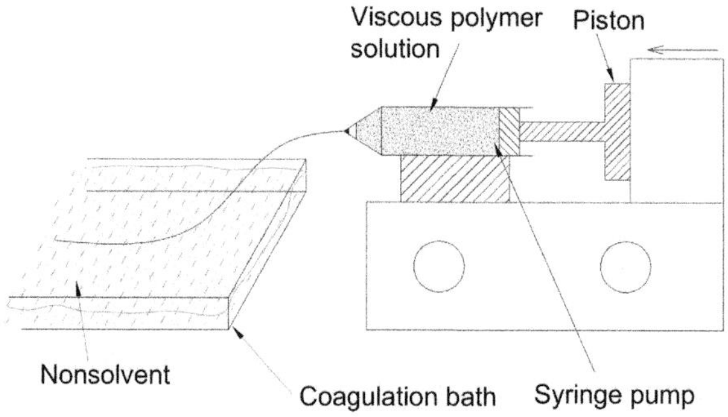

FIGURE 1.37 Motor-driven laboratory scale for wet spinning of fibers.

1.5.2.1.2 Electrospinning

The recent progress in nanotechnology is being extensively applied in diverse biomedical fields and particularly in the generation of scaffolds for tissue engineering, drug delivery mechanism, and enzyme immobilization [153,154]. The tissue engineering has played vital role in the development of biological substitutes that restore, maintain, or improve tissue functions. This is due to the generation of nanofibers produced by electrospinning technique. Therefore, electrospinning is an important process in the era of present modern technology. In this process, an electrical charge is used to draw fine (typically *micro-* to *nano*scale) fibers by choosing suitable polymer solution. This process does not require the use of coagulation bath or high temperatures to produce thin fibers from polymer solution and ensures that no solvent can be carried over into the product. Semicrystalline polymer fibers such as polyethylene, PET, and polypropylene can be produced by this method, which would be otherwise impossible or difficult to produce such fibers using other spinning techniques.

The setup is similar to that employed in conventional spinning process that has a syringe, high voltage supply, and the collector (Figure 1.39). When sufficiently high voltage is applied to a syringe needle, the body of the liquid becomes charged and electrostatic repulsion counteracts the surface tension and the droplet is stretched. At a critical point, a stream of polymer liquid erupts from the surface. The point of eruption is known as the "Taylor cone." If the

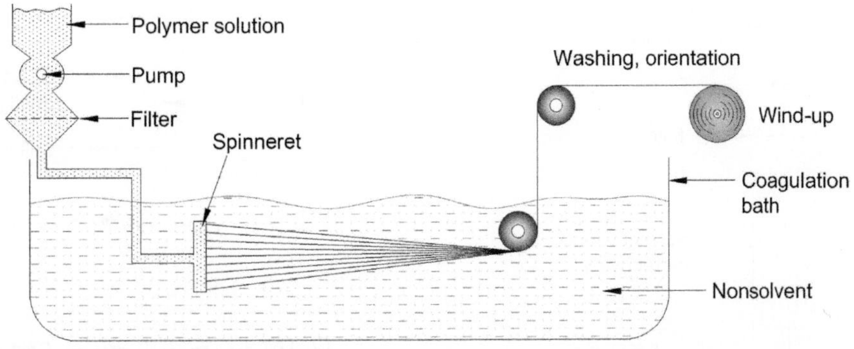

FIGURE 1.38 Large-scale setup for wet spinning of fibers. (For color version of this figure, the reader is referred to the online version of this chapter.)

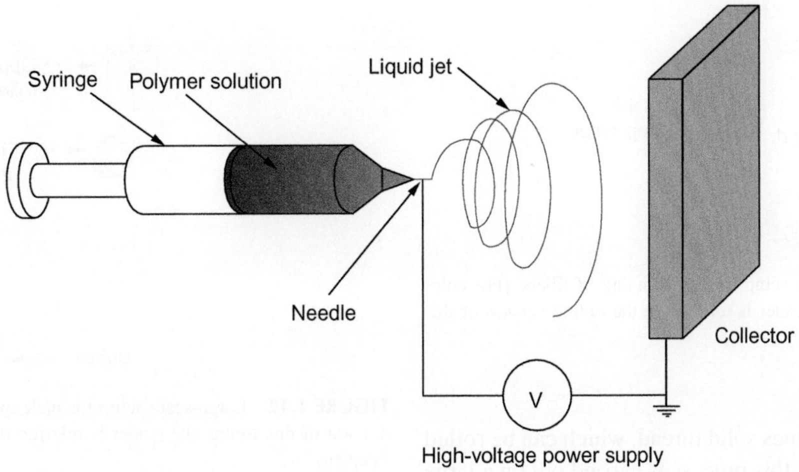

FIGURE 1.39 Electrospinning setup for thin fibers. (For color version of this figure, the reader is referred to the online version of this chapter.)

molecular cohesion of the liquid is sufficiently high, stream breakup does not occur and a charged liquid jet is formed. The jet is elongated by a whipping process caused by electrostatic repulsion initiated at small bends in the fiber, until it is finely deposited on the collector. The elongation and thinning of the fiber resulting from this bending instability leads to the formation of uniform fibers with *nano*meter-scale diameter [155,156]. The thickness of fiber can be varied depending on the properties of polymer, nature of solvent, applied voltage, distance employed between tip of syringe and the collector, etc. [157].

1.5.2.1.3 Dry Spinning

In the dry spinning method, a polymer fiber is formed without using coagulation bath. Instead, hot air is used to dry the fiber, which is extruded from the hypodermic syringe. On a laboratory scale as shown in Figure 1.40, the method is very simple in which the polymer solution is taken into a syringe and pushed down in a vertical cylindrical tube and simultaneously blowing hot air from the bottom to top. When the hot air is blown into the chamber, the flowing

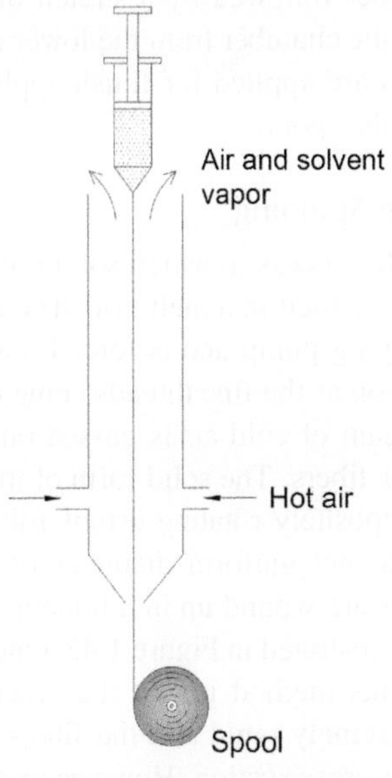

FIGURE 1.40 Laboratory setup for dry spinning of fibers. (For color version of this figure, the reader is referred to the online version of this chapter.)

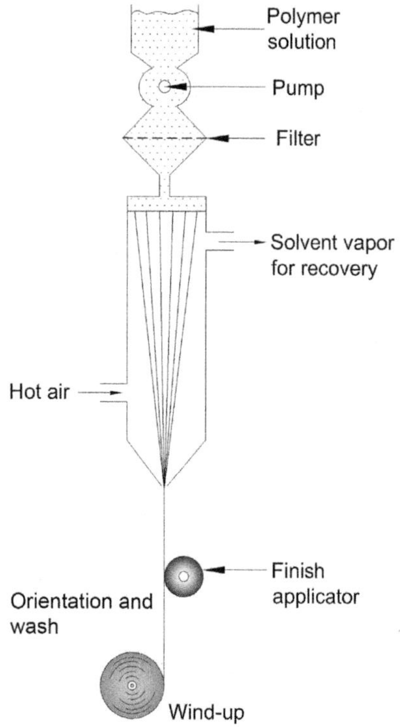

FIGURE 1.41 Large-scale setup for dry spinning of fibers. (For color version of this figure, the reader is referred to the online version of this chapter.)

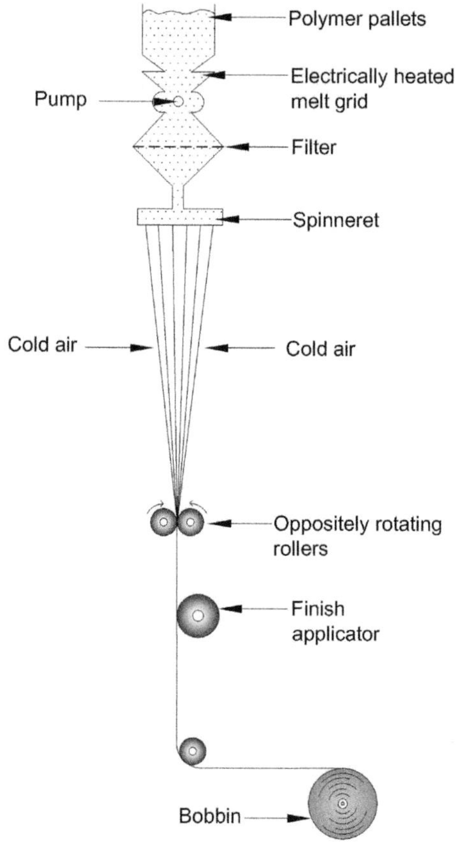

FIGURE 1.42 Large-scale setup for melt spinning of fibers. (For color version of this figure, the reader is referred to the online version of this chapter.)

polymer solution becomes solid thread, which can be rolled with a spool. However, this process is carried out on a large scale by employing the method illustrated in Figure. 1.41. Here, relatively concentrated polymer solution is filtered and pumped through a spinneret. The fibers pass down in a vertical chamber followed by a stream of hot air that is passed through the chamber from the lower end of the tube. The dried fibers are applied for finish applicator and then wound up with the spool.

1.5.2.1.4 Melt Spinning

It is a large-scale process in which solid polymer pellets or chips are heated to melt in a melt grid. The melt solution is filtered by applying pump and is forced down through the spinneret. As soon as the fine threads come out through the spinneret, a stream of cold air is passed on the threads so as to solidify the fibers. The solid form of threads is passed between two oppositely rotating drums followed with finish applicator to get uniform diameter of the fibers and finally the fibers are wound up in a bobbin. The melt spinning process is illustrated in Figure 1.42. One of the greatest advantages of this method is that the process of forming the fibers is extremely rapid and the fibers formed have a uniform circular cross section. However, a disadvantage of this process is that it cannot be applied for the thermally unstable polymers.

1.5.3 Fabrication of Shaped Polymer Objects

Preparation of three-dimensional articles or shaped objects using suitable molds is of great importance in biomedical applications. The applications include artificial heart, kidney, hip, knee, finger, and contact lenses. The molds are made up of either glass or metal. In many situations, the mold-releasing agent is used to release the three-dimensional objects from the molds after the complete curing. There are several molding techniques [26,28,151,152], and out of which, important methods are discussed in the succeeding text.

1.5.3.1 Compression Molding

This molding process is widely used to produce objects from thermosetting materials such as resins. The mold is made up of two halves, one is upper and other is lower. The lower halve usually contains a cavity and the upper halve has a projection, which fits into the cavity when the mold is closed. The gap between the projected upper half and the cavity in the lower one gives the shape of the molded device.

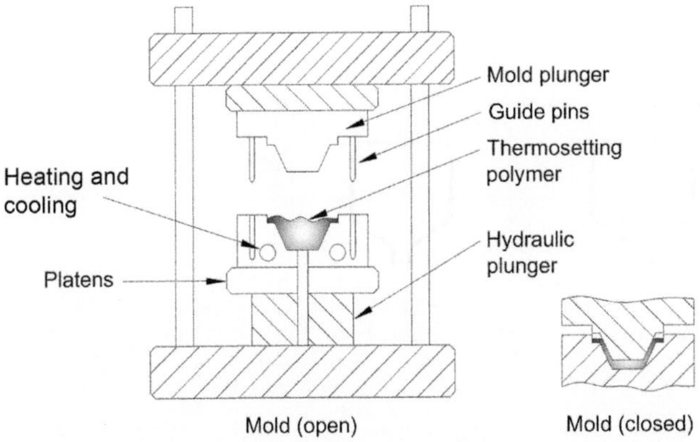

FIGURE 1.43 Preparation of objects by compression mold. (For color version of this figure, the reader is referred to the online version of this chapter.)

During the process of compression molding, the polymer material is subjected in between stationary and movable part of the mold. The heat and pressure are applied simultaneously when the mold is closed so that the material becomes plastic, which flows to fill the mold and becomes homogeneous. Then again, necessary pressure and temperature are applied considerably depending on the thermal and rheological properties of the polymer. Under high pressure, the excess polymer material expelled out as a thin film from the cavity is known as the flash. The molds are held together until all the resin has cured and which may require 30 s to several minutes. The mold is then opened and the object is ejected. In this method, the molds are generally made of chromium-plated metal. The temperature of 150 °C and pressure of about 200 psi are usually employed. This setup is illustrated in Figure 1.43.

1.5.3.2 Injection Molding

It is a high-speed technique used for the fabrication of both thermoplastic and thermosetting polymers. The process is shown in Figure 1.44. The polymer is in the form of powder or liquid resin, which is fed into a hopper and forced into a horizontal cylinder, where it gets softened due to electrical heating plates. The pressure is applied through a hydraulically driven piston to the molten material into a mold, fitted at the end of the cylinder. While moving through a hot zone of the cylinder, a device called *torpedo* helps to speed the plastic material uniformly around the inside wall of the hot cylinder. It can therefore ensure uniform heat distribution on the material. The molten material from the cylinder is then injected through a nozzle into the mold cavity. The polymer after cooling or curing gets solidified and the product of a particular shape can be taken out. This cycle is repeated and the sequence usually takes only about 10-30 s so that the method is quite suitable for bulk production.

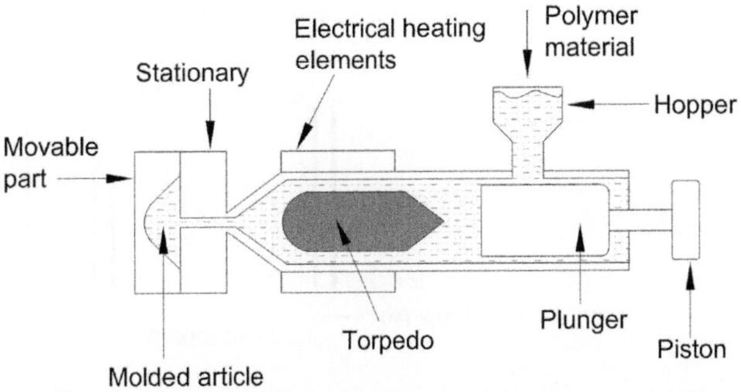

FIGURE 1.44 Preparation of objects by injection mold. (For color version of this figure, the reader is referred to the online version of this chapter.)

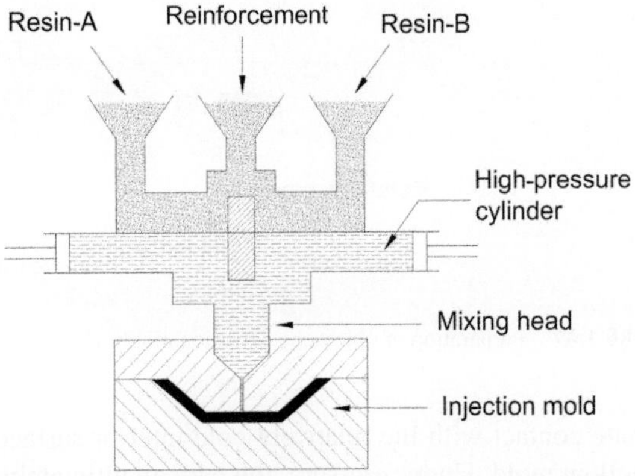

FIGURE 1.45 Preparation of objects by reaction injection mold.

1.5.3.3 Reaction Injection Molding

It is a newer development of injection molding. In this process as shown in Figure 1.45, two monomers (resins) are injected together in the mold. Just before they enter in the mold, a chemical reaction takes place between the two resins. At low temperature, a polymer is formed within less span of time and this process does not require more heat for the molding. This molding method is employed in the manufacture of human body parts.

1.5.3.4 Blow Molding

Most of the hollow biomedical articles are now produced by this technique. Blow molding, in fact, basically belongs to glass industry and it has been also adopted in the polymer industry since it is inexpensive and involves a simple operation. In this method, a piece of molten polymer tube usually called *parison* is properly placed from an extruder in between two halves of a mold. When the two halves of the mold are closed, it pinches or closes one end of the parison and a blowing pin is inserted at the other end. Compressed gas under pressure of about 2 to 100 psi is then injected into a parison through a blowing pin. The hot parison is inflated like a balloon and goes on expanding until it comes in

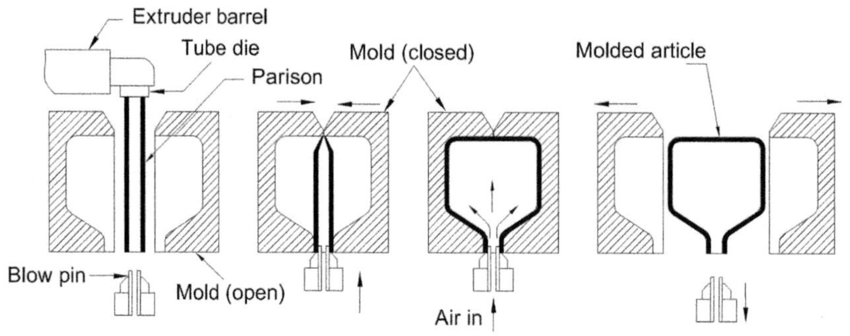

FIGURE 1.46 Preparation of objects by blow mold.

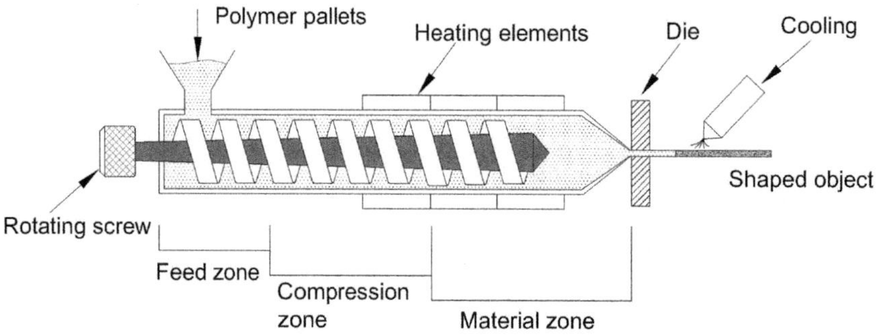

FIGURE 1.47 Preparation of objects by extrusion mold. (For color version of this figure, the reader is referred to the online version of this chapter.)

intimate contact with the relatively cold interior surface of the hollow mold. Under pressure, the parison ultimately assumes the shape of the hollow cavity of the mold. The mold is allowed to cool and rigid article is removed by opening the mold. The operation is repeated in order to get a more number of objects. The typical diagram of blow molding process is shown in Figure 1.46.

1.5.3.5 Extrusion Molding

This is one of the economical methods for producing many common biomaterials in continuous length such as sheets, tubes, rods, and filaments. In this process as illustrated in Figure 1.47, the plastic material is fed into a cylinder through a hopper. The cylinder is having a provision of electrical heating for softening the material. The molten material is further forwarded into a die by means of revolving screw. During its processing from the hopper to the die, the plastic material passes through three distinct zones such as feed zone, compression zone, and material zone. Each zone contributes in its own way to the overall extrusion process. The feed zone receives the material from the hopper and sends it over the compression zone. No heating takes place in the feed zone, but in the compression zone, the material melts due to heat produced by the heating elements and also the material is compressed by operation of the screws. Further, the pasty molten material is then sent to the material section, where it acquires a constant flow rate imparted by the helical flight of the screw. The pressure limits up in this section enables the melt material to enter the die and emerges out of it with the desired profile. The desired profile emerging from the die is quite hot and is subjected to rapid cooling to avoid deshaping. The product formed is then cut to the desired length or wound up in rolls.

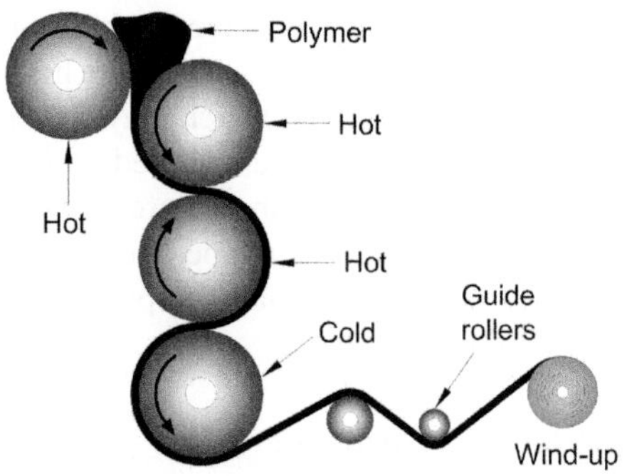

FIGURE 1.48 Preparation sheet/film by calendaring. (For color version of this figure, the reader is referred to the online version of this chapter.)

1.5.4 Calendaring

It is an important method of making film and sheet and is shown in Figure 1.48. In this process, the plastic compound is passed between a series of heated rollers. The sheet emerging from the rollers is cooled by passing through cold rollers. The sheets are finally wound up in rolls. However, the thickness is controlled by a continuation of squeezing and altering the speed of the finishing rolls.

1.6 FUTURE PERSPECTIVES

With respect to the aforementioned discussions, it can be noted that despite the apparent proliferation of biomaterials in the area of medical science, the science and technology of biomaterials is still in its early stages of development. Tremendous opportunities exist and will continue to exist for the penetration of biopolymers in every facet of biomedical field through intensive research and development. The level of research and developmental activity in the general field of biopolymers has been rising dramatically all over the world during the past two decades. This is happening not only because of growing recognition of commercial potentiality of biomaterials but also because of the growing realization of the international scientific and engineering community that biopolymers are capable of playing tremendously beneficial role in the society in general and biomedical science, in particular, for more than what has hitherto been appreciated. This is particularly a significant and welcoming realization for the health and well-being of all humanity. An appreciable example is the biodegradable PCL, which is recently being used as scaffold for the regeneration of lost or defect organ through the development of nanofibers using revisited electrospinning technique. Therefore, the fullest potentiality and practical utilization of all developed biopolymers for application in everyday life and particularly in the field of medical science can be achieved only through sincere efforts and commitment.

1.7 CONCLUSIONS

From the literature, it is concluded that biopolymers play a vital role in meeting the demands of medical science. The methods such as addition, condensation, and metathesis polymerization have been discussed by taking suitable examples. The cross-linked thermoset biopolymers can also be synthesized using vinyl polymerization method. A detailed account on both homogenous and heterogeneous polymerization techniques is explained. An attempt has been made for the compilation of the most recent biopolymers employed in biomedical applications such as tissue engineering, drug release, and dental materials. Biopolymers including PMMA as a main component of contact lens and cosmetic surgery; PLA, PLG, and PLGA in fracture fixation devices; polycyanoacrylates as *super glue*; PCL as a scaffold for bone regeneration; PUs in cardiovascular devices; and polydioxanone in surgical suture and drug delivery have been dealt with in detail. In polymerization techniques, the processing conditions to improve the yield, molecular weight, and purity of the resulting polymers have been discussed. A special attention has been devoted to general properties of biopolymers, synthesis/extraction, and their applications in biomedical science, which could play an important role in developing medical devices. To develop devices depending on the requirements for biomedical applications, fabrication methods that are important have been discussed with illustrations. The processing conditions including the significance of torpedo and viscosity of polymers have also been addressed suitably.

ACKNOWLEDGMENTS

The authors wish to thank the UGC, New Delhi, for providing the financial support under UPE-FAR-I program (Contract No. 14-3/2012 [NS/PE]). The authors are grateful to the Department of Tool Design, NTTF, Dharwad, for neatly designing the polymer processing illustrations.

REFERENCES

[1] J. Jagur-Grodzinski, Biomedical application of functional polymers, React. Funct. Polym. 39 (1999) 99–138.

[2] J.C. Middleton, A.J. Tipton, Synthetic biodegradable polymers as orthopaedic devices, Biomaterials 21 (2000) 2335–2346.

[3] J. Jagur-Grodzinski, Polymers for tissue engineering, medical devices, and regenerative Medicine, Concise general review of recent studies, Polym. Adv. Technol. 17 (2006) 395.

[4] M.E. Gomes, R.L. Reis, Biodegradable polymers and composites in biomedical applications: from catgut to tissue engineering. Part 1 Available systems and their properties, Int. Mater. Rev. 49 (2004) 261–273.

[5] L. Durselen, M. Dauner, H. Hierlemann, H. Planck, L.E. Clases, A. Igantius, Resorbable polymer fibers for ligament augmentation, J. Biomed. Mater. Res. 58 (2001) 666–672.

[6] L.S. Nair, C.T. Laurencin, Polymers as biomaterials for tissue engineering and controlled drug delivery, Adv. Biochem. Eng. Biotechnol. 102 (2006) 47–90.

[7] M.H. Sheridan, L.D. Shera, M.C. Peters, D.J. Mooney, Bioabsorbable polymer scaffolds for tissue engineering capable of sustained growth factor delivery, J. Control. Release 64 (2000) 91–102.

[8] H.C. Kang, M. Lee, Y.H. Bae, Polymeric gene carriers, Crit. Rev. Eukaryot. Gene Expr. 15 (2005) 317–342.

[9] D.G. Anderson, A. Akinc, N. Hossain, R. Langer, Structure/property studies of polymeric gene delivery using a library of poly(bet-amino esters), Mol. Ther. 11 (2005) 426–434.

[10] M. Navarro, A. Michiradi, O. Castano, J.A. Planell, Biomaterials in Orthopedic, J. R. Soc. Interface 5 (2008) 1137–1158.

[11] Q. Lin, Naturally occurring polymer biomaterials, in: S. Donglu (Ed.), Introduction to Biomaterials, World Scientific Publishing Co. Pte. Ltd., Singapore, 2006, pp. 158–171.

[12] K.A. Athanasiou, C.E. Agrawal, F.A. Barber, S.S. Burkhart, Orthopaedic applications for PLA, PGA biodegradable polymers, Arthroscopy 14 (1998) 726–737.

[13] E.A. Wang, V. Rosen, J.S. D'Alessandro, M. Bauduy, P. Cordes, T. Harada, D.I. Isreal, R.M. Hewick, K.M. Kerns, P. LaPan, D. Luxenberg, D. McQuaid, I.K. Moutsatsos, J. Nove, J.M. Wozney, Recombinant human bone morphogenic protein induces bone formation, Proc. Natl. Acad. Sci. U.S.A. 87 (1990) 2220–2224.

[14] N.A. Gharibjanian, W.C. Chua, S. Dhar, T. Scholz, T.Y. Shibuya, G.R. Evans, J.W. Calvert, Release kinetics of polymer bound

bone morphogenetic protein-2 and its effects on the osteogenic expression of MC3T3-E1 osteoprecursor cells, Plast. Reconstr. Surg. 123 (2009) 1169–1177.
[15] M. Azori, Polymeric prodrugs, Crit. Rev. Ther. Drug Carrier Syst. 4 (1987) 39–65.
[16] V.F.M.H. de Groot, Prodrugs: challenges and rewards, in: V.J. Stella, R.T. Borchardt, M.K. Hageman, R. Oliyai, H. Maag, J.W. Tilley (Eds.), Targeting – Cancer – Small Molecules Springer, USA, 2007, pp. 478.
[17] J. Nicholas, S. Mura, D. Brambilla, N. Mackiewicz, P. Coreur, Design functionalization strategies and biomedical applications of targeted biodegradable. Biocompatible polymer based nanocarriers for drug delivery, Chem. Soc. Rev. 42 (2013) 1147–1235.
[18] M. Ramchandani, D. Robinson, In vitro and in vivo release of ciprofloxacin from PLGA 50:50 implants, J. Control. Release 54 (1998) 167–175.
[19] Z. Xia, T. Yoshida, M. Funaoka, Enzymatic degradation of highly phenolic lignin-based polymers (lignophenols), Eur. Polym. J. 39 (2003) 909–914.
[20] R.J. Young, Introduction to Polymers, Chapman & Hall, USA, 1987.
[21] S.E.M. Selke, J.D. Culter, R.J. Hernandez, Plastics Packaging: Properties, Processing, Applications, and Regulations, Hanser Publishers, Munich, 2004, pp. 29.
[22] http://www2.chemistry.msu.edu/faculty/reusch/virttxtjml/polymers.html.
[23] B.D. Ulery, L.S. Nair, C.T. Laurencin, Biomedical applications of biodegradable polymers, J. Polym. Sci. B Polym. Phys. 49 (2011) 832–886.
[24] K.J. Ivin, I.C. Mol, Olefin Metathesis Polymerization, second ed., Academic, New York, 1996.
[25] E.J. Vandenberg, Catalysis in polymer synthesis, in: E.J. Vandenberg, J.C. Salamone (Eds.), ACS Symposium Series 496, American Chemical Society, Washington DC, 1992.
[26] A.J. Peacock, Allison Calhoun, Polymer Chemistry Properties and Applications, Hanser Publishers, Munich, 2006, pp. 36–38.
[27] M.P. Stevens, Polymer Chemistry an Introduction, third ed., Oxford University Press, New York, 1999, pp. 173–175.
[28] P. Bahaddur, N.V. Sastry, Principles of Polymer Science, Narosa Publishing House, New Delhi, 2002, pp. 43–44.
[29] C.G. Pitt, Polycaprolactone and its copolymers, in: M. Chasin, R. Langer (Eds.), Biodegradable Polymers as Drug Delivery Systems, Dekker, New York, 1990, pp. 71–119.
[30] A. Nathan, J. Kohn, Amino acid derived polymers, in: S.W. Shalaby (Ed.), Biomedical Polymers: Designed to Degrade Systems, Hanser Publication, New York, 1994, pp. 117–151.
[31] M.Y. Kariduraganavar, F.J. Davis, G.R. Mitchell, R.H. Olley, Using an additive to control the electrospinning of fibres of poly(ε-caprolactone), Polym. Intl. 59 (2010) 827–835.
[32] R.D. Lundberg, J.V. Koleske, K.B. Wischmann, Lactone polymers. III. Polymerization of ε-caprolactone, J. Polym. Sci. [A1] 7 (1969) 2915–2930.
[33] M. Labet, W. Thielemans, Synthesis of polycaprolactone: a review, Chem. Soc. Rev. 38 (2009) 3484–3504.
[34] C. St. Louis, Drug for Adults is Popular as Children's Remedy, The New York Times, New York, 2012.
[35] J.A. Di Palma, M.V. Cleveland, J. McGowan, J.L. Herrera, A randomized, multicenter comparison of polyethylene glycol laxative and tegaserod in treatment of patients with chronic constipation, Am. J. Gastroenterol. 102 (2007) 1964–1971.
[36] T.L. Krause, G.D. Bittner, Rapid morphological fusion of severed myelinated axons by polyethylene glycol, Proc. Natl. Acad. Sci. U.S.A. 87 (1990) 1471–1475.
[37] S.C. Smolinske, Handbook of Food, Drug, and Cosmetic Excipients, CRC Press, Boca Raton, 1992, p. 287.
[38] J. Kovar, Y. Wang, M.A. Simpson, D.M. Olive, Imaging lymphatics with a variety of near-infrared-labeled optical agents, in: World Molecular Imaging Conference, 2009.
[39] D.E. Corpet, G. Parnaud, M. Delverdier, G. Peiffer, S. Tache, Consistent and fast inhibition of colon carcinogenesis by polyethylene glycol in mice and rats given various carcinogens, Cancer Res. 60 (2000) 3160–3164.
[40] R.B. Borgens, D. Bohnert, Rapid recovery from spinal cord injury after subcutaneously administered polyethylene glycol, J. Neurosci. Res. 66 (2001) 1179–1186.
[41] G. Bittner et al., Melatonin enhances the in vitro and in vivo repair of severed rat sciatic axons, Neurosci. Lett. 376 (2005) 98–101.
[42] A.C. French, A.L. Thompson, B.G. Davis, High purity discrete PEG oligomer crystals allow structural insight, Angew. Chem. Int. Ed. 48 (2009) 1248–1252.
[43] J. Kahovec, R.B. Fox, K. Hatada, Nomenclature of regular single-strand organic polymers, Pure Appl. Chem. 74 (2002) 1921–1956.
[44] Polyurethane.americanchemistry.com/Polyurethanes and Medical Applications.
[45] R.J. Zdrahala, I.J. Zdrahala, Biomedical applications of polyurethanes: a review of past promises, present realities and a vibrant future, J. Biomater. Appl. 14 (2000) 67–90.
[46] P.W. Morgan, Condensation Polymers: By Interfacial and Solution Methods, Wiley-Interscience, New York, 1965, p. 44.
[47] K.N. Cologne, K.N. Bergisch-Gladbach, J.P. Cologne, Process for the preparation of polyurethanes which are dispersible in water, U. S. Patent 4 237 264 (1980).
[48] E.D. Boland, B.D. Coleman, C.P. Barnes, D.G. Simpson, G.E. Wnek, G.L. Bowlin, Electrospinning polydioxanone for biomedical applications, Acta Biomater. 1 (2005) 115–123.
[49] J. Middleton, A. Tipton, Synthetic biodegradable polymers as medical devices, Med. Plast. Biomater. Mag. (March, 1998).
[50] N. Tiberiu, Concepts in biological analysis of resorbable materials in oro-maxillofacial surgery, Rev. Chir. Oro-maxilo-fac. Implantol. 2 (2011) 33–38.
[51] S. Kang, H. Lin, D.E. Day, J.O. Stoffer, Optically transparent polymethyl methacrylate composites made with glass fibers of varying refractive index, J. Mater. Res. 12 (1997) 1091–1101.
[52] J.P. Chen, T. Kodaira, K. Isa, T. Senda, Simultaneous TG-MS studies on polymers derived from monomers with a polar group, J. Mass Spectr. Soc. Jpn. 49 (2001) 41–50.
[53] L. Qiliu, H.D. Wagner, A comparison of the mechanical strength and stiffness of MWNT-PMMA and MWNT-epoxy nanocomposites, Compos. Interfaces 14 (2007) 285–297.
[54] K. Myer, Handbook of Materials Selection, John Wiley & Sons, New York, 2002, p. 341.
[55] R.A. Meyers, Molecular Biology and Biotechnology: A Comprehensive Desk Reference, Wiley-VCH, Germany, 1995, p. 722.
[56] T.J. Kaufmann, M.E. Jensen, G. Ford, L.L. Gill, W.F. Marx, D.F. Kallmes, Cardiovascular effects of polymethylmethacrylate use in percutaneous vertebroplasty, Am. J. Neuroradiol. 23 (2002) 601–604.
[57] D. Blond, V. Barron, M. Ruether, K.P. Ryan, V. Nicolosi, W.J. Blau, J.N. Coleman, Enhancement of modulus, strength, and toughness in poly(methyl methacrylate)-based composites by the incorporation of poly(methyl methacrylate)-functionalized nanotubes. Adv. Funct. Mater. 16 (2006) 1608–1614, doi:10.1002/adfm.200500855.
[58] J.Y. Rho, Young's modulus of trabecular and cortical bone material: ultrasonic and microtensile measurements, J. Biomech. 26 (1993) 111–119.

[59] M.D. Miller, Review of Orthopedics, fourth ed., W.B. Saunders, Philadelphia, 1996, p. 129.
[60] J.A. Moore, Macromolecular Synthesis, John-Wiley and Sons, New York, 1977, pp. 23–25.
[61] D.K. Gilding, A.M. Reed, Biodegradable polymers for use in surgery polyglycolic/poly (lactic acid) homo- and copolymers: 1, Polymer 20 (1979) 1459–1464.
[62] E. Schmitt, Polyglycolic acid in solutions, U.S. Patent 3 737 440, 1973.
[63] P.A. Gunatillake, R. Adhikari, Biodegradable synthetic polymers for tissue engineering, Eur. Cells Mater. 5 (2003) 1–16.
[64] C.E. Lowe, Preparation of high molecular weight polyhydroxyacetic ester, U.S. Patent 2 668 162, 1954.
[65] M. Bero, P. Dobrzynski, J. Kasperczyk, Application of calcium acetylacetonate to the polymerization of glycolide and copolymerization of glycolide with ε-caprolactone and L-lactide, Macromolecules 32 (1999) 4735–4737.
[66] K.M. Stridsberg, M. Ryner, A. Ann-Christine, Controlled ring-opening polymerization: polymers with designed macromolecular architecture, Adv. Polym. Sci. 157 (2002) 41–65.
[67] M. Epple, A detailed characterization of polyglycolide prepared by solid-state polycondensation reaction, Macromol. Chem. Phys. 200 (1999) 2221–2229.
[68] T. A. Masuda et al., Biodegradable plastic composition, U.S. Patent 5 227 415, 1993.
[69] G.L. Fiore, F. Jing, V.G. Young Jr., C.J. Cramer, M.A. Hillmyer, High Tg aliphatic polyesters by the polymerization of spirolactide derivatives, Polym. Chem. 1 (2010) 870–877.
[70] D. Garlotta, A literature review of poly(lactic acid), J. Polym. Environ. 9 (2001) 2.
[71] S. Anders, M. Stolt, Properties of lactic acid based polymers and their correlation with composition, Prog. Polym. Sci. 27 (2002) 1123–1163.
[72] J.C. Middelton, A.J. Tipton, Synthetic biodegradable polymers as orthopedic devices, Biomaterials 21 (2000) 2335–2346.
[73] S. Suzuki, Y. Ikada, Medical applications, in: R. Auras, L.T. Lim, S.E.M. Selke, H. Tsuji (Eds.), Poly(Lactic Acid): Synthesis, Structures, Properties, Processing, and Applications, John Wiley & Sons, New York, 2010, pp. 443–456.
[74] A. Södergård, M. Stolt, Industrial production of high molecular weight poly(lactic acid), in: R. Auras, L.T. Lim, S.E.M. Selke, H. Tsuji (Eds.), Poly(Lactic Acid): Synthesis, Structures, Properties, Processing, and Applications, John Wiley & Sons, New York, 2010, pp. 27–41.
[75] H.R. Kricheldorf, J.J. Michael, New polymer syntheses, Polym. Bull. 9 (1993) 6–7.
[76] M.J. Santander-Ortega, N. Csaba, M.J. Alonso, J.L. Ortega-Vinuesa, D. Bastos-Gonzalez, Stability and physicochemical characteristics of PLGA, PLGA:poloxamer and PLGA: poloxamine blend nanoparticles: a comparative study, Colloids Surf. A 296 (2007) 132–140.
[77] M. Studer, M. Briel, B. Leimenstoll, T.R. Glass, H.C. Bucher, Effect of different antilipidemic agents and diets on mortality: a systematic review, Arch. Intern. Med. 165 (2005) 725–730.
[78] D.A. Vorp, T. Maul, A. Nieponice, Molecular aspects of vascular tissue engineering, Front. Biosci. 10 (2005) 768–789.
[79] F.W. Lichtenthaler, Carbohydrates as organic raw materials, in: G. Bellussi, et al. (Eds.), Ullmann's Encyclopedia of Industrial Chemistry, Wiley-VCH, Weinheim, 2011, pp. 1–33.
[80] Y. Poirier, C. Somerville, L.A. Schechtman, M.M. Satkowski, I. Noda, Synthesis of high-molecular-weight poly([R]-(-)-3-hydroxybutyrate) in transgenic Arabidopsis thaliana plant cells, Int. J. Biol. Macromol. 17 (1995) 7–12.
[81] M. Lemoigne, Produits de dehydration et de polymerisation de l'acide ß-oxobutyrique, Bull. Soc. Chim. Biol. 8 (1926) 770–782.
[82] S. Alexander, Biopolymers, Wiley-VCH, New York, 2002.
[83] N.M. Bikales, H.F. Mark, G. Menges, C.G. Overberger, Encyclopedia of Polymer Science and Engineering, second ed., John Wiley & Sons, Inc., New York, 1985, vol. 1, pp. 275–276, vol. 2, pp. 17–20.
[84] W.H. Brown, C.S. Foote, Organic Chemistry, second ed., Saunders College Publishing, Pennsylvania, 1998, pp. 931–934.
[85] M.A. Ferreira, L.C. Sunback, A biodegradable and biocompatible gecko inspired tissue adhesive, Proc. Natl. Acad. Sci. U.S.A. 105 (2008) 2307–2312.
[86] H.A. Allcock, F.W. Lampe, Contemporary Polymer Chemistry, Prentice-Hall Inc., Englewood Cliffs, New Jersey, 1981, pp. 552–223.
[87] F. Haaf, A. Sanner, F. Straub, Polymers of N-vinylpyrrolidone: synthesis, characterization and uses, Polym. J. 17 (1985) 143–152.
[88] F. Fischer, S. Bauer, Ein Tausendsassa in der Chemie—polyvinylpyrrolidon, Chem. unserer Zeit. 43 (2009) 376–383.
[89] V. Bühler, Excipients for Pharmaceuticals—Povidone, Crospovidone and Copovidone, Springer, Berlin, Heidelberg, New York, 2005, pp. 1–254.
[90] M.D. Santhi Ganesan, Embolized crospovidone (poly[N-vinyl-2-pyrrolidone]) in: the lungs of intravenous drug users, Mod. Pathol. 16 (2003) 286–292.
[91] S.K. Das, S.K. Saha, A. Das, A.K. Halder, S.N. Banerjee, M. Chakraborty, A study of comparison of efficacy and safety of talc and povidone iodine for pleurodesis of malignant pleural effusions, J. Indian Med. Assoc. 106 (2008) 589–590, 592.
[92] P.K. Dutta, M.N.V. Ravikumar, Biomedical applications of chitin, chitosan, and their derivatives, J. Macromol. Sci. C Polym. Rev. 40 (2000) 69–83.
[93] B.G. Kozen, S.J. Kircher, J. Henao, F.S. Godinez, A.S. Johnson, An alternative hemostatic dressing: comparison of CELOX, HemCon, and QuikClot, J. Emerg. Med. 15 (2008) 74–81.
[94] A.E. Pusateri, S.J. McCarthy, K.W. Gregory, R.A. Harris, L. Cardenas, A.T. McManus, C.W. Goodwin Jr., Effect of a chitosan-based hemostatic dressing on blood loss and survival in a model of severe venous hemorrhage and hepatic injury in swine, J. Trauma 4 (2003) 177–182.
[95] T.A. Khan, K.K. Peh, H.S. Ch'ng, Mechanical, bioadhesive strength and biological evaluations of chitosan films for wound dressing, J. Pharm. Pharm. Sci. 3 (2000) 303–311.
[96] C. Hardy, L. Johnson, P. Luksch, Hemostatic material, US Patent, 8106030, 2012.
[97] P. Baldrick, The safety of chitosan as a pharmaceutical excipient, Regul. Toxicol. Pharmacol. 56 (2009) 290–299.
[98] A.S. Pandit, Hemostatic Wound Dressing, U.S. Patent 5836970, 1998.
[99] C. Hardy, A. Darby, G. Eason, Hemostatic Material, U.S. application WO2009130485, 2009.
[100] W. Tiyaboonchai, Chitosan nanoparticles: a promising system for drug delivery, Naresuan Univ. J. 11 (2003) 51–66.
[101] J.H. Park, G. Saravanakumar, K. Kim, I.C. Kwon, Targeted delivery of low molecular drugs using chitosan and its derivatives, Adv. Drug Deliv. Rev. 62 (2010) 28–41.
[102] J. Linden, R. Stoner, K. Knutson, C. Gardner-Hughes, Organic disease control elicitors, Agro Food Ind. Hi-Tech 24 (2000) 12–15.
[103] T. Freier, R. Montenegro, H.S. Koh, M.S. Shoichet, Chitin-based tubes for tissue engineering in the nervous system, Biomaterials 26 (2005) 4624–4632.

[104] P. Mokrejs, F. Langmaier, M. Mladek, D. Janacova, K. Kolomaznik, V. Vasek, Extraction of collagen and gelatine from meat industry by-products for food and non food uses, Waste Manag. Res. 27 (2009) 31–37.
[105] C.G.B. Cole, Gelatin, in: F.J. Francis (Ed.), Encyclopedia of Food Science and Technology, second ed., John Wiley & Sons, New York, 2000, pp. 1183–1188.
[106] Z. Zhang, G. Li, B. Shi, Physicochemical properties of collagen, gelatin and collagen hydrolysate derived from bovine limed split wastes, J. Soc. Leather Technol. Chem. 90 (2012) 23–28.
[107] A.G. Ward, A. Courts, The Science and Technology of Gelatin, New York Academic Press, New York, ISBN: 0-12-735050-0, 1977.
[108] C.R. Smith, Gelatin hand book, J. Am. Chem. Soc. 43 (1921) 1350.
[109] C. Hawkes, Nutrition Labels and Health Claims: The Global Regulatory Environment, World Health Organization, Geneva, ISBN: 92 4 159171 4, 2004.
[110] P. Morganti, S.D. Randazzo, C. Bruno, Effect of gelatin cysteine on hair after a three months treatment, J. Soc. Cosmet. Chem. 33 (1982) 95.
[111] S.D. Randazzo, P. Morganti, The influence of gelatin cysteine supplementation on the amino acids composition of human hair, accepted for presentation on XVI intern. Congress of Dermatology, May 23–28, 1982, Tokyo.
[112] M.G. Mulinos, E.D. Kadison, Effect of gelatin on the vascularity of the finger, Angiology 16 (1965) 170–176, SAGE—Apr 1.
[113] T. Lloyd, M.D. Tyson, The effect of gelatin on fragile finger nails, J. Invest. Dermatol. 14 (1950) 323–325.
[114] M. Jank, Gelatin therapy in onychomycoses, Wein. Med. Wochenschr. 24 (1968) 154–156.
[115] V.L. Campo, F.F. Kawano, D.B. Silva Jr., I. Carvalho, Carrageenans: biological properties, chemical modifications and structural analysis, Carbohydr. Polym. 77 (2009) 167–180.
[116] S. Distantina, Rochmadi, M. Fahrurrozi, Wiratni, Hydrogels based on carrageenan extracted from kappaphycus alvarezii, World Acad. Sci. Eng. Technol. 78 (2013) 981–984.
[117] S. Distantina, Wiratni, M. Fahrurrozi, Rochmadi, Carrageenan properties extracted from eucheuma cottonii, Indonesia, World Acad. Sci. Eng. Technol. 78 (2011) 738–742.
[118] D.A.T. Southgate, Dietary Fibre: Chemical and Biological Aspects, Royal Society of Chemistry, Woodhead Publisher, Norwich, England, 1990, p. 386.
[119] J.K. Tobacman, Review of harmful gastrointestinal effects of carrageenan in animal experiments, Environ. Health Perspect. 109 (2001) 983–994.
[120] J. Watt, R. Marcus, Danger of carrageenan in foods and slimming recipes, The Lancet 317 (1981) 338.
[121] J. Watt, R. Marcus, Harmful effects of carrageenan fed to animals, Cancer Detect. Prev. 4 (1981) 129–134.
[122] http://www.drweil.com/drw/u/QAA401181/Is-Carrageenan-Safe.html.
[123] J.A. Fernández-Romero, J.C. Abraham, A. Rodriguez, L. Kizima, N.J. Pierre, R. Menon, O. Begay, S. Seidor, B.E. Ford, P.I. Gil, J. Peters, D. Katz, M. Robbiani, T.M. Zydowsky, Zinc acetate/carrageenan gels exhibit potent activity *in vivo* against high-dose herpes simplex virus 2 vaginal and rectal challenge, Antimicrob. Agents Chemother. 56 (2012) 358–368.
[124] C.B. Buck, C.D. Thompson, J.N. Roberts, M. Muller, D.R. Lowy, J.T. Schiller, Carrageenan is a potent inhibitor, PLoS Pathog. 2 (671) (2006) 676–677.
[125] E.R. Meier, C. Jawad, M. Weinmüllner, R. Grassauer, A.E. Prieschl-Grassauer, Efficacy and safety of an antiviral Iota-carrageenan nasal spray: a randomized, double-blind, placebo-controlled exploratory study in volunteers with early symptoms of the common cold, Respir. Res. 11 (2010) 108.
[126] J.R.E. Frasher, T.C. Laurent, U.B.G. Laurent, Hyaluronan: its nature, distribution, functions and turnover, J. Intern. Med. 242 (27–33) (1997) 1997.
[127] R. Stern, Hyaluronan catabolism: a new metabolic pathway, Eur. J. Cell Biol. 83 (2004) 317–325.
[128] K. Sugahara, N.B. Schwartz, A. Dorfman, Biosynthesis of hyaluronic acid by streptococcus, J. Biol. Chem. 254 (1979) 6252–6261.
[129] M.R. Wessels, A.E. Moses, J.B. Goldberg, T.J. DiCesare, Hyaluronic acid capsule is a virulence factor for mucoid group A streptococci, Proc. Natl. Acad. Sci. U.S.A. 88 (1991) 8317–8321.
[130] H.M. Schrager, J.G. Rheinwald, M.R. Wessels, Hyaluronic acid capsule and the role of streptococcal entry into keratinocytes in invasive skin infection, J. Clin. Invest. 98 (1996) 1954–1958.
[131] G.B. Carmen, J. Springer, F.K. Kooy, A.M. Lambertus, V.D. Broek, G. Eggink, Production methods for hyaluronan, Int. J. Carbohydr. Chem. (2013) 14 pages, Article ID:624967, http://dx.doi.org/10.1155/2013/624967.
[132] W. Rutjes, P. Jüni, B.R. da Costa, S. Trelle, E. Nüesch, S. Reichenbach, Viscosupplementation for osteoarthritis of the knee: a systematic review and meta-analysis, Ann. Intern. Med. 157 (2012) 180–191.
[133] A. Josefsson, H. Adamo, P. Hammarsten, T. Granfors, P.R. Stattin, L. Egevad, A.E.M. Laurent, P. Wikström, Prostate cancer increases hyaluronan in surrounding nonmalignant stroma and this response is associated with tumor growth and an unfavorable outcome, Am. J. Pathol. 179 (2011) 1961–1968.
[134] P. Gritsenko, O. Ilina, P. Friedl, Interstitial guidance of cancer invasion, J. Pathol. 226 (2) (2012) 185–199.
[135] M.I. De Andrés Santos, A. Velasco-Martín, E. Hernández-Velasco, J. Martín-Gil, F.J. Martín-Gil, Thermal behaviour of aqueous solutions of sodium hyaluronate from different commercial sources, Thermochim Acta 242 (1994) 153–160.
[136] X.Z. Shu, K. Ghosh, Y. Liu, F.S. Palumbo, Y. Luo, R.A.F. Clark, G.D. Prestwich, Attachment and spreading of fibroblast on an RGD peptide-modified injectable hyaluronan hydrogel, J. Biomed. Mater. Res. 68 (2004) 365–375.
[137] S. Shanmugam, R. Manavalan, D. Venkappayya, K. Sundaramoorthy, V.M. Mounisamy, S. Hemalatha, T. Ayyappan, Natural products radians, Annu. Res. Rep. 4 (2005) 478–481.
[138] G.C. Barrére, C.E. Barber, M.J. Daniels, Molecular cloning of genes involved in the production of the extracellular polysaccharide xanthan by *Xanthomonas campestris* pv. *campestris*, Intl. J. Biol. Macromol. 8 (1986) 372–374.
[139] K. Kokate, A.P. Purohit, S.B. Gokhale, Pharmacognosy, 22nd ed., Nirali Prakashan, Pune, India, 2003.
[140] W.C. Evans, Trease and Evans' Pharmacognosy, 14th ed., Harcourt Brace and Co., Asia Pvt. Ltd., Singapore, 1996 196, 208, 209, 213-215, 462, 555.
[141] J.E.F. Reynolds, Martindale, The Extra Pharmacopoeia, 30th ed., The Pharmaceutical Press, London, 1993, 652, 904, 1217, 1221.
[142] K.Y. Lee, D.J. Mooney, Alginate: properties and biomedical applications, Prog. Polym. Sci. 37 (2012) 106–126.
[143] W.R. Gombotz, S.F. Wee, Protein release from alginate matrices, Adv. Drug Deliv. Rev. 31 (1998) 267–285.
[144] O. Smidsrod, G. Skjak-Bræk, Alginate as immobilization matrix for cells, Trend Biotechnol. 8 (1990) 71–78.
[145] E. Clark, H.C. Green, Alginic acid and process of making same, U.S. Patent 2036922, 1936.

[146] M. Rinaudo, Main properties and current applications of some polysaccharides as biomaterials, Polym Int. 57 (2008) 397–430.
[147] Y. Qin, Alginate fibres: an overview of the production processes and applications in wound management, Polym Int. 57 (2008) 171–180.
[148] A.F. Stockwell, S.S. Davis, S.E. Walker, In vitro evaluation of alginate gel systems as sustained release drug delivery systems, J. Control. Release 3 (1986) 167–175.
[149] S.E. Barnett, S.J. Varley, The effects of calcium alginate on wound healing, Ann. R. Coll. Surg. Engl. 69 (1987) 153–155.
[150] Y.B. Kim, G.H. Kim, Collagen/alginate scaffolds comprising core (PCL)–shell (collagen/alginate) struts for hard tissue regeneration: fabrication, characterisation, and cellular activities, J. Mater. Chem. B 1 (2013) 3185–3194.
[151] H.R. Allcock, F.W. Lampe, Contemporary Polymer Chemistry, Prentice-Hall, New Jersey, 1981.
[152] V.R. Gowarikar, N.V. Viswanathan, J. Sreedhar, Polymer Science, New Age International Publishers, New Delhi, 1986.
[153] T.J. Sill, H.A. von Recum, Electrospinning: applications in drug delivery and tissue engineering, Biomaterials 29 (2008) 1989–2006.
[154] Z.G. Wang, L.S. Wan, Z.M. Liu, X.J. Huang, Z.K. Xu, Enzyme immobilization on electrospun polymer nanofibers: an overview, J. Mol. Catal. B: Enzym. 56 (2009) 189–195.
[155] X.J. Huang, Z.K. Xu, L.S.W.C. Innocen, P. Seta, Electrospun nanofibers modified with phospholipid moieties for enzyme immobilization, Macromol. Rapid Commun. 27 (2006) 1341–1345.
[156] S. Ramakrishna, K. Fujihara, W.E. Teo, T.C. Lim, Z. Ma, An Introduction to Electrospinning and Nanofibers, World Scientific Publishing Co. Pte. Ltd., Singapore, 2005, pp. 396.
[157] J. Doshi, D.H. Reneker, Electrospinning process and applications of electrospun fibers, J. Electrostat. 35 (1995) 151–160.

Chapter 2

Hierarchical Characterization of Biomedical Polymers

Meera Parthasarathy, Swaminathan Sethuraman

Centre for Nanotechnology & Advanced Biomaterials (CeNTAB), School of Chemical & Biotechnology, SASTRA University, Thanjavur, India

Chapter Outline

2.1 INTRODUCTION

Polymers are the most versatile class of biomaterials used in a number of biomedical applications like tissue engineering, drug delivery, biosensors, wound healing, and disease diagnosis kits. The versatility of polymers is attributed to the ease of functionalization, processing, and diversity in synthetic approaches. The possibility of altering the chemical functional groups in polymers offers special advantages to biomedical applications such as tunable biocompatibility, controlled release of drug molecules from polymeric carriers, and controlled biodegradation of polymeric implants.

Polymers differ from simple low-molecular-weight compounds in a number of aspects. They are long-chain macromolecules with molecular weight in the range of kilodaltons. All the differences in properties between polymers and simple molecules arise due to the spatial disposition and dynamics of the long molecular chains and the complex, often unpredictable, interactions between the long chains. For example, glucose has a sharp melting point but not starch and cellulose. Starch has alpha glycosidic linkages, while cellulose has beta glycosidic linkages, so starch can be digested by humans but not cellulose. Polymer properties mainly depend on its molecular-weight distribution, average molecular weight, and crystallinity. In addition to these parameters, which can be controlled during synthesis of the polymer, processing method and condition is a major determinant of the properties and functions of the polymeric product. While the dependence of properties on processing conditions endows more tunability, prior knowledge of the same is essential to ensure reproducible results in biomedical applications. For example, properties of PVA hydrogels physically cross-linked by freeze-thaw cycles depend on the number of cycles and duration of freezing and thawing [1]. Incomplete understanding of many polymeric biomaterials has led to critical implant failure [2].

In this context, this chapter discusses recent advances in the characterization of polymeric biomaterials. The discussion focuses on characterization of polymeric products used in biomedical applications and not on molecular-level spectroscopic characterization of polymers.

2.2 THE HIERARCHICAL CHARACTERIZATION APPROACH

Characterization of polymeric biomaterials is more challenging than that of other classes of biomaterials like metals, alloys, glasses, and ceramics, owing to their dynamic

Natural and Synthetic Biomedical Polymers

structure. Method of storage and handling could alter the polymer properties to a significant extent. If a polymer is hygroscopic, it has to be transferred to the characterization chamber without exposure to the atmosphere. Even the surface texture, roughness, or porosity of the hygroscopic polymer implant might change when it is characterized in atmospheric conditions. This, in turn, could affect the interpretation of thrombogenicity and biological fixation of the implant in the host tissue [3]. Polymeric micelles and nanoparticles used in drug delivery have even more dynamic structures and are difficult to be characterized [4]. Sterilization procedure also significantly affects the property of the final polymeric biomaterial. When poly(ethylene glycol) hydrogel drug carriers sterilized by different procedures were analyzed, surface roughness of the hydrogels sterilized using H_2O_2 and gamma radiation was found to be significantly lower than the unsterilized and ethylene oxide-sterilized hydrogels [5]. Gamma radiation and ethylene oxide sterilization are also known to cause fatigue of high-molecular-weight polyethylene implants [6]. Thus, characterization of polymeric biomaterials before and after sterilization is critical before implantation in the host.

Moreover, the surface of a polymeric product is more reactive than its bulk because of the chemical functional groups interacting with the environment. While the high surface reactivity is advantageous in imparting biocompatibility and biodegradability properties to the polymer implant, the same feature imposes critical challenges on characterization. For instance, the properties could be altered during the sample preparation step for a particular characterization technique. *In vivo* characterization of polymeric implants and drug delivery systems is even more challenging due to contributions from adsorbed proteins, inflammatory exudates, and reactive oxygen species coming in contact with the polymer surface. Large-scale characterization of polymeric biomedical products is needed in many cases, which can be tackled by theoretical modeling and simulation to predict the properties.

Complete characterization of a polymeric biomaterial can be achieved by following a hierarchical approach as described in Figure 2.1.

Bulk characterization yields information on the macroscopic properties of the biomaterial such as thermal, mechanical, solubility, optical, and dielectric properties. Surface characterization yields morphological information that is critical for interfacing the implant or drug delivery device with the host tissue. This could be achieved by microscopic and spectroscopic methods. Next in the hierarchy is the characterization of processes such as biodegradation mechanism and kinetics under biomimetic *in vitro* conditions. Cases of implanted device failure need to be assessed by systematic interrogation of explanted medical devices. After knowing the basic characteristics of the biomaterial, real-time investigation of *in vivo* processes plays a major role in the successful journey of an implant.

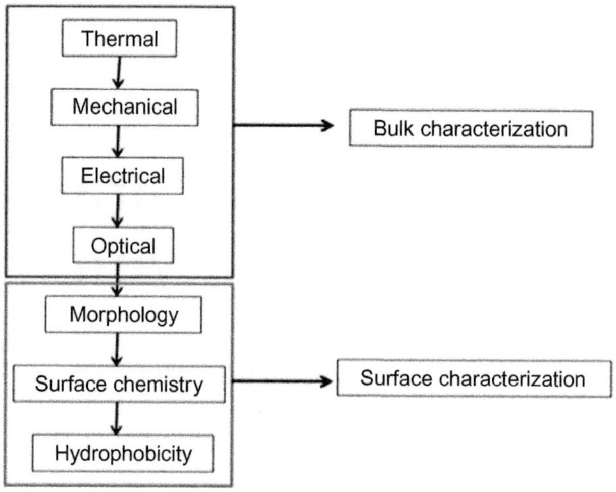

FIGURE 2.1 Complete characterization of a biomedical polymer by the hierarchical approach. (For color version of this figure, the reader is referred to the online version of this chapter.)

2.3 BULK CHARACTERIZATION

2.3.1 Thermal Properties

Thermal characterization of biomedical polymers is mainly aimed at determining their melting temperature, crystallization temperature, and glass transition temperature (T_g). Though these thermal parameters are not directly related to the biocompatibility of the polymers, they are inevitable in deciding the processing conditions to convert a natural or synthetic biomedical polymer into the final implant or device. Differential scanning calorimetry (DSC) is the technique used to determine the thermal parameters mentioned earlier. Sometimes, the polymer processing technique itself induces changes in thermal properties. Electrospinning is a process for preparing for polymer fibers from viscous solutions and melts. The spinning process is reported to reduce the crystallization temperature of poly(lactic acid) (PLA) from 108 to 77 °C and poly(glycolic acid) (PGA) from 68 to 49 °C. DSC data also provide information about the spinnability of PLA/PGA blends as a function of the homopolymer ratio. Blend solution containing 25% PLA and 75% PGA exhibited significantly low crystallinity compared to other ratios and was accompanied by a decrease in spinnability [7]. Additional care should be taken in the case of hygroscopic polymer fibers while doing DSC as the glass transition peak might not be visible in the first heating cycle.

T_g of a polymer is related to its biodegradability. If the T_g of a biomedical polymer is similar to the body temperature, the implanted polymer may be more flexible in the host environment than under *in vitro* conditions, which in turn may accelerate its biodegradation *in vivo* [8]. In addition, the presence of water and other ions in the host environment may also reduce the T_g of implanted polymers like polyesters and accelerate their biodegradation rate.

Another useful thermal characterization technique is thermal compression in which a polymer fabric or biotextile is subjected to different loads at different temperatures. The thickness, pore size, and distribution can be monitored at each condition to prepare ideal scaffolds for tissue engineering. Poly(ethylene terephthalate) (PET) nonwoven fiber scaffolds have been prepared for tissue engineering by thermal compression and simultaneous characterization. Applying pressure near the T_g of the polymer (~70 °C) yielded better control of the pore size distribution and smaller pore sizes, which led to faster and wider proliferation of trophoblast ED_{27} and NIH 3T3 cells on the scaffold [9].

2.3.2 Mechanical Properties

Mechanical strength is one of the major determinants of biocompatibility of polymeric tissue engineering scaffolds and drug delivery systems [10]. In tissue engineering, proper interfacing of the scaffold with the host tissue is ensured only if the mechanical properties of the scaffold match with those of the healing or regenerating tissue. Mechanical parameters of importance in biomedical applications include tensile strength, elastic modulus, compressive strength, dynamic stability, and wear and creep resistance. Influence of these parameters will differ depending on the site of implantation and retention time of the biomaterial. Scaffolds and implants intended for ligament and cartilage tissue engineering are expected to have good tensile strength and directional load-bearing capacity. Whereas scaffolds for bone tissue engineering should have good compressive strength and those used in bone joints/shoulder cuffs should exhibit considerable wear/abrasion resistance, polymers used in vascular grafts and conduits should withstand the pulsatile flow of blood through them. On the other hand, drug delivery carriers should have good deformability and elasticity to control the delivery and bioavailability of drug molecules at the site of infection or injury.

Tensile test is the basic characterization of mechanical properties of a given polymeric biomaterial. Tensile strength is important for polymers used in repairing cartilage and ligament ailments. Desirable values of the parameter can be achieved by altering the degree of crystallinity of the polymer and by controlling the orientation of fibers during electrospinning. For instance, elongation at break of biodegradable scaffolds made of poly(*p*-dioxanone) fibers used for cartilage tissue engineering can be controlled by aligning the fibers along a certain direction using a dynamic collector in electrospinning. A twofold increase in elongation at break was observed between random and aligned fibers [11]. In the case of aligned fiber scaffolds, the direction of applied force during mechanical testing affects the end result drastically because of their anisotropic mechanical behavior. In the previously mentioned report, the elongation at break was more when force was applied perpendicular to the alignment direction than along the alignment direction. Compression modulus of a scaffold should be in the range of 10-1500 MPa for hard tissues and 0.4-350 MPa for soft tissues [10]. Several cases of implant failure have been attributed to the poor mechanical integration of implants with host tissue. Loosening of polymer implants is a major consequence of fatigue and wear. Wear has been frequently observed in implants used in knee joints and hip joints. Fatigue behavior of polymeric implants can be characterized using fracture mechanics, stress/life analysis, and fatigue-wear analysis using multiaxial loading simulated under physiological conditions [12]. The fracture mechanics test and stress/life test are usually performed on polymeric orthopedic implants intended to withstand cyclic stress. Another important mechanical parameter of a polymer scaffold is surface stiffness, which is known to affect cell-biomaterial interactions. Cell differentiation is also known to depend on the stiffness of polymer scaffolds. Undifferentiated mesenchymal stem cells are known to differentiate into neuron-like cells when grown on a polymer scaffold with stiffness of the order 1 kPa and myoblast-like cells over a stiffer polymer surface with stiffness in the range 10-20 kPa [13]. Similarly, elasticity is an important parameter for scaffolds used in skin tissue engineering whereas elasticity with additional toughness is needed for cartilage tissue scaffolds [14].

In vitro mechanical tests should also focus on the dynamic behavior of a given polymer scaffold, which is very crucial for applications like knee/hip joint repair and vascular grafts. Certain polymers like polypropylene show creep behavior, i.e., exhibit dimensional changes under continuous load and cannot be used to make vascular grafts [15].

For this reason, expanded PTFE and certain butyl polyesters are better alternatives than polypropylene for vascular grafts. Certain polymers, instead of exhibiting considerable dimensional changes, become weak with time after implantation in the host environment. For instance, biodegradable poly(L,DL-lactide) cages used in treating spine injury were found to degrade three months earlier than their intended 6-month degradation time when tested in goat lumbar spine due to their time-dependent deterioration of mechanical strength of the polymer material [9]. Variation of mechanical strength during biodegradation is critical in load-bearing orthopedic implants, wherein the degrading implant should gradually transfer the load to the regenerating bone tissue without any stress-shielding issue [12]. Poly(L-lactic acid) (PLLA), which is widely used in fixing small bone fractures but not for large bone fractures as its mechanical properties, degrades with time. If metal poles are used instead, stress shielding will be unavoidable with additional complication of second surgery to remove the metal poles after bone regeneration. Hence, PLLA is blended with poly(hydroxybutyrate-*co*-hydroxyvalerate) (PHBV) to treat major bone fractures, and the composition of the blends is optimized using a mechanical characterization technique called dynamic mechanical analysis [16].

Mechanical characterization techniques used for fixation of tissue engineering scaffolds to the host tissue is equally important as scaffold characterization itself. For engineering vulnerable tissues like the articular cartilage, the method used for fixing the scaffold supporting the autologous chondrocytes needs to be chosen carefully. Three different fixation methods, namely, fibrin glue, transosseous suture, and chondral suture, were mechanically tested by loading PGA and PLGA scaffolds until failure. It was found that chondral-sutured and transosseous-fixed PGA scaffold could be loaded highest until failure compared to fibrin glue-fixed PLGA scaffold [17].

In addition to applications like tissue engineering and sutures, mechanical properties of polymeric biomaterials play a significant role in drug delivery applications. Deformable liposomal carriers have been identified to be more effective for delivering corticosteroid drugs for ear edema than direct application of the drug. The deformable nature of the carrier allows minimum effective dose and increased bioavailability at the site of infection [7]. Similarly, thin and mechanically stable collagen cell carriers allow effective proliferation of cardiomyocytes, wherein the adherent cardiomyocytes could rhythmically contract the whole polymer carrier due to its excellent elasticity [18].

2.3.3 Optical Properties

In addition to the mechanical properties, the three-dimensional microstructure of a polymer scaffold plays a major role in tissue engineering. Quantitative characterization of the pore morphology such as pore size, pore volume, size distribution, tortuosity, and interconnectivity is required to the success of a polymeric tissue engineering scaffold. This is achieved by optical characterization techniques based on light scattering and confocal methods. Confocal microscopy is normally used to assess the depth of penetration of cells seeded on a polymeric scaffold at different time intervals. Resorbable scaffolds of poly(L-lactide-*co*-glycolide) were tested as carriers of osteoblast-like MG-63 cells for bone tissue engineering. The effect of average pore size of the scaffold on the penetration depth of MG-63 cells was studied by confocal microscopy [19]. Another optical technique used for three-dimensional mapping of scaffold microstructure and cell constructs is nuclear magnetic resonance (NMR) imaging.

Nevertheless, NMR imaging suffers from poor lateral resolution (~10 μm) and confocal microscopy is limited by poor penetration depth (~80 μm) and vertical resolution especially in the case of opaque samples [20]. Optical coherence microscopy/optical coherence tomography (OCT) is more advantageous than the earlier-mentioned two optical imaging techniques and is being widely used in characterization of polymer scaffolds and cell constructs. Pore morphology of hydrogel scaffolds made of cyclic acetals has been successfully quantified using OCT [21]. The technique has been extremely useful for *in vivo* characterization of soft contact lenses. For instance, ultra-high-resolution OCT was used to image tear films around different types of contact lenses *in vivo* and demonstrated to be a reliable method to evaluate fitting procedures used to fix contact lenses [22]. Other methods for optical characterization of contact lenses are refractometry (measures refractive index) and visible spectroscopy (measures percentage transmittance that is a measure of transparency) [23]. Spectrofluorimetry has been demonstrated to prove acidification of aqueous humor and corneal epithelium by soft contact lenses in rabbit eyes [24].

2.3.4 Electrical Properties

Conducting polymers capable of conducting ions or electrons are recently recognized as attractive candidates for biomedical applications owing to their electric field-controllable properties. A number of conducting polymers are commercially available or can be easily synthesized by methods such as electrodeposition on surfaces. The most commonly known conducting polymers and their electrical conductivity values are tabulated in Table 2.1.

Nanocomposite scaffolds made of conducting polymers are known to modulate the growth of a number of cell types such as chromaffin cells [26], nerve cells [27], and endothelial cells [28]. They also serve as electrical transducers for therapeutic applications and provide excellent control over drug release kinetics. Electrical characterization of conducting polymer biomaterials is done using four-point probe I-V measurements and cyclic voltammetry using direct current and electrochemical impedance spectroscopy using alternating current. Electrospun PET fibers were coated with PEDOT using vapor-phase polymerization method. The sheet resistance of the coated scaffold was measured by four-point probe method, and SH-SY5Y neuroblastoma cells

TABLE 2.1 Conducting Polymers Commonly Used in Biomedical Applications [25]

Conducting Polymer	Conductivity (S/cm)
Polyacetylene	200-1000
Polyparaphenylene	500
Polyparaphenylene sulfide	3-300
Polyparavinylene	1-1000
Polypyrrole	40-200
Polythiophene	10-100
Polyisothionaphthene	1-50

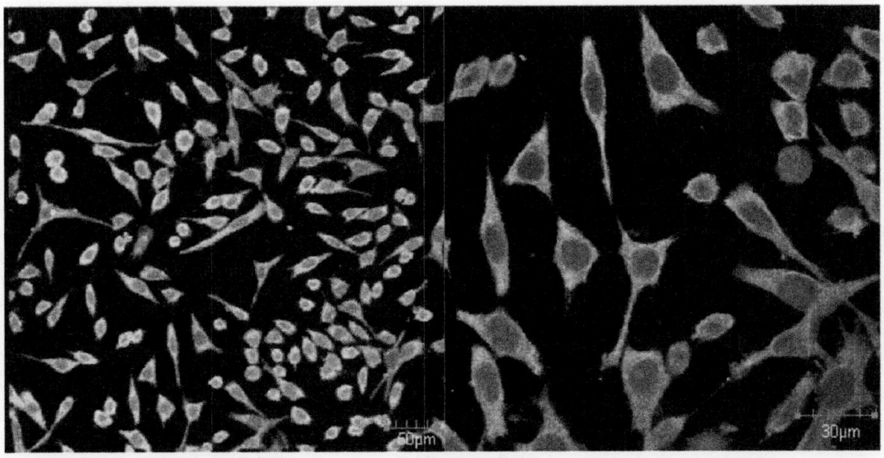

FIGURE 2.2 Fluorescence micrographs of fibroblast cells cultured on PHBV nanofiber scaffolds after 48 h of culture. (For color version of this figure, the reader is referred to the online version of this chapter.)

were cultured on the PEDOT-coated PET fiber scaffolds. Extensive proliferation of the cells was observed on the conducting polymer-coated scaffold. Calcium signaling of the neuroblastoma cells was also controlled by applying voltage of 1.5-3.0 V to the PEDOT-coated scaffold [29]. Dielectric measurement by impedance spectroscopy has been extremely useful in characterizing and controlling the growth of certain cell types on conducting polymer scaffolds. Hepatocytes cultured in highly porous and interconnected alginate and gelatin polymer scaffolds were characterized using alternating current in the frequency range 1 kHz to 2 GHz. Qualitative analysis of the dielectric spectra indicated that the overall survival rate of the hepatocytes was better on alginate scaffold than gelatin scaffold despite both of them having similar pore size distribution and pore interconnectivity. The results of the dielectric spectra were confirmed by cell viability test [30].

Alternating current has also been used to stimulate the growth of PC12 cells on conducting scaffolds made of poly(caprolactone fumarate)-polypyrrole (PCLF-PPy). PC12 cells were cultured on PCLF-PPy scaffolds and alternating current of 10 μA amplitude and 20 Hz frequency was applied to the cell-seeded scaffold everyday for 1 h [31]. Similarly, fluorescent microscope helps in the imaging of fibroblast cells that were cultured on PHBV nanofibers after 48 h of culture (Figure 2.2).

2.4 SURFACE CHARACTERIZATION

The success of a tissue engineering scaffold or a drug carrier depends mainly on its surface properties. The "surface" of a polymeric biomaterial is defined with respect to the biomedical application and the host tissue environment in which the material is intended to perform. Biocompatibility of a polymer implant is directly related to its surface properties. A polymer implant with rough surface accelerates thrombus formation when in contact with blood and is not blood-compatible. On the other hand, textured surfaces are more skin-compatible than nontextured surfaces as they preclude epithelial downgrowth and fibrous encapsulation in subcutaneous implants. Protein adsorption, which is the first stage of host recognition of the implant, depends exclusively on the physical and chemical properties of the implant surface and hence dictates its biocompatibility, inflammatory response, and biological fixation with the host tissue.

The surface of a polymeric biomaterial could be described based on its morphology/topography, hydrophobicity, and chemical functional groups. Accordingly, a number of high-resolution techniques are available today to characterize various aspects of polymer surfaces intended for biomedical applications.

2.4.1 Microscopic Characterization

Microscopic examination of biomaterials has become inevitable in biomedical research owing to the range of information derivable from different types of microscopes available today. Ranging from the conventional compound microscope, we have more advanced optical microscopes with high-end optics and lens materials available to us. Unlike the optical microscopes, which operate on the principle of reflection and refraction of light radiation, microscopes using electron beams offer far better lateral and vertical resolution to enable deeper analysis of implant surfaces. Yet other classes of unconventional microscopes are scanning probe microscopes working on the rastering movement of a sharp solid probe across the sample surface. Different types of light microscopes used in the examination of polymeric biomaterials are listed in Table 2.2.

The resolution of a micrograph is inversely proportional to the wavelength of incident radiation. Hence, one could imagine high-resolution imaging if a radiation having higher

TABLE 2.2 Comparison of Light Microscopic Techniques Used for Characterizing Biomedical Specimens

Type of Light Microscope	Construction/Working Principle	Specimens		Remark
Upright microscope	Specimen placed below the objective lens	Tissue slice (histology work)		High resolution but short working distance
Inverted microscope	Objective lens positioned beneath the specimen	Live cells in transparent tissue culture flasks or well plates		Lower resolution but long working distance, can be integrated with flow cells
Bright field microscope	Based on transmission of light through samples; background appears bright and specimen features appear dark	*Stain* Eosin Hematoxylin Alizarin red Verhoeff's stain	*Information* Cytoplasmic proteins Cell nucleus Calcific deposits on implants Elastin	Unstained cells cannot be seen, immunohistochemistry can be done
Dark field microscope	Oblique illumination of the specimen; based on scattering of light by the specimen; background appears dark and features as bright spots	Specimens with small particulates		
Differential interference contrast microscope	Converts light diffracted by different organelles in the cell into an image	Specimen surfaces can be seen clearly		Shallow depth of focus so more surface-sensitive, cannot be performed in tissue culture plastic wares
Fluorescence microscope	Monitors light emission from fluorophore-tagged lipids, enzymes, membrane components, and antibodies	Fluorophore-tagged specimens		
Confocal laser scanning microscope	A special tool for fluorescence, prevents blurring of images by placing a pinhole at the confocal image plan	Location of proteins, lipids, and cellular components		
Polarization-based microscope	Measures changes in polarization of light caused by ordered arrangements in a sample	Collagen, muscle sarcomeres, remnant synthetic polymers in explanted tissues		Cannot be performed in tissue culture plastic wares since they themselves cause polarization changes

energy than visible light is used in the microscope. Electron microscopes use electron beams instead of light beams to obtain micrographs of the sample. The energy of the primary electron beam ejected from the source can be easily tuned by altering the acceleration voltage. Scanning electron microscopy (SEM) involves shining of a high-energy primary electron beam on the sample and detection of secondary electrons and other radiations emitted from the sample. Analysis of the radiations emitted from the sample provides morphological information and surface elemental composition. As the equipment operates under high vacuum, the specimen should be dry. Electrical conductivity is another prerequisite for SEM analysis; insulating specimens are coated with a very thin layer of platinum or gold by vacuum sputtering prior to imaging. Fixation is needed in the case of proteins, cells, and tissue samples. Morphological information is obtained from the SEM image formed by variations in secondary electron density emitted from the specimen. Morphological information on polymer scaffolds such as pore interconnectivity, alignment, and surface modification can be obtained (Figure 2.3a–c). SEM can also be used to analyze cell-scaffold interactions. Figure 2.3d shows the scanning electron micrograph of fibroblast adhesion on curcumin-loaded PLGA nanofiber scaffolds after 48 h of culture (Figure 2.3d).

Transmission electron microscopy (TEM) has higher resolution than SEM as it uses high-energy electron beam. In this technique, primary electrons transmitted through the specimen are detected, which gives morphological and

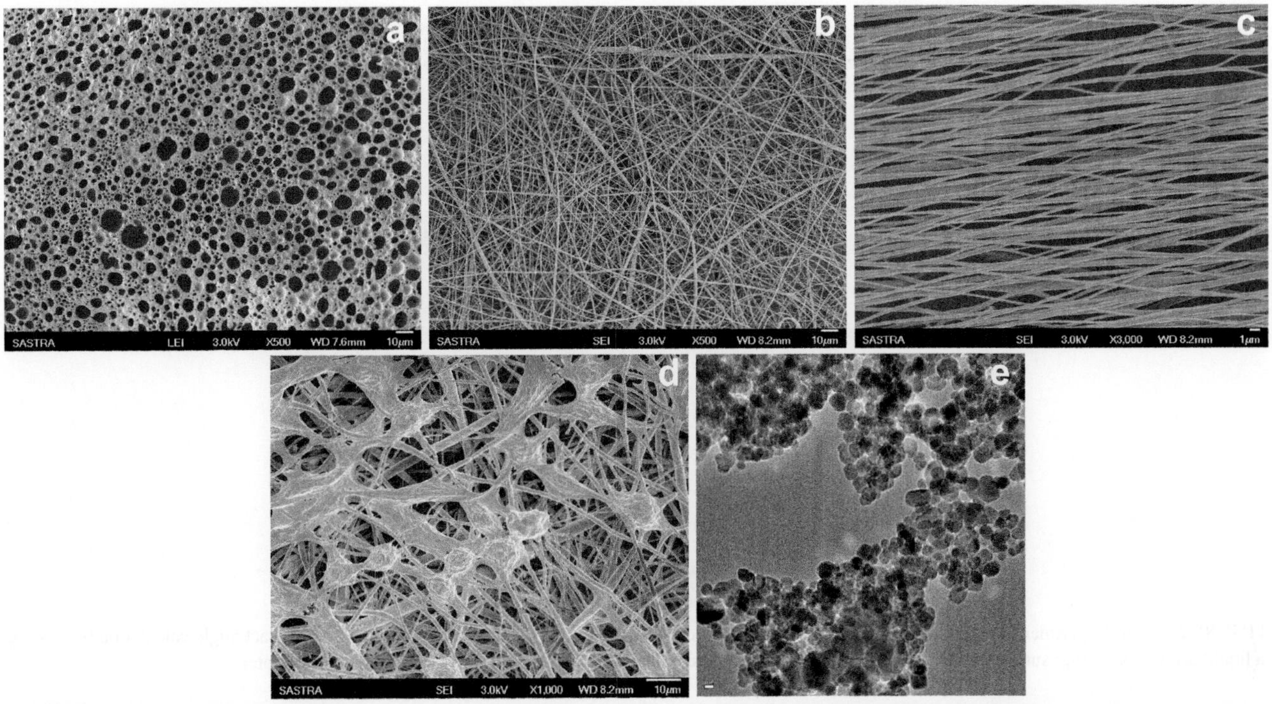

FIGURE 2.3 Scanning electron micrographs of (a) porous PHBV scaffold obtained by particulate leaching, (b) electrospun random PLGA fibers, (c) electrospun radially aligned PHBV fibers [11], (d) fibroblast cells on PLGA curcumin-loaded fibers after 48 h of culture, (e) transmission electron micrographs of iron oxide nanoparticles.

crystallographic information. The specimen should be as thin as possible. Tissue samples are embedded in acrylic resins and sectioned into thin slices using ultramicrotome for TEM analysis since sample thickness should not be more than 1 μm [32]. Figure 2.3e shows the surface morphology of iron oxide nanoparticles, which have been evaluated for theranostic applications.

Scanning tunneling microscopy (STM) and atomic force microscopy (AFM) are scanning probe techniques, in which sharp solid probe scans over the specimen surface and provides topographic information. STM provides information at the atomic level and can identify the arrangement of atoms on sample surface. In this technique, a sharp tungsten tip is scanned very close to the sample surface and tunneling current is measured. The current is inversely proportional to the tip-sample separation. AFM makes use of a silicon nitride probe mounted on a flexible cantilever. The movement of the cantilever is monitored by a laser beam focused on the surface; as the probe scans the sample surface, attractive and repulsive forces between the probe and the surface are measured.

2.4.2 Surface Hydrophobicity

Hydrophobicity refers to the poor wettability of surfaces by water and is quantified using contact angle measurements. The experimental setup consists of a microsyringe to instill fine droplet of the liquid on the polymer specimen surface, fiber optic illumination, telescopic focusing, and computer monitor to record the time-dependent changes. Schematic representation of a goniometer can be found in Figure 2.4a. In this experiment, a drop of water is placed on specimen surface and contact angle is calculated as depicted in Figure 2.4b.

Three methods are usually used to calculate contact angle—Wilhelmy plate method, sessile drop method [33], and captive bubble method [34]. Sessile drop method is the most commonly used method for biomedical polymers. In this method, about 3 μl of a liquid droplet is placed on the polymer surface and images of the drop are acquired about 30 s of equilibration of the drop. Interface energy between the solid sample surface and liquid can also be calculated using the Young's equation:

$$\gamma_{LV} \cos\theta = \gamma_{SV} - \gamma_{SL}$$

where, γ_{LV}, γ_{SV}, γ_{SL} are interfacial tension between liquid-vapor, solid-vapor, and solid-liquid interfaces, respectively, and θ is the contact angle measured.

Hydrophobicity of biomedical polymers influences the biocompatibility depending on the particular application such as tissue engineering, blood contacting devices, and dental implants [35]. Polymers are dynamic structures and can switch their surface functional groups depending on the environment. For example, polymeric biomaterials need to have a hydrophilic surface for most of the applications, so that the cell-adhesive proteins present in the serum will be adsorb and promote cell adhesion and proliferation. This is achieved by surface treatment procedures such as

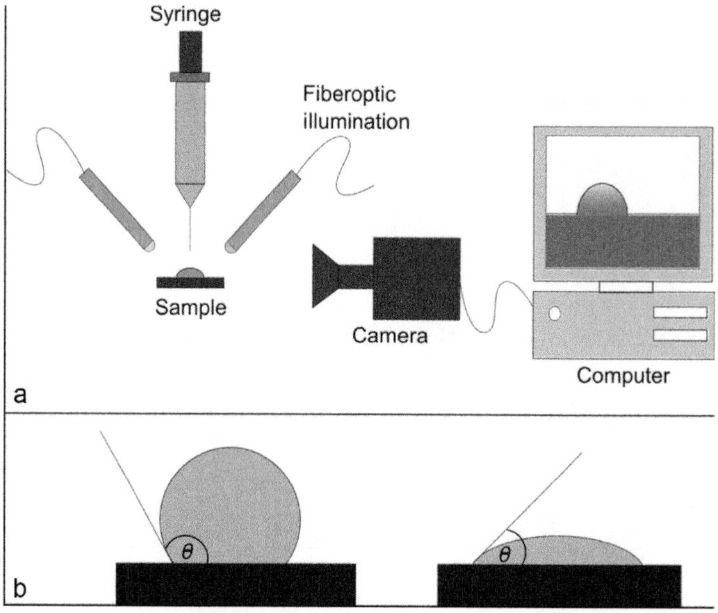

FIGURE 2.4 (a) Experimental setup of a goniometer used for contact angle measurement of solid samples and (b) contact angle calculation by placing a liquid droplet on a solid surface. (For color version of this figure, the reader is referred to the online version of this chapter.)

UV irradiation and plasma treatment. However, it has been observed that the water contact angle of the treated polymer surfaces increases gradually with time, thereby making them hydrophobic and less biocompatible. The phenomenon is called "hydrophobic recovery of polymeric biomaterials" and is being addressed in detail [36].

2.4.3 Spectroscopic Characterization

Surface characterization by spectroscopic techniques yields information on the functional groups and elemental composition on the surface of polymeric biomaterials. The most common spectroscopic tools used for biomedical polymers are X-ray photoelectron spectroscopy (XPS), Auger electron spectroscopy (AES), secondary ion mass spectrometry (SIMS), and Fourier transform infrared spectroscopy (FTIR) (diffuse reflectance and attenuated total internal reflectance modes). Each of these techniques is discussed in the succeeding text.

XPS is used to analyze the elemental composition of polymer surfaces. In this technique, the sample is irradiated with a high-energy monochromatic X-ray and the core level electrons ejected from the sample (called photoelectrons) are detected. The energy of the photoelectrons ejected from the sample depends on the elements present on the sample surface. PGA scaffolds are used in bone tissue engineering, and in order to improve the osteoconduction of the PGA scaffold, hydroxyapatite nanoparticles are coated on the polymer. XPS is a reliable method to verify the deposition of HA nanoparticles on PGA surface [37]. AES is more surface-sensitive than XPS. In this technique, a beam of electrons is incident on the sample, which in turn ejects secondary electrons called Auger electrons. This provides elemental composition and enables depth profiling to verify cross-sectional composition changes.

XPS and AES cannot detect lighter elements whereas SIMS can detect even hydrogen and high-mass fragments with high sensitivity. The sample bombarded by high-energy primary ion beam ejects secondary ions, which are detected based on their mass-to-charge ratio. Mass spectrometric techniques are known for their high sensitivity and accuracy, as they do not require any histochemical labeling. For instance, time-of-flight SIMS has been applied successfully to monitor the activity of macrophages on polymer scaffolds after implantation in the host tissue [38]. SIMS has also been demonstrated successfully to monitor biodegradation mechanisms of polymers by following their hydrolytic oligomeric fragments [39].

FTIR is another surface-sensitive spectroscopic tool to analyze biomedical polymers since sample preparation is very simple for this technique. FTIR spectra should be recorded in reflection mode instead of transmission mode in order to analyze the functional groups present on polymer surfaces. ATR-FTIR has been applied to study *in vitro* mineralization of porous starch scaffolds cultured in bone marrow stromal cells harvested from Wistar rats. Mineral deposition in *in vitro* cultures is usually followed by von Kossa stain or Alizarin red stain or by calcium uptake. These methods provide erroneous results because the scaffold matrix itself can take up some calcium from the medium. ATR-FTIR is devoid of the limitation and provides reliable information on the mineralization process. In the cited report, evidences for protein matrix formation on the scaffold and mineralized extracellular matrix were obtained by ATR-FTIR analysis [40].

2.5 FUTURE PROSPECTS

Research in the area of nanotechnology has been rapidly advancing that we find that the tools either lack sensitivity or resolution that is required to effectively characterize very low signals. Characterization of the various properties such as topography, morphology, mechanical, and porosity of biomaterials before their use is very critical and important so that we can predict their behavior *in vivo*. However, characterization tools have improved over the past couple of decades, and collectively, we have been able to image and understand not only the surface of materials but also their properties. For example, the atomic force microscope provides atomic-scale surface data, while the scanning electron microscope provides micro- and nanoscale surface data, and the nanoscale data about the internal structure of materials can be obtained from transmission electron microscope. With this information and with data on the mechanical properties, wettability, porosity, etc., we would be able to understand the surfaces of materials and how they would behave *in vitro* and *in vivo*.

REFERENCES

[1] N.E. Vrana, A.O. Grady, E. Kay, P.A. Cahill, G.B. Mcguinness, Cell encapsulation within PVA-based hydrogels via freeze-thawing: a one-step scaffold formation and cell storage technique, J. Tissue Eng. Regen. Med. 3 (2009) 567–572.

[2] B.D. Ratner, A.S. Hoffmann, F.J. Schoen, J.E. Lemons, Biomaterials Science: An Introduction to Materials in Medicine, second ed., 2004.

[3] W.A. Hanna, F.E. Gharib, I.I. Marhoon, Characterization of ceramic filled polymer matrix composite used for biomedical applications, J. Miner. Mater. Character. Eng. 10 (2011) 1167–1178.

[4] K. Yasugi, Y. Nagasaki, M. Kato, K. Kataoka, Preparation and characterization of polymeric micelles from poly(ethylene glycol)-poly(D, L-Lactide) block copolymers as potential drug carrier, J. Control. Release 62 (1999) 89–100.

[5] D. Kanjickal, S. Lopina, M.M. Chapino, S. Schmidt, D. Donovan, Effects of sterilization on poly(ethylene glycol) hydrogels, J. Biomed. Mater. Res. A 87A (3) (2008) 608–617.

[6] M.D. Ries, K. Weaver, R.M. Rose, J. Gunther, W. Sauer, N. Seals, Fatigue strength of polyethylene after sterilization by gamma irradiation or ethylene oxide, Curr. Orthop. Pract. 333 (1996) 87–95.

[7] T.H. Smit, T.A.P. Engels, P.I.J.M. Wuisman, L.E. Govaert, Time-dependent mechanical strength of poly(L, DL-lactide): shedding light on the premature failure of degradable spinal cages, Spine 33 (2008) 14–18.

[8] T. Schmidt, S. Stachon, A. Mack, M. Rohde, L. Just, Evaluation of a thin and mechanically stable collagen cell carrier, Tissue Eng. C 17 (2011) 1161–1170.

[9] L.I. Ramdhanie, S.R. Aubuchon, E.D. Boland, D.C. Knapp, C.P. Barnes, D.G. Simpson, G.E. Wnek, G.L. Bowlin, Thermal and mechanical characterization of electrospun blends of poly(lactic acid) and poly(glycolic acid), Polym. J. 38 (2006) 1137–1145.

[10] S.J. Hollister, Porous scaffold design for tissue engineering, Nat. Mater. 4 (2005) 518–524.

[11] T. Schneider, B. Kohl, T. Sauter, K. Kratz, A. Lendlein, W. Ertel, G. S-Tanzil, Influence of fiber orientation in electrospun polymer scaffolds on viability, adhesion and differentiation of articular chondrocytes, Clin. Hemorheol. Microcirc. 52 (2012) 325–336.

[12] S.H. Teoh, Fatigue of biomaterials, Int. J. Fatigue 22 (2000) 825–837.

[13] J.A.S. Herrera, E.R. Romo, Cell-Biomaterial mechanical interaction in the framework of tissue engineering: insights, computational modelling and perspectives, Int. J. Mol. Sci. 12 (2011) 8217–8244.

[14] C.G. Jeong, S.J. Hollister, Mechanical, permeability and degradation properties of 3D designed poly(1,8-octanediol-co-citrate) scaffolds for soft tissue engineering, J. Biomed. Mater. Res. B Appl. Biomater. (2010) 142–149.

[15] S.L. Weinberg, Biomedical Device Consultants Laboratory data, 1998.

[16] B.M.P. Ferreira, C.A.C. Zavaglia, E.A.R. Duek, Thermal, morphologic and mechanical characterization of poly(L-Lactic acid) and poly(hydroxyl butyrate-co-valerate) blends, Braz. J. Biomed. Eng. 19 (2003) 21–27.

[17] S. Knecht, C. Erggelet, M. Endres, M. Sittinger, C. Kaps, E. Stussi, Mechanical testing of fixation techniques for scaffold-based tissue engineering grafts, J. Biomed. Mater. Res. B Appl. Biomater. 83 (2007) 50–57.

[18] J.C. Middleton, A.J. Tipton, Synthetic biodegradable polymers as orthopaedic devices, Biomaterials 21 (2000) 2335–2346.

[19] E. Pamula, E. Filova, L. Bacakova, V. Lisa, D. Adamczyk, Resorbable polymer scaffolds for bone tissue engineering: the influence of their microstructure on the growth of human osteoblast-like MG 63 cells, J. Biomed. Mater. Res. A 89 (2009) 432–443.

[20] M.T. Cicerone, J.P. Dunkers, N.R. Washburn, F.A. Landis, J.A. Copper, Optical coherence microscopy for in-situ monitoring of cell growth in scaffold constructs, in: 7th World Biomaterials Congress, 2004, pp. 584.

[21] C.W. Chen, M.W. Betz, J.P. Fisher, A.B.S. Paek, Y. Chen, Macroporous hydrogel scaffolds and their characterization by optical coherence tomography, Tissue Eng. Part C Methods 17 (2011) 101–112.

[22] J. Wang, S. Jiao, M. Ruggeri, M.A. Shousha, Q. Chen, In situ visualization of tears on contact lenses using ultra high resolution optical coherence tomography, Eye Contact Lens 35 (2009) 44–49.

[23] V. Sankar, T. Suresh Kumar, K. Panduranga Rao, Preparation, characterization and fabrication of intraocular lenses from photo-initiated polymerized poly(methyl methacrylate), Trends Biomater. Artif. Organs 17 (2004) 24–30.

[24] C. Giasson, J.A. Bonanno, Corneal epithelial and aqueous humour acidification during in vivo contact lens wear in rabbit eyes, Invest. Ophthalmol. Vis. Sci. 35 (1994) 851–861.

[25] M. Mozafari, M. Mehraien, D. Vashaee, L. Tayebi, Electroconductive nanocomposite scaffolds: a new strategy into tissue engineering and regenerative medicine, in: F. Ebrahimi (Ed.), Nanocomposites—New Trends and Developments, InTech Open Publishers, 2012, Chapter 14.

[26] A. Kotwal, C.E. Schmidt, Biomaterials 22 (2001) 1055.

[27] R.F. Valentini, T.G. Vargo, A. Gardellajr, P. Aebischer, Biomaterials 13 (1992) 193.

[28] B. Garner, A. Georgevich, A.J. Hodgson, L. Liu, G.G.J. Wallace, Biomed. Mater. Res. 44 (1999) 121.

[29] M.H. Bolin, K. Svennersten, X. Wang, I.S. Chronakis, A.R. Dalhfors, E.W.H. Jager, M. Berggrenn, Nano-fiber scaffold electrodes based on PEDOT for cell stimulation, Sensors Actuat. B 142 (2009) 451–456.

[30] M. Massimi, A. Stampella, L.C. Deyirgiliis, G. Rizzitelli, A. Barbetta, M. Dentini, C. Cametti, Colloids Surf. B Biointerfaces 102 (2013) 700–707.
[31] P. Moroder, H. Wang, T. Ruesink, L. Lu, A.J. Windebank, M.J. Yaszemski, M.B. Runge, Material properties and electrical stimulation regimes through polycaprolactone fumarate—polypyrrole scaffolds as potential conductive nerve conduits, Acta Biomater. 7 (2011) 944–953.
[32] K. Merrett, R.M. Cornelius, W.G. Mcclung, L.D. Unsworth, H. Sheardown, Surface analysis methods for characterizing polymeric biomaterials, J. Biomater. Sci. Polym. Ed. 13 (2002) 593–621.
[33] T. Matsunaga, Y. Ikada, J. Colloid Interface Sci. 84 (1981) 8.
[34] D. Li, A.W.J. Neumann, J. Colloid Interface Sci. 148 (1992) 190.
[35] K.L. Menzies, L. Jones, The impact of contact angle on the biocompatibility of biomaterials, Optom. Vis. Sci. 87 (2010) 387–399.
[36] C.O. Connell, R. Sherlock, M.D. Ball, B.A. Kiss, U. Prendergast, T.J. Glynn, Investigation of the hydrophobic recovery of various polymeric biomaterials after 172 nm UV treatment using contact angle, surface free energy and XPS measurements, Appl. Surf. Sci. 255 (2009) 4405–4413.
[37] H.S. Yang, J. Park, W.G. La, H. Jang, M. Lee, B.S. Kim, 3,4-Dihydroxyphenylalanine-assisted hydroxyapatite nanoparticle coating on polymer scaffolds for efficient osteoconduction, Tissue Eng. Part C Methods 18 (2012) 245–251.
[38] L.A. Klerk, P.Y. Dankers, E.R. Popa, A.W. Bosman, M.E. Sanders, K.A. Reedquist, R.M. Heeren, TOF-secondary ion mass spectrometry imaging of polymer scaffolds with surrounding tissue after in vivo implantation, Anal. Chem. 82 (2010) 4337–4343.
[39] J.W. Lee, J.A. Gardella, Macromolecules 34 (2001) 3928.
[40] M.E. Gomes, H.L. Holtorf, R.L. Reis, A.G. Mikos, Influence of the porosity of starch-based fiber mesh scaffolds on the proliferation and osteogenic differentiation of bone marrow stromal cells cultured in a flow perfusion bioreactor, Tissue Eng. 12 (2006) 801–809.

Chapter 3

Proteins and Poly(Amino Acids)

Tarun Saxena*, Lohitash Karumbaiah*, Chandra M. Valmikinathan†
*Department of Biomedical Engineering, Georgia Institute of Technology, Atlanta, Georgia, USA
†Global Surgery Group, Johnson and Johnson, Somerville, New Jersey, USA

Chapter Outline

3.1 INTRODUCTION

The use of implantable materials for tissue repair and regeneration and drug delivery applications is a reality, owing to the advances in tissue engineering, bioengineering, and biomaterials science. Typical examples include artificial organs [1–3], implantable drug-releasing modules [4–7], and conduits that support regeneration [8–11]. A key challenge for these implantable devices/materials is to maintain functionality over long periods of time. However, limited functionality often results due to implantation injury, lack of biocompatibility, and the foreign body response at the host/material interface. The immune response to an implanted material determines the biocompatibility of the material [12–15]. Therefore, naturally occurring biomaterials and their derivatives offer a viable alternative over synthetic materials, as they offer superior biocompatibility that improves the host/material interactions, thereby improving device functionality and longevity [16–18].

One of the key guideposts for biomaterial design is the extracellular matrix (ECM). The ECM of cells is a complex network on biomacromolecules including proteins and polysaccharides. Two main classes of biomacromolecules are found in the ECM of tissues: (1) glycosaminoglycans, which are polysaccharides and (2) fibrous proteins such as collagens, laminin, elastin, and fibronectin [19]. All ECMs provide structural support and serve as a reservoir of growth factors and cytokines and modulate the presentation, synthesis, and degradation of these factors to the surrounding cells. The composition and organization of the ECM changes according to the tissue it consists of and determines the tissue physical properties. Thus, providing structural support and modulating the activity of growth factors spatiotemporally in cellular microenvironments makes the ECM an ideal scaffold for tissue repair and reconstruction and the gold standard for tissue regeneration strategies. These strategies employ biocompatible materials that seek to provide adequate mechanical and structural support, actively guide and control cell fate, and supply biological signals via trophic factor sequestration. Further, these materials have to offer tunable biodegradability as well as provide control over the sequestration and delivery of specific bioactive factors to enhance and guide the regeneration process over long time periods [3,6,14,16,20–26]. Thus, one paradigm of approaching tissue regeneration using materials-based strategies is to view the material as a structural scaffold that also offers controlled release of bioactive molecules such as proteins and drugs.

The focus of this review is drug delivery and tissue regeneration using natural biomaterials consisting of proteins and polymer chains of naturally occurring amino acids, or poly(amino acids) (PAAs). PAAs or polypeptides are biopolymers made from repeating units of amino acids. They differ chiefly from proteins in that they contain only one type of amino acid (monomer) and are polydisperse, whereas proteins are assemblies of various amino acids and are monodisperse [12]. In both proteins and PAAs, the amino acid side chains offer sites for attachment of various moieties that can modify the physical and biochemical properties of the original biopolymer [6,25]. Proteins such as collagen, elastin, fibroin (silk), and PAAs such as poly-L-lysine), poly(glutamic acid), and poly(aspartic acid), which have been discussed in this chapter, have been investigated as potential scaffolds for tissue engineering applications [6,12,25]. PAAs are prone to enzymatic hydrolysis and allow for tunability in degradation rates via modification of their side chains. Further, the products of enzymatic hydrolytic degradation are readily eliminated by virtue of being naturally occurring metabolites, making them ideal candidates for tissue engineering applications [12,27,28]. This book chapter concisely reviews some of the naturally occurring ECM proteins and PAAs used in tissue engineering applications for controlled drug release and tissue regeneration of various organ systems. In this review, we focus on approaches that aim to create fully biological materials without the use of synthetic materials for support. Further, while there is a large body of work in these areas, we have focused our attention on advances that have been made in the last decade with the intent to provide a concise update of recent applications of protein- and PAA-based biomaterials strategies. We do not intend to neglect the early pioneering contributions of various researchers. The reader is encouraged to peruse the cited references (which include several excellent reviews) for in-depth understanding of the various experimental and manufacturing strategies.

3.2 FIBRIN-BASED BIOMATERIALS

Fibrin is a polymer of choice for tissue engineering applications due to its porosity [29], biodegradability [30], and ability to provide mechanical support [31], promote cell attachment, proliferation [32], and angiogenesis [33], all of which are desirable properties for any biomaterial-based tissue engineering strategies. Fibrin-based hydrogels, glues, sprays, and microbeads have been used to evaluate cell adhesion, proliferation, and differentiation *in vitro* and for the repair and regeneration of a host of tissues and organs *in vivo*.

Fibrin is a fibrous protein monomer derived from the thrombin-mediated cleavage of fibrinopeptide A and fibrinopeptide B, which belong to the Aα and Bβ chains of fibrinogen [34,35]. Fibrin monomers thus formed are subsequently cross-linked via covalent bonding of lysine and glutamine residues of individual γ-chains by the transglutaminase, factor XIIIa, producing a protease-resistant fibrin mesh [36–39]. Fibrin's role as the primary scaffolding material surrounding platelets in blood clots prompted its use as a sealant, as well as a hemostatic, and therapeutic agent in a number of applications involving cardiovascular [40–42], liver [43,44], kidney [45], neurosurgical [46,47], orthopedic and periodontal [48,49], wound healing [50], and cosmetic- and reconstructive-related [51,52] surgical procedures, among others. Although there are a number of clinical applications of fibrin-based sealants, this section will briefly discuss the use of fibrin-based materials for tissue engineering applications. For a more in-depth review of the tissue engineering applications of fibrin, the reader is referred to Ahmed *et al.* [35].

The ability of a target biomaterial to control cell adhesion, proliferation, and differentiation into the desired phenotype for regenerative and repair outcomes *in vivo* begins with the evaluation of these properties *in vitro*. Fibrin-based hydrogels prepared from porcine fibrinogen and thrombin have been shown to support encapsulation and growth of chondroprogenitor cells that could be used for the replacement and repair of abnormal cartilage. Encapsulated cells were shown to eventually produce matrix metalloproteinases, which induced breakdown of the fibrin gels eventually [53]. Fibrin hydrogels can be further cross-linked with ECM in order to prevent their rapid degradation, to enhance cell adhesion, and to provide bioactive surfaces for growth factor binding and release. These strategies have been successfully used to influence neurite extension from chicken dorsal root ganglia *in vitro* [36,54]. Other approaches to overcome the rapid degradation of fibrin hydrogels involve the use of fibrin microbeads (FMBs). Using FMBs, high yields of purified bone marrow-derived mesenchymal stem cells (BMSCs) and MSCs from human peripheral blood were achieved [55,56].

Fibrin glues, microbeads, and hydrogels have also found extensive use in several *in vivo* applications. Dog MSCs containing platelet-rich plasma and fibrin glue when simultaneously administered along with dental implants significantly enhanced bone formation when compared to controls [57]. In other studies, bone morphogenetic protein-2 (BMP-2) was modified to bind fibrin glue and implanted in a rat cranial defect, which resulted in increased bone formation [58]. Platelet-rich fibrin glue containing BMSCs and BMP-2 when injected into nude mice induced the formation of bone tissue after 12 weeks [59]. More recently, animals subjected to a middle cerebral artery occlusion when treated with induced pluripotent stem cells (iPS) mixed with fibrin glue showed reduced infarct sizes and improved motor function [60]. Fibrin bead implants seeded with perichondrial cells when implanted into physeal defects of rabbits demonstrated longer tibial length and reduced deformities when compared to the untreated contralateral tibiae [61]. Fibrin cable formation is a key determinant of the nerve regeneration following peripheral nerve injuries. Therefore, the use of fibrin as a scaffolding material is a rational strategy being employed to facilitate nerve repair and regeneration after peripheral nerve injuries. Fibrin glue containing the neurotrophic factors has been reported to facilitate nerve repair and hind limb function in paraplegic adult rats by mediating regeneration of the injured spinal cord [62,63]. Fibrin hydrogel containing nerve guides when implanted in a rat sciatic nerve defect induced nerve regeneration and superior myelinated axon numbers when compared to other nerve guides [64]. Fibrin-based hydrogels have also been used for the repair of damaged cardiac tissue. Adult rats implanted around the femoral artery with silicon tubing filled with fibrin gel showed the formation of myocardial tissue 3 weeks postimplantation [65]. Since fibrin is a single protein lacking the complexity of ECM required for advanced tissue engineering applications, more advanced tissue engineering strategies demand the use of a combination of cell transplantation, ECM presentation, and biocompatible scaffolding materials for successful outcomes. More recently, Godier-Furnemont *et al.* [66] demonstrated that cell-matrix composites consisting of fibrin hydrogels encapsulated human mesenchymal progenitor cells and decellularized sheets of human myocardium when implanted into a nude rat model of cardiac infarction, induced enhanced vascular network formation in the infarct bed and recovery of left ventricular systolic activity.

There is extensive evidence to support the use of fibrin as a biomaterial of choice for a range of *in vitro* and *in vivo* applications. It lends itself to be modified to suit the application with precise control over material architecture and structure. The ability to control degradation by coupling ECM proteins and growth factors to fibrin-based biomaterials enables control over cell adhesion, proliferation, migration, and cell-cell interactions and makes it an ideal scaffolding material for *in vivo* cell transplantation and 3D tissue engineering strategies.

3.3 ELASTIN-BASED BIOMATERIALS

Elastin is the second most common structural component in the ECM, second only to the family of collagens [67]. It is the major component of the elastic matrix in mammalian tissues, which provides the mechanical properties and elasticity to the matrix [68,69]. In the succeeding text, we discuss some critical factors in the choice and application of elastin as a biomaterial including synthesis and production of elastin, biomechanical and biochemical roles of elastin, and current technologies for three-dimensional (3D) matrices produced using elastin.

The ECM of several tissues, including vascular and connective tissues, consists primarily of a network of elastic fibers. These fibers provide the mechanical properties associated with the tissues and their ability to elongate and recoil after stretch [70–72]. Elastic fibers are composed of cross-linking elastin protein and surrounded by a variety of fibrous proteins such as fibrillin [73]. The process of new elastin fiber formation is called elastogenesis [74]. Typically, *in vivo*, this process involves assembly of tropoelastin, the monomer unit that cross-links to form the polymer, elastin [75,76]. During the process of elastogenesis, a critical enzyme called lysine 6-oxidase [77] or, otherwise, lysyl oxidase plays a significant role in the cross-linking process [78]. Lysyl oxidase is a cuproenzyme, which functions in their presence of copper as a chelating agent, stabilizes the ECM, and catalyzes the cross-linking of elastin [79]. During prenatal development, elastogenic cells, such as vascular endothelial cells, smooth muscle cells, and skeletal muscle cells, synthesize tropoelastin [80], which is further regulated to maturation and helps in synthesis of 3D strands of tropoelastin [81]. This further associates with elastin-binding protein (EBP) and is secreted to the cell surface for further polymerization to form elastin strands [82]. Once at the surface, the elastin-binding peptide dissociates from the tropoelastin, which is mediated by competitive inhibition from galactosides [83,84]. Meanwhile, the secreted tropoelastin is deposited onto the ECM and forms microfibrils, which further mature into stable elastic fibers. This process of cross-linking to yield insoluble elastin fiber is catalyzed by lysyl oxidase, as described earlier [79].

The location and the role of the organ or tissues define the arrangement of elastin in the ECM. For instance, in the arterial walls, elastin 3D structure is composed of thin laminar architectures and arranged radially and imparts extreme elasticity, necessary for constant cyclic expansion and contraction to regulate consistent blood flow across them [85]. On the other hand, in the lungs, elastin exists as a lattice and spreads across the surface of the organ, allowing for continuous expansion and contraction during breathing [86]. In the dermis, elastin plays a different role, where it acts a mechanical barrier to maintain stiffness and prevents wrinkling and maintains tautness to act as a barrier layer [87].

Dysfunction in elastin or mutations in the gene that produces tropoelastin could have tremendous consequences in synthesis, properties, and functions of the tropoelastin and eventually elastin [88]. As elastin plays a critical role in the structural integrity of several tissues including skin, blood vessels, and other organs, dysfunctions in its structures could lead to scarring, unusual wound contractions, loss of elasticity, leaky blood vessels, and loss of elastic properties of the skin [89]. More significant implications can arise from elastin dysfunction in the organs and vasculature. Apart from structural roles, elastin in the ECM also plays significant roles in the maintenance of SMC populations and controlling migration. For instance, in the vasculature, elastin misfolding and insufficient elastin formation results in severe, debilitating diseases such as aortic stenosis or stiffer blood vessels and obstruction of the aorta due to unregulated smooth muscle cell proliferation [90,91]. Therefore, it is important to understand the properties of elastin for application in most suitable approach as biomaterials or scaffolds.

Most native elastic tissues consist entirely of elastin. As one would imagine from the name, the elastic properties imbued on the tissues arise from the aligned elastin fibrils in the tissue. The elastic modulus of the fibers varies largely depending on their location and function. In some cases, elastin solely acts as a barrier layer (skin) versus in other where it is present to serve in resisting infinite expansion of tissues and in other where it is present as a signaling material to regulate cell functions [92]. For instance, in blood vessels, the modulus typically ranges from 50 to 200 kPa, where the expansion is radial and is limited. On the other hand, in the heart, the modulus peaks to 300-600 kPa, due to larger volume flux and restricted regenerative capacities, in case ECM damage occurs [93]. The variations in the elasticity are affected entirely by number of tropoelastin monomers and the number of coils or strands of tropoelastin that might be present in each fiber [81,94]. It is also interesting to note that the elastic mechanism is believed to be entropically driven, wherein the extension of the protein fibers results in a more ordered 3D structural organization, and upon constriction or recoil, a disorderedly state is attained [95].

In other cases, where elastin serves in ECM to control cell and matrix function, elastin has lesser tropoelastin chains and, along with collagen, plays very important roles in the regulation of cells, especially in elastic tissues such as blood vessels [96]. Several studies have shown the roles of elastin during arterial morphogenesis, via knockout experiments. Significant differences in SMC proliferation were observed in cases where tropoelastin was knocked out [96,97]. In another study, elastin knockout mitigated SMC proliferation and the effect was shown to be dose-dependent [98]. Similar effects have been demonstrated in other studies with endothelial cells [99]. Cell proliferation and differentiation was reduced due to elastin knockout. Several other studies have also shown very similar effects with dermal fibroblasts, where elastin plays a very important role in the matrix as a defense mechanism as described earlier. Dermal fibroblasts failed to attach and proliferate on the elastin deprived matrices [89,100].

Several surface-bound receptors on cells, including EBP, have been studied to evaluate cell-matrix interactions, especially with elastin-derived matrices. Several studies have described the role of EBP elastin-binding interactions and downstream effects including cell proliferation, chemotaxis, and changes in cell morphology for several cell types including smooth muscle cells, endothelial cells, fibroblasts, and MSCs [101–104].

ECM interactions of cells with elastin define several cell process parameters such as migration, focal adhesion formation, cell proliferation, and phenotypic differentiation (in stem cells) [105–108]. Integrin $\alpha v\beta 3$ plays very important role on the cell surface during interactions with elastin in the matrix [100,109]. In another study, the chemotoxic effect of elastin was described elegantly. Indik *et al.* [102] showed that elastin-containing extracellular matrices attracted more SMCs and endothelial cells [101,110], as compared to other surfaces including collagen. This fact has been attributed to the cell-surface receptor interactions with elastin via integrins.

Elastin-containing tissues are prevalent in the human anatomy, especially in the vascular tissues, cardiovascular tissues, dermis, and muscles [111]. Contemporary approaches to replace elastin-based extracellular matrices employ the application of synthetic biomaterials such as polytetrafluoroethylene and polyethylene terephthalate (Dacron) [112–115]. However, these materials are poor substitutes due to mechanical mismatch and inability to be remodeled and more importantly are not biodegradable, *in vivo* [116]. Furthermore, their interactions with native ECM and associated cell types cannot be mimicked appropriately, leading to increased thrombogenicity and therefore high failure rates, postimplantation [117]. In order to overcome this limitation, several studies have detailed the fabrication of pure cross-linked elastin- or tropoelastin-based biomaterials or by surface coating PTFE or other grafts described earlier with elastin mimetic peptides or the whole protein itself [118–120]. Cross-linked 3D elastin-based scaffolds or matrices are traditionally fabricated using two approaches, one by cross-linking of tropoelastin- or elastin-like peptides with appropriate enzymes or chemical moieties or from decellularized native tissues from allogeneic or xenogeneic sources. More recently, synthetic peptides designed to mimic elastin's biological properties that contain appropriate end groups for cross-linking and forming hydrogels or nanofibrous materials have also been developed [121–123].

Tropoelastin is the precursor to elastin *in vivo* during matrix development and is the most common approach to generate elastin-based constructs for regenerative medicine. However, obtaining significant amounts of tropoelastin from tissues is challenging [124]. Recombinant human tropoelastin (rhTE), expressed by *Escherichia coli*, is the most common source of tropoelastin [124,125]. The recombinant tropoelastin exhibits all critical properties exhibited by native tropoelastin, including the coacervation ability at physiological conditions and availability of end groups for cross-linking to form insoluble elastin fibers or matrices [126,127]. Functionally, rhTE showed enhanced endothelial cell attachment and proliferation [89,100] and as a surface coating demonstrated improved biocompatibility of implanted devices [110].

Hydrogel matrices fabricated by physical cross-linking of recombinant TE have been used diversely for regenerative medicine applications [128–130]. Several chemical cross-linking approaches have been evaluated over the last decade to produce matrices from rhTE, including chemical cross-linking [122], enzyme treatment [93], or pH-induced [131] gelation. The hydrogels hence formed demonstrated similar mechanical properties to native elastin and supported attachment and proliferation of dermal fibroblasts [122,128,130]. Newer approaches to generate porous elastin matrices have been developed employing supercritical carbon dioxide with and without glycosaminoglycans to generate living dermal matrices [132].

Understanding of properties, binding sites, and other key peptides in the elastin sequence has led to the development of synthetic elastin-based peptides, which could self-assemble to form fibrous hydrogel-like matrices [106,123]. Applications of these hydrogel-based matrices can be found widely in regenerative medicine and drug delivery applications, especially in cardiac or vascular tissue engineering [71,75,123,133]. 3D matrices formed with elastin-based peptides have demonstrated increased cell attachment and proliferation and, upon matrix formation and differentiation, form high-density elastic fibers, which are ideal for tissue regeneration [72,121]. As described earlier, synthetic grafts could provide mechanical properties necessary for tissue replacements, but do not provide ideal matrices for enhancing cell interactions. In order to enhance biocompatibility and biological properties of synthetic matrices, elastin-based peptides have been used as a coating material [70] (Figure 3.1).

The process of decellularization traditionally involves the removal of cells and cellular components, leaving behind an intact matrix. Decellularized materials have been used in the clinic widely and have received tremendous interest owing to its ability to induce a favorable cell response and migration, leading to better regeneration of damaged tissues. This has been attributed to the nature of the ECM and the cues it provides necessary to induce endogenous cell migration. Decellularized tissues primarily composed of elastin have been previously developed and have been used in the regeneration of tissues such as blood vessels, heart valves, bladder skin, and lungs [71,135].

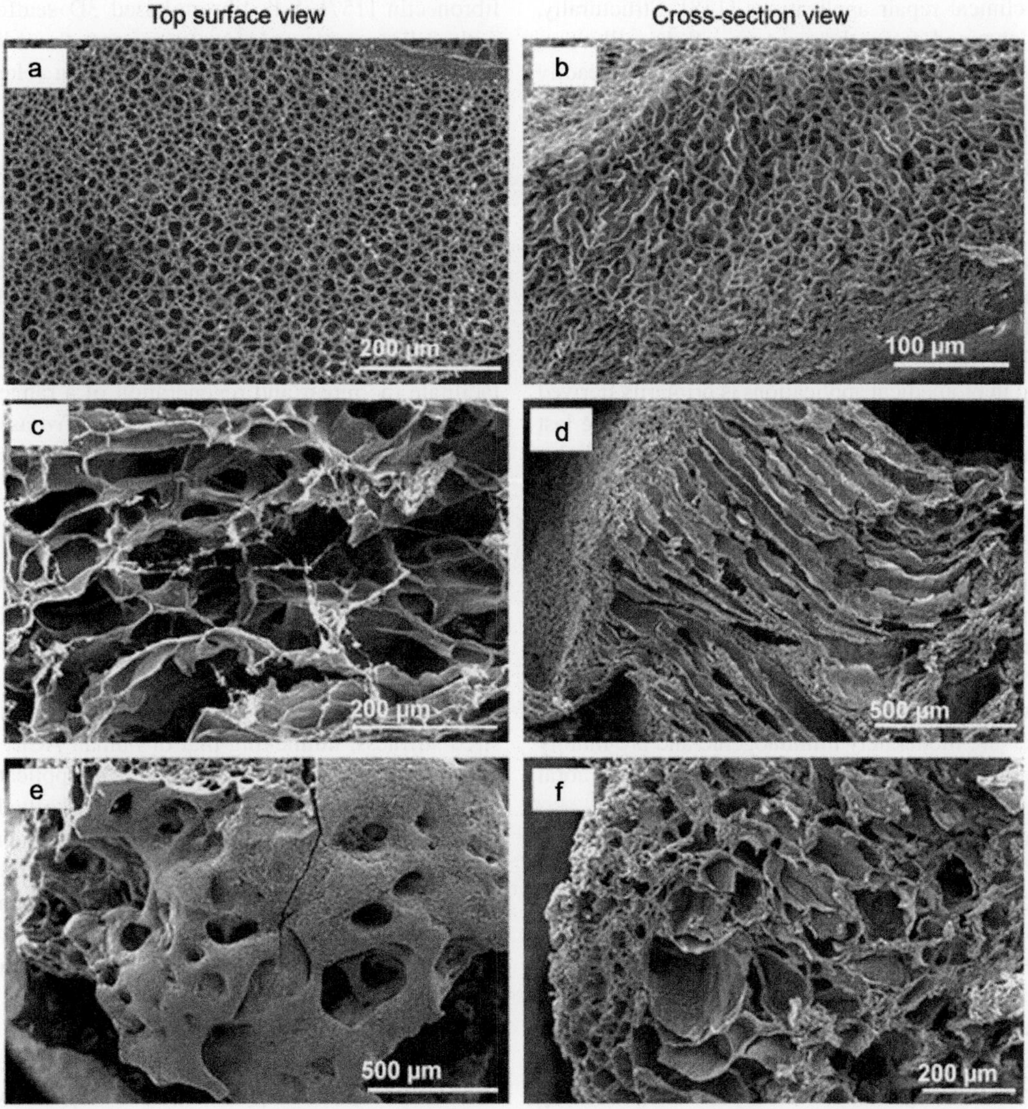

FIGURE 3.1 Scanning electron micrograph (SEM) images of α-elastin hydrogels fabricated at atmospheric pressure using 2% HMDI (a, b), 60 bar CO_2 pressure using 2% HMDI (c, d), 60 bar CO_2 pressure using 5% HMDI (e, f). With permissions from Annabi *et al.* (Elsevier) [134].

Furthermore, enhanced grafts composed of elastin only have been developed by digesting the remaining collagen in the matrices with collagenases to retain intact elastin. Studies by Chuang *et al.* have shown that such elastin-rich matrices have sufficient mechanical properties and can support fibroblast proliferation and migration [136]. Dermal matrices derived from acellular porcine dermis, which contains a significant amount of elastin (up to 30%), have shown enhanced potential to vascularize and superior regeneration in a rat excision model [137].

3.4 SILK-BASED BIOMATERIALS

Silk fibroin is a fibrous protein secreted by the silkworm *Bombyx mori*, as well as by a number of different species of spiders. Used originally as a suture to facilitate wound approximation and/or ligation, silk has since been used for a range of clinical repair applications [138]. Structurally, insect silk obtained from the cocoons of the silkworm *B. mori* is composed of two distinct proteins: the readily water-soluble sericin and fibroin, which was first shown to be dissolvable in aqueous inorganic salt solutions by von Weimarn [139,140]. The solubilized fibroin could then be neutralized and dialyzed, resulting in a water-soluble form of fibroin [141]. Solubilized fibroin thus obtained consists of segments primarily in the α-form, which are unstable and readily transition to the β-form, rendering the protein insoluble and giving it its fibrous structure and mechanical strength [140].

Silk used in biomedical applications is primarily derived from *B. mori*. Silk has been extensively used over the last 100 years as the suture material of choice for wound closure [142], owing to its optimal strength, biodegradability, elasticity, and minimal tissue reactivity. The induction of allergic and inflammatory reactions in virgin silk suture implanted patients [143] has been attributed to the presence of sericin in these preparations [144–146]. These adverse reactions prompted the removal of sericin and coating of sericin-free silk fibroin with waxes or silicone. The resulting degummed and coated silk sutures, referred to as "black braided silk," are moderately immunogenic and are mainly used in surgical applications such as eye, lip, and intraoral surgeries [147].

Silk is also increasingly being used in the form of fiber ropes [138], nonwoven mats [148], films [149], porous 3D matrices [150–152], and gels [153] for a host of tissue engineering applications. A majority of cells require cell-surface receptor interactions with the underlying ECM to adhere, proliferate, and differentiate. In addition to receptor-ECM interactions, cells in culture also rely on nonspecific surface interactions involving charge and polarity to adhere [154]. Silk fibroin matrices, gels, and films have been previously shown to support cellular adherence and growth of a variety of different cell types either in their native state or as ECM-coated substrates *in vitro*. Silk fibroin films loaded with nerve growth factor (NGF) were previously shown to support adherence and neurite outgrowth from PC12 cells when used in nerve conduits [155]. Silk fibroin films decorated with the RGD integrin recognition sequence and parathyroid hormone amino acids reportedly demonstrated osteoblast-like cell formation along with the expression of bone matrix-inducing factors, demonstrating their ability to induce bone formation [156]. Silk fibroin rope matrices and films functionalized with RGD peptides also showed greater attachment and proliferation of human bone marrow stromal cells (BMSCs) and anterior cruciate ligament (ACL) fibroblasts, along with the significant upregulation of collagen type-1 and decorin transcripts when compared to unmodified matrices and films. Oriented growth of human mesenchymal stem cell cultures was also achieved by functionalizing electrospun silk fibroin aligned fibers with fibronectin [157]. Silk fibroin-based 3D scaffolds coated with collagen were shown to support the adhesion and growth of liver hepatocellular carcinoma (HepG2) cells, owing to their uniform pore distribution and hydrophilicity [158]. Nonwoven silk fibroin nets without any ECM coatings were also shown to facilitate the adherence and growth of human endothelial, epithelial, glial, keratinocyte, and osteoblast cells for at least 7 weeks [159]. These studies along with several others reviewed elsewhere [160] demonstrate the biocompatibility and use of silk-based materials as a scaffolding biomaterial for clinical applications.

Silk-based biomaterials have also been used *in vivo* to facilitate the repair and regeneration of a range of defects including for bone and ACL repair. *In vivo* assessment of the inflammatory response to implanted unmodified and RGD-modified silk constructs indicated that they induced a more muted immune response when compared to collagen and polylactic acid (PLA) films [161]. Previous studies using collagen fibers and collagen-PLA composite fibers in rabbit ACL defects showed inconclusive results wherein significant degradation of the ACL prosthesis was observed [138,162,163]. The slow degradation times of silk coupled with the ability to tune silk-based biomaterials to the desired stiffness, mimicking that of human ACL, make it an optimal material for use in these sorts of applications [164]. The *in vivo* implantation of MSC-laden knitted silk mesh scaffolds in rabbit and pig ACL defects showed remarkable ligament regeneration, maintenance of tensile load, and ACL-bone insertion characteristics similar to that of the native ligament. Uniform distribution of MSCs throughout the regenerated ligament and expression of collagen and tenascin ECM components were also observed [165,166]. More recent work using low-crystallinity hydroxyapatite-silk scaffolds showed encouraging bone/ligament regeneration as assessed by histology and microcomputed tomography over a period of 4 months [167]. Porous silk fibroin scaffolds have been used to engineer bone tissue in a bioreactor using human BMSCs cultured for a period of 5 weeks. The resulting bone tissue that was implanted into calvarial critical-sized defects in mice resulted in advanced bone formation when compared to silk scaffolds alone or scaffolds laden with BMSCs within a period of 5 weeks [168]. Injectable silk fibroin hydrogels have also been used to study bone formation both *in vitro* and *in vivo* [169]. When compared to the biodegradable polymer poly(D,L lactide-co-glycolide) (PLGA), injectable silk hydrogels facilitated significantly higher bone formation rate, trabecular volume, thickness, and significantly lower trabecular separation when assessed in a rabbit distal femur critical-sized defect.

Insect and spider silks are natural biopolymers whose molecular structure enables their use in applications requiring exceptional strength and flexibility of the material. These traits along with their biocompatibility, biodegradability, and the ability to produce large amounts of the material make the use of silk and silk-based biomaterials a rational choice for a host of tissue engineering applications.

3.5 COLLAGEN-BASED BIOMATERIALS

Collagens are the most abundant proteins of the ECM in vertebrates and play a dominant role in maintaining the biological and structural integrity of ECM [19]. To this date, 29 different collagens have been found, classified based on their structure. The primary structure of a collagen is a triple-stranded helix consisting of collagen polypeptide-alpha chains. Collagens have a simple repetitive [Gly-Xaa-Yaa]$_n$ sequence motif, where X and Y are often proline and hydroxyproline, that can be used as a building block for self-assembly into complex hierarchical structures such as fibrils, fibers, and bundles [19,170–172]. Collagens are rich in proline and glycine and usually form strong, long, cross-linked fibrils. Collagen fibrils have high resistance to tensile forces and hence are major constituents of connective tissue, skin, tendons, blood vessels, cornea, cartilage, and bone. Collagen type I is abundant in vertebrate tissues and collagen is usually extracted from the skin and tendon of the source. During the past decade, collagen-based biomaterials have been used for applications ranging from injectable collagen matrices for drug and protein delivery to bone regeneration scaffolds [170,173–176]. Collagen-based materials are defined as materials that contain collagen that has undergone some processing, for example, cross-linking or reinforcement with a different material. Collagen is highly versatile in terms of the physical forms collagen-based materials can be fabricated into such as injectable gels, films, meshes, and fibers for biomedical applications including 3D cell culture [176–181], bone and cartilage repair [182–184], vascular and cardiovascular grafts [185–188], skin grafts [189–192], corneal defects [193–195], peripheral nerve repair [8,10,11,196,197], spinal cord and traumatic brain injury repair [198–205], wound dressings [191,206–209], neural interfacing [210], and various other applications [211–219]. Several FDA-approved collagen-based biomaterials are available for the aforementioned applications [6,25,184,220]. Collagen-based biomaterials are fabricated using two fundamental techniques. The first technique involves decellularizing a tissue or matrix that contains collagen, thereby preserving the original tissue shape and ECM structure. Acellularized matrices are produced using physical, chemical, and enzymatic methods such as snap freezing, detergent washing, and trypsinization, respectively [174,175,221], generally used in combination. The second technique involves extraction, purification, and polymerization of collagen from synthetic, plant-based, or animal sources to form functional scaffolds. The reader is encouraged to read the many excellent reviews cited here for an in-depth understanding of the various processing techniques and applications of collagen-based biomaterials. Following these extraction methods, these collagen-based biomaterials often involve excessive pre- and postprocessing steps, such as cross-linking, to form the final material [170,172–175,221]. This processing is necessary because the extraction methods usually lead to a mechanically and thermally unstable product [222,223]. As a result, this processing invariably changes the normal signaling achieved by collagens due to changes in primary structure, epitope display, and different peptide fragments that could be released upon degradation of the materials [171,174,175,224,225]. While collagen-based materials have been used in various biomedical applications, pure forms of collagens, which can provide greater physiological and biological integration, have not been viable. There are two major issues with using pure forms of collagen for *in vivo* applications. First, only limited supplies of the protein in pure form are available and are predominantly from animal (bovine) tissues. This animal origin is associated with significant safety concerns due to immunogenicity arising from the source-donor species mismatch and the danger of potential carryover of animal-derived diseases. While other limited sources of animal collagens are also available commercially, there are economic and structural limitations such as high cost of production; structural diversity of acellular collagens; and the accompanying immunologic consequences, yield, and availability of these sources. The second issue involves challenges associated with *ex vivo* engineering of collagens, as it involves complex processing and assembly into functional biomaterials that retain the structural and physiological complexity of natural collagens, and most collagen mimics or synthetic collagens lack the higher-order supramolecular structure of native collagens [170–172,174,175,221,224,226]. An area of intense research focus is recombinant production of collagen in systems such as plants, yeast, bacteria, and *in vitro* cell culture. For drug release applications and tissue repair, pure collagen gels degrade rapidly due to their high hydrophilicity that leads to swelling and more rapid drug release and disintegration compared with synthetic polymers. Increasing the amount of collagen can make the gels stronger but leads to nonphysiological concentrations of collagen that causes poor physiological integration as the diffusion of growth factors is hindered.

Finally, collagen gels have variable enzymatic degradation rates leading to poor drug release and lack of integration with host tissues [176,212]. Overall, although extensively researched and holding tremendous biomedical potential, collagen and collagen-based biomaterials will continue to face significant limitations until alternative biological sources can be found, or recombinant and synthetic production [193,227–230] of pure collagen is accomplished.

3.6 POLY(GLUTAMIC ACID)-BASED BIOMATERIALS

Among the three different poly(amino acids), poly(glutamic acid) (PGA) is the most abundant and widely used. The reasons for its wide applications can be attributed to its safety profile, degradability to nontoxic components, and its edibility. PGA has been obtained in nature as well as can be synthesized in several ways. Some of the most common natural and synthetic pathways have been discussed in the succeeding text.

Several classes of bacteria synthesize poly(glutamic acid). The primary reason of PGA in bacterium in nature is for protection against hostile environment. The most common reasons for synthesis of capsule or PGA-based matrix are for protection against antibiotics, viruses, or phage and for protection during their senescence period. Some of the most common bacterial strains that produce PGA are *Bacillus licheniformis*, *Bacillus subtilis*, *Bacillus megaterium*, *Bacillus pumilus*, *Bacillus mojavensis*, and *Bacillus amyloliquefaciens*; however, *Bacillus licheniformis* and *Bacillus subtilis* have been most commonly used in industrial-scale production of PGA, via fermentation [231]. The diverse physiological properties and functions can be attributed to the species and the environmental conditions, where they were raised. Primarily, PGA serves as a source of glutamate for bacteria during the latent phase [232,233]. PGA also serves as a mode of protection. For example, in the case of highly pathogenic bacteria such as *Staphylococcus epidermidis*, PGA is synthesized and secreted on the surface to enable them to escape phagocytosis [234]. Also, PGA has been associated with antibody resistance. For example, the outer capsule synthesized by *B. anthracis* is capable of preventing antibodies from penetrating the bacterial walls or colonies [235,236]. Recent studies have led to better understanding of PGA synthesis, factors affecting quality of PGA, etc., and have been employed for successful biosynthetic production of PGA.

Several lines of evidence suggest that biosynthesis of PGA is carried out in two steps. First, synthesis of L- and D-glutamic acid occurs, and in the second, it is polymerized to form PGA. Shih *et al.* have described this entire process in details. This review also discusses briefly several parameters, enzymes, and precursors necessary during the synthesis and polymerization process [237]. Over the last decade, cloning of PGA biosynthesis genes for PGA production has drawn tremendous attention. Sung *et al.* have recently reported the cloning and expression of three specific genes, encoding a PGA synthetic system in *E. coli* [238]. However, this approach secreted PGA that was extracellular, but the genetically modified strain was found to produce sufficient PGA in both L-glutamic acid- and DL-glutamic acid-rich media and in large quantities.

Several bacterial strains have been studied for development of suitable fermentation process and scale up of PGA synthesis [231,238]. Since PGA primarily serves as a capsule or as ECM for bacterium, it is viscous. Therefore, fermentation-based approach of PGA is quite challenging and needs very strict control on the factors such as temperature, nutrient conditions, and humidity; as one would imagine, one of the most important ingredients in the synthesis of PGA is glutamic acid. However, there are several strains of bacteria that synthesize PGA without glutamic acid as the starting material. Furthermore, apart from glutamic acid, several other factors such as nutrients rich in nitrogen sources, medium pH, and aeration affect the quality and efficiency of PGA production.

It has been well established that L-glutamic acid plays a very significant role in the PGA production process. The conversion rate of L-glutamic acid to PGA is an important factor that determines the effective concentration of L-glutamic acid in the medium. Several studies suggest using a range 20-30 g/l of L-glutamic acid for the production of PGA [239,240]. Also, several studies have shown that L-glutamic acid is produced in the tricarboxylic acid cycle (TCA) and is then polymerized to PGA [241,242]. Therefore, citric acid is a critical component needed for successful synthesis of PGA. Furthermore, a source of carbon, primarily in the form of sugars such as glucose, is necessary for successful production of glutamic acid. Studies by Ko *et al.* suggested that PGA synthesis by a commercially available class of bacteria, *B. licheniformis*, occurs by conversion of glucose to acetyl-CoA and the TCA cycle intermediates, which then forms L-glutamic acid [242]. Also, several enzymes such as polyglutamyl synthetase that catalyzes the polymerization of glutamic acid to PGA are a vital cog in the production process [243].

Some of the most interesting properties of PGA that make it an ideal material for biological applications are its molecular weight, viscosity, biodegradability, and most importantly its vast availability[237,240,244]. It has found profound applications in the areas of development of scaffolds, drug and gene delivery, and tissue adhesive and sealant and coatings for several applications. Some of the most relevant properties of PGA that make it a unique material for such applications are described in the following sections.

One of the most critical properties of PGA for the application in regenerative medicine and drug delivery is its molecular weight. Higher molecular weights are associated with more surface charge, higher viscosity, and reduced normal tissue penetration, allowing for applications in tumor drug delivery, which has been discussed in detail in the second half of this chapter. Using GPC, the estimated molecular weight of biosynthesized PGA is in the range of 105-106 Da [237]. However, the molecular weight of PGA is variable and dependent on the bacterial strains, medium components [245], and culture conditions [239] as described earlier. Furthermore, it is challenging to obtain a narrow range of molecular weight distribution by biosynthetic methods, primarily due to its structural complexity and variations in strains and synthetic outputs [240,246].

The other critical aspect of using PGA as a starting material for biomedical applications is its degradability. The most common approaches for degradation have been either enzymatic or by the application of alkaline conditions or heat. Troy *et al.* first reported the biodegradation of PGA in the presence of depolymerase enzyme [247]. It has been identified that the enzymes are typically located in the cytoplasm of the cells or in some cases bound to the cell surface, allowing for matrix degradation and migration [248].

It has also been established that alkaline hydrolysis under mild conditions degrades microbial PGA. However, since this method is very nonspecific, tight control over the degradation process and molecular weights obtained at the end of the process cannot be controlled [249]. Also, this method could potentially lead to denatured product, if downstream applications are the goal of the application. Therefore, it is critical that neutral pH conditions are evaluated. Goto *et al.* showed that at pH 7.0 with heating at 80-120 °C, random degradation or chain cleavage can occur leading to reduced molecular weights and degradation [250]. It was also reported that the rate of hydrolysis or degradation was directly proportional to temperature. Also, the size or molecular weights were inversely proportional to the temperature. Therefore, this method afforded tighter control over the degradation process as compared to alkaline hydrolysis.

Naturally synthesized PGA has been used widely as food supplements and in cosmetics and pharmaceuticals. The toxicity of g-PGA on the human β-cell line EHRB and on mice has been evaluated. *In vitro* evaluation of PGA with cell lines at significantly high doses (up to 100 g/l) showed low to no toxic effects. Nevertheless, at the same or higher concentrations, no significant toxic or inflammatory effects were observed in mice [238].

However, recently, some studies have reported acute toxicity of PGA, primarily as an effect of its molecular weights. PGA obtained from *B. subtilis*, typically used in cosmetic applications, showed an inflammatory effect at molecular weights less than 2000 kDa. However, the dose was not lethal and no significant changes were observed in the health profile, including weight loss, fluid intake, and food intake. The results were very deemed minor, with effects primarily local to the site of injections [251]. Therefore, it can be concluded that low dosages or at higher molecular weights, PGA is a very safe material for biomedical applications.

It has been reported in several studies that molecules or drugs larger than 30 kDa cannot pass through normal capillaries present in tissues [237,245,248,252]. On the other hand, tumors have a significantly enhanced permeability due to the leaky vasculature and underdeveloped blood vessels, hence allowing for better drug penetration into the tissue. Considering this effect, the molecular weight of PGA is a very critical to design suitable drug carrier, especially to specifically target and differentiate normal tissues from tumor tissues [244]. As discussed earlier, typical molecular weight of PGA is about 40 kDa. However, the molecular weights can be reduced by controlled degradation for applicability to normal tissues [253].

Some of the most significant properties that PGA possesses are surface charge, degradability, and biocompatibility, which ideally place them as material of choice for drug delivery. Also, the carboxylic groups provide an ideal surface for further functionalization, including addition of polyethylene glycol(PEG) for enhancing residence time during circulation [254]. Also, the surface charge and the functionalization ability offers enhanced binding capacity to drugs, by either physical or chemical means [255,256]. This leads to enhanced drug loading, controlled release over longer durations, and increased delivery to targeted sites, with appropriately designed targeting moieties [244,254,255]. There are several studies describing the development of PGA-drug conjugates for tumor therapeutics, some of which are discussed in the succeeding text.

The most common of the drug nanoparticle conjugate is paclitaxel-loaded PGA system, typically prepared by chemical binding (covalent) of paclitaxel to PGA [257,258]. The conjugate typically showed enhanced efficacy, improved drug pharmacokinetic profile, and, more importantly, increased hydrophility of the carrier matrix, therefore affording enhanced distribution and systemic delivery potential [259]. Furthermore, Li *et al.* showed that in mice, tumor xenografts exhibited significantly higher tumor control and death, as compared to paclitaxel delivered systemically [259], and complete tumor regressions were exhibited in ovarian and breast cancer studies with murine models, upon administration of significantly high doses of paclitaxel-bound PGA particles, with a single dose, delivered systemically. What was more remarkable was the uptake of the particles by tumors was at least fivefold higher than controls. A preclinical study in animal tumor models demonstrated

PGA-tagged paclitaxel was more effective than systemically delivered paclitaxel and more effective than standard paclitaxel [260]. In clinical trials, PGA-bound paclitaxel showed enhanced pharmacokinetic stability as compared to intravenously delivered paclitaxel alone. Also, significantly fewer side effects were observed, leading to the belief that it could stabilize tumor progression and help as a combination method for treatment and maintenance of tumors, along with radiation therapy. Singer *et al.* reported that the radiation dosage could significantly be reduced (53.9-7.5 Gy), therefore minimizing side effects and exposure to normal cells [258].

PGA has also been used successfully as an adjuvant for protein delivery [261,262]. Studies by Sung *et al.* showed that a combination of PGA and recombinant transmissible gastroenteritis virus nucleocapsid protein in rabbits produced considerably higher titers of antibodies than controls [263]. More importantly, the study also showed that the molecular weight of PGA had a significant effect on the antibody titer [251].

Nanoparticles fabricated from PGA have also shown some tremendous promise as viral carriers for tumor vaccines. Earlier studies have shown that they can deliver antigens at significantly higher amounts to antigen-presenting cells as controls. Enhanced downstream effect and increased immune response were observed in antigen-specific cytotoxic T lymphocytes [264–266]. Similar nanoparticles have been synthesized with PGA in the backbone [267]. The surface charge and the hydrophilicity of the PGA are some of the characteristics that have been exploited in such applications [268]. For example, a novel biodegradable diblock copolymer composed of polylactide-co-glycolide (PLGA) and PGA has been synthesized by Yang *et al.* [269]. The copolymer demonstrated pH response and nanosphere aggregates hence formed could be tailored between a disorganized form, micelles, semivesicles with thick walls, and vesicles, in a pH-dependent manner. Such a pH-dependent self-assembly process is promising for drug delivery and bioapplications [270] (Figure 3.2).

DNA delivery has been a significant challenge over the last decade. Therapeutic efficacy with significant endonuclear delivery has restricted the DNA-based therapeutic applications. Previously, Dekie *et al.* showed the development of PGA-based gene delivery vectors [271]. The complexes formed between the DNA and the PGA were shown to be extremely stable and were shown to successfully transfect target cells.

Arresting blood loss, sealing tissues after injury, and preventing tissue adhesion are some of the critical areas of interest for applications of biomaterials. One of the most commonly used sealant for hemostatic applications is fibrin, but it has very minimal tissue adhesion and can lead to inflammatory cascades and thrombi formation leading to blood vessel blockage [272]. PGA along with collagen and gelatin has shown excellent potential as a hemostasis agent and possesses better tissue adhesion, leading to extended application [273]. Preclinical studies with gelatin-bound PGA foams or hydrogels showed enhanced hemostasis, without the side effects, such as inflammatory response, which is associated with animal-derived fibrin. It was shown that the

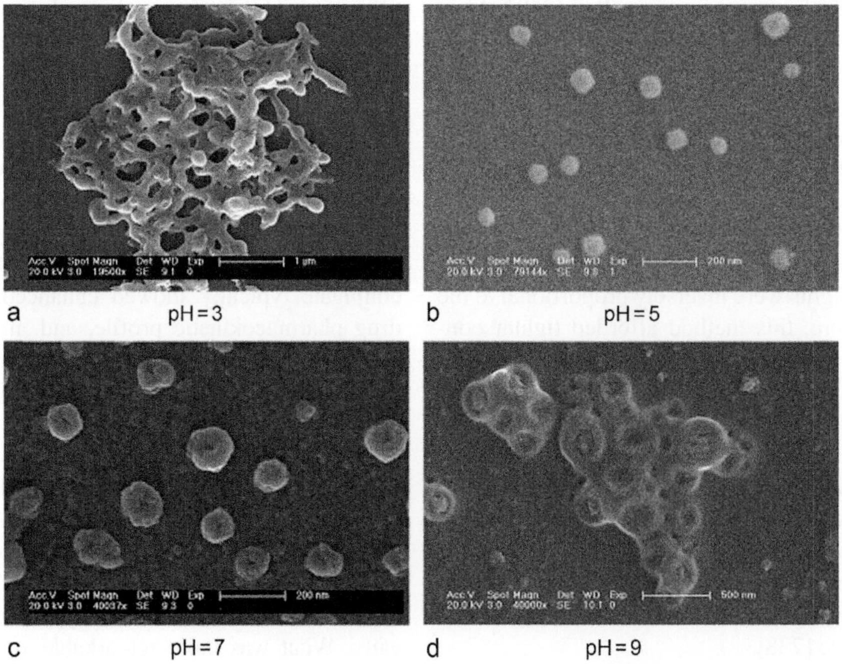

FIGURE 3.2 Scanning electron micrograph (SEM) images of PGA nanoparticles formed at different pH values. With permission from Yang *et al.* (Elsevier) [270].

gelatin PGA hydrogel gel was rapidly curable, comparable to fibrin-based matrices, and was significantly better in function as a sealant for lung air leaks [273].

Recently, PGA has been one of the candidate materials for tissue engineering applications [274–276]. Some of the critical properties of tissue-engineered matrices are the biocompatibility, biodegradability, and the nature of the degradation products, which all work in favor of PGA [6]. PGA has been previously formulated into several forms to make scaffolds, including, nanofibers [277,278], hydrogel [279,280], self-assembled layers [280–282], and nanoparticles for surface functionalization [283]. It has also been used to form blends along with natural polymers such as chitosan [284], collagen [285], and gelatin [286], in order to enhance surface properties, charge, and biocompatibility. Also, as described earlier, the available carboxylic acid surface groups provide ideal substrates for functionalization. Previously, Gentilini *et al.* showed the synthesis of water-resistant formulations of PGA by functionalizing it with ethyl, propyl, or benzyl groups, via esterification [287]. They further showed the formulation of the polymers into nanofibrous scaffolds. In their studies, the functionalized PGA had no cytotoxic effects. Furthermore, benzyl-functionalized PGA showed enhanced cell attachment and proliferation as compared to ethyl- or propyl-functionalized matrices.

PGA-based matrices have also been used in nerve tissue engineering. Kuo *et al.* showed that hydrogels formed by blends of alginate, gamma-PGA, and surface-bound NGF showed enhanced neural differentiation from induced pluripotent stem (iPS) cells [288]. Furthermore, they showed that, with increasing weight ratios of alginate to PGA, the pore diameter increased. PGA-based two dimensional (2D) assays have also been developed to understand cell material interactions for neural applications. Song *et al.* demonstrated a novel method for preparing 2D micropatterns composed of guidance cues for neuronal outgrowth. In their study, they employed an oxidizing potential to an indium tin oxide substrate in the presence of pyrrole and PGA in solution, to electrochemically deposit a matrix consisting of polypyrrole doped with poly(glutamic acid) [289]. The carboxylic group on the PGA was further functionalized with laminin or poly-L-lysine, mediated by carbodiimide coupling. The results showed that cells preferentially adhered to the micropatterned regions where the proteins were attached.

In another study, Goncalves *et al.* showed a method to enhance stem cell recruitment by incorporation and release of stromal-derived factor 1 (SDF1α) from self-assembled matrices composed of PGA and chitosan [282]. Chitosan and γ-PGA, being oppositely charged, could self-assemble via electrostatic interactions to form stable surface coatings, such as polyelectrolyte multilayer films with thicknesses in the range of 120 nm. They showed the incorporation of SDF1α into the matrix and the controlled release of the chemokine over 120 h. Across the 5 days, the authors showed that MSCs could be recruited to the matrix successfully. Such approaches could enhance therapeutic efficacy by recruiting appropriate cells to injured regions, thereby enhancing the quality of wound healing [282].

3D hydrogels are of tremendous interest for tissue engineering applications. Several methods have been previously reported for the synthesis of PGA-based hydrogels [290–292]. It has also been reported earlier that sulfonated γ-PGA mimics heparin and hence can be used as an anticoagulant and for controlled release applications in tissue engineering [293]. In a significant study, Matsusaki *et al.* showed the formation of semi-interpenetrating network hydrogels composed of sulfonated PGA and regular PGA that provided enhanced FGF2 stability and release [274,293,294]. Furthermore, functional significance was demonstrated using fibroblasts, where they proliferated in the presence of FGF but in the absence of serum. In a similar study, Hsieh *et al.* showed a blend of chitosan and PGA showed enhanced binding and controlled release of BMP-2 [284].

3.7 CYANOPHYCIN AND POLY(ASPARTIC ACID)-BASED BIOMATERIALS

Cyanophycin, also referred to as cyanophycin granule polypeptide, is a naturally occurring PAA. It consists of equimolar amounts of aspartic acid and arginine with a poly(aspartic acid) backbone with arginine residues linked to the β-carboxyl group of each aspartate by its α-amino group, making it a highly polydisperse polymer. Cyanophycin is produced in many cyanobacteria as a temporary nitrogen reserve material. The polymerization reaction is catalyzed by the enzyme cyanophycin synthetase [295–299]. Cyanophycin is of biomedical interest because purified cyanophycin can be chemically converted into a polymer with a reduced arginine content, which might be used like poly(aspartic acid) as a biodegradable substitute for synthetic polyacrylates. However, bacterial production of cyanophycin on a large scale is severely impeded by the low yield and the highly complicated fermentation process, and the search is on for newer methods for large-scale production [295–300] (Figure 3.3).

Further, the material and physical properties of cyanophycin have not been well characterized, and hence, studies on the applications of cyanophycin as a biomaterial have been limited. A synthetic PAA similar to cyanophycin that has found biomedical applications is poly(aspartic acid). Poly(aspartic acid) is synthesized using various schemes, the most common being thermal polymerization of aspartic acid and polycondensation of carboxyanhydrides. Poly(aspartic acid) has also been found to undergo biodegradation by lysosomal enzymes [302–305]. The

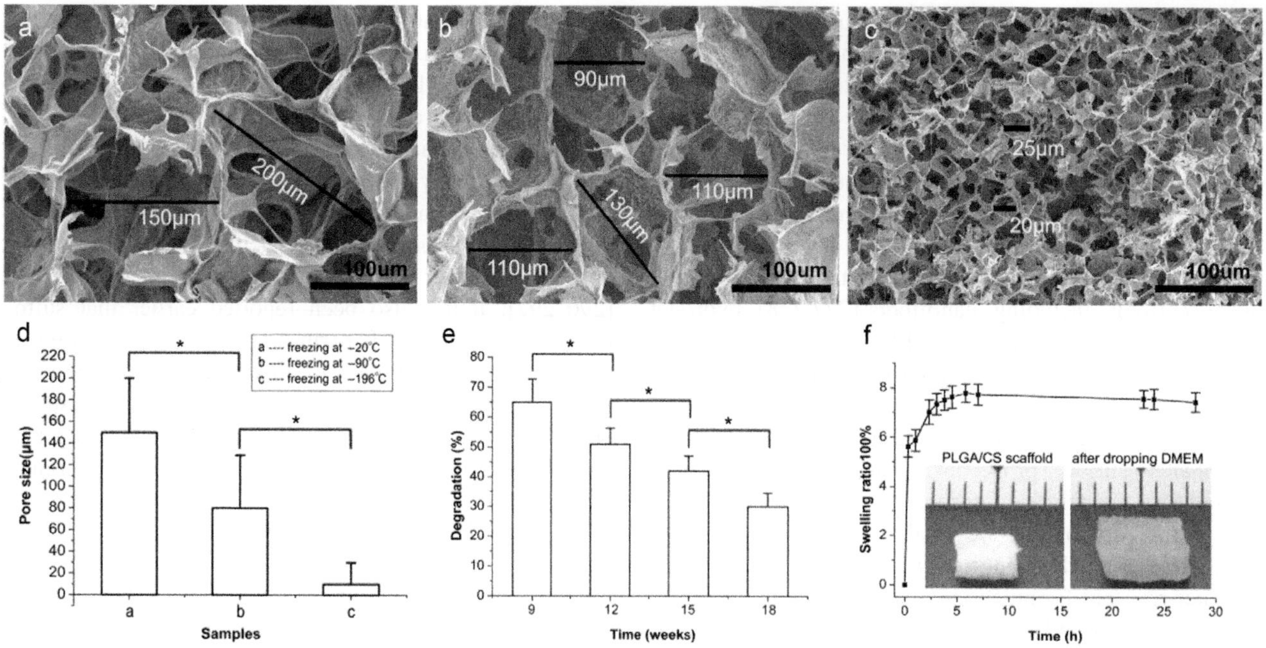

FIGURE 3.3 SEM images of PLGA/CHI porous complex scaffolds at different freezing temperatures: (a) −20 °C; (b) −90 °C; (c) −196 °C. (d) Pore size distribution. (e) Degradation of PLGA/CHI (1:1, 2% solid content, freezing temperature −20 °C) in PBS containing lysozyme. (f) Swelling behavior of PLGA/CHI (1:1, 2%solid content, freezing temperature −20 °C) in DMEM. With permission from Zhang *et al.* (Elsevier) [301].

polymer has high functionality, and several chemically modified forms of PAA are being developed as potential biomaterials. The biodegradability and the high density of negatively charged carboxylic groups makes poly(aspartic acid) an ideal substitute for synthetic nonbiodegradable molecules. PAA-based hydrogels [306–314] and block copolymers with aspartic acid and other synthetic biodegradable polymeric moieties including PLA, PEG, and polycaprolactone have been developed to form micellar nanostructures, and nanoparticles for use as programmable drug delivery vehicles [315–325], and as tissue adhesives [326]. For example, Ponta and Bae [324] recently demonstrated controlled release of anticancer drugs using a block copolymer system of PAA and PEG that formed nanoassemblies of polymer micelles. The controlled release was obtained as the micelles were engineered to have pH-sensitive degradability. Many of these systems are currently undergoing late-stage clinical trials [6,25,305,327,328]. Block copolymers based on poly(aspartic acid) have also been used as a coating on nanoparticles for magnetic resonance imaging [329] of tumors, and in general, nanoparticles based on PEG-poly(aspartic acid) block copolymers exhibit long circulation times [330], making them ideal imaging and drug delivery agents [331]. Recently, Yang *et al.* [325] took advantage of these properties and developed a nanoparticle system based on poly(aspartic acid) to develop a multifunctional nanoparticle that enabled *in vivo* imaging of tumors and deliver anticancer agents as well. To do this, Yang *et al.* created multifunctional poly(aspartic acid) nanoparticles containing iron oxide nanocrystals that enable imaging and doxorubicin that is a potent chemotherapeutic. Kakizawa *et al.* have developed poly(aspartic acid)-based core-shell nanoparticles for nonviral plasmid DNA delivery. This system consists of the DNA as part of the core and the PEG-poly(aspartic acid) as the shell of the nonviral delivery system. Kakizawa *et al.* showed that these systems have high transfection efficiency and low toxicity *in vitro*. In conclusion, cyanophycin and poly(aspartic acid) are highly versatile in terms of their ability to complex with other biomoieties and provide control over their degradability that makes them ideal drug delivery vehicles. As newer and facile sources of cyanophycin become available, its potential as a biomaterial will be realized.

3.8 POLY-L-LYSINE-BASED BIOMATERIALS

ε-Poly-L-lysine is a homo-poly-amino acid polymer formed by the amide linkage between the ε-amino carbon and the carbonyl group of the "L" chiral form of lysine. Originally discovered as 25-35-residue-long polymers of L-lysine residues in the cell culture filtrates of *Streptomyces albulus* 346 yeast cultures [332], ε-poly-L-lysine (ε-PLL) eventually found use as antimicrobial, antibacteriophage, and food-preserving agents [333,334]. α-PLL is an isoform that differs from ε-poly-L-lysine due to the presence of the amide bond between the α-amino carbon and the carbonyl group of PLL. PLL has found widespread use in biomedical applications in nanoparticles, liposome, and coatings for drug delivery agents; as coatings for enzyme-linked immunosorbent assay and tissue culture surface coatings; and as hydrogels for neural stem cell differentiation and central nervous system repair strategies [335]. Polymeric micelles from PEG-PLL block copolymer have also been developed for drug and gene delivery [331].

The availability of a range of PLL molecular weights and reactive groups and the presence of an overall positive charge have facilitated its popular use in biomedical applications as a surface coating for tissue culture [336] and to decorate tissue engineering scaffolds and drug delivery agents. The presence of free amines on the PLL molecule allows for a variety of biomoieties such as proteins, sugars, peptides, and antibodies to be conjugated to it via simple conjugation chemistry. This allows for PLL to be used as a drug carrier for pharmaceuticals [337] and gene therapy [338–340] using nanoparticles and liposomal forms of PLL. As an example, Guo *et al.* used pH-responsive PEG-PLL-based siRNA complexes to target prostate tumors in mice. They showed that PEG-PLL polymers formed stable complexes with siRNA and demonstrated intracellular trafficking with membrane internalization and reporter gene knockdown, leading to a reduction in tumor volume [340]. Sasaki *et al.* [337] developed PLL-based liposomes with high biocompatibility and low toxicity to enhance the efficiency of eye drop drug delivery to the retina. Sasaki *et al.* determined physicochemical properties of liposomes that are suitable for drug delivery to the eye and showed that liposome surface modification with PLL significantly increased delivery of a model drug. PLL is commonly used as a coating material in cell culture studies, especially on substrates such as glass and silicone to improve cell attachment [336,341,342], and PLL-based hydrogels [343,344] are being used in cell culture studies and hold potential for *in vivo* applications. Recently, Cai *et al.* [344] developed a photopolymerizable PLL gel and used it to modify PEG diacrylate hydrogels for nerve cell culture. Cai *et al.* demonstrated that compared with their non-PLL-grafted gels, the PLL-grafted hydrogels greatly enhanced neuronal cell survival, proliferation, and neurite growth. Further, these gels also promoted neural progenitor cell proliferation and differentiation. Overall, PLL has found numerous applications *in vitro* and holds tremendous potential *in vivo* as a drug delivery module and enabling tissue engineering strategies.

3.9 CONCLUSIONS AND FUTURE WORK

Materials that can mimic the ECM are the gold standard for tissue engineering applications. Both PAAs and proteins lend versatility to tissue engineering applications as they contain multiple attachment sites for exogenous drugs, as well as cells and other biomolecules. It is imperative that the deficit being treated should dictate the biomaterial of choice for the tissue engineering strategy being applied, as placing naturally occurring materials in their nonnatural niches may lead to nonfunctional or incomplete regeneration. Naturally occurring PAAs are excellent candidates as drug release modules since they offer multiple sites for drug attachment but, more importantly, are biodegradable with products that do not cause unwarranted side effects. Naturally occurring proteins are excellent as scaffolds and can also act as reservoirs of bioactive molecules, making them an automatic choice for many tissue engineering applications. As the field of tissue engineering matures, the demand for ECM and ECM mimics will increase. As is evident from the discussions in this review, that while naturally occurring proteins and PAAs hold great promise, substantial challenges are present in the form of the need for nonimmunogenic ECM, large-scale production, and possessing hierarchical structures that mimic the ECM specific to the tissue engineering application. Fortunately, tremendous research efforts are under way to produce recombinant and humanized versions of many of these substrates. Ultimately, substantial progress in biomaterial design has been made and will continue, as collaborative research between engineers, clinicians, and scientists in the life sciences continues.

REFERENCES

[1] A. Atala, S.B. Bauer, S. Soker, J.J. Yoo, A.B. Retik, Tissue-engineered autologous bladders for patients needing cystoplasty, The Lancet 367 (2006) 1241–1246.

[2] T.W. Gilbert, A.M. Stewart-Akers, A. Simmons-Byrd, S.F. Badylak, Degradation and remodeling of small intestinal submucosa in canine Achilles tendon repair, J. Bone Jt. Surg. 89 (2007) 621–630.

[3] I.V. Yannas, Emerging rules for inducing organ regeneration, Biomaterials 34 (2013) 321–330.

[4] R. Langer, Biomaterials in drug delivery and tissue engineering: one laboratory's experience, Acc. Chem. Res. 33 (2000) 94–101.

[5] R. Langer, N.A. Peppas, Advances in biomaterials, drug delivery, and bionanotechnology, AIChE J. 49 (2003) 2990–3006.

[6] L.S. Nair, C.T. Laurencin, Polymers as biomaterials for tissue engineering and controlled drug delivery, Adv. Biochem. Eng. Biotechnol. 102 (2006) 47–90.

[7] G.D. Prestwich, Y. Luo, Novel biomaterials for drug delivery, Expert Opin. Ther. Pat. 11 (2001) 1395–1410.

[8] V. Mukhatyar, L. Karumbaiah, J. Yeh, R. Bellamkonda, Tissue engineering strategies designed to realize the endogenous regenerative potential of peripheral nerves, Adv. Mater. 21 (2009) 4670–4679.

[9] L. Karumbaiah, R. Bellamkonda, Neural Tissue Engineering. Neural Engineering, Springer Science+Business Media, New York, 2013, pp. 765–794.

[10] B.D. Bushnell, A.D. McWilliams, G.B. Whitener, T.M. Messer, Early clinical experience with collagen nerve tubes in digital nerve repair, J. Hand Surg. 33 (2008) 1081–1087.

[11] A.R. Nectow, K.G. Marra, D.L. Kaplan, Biomaterials for the development of peripheral nerve guidance conduits, Tissue Eng. Part B Rev. 18 (2011) 40–50.

[12] B.D. Ratner, A.S. Hoffman, F.J. Schoen, J.E. Lemons, Biomaterials Science: An Introduction to Materials in Medicine, Academic Press, San Diego, California, 2004, pp. 162–164.

[13] J.S. Temenoff, A.G. Mikos, Biomaterials: The Intersection of Biology and Materials Science, Pearson/Prentice Hall, New Jersey, 2008.

[14] J.M. Morais, F. Papadimitrakopoulos, D.J. Burgess, Biomaterials/tissue interactions: possible solutions to overcome foreign body response, AAPS J. 12 (2010) 188–196.

[15] S. Franz, S. Rammelt, D. Scharnweber, J.C. Simon, Immune responses to implants—a review of the implications for the design of immunomodulatory biomaterials, Biomaterials 32 (2011) 6692–6709.

[16] M. Lutolf, J. Hubbell, Synthetic biomaterials as instructive extracellular microenvironments for morphogenesis in tissue engineering, Nat. Biotechnol. 23 (2005) 47–55.

[17] R. Langer, D.A. Tirrell, Designing materials for biology and medicine, Nature 428 (2004) 487–492.

[18] L.R. Brown, Biomaterials in their role in creating new approaches for the delivery of drugs, proteins, nucleic acids, and mammalian cells, Drug Discov. Eval. Saf. Pharmacokinetic Assays (2013) 677–690.

[19] B. Alberts, A. Johnson, J. Lewis, M. Raff, K. Roberts, P. Walter, Molecular Biology of the Cell, Garland Science, New York, 2002, There is no corresponding record for this reference. 1997.

[20] S.F. Badylak, The extracellular matrix as a scaffold for tissue reconstruction, in: Seminars in Cell & Developmental Biology, Elsevier, New York, 2002, pp. 377–383.

[21] S.J. Hollister, Porous scaffold design for tissue engineering, Nat. Mater. 4 (2005) 518–524.

[22] J. Hodde, Extracellular matrix as a bioactive material for soft tissue reconstruction, ANZ J. Surg. 76 (2006) 1096–1100.

[23] S.F. Badylak, The extracellular matrix as a biologic scaffold material, Biomaterials 28 (2007) 3587–3593.

[24] S.F. Badylak, D.O. Freytes, T.W. Gilbert, Extracellular matrix as a biological scaffold material: structure and function, Acta Biomater. 5 (2009) 1–13.

[25] B.D. Ulery, L.S. Nair, C.T. Laurencin, Biomedical applications of biodegradable polymers, J. Polym. Sci. B 49 (2011) 832–864.

[26] D.W. Hutmacher, Biomaterials offer cancer research the third dimension, Nat. Mater. 9 (2010) 90–93.

[27] A.J. Domb, Biodegradable polymers derived from amino acids, Biomaterials 11 (1990) 686–689.

[28] W. Khan, S. Muthupandian, S. Farah, N. Kumar, A.J. Domb, Biodegradable polymers derived from amino acids, Macromol. Biosci. 11 (2011) 1625–1636.

[29] W. Bensaïd, J. Triffitt, C. Blanchat, K. Oudina, L. Sedel, H. Petite, A biodegradable fibrin scaffold for mesenchymal stem cell transplantation, Biomaterials 24 (2003) 2497–2502.

[30] R. Gorodetsky, R.A. Clark, J. An, J. Gailit, L. Levdansky, A. Vexler, et al., Fibrin microbeads (FMB) as biodegradable carriers for culturing cells and for accelerating wound healing, J. Invest. Dermatol. 112 (1999) 866–872.

[31] F. Rosso, G. Marino, A. Giordano, M. Barbarisi, D. Parmeggiani, A. Barbarisi, Smart materials as scaffolds for tissue engineering, J. Cell. Physiol. 203 (2005) 465–470.

[32] C.A. Bootle-Wilbraham, S. Tazzyman, W.D. Thompson, C.M. Stirk, C.E. Lewis, Fibrin fragment E stimulates the proliferation, migration and differentiation of human microvascular endothelial cells in vitro, Angiogenesis 4 (2001) 269–275.

[33] H. Hall, T. Baechi, J.A. Hubbell, Molecular properties of fibrin-based matrices for promotion of angiogenesis in vitro, Microvasc. Res. 62 (2001) 315–326.

[34] M.W. Mosesson, Fibrinogen and fibrin structure and functions, J. Thromb. Haemost. 3 (2005) 1894–1904.

[35] T.A. Ahmed, E.V. Dare, M. Hincke, Fibrin: a versatile scaffold for tissue engineering applications, Tissue Eng. Part B Rev. 14 (2008) 199–215.

[36] J.C. Schense, J.A. Hubbell, Cross-linking exogenous bifunctional peptides into fibrin gels with factor XIIIa, Bioconjug. Chem. 10 (1999) 75–81.

[37] M.W. Mosesson, K.R. Siebenlist, J.F. Hainfeld, J.S. Wall, The covalent structure of factor XIIIa crosslinked fibrinogen fibrils, J. Struct. Biol. 115 (1995) 88–101.

[38] K.R. Siebenlist, D.A. Meh, M.W. Mosesson, Protransglutaminase (factor XIII) mediated crosslinking of fibrinogen and fibrin, Thromb. Haemost. 86 (2001) 1221–1228.

[39] M.W. Mosesson, K.R. Siebenlist, I. Hernandez, J.S. Wall, J.F. Hainfeld, Fibrinogen assembly and crosslinking on a fibrin fragment E template, Thromb. Haemost. 87 (2002) 651–658.

[40] J. Rousou, S. Levitsky, L. Gonzalez-Lavin, D. Cosgrove, D. Magilligan, C. Weldon, et al., Randomized clinical trial of fibrin sealant in patients undergoing resternotomy or reoperation after cardiac operations. A multicenter study, J. Thorac. Cardiovasc. Surg. 97 (1989) 194–203.

[41] E. Wolner, Fibrin gluing in cardiovascular surgery, Thorac. Cardiovasc. Surg. 30 (2008) 236–237.

[42] W.D. Spotnitz, M.S. Dalton, J.W. Baker, S.P. Nolan, Reduction of perioperative hemorrhage by anterior mediastinal spray application of fibrin glue during cardiac operations, Ann. Thorac. Surg. 44 (1987) 529–531.

[43] A. Frilling, G.A. Stavrou, H.J. Mischinger, B. de Hemptinne, M. Rokkjaer, J. Klempnauer, et al., Effectiveness of a new carrier-bound fibrin sealant versus argon beamer as haemostatic agent during liver resection: a randomised prospective trial, Langenbecks Arch. Surg. 390 (2005) 114–120.

[44] H. Ding, J. Zhou, X. Zheng, P. Ye, Q. Chen, J. Yuan, et al., Systematic review and meta-analysis of application of fibrin sealant after liver resection, Curr. Med. Res. Opin. 29 (2013) 387–394.

[45] R.S. Pruthi, J. Chun, M. Richman, The use of a fibrin tissue sealant during laparoscopic partial nephrectomy, BJU Int. 93 (2004) 813–817.

[46] C.I. Shaffrey, W.D. Spotnitz, M.E. Shaffrey, J.A. Jane, Neurosurgical applications of fibrin glue: augmentation of dural closure in 134 patients, Neurosurgery 26 (1990) 207–210.

[47] L.M. Cavallo, D. Solari, T. Somma, D. Savic, P. Cappabianca, The awake endoscope-guided sealant technique with fibrin glue in the treatment of post-operative CSF leak after extended transsphenoidal surgery, Technical note, World neurosurgery, (2013).

[48] E. Anitua, M. Sanchez, A.T. Nurden, P. Nurden, G. Orive, I. Andia, New insights into and novel applications for platelet-rich fibrin therapies, Trends Biotechnol. 24 (2006) 227–234.

[49] O. Levy, U. Martinowitz, A. Oran, C. Tauber, H. Horoszowski, The use of fibrin tissue adhesive to reduce blood loss and the need for blood transfusion after total knee arthroplasty. A prospective, randomized, multicenter study, J. Bone Jt. Surg. 81 (1999) 1580–1588.

[50] L.J. Currie, J.R. Sharpe, R. Martin, The use of fibrin glue in skin grafts and tissue-engineered skin replacements: a review, Plast. Reconstr. Surg. 108 (2001) 1713–1726.

[51] T.R. Hester Jr., Z.E. Gerut, J.R. Shire, D.B. Nguyen, A.H. Chen, J. Diamond, et al., Exploratory, randomized, controlled, phase 2 study to evaluate the safety and efficacy of adjuvant fibrin sealant

VH S/D 4 S-Apr (ARTISS) in patients undergoing rhytidectomy, Aesthet. Surg. J. 33 (2013) 323–333.
[52] T.R. Hester Jr., J.R. Shire, D.B. Nguyen, Z.E. Gerut, A.H. Chen, J. Diamond, et al., Randomized, controlled, phase 3 study to evaluate the safety and efficacy of fibrin sealant VH S/D 4 s-apr (Artiss) to improve tissue adherence in subjects undergoing rhytidectomy, Aesthet. Surg. J. 33 (2013) 487–496.
[53] T.A. Ahmed, M. Griffith, M. Hincke, Characterization and inhibition of fibrin hydrogel-degrading enzymes during development of tissue engineering scaffolds, Tissue Eng. 13 (2007) 1469–1477.
[54] S.E. Sakiyama-Elbert, J.A. Hubbell, Development of fibrin derivatives for controlled release of heparin-binding growth factors, J. Control. Release 65 (2000) 389–402.
[55] R. Rivkin, A. Ben-Ari, I. Kassis, L. Zangi, E. Gaberman, L. Levdansky, et al., High-yield isolation, expansion, and differentiation of murine bone marrow-derived mesenchymal stem cells using fibrin microbeads (FMB), Cloning Stem Cells 9 (2007) 157–175.
[56] I. Kassis, L. Zangi, R. Rivkin, L. Levdansky, S. Samuel, G. Marx, et al., Isolation of mesenchymal stem cells from G-CSF-mobilized human peripheral blood using fibrin microbeads, Bone Marrow Transplant. 37 (2006) 967–976.
[57] K. Ito, Y. Yamada, T. Naiki, M. Ueda, Simultaneous implant placement and bone regeneration around dental implants using tissue-engineered bone with fibrin glue, mesenchymal stem cells and platelet-rich plasma, Clin. Oral Implants Res. 17 (2006) 579–586.
[58] H. Schmoekel, J.C. Schense, F.E. Weber, K.W. Gratz, D. Gnagi, R. Muller, et al., Bone healing in the rat and dog with nonglycosylated BMP-2 demonstrating low solubility in fibrin matrices, J. Orthop. Res. 22 (2004) 376–381.
[59] S.J. Zhu, B.H. Choi, J.H. Jung, S.H. Lee, J.Y. Huh, T.M. You, et al., A comparative histologic analysis of tissue-engineered bone using platelet-rich plasma and platelet-enriched fibrin glue, Oral Surg. Oral Med. Oral Pathol. Oral Radiol. Endod. 102 (2006) 175–179.
[60] S.J. Chen, C.M. Chang, S.K. Tsai, Y.L. Chang, S.J. Chou, S.S. Huang, et al., Functional improvement of focal cerebral ischemia injury by subdural transplantation of induced pluripotent stem cells with fibrin glue, Stem Cells Dev. 19 (2010) 1757–1767.
[61] H. Schmoekel, J.C. Schense, F.E. Weber, K.W. Gratz, D. Gnagi, R. Muller, et al., Bone healing in the rat and dog with nonglycosylated BMP-2 demonstrating low solubility in fibrin matrices, J. Orthopaed. Res. 22 (2004) 376–381.
[62] H. Cheng, Y. Cao, L. Olson, Spinal cord repair in adult paraplegic rats: partial restoration of hind limb function, Science 273 (1996) 510–513.
[63] K. Iwaya, K. Mizoi, A. Tessler, Y. Itoh, Neurotrophic agents in fibrin glue mediate adult dorsal root regeneration into spinal cord, Neurosurgery 44 (1999) 589–595.
[64] K. Nakayama, K. Takakuda, Y. Koyama, S. Itoh, W. Wang, T. Mukai, et al., Enhancement of peripheral nerve regeneration using bioabsorbable polymer tubes packed with fibrin gel, Artif. Organs 31 (2007) 500–508.
[65] R.K. Birla, G.H. Borschel, R.G. Dennis, D.L. Brown, Myocardial engineering in vivo: formation and characterization of contractile, vascularized three-dimensional cardiac tissue, Tissue Eng. 11 (2005) 803–813.
[66] A.F. Godier-Furnemont, T.P. Martens, M.S. Koeckert, L. Wan, J. Parks, K. Arai, et al., Composite scaffold provides a cell delivery platform for cardiovascular repair, Proc. Natl. Acad. Sci. U.S.A. 108 (2011) 7974–7979.
[67] L. Nivison-Smith, A. Weiss, Elastin based constructs, 2011.
[68] G. Wood, Some tensile properties of elastic tissue, Biochim. Biophys. Acta 15 (1954) 311–324.
[69] I. Vesely, The role of elastin in aortic valve mechanics, J. Biomech. 31 (1997) 115–123.
[70] J.F. Almine, D.V. Bax, S.M. Mithieux, L. Nivison-Smith, J. Rnjak, A. Waterhouse, et al., Elastin-based materials, Chem. Soc. Rev. 39 (2010) 3371–3379.
[71] W.F. Daamen, J.H. Veerkamp, J.C. van Hest, T.H. van Kuppevelt, Elastin as a biomaterial for tissue engineering, Biomaterials 28 (2007) 4378–4398.
[72] F.W. Keeley, C.M. Bellingham, K.A. Woodhouse, Elastin as a self-organizing biomaterial: use of recombinantly expressed human elastin polypeptides as a model for investigations of structure and self-assembly of elastin, Philos. Trans. R. Soc. Lond. B Biol. Sci. 357 (2002) 185–189.
[73] Y. Yamauchi, E. Tsuruga, K. Nakashima, Y. Sawa, H. Ishikawa, Fibulin-4 and-5, but not fibulin-2, are associated with tropoelastin deposition in elastin-producing cell culture, Acta histochem. cytochem. 43 (2010) 131.
[74] T. Nakamura, P.R. Lozano, Y. Ikeda, Y. Iwanaga, A. Hinek, S. Minamisawa, et al., Fibulin-5/DANCE is essential for elastogenesis in vivo, Nature 415 (2002) 171–175.
[75] S.M. Mithieux, A.S. Weiss, Elastin, Adv. Protein Chem. 70 (2005) 437–461.
[76] J. Rosenbloom, W.R. Abrams, R. Mecham, Extracellular matrix 4: the elastic fiber, FASEB J. 7 (1993) 1208–1218.
[77] D.R. Eyre, M.A. Paz, P.M. Gallop, Cross-linking in collagen and elastin, Annu. Rev. Biochem. 53 (1984) 717–748.
[78] P.J. Stone, S.M. Morris, S. Griffin, S. Mithieux, A.S. Weiss, Building elastin. Incorporation of recombinant human tropoelastin into extracellular matrices using nonelastogenic rat-1 fibroblasts as a source for lysyl oxidase, Am. J. Respir. Cell Mol. Biol. 24 (2001) 733.
[79] R.B. Rucker, T. Kosonen, M.S. Clegg, A.E. Mitchell, B.R. Rucker, J.Y. Uriu-Hare, et al., Copper, lysyl oxidase, and extracellular matrix protein cross-linking, Am. J. Clin. Nutr. 67 (1998) 996S–1002S.
[80] J. Uitto, A.M. Christiano, V.M. Kahari, M.M. Bashir, J. Rosenbloom, Molecular biology and pathology of human elastin, Biochem. Soc. Trans. 19 (1991) 824–829.
[81] C. Baldock, A.F. Oberhauser, L. Ma, D. Lammie, V. Siegler, S.M. Mithieux, et al., Shape of tropoelastin, the highly extensible protein that controls human tissue elasticity, Proc. Natl. Acad. Sci. U.S.A. 108 (2011) 4322–4327.
[82] A. Hinek, F.W. Keeley, J. Callahan, Recycling of the 67-kDa elastin binding protein in arterial myocytes is imperative for secretion of tropoelastin, Exp. Cell Res. 220 (1995) 312–324.
[83] R.P. Mecham, L. Whitehouse, M. Hay, A. Hinek, M.P. Sheetz, Ligand affinity of the 67-kD elastin/laminin binding protein is modulated by the protein's lectin domain: visualization of elastin/laminin-receptor complexes with gold-tagged ligands, J. Cell Biol. 113 (1991) 187–194.
[84] R.P. Mecham, Elastin synthesis and fiber assembly, Ann. N. Y. Acad. Sci. 624 (1991) 137–146.
[85] S. Glagov, R. Vito, D.P. Giddens, C.K. Zarins, Micro-architecture and composition of artery walls: relationship to location, diameter and the distribution of mechanical stress, J. Hypertens. Suppl. 10 (1992) S101–S104.
[86] R. Mercer, J. Crapo, Spatial distribution of collagen and elastin fibers in the lungs, J. Appl. Physiol. 69 (1990) 756–765.

[87] C. Frances, L. Robert, Elastin and elastic fibers in normal and pathologic skin, Int. J. Dermatol. 23 (1984) 166–179.
[88] M. Tassabehji, K. Metcalfe, J. Hurst, G.S. Ashcroft, C. Kielty, C. Wilmot, et al., An elastin gene mutation producing abnormal tropoelastin and abnormal elastic fibres in a patient with autosomal dominant cutis laxa, Hum. Mol. Genet. 7 (1998) 1021–1028.
[89] J. Rnjak, S.G. Wise, S.M. Mithieux, A.S. Weiss, Severe burn injuries and the role of elastin in the design of dermal substitutes, Tissue Eng. Part B Rev. 17 (2011) 81–91.
[90] A.K. Ewart, W. Jin, D. Atkinson, C.A. Morris, M.T. Keating, Supravalvular aortic stenosis associated with a deletion disrupting the elastin gene, J. Clin. Invest. 93 (1994) 1071.
[91] M.E. Curran, D.L. Atkinson, A.K. Ewart, C.A. Morris, M.F. Leppert, M.T. Keating, The elastin gene is disrupted by a translocation associated with supravalvular aortic stenosis, Cell 73 (1993) 159–168.
[92] M. Lillie, J. Gosline, The effects of hydration on the dynamic mechanical properties of elastin, Biopolymers 29 (1990) 1147–1160.
[93] S.M. Mithieux, S.G. Wise, M.J. Raftery, B. Starcher, A.S. Weiss, A model two-component system for studying the architecture of elastin assembly in vitro, J. Struct. Biol. 149 (2005) 282–289.
[94] J. Holst, S. Watson, M.S. Lord, S.S. Eamegdool, D.V. Bax, L.B. Nivison-Smith, et al., Substrate elasticity provides mechanical signals for the expansion of hemopoietic stem and progenitor cells, Nat. Biotechnol. 28 (2010) 1123–1128.
[95] B. Vrhovski, A.S. Weiss, Biochemistry of tropoelastin, Eur. J. Biochem. 258 (1998) 1–18.
[96] B.S. Brooke, A. Bayes-Genis, D.Y. Li, New insights into elastin and vascular disease, Trends in cardiovasc. Med. 13 (2003) 176–181.
[97] D.Y. Li, B. Brooke, E.C. Davis, R.P. Mecham, L.K. Sorensen, B.B. Boak, et al., Elastin is an essential determinant of arterial morphogenesis, Nature 393 (1998) 276–280.
[98] S. Ito, S. Ishimaru, S.E. Wilson, Effect of coacervated α-elastin on proliferation of vascular smooth muscle and endothelial cells, Angiology 49 (1998) 289–297.
[99] M.R. Williamson, A. Shuttleworth, A.E. Canfield, R.A. Black, C.M. Kielty, The role of endothelial cell attachment to elastic fibre molecules in the enhancement of monolayer formation and retention, and the inhibition of smooth muscle cell recruitment, Biomaterials 28 (2007) 5307–5318.
[100] D.V. Bax, U.R. Rodgers, M.M. Bilek, A.S. Weiss, Cell adhesion to tropoelastin is mediated via the C-terminal GRKRK motif and integrin αVβ3, J. Biol. Chem. 284 (2009) 28616–28623.
[101] R.M. Senior, G.L. Griffin, R.P. Mecham, Chemotactic activity of elastin-derived peptides, J. Clin. Invest. 66 (1980) 859–862.
[102] Z. Indik, W.R. Abrams, U. Kucich, C.W. Gibson, R.P. Mecham, J. Rosenbloom, Production of recombinant human tropoelastin: characterization and demonstration of immunologic and chemotactic activity, Arch. Biochem. Biophys. 280 (1990) 80–86.
[103] S. Mochizuki, B. Brassart, A. Hinek, Signaling pathways transduced through the elastin receptor facilitate proliferation of arterial smooth muscle cells, J. Biol. Chem. 277 (2002) 44854–44863.
[104] S. Kamisato, Y. Uemura, N. Takami, K. Okamoto, Involvement of intracellular cyclic GMP and cyclic GMP-dependent protein kinase in alpha-elastin-induced macrophage chemotaxis, J. Biochem. 121 (1997) 862–867.
[105] W. Hornebeck, J.M. Tixier, L. Robert, Inducible adhesion of mesenchymal cells to elastic fibers: elastonectin, Proc. Natl. Acad. Sci. U.S.A. 83 (1986) 5517–5520.
[106] S.K. Karnik, J.D. Wythe, L. Sorensen, B.S. Brooke, L.D. Urness, D.Y. Li, Elastin induces myofibrillogenesis via a specific domain, VGVAPG, Matrix Biol. 22 (2003) 409–425.
[107] S.K. Karnik, B.S. Brooke, A. Bayes-Genis, L. Sorensen, J.D. Wythe, R.S. Schwartz, et al., A critical role for elastin signaling in vascular morphogenesis and disease, Development 130 (2003) 411–423.
[108] K. Akhtar, T.J. Broekelmann, H. Song, J. Turk, T.J. Brett, R.P. Mecham, et al., Oxidative modifications of the C-terminal domain of tropoelastin prevent cell binding, J. Biol. Chem. 286 (2011) 13574–13582.
[109] U. Rodgers, A.S. Weiss, Integrin αvβ3 binds a unique non-RGD site near the C-terminus of human tropoelastin, Biochimie 86 (2004) 173–178.
[110] B.D. Wilson, C.C. Gibson, L.K. Sorensen, M.Y. Guilhermier, M. Clinger, L.L. Kelley, et al., Novel approach for endothelializing vascular devices: understanding and exploiting elastin–endothelial interactions, Ann. Biomed. Eng. 39 (2011) 337–346.
[111] D. Lloyd-Jones, R.J. Adams, T.M. Brown, M. Carnethon, S. Dai, G. De Simone, et al., Heart disease and stroke statistics—2010 update. A report from the American Heart Association, Circulation 121 (2010) e46–e215.
[112] S.P. Hoerstrup, G. Zünd, R. Sodian, A.M. Schnell, J. Grünenfelder, M.I. Turina, Tissue engineering of small caliber vascular grafts, Eur. J. Cardiothorac. Surg. 20 (2001) 164–169.
[113] H. Verhagen, J. Blankensteijn, P.G. de Groot, G. Heijnen-Snyder, A. Pronk, T.M. Vroom, et al., In vivo experiments with mesothelial cell seeded ePTFE vascular grafts, Eur. J. Vasc. Endovasc. Surg. 15 (1998) 489–496.
[114] D.Y. Tseng, E.R. Edelman, Effects of amide and amine plasma-treated ePTFE vascular grafts on endothelial cell lining in an artificial circulatory system, J. Biomed. Mater. Res. 42 (1998) 188–198.
[115] P. Ramires, L. Mirenghi, A. Romano, F. Palumbo, G. Nicolardi, Plasma-treated PET surfaces improve the biocompatibility of human endothelial cells, J. Biomed. Mater. Res. 51 (2000) 535–539.
[116] R.J. Zdrahala, Small caliber vascular grafts. Part I: state of the art, J. Biomater. Appl. 10 (1996) 309–329.
[117] J. Chlupáč, E. Filova, L. Bačáková, Blood vessel replacement: 50 years of development and tissue engineering paradigms in vascular surgery, Physiol. Res. 58 (2009) S119.
[118] S.W. Jordan, C.A. Haller, R.E. Sallach, R.P. Apkarian, S.R. Hanson, E.L. Chaikof, The effect of a recombinant elastin-mimetic coating of an ePTFE prosthesis on acute thrombogenicity in a baboon arteriovenous shunt, Biomaterials 28 (2007) 1191–1197.
[119] M. Byrom, S. Wise, P. Bannon, A. Weiss, M. Ng, Elastin-coated ePTFE vascular conduit, Heart Lung Circ. 17 (2008) S19.
[120] M. Byrom, S. Wise, A. Waterhouse, P. Bannon, A. Weiss, M. Bilek, et al., Plasma treatment of ePTFE for covalent attachment of human elastin, and its effects on endothelialisation, Heart Lung Circ. 18 (2009) S70.
[121] K. Trabbic-Carlson, L.A. Setton, A. Chilkoti, Swelling and mechanical behaviors of chemically cross-linked hydrogels of elastin-like polypeptides, Biomacromolecules 4 (2003) 572–580.
[122] S.M. Mithieux, J.E. Rasko, A.S. Weiss, Synthetic elastin hydrogels derived from massive elastic assemblies of self-organized human protein monomers, Biomaterials 25 (2004) 4921–4927.
[123] J. Kopeček, J. Yang, Peptide-directed self-assembly of hydrogels, Acta Biomater. 5 (2009) 805.
[124] S.L. Martin, B. Vrhovski, A.S. Weiss, Total synthesis and expression in *Escherichia coli* of a gene encoding human tropoelastin, Gene 154 (1995) 159–166.

[125] D.A. Tirrell, M.J. Fournier, T.L. Mason, Protein engineering for materials applications, Curr. Opin. Struct. Biol. 1 (1991) 638–641.
[126] B. Vrhovski, S. Jensen, A.S. Weiss, Coacervation characteristics of recombinant human tropoelastin, Eur. J. Biochem. 250 (1997) 92–98.
[127] L.D. Muiznieks, S.A. Jensen, A.S. Weiss, Structural changes and facilitated association of tropoelastin, Arch. Biochem. Biophys. 410 (2003) 317–323.
[128] N. Annabi, S.M. Mithieux, A.S. Weiss, F. Dehghani, Cross-linked open-pore elastic hydrogels based on tropoelastin, elastin and high pressure CO_2, Biomaterials 31 (2010) 1655–1665.
[129] S.G. Wise, S.M. Mithieux, A.S. Weiss, Engineered tropoelastin and elastin-based biomaterials, Adv. Protein Chem. Struct. Biol. 78 (2009) 1–24.
[130] N. Annabi, S.M. Mithieux, A.S. Weiss, F. Dehghani, The fabrication of elastin-based hydrogels using high pressure CO_2, Biomaterials 30 (2009) 1–7.
[131] S.M. Mithieux, Y. Tu, E. Korkmaz, F. Braet, A.S. Weiss, In situ polymerization of tropoelastin in the absence of chemical cross-linking, Biomaterials 30 (2009) 431–435.
[132] Y. Tu, S.M. Mithieux, N. Annabi, E.A. Boughton, A.S. Weiss, Synthetic elastin hydrogels that are coblended with heparin display substantial swelling, increased porosity, and improved cell penetration, J. Biomed. Mater. Res. A 95 (2010) 1215–1222.
[133] L. Huang, R.A. McMillan, R.P. Apkarian, B. Pourdeyhimi, V.P. Conticello, E.L. Chaikof, Generation of synthetic elastin-mimetic small diameter fibers and fiber networks, Macromolecules 33 (2000) 2989–2997.
[134] N. Annabi, S.M. Mithieux, E.A. Boughton, A.J. Ruys, A.S. Weiss, F. Dehghani, Synthesis of highly porous crosslinked elastin hydrogels and their interaction with fibroblasts *in vitro*, Biomaterials 30 (2009) 4550–4557.
[135] A.P. Price, K.A. England, A.M. Matson, B.R. Blazar, A. Panoskaltsis-Mortari, Development of a decellularized lung bioreactor system for bioengineering the lung: the matrix reloaded, Tissue Eng. Part A 16 (2010) 2581–2591.
[136] T.-H. Chuang, C. Stabler, A. Simionescu, D.T. Simionescu, Polyphenol-stabilized tubular elastin scaffolds for tissue engineered vascular grafts, Tissue Eng. Part A 15 (2009) 2837–2851.
[137] B. Hafemann, S. Ensslen, C. Erdmann, R. Niedballa, A. Zühlke, K. Ghofrani, et al., Use of a collagen/elastin-membrane for the tissue engineering of dermis, Burns 25 (1999) 373–384.
[138] G.H. Altman, F. Diaz, C. Jakuba, T. Calabro, R.L. Horan, J. Chen, et al., Silk-based biomaterials, Biomaterials 24 (2003) 401–416.
[139] P.P. von Weimarn, Generally applicable methods of bringing fibroin, chitin, casein and similar substances into plastic state and the state of colloidal solution by means of concentrated watery solutions of slightly soluble and heavily hydrated salts, Kolloid-Zeitschrift 40 (1926) 120–122.
[140] E.J. Ambrose, C.H. Bamford, A. Elliott, W.E. Hanby, Water soluble silk; an alpha-protein, Nature 167 (1951) 264–265.
[141] D. Coleman, F.O. Howitt, Studies on silk proteins. I. The properties and constitution of fibroin. The conversion of fibroin into a water-soluble form and its bearing on the phenomenon of denaturation, Proc. R. Soc. Med. 134 (1947) 544.
[142] R.L. Moy, A. Lee, A. Zalka, Commonly used suture materials in skin surgery, Am. Fam. Physician 44 (1991) 2123–2128.
[143] H.K. Soong, K.R. Kenyon, Adverse reactions to virgin silk sutures in cataract surgery, Ophthalmology 91 (1984) 479–483.
[144] M. Dewair, X. Baur, K. Ziegler, Use of immunoblot technique for detection of human IgE and IgG antibodies to individual silk proteins, J. Allergy Clin. Immunol. 76 (1985) 537–542.
[145] C.M. Wen, S.T. Ye, L.X. Zhou, Y. Yu, Silk-induced asthma in children: a report of 64 cases, Ann. Allergy 65 (1990) 375–378.
[146] W. Zaoming, R. Codina, E. Fernandez-Caldas, R.F. Lockey, Partial characterization of the silk allergens in mulberry silk extract, J. Investig. Allergol. Clin. Immunol. 6 (1996) 237–241.
[147] F.G. Omenetto, D.L. Kaplan, New opportunities for an ancient material, Science 329 (2010) 528–531.
[148] I. Dal Pra, G. Freddi, J. Minic, A. Chiarini, U. Armato, De novo engineering of reticular connective tissue in vivo by silk fibroin nonwoven materials, Biomaterials 26 (2005) 1987–1999.
[149] H.J. Jin, J. Park, R. Valluzzi, P. Cebe, D.L. Kaplan, Biomaterial films of Bombyx mori silk fibroin with poly(ethylene oxide), Biomacromolecules 5 (2004) 711–717.
[150] R. Nazarov, H.J. Jin, D.L. Kaplan, Porous 3-D scaffolds from regenerated silk fibroin, Biomacromolecules 5 (2004) 718–726.
[151] Y. Tamada, New process to form a silk fibroin porous 3-D structure, Biomacromolecules 6 (2005) 3100–3106.
[152] H.J. Kim, H.S. Kim, A. Matsumoto, I.J. Chin, H.J. Jin, D.L. Kaplan, Processing windows for forming silk fibroin biomaterials into a 3D porous matrix, Aust. J.Chem. 58 (2005) 716–720.
[153] S. Rammensee, D. Huemmerich, K.D. Hermanson, T. Scheibel, A.R. Bausch, Rheological characterization of hydrogels formed by recombinantly produced spider silk, Appl. Phys.A Mater. Sci. Process. 82 (2006) 261–264.
[154] A.J. Campillo-Fernandez, R.E. Unger, K. Peters, S. Halstenberg, M. Santos, M. Salmeron Sanchez, et al., Analysis of the biological response of endothelial and fibroblast cells cultured on synthetic scaffolds with various hydrophilic/hydrophobic ratios: influence of fibronectin adsorption and conformation, Tissue Eng. Part A 15 (2009) 1331–1341.
[155] L. Uebersax, M. Mattotti, M. Papaloizos, H.P. Merkle, B. Gander, L. Meinel, Silk fibroin matrices for the controlled release of nerve growth factor (NGF), Biomaterials 28 (2007) 4449–4460.
[156] S. Sofia, M.B. McCarthy, G. Gronowicz, D.L. Kaplan, Functionalized silk-based biomaterials for bone formation, J. Biomed. Mater. Res. 54 (2001) 139–148.
[157] A.J. Meinel, K.E. Kubow, E. Klotzsch, M. Garcia-Fuentes, M.L. Smith, V. Vogel, et al., Optimization strategies for electrospun silk fibroin tissue engineering scaffolds, Biomaterials 30 (2009) 3058–3067.
[158] Q.A. Lv, Q.L. Feng, K. Hu, F.Z. Cui, Three-dimensional fibroin/collagen scaffolds derived from aqueous solution and the use for HepG2 culture, Polymer 46 (2005) 12662–12669.
[159] R.E. Unger, M. Wolf, K. Peters, A. Motta, C. Migliaresi, C.J. Kirkpatrick, Growth of human cells on a non-woven silk fibroin net: a potential for use in tissue engineering, Biomaterials 25 (2004) 1069–1075.
[160] C. Vepari, D.L. Kaplan, Silk as a biomaterial, Prog. Polym. Sci. 32 (2007) 991–1007.
[161] L. Meinel, S. Hofmann, V. Karageorgiou, C. Kirker-Head, J. McCool, G. Gronowicz, et al., The inflammatory responses to silk films in vitro and in vivo, Biomaterials 26 (2005) 147–155.
[162] M.G. Dunn, A.J. Tria, Y.P. Kato, J.R. Bechler, R.S. Ochner, J.P. Zawadsky, et al., Anterior cruciate ligament reconstruction using a composite collagenous prosthesis—a biomechanical and histologic-study in rabbits, Am. J. Sports Med. 20 (1992) 507–575.
[163] M.G. Dunn, L.D. Bellincampi, A.J. Tria, J.P. Zawadsky, Preliminary development of a collagen-PLA composite for ACL reconstruction, J. Appl. Polym. Sci. 63 (1997) 1423–1428.

[164] G.H. Altman, R.L. Horan, H.H. Lu, J. Moreau, I. Martin, J.C. Richmond, et al., Silk matrix for tissue engineered anterior cruciate ligaments, Biomaterials 23 (2002) 4131–4141.
[165] H.B. Fan, H.F. Liu, E.J.W. Wong, S.L. Toh, J.C.H. Goh, In vivo study of anterior cruciate ligament regeneration using mesenchymal stem cells and silk scaffold, Biomaterials 29 (2008) 3324–3337.
[166] H.B. Fan, H.F. Liu, S.L. Toh, J.C.H. Goh, Anterior cruciate ligament regeneration using mesenchymal stem cells and silk scaffold in large animal model, Biomaterials 30 (2009) 4967–4977.
[167] P. Shi, T.K. Teh, S.L. Toh, J.C. Goh, Variation of the effect of calcium phosphate enhancement of implanted silk fibroin ligament bone integration, Biomaterials 34 (2013) 5947–5957.
[168] L. Meinel, R. Fajardo, S. Hofmann, R. Langer, J. Chen, B. Snyder, et al., Silk implants for the healing of critical size bone defects, Bone 37 (2005) 688–698.
[169] M. Fini, A. Motta, P. Torricelli, G. Giavaresi, N. Nicoli Aldini, M. Tschon, et al., The healing of confined critical size cancellous defects in the presence of silk fibroin hydrogel, Biomaterials 26 (2005) 3527–3536.
[170] P. Balasubramanian, M.P. Prabhakaran, M. Sireesha, S. Ramakrishna, Collagen in human tissues: structure, function, and biomedical implications from a tissue engineering perspective, Adv. Pol. Sci. 251 (2013) 173–206.
[171] D.L.P. Kaplan, B. Brodksky, Shining light on collagen: expressing collagen in plants, Tissue Eng. Part A 19 (2013) 1499–1501.
[172] M.D. Shoulders, R.T. Raines, Collagen structure and stability, Annu. Rev. Biochem. 78 (2009) 929.
[173] L. Cen, W. Liu, L. Cui, W. Zhang, Y. Cao, Collagen tissue engineering: development of novel biomaterials and applications, Pediatr. Res. 63 (2008) 492–496.
[174] W. Friess, Collagen–biomaterial for drug delivery, Eur. J. Pharm. Biopharm. 45 (1998) 113–136.
[175] R. Parenteau-Bareil, R. Gauvin, F. Berthod, Collagen-based biomaterials for tissue engineering applications, Materials 3 (2010) 1863–1887.
[176] D.G. Wallace, J. Rosenblatt, Collagen gel systems for sustained delivery and tissue engineering, Adv. Drug Deliv. Rev. 55 (2003) 1631–1649.
[177] M.J. Blewitt, R.K. Willits, The effect of soluble peptide sequences on neurite extension on 2D collagen substrates and within 3D collagen gels, Ann. Biomed. Eng. 35 (2007) 2159–2167.
[178] H.G. Sundararaghavan, G.A. Monteiro, B.L. Firestein, D.I. Shreiber, Neurite growth in 3D collagen gels with gradients of mechanical properties, Biotechnol. Bioeng. 102 (2009) 632–643.
[179] W. Ma, W. Fitzgerald, Q.-Y. Liu, T. O'shaughnessy, D. Maric, H. Lin, et al., CNS stem and progenitor cell differentiation into functional neuronal circuits in three-dimensional collagen gels, Exp. Neurol. 190 (2004) 276–288.
[180] C. Shi, Q. Li, Y. Zhao, W. Chen, B. Chen, Z. Xiao, et al., Stem-cell-capturing collagen scaffold promotes cardiac tissue regeneration, Biomaterials 32 (2011) 2508–2515.
[181] Y.-l. Yang, S. Motte, L.J. Kaufman, Pore size variable type I collagen gels and their interaction with glioma cells, Biomaterials 31 (2010) 5678–5688.
[182] T. Matsuno, T. Nakamura, K. Kuremoto, S. Notazawa, T. Nakahara, Y. Hashimoto, et al., Development of beta-tricalcium phosphate/collagen sponge composite for bone regeneration, Dent. Mater. J. 25 (2006) 138–144.
[183] S.M. Oliveira, I.F. Amaral, M.A. Barbosa, C.C. Teixeira, Engineering endochondral bone: in vitro studies, Tissue Eng. Part A 15 (2009) 625–634.
[184] A. Aravamudhan, D.M. Ramos, J. Nip, M.D. Harmon, R. James, M. Deng, et al., Cellulose and collagen derived micro-nano structured scaffolds for bone tissue engineering, J. Biomed. Nanotechnol. 9 (2013) 719–731.
[185] C.L. Cummings, D. Gawlitta, R.M. Nerem, J.P. Stegemann, Properties of engineered vascular constructs made from collagen, fibrin, and collagen–fibrin mixtures, Biomaterials 25 (2004) 3699–3706.
[186] W. He, T. Yong, W.E. Teo, Z. Ma, S. Ramakrishna, Fabrication and endothelialization of collagen-blended biodegradable polymer nanofibers: potential vascular graft for blood vessel tissue engineering, Tissue Eng. 11 (2005) 1574–1588.
[187] S.I. Jeong, S.Y. Kim, S.K. Cho, M.S. Chong, K.S. Kim, H. Kim, et al., Tissue-engineered vascular grafts composed of marine collagen and PLGA fibers using pulsatile perfusion bioreactors, Biomaterials 28 (2007) 1115–1122.
[188] J.P. Stegemann, S.N. Kaszuba, S.L. Rowe, Review: advances in vascular tissue engineering using protein-based biomaterials, Tissue Eng. 13 (2007) 2601–2613.
[189] N.-T. Dai, M. Williamson, N. Khammo, E. Adams, A. Coombes, Composite cell support membranes based on collagen and polycaprolactone for tissue engineering of skin, Biomaterials 25 (2004) 4263–4271.
[190] M.B. Klein, L.H. Engrav, J.H. Holmes, J.B. Friedrich, B.A. Costa, S. Honari, et al., Management of facial burns with a collagen/glycosaminoglycan skin substitute—prospective experience with 12 consecutive patients with large, deep facial burns, Burns 31 (2005) 257–261.
[191] H.M. Powell, D.M. Supp, S.T. Boyce, Influence of electrospun collagen on wound contraction of engineered skin substitutes, Biomaterials 29 (2008) 834–843.
[192] W. Haslik, L.-P. Kamolz, G. Nathschläger, H. Andel, G. Meissl, M. Frey, First experiences with the collagen-elastin matrix Matriderm® as a dermal substitute in severe burn injuries of the hand, Burns 33 (2007) 364–368.
[193] W. Liu, K. Merrett, M. Griffith, P. Fagerholm, S. Dravida, B. Heyne, et al., Recombinant human collagen for tissue engineered corneal substitutes, Biomaterials 29 (2008) 1147–1158.
[194] Y. Liu, L. Gan, D.J. Carlsson, P. Fagerholm, N. Lagali, M.A. Watsky, et al., A simple, cross-linked collagen tissue substitute for corneal implantation, Invest. Ophthalmol. Vis. Sci. 47 (2006) 1869–1875.
[195] J. Torbet, M. Malbouyres, N. Builles, V. Justin, M. Roulet, O. Damour, et al., Orthogonal scaffold of magnetically aligned collagen lamellae for corneal stroma reconstruction, Biomaterials 28 (2007) 4268–4276.
[196] O. Alluin, C. Wittmann, T. Marqueste, J.-F. Chabas, S. Garcia, M.-N. Lavaut, et al., Functional recovery after peripheral nerve injury and implantation of a collagen guide, Biomaterials 30 (2009) 363–373.
[197] J.-S. Yi, H.-J. Lee, H.-J. Lee, I.-W. Lee, J.-H. Yang, Rat peripheral nerve regeneration using nerve guidance channel by porcine small intestinal submucosa, J. Korean Neurosurg. Soc. 53 (2013) 65–71.
[198] K. Fukushima, M. Enomoto, S. Tomizawa, M. Takahashi, Y. Wakabayashi, S. Itoh, et al., The axonal regeneration across a honeycomb collagen sponge applied to the transected spinal cord, J. Med. Dent. Sci. 55 (2008) 71–79.
[199] Q. Han, W. Jin, Z. Xiao, H. Ni, J. Wang, J. Kong, et al., The promotion of neural regeneration in an extreme rat spinal cord injury

model using a collagen scaffold containing a collagen binding neuroprotective protein and an EGFR neutralizing antibody, Biomaterials 31 (2010) 9212–9220.

[200] X. Li, Z. Xiao, J. Han, L. Chen, H. Xiao, F. Ma, et al., Promotion of neuronal differentiation of neural progenitor cells by using EGFR antibody functionalized collagen scaffolds for spinal cord injury repair, Biomaterials 34 (2013) 5107–5116.

[201] T. Liu, J. Xu, B.P. Chan, S.Y. Chew, Sustained release of neurotrophin-3 and chondroitinase ABC from electrospun collagen nanofiber scaffold for spinal cord injury repair, J. Biomed. Mater. Res. A 100 (2012) 236–242.

[202] L. Yao, G.C. de Ruiter, H. Wang, A.M. Knight, R.J. Spinner, M.J. Yaszemski, et al., Controlling dispersion of axonal regeneration using a multichannel collagen nerve conduit, Biomaterials 31 (2010) 5789–5797.

[203] R.H. Cholas, H.-P. Hsu, M. Spector, The reparative response to cross-linked collagen-based scaffolds in a rat spinal cord gap model, Biomaterials 33 (2012) 2050–2059.

[204] C. Qu, Y. Xiong, A. Mahmood, D.L. Kaplan, A. Goussev, R. Ning, et al., Treatment of traumatic brain injury in mice with bone marrow stromal cell-impregnated collagen scaffolds, J. Neurosurg. 111 (2009) 658.

[205] T. Saxena, R.J. Gilbert, B.S. Pai, R.V. Bellamkonda, Biomedical strategies for axonal regeneration, in: eLS, 2011.

[206] J.-P. Chen, G.-Y. Chang, J.-K. Chen, Electrospun collagen/chitosan nanofibrous membrane as wound dressing, Colloids Surf. A Physicochem. Eng. Asp. 313 (2008) 183–188.

[207] R.-N. Chen, G.-M. Wang, C.-H. Chen, H.-O. Ho, M.-T. Sheu, Development of N,O-(carboxymethyl) chitosan/collagen matrixes as a wound dressing, Biomacromolecules 7 (2006) 1058–1064.

[208] S.-J. Liu, Y.-C. Kau, C.-Y. Chou, J.-K. Chen, R.-C. Wu, W.-L. Yeh, Electrospun PLGA/collagen nanofibrous membrane as early-stage wound dressing, J. Membr. Sci. 355 (2010) 53–59.

[209] P. Soltysiak, M.E. Hollwarth, A.K. Saxena, Comparison of suturing techniques in the formation of collagen scaffold tubes for composite tubular organ tissue engineering, Biomed. Mater. Eng. 20 (2010) 1–11.

[210] X. Liu, Z. Yue, M.J. Higgins, G.G. Wallace, Conducting polymers with immobilised fibrillar collagen for enhanced neural interfacing, Biomaterials 32 (2011) 7309–7317.

[211] Y. Liu, S. Bharadwaj, S.J. Lee, A. Atala, Y. Zhang, Optimization of a natural collagen scaffold to aid cell–matrix penetration for urologic tissue engineering, Biomaterials 30 (2009) 3865–3873.

[212] M.G. Albu, I. Titorencu, M.V. Ghica, Collagen-Based Drug Delivery Systems for Tissue Engineering, in: Biomaterials Applications for Nanomedicine, Intech Open Acces Publisher, Rijeka, 2011, pp. 339–342.

[213] W. Zheng, W. Zhang, X. Jiang, Biomimetic collagen nanofibrous materials for bone tissue engineering, Adv. Eng. Mater. 12 (2010) B451–B466.

[214] G.M. Cunniffe, F.J. O'Brien, Collagen scaffolds for orthopedic regenerative medicine, JOM 63 (2011) 66–73.

[215] M.M.G. Guille, C. Helary, S. Vigier, N. Nassif, Dense fibrillar collagen matrices for tissue repair, Soft Matter 6 (2010) 4963–4967.

[216] A. Gaspar, L. Moldovan, D. Constantin, A. Stanciuc, P.S. Boeti, I. Efrimescu, Collagen-based scaffolds for skin tissue engineering, J. Med. Life 4 (2011) 172.

[217] S. Kew, J. Gwynne, D. Enea, M. Abu-Rub, A. Pandit, D. Zeugolis, et al., Regeneration and repair of tendon and ligament tissue using collagen fibre biomaterials, Acta Biomater. 7 (2011) 3237–3247.

[218] D.N. Woolfson, Building fibrous biomaterials from α-helical and collagen-like coiled-coil peptides, Pept. Sci. 94 (2010) 118–127.

[219] C.H. Lee, A. Singla, Y. Lee, Biomedical applications of collagen, Int. J. Pharm. 221 (2001) 1–22.

[220] R.T. Hansen, G. Choi, E. Bryk, V. Vigorita, The human knee meniscus: a review with special focus on the collagen meniscal implant, J. Long Term Eff. Med. Implants 21 (2011) 321–337.

[221] S.M. Krane, The importance of proline residues in the structure, stability and susceptibility to proteolytic degradation of collagens, Amino acids 35 (2008) 703–710.

[222] L. He, C. Mu, J. Shi, Q. Zhang, B. Shi, W. Lin, Modification of collagen with a natural cross-linker, procyanidin, Int. J. Biol. Macromol. 48 (2011) 354–359.

[223] V. Charulatha, A. Rajaram, Influence of different crosslinking treatments on the physical properties of collagen membranes, Biomaterials 24 (2003) 759–767.

[224] R.A. Brown, Direct collagen-material engineering for tissue fabrication, Tissue Eng. Part A 19 (13–14) (2013) 1495–1498.

[225] D.I. Zeugolis, S.T. Khew, E.S. Yew, A.K. Ekaputra, Y.W. Tong, L.-Y.L. Yung, et al., Electro-spinning of pure collagen nano-fibres—just an expensive way to make gelatin? Biomaterials 29 (2008) 2293–2305.

[226] P. Angele, J. Abke, R. Kujat, H. Faltermeier, D. Schumann, M. Nerlich, et al., Influence of different collagen species on physico-chemical properties of crosslinked collagen matrices, Biomaterials 25 (2004) 2831–2841.

[227] D. Olsen, C. Yang, M. Bodo, R. Chang, S. Leigh, J. Baez, et al., Recombinant collagen and gelatin for drug delivery, Adv. Drug Deliv. Rev. 55 (2003) 1547–1567.

[228] C. Yang, P.J. Hillas, J.A. Báez, M. Nokelainen, J. Balan, J. Tang, et al., The application of recombinant human collagen in tissue engineering, BioDrugs 18 (2004) 103–119.

[229] J. Báez, D. Olsen, J.W. Polarek, Recombinant microbial systems for the production of human collagen and gelatin, Appl. Microbiol. Biotechnol. 69 (2005) 245–252.

[230] S. Browne, D.I. Zeugolis, A. Pandit, Collagen: finding a solution for the source, Tissue Eng. Part A 19 (2013) 1491–1494.

[231] F.F. Hezayen, B.H. Rehm, B.J. Tindall, A. Steinbuchel, Transfer of Natrialba asiatica B1T to Natrialba taiwanensis sp. nov. and description of Natrialba aegyptiaca sp. nov., a novel extremely halophilic, aerobic, non-pigmented member of the Archaea from Egypt that produces extracellular poly(glutamic acid), Int. J. Syst. Evol. Microbiol. 51 (2001) 1133–1142.

[232] K. Kimura, L.S. Tran, I. Uchida, Y. Itoh, Characterization of Bacillus subtilis gamma-glutamyltransferase and its involvement in the degradation of capsule poly-gamma-glutamate, Microbiology 150 (2004) 4115–4123.

[233] K. Kimura, L.S. Tran, Y. Itoh, Roles and regulation of the glutamate racemase isogenes, racE and yrpC, in Bacillus subtilis, Microbiology 150 (2004) 2911–2920.

[234] S. Kocianova, C. Vuong, Y. Yao, J.M. Voyich, E.R. Fischer, F.R. DeLeo, et al., Key role of poly-gamma-DL-glutamic acid in immune evasion and virulence of Staphylococcus epidermidis, J. Clin. Invest. 115 (2005) 688–694.

[235] H. Smith, H.T. Zwartouw, The polysaccharide from Bacillus anthracis grown in vivo, Biochem. J. 63 (1956) 447–453.

[236] H.T. Zwartouw, H. Smith, Polyglutamic acid from Bacillus anthracis grown in vivo; structure and aggressin activity, Biochem. J. 63 (1956) 437–442.

[237] I.L. Shih, Y.T. Van, The production of poly-(gamma-glutamic acid) from microorganisms and its various applications, Bioresour. Technol. 79 (2001) 207–225.

[238] M.H. Sung, C. Park, C.J. Kim, H. Poo, K. Soda, M. Ashiuchi, Natural and edible biopolymer poly-gamma-glutamic acid: synthesis, production, and applications, Chem. Rec. 5 (2005) 352–366.

[239] A.M. Cromwick, G.A. Birrer, R.A. Gross, Effects of pH and aeration on gamma-poly(glutamic acid) formation by Bacillus licheniformis in controlled batch fermentor cultures, Biotechnol. Bioeng. 50 (1996) 222–227.

[240] J.M. Buescher, A. Margaritis, Microbial biosynthesis of polyglutamic acid biopolymer and applications in the biopharmaceutical, biomedical and food industries, Crit. Rev. Biotechnol. 27 (2007) 1–19.

[241] I.B. Bajaj, R.S. Singhal, Enhanced production of poly (gamma-glutamic acid) from Bacillus licheniformis NCIM 2324 by using metabolic precursors, Appl. Biochem. Biotechnol. 159 (2009) 133–141.

[242] Y.H. Ko, R.A. Gross, Effects of glucose and glycerol on gamma-poly(glutamic acid) formation by Bacillus licheniformis ATCC 9945a, Biotechnol. Bioeng. 57 (1998) 430–437.

[243] F.A. Troy, Chemistry and biosynthesis of the poly(-D-glutamyl) capsule in Bacillus licheniformis. I. Properties of the membrane-mediated biosynthetic reaction, J. Biol. Chem. 248 (1973) 305–315.

[244] K. Hoste, E. Schacht, L. Seymour, New derivatives of polyglutamic acid as drug carrier systems, J. Control. Release 64 (2000) 53–61.

[245] Q. Wu, H. Xu, N. Shi, J. Yao, S. Li, P. Ouyang, Improvement of poly(gamma-glutamic acid) biosynthesis and redistribution of metabolic flux with the presence of different additives in Bacillus subtilis CGMCC 0833, Appl. Microbiol. Biotechnol. 79 (2008) 527–535.

[246] A. Yaron, A. Berger, Multi-chain polyamino acids containing glutamic acid, aspartic acid and proline, Biochim. Biophys. Acta 107 (1965) 307–332.

[247] F.A. Troy, Chemistry and biosynthesis of the poly(-D-glutamyl) capsule in Bacillus licheniformis. II. Characterization and structural properties of the enzymatically synthesized polymer, J. Biol. Chem. 248 (1973) 316–324.

[248] G.A. Birrer, A.M. Cromwick, R.A. Gross, Gamma-poly(glutamic acid) formation by Bacillus licheniformis 9945a: physiological and biochemical studies, Int. J. Biol. Macromol. 16 (1994) 265–275.

[249] H. Kubota, Y. Nambu, T. Endo, Alkaline hydrolysis of poly (γ-glutamic acid) produced by microorganism, J. Polym. Sci. A Polym. Chem. 34 (1996) 1347–1351.

[250] A. Goto, M. Kunioka, Biosynthesis and hydrolysis of poly (γ-glutamic acid) from Bacillus subtilis IFO3335, Biosci. Biotechnol. Biochem. 56 (1992) 1031–1035.

[251] I. Bajaj, R. Singhal, Poly (glutamic acid)–an emerging biopolymer of commercial interest, Bioresour. Technol. 102 (2011) 5551–5561.

[252] I.L. Shih, Y.T. Van, L.C. Yeh, H.G. Lin, Y.N. Chang, Production of a biopolymer flocculant from Bacillus licheniformis and its flocculation properties, Bioresour. Technol. 78 (2001) 267–272.

[253] A. Richard, A. Margaritis, Kinetics of molecular weight reduction of poly (glutamic acid) by in situ depolymerization in cell-free broth of *Bacillus subtilis*, Biochem. Eng. J. 30 (2006) 303–307.

[254] J. Vega, S. Ke, Z. Fan, S. Wallace, C. Charsangavej, C. Li, Targeting doxorubicin to epidermal growth factor receptors by site-specific conjugation of C225 to poly (L-glutamic acid) through a polyethylene glycol spacer, Pharm. Res. 20 (2003) 826–832.

[255] H.-F. Liang, T.-F. Yang, C.-T. Huang, M.-C. Chen, H.-W. Sung, Preparation of nanoparticles composed of poly (γ-glutamic acid)-poly (lactide) block copolymers and evaluation of their uptake by HepG2 cells, J. Control. Release 105 (2005) 213–225.

[256] J.W. Singer, P. Vries, R. Bhatt, J. Tulinsky, P. Klein, C. Li, et al., Conjugation of camptothecins to poly-(l-glutamic acid), Ann. N. Y. Acad. Sci. 922 (2000) 136–150.

[257] E. Oldham, C. Li, S. Ke, S. Wallace, P. Huang, Comparison of action of paclitaxel and poly (L-glutamic acid)-paclitaxel conjugate in human breast cancer cells, Int. J. Oncol. 16 (2000) 125–132.

[258] J.W. Singer, B. Baker, P. de Vries, A. Kumar, S. Shaffer, E. Vawter, et al., Poly-(l)-glutamic acid-paclitaxel (CT-2103)[XYOTAX™], a biodegradable polymeric drug conjugate, Polymer Drugs in the Clinical Stage, Springer Science+Business Media, New York, 2004, p. 81–99.

[259] C. Li, D.-F. Yu, R.A. Newman, F. Cabral, L.C. Stephens, N. Hunter, et al., Complete regression of well-established tumors using a novel water-soluble poly (L-glutamic acid)-paclitaxel conjugate, Cancer Res. 58 (1998) 2404–2409.

[260] Y. Matsumura, Poly (amino acid) micelle nanocarriers in preclinical and clinical studies, Adv. Drug Deliv. Rev. 60 (2008) 899–914.

[261] S. Okamoto, M. Matsuura, T. Akagi, M. Akashi, T. Tanimoto, T. Ishikawa, et al., Poly (γ-glutamic acid) nano-particles combined with mucosal influenza virus hemagglutinin vaccine protects against influenza virus infection in mice, Vaccine 27 (2009) 5896–5905.

[262] T. Akagi, X. Wang, T. Uto, M. Baba, M. Akashi, Protein direct delivery to dendritic cells using nanoparticles based on amphiphilic poly (amino acid) derivatives, Biomaterials 28 (2007) 3427–3436.

[263] J.S.L. Ji Youn Kim, Ha Ryoung Poo, Moon Hee Sung, Composition for adjuvant containing poly-gamma-glutamic acid, in: WO (Ed.), Bioleaders Corporation, S. Korea.

[264] X. Wang, Y. Wang, Z.G. Chen, D.M. Shin, Advances of cancer therapy by nanotechnology, Cancer Res. Treat. 41 (2009) 1–11.

[265] X. Wang, T. Uto, T. Akagi, M. Akashi, M. Baba, Poly (γ-glutamic acid) nanoparticles as an efficient antigen delivery and adjuvant system: potential for an AIDS vaccine, J. Med. Virol. 80 (2008) 11–19.

[266] T. Uto, X. Wang, K. Sato, M. Haraguchi, T. Akagi, M. Akashi, et al., Targeting of antigen to dendritic cells with poly (γ-glutamic acid) nanoparticles induces antigen-specific humoral and cellular immunity, J. Immunol. 178 (2007) 2979–2986.

[267] G. Réthoré, A. Mathew, H. Naik, A. Pandit, Preparation of chitosan/polyglutamic acid spheres based on the use of polystyrene template as a nonviral gene carrier, Tissue Eng. Part C Methods 15 (2009) 605–613.

[268] R. Duncan, Polymer conjugates as anticancer nanomedicines, Nat. Rev. Cancer 6 (2006) 688–701.

[269] Y.Z. Jing Yang, Z. Wenling, S. Cunxia, Y. Mei, Locally infused gene containing nanoparticles to inhibit rabbit intimal hyperplasia, J. Control. Release 152 (2011) 255–256.

[270] Y. Yang, J. Cai, X. Zhuang, Z. Guo, X. Jing, X. Chen, pH-dependent self-assembly of amphiphilic poly (l-glutamic acid)-block-poly (lactic-co-glycolic acid) copolymers, Polymer 51 (2010) 2676–2682.

[271] L. Dekie, V. Toncheva, P. Dubruel, E.H. Schacht, L. Barrett, L.W. Seymour, Poly-L-glutamic acid derivatives as vectors for gene therapy, J. Control. Release 65 (2000) 187–202.

[272] D. Thompson, N. Letassy, G. Thompson, Fibrin glue: a review of its preparation, efficacy, and adverse effects as a topical hemostat, Ann. Pharmacother. 22 (1988) 946–952.

[273] Y. Otani, Y. Tabata, Y. Ikada, Hemostatic capability of rapidly curable glues from gelatin, poly (L-glutamic acid), and carbodiimide, Biomaterials 19 (1998) 2091–2098.

[274] M. Matsusaki, M. Akashi, Novel functional biodegradable polymer IV: pH-sensitive controlled release of fibroblast growth factor-2 from a poly (γ-glutamic acid)-sulfonate matrix for tissue engineering, Biomacromolecules 6 (2005) 3351–3356.

[275] K.-Y. Chang, L.-W. Cheng, G.-H. Ho, Y.-P. Huang, Y.-D. Lee, Fabrication and characterization of poly (γ-glutamic acid)-graft-chondroitin sulfate/polycaprolactone porous scaffolds for cartilage tissue engineering, Acta Biomater. 5 (2009) 1937–1947.

[276] R.M. Day, A.R. Boccaccini, S. Shurey, J.A. Roether, A. Forbes, L.L. Hench, et al., Assessment of polyglycolic acid mesh and bioactive glass for soft-tissue engineering scaffolds, Biomaterials 25 (2004) 5857–5866.

[277] S. Wang, X. Cao, M. Shen, R. Guo, I. Bányai, X. Shi, Fabrication and morphology control of electrospun poly (γ-glutamic acid) nanofibers for biomedical applications, Colloids Surf. B Biointerfaces 89 (2012) 254–264.

[278] J.-w. Huang, M.-t. Fan Chiang, K.-y. Chang, Process of making water-insoluble polyglutamic acid fibers, US Patent 8,273,278 (2012).

[279] A. Sugino, T. Miyazaki, C. Ohtsuki, Apatite-forming ability of polyglutamic acid hydrogels in a body-simulating environment, J. Mater. Sci. Mater. Med. 19 (2008) 2269–2274.

[280] K. Fan, D. Gonzales, M. Sevoian, Hydrolytic and enzymatic degradation of poly (γ-glutamic acid) hydrogels and their application in slow-release systems for proteins, J. Environ. Polym. Degradation 4 (1996) 253–260.

[281] A.P. Johnston, C. Cortez, A.S. Angelatos, F. Caruso, Layer-by-layer engineered capsules and their applications, Curr. Opin. Colloid Interface Sci. 11 (2006) 203–209.

[282] R.M. Goncalves, J.C. Antunes, M.A. Barbosa, Mesenchymal stem cell recruitment by stromal derived factor-1-delivery systems based on chitosan/poly (γ-glutamic acid) polyelectrolyte complexes, Eur. Cell. Mater. 23 (2012) 249–260.

[283] B.C. Dash, G. Réthoré, M. Monaghan, K. Fitzgerald, W. Gallagher, A. Pandit, The influence of size and charge of chitosan/polyglutamic acid hollow spheres on cellular internalization, viability and blood compatibility, Biomaterials 31 (2010) 8188–8197.

[284] C.-Y. Hsieh, S.-P. Tsai, D.-M. Wang, Y.-N. Chang, H.-J. Hsieh, Preparation of γ-PGA/chitosan composite tissue engineering matrices, Biomaterials 26 (2005) 5617–5623.

[285] T. Sekine, T. Nakamura, Y. Shimizu, H. Ueda, K. Matsumoto, Y. Takimoto, et al., A new type of surgical adhesive made from porcine collagen and polyglutamic acid, J. Biomed. Mater. Res. 54 (2001) 305–310.

[286] H. Layman, M.-G. Spiga, T. Brooks, S. Pham, K.A. Webster, F.M. Andreopoulos, The effect of the controlled release of basic fibroblast growth factor from ionic gelatin-based hydrogels on angiogenesis in a murine critical limb ischemic model, Biomaterials 28 (2007) 2646–2654.

[287] C. Gentilini, Y. Dong, J.R. May, S. Goldoni, D.E. Clarke, B.H. Lee, et al., Functionalized poly (γ-glutamic acid) fibrous scaffolds for tissue engineering, Adv. Healthc. Mater. 1 (2012) 308–315.

[288] Y.-C. Kuo, C.-C. Wang, Guided differentiation of induced pluripotent stem cells into neuronal lineage in alginate–chitosan–gelatin hydrogels with surface neuron growth factor, Colloids Surf. B Biointerfaces 104 (2013) 194–199.

[289] H.-K. Song, B. Toste, K. Ahmann, D. Hoffman-Kim, G. Palmore, Micropatterns of positive guidance cues anchored to polypyrrole doped with polyglutamic acid: a new platform for characterizing neurite extension in complex environments, Biomaterials 27 (2006) 473–484.

[290] Z. Yang, Y. Zhang, P. Markland, V.C. Yang, Poly (glutamic acid) poly (ethylene glycol) hydrogels prepared by photoinduced polymerization: synthesis, characterization, and preliminary release studies of protein drugs, J. Biomed. Mater. Res. 62 (2002) 14–21.

[291] M. Kunioka, H.J. Choi, Preparation conditions and swelling equilibria of biodegradable hydrogels prepared from microbial poly (γ-glutamic acid) and poly (ε-lysine), J. Environ. Polym. Degradation 4 (1996) 123–129.

[292] G.-h. Ho, T.-h. Yang, K.-h. Yang, Stable biodegradable, water absorbing gamma-polyglutamic acid hydrogel, EP Patent 1,550,469 (2006).

[293] M. Matsusaki, T. Serizawa, A. Kishida, T. Endo, M. Akashi, Novel functional biodegradable polymer: synthesis and anticoagulant activity of poly(gamma-glutamic acid)sulfonate (gamma-PGA-sulfonate), Bioconjug. Chem. 13 (2002) 23–28.

[294] M. Matsusaki, T. Serizawa, A. Kishida, M. Akashi, Novel functional biodegradable polymer. III. The construction of poly (γ-glutamic acid)-sulfonate hydrogel with fibroblast growth factor-2 activity, J. Biomed. Mater. Res. Part A 73 (2005) 485–491.

[295] H. Mooibroek, N. Oosterhuis, M. Giuseppin, M. Toonen, H. Franssen, E. Scott, et al., Assessment of technological options and economical feasibility for cyanophycin biopolymer and high-value amino acid production, Appl. Microbiol. Biotechnol. 77 (2007) 257–267.

[296] F. Oppermann-Sanio, A. Steinbüchel, Occurrence, functions and biosynthesis of polyamides in microorganisms and biotechnological production, Naturwissenschaften 89 (2002) 11–22.

[297] F.B. Oppermann-Sanio, A. Steinbüchel, Cyanophycin, Biopolymers Online (2003).

[298] A. Sallam, A. Kast, S. Przybilla, T. Meiswinkel, A. Steinbuchel, Biotechnological process for production of beta-dipeptides from cyanophycin on a technical scale and its optimization, Appl. Environ. Microbiol. 75 (2009) 29–38.

[299] A. Sallam, A. Steinbüchel, Dipeptides in nutrition and therapy: cyanophycin-derived dipeptides as natural alternatives and their biotechnological production, Appl. Microbiol. Biotechnol. 87 (2010) 815–828.

[300] K.M. Frey, F.B. Oppermann-Sanio, H. Schmidt, A. Steinbuchel, Technical-scale production of cyanophycin with recombinant strains of Escherichia coli, Appl. Environ. Microbiol. 68 (2002) 3377–3384.

[301] K. Zhang, Y. Zhang, S. Yan, L. Gong, J. Wang, X. Chen, et al., Repair of an articular cartilage defect using adipose-derived stem cells loaded on a polyelectrolyte complex scaffold based on poly (L-glutamic acid) and chitosan, Acta Biomater. 9 (7) (2013) 7276–7288.

[302] L. Ni, A. Chiriac, C. Popescu, I. NEAM, Possibilities for poly (aspartic acid) preparation as biodegradable compound, J. Optoelectronics Adv. Mater. 8 (2006) 663–666.

[303] S. Roweton, S. Huang, G. Swift, Poly (aspartic acid): synthesis, biodegradation, and current applications, J. Environ. Polym. Degradation 5 (1997) 175–181.

[304] S.M. Thombre, B.D. Sarwade, Synthesis and biodegradability of polyaspartic acid: a critical review, J. Macromol. Sci. A 42 (2005) 1299–1315.

[305] J.V. Gonzalez-Aramundiz, M.V. Lozano, A. Sousa-Herves, E. Fernandez-Megia, N. Csaba, Polypeptides and polyaminoacids in drug delivery, Expert Opin. Drug Deliv. 9 (2012) 183–201.

[306] B. Gyarmati, B. Vajna, Á. Némethy, K. László, A. Szilágyi, Redox- and pH-responsive cysteamine-modified poly (aspartic acid) showing a reversible sol–gel transition, Macromol. Biosci. 13 (5) (2013) 633–640.

[307] C. Liu, Y. Chen, J. Chen, Synthesis and characteristics of pH-sensitive semi-interpenetrating polymer network hydrogels based on konjac glucomannan and poly (aspartic acid) for in vitro drug delivery, Carbohydr. Polym. 79 (2010) 500–506.
[308] C. Lu, X. Wang, G. Wu, J. Wang, Y. Wang, H. Gao, et al., An injectable and biodegradable hydrogel based on poly (alpha,beta-aspartic acid) derivatives for localized drug delivery, J. Biomed. Mater. Res. Part A (2013), doi:10.1002/jbm.a.34725.
[309] S.K. Min, S.H. Kim, J.-H. Kim, Preparation and swelling behavior of biodegradable poly (aspartic acid)-based hydrogel, J. Ind. Eng. Chem. Seoul 6 (2000) 276–279.
[310] N.M. Oh, K.T. Oh, Y.S. Youn, D.-K. Lee, K.-H. Cha, D.H. Lee, et al., Poly (L-aspartic acid) nanogels for lysosome-selective antitumor drug delivery, Colloids Surf. B Biointerfaces 101 (2013) 298–306.
[311] S. Umeda, H. Nakade, T. Kakuchi, Preparation of superabsorbent hydrogels from poly (aspartic acid) by chemical crosslinking, Polym. Bull. 67 (2011) 1285–1292.
[312] Y. Wang, M. Xue, J. Wei, C. Li, R. Zhang, H. Cao, et al., Novel solvent-free synthesis and modification of polyaspartic acid hydrogel, RSC Adv. 2 (2012) 11592–11600.
[313] J. Yang, F. Wang, T. Tan, Degradation behavior of hydrogel based on crosslinked poly (aspartic acid), J. Appl. Polym. Sci. 117 (2010) 178–185.
[314] J.-H. Kim, J.H. Lee, S.-W. Yoon, Preparation and swelling behavior of biodegradable superabsorbent gels based on polyaspartic acid, J. Ind. Eng. Chem. Seoul 8 (2002) 138–142.
[315] H. Arimura, Y. Ohya, T. Ouchi, Formation of core-shell type biodegradable polymeric micelles from amphiphilic poly(aspartic acid)-block-polylactide diblock copolymer, Biomacromolecules 6 (2005) 720–725.
[316] K. Cai, K. Yao, X. Hou, Y. Wang, Y. Hou, Z. Yang, et al., Improvement of the functions of osteoblasts seeded on modified poly(D,L-lactic acid) with poly(aspartic acid), J. Biomed. Mater. Res. 62 (2002) 283–291.
[317] E.J. Cha, J.E. Kim, C.H. Ahn, Stabilized polymeric micelles by electrostatic interactions for drug delivery system, Eur. J. Pharm. Sci. 38 (2009) 341–346.
[318] J. Elisseeff, K. Anseth, R. Langer, J.S. Hrkach, Synthesis and characterization of photo-cross-linked polymers based on poly (L-lactic acid-co-L-aspartic acid), Macromolecules 30 (1997) 2182–2184.
[319] Y. Kakizawa, S. Furukawa, K. Kataoka, Block copolymer-coated calcium phosphate nanoparticles sensing intracellular environment for oligodeoxynucleotide and siRNA delivery, J. Control. Release 97 (2004) 345–356.
[320] O. Karal-Yilmaz, N. Kayaman-Apohan, Z. Misirli, K. Baysal, B.M. Baysal, Synthesis and characterization of poly(L-lactic acid-co-ethylene oxide-co-aspartic acid) and its interaction with cells, J. Mater. Sci. Mater. Med. 17 (2006) 213–227.
[321] J.-H. Kim, C.M. Son, Y.S. Jeon, W.-S. Choe, Synthesis and characterization of poly (aspartic acid) derivatives conjugated with various amino acids, J. Polym. Res. 18 (2011) 881–890.
[322] S. Kim, C.M. Son, Y.S. Jeon, J.H. Kim, Characterizations of novel poly (aspartic acid). Derivatives conjugated with γ-amino butyric acid (GABA) as the bioactive molecule, Bull. Korean Chem. Soc. 30 (2009) 3025–3030.
[323] J. Pan, L. Ma, B. Li, Y. Li, L. Guo, Novel dendritic naproxen prodrugs with poly (aspartic acid) oligopeptide: synthesis and hydroxyapatite binding in vitro, Synth. Commun. 42 (2012) 3441–3454.
[324] A. Ponta, Y. Bae, PEG-poly(amino acid) block copolymer micelles for tunable drug release, Pharm. Res. 27 (2010) 2330–2342.
[325] H.-M. Yang, B.C. Oh, J.H. Kim, T. Ahn, H.-S. Nam, C.W. Park, et al., Multifunctional poly (aspartic acid) nanoparticles containing iron oxide nanocrystals and doxorubicin for simultaneous cancer diagnosis and therapy, Colloids Surf. A Physicochem. Eng. Asp. 391 (2011) 208–215.
[326] J.J. Hwang, S.I. Stupp, Poly (amino acid) bioadhesives for tissue repair, J. Biomater. Sci. Polym. Ed. 11 (2000) 1023–1038.
[327] T. Hamaguchi, K. Kato, H. Yasui, C. Morizane, M. Ikeda, H. Ueno, et al., A phase I and pharmacokinetic study of NK105, a paclitaxel-incorporating micellar nanoparticle formulation, Br. J. Cancer 97 (2007) 170–176.
[328] Y. Matsumura, T. Hamaguchi, T. Ura, K. Muro, Y. Yamada, Y. Shimada, et al., Phase I clinical trial and pharmacokinetic evaluation of NK911, a micelle-encapsulated doxorubicin, Br. J. Cancer 91 (2004) 1775–1781.
[329] M. Kumagai, Y. Imai, T. Nakamura, Y. Yamasaki, M. Sekino, S. Ueno, et al., Iron hydroxide nanoparticles coated with poly (ethylene glycol)-poly (aspartic acid) block copolymer as novel magnetic resonance contrast agents for *in vivo* cancer imaging, Colloids Surf. B Biointerfaces 56 (2007) 174–181.
[330] M. Yokoyama, T. Okano, Y. Sakurai, H. Ekimoto, C. Shibazaki, K. Kataoka, Toxicity and antitumor activity against solid tumors of micelle-forming polymeric anticancer drug and its extremely long circulation in blood, Cancer Res. 51 (1991) 3229–3236.
[331] K. Osada, R.J. Christie, K. Kataoka, Polymeric micelles from poly (ethylene glycol)–poly (amino acid) block copolymer for drug and gene delivery, J. R. Soc. Interface 6 (2009) S325–S339.
[332] S. Shima, H. Sakai, Polylysine produced by streptomyces, Agric. Biol. Chem. 41 (1977) 1807–1809.
[333] S. Shima, H. Matsuoka, T. Iwamoto, H. Sakai, Antimicrobial action of epsilon-poly-L-lysine, J. Antibiot. 37 (1984) 1449–1455.
[334] T. Yoshida, T. Nagasawa, Epsilon-poly-L-lysine: microbial production, biodegradation and application potential, Appl. Microbiol. Biotechnol. 62 (2003) 21–26.
[335] S.C. Shukla, A. Singh, A.K. Pandey, A. Mishra, Review on production and medical applications of epsilon-polylysine, Biochem. Eng. J. 65 (2012) 70–81.
[336] L. Karumbaiah, S. Anand, R. Thazhath, Y. Zhong, R.J. McKeon, R.V. Bellamkonda, Targeted downregulation of N-acetylgalactosamine 4-sulfate 6-O-sulfotransferase significantly mitigates chondroitin sulfate proteoglycan-mediated inhibition, Glia 59 (2011) 981–996.
[337] H. Sasaki, K. Karasawa, K. Hironaka, K. Tahara, Y. Tozuka, H. Takeuchi, Retinal drug delivery using eyedrop preparations of poly-l-lysine-modified liposomes, Eur. J. Pharm. Biopharm. 83 (2013) 364–369.
[338] X. Zhang, M. Oulad-Abdelghani, A.N. Zelkin, Y. Wang, Y. Haîkel, D. Mainard, et al., Poly (L-lysine) nanostructured particles for gene delivery and hormone stimulation, Biomaterials 31 (2010) 1699–1706.
[339] Y. Liu, J. Li, K. Shao, R. Huang, L. Ye, J. Lou, et al., A leptin derived 30-amino-acid peptide modified pegylated poly-L-lysine dendrigraft for brain targeted gene delivery, Biomaterials 31 (2010) 5246–5257.
[340] J. Guo, W.P. Cheng, J. Gu, C. Ding, X. Qu, Z. Yang, et al., Systemic delivery of therapeutic small interfering RNA using a pH-triggered amphiphilic poly-l-lysine nanocarrier to suppress prostate cancer growth in mice, Eur. J. Pharm. Sci. 45 (2012) 521–532.
[341] L. Karumbaiah, S.E. Norman, N.B. Rajan, S. Anand, T. Saxena, M. Betancur, R. Patkar, R.V. Bellamkonda, H. Wallace, The

upregulation of specific interleukin (IL) receptor antagonists and paradoxical enhancement of neuronal apoptosis due to electrode induced strain and brain micromotion, Biomaterials 33 (26) (2012) 5983–5996.

[342] S.S. Rao, N. Han, J.O. Winter, Polylysine-modified PEG-based hydrogels to enhance the neuro–electrode interface, J. Biomater. Sci. Polym. Ed. 22 (2011) 611–625.

[343] S.R. Hynes, M.F. Rauch, J.P. Bertram, E.B. Lavik, A library of tunable poly (ethylene glycol)/poly (L-lysine) hydrogels to investigate the material cues that influence neural stem cell differentiation, J. Biomed. Mater. Res. A 89 (2009) 499–509.

[344] L. Cai, J. Lu, V. Sheen, S. Wang, Promoting nerve cell functions on hydrogels grafted with poly (L-lysine), Biomacromolecules 13 (2012) 342–349.

Chapter 4

Natural Polymers: Polysaccharides and Their Derivatives for Biomedical Applications

Aja Aravamudhan*,†,‡, Daisy M. Ramos*,†,‡,§, Ahmed A. Nada*,†,‡, Sangamesh G. Kumbar*,†,‡,§

*Institute for Regenerative Engineering, University of Connecticut Health Center, Farmington, Connecticut, USA

†Raymond and Beverly Sackler Center for Biological, Physical and Engineering Sciences, University of Connecticut Health Center, Farmington, Connecticut, USA

‡Department of Orthopaedic Surgery, University of Connecticut Health Center, Farmington, Connecticut, USA

§Department of Chemical, Materials and Biomedical Engineering, University of Connecticut, Farmington, Connecticut, USA

Chapter Outline

4.1 INTRODUCTION

Polymers of both natural and synthetic origin have been used for a variety of biomedical applications. Polysaccharides, proteins, and polyesters derived from both plant and animal kingdoms constitute the family of natural polymers. Several of these polymers are part of our diet and have been used in a variety of human applications in pharmaceutical excipients, prosthetics, drug delivery, and imaging applications. These polymers are known to be recognized by the biological environment and channeled into metabolic degradation. Due to the similarity that natural polymers share with the extracellular matrix (ECM) components, these materials

Natural and Synthetic Biomedical Polymers

may also avoid the stimulation of chronic immunological reactions and toxicity, often detected with synthetic polymers [1].

Polysaccharides consist of monosaccharides (sugars) linked together by *O*-glycosidic linkages. Differences in the monosaccharide composition, linkage types and patterns, chain shapes, and molecular weight dictate their physical properties, such as solubility, viscosity, gelling potential, and/or surface and interfacial properties. Polysaccharides are derived from renewable resources, like plants, animals, and microorganisms, and are therefore widely distributed in nature. In addition, polysaccharides perform different physiological functions and hence have great potential applications in the fields of tissue engineering and regenerative medicine [1] (Figure 4.1).

There are hundreds of known polysaccharides. A list of polysaccharides from varying sources is given below:

1. Examples of polysaccharides from higher plants include starch, cellulose, and exudate gums like arabinogalactan, guar gum, and gum arabic.
2. Examples of algal polysaccharides: alginates, galactans, and carrageenan.
3. Examples of polysaccharides from animals: chitin, chitosan, glycosaminoglycans (GAGs), and hyaluronic acid (HA).
4. Examples of polysaccharides from microorganisms: dextran, gellan gum, pullulan, xanthan gum, and bacterial cellulose.

Their monomer composition and biological source provides these polysaccharides with different sets of physicochemical properties. Often, polymers of natural origin have limitations in terms of their solubility and industrially acceptable processability factors such as high temperature of melting, which are commonly applied to synthetic polymers. For instance, the majority of polysaccharides are water-soluble and oxidize at elevated temperatures beyond their melting point. These limitations have to be overcome prior to designing any products using polysaccharides. For instance, techniques to cross-link polymer chains have been developed to stabilize polysaccharide structures in order to give structural stability in aqueous environments [2,3]. Polysaccharide chitin in its native form cannot be produced into desired sizes and shapes due to its inability to dissolve in most common industrial solvents. Thus, chitosan, the deacetylated form of chitin, was produced and applied widely instead of native chitin itself, as chitosan is a water-soluble polymer at low pH. Due to its properties, chitosan is widely used for biomedical applications [4]. The following sections will summarize these efforts in the context of biomedical applications (as shown in Figure 4.1)

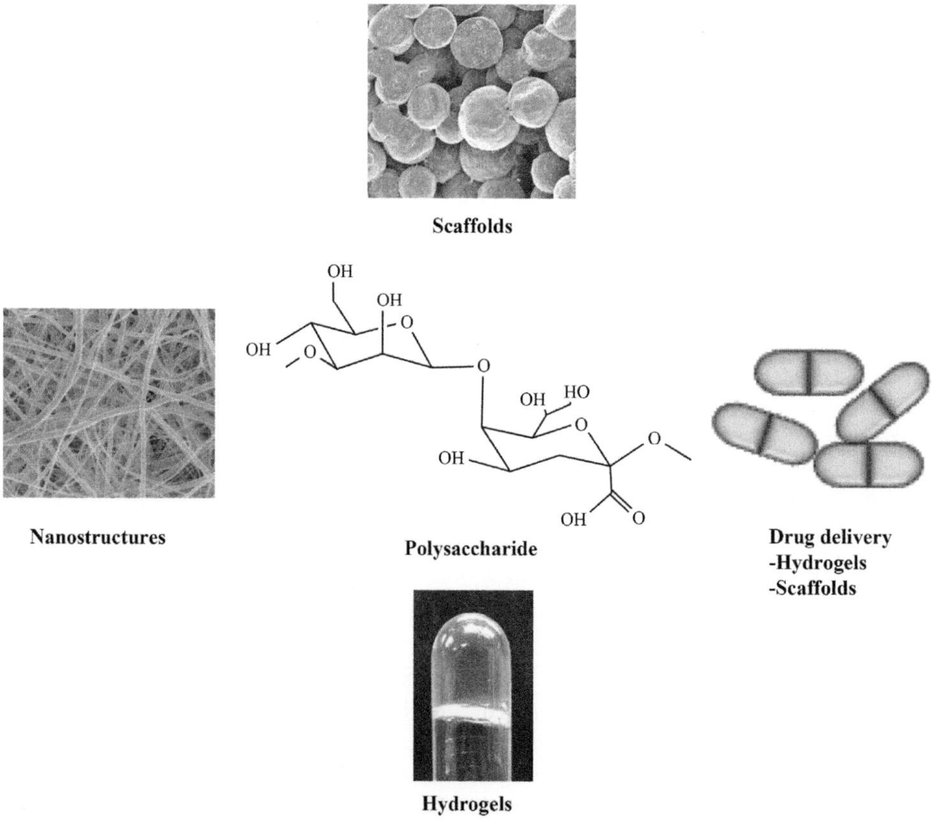

FIGURE 4.1 The use of polysaccharides in biomedical applications.

of polysaccharides, invoking native polysaccharides, semi-synthetic polysaccharide derivatives, and their blends with other synthetic polymers.

4.2 HYALURONIC ACID

4.2.1 Chemical Structure, Properties, and Sources

Chemically, HA is a linear polysaccharide made up of D-glucuronic acid and *N*-acetyl-D-glucosamine that are linked to one another by a β-(1→3) linkage. There could be 250-25,000 such basic disaccharide units in a polymer chain of HA, connected by β-(1→4) linkage (Figure 4.2). The disaccharide units of HA are extended, forming a rigid molecule whose many repelling anionic groups bind cations and water molecules. In solution, hyaluronate occupies a volume approximately 1000 times than in its dry state. Hyaluronate solutions exhibit clear viscoelastic properties that make them excellent biological absorbers and lubricants. These properties also attribute to its preferred form of fabrication into hydrogels. Because of its hydration properties, HA has the ability to bear compressive loads *in vivo* and provide lubrication at the same time. *In vitro*, HA has been shown to facilitate cell migration and pericellular matrix formation [5].

Biologically, HA is an important GAG component of connective tissue, synovial fluid (the fluid that lubricates joints), and the vitreous humor of the eye in mammals [6]. The biological roles of HA are widespread and widely appreciated [7–14]. They range from development, angiogenesis, cellular migration, and receptor-mediated signaling through receptor CD44 and receptor for HA-mediated motility in ECM remodeling and mediation of inflammatory responses [15,16]. HA chain length plays an essential role in the biological functions elicited in native and hence the engineered tissues. Therefore, the molecular weight of HA is an important consideration for the response elicited. For instance, while the low-molecular-weight HA (less than 3.5×10^4 Da) is known to be involved in cytokine activity implicated in inflammatory responses [17], the higher-molecular-weight HA (above 2×10^5 Da) is known to inhibit cell proliferation [18]. Similarly, (1-4 kDa) smaller fragments of HA have a positive effect in promoting vascularization during injury, whereas the (1-9 kDa) large fragment showed no significant effects [18,19]. Not only is the usage of HA at the correct molecular weight and chain length, but also the hyaluronidase (HAS) isozyme (Figure 4.3) that is responsible for degradation of HA in the tissue determines the chain length of the degradation product. Therefore, the enzymes in the tissue where the HA material is to be implanted are a factor for consideration while using the material in tissue engineering [20].

4.2.2 Attempts Made in Tissue Engineering and Drug Delivery

4.2.2.1 HA Alone

HA is highly water-soluble at room temperature and at acidic pH values and exhibits high rates of turnover *in vivo* (half-life varies from only minutes in the blood to weeks in cartilage) [7,21,22]. These properties pose challenges for the material's integrity *in vivo*. Therefore, the usage of HA in its native form in tissue engineering and drug delivery applications is pretty limited. Several cross-linking methods are used to increase the stability of HA in such applications. Some of them are as follows: water-soluble carbodiimide cross-linking, polyvalent hydrazide cross-linking, divinyl sulfone (DVS) cross-linking, disulfide cross-linking, and photo-cross-linking hydrogels through glycidyl methacrylate-HA conjugates [20]. Hence, these techniques of covalent cross-linking provide the opportunity to combine HA with more mechanically stronger polymers.

4.2.2.2 HA Derivatives and Combinations with Other Polymers

The early usage of HA was in ophthalmic drug delivery systems where it provided an ideal matrix for covalent

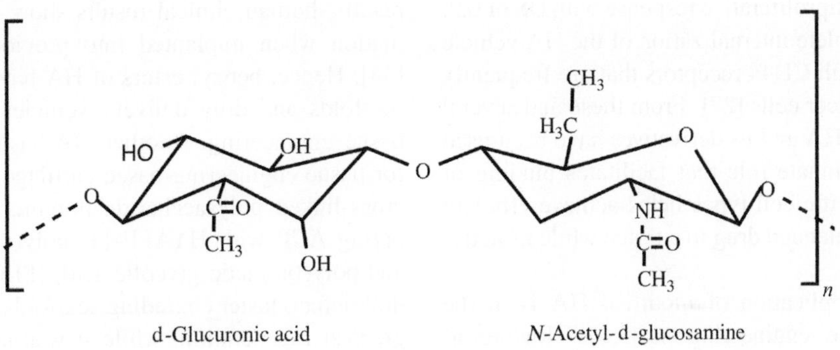

FIGURE 4.2 The structure of hyaluronic acid (HA).

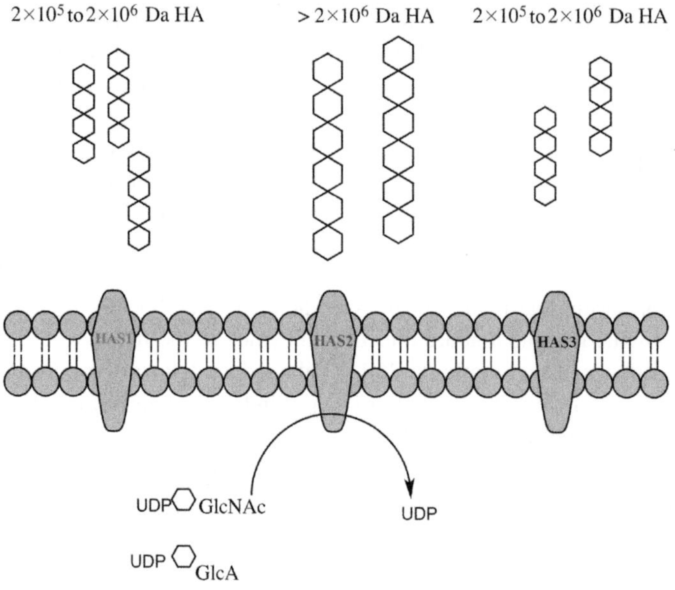

FIGURE 4.3 The three HAS isozymes produce distinct chain lengths of HA. HAS1 and HAS3 can produce chains of 2×10^5 to 2×10^6 Da, while HAS2 produces HA of chain length greater than 2×10^5 Da. (For color version of this figure, the reader is referred to the online version of this chapter.) Modified from ref. [23].

attachment of drugs and showed as much as twice the retention in contrast to free drug (methylprednisolone esters of HA) [24,25]. Different formulations such as gels, solutions, and hydrogels with several model drugs such as pilocarpine and tropicamide showed that HA ester systems were effective ophthalmic drug delivery systems [25]. HA has also been employed in liposomal dermal drug delivery. HA was conjugated to the surface of liposomes by carbodiimide cross-linking of its surface carboxyl residues to the amine residues on the liposome. The epidermal growth factor showed an encapsulation efficiency of >87% in these HA-conjugated liposomes. Avid binding of these HA-conjugated liposomes to a cellular monolayer in culture was seen that did not occur with the unmodified liposomes [26]. Sodium butyrate used as an antiproliferative drug in treatment of cancer has an extremely short half-life of 5 min *in vivo*. In order to bypass this constraint, butyric ester derivatives of HA were synthesized by stepwise chemical treatment of HA. The degree of substitution (DS) varied from 0.1 to 2.24 (1.8-28.4%). MCF breast cancer cell lines showed maximum antiproliferative response with DS of 0.2. It was seen that complete internalization of the HA vehicle occurred in 2 h through CD44 receptors that are frequently overexpressed on cancer cells [27]. From these and several other systems where HA and its derivatives have been used to deliver drugs, its innate role that facilitates binding of HA to receptors/specific cell types helps achieve efficient targeted delivery of intended drug to a tissue while its activity is preserved.

Another major application of modified HA is in the formulation of tissue engineering scaffolds. Radiation-mediated cross-linked HA networks have been used as scaffolds for cell growth with positive outcomes [28]. HA modified with methacrylic anhydride (MA) was photopolymerized to produce HA-MA hydrogels containing porcine chondrocytes [29]. The chondrocytes within the HA scaffolds were viable and were able to produce neo-cartilage within the porous networks. Photopolymerizable HA-MA hydrogels have also been used in heart valve applications in which the HA hydrogels were designed to mimic the cardiac ECM from which the heart valves develop [30]. HA has been combined with other natural/synthetic polymers to produce scaffolds. For example, HA has been combined with polypyrrole to create a multifunctional copolymer [31,32]. When implanted into rats, there was a marked early increase in local vascularization [32]. On the other hand, it was observed that polypyrrole-HA polymers were subsequently sulfonated, which was shown to significantly reduce platelet and cell adhesion [31].

Benzyl derivatives of HA (Hyalograft C and HYAFF-11) are used as polymeric scaffolds for tissue engineering of cartilage [33]. Though laboratory tests have given mixed results, human clinical results show normal cartilage formation when implanted into previously damaged tissue [34]. Hence, benzyl esters of HA have a great potential as scaffolds and drug delivery vehicles for chondrocytes in tissue engineering. Another HA-based scaffold examined for tissue engineering-based cartilage regeneration is auto-cross-linked polysaccharide polymer (ACP). While comparing ACP with HYAFF-11, poly(L-lactic acid) (PLLA) and poly(DL-lactic-glycolic acid) (PLGA) in an osteochondral defect, faster degrading scaffolds of ACP, PLAGA had greatest regeneration, while it was slower in HYAFF-11, PLLA that had slower degradation rates. These *in vivo* data revealed that the scaffolds degraded within 4 months and were able to repair the osteochondral lesions, again emphasizing the selection of the type of HA and the consideration of HAS system in the tissue that would degrade it [35,36]. HYAFF-11 scaffolds have shown very positive effects as scaffolds for engineering of vascular and hepatic tissue and showed great ability for maintenance of cell phenotype, indicating that it could be used as scaffolds for many tissue engineering applications [37,38].

Combining chitosan and high-molecular-weight HA (2.4×10^6 Da) in a three-dimensional copolymer system promoted chondrocyte adhesion. The production of aggrecan and the native rounded morphology of seeded chondrocytes increased with the concentration of added HA and the scaffolds with HA performed better than scaffolds of chitosan alone [39]. Incorporation of chondrocytes and HA into orthopedic implants by cross-linking HA to amine-terminated PLGA-poly(ethylene glycol) (PEG) scaffolds allowed the attachment and proliferation of donor chondrocytes. It was also observed that collagen II expression, a marker of healthy cartilage phenotype, and DNA synthesis were significantly increased in polymers that incorporated HA [40]. Combining HA with DVS, and cross-linking with ultraviolet light, created suitable surfaces for cell adhesion. When a DVS-cross-linked HA scaffold was dehydrated before seeding with smooth muscle cells, gels were more porous and conducive to cell migration and infiltration, but did not lose their nonimmunogenic properties [41–43]. Follow-up studies conducted on the same material showed that smooth muscle cells increased the synthesis of ECM components elastin and collagen of aortic valve tissues, over cells cultured on tissue culture plastic; this synthesis was controlled by HA fragment size and dose [42,44]. Such studies have shown that HA is a suitable material for vascular and cardiac tissue engineering. HA was also introduced directly to engineered structures exogenously. For example, collagen matricides implanted in rabbits have shown greater numbers of chondrocytes (1.5 times control values) after the addition of soluble HA. The addition of HA fragments increased the amounts of proteoglycans, a desirable component of remodeling [45]. However, small HA oligosaccharides (4-16 disaccharides), on the other hand, can prevent cell proliferation *in vivo* [46]. These findings could have implications for improving the potency of implanted small-diameter vascular grafts and preventing stenotic lesions after graft implantation in cardiovascular tissue engineering.

With knee osteoarthritis (OA) patients increasing to a staggering 19 million in 2010, treatment by visco-supplementation has become popular. Visco-supplementation refers to the concept of synovial fluid replacement with intra-articular injections of hyaluronan mainly for the relief of pain associated with OA [47,48]. Intra-articular injections of hyaluronan ameliorate pain and function, generally for up to 3 months with no serious adverse events. In the United States, the first single-injection hyaluronan for visco-supplementation, Synvisc-One®, was approved in early 2009. The global market for dermal fillers is constantly increasing with 100 different dermal fillers on the market for aesthetic plastic surgery and about half of them are based on hyaluronan [49]. Thus, it can be summed up that HA has already stepped into clinical treatments in tissue engineering and will gain greater importance in the future.

4.2.3 Promises and Challenges with HA

HA has had a profound impact on the field of tissue engineering. The incorporation of HA into biomaterials and scaffolds has yielded a new class of biocompatible, controllable, and readily degradable materials. These new scaffolds have been tested with multiple types of cells and have been shown to promote beneficial remodeling of engineered tissues, as well as the gross preservation of cell phenotypes. However, harnessing the endogenous activity of the hyaluronan synthases to stimulate endogenous HA production is yet another useful strategy for the future. Recent research developments regarding HA as a molecular delivery vehicle for pharmacological and oncological applications, as well as in the orthopedic and cardiovascular arenas, have the potential to transform the clinical future of tissue engineering.

4.3 CHONDROITIN SULFATE

4.3.1 Chemical Structure, Properties, and Sources

Chemically, chondroitin sulfate (CS) is a sulfated GAG derivative. Alternating units of *N*-acetylgalactosamine and glucuronic acid are the monomers that constitute the basic unit of CS (Figure 4.4). Hundreds of these alternating units could be present together in a CS polymer chain. CS is a GAG seen associated with proteins in living systems termed proteoglycan. CS is classified according to the position of sulfation of the monomer unit. The types of CS and their chemical structure are listed in Table 4.1.

Biologically, CS is a major structural component of the ECM of several tissues of the body, like cartilage. The biological function and efficacy of CS seems to be highly

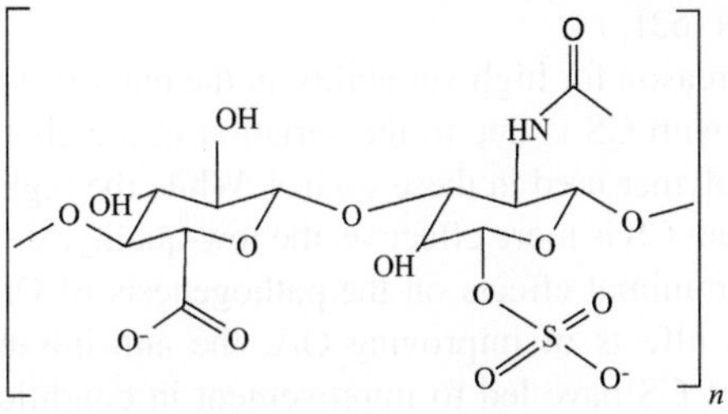

FIGURE 4.4 The structure of chondroitin sulfate (CS).

TABLE 4.1 Types of Chondroitin Sulfate (CS)

S. No.	Name (Type)	Synthetic Name (Sulfation of GalNAc)
1.	Chondroitin sulfate A	Chondroitin-4-sulfate
2.	Chondroitin sulfate C	Chondroitin-6-sulfate
3.	Chondroitin sulfate D	Chondroitin-2,6-sulfate
4.	Chondroitin sulfate E	Chondroitin-4,6-sulfate

dependent on the chain length of the polymer. For instance, CS extracted from the trachea is usually shorter ca. 20-25 kDa and considered to be of lower quality. On the other hand, CS polymer extracted from shark is considered more bioactive and of higher quality with longer chain length ca. 50-80 kDa. Even CS extracted from a single source could have variable polymer chain length [50]. Hence, the type and chain length of CS is likely to determine its biological functions, such as interaction with growth factors and proteins in GAG complexes of the ECM, and its ability to influence cellular function [51,52]. CS is also an indispensible component of tissues for maintaining their mechanical properties. The resistance of cartilage to compression is attributed to the tightly packed, charged sulfate groups of CS. This leads to osmotic water retention and swelling of cartilage and hence endows it with the weight-bearing mechanical properties [53]. Apart from its structural role, CS has also been an important player in basic biological processes including cell division and development of the nervous system. Such biological roles are mediated by the binding of CS to growth factors and cytokines and regulating signaling pathways in neurons [54,55]. Age-related changes in sulfation of the CS chains indicate their fundamental biochemical role in tissues [56] during age-related pathological changes. Loss of CS has been implicated in pathological conditions. For example, OA in cartilage is seen to occur due to the loss of CS leading to cartilage degeneration. Many studies on the impact of CS administration in patients suffering from OA along with *in vitro* results suggest that CS could improve pathology of OA through promoting proteoglycans synthesis, usually lost during cartilage degeneration [57], inhibiting elastase [58–60] and cathepsin G activity [61] and reducing gene expression for a number of proteolytic enzymes [62].

The reason for high variability in the outcomes of treating OA with CS is due to the variation in the chain length of the polymer used in these studies. While the high-quality, long-chain CS is more effective, the low-quality, short-chain CS has minimal effects on the pathogenesis of OA. Apart from its effects of improving OA, the anti-inflammatory effects of CS have led to improvement in conditions such as psoriasis and inflammatory bowel disease. These effects seem to be related to the inhibition of cytokines such as TNF-α [63], and IL-1β-induced translocation of NF-κB [64]. However, most of the effects are dependent on the type of CS determined by degree and position of sulfation, as well as the chain length of the polymer [65].

4.3.2 Attempts Made in Tissue Engineering and Drug Delivery

4.3.2.1 CS Alone

CS is often administered orally as it is seen to improve joint-related pathologies [50]. The intact polymer is often consumed, and several studies have examined the levels of the CS in blood plasma after consumption [65,66] in humans, mice, and horses. The degree of sulfation and chain length of the polymer were seen to determine the rate of uptake, retention, and clearance from these biological systems. For instance, tracheal CS with lower molecular weight is absorbed quickly within 1-5 h and reached a peak plasma concentration at around 10 h. On the other hand, shark CS of high molecular weight is seen to have a slower rate of uptake at around 8.7 h and was retained in the system for as long as 16 h [66]. It was also observed that desulfated chondroitin had quick uptake rate (15 min) and was also cleared within 3 h [67]. Further, *in vitro* experiments showed that there was a direct correlation between the polymer size and the rate at which it crossed the gut wall [68]. This in turn reflected as the rate of uptake and retention of the polymer in the biological system. Therefore, smaller CS polymer is taken up and cleared quickly, while the larger CS polymer takes longer time to be absorbed and cleared from a system.

4.3.2.2 CS Derivatives and Combination with Other Polymers

CS is a water-soluble polymer. Early attempts of using CS as a vehicle for colonic drug delivery cross-linked the polymer to different degrees to achieve a biodegradable system for controlled release of model drugs, indomethacin. A linear correlation between degree of cross-linking and rate of drug release emerged, indicating that the level of cross-linking of CS can be used to regulate the kinetics of drug delivery [69,70]. Still, cross-linking of CS is used as a strategy to decrease the dissolution of CS in water. For example, treatment of water-soluble CS polymer with different proportions of trisodium trimetaphosphate achieved cross-linking of CS and reduced solubility for the usage of the polymer in drug delivery applications [71].

Mostly due to its highly water-soluble nature and lack of mechanical stability, CS has been modified with or used along with other polymers for drug delivery and tissue engineering applications. Porous sponge of CS-chitosan was used for the delivery of platelet-derived growth factor

(PDGF), aimed at achieving greater bone regeneration. Aqueous CS-chitosan solution was subjected to freeze-drying followed by cross-linking to form a porous sponge with 100-150 μm pore size. PDGF-BB was incorporated into the CS-chitosan sponge by soaking CS-chitosan sponge into the PDGF-BB solution. The amount of CS could act as a factor to control the release of PDGF-BB from the sponge [72]. It was also seen that the presence of CS increased the osteoconductive characteristics of the material. Hence, the bioactive CS could be used in combination with other polymers in the formulation of materials for delivery of growth factors.

As CS has biological roles that are advantageous and is a water-soluble and enzymatically biodegradable polysaccharide, it has been used in combination with materials such as collagen in preparation of scaffold matrices. Addition of CS to collagen type I matrix was achieved by using 1-ethyl-3-(3-dimethyl aminopropyl) carbodiimide as a cross-linking agent. The CS-bound matrix had increased water-binding capacity, with decreased tensile strength and temperature of denaturation [72]. Apart from imparting advantageous biological properties to the scaffold, CS also presented the opportunity to control the scaffold's mechanical and degradation characteristics.

As explained in the previous section, CS is an important component of joint ECM and known to play crucial roles in development and amelioration of joint/cartilage pathologies. Hence, it is used extensively in osteochondral tissue engineering. Here again, a number of CS variants such as modified CS or CS in combination with other natural and/or synthetic polymers have been utilized for the formulation of scaffolds. Scaffolds of different formats were utilized. Both fibrous spongy scaffolds and hydrogel scaffolds have been formulated utilizing CS. While employing CS in combination with collagen type I in the form of fibrous sponge matrix, a clear advantage in chondrocyte proliferation and phenotype maintenance was seen [73]. Similarly, sponges with a pore diameter of 180 μm were synthesized using natural polymers gelatin, CS, and HA together. When these scaffolds were seeded with chondrocytes, they maintained their morphology for up to 5 weeks and showed higher levels of aggrecan production than scaffolds that did not contain CS and HA components. These *in vitro* studies clearly point to the potential of scaffolds incorporating CS for cartilage tissue engineering [74,75].

A number of hydrogel scaffolds have also been formulated using CS for cartilage. Chondrocytes have a rounded morphology and this is seen to be best preserved in hydrogels, and therefore, hydrogel scaffolds are an ideal option for cartilage regeneration. Polymers such as PEG, polyvinyl alcohol (PVA), and poly(3-hydroxybutyrate) (PHB) have been used with CS to formulate hydrogels. While PVA hydrogels alone could not support cell adhesion, the addition of CS lowered the extreme hydrophilicity of the hydrogel and facilitated greater cell attachment in the scaffold [76]. When CS containing collagen hydrogel was conjugated to a fabric of PHB, it showed greater osteogenic potential than the parent fabric, on the seeded osteoblasts [77]. The presence of CS in PEG hydrogels promoted the chondrogenic differentiation of seeded bone marrow-derived mesenchymal stem cells, but did not allow them to proceed onto a hypertrophic state. This was an optimal outcome suitable for chondrocyte differentiation, to achieve cartilage regeneration [78]. In a recent study, the polysaccharide backbone was modified with methacrylate and aldehyde groups to form an adhesive gel. This modification led to better integration of the hydrogel with the implanted tissue *in vivo*, as seen in a rat model. This was due to greater scaffold-protein interaction brought about by the addition of modified CS [79]. Efforts have been made to deliver CS using scaffolding principles to joints affected by OA. For instance, a scaffold with 40% chitosan and 60% CS was designed to serve as a carrier of CS proved effective [80]. Hence, CS by itself after certain modifications and in combination with other synthetic and natural polymers has a great potential in scaffolding and drug delivery for osteochondral regeneration.

4.3.3 Promises and Challenges with CS

There have been many promising outcomes with CS as a drug and as a matrix for tissue engineering, both *in vitro* and *in vivo*. However, clinical outcomes have been inconsistent and have variable results. Better experimental design and ways to determine the polymer length and level of sulfation would be crucial to determining the fate of exogenously administered CS. Better experimental design and a greater understanding of the biological role of CS are important factors for determining the polymer's role and the most suitable type of CS polymer that lend the desirable biological outcomes.

4.4 CHITIN AND CHITOSAN

4.4.1 Chemical Structure, Properties, and Sources

Chitin is a structural polysaccharide found in nature. Chemically, chitin is made monomer units of 2-acetamido-2-deoxy-b-D-glucose connected through β (1→4) linkages (Figure 4.5). The C2 position in the glucose ring in the monomers has an acetamido group. The N-deacetylation of this chitin leads to the formation of chitosan. The degree of conversion of the acetamido group to amine group is never really complete. This is given by the degree of deacetylation (DD) of the chitosan. The DD of chitosan can vary from 30% to 95%. This conversion of chitin to chitosan renders the material more readily soluble and processable for various applications (Figure 4.6). Though crystalline chitosan is insoluble in aqueous solution at a pH > 7.0, in dilute acid where pH < 6.0, the positively charged amino group facilitates its solubility [82,83]. There are three reactive functional groups in chitosan—the amino group at C2 and primary and secondary –OH groups at C3 and C6 positions, respectively.

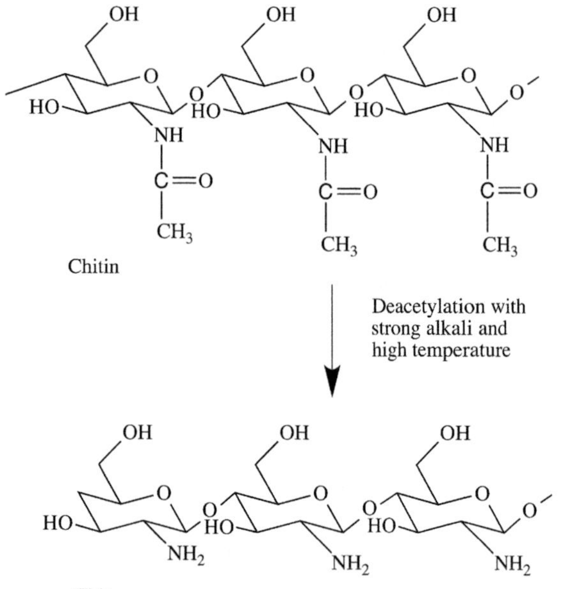

FIGURE 4.5 Structure of chitin and chitosan.

These reactive groups allow for chemical modification of chitosan such as covalent and ionic modifications.

Chitosan is one of the highly studied and utilized polysaccharide for tissue engineering. A process called "internal bubbling process" can be used to form porous structures from chitosan by freezing and lyophilizing a solution of chitosan. The addition of $CaCO_3$ to chitosan is used in this process to create porous chitosan gels [84]. This processability of chitosan is attributed to its cationic nature. The cationic property also leads to interaction of the material with negatively charged small molecules and proteins in biological systems. This is structurally very similar to native GAGs and hence plays crucial role in stimulating favorable responses in biological systems. Chitosan's mechanical properties are determined by pore sizes, the molecular weight, and crystallinity of the polymer. High-porosity, lower molecular weight, and less crystalline polymers are mechanically less competent and vice versa. Apart from these properties, chitosan is biodegradable. The acetylated residues of chitosan are targeted by lysozyme *in vivo* and this seems to be the major mechanism of chitosan degradation. Therefore, DD and crystallinity of the polymer are inversely related to degradation. Thus, the higher DD (>85%), the more crystalline the chitosan polymer and the slower its degradation in the body [85–87].

Chitosan is known to have antibacterial properties that [88] are attributed to the attack of negatively charged

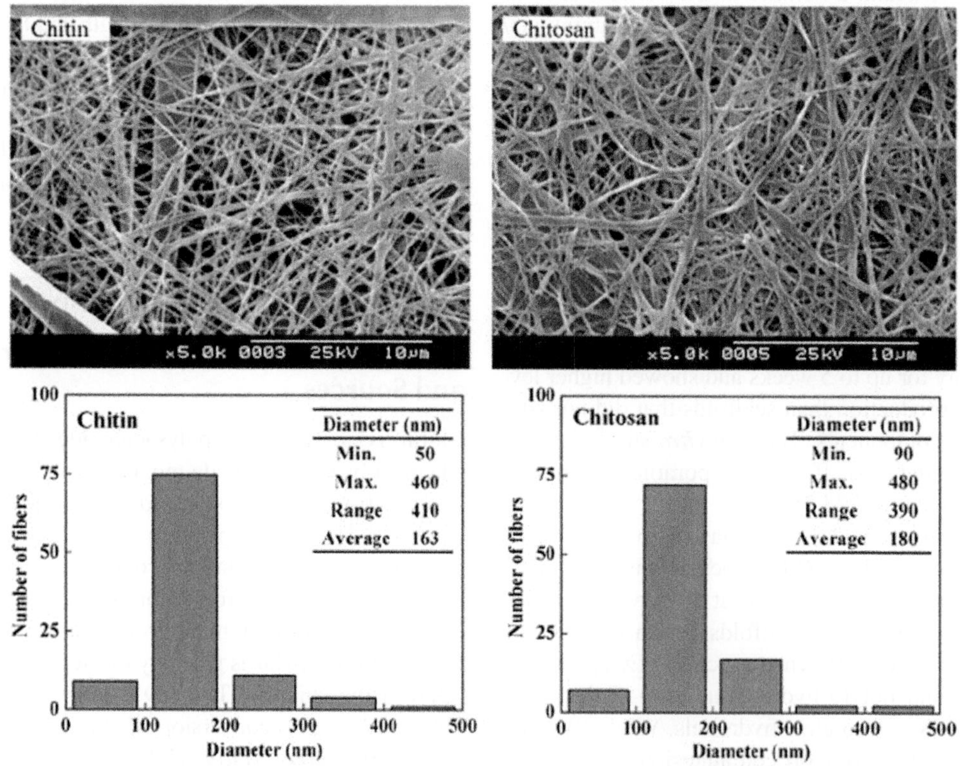

FIGURE 4.6 Chitin (on left) and chitosan after deacetylation (on right) nanofibers [73].

groups on the cell wall by the positively charged chitosan polymer. This leads to lysis of bacterial cell wall and hence its bactericidal activity. Inhibition of bacterial growth by chitosan is also attributed to its binding of bacterial DNA and interference of bacterial transcription [89]. Chitosan is also seen to have minimal immune rejection. Chitin is the major source for chitosan (Figure 4.6). It is the most abundant polymer to undergo biosynthesis, next to cellulose. It is a constituent of the exoskeleton in animals, like crustaceans, mollusks, and insects. It is also a polymer found in the cell wall of certain fungi. Most of the polymer used for commercial source comes as a by-product of the fishery industry [90].

4.4.2 Attempts Made in Tissue Engineering and Drug Delivery

4.4.2.1 Chitosan Alone

The various biological properties of chitosan have been discussed earlier. Chitosan has also been used as a dietary supplement. It is seen to lower low-density cholesterol and is helpful in weight loss [91]. Apart from its direct consumption, chitosan is used in drug formulations of different types such as microparticles, liposomes, granules, and gels for oral and parenteral drug delivery. In most of these applications, chitosan is physically or chemically cross-linked to obtain stability. The degree of cross-linking and drug loading are parameters used to control drug delivery [92,93]. In tissue engineering, chitosan alone was used initially. Since the mechanical and dissolution properties of the polymer make it tough to work with and formulate scaffolds, chemical modifications and their combinations with other natural and synthetic polymers are at present the more popular strategy. Common strategies are discussed in the following section.

4.4.2.2 Chitosan Derivatives and Combination with Other Polymers

Some of the limitations chitosan suffer from are those of insolubility at neutral pH and high water absorption by the polymer at a rapid rate. These factors pose problems of processability and also lead to rapid drug release from chitosan. Hence, chitosan is modified to overcome these limitations. Most modifications are brought about by reactions with the amine or hydroxyl groups of the glucosamine unit in chitosan. Using the reactive amine group, a number of modifying reactions are carried out. A simple example is a reaction in which an aldehyde functional group reacts with $-NH_2$ group of chitosan by reductive amination [94]. The introduction of *N*-cyanoethyl groups into the side chain of glucosamine in chitosan is a good example of this process. This reaction produced some cross-linking through a reaction between the nitrile group and the amine group of chitosan [95]. Examples of covalent modifications of chitosan include acylation and quarternization. When two oppositely charged polymers (a polycation and a polyanion), in a solution phase, separate out in a solution, a dense polymer phase called coacervate and a supernatant with low polymer content separate out. This process is termed polyelectrolyte complex formation. Polyelectrolyte complex formation has been used in a number of chitosan drug delivery systems [96,97] where controlled release of the loaded drug was desired. Modified chitosan and its blends with other polymers have been used in different formats. Nanoparticles and microparticles of chitosan and its derivatives have been formulated using techniques such as emulsification/solvent evaporation [81], spray drying [98], ionotropic gelation and coacervation [99], emulsion cross-linking [100], and sieving [101]. Thin films have been produced using solution casting, while cross-linking and gelation processes have been applied to produce hydrogels of chitosan for drug delivery [102]. Chitosan drug delivery vehicles in the form of tablets and gels are applied in dental, buccal, gastrointestinal, colon-specific, and gene delivery applications due to their favorable biological properties [102]. In tissue engineering, chitosan had been used mostly in minimally modified forms. The focus presently has shifted to improving the properties by introducing chemical modifications to form derivatives of chitosan for specific tissue regeneration purposes. Some of them are listed below.

4.4.2.2.1 Introduction of Sugars

Synthesis of chitosan bound to sugar has many applications in drug delivery and tissue engineering. This is due to the fact that cells, viruses, and bacteria recognize these sugar moieties and hence render these polymers good agents for targeting several target components in tissue engineering. For example, galactosylated chitosan worked as a good ECM for hepatocytes [103]. Specific antigen presenting B cells were recognized by mannosylated chitosan [104].

4.4.2.2.2 Graft Polymerization

Chemical grafting of chitosan can be used to functionalize chitosan and obtain important derivatives. Ceric ion, Fenton's reagent, gamma irradiation, various radicals, and ring opening reactions are the various routes used to achieve graft polymerization of chitosan [105]. Cell morphology and function were controlled by chitosan graft polymerized onto poly(L-lactide) (PLA) by plasma coupling reaction [106]. Further cooperative complementation through graft copolymerization or blend with poly (α-hydroxy acids) using a photosensitive cross-linking agent led to attachment of chitosan onto PLA films. These films showed improved cell attachment [107]. On the other hand, copolymerization of chitosan with heparin inhibited platelet adhesion [108]. Therefore, graft polymerization can help modulate chitosan's properties to elicit a desired cellular response.

4.4.2.2.3 Immobilization of Specific Sequences

Specific amino acid sequences promote cell adhesion. The most commonly used sequence is the RGDs from adhesion proteins. Photo-cross-linking RGD peptides to chitosan improved the adhesion of human endothelial cells, compared to unmodified chitosan scaffolds [109]. In another approach, the –COOH group of amino acids such as lysine, arginine, aspartate, and phenylalanine reacts to the $-NH_2$ group of chitosan. These functionalized chitosan polymers were entrapped onto the surface of PLA to improve cellular responses [110].

4.4.2.2.4 Production of Nanofibers

Nanofibers mimic the structure of natural ECM closely. Hence, enhanced cellular responses are achieved on electrospun nanofiber scaffolds. Chitosan nanofibers ranging from several down to a few nanometers have been produced by electrospinning technique [111,112].

4.4.2.2.5 Thermal Gelation

Thermal gelling is a technique of injecting a polymeric aqueous solution while keeping the temperature above the polymer's sol-gel transition temperature and allowing the polymer to form a gel as it reaches the body temperature. A thermal gelling chitosan polymer was formed by neutralizing highly deacetylated chitosan solution with glycerol phosphate (GP). Chitosan remained in solution at physiological pH. Chitosan/GP solution gelled at body temperature and hence was an attractive, injectable hydrogel drug delivery system local delivery of antineoplastic drugs like paclitaxel [113]. Derivatives of chitosan have been used in skin, bone, cartilage, and liver tissue engineering. A detailed description of chitosan is given elsewhere in this book.

4.4.3 Promises and Challenges with Chitosan in Tissue Engineering

Chitin and its derivatives have a number of applications in drug delivery and tissue engineering. While being an abundantly available polymer, chitosan is also biodegradable inside the body, mostly by the enzymatic activity of lysozyme. However, chitosan/chitin by itself lacks good mechanical properties required in certain structural applications. Chemical modifications of chitosan and blending the polymer with other natural and synthetic polymers are done to overcome this limitation. On the whole, chitosan has a huge potential in tissue engineering and drug delivery applications.

4.5 ALGINIC ACID

4.5.1 Chemical Structure, Properties, and Sources

Alginic acid is an anionic, strictly linear (unbranched) copolymer of mannuronic acid (M block) and guluronic acid (G block) units arranged in an irregular pattern of varying proportions of GG, MG, and MM blocks [114] (Figure 4.7). C5 epimerization and flipping of sugar ring to $^1C_4$ position [115] is seen to occur for steric stability in the polymer. The M residues are linked at $^4C_1$ (diequatorial links). The G residues are the C5 epimers of M. These G residues are linked at $^1C_4$ with diaxial links [116]. The G and M blocks are present as similar or strictly alternating (GG, MM, or GM). Due to the diaxial linkage, G blocks (GG) are stiffer than alternating blocks (GM) and hence more soluble at lower pH. The content of G in alginates varies from 40% to 70%, depending on the source, and determines the quality of the alginate polymer. The molecular weight of alginate can vary widely between 50 and 100,000 kDa. It is generally seen that alginates with high G block content are highly suitable for biomedical application due to the ease of processability and low immunogenicity in the body. Hence, the content of G and M blocks is a crucial factor that determines the properties and applications of the resultant alginate [115].

Alginates are polysaccharides produced by a wide variety of brown seaweeds (*Laminaria* sp., *Macrocystis* sp., *Lessonia* sp., etc.). Additionally, bacteria also synthesize alginates and these can be used as tools to tailor alginate production, by understanding the biosynthesis of the polymer in these bacteria. A family of enzymes termed mannuronan-C5-epimerases convert M into G at the polymer level. By genetically selecting and engineering *Pseudomonas* strains that contain only a single epimerase for the production of high G containing alginates has been possible. Using such strategies, alginates with

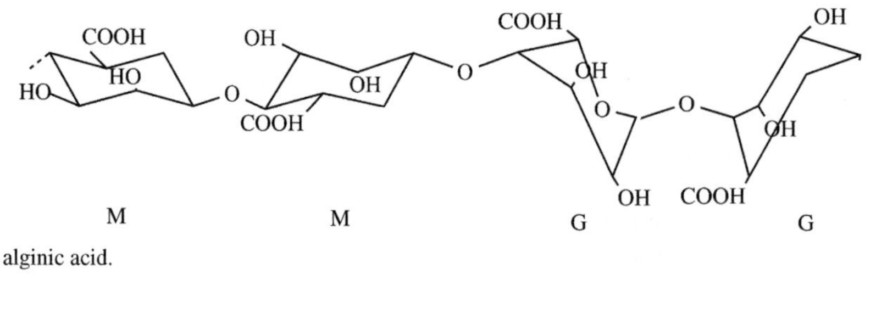

FIGURE 4.7 Structure of alginic acid.

up to 90% G content and extremely long G blocks have been produced [117–119].

Though such strategies are useful to engineer alginates, most of the alginates extracted for large-scale applications originate from natural sources such as seaweeds. The quality is determined by the species and even the seasonal variations. The alginate could contain from 10% to 70% G. Techniques of separation such as fractionation and precipitation in calcium can help separate the G block- and M block-rich alginates. The molecular weight of alginate is a critical factor to influence its viscosity in solution, besides the concentration of the polymer. The most important property of alginate is its ability to gel in the presence of cations (like Ca^{2+} and Ba^{2+}) (Figure 4.8). The carboxylic acid groups of sugars in G blocks of adjacent polymer chains cross-link with multivalent cations to form a gel. Factors that influence the stiffness of the gel are molecular weight

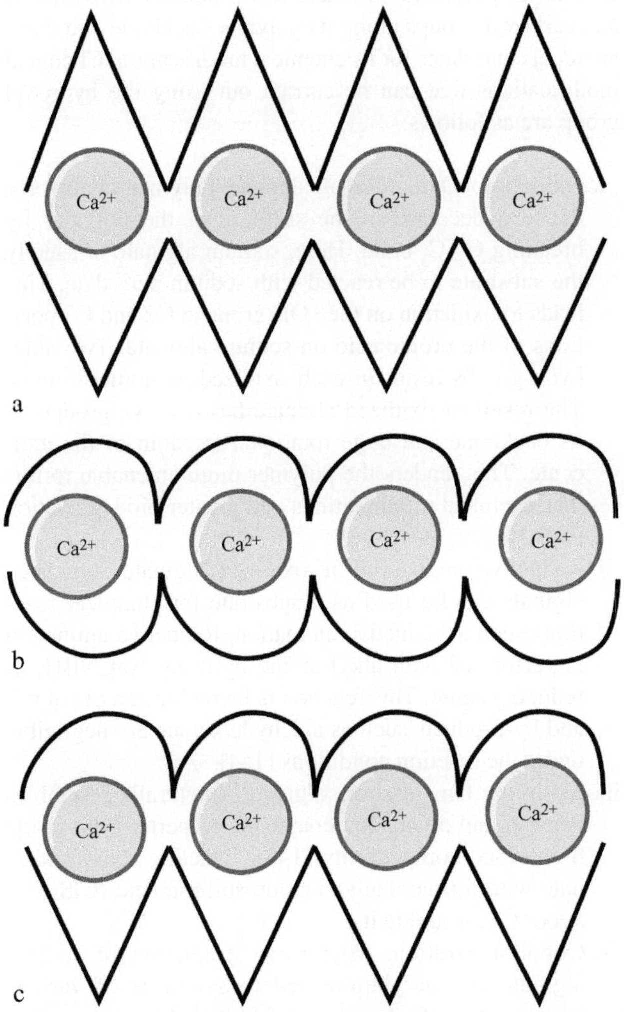

FIGURE 4.8 Gelation of alginic acid in the presence of Ca^{2+}. Possible junctions: (a) GG/GG junctions, (b) MG/MG junctions, and (c) mixed GG/MG, with Ca^{2+}. (For color version of this figure, the reader is referred to the online version of this chapter.)

distribution of the alginate polymer (dependent on M/G ratio) and the stoichiometry of alginate with the chelating cation [120,121].

4.5.2 Attempts Made in Tissue Engineering and Drug Delivery

4.5.2.1 Alginate Alone

The usage of polymers in biomedical applications such as drug delivery and tissue engineering is highly dependent on its ability to degrade in the body [122]. There are enzymes that can degrade alginate in the human body. The mechanism of degradation of alginate in the body happens in multiple ways as described below:

1. Disintegration of the alginate material by exchange of gelling calcium ion with sodium.
2. Acid hydrolysis and alkali hydrolysis. At a physiological pH of 7.4, β-elimination by alkali hydrolysis is the predominant mechanism of alginate polymer length reduction. Oxidation of the gels, by agents such as peroxides [123], not only helps hasten the process but also weakens the gel, in a ring opening reaction of the polymer.
3. Degradation by reactive oxygen species. Most polymers including alginates undergo "free-radical depolymerization" or "oxidative-reductive depolymerization." Water molecules or molecular oxygen generates free radicals in living systems. Exposure of polymer to gamma radiation increases this process and can be used to enhance the rate of degradation [124].

Thus, the major disadvantages of alginate are the lack of enzymatic degradation and its inert nature that makes it nonadherent for cells. To overcome these limitations, alginate is widely used as its derivative in combination with other polymers in drug delivery and tissue engineering applications. Alginate by itself is used widely in many industries. It is used as a stabilizer and emulsifier in food industry, as it interacts with proteins, fats, and fibers. Alginate-pectin mixtures are used as gelling agents independent of sugar content in foods. Hence, alginate is used in many low-calorie substitute foods. The high hydrophilicity of alginates renders the material biocompatible and nonimmunogenic. Therefore, it is used widely in pharmaceutical industry as drug excipient [116], as dental impression material [125], and as a material for wound dressing [126].

4.5.2.2 Alginate Derivatives and Combinations with Other Polymers

In general, alginates are formed into gels either by using multivalent cations or by covalent cross-linking. These modifications will be discussed further, here. The major purpose in subjecting alginates to chemical and physical modifications is to tailor their physical properties such as degradation, mechanical strength, and biological properties such as enhanced interaction with cells. Mechanical properties like stiffness and strength of alginate gels can be controlled by physical factors. Concentration of polymer [127] and its molecular weight [121] can be used to determine the density of polymer solution for formulation of gels. Increase of both these factors leads to higher viscosity of the polymer solution and hence a stiffer and mechanically strong gel. Cationic poly(ethyleneimine) (PEI) [128] addition leads to improvement of alginate gel's mechanical properties. High-molecular-weight PEI increases the resistance of the gel to de-cross-linking agents and thus improves the gel's stiffness. Gelling conditions such as temperature, type, and concentration of cross-linker also affect the mechanical properties of alginate gels. Low-temperature cross-linking [129] leads to slow cross-linking, due to reduced rate of calcium ion diffusion. This results in the formation of gels with enhanced mechanical properties. Apart from these factors, the presence of cells in the gels is also shown to improve its mechanical strength [130].

As discussed earlier in the previous section, there are many methods used to control the degradation of alginates. Gama irradiation [131] and partial oxidation [132] can be used to reduce the molecular weight of polymer and accelerate degradation rates. These techniques also affect the mechanical strength of alginate gels. Gels with bimodal molecular weight distribution have been formulated with one molecular weight polymer being oxidized and the other left untreated. This approach also accelerated gel degradation rate. In gels that are covalently cross-linked, the linker density [133] determines the rate of degradation as well as the strength of the gel. As alginates are not conducive to cell adhesion, covalent modification of the polymer, coupling whole (like fibronectin and collagen) [134,135] or parts of cell adhesion molecules (like RGD peptides), is a popular strategy used to increase cellular responses. Here, the concentration [136] of these adhesion molecules and the composition of the gel (M/G ratio) determine the effectiveness of the gel in inducing cell adhesion.

Alginate-based hydrogels have been used as drug delivery vehicles for low-molecular-weight (small) molecule drugs as well as proteins such as growth factors. Drug-alginate interactions play a crucial role in determining the rate of drug release from alginate gel matrices. Drug release rate can be completely controlled by charge polarity (hydrophilic molecules will be released quickly and hydrophobic ones more slowly), when there is no chemical interaction between the drug and the alginate matrix. Carbodiimide chemistry is used to link hydrophilic drugs [137] to alginate matrices to delay their release. In such a scenario, the rate of drug release is determined by polymer degradation. Here, a linker such as AAD could be used to attach the drug to alginate and the rate of release would be controlled by the concentration of the spacer. Ionic complexes can also be used to attach drugs to alginate [138]. Proteins such as growth factors have been successfully delivered by alginate gel systems. It was seen that such deliver systems preserved the bioactivity of the factors. Hence, angiogenic growth factors like β-fibroblast growth factor [139] and vascular endothelial growth factor (VEGF) loaded into alginate beads were successful in inducing angiogenesis to a greater extent than free administration of the growth factors. Ionic complexes were used to link VEGF [140,141] to alginate matrix and the dissolution of this complex along with diffusion acted to control the release of active VEGF from the gels over several weeks *in vivo*.

In most tissue engineering applications, alginates with modified physical, chemical, and biological properties are desirable. These characteristics can be achieved by chemically modifying alginate to form its derivatives and blends with other polymers. Alginate has a number of hydroxyl and carboxyl groups along its polymer backbone and these are ideal candidates for its chemical modification. Chemical modifications that can be carried out using the hydroxyl group are as follows:

i. Oxidation: Oxidation of alginate polymer chain produced a decrease in the stiffness of the polymer by breaking C_2–C_3 bond. Here, sodium alginate is usually the substrate to be reacted with sodium periodate. This leads to oxidation on the –OH group at C2 and C3 positions of the uronic acid on sodium alginate. Two aldehyde groups result in each oxidized monomeric units. The resultant oxidized alginate has reactive groups on its backbone and large rotational freedom of the molecule. This renders the polymer more amenable to further chemical modifications and greater biodegradation [142,143].

ii. Reductive amination of oxidized alginate: Oxidized alginate can be used as a substrate for chemical reactions such as reductive amination. Reductive amination is performed with alkyl amine by using $NaCNBH_3$ as reducing agent. This reaction is favorable at a pH of 6-7 and by-products such as aldehyde/ketone are negligible under the reaction conditions [144].

iii. Sulfation: On sulfation, alginate structurally resembles heparin and attains anticoagulant properties, alongside high blood compatibility [145]. Reacting sodium alginate with formamide and chlorosulfuric acid ($ClSO_3H$) at 60 °C can sulfate it.

iv. Copolymerization: Microwave irradiation of sodium alginate and acrylamide led to synthesis of various grades of grafted polymers [146]. Alginate-*g*-vinyl sulfonic acid prepared by employing potassium peroxydiphosphate/thiourea redox system has also been reported [147]. These synthesized graft copolymers exhibit better results for swelling, metal ion uptake, and resistance to biodegradability in comparison to parent alginates themselves.

v. Linking cyclodextrins: α-Cyclodextrin can be covalently linked to alginate. This reaction can be targeted to the hydroxyl groups of the alginate by cyanogen bromide (CNBr) method to prevent reaction at the carboxyl group. This specificity was necessary to form the calcium-alginate beads that have great potential in encapsulating bacteria for environmental remediation [148].

Chemical modifications that can be carried out using the carboxyl group are as follows:

i. Esterification: Alkyl group is attached to a molecule during esterification. Addition of alkyl group to the backbone of alginate results in increasing the hydrophobicity of alginate. Alginate can be modified by direct esterification using several alcohols in the presence of catalyst. The alcohol is present in excess to ensure that the equilibrium is in favor of product formation. Propylene glycol ester of alginate was obtained by esterification of alginate with propylene oxide. This is a commercially useful derivative of alginate [149].

ii. Ugi reaction: Hydrophobically modified alginate can be prepared by the Ugi multicomponent condensation reaction. The Ugi reaction is a multicomponent reaction in organic chemistry involving a ketone or aldehyde, an amine, an isocyanide, and a carboxylic acid to form a bis-amide [150].

iii. Amidation: In amidation reaction, a coupling agent, 1-ethyl-3-(3-dimethylaminopropyl) carbodiimide hydrochloride, is reacted with alginate to form amide linkages between amine-containing molecules and the carboxylate moieties on the alginate polymer backbone. This results in hydrophobic modification of the alginate [151].

A general outline of reactions used to modify alginate for biomedical applications was summed up earlier. Though mostly hydrogel of alginate is the most popular form of application of the polymer, other formats have also been formulated. Alginate foams [152], fibers [153,154], and nanofibers [155] are other forms in which alginates are fabricated [115]. Bioartificial pancreas, bone [156], vasculature, and liver [157] are some of the tissue-engineered organs where alginate materials have been used successfully.

4.5.3 Promises and Challenges with Alginates in Tissue Engineering

Alginates are a versatile class of polysaccharides that present a great tool as materials for tissue engineering. They have been formulated as gels, microspheres, foams, and fibers in tissue engineering and for delivery of drugs and biological factors. Some constraints posed by the material are its nonenzymatic degradation in the human body and its extremely hydrophilic nature that discourage cell anchorage. However, these limitations have been overcome by subjecting its −OH and −COOH functionalities to chemical modifications. Understanding the biosynthetic pathways leading to alginate biosynthesis in bacterial systems has been very useful in tailoring their polymer chain composition and the molecular weight of the polymer. Yet, more work needs to be done in improving these systems. Alginate hence serves as a low-cost biopolymer that is a good tool in biomedical engineering.

4.6 CELLULOSE

4.6.1 Chemical Structure, Properties, and Sources

Cellulose, the "sugar of plant cell wall," is the most abundant biopolymer in the biosphere.

The basic monomer unit of cellulose is β-D-anhydroglucopyranose. These units are joined together covalently by acetal functions between the equatorial group of the C4 carbon atom and the C1 carbon atom (β-1, 4-glycosidic bonds). These β-1,4-glycosidic linkages bestow cellulose with its resistance to chemical/enzymatic attack [158] (Figure 4.9). Therefore, cellulose is a linear-chain polymer with a large number of hydroxyl groups (three −OH groups per anhydrous AGU unit). This linear structure can be extended to molecules containing 1000-1500 β-D-glucose monomer units, in a cellulose polymer chain. This chain length of cellulose is expressed in terms of number of constituent AGUs, termed degree of polymerization. The degree of linearity and the presence of extensive −OH groups throughout the cellulose chain are responsible for formation of inter- and intramolecular hydrogen bonds throughout the polymer chain. This causes cellulose chains to organize in parallel arrangements into crystallites and crystallite strands, the basic elements of the supramolecular structure of the cellulose fibrils and the cellulose fibers. This arrangement of fibers in the polymer is termed its supramolecular structure and it in turn influences its physical and chemical properties. Cellulose is known to exist in at

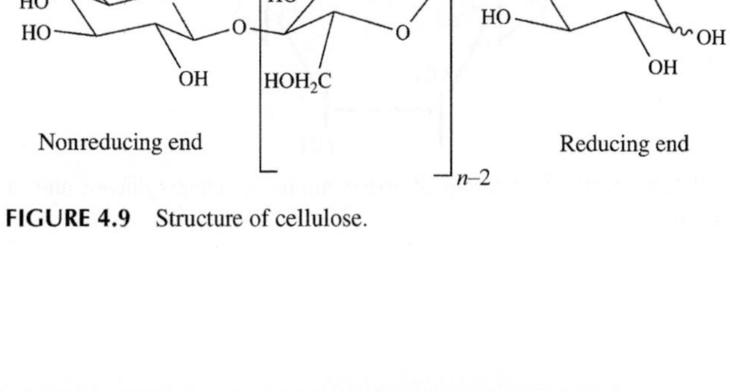

FIGURE 4.9 Structure of cellulose.

least five allomorphic forms. Cellulose I is the form found in nature. Cellulose may occur in other crystal structures denoted celluloses II, III, and IV. Cellulose II is the most stable structure of technical relevance. This structure can be formed from cellulose I by treatment with an aqueous solution of sodium hydroxide. This leads to regeneration of native cellulose from solutions of semi-stable derivatives. This crystalline structure is modified from cellulose I. A parallel chain arrangement of cellulose in cellulose I form undergoes a change to form cellulose II. This renders the cellulose II more accessible to chemical treatments and hence more reactive. A fringe fibrillar model is used to describe the microfibrillar structure of cellulose polymer. It describes the structure to be made of crystalline regions of varying dimensions called crystallites and noncrystalline regions. This structure explained the partial crystallinity and reactivity of cellulose in relation to its microfibrillar structure [159].

Cellulose behaves as an active chemical due to the three hydroxyl groups in each glucose unit. The hydroxyl groups at the second and third positions behave as secondary alcohols, while the hydroxyl group at the sixth position acts as a primary alcohol (numbering as shown in Figure 4.10). These –OH groups are responsible for the reactivity of cellulose. DS is the term used to indicate the average number of –OH groups substituted in an anhydroglucose unit of a cellulose molecule. That is, a DS of 3 indicates that all the three –OH groups have been substituted in the anhydroglucose units of the cellulose derivative. In general, the relative reactivity of the hydroxyl groups can be expressed as $OH\text{-}C_6 \gg OH\text{-}C_2 > OH\text{-}C_3$ [160].

Cellulose, like the polysaccharides above, has certain drawbacks. These include poor solubility in common solvents, poor crease resistance, poor dimensional stability, lack of thermoplasticity, high hydrophilicity, and lack of antimicrobial properties. To overcome such drawbacks, the controlled physical and/or chemical modification of the cellulose structure is essential [160]. Introduction of functional groups into cellulose can alleviate these problems while maintaining the desirable intrinsic properties of cellulose. Apart from the conventional plant source, cellulose is also obtained from bacteria, termed bacterial cellulose.

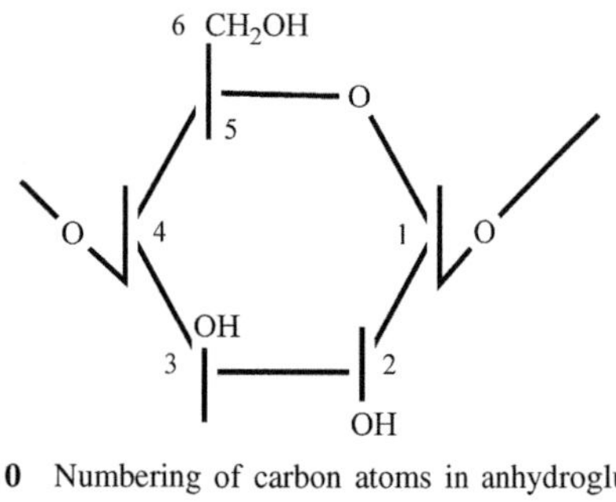

FIGURE 4.10 Numbering of carbon atoms in anhydroglucose unit of cellulose.

Understanding and engineering these biological systems has opened further doors for bringing in desired modifications into cellulose.

4.6.2 Attempts Made in Tissue Engineering and Drug Delivery

4.6.2.1 Cellulose Alone

As cellulose in its native form has extensive hydrogen bonds, it is not very processable. Most of the early attempts resorted to viscose process of regenerating (restoring cellulose structure back) cellulose from its derivatives. Regenerated cellulose has been used as a matrix for wound dressing. Early results indicated that implanted viscose cellulose sponges led to increased granulation tissue formation over a period of time and that the pore structure of the scaffold could influence cell infiltration within a certain limit [161]. On the other hand, it was seen that a lower content of cellulose and lower pore diameter induced greater tissue invasion of the implant, and later proved to be a matrix conducive for bone formation in a rat model [162]. However, cellulose sponges were seen to be slow-degrading matrices taking up to 60 weeks for them to be degraded [163]. Cellulose also was used successfully as an enzyme carrier by dissolving cellulose in ionic liquid and regenerating it. Usage of a hydrophobic ionic liquid preserved the enzymatic activity to a greater extent [164]. While biodegradation remains a challenge for the absorption of cellulose, it was seen that treatments that markedly reduced crystallinity led to degradation as well as high biocompatibility of regenerated cellulose [165]. In recent years, microbial cellulose (MC) produced by bacterial species such as *Acetobacter xylinum* has gained much attention as a material for use in biomedical appliances. Though chemically similar to plant cellulose, MC has a microfibrillar and nanostructured arrangement that enables higher water retention by the material. This property is conducive in its application in wound dressing, production of vascular conduits, etc. [166]. This cellulose was also seen to have a high degree of biocompatibility [167].

4.6.2.2 Cellulose Derivatives and Combination with Other Polymers

The supramolecular structure of cellulose and the extensive hydrogen-bonded chemical structure of cellulose render it insoluble in water and organic compounds. Therefore, most of the reactions involving cellulose are carried out in solid or swollen state as heterogeneous reactions. Here, the limiting factors for the heterogenous reaction are the breakage of hydrogen bonds (by alkaline treatment) and degree of interaction with the reaction media (by swelling). Therefore, specific solvents that disrupt the hydrogen bonds in cellulose such as *N,N*-dimethylacetamide/lithium chloride (DMA/LiCl) and dimethyl sulfoxide/tetrabutylammonium fluoride (DMSO/TBAF) [168] are used widely for the purpose. Hence, even though using such strategies, cellulose alone has been employed for different purposes as a material, its derivatives are easier to work with as they overcome the limitations posed by cellulose. The following section gives an overview of derivatives of cellulose and its combination with other polymers:

4.6.2.2.1 Cellulose Esters

Esters of cellulose with interesting properties such as bioactivity and thermal and dissolution behavior can be obtained by esterification of cellulose with nitric acid in the presence of sulfuric acid, phosphoric acid, or acetic acid. Commercially important cellulose esters are cellulose acetate, cellulose acetate propionate, and cellulose acetate butyrate. Cellulose esters of aliphatic, aromatic, bulky, and functionalized carboxylic acids can be synthesized through the activation of free acids *in situ* with tosyl chloride, *N,N′*-carbonyldiimidazole, and iminium chloride under homogeneous acylation with DMA/LiCl or DMSO/TBAF. A wide range of cellulose esters that vary in their DS, various substituent distributions, and several desirable properties can be obtained through these reactions. Recently, a number of enzymes that degrade cellulose esters have been reported. Some of them are acetyl esterases, carbohydrate esterase (CE) family 1, and esterases of the CE 5 [169–172] family.

Cellulose esters have been put to use in many biomedical applications. Hemodialysis membranes used in purification of blood, for patients with renal failure have employed melt-spun cellulose diacetate membrane. These membranes have been produced by Altin (company) and used successfully. They were advantageous and had less toxicity than synthetic polymer-based membranes [173]. Cellulose acetate is seen as a preferred material for the fabrication of blood filtration devices that are used to separate a fraction of the blood such as red blood cells and leukocytes [174,175]. Cellulose acetate and regenerated cellulose fibrous matrices have been successful in supporting the growth of cardiac myocytes and present a potential scaffold platform for cardiac regeneration [176]. Our lab has been successful in the formulation of mechanically competent cellulose acetate- and ethyl cellulose-based scaffolds [177]. Further, these scaffolds could maintain the growth of osteoblasts (bone cells), *in vitro* [178] (Figure 4.11), indicating their potential application as scaffolds for bone regeneration. Hence, cellulose esters have a huge potential in tissue engineering and drug delivery applications.

4.6.2.2.2 Cellulose Ethers

Carboxymethyl cellulose (CMC) is the major cellulose ether. By activating the noncrystalline regions of cellulose, selective regions of alkylating reagents can attack the cellulose. This is termed the concept of reactive structure fractions and is used widely for the production of CMC. Another route for carrying out the same reaction is by derivatization of cellulose in reactive microstructures, formed by induced phase separation. This process involves the usage of NaOH in anhydrous state in combination with solvents like DMA/LiCl. These CMC products have a distribution of substituents that deviate significantly from statistical prediction of the product theoretically.

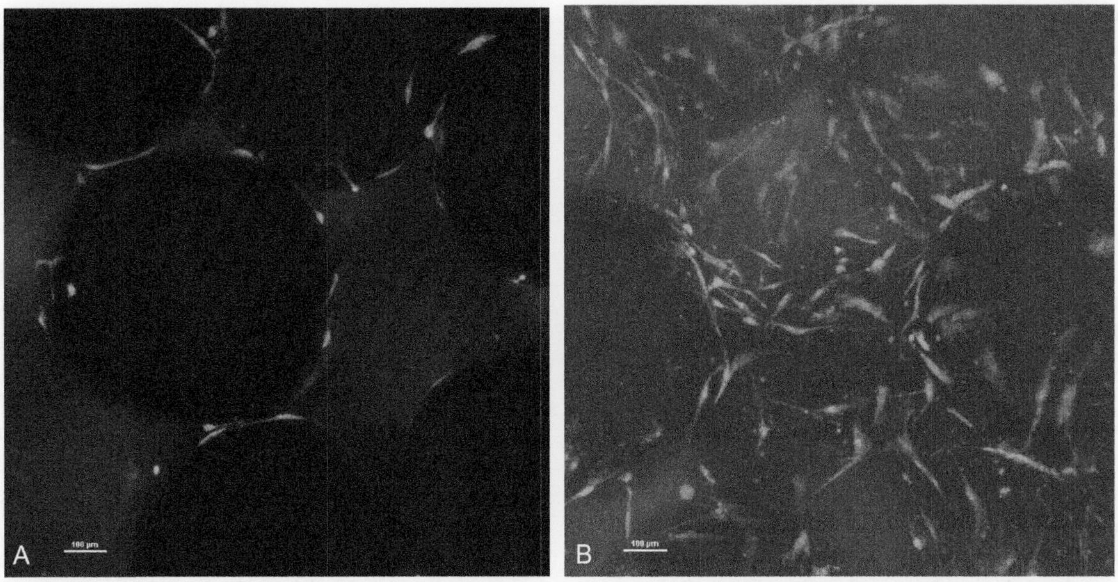

FIGURE 4.11 Cellular viability of osteoblasts on (a) cellulose acetate scaffolds and (b) cellulose acetate collagen scaffolds. The green cells are viable and the red ones are nonviable (Live/Dead staining) [170]. (For interpretation of the references to color in this figure legend, the reader is referred0 to the online version of this chapter.)

CMC is used in several drug delivery and tissue engineering purposes. The release of apomorphine, a drug used to regulate motor responses in Parkinson's disease, was successfully incorporated into CMC powder formulation and exhibited a sustained nasal release, and performed better than starch-based delivery vehicle [179]. Sodium CMC has been used successfully in gastrointestinal drug delivery [180]. Hence, CMC is seen as a successful drug delivery system for mucosal tissue [181]. Apart from drug delivery, CMC is useful as a scaffold in tissue engineering. CMC hydrogels having pH-dependent swelling characteristics were capable of releasing entrapped drug at the right pH present in the tissue of interest and showed great potential as a wound dressing material [182]. CMC hydrogels could be used for encapsulating cells of nucleus pulposis and hence are a potential replacement for intervertebral disk degeneration [183]. CMC has been combined with chitosan [184] and hydroxyapatite [185] for bone and dental regeneration purposes too.

4.6.2.2.3 Sillyl Cellulose

Silyl ethers of cellulose are characterized by a remarkable increase in thermal stability, lipophilic behavior, and a lack of hydrogen bonds. They can be used as selective protecting groups in organic synthesis, due to the simple cleavage of the silyl ethers under acidic conditions or through nucleophilic attack. Therefore, the silyl ethers of cellulose are very attractive for engineering polysaccharide chemistry [186]. The silylation of polar protic –OH groups of cellulose with chlorosilanes and silazanes leads to these silyl ethers. The degree (DS) and position of silyl substitution is determined by the reaction condition. All the three –OH groups will be substituted when trimethylsilylation with hexamethyldisilazane in liquid ammonia is used in the reaction. Dissolution of cellulose in DMA/LiCl (homogeneous reaction) makes the –OH groups more accessible. Following this, if the synthesis takes place in the presence of imidazole, the bulky silylation reagent thexyldimethylchlorosilane leads to complete silylation at O_6 and O_2 (DS value=2.0). Here, the primary and the most reactive secondary –OH groups are converted. If silyl ether formation starts with the same reagent in cellulose suspension in aprotic dipolar media like *N*-methylpyrrolidone, which contain gaseous ammonia, silylation of all primary C_6-OH groups takes place. This state does not permit any further reaction of the secondary hydroxy groups [187].

4.6.2.2.4 Cellulose Sulfonates

The most frequently synthesized and used cellulose sulfonates are the *p*-toluenesulfonates (tosylates), methanesulfonates (mesylates), *p*-bromobenzenesulfonates (brosylates), and trifluoromethanesulfonates (triflates). The synthesis of sulfonates through simple esterification of the –OH groups of cellulose with the corresponding sulfonic acid chlorides or anhydride is a way to attach nucleofuge groups to cellulose [188]. By varying the solvent and reaction conditions, the DS of the polymer can be controlled. For instance, at temperatures of 7 °C, cellulose tosylate with a maximum DS of 2.3 can be formed with tosyl chloride in the presence of triethylamine. The DS of sulfonated cellulose can also be controlled by the molar ratio of tosyl chloride to glucose units of the cellulose.

4.6.2.2.5 Aminocellulose

Aminocellulose is an aminodeoxy derivative bearing the nitrogen function directly on the cellulose skeleton. These are useful in the immobilization of enzymes and other proteins, by having a specific structural design based on cellulose tosylates. Aminodeoxycellulose is synthesized with corresponding halogen derivatives and sulfonates as starting materials. PDA cellulose is a material used successfully for immobilization of enzymes like oxidoreductases, glucose oxidases, and peroxidases using glutaraldehyde. Diazo coupling and redox coupling have also been utilized to link ascorbates and dyes. Here too, the solvent composition and reaction conditions can determine the reaction chemistry (S_N2) and hence the DS of the product formed [189].

4.6.2.2.6 Resinification of Cellulose

Resinification of cellulose can impart crease resistance termed "durable press" properties to cellulose. The reaction of cellulose with bi- or polyfunctional compounds leads to formation of cross-linked cellulose matrix, called resinified cellulose [158].

4.6.2.2.7 Graft Polymerization of Cellulose

A new approach to modification of cellulose is by graft polymerization. A graft copolymer generally consists of a long sequence of one monomer, referred to as the backbone polymer (main chain) (cellulose in this case) with one or more branches (grafts) of long sequences of a different monomer [190]. Graft copolymerization permits the combination of the best properties of two or more polymers in one physical unit [160]. The aim with cellulose graft polymerization is to retain the inherent properties of cellulose and incorporate qualities from the polymer grafted onto it. Depending on the nature of the grafted polymer, properties such as dimensional stability, resistance to abrasion and wear, wrinkle recovery, oil and water repellence, elasticity, sorbancy, ion exchange capabilities, temperature responsiveness, thermal resistance, and resistance to microbiological attack can be incorporated into cellulose [191–195].

The methods for graft polymerization of cellulose can be generally classified into three major groups such as (i) free-radical polymerization, (ii) ionic and ring opening polymerization, and (iii) living radical polymerization.

Strategies used in cellulose graft polymerization can be divided into three categories:

(i) The "grafting to" approach—here, the functional preformed polymer with its reactive end-group is coupled with the functional groups located on the backbone of cellulose, the major polymer. The inherent weakness of this approach is a hindrance to diffusion caused by crowding of polymer chains on the surface [196].

(ii) The "grafting from" approach—here, the growth of polymer chains occurs from initiating sites on the cellulose backbone. This is the most commonly used approach. Easy access of the reactive groups to the chain end of the growing polymer is achieved in "grafting from" approach. This makes it possible to attain high graft density [196].

(iii) The "grafting through" approach—in this approach, a vinyl macromonomer of cellulose is copolymerized with a low-molecular-weight comonomer. Though this approach is more convenient, a cellulose-derived macromonomer has to be synthesized. This poses a limitation to this technique [196].

4.6.3 Promises and Challenges with Cellulose

Cellulose and other polysaccharides discussed earlier have been gaining importance as polymeric materials. Increasing the knowledge of organic, polymer chemistry and the chemistry of low-molecular-weight polysaccharides can greatly help us understand more about the chemistry of cellulose and control the different processing techniques for cellulose to a greater extent. Having a multidisciplinary approach will further help us utilize the polymer more in biomedical applications. The development of derivatives and grafted polymers of cellulose has been an important step toward the utilization of cellulose, which is considered as a renewable resource. Processes such as lyocell processing of cellulose are environment-friendly techniques and promise a safer polymer processing technology. New insights are still being obtained on the process of wood pulping and biosynthetic pathways in cellulose synthesis. By using this knowledge, engineering cellulose and utilizing bacteria for production of the polymer are the advances we could expect in the future.

4.7 CONCLUSIONS

Polymers derived from plants and animal kingdoms have been widely researched as biomaterials for a variety of biomedical applications including drug delivery and regenerative medicine. These polymers have biochemical similarity with human ECM components and hence are readily accepted by the body. Additionally, these polymers inherit several advantages including natural abundance, relative ease of isolation, and room for chemical modification to meet the technological needs. In addition, these polymers undergo enzymatic and/or hydrolytic degradation in the biological environment with body-friendly degradation by-products. Natural polymers include the list of polysaccharides (carbohydrates) and animal-derived proteins. Polysaccharides are an important class of biomaterials with significant research interest for a variety of drug delivery and tissue engineering applications due to their assured biocompatibility and bioactivity. Polysaccharides are often isolated and purified from renewable sources including plants, animals, and microorganisms. Essentially, these polymers have structural similarities, chemical versatilities, and biological performance similar to ECM components, which often mitigate issues associated with biomaterial toxicity and host immune responses. The building blocks of carbohydrate monosaccharide are joined together by *O*-glycosidic linkages to form a polysaccharide chain. Polysaccharides offer a diverse set of physicochemical properties based on the monosaccharide that constitutes the chain, its composition, and source. The popular list of polysaccharides used for a variety of biomedical applications includes cellulose, chitin/chitosan, starch, alginates, HAs, pullulan, guar gum, xanthan gum, and GAGs. In spite of many merits as biomaterials, these polysaccharides suffer from various drawbacks including variations in material properties based on the source, microbial contamination, uncontrolled water uptake, poor mechanical strength, and unpredictable degradation pattern. These inconsistencies have limited their usage and biomedical application-related technology development.

Numerous synthetic polymers with well-defined mechanical and degradation properties have been developed to meet the technological needs in the biomedical applications. However, these polymers from the biological standpoint lack much-desired bioactivity and biocompatibility and may cause toxicity and immune response. Polysaccharide structure offers freely available hydroxyl and amine functionalities that make it possible to alter its physicochemical properties by chemically modifying polysaccharide structure. For instance, grafting synthetic monomers on the polysaccharide chain offers an easy way to control polymer solubility in desired solvents, water uptake, and degradation. These semi-synthetic polymers offer best features of the both natural and synthetic polymers. Various cross-linking techniques to restrict the polysaccharide chain movement to control their water uptake, degradation, and mechanical properties have also been developed. Polysaccharide-based porous scaffolds, fiber matrices, hydrogels, and micro- and nanoparticles have been developed for a variety of tissue regeneration and drug delivery applications. In the recent years, glycochemistry has gained research momentum for understanding carbohydrate biological functions and development of carbohydrate-based drugs and vaccines. Engineered carbohydrate-based polymeric structures may serve as an alternative material platform for a variety of regenerative medicine and drug delivery applications.

A new nonpetroleum-based biomaterial platform to meet the versatile needs in biological science and biomedical engineering could be achieved by collaborative efforts between academia, government, and industry partnership. The collaborative efforts should include bringing scientists working in different disciplines of chemistry, biology, polymers, materials sciences, and engineering to work toward these activities. These collaborative efforts could lead to the development of a methodology for synthesis of natural polymer-based semi-synthetic polymers and provide a greater depth of understanding of carbohydrate biological functions, polymer structure, material properties, degradation, and mechanical properties. Further, the development of modeling tools to predict structure, property and biological activity of carbohydrates for biomedical applications is a step in this direction. The goal of the new initiatives should focus on the development of natural polymer-based orthopedic fixation devices, biomedical implants, drug delivery vehicles, carbohydrate-based drugs, hydrogels, surfactants, coagulants, and absorbents for a variety of biomedical applications. The research activities in this area could generate commercially available technologies and products from the renewable resources and contribute immensely toward economic development.

ACKNOWLEDGMENTS

The authors gratefully acknowledge funding from the Raymond and Beverly Sackler Center for Biomedical, Biological, Physical and Engineering Sciences. Authors also acknowledge the funding from National Science Foundation (IIP-1311907 and EFRI-1332329) and Early Career Translational Research Award in Biomedical Engineering from Wallace H. Coulter Foundation.

REFERENCES

[1] J.F. Mano, G.A. Silva, H.S. Azevedo, P.B. Malafaya, R.A. Sousa, S.S. Silva, et al., Natural origin biodegradable systems in tissue engineering and regenerative medicine: present status and some moving trends, J. R. Soc. Interface 4 (17) (2007) 999–1030.

[2] G.G. d'Ayala, M. Malinconico, P. Laurienzo, Marine derived polysaccharides for biomedical applications: chemical modification approaches, Molecules 13 (9) (2008) 2069–2106.

[3] L. Klouda, A.G. Mikos, Thermoresponsive hydrogels in biomedical applications, Eur. J. Pharm. Biopharm. 68 (1) (2008) 34–45.

[4] S. Hirano, Chitin and Chitosan, Cancer Chemotherapy to Ceramic Colorants, Ullmann's Encyclopedia of Industrial Chemistry, VCH Verlagsgesellschaft, Weinheim - Deerfield Beach - Basel, A5 59, 1986, p. 898.

[5] S.P. Evanko, J.C. Angello, T.N. Wight, Formation of hyaluronan- and versican-rich pericellular matrix is required for proliferation and migration of vascular smooth muscle cells, Arterioscler. Thromb. Vasc. Biol. 19 (4) (1999) 1004–1013.

[6] J.L. Drury, D.J. Mooney, Hydrogels for tissue engineering: scaffold design variables and applications, Biomaterials 24 (24) (2003) 4337–4351.

[7] T.C. Laurent, J.R. Fraser, Hyaluronan, FASEB J. 6 (7) (1992) 2397–2404.

[8] T.C. Laurent, U.B. Laurent, J.R. Fraser, The structure and function of hyaluronan: an overview, Immunol. Cell Biol. 74 (2) (1996) A1–A7.

[9] P.H. Weigel, V.C. Hascall, M. Tammi, Hyaluronan synthases, J. Biol. Chem. 272 (22) (1997) 13997–14000.

[10] J.R. Fraser, T.C. Laurent, U.B. Laurent, Hyaluronan: its nature, distribution, functions and turnover, J. Intern. Med. 242 (1) (1997) 27–33.

[11] K.P. Vercruysse, G.D. Prestwich, Hyaluronate derivatives in drug delivery, Crit. Rev. Ther. Drug Carrier Syst. 15 (5) (1998) 513–555.

[12] B.P. Toole, Hyaluronan is not just a goo! J. Clin. Invest. 106 (3) (2000) 335–336.

[13] T.D. Camenisch, J.A. McDonald, Hyaluronan: is bigger better? Am. J. Respir. Cell Mol. Biol. 23 (4) (2000) 431–433.

[14] J.A. McDonald, T.D. Camenisch, Hyaluronan: genetic insights into the complex biology of a simple polysaccharide, Glycoconj. J. 19 (4–5) (2002) 331–339.

[15] C.M. McKee, M.B. Penno, M. Cowman, M.D. Burdick, R.M. Strieter, C. Bao, P.W. Noble, Hyaluronan (HA) fragments induce chemokine gene expression in alveolar macrophages. The role of HA size and CD44, J. Clin. Invest. 98 (10) (1996) 2403–2413.

[16] O. Ishida, Y. Tanaka, I. Morimoto, M. Takigawa, S. Eto, Chondrocytes are regulated by cellular adhesion through CD44 and hyaluronic acid pathway, J. Bone Miner. Res. 12 (10) (1997) 1657–1663.

[17] J. Hodge-Dufour, P.W. Noble, M.R. Horton, C. Bao, M. Wysoka, M.D. Burdick, et al., Induction of IL-12 and chemokines by hyaluronan requires adhesion-dependent priming of resident but not elicited macrophages, J. Immunol. 159 (5) (1997) 2492–2500.

[18] D. West, S. Kumar, Hyaluronan and angiogenesis, Biol. Hyaluronan 143 (1989) 187–207.

[19] V.C. Lees, T. Fan, D.C. West, Angiogenesis in a delayed revascularization model is accelerated by angiogenic oligosaccharides of hyaluronan, Lab. Invest. 73 (2) (1995) 259–266.

[20] G.D. Prestwich, D.M. Marecak, J.F. Marecek, K.P. Vercruysse, M.R. Ziebell, Controlled chemical modification of hyaluronic acid: synthesis, applications, and biodegradation of hydrazide derivatives, J. Control. Release 53 (1) (1998) 93–103.

[21] S. Eriksson, J.R.E. Fraser, T.C. Laurent, H. Pertoft, B. Smedsrød, Endothelial cells are a site of uptake and degradation of hyaluronic acid in the liver, Exp. Cell Res. 144 (1) (1983) 223–228.

[22] R. Tammi, K. Rilla, J.P. Pienimäki, D.K. MacCallum, M. Hogg, M. Luukkonen, et al., Hyaluronan enters keratinocytes by a novel endocytic route for catabolism, J. Biol. Chem. 276 (37) (2001) 35111–35122.

[23] D.D. Allison, K.J. Grande-Allen, Review. Hyaluronan: a powerful tissue engineering tool, Tissue Eng. 12 (8) (2006) 2131–2140.

[24] K. Kyyrönen, L. Hume, L. Benedetti, A. Urtti, E. Topp, V. Stella, Methylprednisolone esters of hyaluronic acid in ophthalmic drug delivery: in vitro and in vivo release studies, Int. J. Pharm. 80 (1) (1992) 161–169.

[25] M.F. Saettone, P. Chetoni, M. Tilde Torracca, S. Burgalassi, B. Giannaccini, Evaluation of muco-adhesive properties and in vivo activity of ophthalmic vehicles based on hyaluronic acid, Int. J. Pharm. 51 (3) (1989) 203–212.
[26] N. Yerushalmi, A. Arad, R. Margalit, Molecular and cellular studies of hyaluronic acid-modified liposomes as bioadhesive carriers for topical drug delivery in wound healing, Arch. Biochem. Biophys. 313 (2) (1994) 267–273.
[27] D. Coradini, C. Pellizzaro, G. Miglierini, M.G. Daidone, A. Perbellini, Hyaluronic acid as drug delivery for sodium butyrate: improvement of the anti-proliferative activity on a breast-cancer cell line, Int. J. Cancer 81 (3) (1999) 411–416.
[28] T. Segura, B.C. Anderson, P.H. Chung, R.E. Webber, K.R. Shull, L.D. Shea, Crosslinked hyaluronic acid hydrogels: a strategy to functionalize and pattern, Biomaterials 26 (4) (2005) 359–371.
[29] J.A. Burdick, C. Chung, X. Jia, M.A. Randolph, R. Langer, Controlled degradation and mechanical behavior of photopolymerized hyaluronic acid networks, Biomacromolecules 6 (1) (2005) 386–391.
[30] B.P. Toole, Hyaluronan in morphogenesis, Semin. Cell Dev. Biol. 12 (2) (2001) 79–87.
[31] L. Cen, K.G. Neoh, Y. Li, E.T. Kang, Assessment of in vitro bioactivity of hyaluronic acid and sulfated hyaluronic acid functionalized electroactive polymer, Biomacromolecules 5 (6) (2004) 2238–2246.
[32] J.H. Collier, J.P. Camp, T.W. Hudson, C.E. Schmidt, Synthesis and characterization of polypyrrole-hyaluronic acid composite biomaterials for tissue engineering applications, J. Biomed. Mater. Res. 50 (4) (2000) 574–584.
[33] D. Campoccia, P. Doherty, M. Radice, P. Brun, G. Abatangelo, D.F. Williams, Semisynthetic resorbable materials from hyaluronan esterification, Biomaterials 19 (23) (1998) 2101–2127.
[34] A. Pavesio, G. Abatangelo, A. Borrione, D. Brocchetta, A.P. Hollander, E. Kon, et al., Hyaluronan-based scaffolds (Hyalograft C) in the treatment of knee cartilage defects: preliminary clinical findings, Novartis Found. Symp. 249 (2003) 203–217, discussion 229–233, 234–238, 239–241.
[35] M. Radice, P. Brun, R. Cortivo, R. Scapinelli, C. Battaliard, G. Abatangelo, Hyaluronan-based biopolymers as delivery vehicles for bone-marrow-derived mesenchymal progenitors, J. Biomed. Mater. Res. 50 (2) (2000) 101–109.
[36] L.A. Solchaga, J.S. Temenoff, J. Gao, A.G. Mikos, A.I. Caplan, V.M. Goldberg, Repair of osteochondral defects with hyaluronan- and polyester-based scaffolds, Osteoarthr. Cartil. 13 (4) (2005) 297–309.
[37] K. Hemmrich, D. von Heimburg, R. Rendchen, C. Di Bartolo, E. Milella, N. Pallua, Implantation of preadipocyte-loaded hyaluronic acid-based scaffolds into nude mice to evaluate potential for soft tissue engineering, Biomaterials 26 (34) (2005) 7025–7037.
[38] B. Zavan, P. Brun, V. Vindigni, A. Amadori, W. Habeler, P. Pontisso, et al., Extracellular matrix-enriched polymeric scaffolds as a substrate for hepatocyte cultures: in vitro and in vivo studies, Biomaterials 26 (34) (2005) 7038–7045.
[39] S. Yamane, N. Iwasaki, T. Majima, T. Funakoshi, T. Masuko, K. Harada, et al., Feasibility of chitosan-based hyaluronic acid hybrid biomaterial for a novel scaffold in cartilage tissue engineering, Biomaterials 26 (6) (2005) 611–619.
[40] H.S. Yoo, E.A. Lee, J.J. Yoon, T.G. Park, Hyaluronic acid modified biodegradable scaffolds for cartilage tissue engineering, Biomaterials 26 (14) (2005) 1925–1933.
[41] L.P. Amarnath, A. Srinivas, A. Ramamurthi, In vitro hemocompatibility testing of UV-modified hyaluronan hydrogels, Biomaterials 27 (8) (2006) 1416–1424.
[42] A. Ramamurthi, I. Vesely, Evaluation of the matrix-synthesis potential of crosslinked hyaluronan gels for tissue engineering of aortic heart valves, Biomaterials 26 (9) (2005) 999–1010.
[43] A. Ramamurthi, I. Vesely, Ultraviolet light-induced modification of crosslinked hyaluronan gels, J. Biomed. Mater. Res. A 66 (2) (2003) 317–329.
[44] B. Joddar, A. Ramamurthi, Fragment size-and dose-specific effects of hyaluronan on matrix synthesis by vascular smooth muscle cells, Biomaterials 27 (15) (2006) 2994–3004.
[45] R. Ohri, S.K. Hahn, A.S. Hoffman, P.S. Stayton, C.M. Giachelli, Hyaluronic acid grafting mitigates calcification of glutaraldehyde-fixed bovine pericardium, J. Biomed. Mater. Res. A 70 (2) (2004) 328–334.
[46] A. Chajara, M. Raoudi, B. Delpech, H. Levesque, Inhibition of arterial cells proliferation in vivo in injured arteries by hyaluronan fragments, Atherosclerosis 171 (1) (2003) 15–19.
[47] E.A. Balazs, J.L. Denlinger, Viscosupplementation: a new concept in the treatment of osteoarthritis, J. Rheumatol. Suppl. 39 (1993) 3.
[48] K.W. Marshall, Intra-articular hyaluronan therapy, Curr. Opin. Rheumatol. 12 (5) (2000) 468–474.
[49] L. Liu, Y. Liu, J. Li, G. Du, J. Chen, Microbial production of hyaluronic acid: current state, challenges, and perspectives, Microb. Cell Fact. 10 (1) (2011) 99.
[50] R.M. Lauder, Chondroitin sulphate: a complex molecule with potential impacts on a wide range of biological systems, Complement. Ther. Med. 17 (1) (2009) 56–62.
[51] C.D. Nandini, N. Itoh, K. Sugahara, Novel 70-kDa chondroitin sulfate/dermatan sulfate hybrid chains with a unique heterogeneous sulfation pattern from shark skin, which exhibit neuritogenic activity and binding activities for growth factors and neurotrophic factors, J. Biol. Chem. 280 (6) (2005) 4058–4069.
[52] C.D. Nandini, T. Mikami, M. Ohta, N. Itoh, F. Akiyama-Nambu, K. Sugahara, Structural and functional characterization of oversulfated chondroitin sulfate/dermatan sulfate hybrid chains from the notochord of hagfish. neuritogenic and binding activities for growth factors and neurotrophic factors, J. Biol. Chem. 279 (49) (2004) 50799–50809.
[53] T.E. Hardingham, A.J. Fosang, Proteoglycans: many forms and many functions, FASEB J. 6 (3) (1992) 861–870.
[54] C.D. Nandini, K. Sugahara, Role of the sulfation pattern of chondroitin sulfate in its biological activities and in the binding of growth factors, Adv. Pharm. 53 (2006) 253–279.
[55] K. Sugahara, T. Mikami, Chondroitin/dermatan sulfate in the central nervous system, Curr. Opin. Struct. Biol. 17 (5) (2007) 536–545.
[56] R.M. Lauder, T.N. Huckerby, G.M. Brown, M.T. Bayliss, I.A. Nieduszynski, Age-related changes in the sulphation of the chondroitin sulphate linkage region from human articular cartilage aggrecan, Biochem. J. 358 (Pt. 2) (2001) 523–528.
[57] H. Nagase, M. Kashiwagi, Aggrecanases and cartilage matrix degradation, Arthritis Res. Ther. 5 (2) (2003) 94–103.
[58] L. Antonilli, E. Paroli, Role of the oligosaccharide inner core in the inhibition of human leukocyte elastase by chondroitin sulfates, Int. J. Clin. Pharmacol. Res. 13 (Suppl.) (1993) 11–17.

[59] A. Baici, P. Bradamante, Interaction between human leukocyte elastase and chondroitin sulfate, Chem. Biol. Interact. 51 (1) (1984) 1–11.
[60] N. Volpi, Inhibition of human leukocyte elastase activity by chondroitin sulfates, Chem. Biol. Interact. 105 (3) (1997) 157–167.
[61] E.J. Campbell, C.A. Owen, The sulfate groups of chondroitin sulfate- and heparan sulfate-containing proteoglycans in neutrophil plasma membranes are novel binding sites for human leukocyte elastase and cathepsin G, J. Biol. Chem. 282 (19) (2007) 14645–14654.
[62] F. Legendre, C. Bauge, R. Roche, A.S. Saurel, J.P. Pujol, Chondroitin sulfate modulation of matrix and inflammatory gene expression in IL-1beta-stimulated chondrocytes-study in hypoxic alginate bead cultures, Osteoarthr. Cartil. 16 (1) (2008) 105–114.
[63] S.Y. Cho, J.S. Sim, C.S. Jeong, S.Y. Chang, D.W. Choi, T. Toida, et al., Effects of low molecular weight chondroitin sulfate on type II collagen-induced arthritis in DBA/1J mice, Biol. Pharm. Bull. 27 (1) (2004) 47–51.
[64] C.X. Xu, H. Jin, Y.S. Chung, J.Y. Shin, M.A. Woo, K.H. Lee, et al., Chondroitin sulfate extracted from the Styela clava tunic suppresses TNF-alpha-induced expression of inflammatory factors, VCAM-1 and iNOS by blocking akt/NF-kappaB signal in JB6 cells, Cancer Lett. 264 (1) (2008) 93–100.
[65] J. Du, N. Eddington, Determination of the chondroitin sulfate disaccharides in dog and horse plasma by HPLC using chondroitinase digestion, precolumn derivatization, and fluorescence detection, Anal. Biochem. 306 (2) (2002) 252–258.
[66] N. Volpi, Oral absorption and bioavailability of ichthyic origin chondroitin sulfate in healthy male volunteers, Osteoarthr. Cartil. 11 (6) (2003) 433–441.
[67] S. Kusano, A. Ootani, S. Sakai, N. Igarashi, A. Takeguchi, H. Toyoda, et al., HPLC determination of chondrosine in mouse blood plasma after intravenous or oral dose, Biol. Pharm. Bull. 30 (8) (2007) 1365–1368.
[68] L. Barthe, J. Woodley, M. Lavit, C. Przybylski, C. Philibert, G. Houin, In vitro intestinal degradation and absorption of chondroitin sulfate, a glycosaminoglycan drug, Arzneimittelforschung 54 (5) (2011) 286–292.
[69] A. Rubinstein, D. Nakar, A. Sintov, Chondroitin sulfate: a potential biodegradable carrier for colon-specific drug delivery, Int. J. Pharm. 84 (2) (1992) 141–150.
[70] A. Sintov, N. Di-Capua, A. Rubinstein, Cross-linked chondroitin sulphate: characterization for drug delivery purposes, Biomaterials 16 (6) (1995) 473–478.
[71] O.A. Cavalcanti, C.C. da SILVA, E.A.G. Pineda, A.A.W. Hechenleitner, Synthesis and characterization of phosphated cross-linked chondroitin sulfate: potential ingredient for specific drug delivery, Acta Farm. Bonaerense 24 (2) (2005) 234.
[72] J. Pieper, A. Oosterhof, P. Dijkstra, J. Veerkamp, T. Van Kuppevelt, Preparation and characterization of porous crosslinked collagenous matrices containing bioavailable chondroitin sulphate, Biomaterials 20 (9) (1999) 847–858.
[73] J.L.C. van Susante, J. Pieper, P. Buma, T.H. van Kuppevelt, H. van Beuningen, P.M. van der Kraan, et al., Linkage of chondroitin-sulfate to type I collagen scaffolds stimulates the bioactivity of seeded chondrocytes in vitro, Biomaterials 22 (17) (2001) 2359–2369.
[74] C. Chang, H. Liu, C. Lin, C. Chou, F. Lin, Gelatin-chondroitin-hyaluronan tri-copolymer scaffold for cartilage tissue engineering, Biomaterials 24 (26) (2003) 4853–4858.
[75] C.S. Ko, J.P. Huang, C.W. Huang, I. Chu, Type II collagen-chondroitin sulfate-hyaluronan scaffold cross-linked by genipin for cartilage tissue engineering, J. Biosci. Bioeng. 107 (2) (2009) 177–182.
[76] C.T. Lee, P.H. Kung, Y.D. Lee, Preparation of poly (vinyl alcohol)-chondroitin sulfate hydrogel as matrices in tissue engineering, Carbohydr. Polym. 61 (3) (2005) 348–354.
[77] M. Wollenweber, H. Domaschke, T. Hanke, S. Boxberger, G. Schmack, K. Gliesche, et al., Mimicked bioartificial matrix containing chondroitin sulphate on a textile scaffold of poly (3-hydroxybutyrate) alters the differentiation of adult human mesenchymal stem cells, Tissue Eng. 12 (2) (2006) 345–359.
[78] S. Varghese, N.S. Hwang, A.C. Canver, P. Theprungsirikul, D.W. Lin, J. Elisseeff, Chondroitin sulfate based niches for chondrogenic differentiation of mesenchymal stem cells, Matrix Biol. 27 (1) (2008) 12–21.
[79] D.A. Wang, S. Varghese, B. Sharma, I. Strehin, S. Fermanian, J. Gorham, et al., Multifunctional chondroitin sulphate for cartilage tissue–biomaterial integration, Nat. Mater. 6 (5) (2007) 385–392.
[80] J.F. Piai, A.F. Rubira, E.C. Muniz, Self-assembly of a swollen chitosan/chondroitin sulfate hydrogel by outward diffusion of the chondroitin sulfate chains, Acta Biomater. 5 (7) (2009) 2601–2609.
[81] J. Kim, B. Kwak, H. Shim, Y. Lee, H. Baik, M. Lee, et al., Preparation of doxorubicin-containing chitosan microspheres for transcatheter arterial chemoembolization of hepatocellular carcinoma, J. Microencapsul. 24 (5) (2007) 408–419.
[82] M. Dornish, D. Kaplan, Ø. Skaugrud, Standards and guidelines for biopolymers in tissue-engineered medical products, Ann. N.Y. Acad. Sci. 944 (1) (2001) 388–397.
[83] P.J. VandeVord, H.W. Matthew, S.P. DeSilva, L. Mayton, B. Wu, P.H. Wooley, Evaluation of the biocompatibility of a chitosan scaffold in mice, J. Biomed. Mater. Res. 59 (3) (2002) 585–590.
[84] K.S. Chow, E. Khor, Novel fabrication of open-pore chitin matrixes, Biomacromolecules 1 (1) (2000) 61–67.
[85] K. Kamiyama, H. Onishi, Y. Machida, Biodisposition characteristics of N-succinyl-chitosan and glycol-chitosan in normal and tumor-bearing mice, Biol. Pharm. Bull. 22 (2) (1999) 179–186.
[86] K.Y. Lee, W.S. Ha, W.H. Park, Blood compatibility and biodegradability of partially N-acylated chitosan derivatives, Biomaterials 16 (16) (1995) 1211–1216.
[87] G. Paradossi, E. Chiessi, M. Venanzi, B. Pispisa, A. Palleschi, Branched-chain analogues of linear polysaccharides: a spectroscopic and conformational investigation of chitosan derivatives, Int. J. Biol. Macromol. 14 (2) (1992) 73–80.
[88] Y. Chen, Y. Chung, L. Woan Wang, K. Chen, S. Li, Antibacterial properties of chitosan in waterborne pathogen, J. Environ. Sci. Health A 37 (7) (2002) 1379–1390.
[89] S. Hu, C. Jou, M. Yang, Protein adsorption, fibroblast activity and antibacterial properties of poly (3-hydroxybutyric acid-co-3-hydroxyvaleric acid) grafted with chitosan and chitooligosaccharide after immobilized with hyaluronic acid, Biomaterials 24 (16) (2003) 2685–2693.
[90] M.N.V. Ravi Kumar, A review of chitin and chitosan applications, React. Funct. Polym. 46 (1) (2000) 1–27.
[91] C.N. Mhurchu, C. Dunshea-Mooij, D. Bennett, A. Rodgers, Effect of chitosan on weight loss in overweight and obese individuals: a systematic review of randomized controlled trials, Obes. Rev. 6 (1) (2005) 35–42.
[92] O. Felt, P. Buri, R. Gurny, Chitosan: a unique polysaccharide for drug delivery, Drug Dev. Ind. Pharm. 24 (11) (1998) 979–993.

[93] M.P. Patel, R.R. Patel, J.K. Patel, Chitosan mediated targeted drug delivery system: a review, J. Pharm. Pharm. Sci. 13 (4) (2010) 536–557.
[94] S.G. Kumbar, A.R. Kulkarni, T.M. Aminabhavi, Crosslinked chitosan microspheres for encapsulation of diclofenac sodium: effect of crosslinking agent, J. Microencapsul. 19 (2) (2002) 173–180.
[95] V. Mourya, N.N. Inamdar, Trimethyl chitosan and its applications in drug delivery, J. Mater. Sci. Mater. Med. 20 (5) (2009) 1057–1079.
[96] M. Chen, H. Wong, K. Lin, H. Chen, S. Wey, K. Sonaje, et al., The characteristics, biodistribution and bioavailability of a chitosan-based nanoparticulate system for the oral delivery of heparin, Biomaterials 30 (34) (2009) 6629–6637.
[97] Y. Lin, F. Mi, C. Chen, W. Chang, S. Peng, H. Liang, et al., Preparation and characterization of nanoparticles shelled with chitosan for oral insulin delivery, Biomacromolecules 8 (1) (2007) 146–152.
[98] N. Arya, S. Chakraborty, N. Dube, D.S. Katti, Electrospraying: a facile technique for synthesis of chitosan-based micro/nanospheres for drug delivery applications, J. Biomed. Mater. Res. B 88 (1) (2009) 17–31.
[99] Y. Lin, C. Chung, C. Chen, H. Liang, S. Chen, H. Sung, Preparation of nanoparticles composed of chitosan/poly-γ-glutamic acid and evaluation of their permeability through caco-2 cells, Biomacromolecules 6 (2) (2005) 1104–1112.
[100] S.W. Shalaby, J.A. DuBose, M. Shalaby, Chitosan Based Systems, Absorbable and Biodegradable Polymers, vol. 6, CRC Press, Boca Raton, 2004, pp. 77–89.
[101] S.A. Agnihotri, T.M. Aminabhavi, Chitosan nanoparticles for prolonged delivery of timolol maleate, Drug Dev. Ind. Pharm. 33 (11) (2007) 1254–1262.
[102] T. Sonia, C.P. Sharma, Chitosan and its derivatives for drug delivery perspective, anonymous Chitosan for biomaterials I, J. Appl. Polym. Sci., Springer-Verlag Berlin Heidelberg, 119 (2011) 2902–2910.
[103] S. Seo, I. Park, M. Yoo, M. Shirakawa, T. Akaike, C. Cho, Xyloglucan as a synthetic extracellular matrix for hepatocyte attachment, J. Biomater. Sci. Polym. Ed. 15 (11) (2004) 1375–1387.
[104] A. Gamian, M. Chomik, C.A. Laferrière, R. Roy, Inhibition of influenza A virus hemagglutinin and induction of interferon by synthetic sialylated glycoconjugates, Can. J. Microbiol. 37 (3) (1991) 233–237.
[105] D.W. Jenkins, S.M. Hudson, Review of vinyl graft copolymerization featuring recent advances toward controlled radical-based reactions and illustrated with chitin/chitosan trunk polymers, Chem. Rev. 101 (11) (2001) 3245–3274.
[106] Z. Ding, J. Chen, S. Gao, J. Chang, J. Zhang, E. Kang, Immobilization of chitosan onto poly-L-lactic acid film surface by plasma graft polymerization to control the morphology of fibroblast and liver cells, Biomaterials 25 (6) (2004) 1059–1067.
[107] A. Zhu, M. Zhang, J. Wu, J. Shen, Covalent immobilization of chitosan/heparin complex with a photosensitive hetero-bifunctional crosslinking reagent on PLA surface, Biomaterials 23 (23) (2002) 4657–4665.
[108] S. Mao, X. Shuai, F. Unger, M. Wittmar, X. Xie, T. Kissel, Synthesis, characterization and cytotoxicity of poly (ethylene glycol)-graft-trimethyl chitosan block copolymers, Biomaterials 26 (32) (2005) 6343–6356.
[109] E. Khor, L.Y. Lim, Implantable applications of chitin and chitosan, Biomaterials 24 (13) (2003) 2339–2349.
[110] T. Chung, Y. Lu, S. Wang, Y. Lin, S. Chu, Growth of human endothelial cells on photochemically grafted Gly–Arg–Gly–Asp (GRGD) chitosans, Biomaterials 23 (24) (2002) 4803–4809.
[111] N. Bhattarai, D. Edmondson, O. Veiseh, F.A. Matsen, M. Zhang, Electrospun chitosan-based nanofibers and their cellular compatibility, Biomaterials 26 (31) (2005) 6176–6184.
[112] L. Li, Y. Hsieh, Chitosan bicomponent nanofibers and nanoporous fibers, Carbohydr. Res. 341 (3) (2006) 374–381.
[113] E. Ruel-Gariépy, J. Leroux, In situ-forming hydrogels—review of temperature-sensitive systems, Eur. J. Pharm. Biopharm. 58 (2) (2004) 409–426.
[114] T. Matsumoto, M. Kawai, T. Masuda, Influence of concentration and mannuronate/guluronate [correction of gluronate] ratio on steady flow properties of alginate aqueous systems, Biorheology 29 (4) (1992) 411–417.
[115] B.E. Christensena, Alginates as biomaterials in tissue engineering, Carbohydr. Chem. 37 (2011) 227–258.
[116] A.D. Augst, H.J. Kong, D.J. Mooney, Alginate hydrogels as biomaterials, Macromol. Biosci. 6 (8) (2006) 623–633.
[117] S. Holtan, P. Bruheim, G. Skjåk-Bræk, Mode of action and subsite studies of the guluronan block-forming mannuronan C-5 epimerases AlgE1 and AlgE6, Biochem. J. 395 (Pt. 2) (2006) 319.
[118] Y.A. Mørch, S. Holtan, I. Donati, B.L. Strand, G. Skjåk-Bræk, Mechanical properties of C-5 epimerized alginates, Biomacromolecules 9 (9) (2008) 2360–2368.
[119] Ý. Mørch, I. Donati, B.L. Strand, G. Skjåk-Bræk, Molecular engineering as an approach to design new functional properties of alginate, Biomacromolecules 8 (9) (2007) 2809–2814.
[120] I. Donati, P. Sergio, Material Properties of Alginates, Alginates: Biology and Applications, vol. 13, Springer, Berlin Heidelberg, 2009, pp. 1–53.
[121] H.J. Kong, M.K. Smith, D.J. Mooney, Designing alginate hydrogels to maintain viability of immobilized cells, Biomaterials 24 (22) (2003) 4023–4029.
[122] H. Holme, H. Foros, H. Pettersen, M. Dornish, O. Smidsrød, Thermal depolymerization of chitosan chloride, Carbohydr. Polym. 46 (3) (2001) 287–294.
[123] H.K. Holme, K. Lindmo, A. Kristiansen, O. Smidsrød, Thermal depolymerization of alginate in the solid state, Carbohydr. Polym. 54 (4) (2003) 431–438.
[124] O. Jeon, K.H. Bouhadir, J.M. Mansour, E. Alsberg, Photocrosslinked alginate hydrogels with tunable biodegradation rates and mechanical properties, Biomaterials 30 (14) (2009) 2724–2734.
[125] M. Ashley, A. McCullagh, C. Sweet, Making a good impression: (A 'how to' paper on dental alginate), Dent. Update 32 (3) (2005) 169–170, 172, 174–175.
[126] I.R. Matthew, R.M. Browne, J.W. Frame, B.G. Millar, Subperiosteal behaviour of alginate and cellulose wound dressing materials, Biomaterials 16 (4) (1995) 275–278.
[127] M.A. LeRoux, F. Guilak, L.A. Setton, Compressive and shear properties of alginate gel: effects of sodium ions and alginate concentration, J. Biomed. Mater. Res. 47 (1) (1999) 46–53.
[128] H.J. Kong, D.J. Mooney, The effects of poly (ethyleneimine)(PEI) molecular weight on reinforcement of alginate hydrogels, Cell Transplant. 12 (7) (2003) 779–785.
[129] J.L. Drury, R.G. Dennis, D.J. Mooney, The tensile properties of alginate hydrogels, Biomaterials 25 (16) (2004) 3187–3199.
[130] D.F. Emerich, C. Halberstadt, C. Thanos, Role of nanobiotechnology in cell-based nanomedicine: a concise review, J. Biomed. Nanotechnol. 3 (3) (2007) 235–244.
[131] E. Alsberg, H. Kong, Y. Hirano, M. Smith, A. Albeiruti, D. Mooney, Regulating bone formation via controlled scaffold degradation, J. Dent. Res. 82 (11) (2003) 903–908.

[132] K.Y. Lee, K.H. Bouhadir, D.J. Mooney, Evaluation of chain stiffness of partially oxidized polyguluronate, Biomacromolecules 3 (6) (2002) 1129–1134.
[133] K.Y. Lee, K.H. Bouhadir, D.J. Mooney, Degradation behavior of covalently cross-linked poly (aldehyde guluronate) hydrogels, Macromolecules 33 (1) (2000) 97–101.
[134] A. Mosahebi, M. Wiberg, G. Terenghi, Addition of fibronectin to alginate matrix improves peripheral nerve regeneration in tissue-engineered conduits, Tissue Eng. 9 (2) (2003) 209–218.
[135] P. Prang, R. Müller, A. Eljaouhari, K. Heckmann, W. Kunz, T. Weber, et al., The promotion of oriented axonal regrowth in the injured spinal cord by alginate-based anisotropic capillary hydrogels, Biomaterials 27 (19) (2006) 3560–3569.
[136] R. Pasqualini, E. Koivunen, E. Ruoslahti, αv integrins as receptors for tumor targeting by circulating ligands, Nat. Biotechnol. 15 (6) (1997) 542–546.
[137] K.H. Bouhadir, G.M. Kruger, K.Y. Lee, D.J. Mooney, Sustained and controlled release of daunomycin from cross-linked poly (aldehyde guluronate) hydrogels, J. Pharm. Sci. 89 (7) (2000) 910–919.
[138] K.H. Bouhadir, E. Alsberg, D.J. Mooney, Hydrogels for combination delivery of antineoplastic agents, Biomaterials 22 (19) (2001) 2625–2633.
[139] R.J. Laham, F.W. Sellke, E.R. Edelman, J.D. Pearlman, J.A. Ware, D.L. Brown, et al., Local perivascular delivery of basic fibroblast growth factor in patients undergoing coronary bypass surgery results of a phase I randomized, double-blind, placebo-controlled trial, Circulation 100 (18) (1999) 1865–1871.
[140] K. Lee, J. Yoon, J. Lee, S. Kim, H. Jung, S. Kim, et al., Sustained release of vascular endothelial growth factor from calcium-induced alginate hydrogels reinforced by heparin and chitosan, Transplant. Proc. 36 (8) (2004) 2464–2465.
[141] Q. Sun, R.R. Chen, Y. Shen, D.J. Mooney, S. Rajagopalan, P.M. Grossman, Sustained vascular endothelial growth factor delivery enhances angiogenesis and perfusion in ischemic hind limb, Pharm. Res. 22 (7) (2005) 1110–1116.
[142] C. Gomez, M. Rinaudo, M. Villar, Oxidation of sodium alginate and characterization of the oxidized derivatives, Carbohydr. Polym. 67 (3) (2007) 296–304.
[143] S. He, M. Zhang, Z. Geng, Y. Yin, K. Yao, Preparation and characterization of partially oxidized sodium alginate, Chin. J. Appl. Chem. 22 (9) (2005) 1007.
[144] F. Mi, H. Sung, S. Shyu, Drug release from chitosan–alginate complex beads reinforced by a naturally occurring cross-linking agent, Carbohydr. Polym. 48 (1) (2002) 61–72.
[145] S. Alban, A. Schauerte, G. Franz, Anticoagulant sulfated polysaccharides: part I. synthesis and structure-activity relationships of new pullulan sulfates, Carbohydr. Polym. 47 (3) (2002) 267–276.
[146] G. Sen, R.P. Singh, S. Pal, Microwave-initiated synthesis of polyacrylamide grafted sodium alginate: synthesis and characterization, J. Appl. Polym. Sci. 115 (1) (2010) 63–71.
[147] A. Sand, M. Yadav, K. Behari, Synthesis and characterization of alginate-g-vinyl sulfonic acid with a potassium peroxydiphosphate/thiourea system, J. Appl. Polym. Sci. 118 (6) (2010) 3685–3694.
[148] W. Pluemsab, N. Sakairi, T. Furuike, Synthesis and inclusion property of α-cyclodextrin-linked alginate, Polymer 46 (23) (2005) 9778–9783.
[149] M. Carré, C. Delestre, P. Hubert, E. Dellacherie, Covalent coupling of a short polyether on sodium alginate: synthesis and characterization of the resulting amphiphilic derivative, Carbohydr. Polym. 16 (4) (1991) 367–379.
[150] I. Ugi, The α-addition of immonium ions and anions to isonitriles accompanied by secondary reactions, Angew. Chem. Int. Ed. Engl. 1 (1) (1962) 8–21.
[151] C. Galant, A. Kjøniksen, G.T. Nguyen, K.D. Knudsen, B. Nyström, Altering associations in aqueous solutions of a hydrophobically modified alginate in the presence of β-cyclodextrin monomers, J. Phys. Chem. B 110 (1) (2006) 190–195.
[152] B.A. Justice, N.A. Badr, R.A. Felder, 3D cell culture opens new dimensions in cell-based assays, Drug Discov. Today 14 (1) (2009) 102–107.
[153] K.H. Lee, S.J. Shin, Y. Park, S. Lee, Synthesis of cell-laden alginate hollow fibers using microfluidic chips and microvascularized tissue-engineering applications, Small 5 (11) (2009) 1264–1268.
[154] S. Shin, J. Park, J. Lee, H. Park, Y. Park, K. Lee, et al., "On the fly" continuous generation of alginate fibers using a microfluidic device, Langmuir 23 (17) (2007) 9104–9108.
[155] C.A. Bonino, M.D. Krebs, C.D. Saquing, S.I. Jeong, K.L. Shearer, E. Alsberg, et al., Electrospinning alginate-based nanofibers: from blends to crosslinked low molecular weight alginate-only systems, Carbohydr. Polym. 85 (1) (2011) 111–119.
[156] M. Bongio, Jeroen J.J.P. van den Beucken, S.C. Leeuwenburgh, J.A. Jansen, Development of bone substitute materials: from 'biocompatible' to 'instructive', J. Mater. Chem. 20 (40) (2010) 8747–8759.
[157] M. Dvir-Ginzberg, T. Elkayam, S. Cohen, Induced differentiation and maturation of newborn liver cells into functional hepatic tissue in macroporous alginate scaffolds, FASEB J. 22 (5) (2008) 1440–1449.
[158] D. Roy, M. Semsarilar, J.T. Guthrie, S. Perrier, Cellulose modification by polymer grafting: a review, Chem. Soc. Rev. 38 (7) (2009) 2046–2064.
[159] H. Fink, D. Hofmann, B. Philipp, Some aspects of lateral chain order in cellulosics from X-ray scattering, Cellulose 2 (1) (1995) 51–70.
[160] A. Hebeish, J.T. Guthrie, Industrial Application of Cellulose Graft Copolymers, The Chemistry and Technology of Cellulosic Copolymers, Springer, Berlin Heidelberg, 1981, pp. 326–342.
[161] O. Pajulo, J. Viljanto, T. Hurme, P. Saukko, B. Lönnberg, K. Lönnqvist, Viscose cellulose sponge as an implantable matrix: changes in the structure increase the production of granulation tissue, J. Biomed. Mater. Res. 32 (3) (1996) 439–446.
[162] M. Martson, J. Viljanto, T. Hurme, P. Saukko, Biocompatibility of cellulose sponge with bone, Eur. Surg. Res. 30 (6) (1998) 426–432.
[163] M. Martson, J. Viljanto, T. Hurme, P. Laippala, P. Saukko, Is cellulose sponge degradable or stable as implantation material? An in vivo subcutaneous study in the rat, Biomaterials 20 (21) (1999) 1989–1995.
[164] M.B. Turner, S.K. Spear, J.D. Holbrey, R.D. Rogers, Production of bioactive cellulose films reconstituted from ionic liquids, Biomacromolecules 5 (4) (2004) 1379–1384.
[165] T. Miyamoto, S. Takahashi, H. Ito, H. Inagaki, Y. Noishiki, Tissue biocompatibility of cellulose and its derivatives, J. Biomed. Mater. Res. 23 (1) (1989) 125–133.
[166] W.K. Czaja, D.J. Young, M. Kawecki, R.M. Brown, The future prospects of microbial cellulose in biomedical applications, Biomacromolecules 8 (1) (2007) 1–12.
[167] G. Helenius, H. Bäckdahl, A. Bodin, U. Nannmark, P. Gatenholm, B. Risberg, In vivo biocompatibility of bacterial cellulose, J. Biomed. Mater. Res. A 76 (2) (2006) 431–438.
[168] G.T. Ciacco, T.F. Liebert, E. Frollini, T.J. Heinze, Application of the solvent dimethyl sulfoxide/tetrabutyl-ammonium fluoride trihydrate as reaction medium for the homogeneous acylation of sisal cellulose, Cellulose 10 (2) (2003) 125–132.

[169] C. Altaner, J. Puls, B. Saake, Enzyme aided analysis of the substituent distribution along the chain of cellulose acetates regioselectively modified by the action of an Aspergillus niger acetylesterase, Cellulose 10 (4) (2003) 391–395.

[170] T. Heinze, T. Liebert, Unconventional methods in cellulose functionalization, Prog. Polym. Sci. 26 (9) (2001) 1689–1762.

[171] T. Heinze, T.F. Liebert, K.S. Pfeiffer, M.A. Hussain, Unconventional cellulose esters: synthesis, characterization and structure–property relations, Cellulose 10 (3) (2003) 283–296.

[172] S. Lee, C. Altaner, J. Puls, B. Saake, Determination of the substituent distribution along cellulose acetate chains as revealed by enzymatic and chemical methods, Carbohydr. Polym. 54 (3) (2003) 353–362.

[173] A. Althin, B. Fernandez, R. Elsen, K. Ruzius, L. Silva, G. Washington, High-flux Hollow-Fiber Membrane with Enhanced Transport Capability and Process for Making Same, EP0598690, 1998.

[174] K.J. Edgar, C.M. Buchanan, J.S. Debenham, P.A. Rundquist, B.D. Seiler, M.C. Shelton, et al., Advances in cellulose ester performance and application, Prog. Polym. Sci. 26 (9) (2001) 1605–1688.

[175] S. Sternberg, D.R. Lynn, Methods for Correlating Average Fiber Diameter with Performance in Complex Filtration Media, US 6032807 A, 2000.

[176] E. Entcheva, H. Bien, L. Yin, C.Y. Chung, M. Farrell, Y. Kostov, Functional cardiac cell constructs on cellulose-based scaffolding, Biomaterials 25 (26) (2004) 5753–5762.

[177] S.G. Kumbar, C.T. Laurencin, Natural Polymer-Based Porous Orthopedic Fixation Screw for Bone Repair and Regeneration, 20110208190 (2011).

[178] A. Aravamudhan, D.M. Ramos, J. Nip, M.D. Harmon, R. James, M. Deng, et al., Cellulose and collagen derived micro-nano structured scaffolds for bone tissue engineering, J. Biomed. Nanotechnol. 9 (4) (2013) 719–731.

[179] M.I. Ugwoke, R.U. Agu, H. Vanbilloen, J. Baetens, P. Augustijns, N. Verbeke, et al., Scintigraphic evaluation in rabbits of nasal drug delivery systems based on carbopol and carboxymethylcellulose, J. Control. Release 68 (2) (2000) 207–214.

[180] R. Chen, H. Ho, C. Yu, M. Sheu, Development of swelling/floating gastroretentive drug delivery system based on a combination of hydroxyethyl cellulose and sodium carboxymethyl cellulose for losartan and its clinical relevance in healthy volunteers with CYP2C9 polymorphism, Eur. J. Pharm. Sci. 39 (1) (2010) 82–89.

[181] A.H. Shojaei, Buccal mucosa as a route for systemic drug delivery: a review, J. Pharm. Pharm. Sci. 1 (1) (1998) 15–30.

[182] K. Pal, A. Banthia, D. Majumdar, Development of carboxymethyl cellulose acrylate for various biomedical applications, Biomed. Mater. 1 (2) (2006) 85.

[183] A.T. Reza, S.B. Nicoll, Characterization of novel photocrosslinked carboxymethylcellulose hydrogels for encapsulation of nucleus pulposus cells, Acta Biomater. 6 (1) (2010) 179–186.

[184] H. Chen, M. Fan, Novel thermally sensitive pH-dependent chitosan/carboxymethyl cellulose hydrogels, J. Bioact. Compat. Polym. 23 (1) (2008) 38–48.

[185] J. Liuyun, L. Yubao, X. Chengdong, Preparation and biological properties of a novel composite scaffold of nano-hydroxyapatite/chitosan/carboxymethyl cellulose for bone tissue engineering, J. Biomed. Sci. 16 (2009) 65.

[186] P.G. Wuts, T.W. Greene, Greene's Protective Groups in Organic Synthesis, Hoboken, Wiley-Interscience, N.J, 2006, Internet resource.

[187] A.L. Schwan, M.L. Kalin, K.E. Vajda, T. Xiang, D. Brillon, Oxidative fragmentations of selected 1-alkenyl sulfoxides. Chemical and spectroscopic evidence for 1-alkenesulfinyl chlorides, Tetrahedron Lett. 37 (14) (1996) 2345–2348.

[188] K. Rahn, M. Diamantoglou, D. Klemm, H. Berghmans, T. Heinze, Homogeneous synthesis of cellulose p-toluenesulfonates in N,N-dimethylacetamide/LiCl solvent system, Die Angew. Makromolek. Chem. 238 (1) (1996) 143–163.

[189] D. Klemm, B. Heublein, H.P. Fink, A. Bohn, Cellulose: fascinating biopolymer and sustainable raw material, Angew. Chem. Int. Ed. Engl. 44 (22) (2005) 3358–3393.

[190] H. Krässig, V. Stannett, Graft Co-Polymerization to Cellulose and Its Derivatives, Anonymous Fortschritte Der Hochpolymeren-Forschung, vol. 4(2), Springer, 1965, pp. 111–156.

[191] G.S. Chauhan, S. Mahajan, L.K. Guleria, Polymers from renewable resources: sorption of Cu^{2+} ions by cellulose graft copolymers, Desalination 130 (1) (2000) 85–88.

[192] K. El-Salmawi, M. Zaid, S. Ibraheim, A. El-Naggar, A. Zahran, Sorption of dye wastes by poly (vinyl alcohol)/poly (carboxymethyl cellulose) blend grafted through a radiation method, J. Appl. Polym. Sci. 82 (1) (2001) 136–142.

[193] K. Gupta, K. Khandekar, Temperature-responsive cellulose by ceric (IV) ion-initiated graft copolymerization of N-isopropylacrylamide, Biomacromolecules 4 (3) (2003) 758–765.

[194] S. Vitta, E. Stahel, V. Stannett, The preparation and properties of acrylic and methacrylic acid grafted cellulose prepared by ceric ion initiation. Part I. Preparation of the grafted cellulose, J. Macromol. Sci. 22 (5–7) (1985) 579–590.

[195] A. Waly, F. Abdel-Mohdy, A. Aly, A. Hebeish, Synthesis and characterization of cellulose ion exchanger. II. Pilot scale and utilization in dye–heavy metal removal, J. Appl. Polym. Sci. 68 (13) (1998) 2151–2157.

[196] G. Odian, Radical Chain Polymerization, Principles of Polymerization, fourth ed., John Wiley & Sons, Inc., Hoboken, NJ, USA, 2004, pp. 198–349.

Chapter 5

Chitosan as a Biomaterial: Structure, Properties, and Applications in Tissue Engineering and Drug Delivery

Tao Jiang*,†,‡, Roshan James§, Sangamesh G. Kumbar†,‡,§,¶, Cato T. Laurencin†,‡,§,¶

*Department of Medicine, University of Connecticut Health Center, Farmington, Connecticut, USA

†Institute for Regenerative Engineering, University of Connecticut Health Center, Farmington, Connecticut, USA

‡Raymond and Beverly Sackler Center for Biological, Physical and Engineering Sciences, University of Connecticut Health Center, Farmington, Connecticut, USA

§Department of Orthopaedic Surgery, University of Connecticut Health Center, Farmington, Connecticut, USA

¶Department of Chemical, Materials and Biomolecular Engineering, University of Connecticut, Storrs, Connecticut, USA

Chapter Outline

5.1 INTRODUCTION

Chitin and chitosan belong to the polysaccharide family and are commonly found in the exoskeleton of crustaceans. In crabs and shrimps, chitin functions as structural components providing strength and protection to the organisms. Ideally, chitin is a linear polysaccharide consisting of β-(1-4)-2-acetamido-2-deoxy-D-glucopyranose repeating units where the amine groups are entirely acetylated. The ideal structure of chitosan, the primary derivative of chitin, is comprised of linear β-(1-4)-2-amino-2-deoxy-D-glucopyranose repeating units where the *N*-acetylglucosamine residues in chitin macromolecular chain are fully deacetylated to become *N*-glucosamine residues. In general, chitosan occurs as a copolymer of *N*-acetylglucosamine and *N*-glucosamine units randomly or block distributed throughout the biopolymer chain. The percentage of *N*-glucosamine units is also defined as the degree of deacetylation (DDA) of chitosan. Although chitin was discovered in the early 1800s, it has not received much research interest as compared to its structurally similar biopolymer cellulose; partially because chitin has a rigid structure and is impregnable to common solvents. The discovery of chitosan was made by the French physiologist Charles Rouget in 1859 [1]. As compared to its parent polymer chitin, chitosan can be processed into different forms at much milder conditions due to its solubility in dilute acid solutions, making chitosan a more attractive biopolymer for a variety of applications. For example, chitosan is particularly useful in color removal from textile mill effluents due to the protonation of its amine groups and their affinity to bind negatively charged dye molecules [2]. Other important applications of chitosan include heavy metal removal from industrial wastewater due to the chelating capability of chitosan [3] and cosmetic application attributing to its solubility and antimicrobial and antifungal properties [4].

Chitosan has been used extensively in biomedical applications. In ophthalmology, chitosan has been explored as a material for making contact lenses due to its film-forming ability, optical clarity, and mechanical stability [5]. In wound healing, a great amount of research has investigated the use of different chitosan formulations for skin wound closure in animal models [6,7]. Chitosan is the active component of a number of wound-dressing products on the United States market such as Tegasorb™, Chito-Seal®, HemCon™, and Syvek Patch® [8] and used in the battlefield attributing to the hemostatic and antimicrobial properties of chitosan. In the late 1980s, tissue engineering emerged as a new interdisciplinary field that applies the principles of engineering and the life sciences toward the development of biological substitutes that restore, maintain, or improve tissue function [9]. Since then, there has been an explosion of research on evaluating a broad range of polymeric, ceramic, and metallic biomaterials for regenerating tissues. As a naturally occurring biopolymer, chitosan has attracted much research interest in tissue engineering applications because of its biocompatibility, biodegradability, antimicrobial properties, and functionality [10]. Furthermore, due to the cationic nature of chitosan molecules and the abundant functional amine and hydroxyl groups on the molecular chain, chitosan has been an excellent candidate as a biomaterial for carrying proteins and other active molecules through physical or chemical means [11]. These chitosan-based systems have been extensively investigated for delivering biopharmaceuticals.

In this chapter, we first discuss the chemical and physical properties of chitosan, including the synthesis, modification, molecular structure, characterization, and structure-property relationship. Second, we review the topics of biocompatibility, biodegradability, and antimicrobial activity of chitosan. These properties make chitosan a potential biomaterial for many biomedical applications. Last, we review the research on using chitosan-based biomaterials and systems for tissue engineering and drug delivery applications.

5.2 CHITOSAN CHEMISTRY

5.2.1 Synthesis

Chitosan is synthesized by deacetylation of chitin. The deacetylation process is relatively harsh and involves the removal of acetyl groups from the chitin molecular chain by treating with concentrated NaOH in a prolonged period of time, leaving behind complete amine groups ($-NH_2$). In order to avoid undesirable side reactions such as depolymerization and generation of reactive species, deacetylation of chitin is usually carried out under the protection of nitrogen or by addition of sodium borohydride to the NaOH solution. The versatility of chitosan depends largely on the high degree of chemically reactive amine group [12]. Several techniques are reported for the alkali deacetylation of chitin. The DDA increases with reaction temperature (70-150 °C), NaOH concentration (10-60%), and solution to chitin ratio (10-20%) [13]. By fixing the solid/liquid ratio of chitin/NaOH (1:25 weight to volume ratio), Laurencin *et al.* [14] demonstrated that the DDA of chitosan could be controlled by changing the reaction pressure and NaOH concentration. At the NaOH concentration of 60%, a high reaction pressure of 50 psi increased the DDA of chitosan compared to a lower pressure of 25 psi. With the reaction pressure fixed at 50 psi, a higher NaOH concentration (70%) reduced the DDA, which could be due to the detrimental effect of high alkali concentration on the backbone of chitosan molecular chains. In a separate study, autoclaving conditions (15 psi/121 °C) were applied with 45% NaOH and a solid/liquid ratio of 1:15 or 1:10 [15]. It was shown that a DDA of 90.4% was achieved and the DDA was independent of the solid/liquid ratio. In addition, promoters were reported to reduce the steps needed for *N*-acetyl group removal [16,17]. One challenge in chitosan synthesis was to prepare fully deacetylated chitosan with minimal depolymerization. This challenge has been overcome by repeatedly treating chitosan with concentrated NaOH solution at high temperatures, leading to the production of chitosan with 100% DDA [18,19].

Apart from DDA, the distribution of the *N*-acetylglucosamine and *N*-glucosamine units in chitosan molecular chain is of importance. Kurita *et al.* [20] investigated the processes of alkaline hydrolyses of chitin under both heterogeneous and homogeneous conditions. The deacetylation under heterogeneous conditions proceeded preferentially from the amorphous regions to give block-type copolymers of *N*-acetylglucosamine and *N*-glucosamine units. On the other hand, the deacetylation under homogeneous conditions proceeded in both crystalline and amorphous regions indiscriminately, leading to the formation of random-type copolymers. In a not only seemingly opposite but also successful approach, Aiba [21] synthesized random copolymer of the two units by *N*-acetylating highly deacetylated chitosan under homogeneous condition and block copolymer by *N*-acetylating moderately deacetylated chitosan under heterogeneous condition.

5.2.2 Modification

One of the most intriguing features of chitosan is the presence of abundant reactive amine and hydroxyl groups on its backbone. The functionality of the biopolymer, due to the two hydroxyl groups (C3 and C6 positions) and an amine group (C2 position) of each repeating unit, makes chitosan versatile for chemical modification. The preparation of various types of chitosan derivatives via chemical modifications has been thoroughly reviewed in the literature [22,23].

In this chapter, the chemical modification strategies that lead to the formation of bioactive chitosan derivatives for tissue engineering applications are of particular interest. In this regard, two strategies have been most commonly used. The first approach involves chemical modification of chitosan to alter its physicochemical properties through the introduction of certain structural moieties to the polymer chain. Biomolecules could be physically adsorbed or entrapped on chemically modified chitosan surface or within chitosan substrates. For example, chitosan can be converted into water-soluble form by introducing acidic functionalities via modification with lactobionic acid [24]. The modified chitosan in combination with heparin hydrogels was used to immobilize fibroblast growth factor-2 (FGF-2) to attain neovascularization and fibrous tissue formation in a mouse model [24]. Chitosan was further modified with multiple functionalities such as lactose and azide via sequential condensation reaction at the chitosan amine sites [25]. The resultant polymer can be photo-cross-linked, avoiding the use of toxic cross-linking agents during scaffold fabrication and biomolecules incorporation. These chitosan derivatives show great potential in incorporating growth factors under mild conditions for a variety of biomedical applications such as wound healing and controlled angiogenesis. In the second approach, chemical modification is conducted by covalently binding specific biomolecules such as oligopeptides or proteins to chitosan. For example, oligopeptide sequences of interest can be directly bound to chitosan through reactions between carboxyl groups on the peptide and amine groups on chitosan forming imide bonds [26]. This reaction often involves the use of carbodiimide such as 1-ethyl-3-(3-dimethylaminopropyl) carbodiimide (EDC) and *N*-hydroxysuccinimide (NHS) to activate the carboxyl groups on peptides and decrease side reactions. Another approach is to prepare a reactive chitosan intermediate such as thiolated chitosan, carboxylated chitosan, or maleimide-activated chitosan through the amine groups [27–31]. These active chitosan intermediates can further react with sulfhydryl or amine groups of some peptide sequences or whole protein molecules enabling covalent immobilization. Other chemical modification methods of chitosan through its hydroxyl groups for biological purposes have also been reported. For example, carboxylation can also be achieved through the hydroxyl groups on chitosan followed by imide-bonding formation with peptides [32]. In addition, photochemical reactions can occur between the hydroxyl groups on chitosan and azido derivative of peptides to form covalent bonding [33,34]. All the aforementioned modification methods take advantage of the amine and hydroxyl functionalities of chitosan, to develop biologically active chitosan with cell recognition and specificities for various tissue engineering and regenerative medicine applications. Table 5.1 summarizes some major chemical modifications of chitosan reported in the literature toward the synthesis of active chitosan derivatives for coupling with peptides or proteins.

5.3 CHITOSAN PHYSICS

5.3.1 Physical Properties and Characterization

The two most important structural parameters that affect the properties of chitosan are the DDA and molecular weight. These two factors together determine the physical, chemical, and biological properties of the biopolymer. As for any other polymer, these properties are an average of the contributions of all polymer chains in the sample.

5.3.1.1 Degree of Deacetylation

The DDA of chitosan is an important parameter that determines many physicochemical and biological properties of the biopolymer such as solubility, biodegradability, and biocompatibility. As mentioned earlier in this chapter, the DDA of chitosan can be defined as

$$\mathrm{DDA} = \frac{n(\mathrm{GlcN})}{n(\mathrm{GlcN}) + n(\mathrm{GlcNAc})} \times 100\% \qquad (5.1)$$

where n(GlcN) is the average number of *N*-glucosamine units and n(GlcNAc) is the average number of *N*-acetylglucosamine units.

Many methods have been developed to measure the DDA including conductometric titration [36], potentiometric titration [37], elemental analysis [36], and acidic or enzymatic hydrolysis followed by colorimetric assay or high-performance liquid chromatography (HPLC) [38,39], pyrolysis-gas chromatography [40], and spectroscopic methods such as nuclear magnetic resonance spectroscopy

TABLE 5.1 Selected Chemical Modification of Chitosan Toward Bioactive Intermediates of Chitosan Derivatives for Coupling with Peptides or Proteins

Chitosan and Reactive Functional Group[a]	Coupling Agent	Modified Chitosan	Major Reacting Groups on Peptides or Proteins
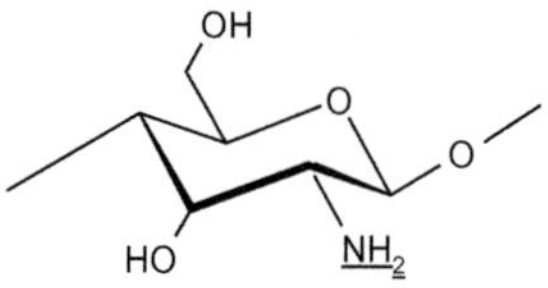	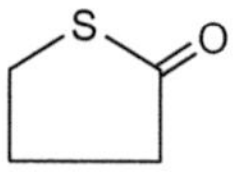	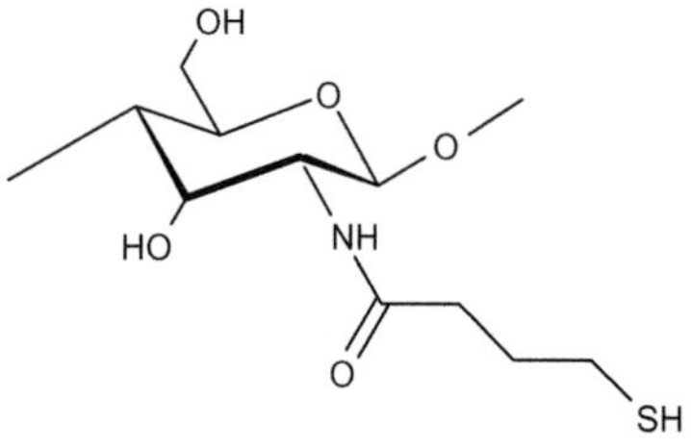	–SH
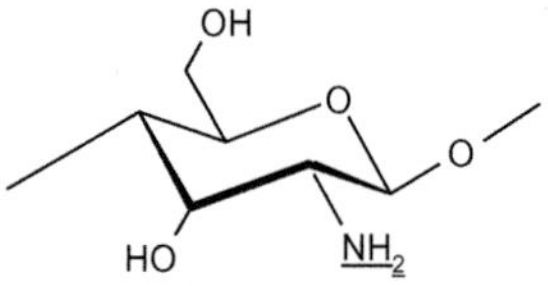	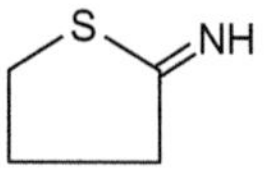	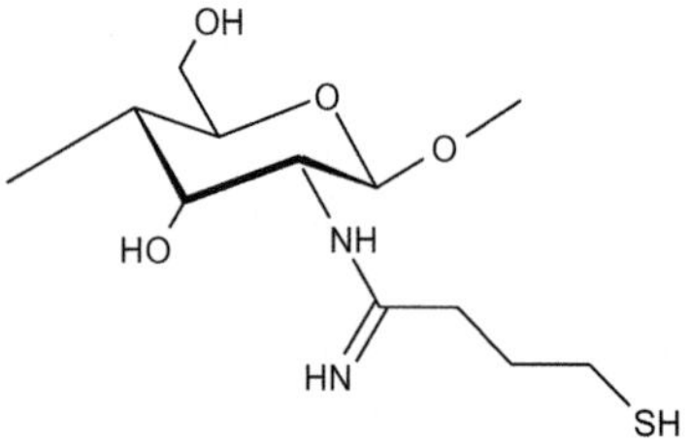	–SH
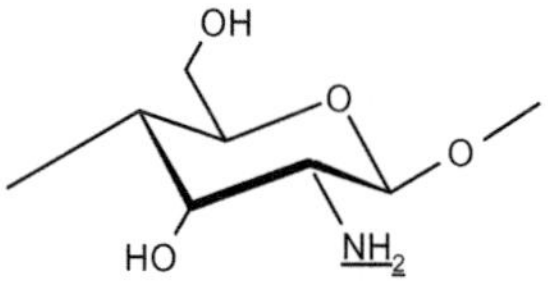	HS-CH_2COOH	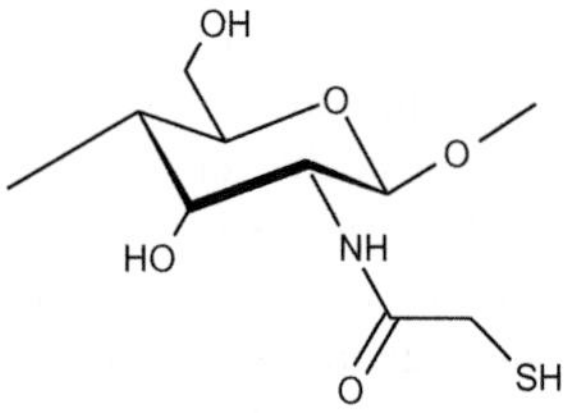	–SH
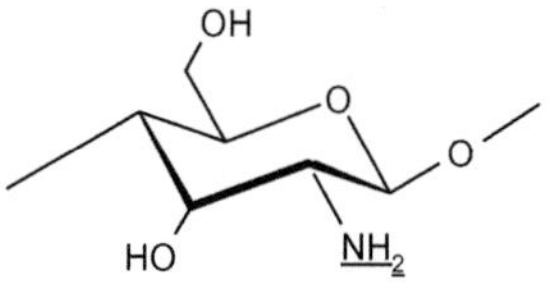	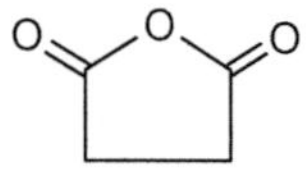	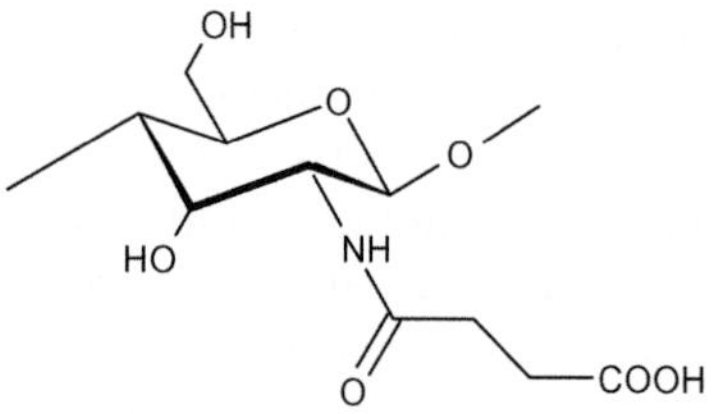	–NH_2
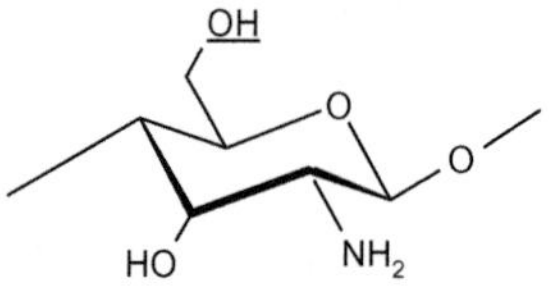	Cl-CH_2COOH	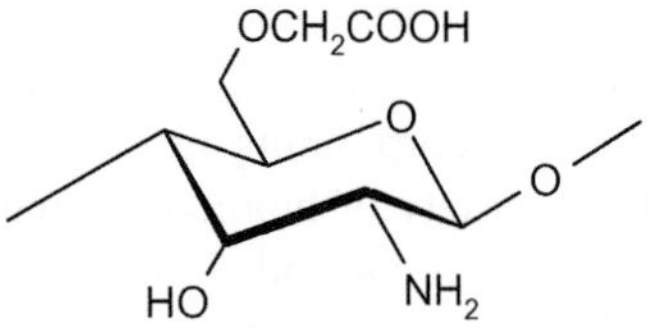	–NH_2
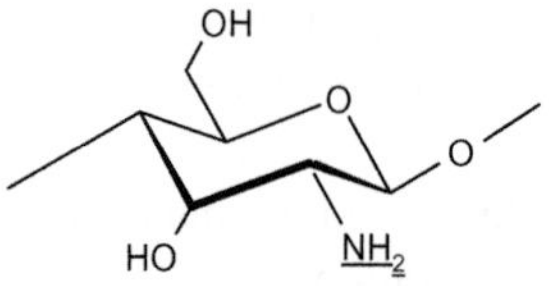	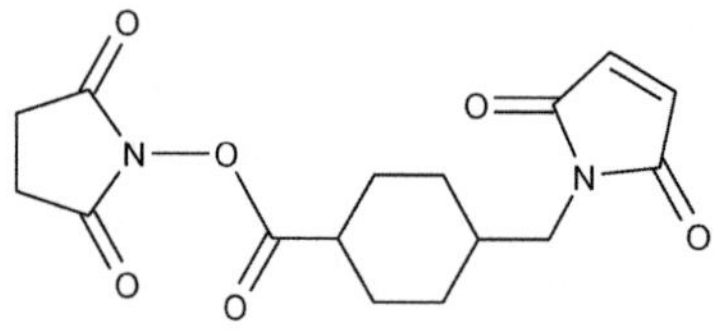	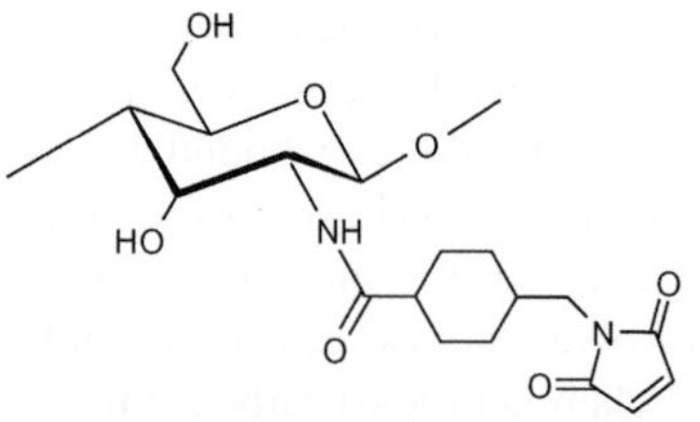	–SH

[a]*Underlined group represents the reacting group.*

(^{1}H NMR, ^{13}C NMR, and ^{15}N NMR) [41–43], infrared spectroscopy (IR) [44,45], ultraviolet spectroscopy (UV) [46,47], and X-ray powder diffraction [48].

NMR spectroscopy is arguably the most precise method to measure the DDA of chitosan. In ^{1}H liquid-state NMR spectroscopy, a dilute solution of the chitosan sample is usually prepared in a deuterated aqueous acid such as CD_3COOD/D_2O or DCl/D_2O and the spectra between 0 and 10 ppm are recorded. Using ^{1}H liquid-state NMR spectroscopy, chitosan with DDA ranging in 40-100% can be accurately determined [49]. In contrast, ^{13}C and ^{15}N NMR can be performed on solid-state chitosan samples and do not require extra sample preparation. In addition, ^{13}C and ^{15}N solid-state NMR can be applied to chitin and chitosan in the whole range of DDA from 0% to 100% [49]. The DDA of chitosan by different types of NMR can be calculated based on the ratio of I_P/I_R, where I_P is integral of the probe signal and I_R is the integral of reference signal [41]. Though NMR spectroscopy is accurate in measuring the DDA of chitosan, it is associated with high operational costs and may not be best for routine measurements of DDA.

IR spectroscopy is a relatively fast and cheap technique to determine the DDA of chitosan and probably the most extensively studied method. In IR spectroscopy, solid-state chitosan is mixed with KBr to form a thin disc and the IR spectrum in the range of 500-4000 cm^{-1} is recorded. Different absorption ratios such as A_{1560}/A_{2875}, A_{1655}/A_{2875}, A_{1655}/A_{3450}, A_{1320}/A_{3450}, A_{1655}/A_{1070}, A_{1655}/A_{1030}, A_{1560}/A_{1070}, A_{1560}/A_{1130}, A_{1560}/A_{1160}, A_{1560}/A_{897}, and A_{1320}/A_{1420} have been used to determine the DDA [49,50]. Each ratio has its own limitations, particular with respect to the DDA range for which they are suitable. To date, the preferred ratio in the literature is A_{1655}/A_{3450}, where the adsorption at 1655 cm^{-1} corresponds to amide I band and the adsorption at 3450 cm^{-1} corresponds to the hydroxyl group (Figure 5.1). The A_{1655}/A_{3450} ratio method together with the originally proposed baseline A for the amide I band gives DDA in good agreement with other methods such as the titrimetric method particularly for chitosans with DDA less than 80%. For chitosans with higher DDA, using baseline A gives inaccurate measurements. For example, with fully deacetylated chitosan in which the amide I band is absent, the use of this baseline still leads to a percent acetylation value of 3-5% [52]. Miya *et al.* [51] proposed an alternative baseline B for samples with DDA greater than 90%. Using the adjusted baseline B, the new equation for calculating DDA is

$$\mathrm{DDA} = \left[100 - \left(A_{1655} / A_{3450}\right) \times 115\right]\% \qquad (5.2)$$

This revision eliminates the inaccuracy of this ratio at very high level of deacetylation and is in good agreement with the dye method over a wide range of DDA [53].

5.3.1.2 Molecular Weight

The molecular weight of chitosan, like that of any other polymer, is an important characteristic that determines many physical properties of the polymer. There are three most common methods to calculate polymer molecular weights, namely, the weight average molecular weight (M_w), the number average molecular weight (M_n), and the viscosity average

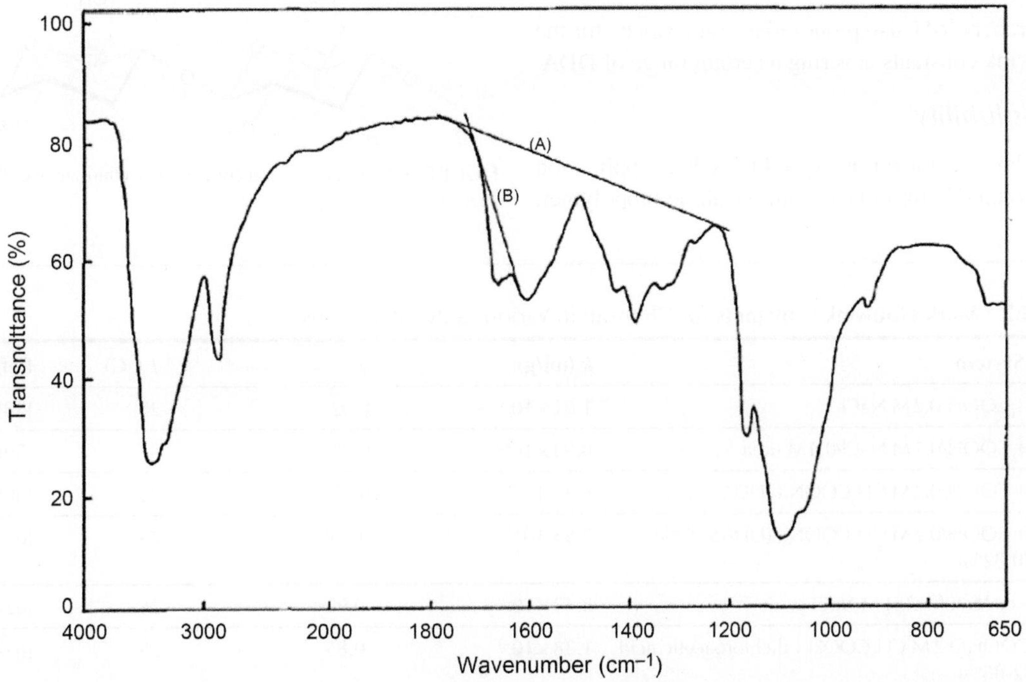

FIGURE 5.1 The IR spectrum of chitosan and the assignment of baseline of amide I absorbance band for the calculation of A_{1655}/A_{3450} ratio. Line (a) is the originally used baseline and line (b) is the revised baseline. Reprinted with permission from Ref. [51], copyright © 1980 Elsevier.

molecular weight (M_v). Polydispersity is then defined as the ratio of M_w/M_n. The methods to determine the molecular weight of chitosan have been investigated by many researchers. Muzzarelli *et al.* [54] used laser light scattering to determine the absolute M_w of a number of chitosans. Gel permeation chromatography [55], HPLC [56], and size-exclusion chromatography [56,57] are relative methods that have been adopted to measure the molecular weight of chitosan.

The most popular method to measure the molecular weight of chitosan is viscometry, which is based on the Mark-Houwink equation

$$[\eta] = k \bullet M^{\alpha} \tag{5.3}$$

where $[\eta]$ is the intrinsic viscosity of the solution, M is the viscosity average molecular weight (M_v), and k and α are constants that are dependent on the polymer and the solvent system used and the temperature. Various reports have tested different solvent systems for measuring the M_v of chitosan and the Mark-Houwink constants for those systems are summarized in Table 5.2 [58–63].

It is widely accepted that the k and α values for chitosan depend on its DDA. Several studies have measured the k and α values for chitosans with different DDA using the same solvent system [60,64]. Wang *et al.* determined k and α for chitosans with four different DDA and further generalized the relationships between k and DDA and α and DDA where DDA is expressed as percentage:

$$k = 1.64 \times 10^{-30} \times \text{DDA}^{14.0} \tag{5.4}$$

$$\alpha = -1.02 \times 10^{-2} \times \text{DDA} + 1.82 \tag{5.5}$$

Other researchers [61] also proposed average values for the Mark-Houwink constants covering a certain range of DDA.

5.3.1.3 Solubility

Solubility plays a critical role in the biomedical application of a biomaterial. Chitosan is a semicrystalline biopolymer, resulting from strong inter- and intramolecular hydrogen bonds. Chitosan is readily soluble in dilute acidic solutions at pH < 6 but insoluble in organic solvents. Organic acids such as acetic, formic, and lactic acid and inorganic acid such as hydrochloric acid can dissolve chitosan. The pK_a value of the primary amine groups of chitosan is approximately 6.3; therefore, under acidic conditions of pH below 6, the amine groups are protonated (Figure 5.2), leading to the repulsion between positively charged macromolecular chains that allows water molecules to diffuse in and solvate the polymer.

Crystallinity, DDA, the distribution of the acetyl groups, and molecular weight are among the main factors that influence the solubility of chitosan; others include deacetylation conditions during chitosan synthesis such as the temperature, alkaline concentration, and ratio of chitin to alkali solution. Reducing the crystallinity by partial reacetylation or physical disruption by the addition of urea and guanidine hydrochloride improves the solubility of chitosan [65,66]. In addition, the solubility of chitosan in acidic solution increases with the decrease of the molecular weight [67].

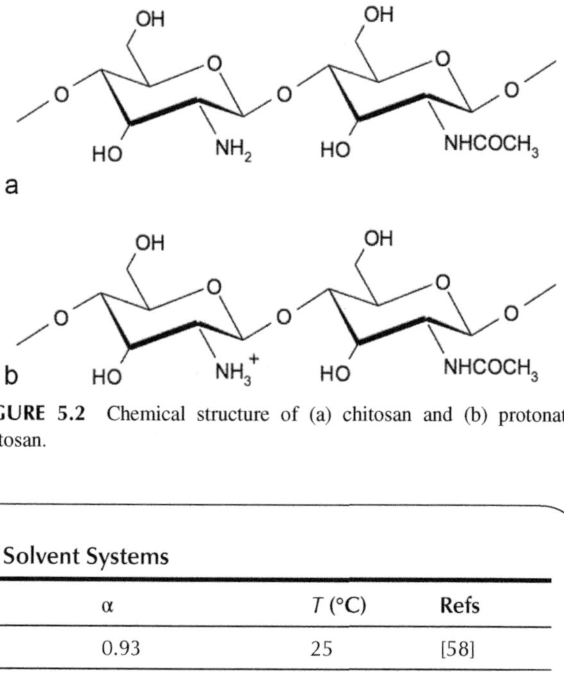

FIGURE 5.2 Chemical structure of (a) chitosan and (b) protonated chitosan.

TABLE 5.2 Mark-Houwink Constants for Chitosan in Various Solvent Systems

Solvent System	*k* (ml/g)	α	*T* (°C)	Refs
0.1 M CH_3COOH/0.2 M NaCl	1.81×10^{-3}	0.93	25	[58]
0.2 M CH_3COOH/0.1 M NaCl/0.4 M urea	8.93×10^{-2}	0.71	–	[59]
0.3 M CH_3COOH/0.2 M CH_3COONa (DDA = 98%)	8.2×10^{-2}	0.76	25	[60]
0.3 M CH_3COOH/0.2 M CH_3COONa (DDA > 97%) (DDA = 80-82%)	7.9×10^{-2}	0.796	25	[61]
0.02 M acetate buffer/0.1 M NaCl	8.43×10^{-2}	0.92	25	[62]
2% CH_3COOH/0.2 M CH_3COOH + dichloroacetic acid (DDA = 82-88%)	1.38×10^{-2}	0.85	25	[63]

Given the fact that chitosan will precipitate from solutions once the pH reaches above 6, the application of chitosan in physiological condition is severely limited. Therefore, strategies that may improve the solubility over a broader pH range have been enthusiastically investigated. Both chemical substitution and graft copolymerization have been successfully used to prepare water-soluble chitosan derivatives and copolymers by introducing water-soluble entities, hydrophilic moieties, bulky and hydrocarbon groups, etc. [22,68]. For example, chemical reactions such as acylation, alkylation, PEGylation, hydroxyalkylation, and carboxyalkylation have been used to improve the water solubility of chitosan, resulting in such derivatives as acyl-chitosan, *N*-alkyl-chitosan, PEG-chitosan, hydroxyalkyl-chitosan, and carboxyalkyl-chitosan [22]. These water-soluble forms of chitosan derivatives have found tremendous applications in wound healing [69,70], drug delivery [71–73], and tissue engineering [74,75].

5.3.2 Structure and Property Relationship

The physical and biological properties of chitosan are largely dependent on two major structural parameters, i.e., molecular weight (M_w) and DDA, among a few others. Considering that the sources and synthesis techniques also have great impacts on the material properties, chitosans with various DDA and Mw are often prepared by selective N-acetylation of one single stock of chitosan in order to perform fair comparisons of the properties. Obviously, the solubility of chitin/chitosan is determined by DDA, M_w, and crystallinity as discussed earlier. Crystallinity itself is also dependent on DDA with maximal crystallinity for both chitin (i.e., 0% deacetylated) and fully deacetylated chitosan (i.e., 100% deacetylated) due to high chemical regularity in these materials [76]. For chitosans (when DDA > 50%), the crystallinity of the polymer increases with the increase in DDA because of higher chemical regularity and more flexible polymer chains. DDA, M_w, and crystallinity in combination also control the biodegradability of chitosan. Freier *et al.* [77] investigated the *in vitro* degradation of chitin/chitosan with a broad range of DDA from approximately 0% to 99.5% with or without lysozyme. As shown in Figure 5.3, the degradation rate for chitosan with a DDA of approximately 50% is the fastest among all the chitosans. The degradation rate decreases with increase in DDA (if DDA is greater than 50%, i.e., in the "chitosan" range) and also decreases with decrease in DDA (if DDA is less than 50%, i.e., in the "chitin" range) [77]. Due to the fact that chitosan is water-soluble when the DDA is close to 50% [78], it is anticipated that the solubility of low and medium molar mass fragments cleaved during degradation is significantly higher in chitosans having intermediate DDA, which explains the enhanced degradation rate of samples with approximately 50% DDA.

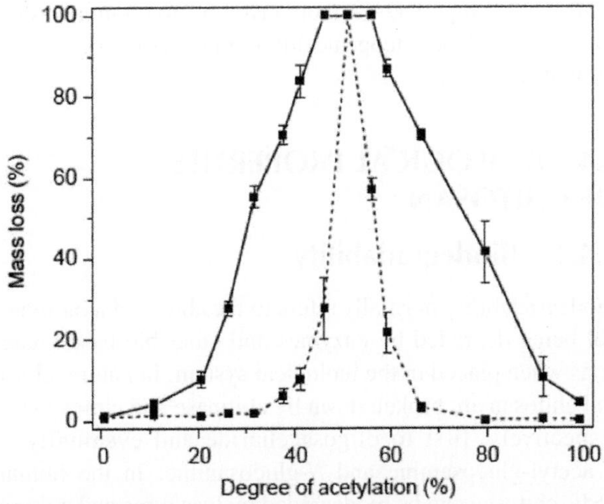

FIGURE 5.3 Influence of the degree of deacetylation (DDA) on the degradation rate of chitosan. Reprinted with permission from Ref. [77], copyright © 2005 Elsevier.

The mechanical properties of chitosan are also affected by M_w and DDA. Several studies investigated the relationship between the mechanical properties of chitosan films and the M_w and DDA of the chitosan used [56,79,80]. Although absolute results from different research groups are difficult to be directly compared due to the different chitosan materials used in these studies, in general, higher molecular weight leads to better mechanical properties as evidenced by the enhanced tensile strength. It is believed that increased molecular entanglement in high-molecular-weight chitosan is attributed to the improvement of mechanical properties. Increase in DDA also has a positive effect on the tensile strength of chitosan films [79,80], the reason being higher DDA in the chitosan range (when DDA > 50%) resulted in higher crystallinity.

Furthermore, the DDA of chitosan plays an important role in the biological properties of the biopolymer. For example, neural cells show better affinity to and greater cell viability on chitosan substrates with higher DDA [77,80]. These findings suggest that the interactions between the positively charged amine groups and negatively charged cell membrane surfaces contribute to cell attachment and proliferation. Similarly, the DDA of chitosan affects the proliferation of other anchorage-dependent cells such as keratinocytes, fibroblasts, and osteoblasts have also been reported, suggesting that these interactions are not cell-specific [81,82].

Due to the fact that the physicochemical and biological properties are greatly influenced by chitosan structural parameters, controlling M_w and DDA could be a useful tool to design finely tuned chitosan-based devices for biomedical applications. For example, in drug delivery, chitosan with high M_w and high DDA is used to improve drug encapsulation and reduce the release rate [12,83]. In tissue engineering, chitosan with high DDA is more favorable to promote proliferation and induce osteogenic differentiation of bone marrow stromal cells [84].

5.4 BIOLOGICAL PROPERTIES OF CHITOSAN

5.4.1 Biodegradability

Biodegradability normally refers to the ability of a biomaterial being degraded by enzymes and other bio-based reactions when placed in the biological system. In nature, chitin and chitosan are broken down by chitinase and chitosanase, respectively, first to oligosaccharide and eventually to *N*-acetyl-glucosamine and *N*-glucosamine. In the human body, chitosan can be biodegraded by lysozyme and colonic bacterial enzyme. In addition, several human chitinases [85], glucosidase [86] and proteases [87] have been identified to have enzymatic activities and be able to degrade chitosan to varying degrees.

Several studies have demonstrated that the *in vitro* degradation rate is maximal at approximately 50% DDA due to enhanced hydrophilicity and easy enzyme accessibility at such a DDA [77,88]. Increasing DDA (when DDA > 50%) reduces the degradation rate because of increased crystallinity and reduction in the number of *N*-acetylglucosamine groups. Highly deacetylated chitosans exhibit minimal degradation [88]. The molecular weight of chitosan plays a role in the degradation rate. Maintaining the DDA unchanged, chitosans of higher molecular weight degrade more slowly than those having lower molecular weights [86].

Chitosan biodegradation *in vivo* has been evaluated using different approaches. Since chitosan is commonly used as a food additive, the biodegradation of chitosan after oral administration has been investigated. Research has shown that chitosans with relatively low DDA (~50%) are highly biodegradable and easily excreted in urine, showing no accumulation in the body [89]. Similar to the trend demonstrated *in vitro*, the *in vivo* biodegradation of chitosan also depends strongly on the molecular weight. As the molecular weight decreases, the absorption of chitosan by intestine increases [90]. Chitosan biodegradation has also been evaluated by implantation in different animal models. In one study [91], chitosan films were implanted subcutaneously in rats. Results showed similar influence of chitosan DDA on the *in vivo* biodegradation rate as we have seen *in vitro*, i.e., the higher the DDA, the slower the biodegradation. In another study [92], chitosan fibrous scaffolds with different DDA were implanted between two nerve stumps of the rat sciatic nerve gaps. Highly deacetylated chitosan with DDA of 92.3% was resistant to biodegradation; however, chitin and chitosan with lower DDA were readily biodegradable. These findings suggest that the biodegradation of chitosan is highly controllable. Structural parameters such as molecular weight, DDA, and crystallinity could be used alone or in combination to finely tune the degradation rate for specific applications.

5.4.2 Biocompatibility

Biocompatibility is concerned with the performance of biomaterials and has been extensively used in the field of biomaterial science. Historically, it is defined as "the ability of a biomaterial to perform with an appropriate host response in a specific application" [93]. This definition places an emphasis on that biocompatibility is a relative concept concerning with a specific biomaterial for a specific application. More recently, Williams redefined the biocompatibility of a scaffold or matrix for a tissue engineering product as "the ability to perform as a substrate that will support the appropriate cellular activity, including the facilitation of molecular and mechanical signaling systems, in order to optimize tissue regeneration, without eliciting any undesirable local or systemic responses in the eventual host" [94]. This revised definition underlines the importance that a biomaterial or a device shall not produce any clinically significant adverse effects in the host. As a more concrete reflection to the point, ISO 10993 guidelines represent the series of standards for evaluating the biocompatibility of a biomaterial or a medical device prior to a clinical study.

The biocompatibility of chitosan has been investigated and reported over the past several decades. A great number of studies have been performed *in vitro* to evaluate the biocompatibility of chitosan and chitosan-based systems using a variety of cell types such as fibroblasts [95,96], osteoblasts [97,98], chondrocytes [98], endothelial cells [34], neural cells [99], and hepatocytes [100,101]. Results have shown that chitosan is nontoxic and can support these types of cells to adhere and proliferate, which suggests that chitosan is compatible with these cell types.

The *in vivo* performance of chitosan-based matrices has also been evaluated using a variety of animal models particularly in subcutaneous models using mice or rats. VandeVord *et al.* [102] implanted a tubular chitosan scaffold into mice subcutaneously. It was found that chitosan did not illicit pathological inflammatory responses. In addition, there was no evidence of infection or endotoxin, and low incidence of chitosan-specific immune response, and minimal sign of inflammatory reaction to the implanted materials. Wang *et al.* [103] tested the biocompatibility of a porous chitosan-based fiber-reinforced conduit in a rat subcutaneous model. The conduit was shown to illicit only mild tissue response and was compatible with the surrounding tissue. Tissue-specific evaluation has also been investigated. For example, Bumgardner *et al.* [104] implanted chitosan-coated titanium pins in the tibia of New Zealand white rabbits. Histological evaluations of tissues in contact with the chitosan-coated pins indicated minimal inflammatory response and a typical healing sequence of fibrous, woven bone formation, followed by development of lamellar bone. The DDA of chitosan has shown some influence on the biological performance of the material both *in vitro* and *in vivo*. Evidence suggests that chitosan with higher DDA supports better cell growth [105], yields lesser inflammatory reaction [106], and thus is more biocompatible.

5.4.3 Antimicrobial Activity

One intriguing feature of chitosan is its antimicrobial activity. Over the past three decades, numerous studies have demonstrated broad-spectrum antimicrobial activities of chitosan against Gram-positive and Gram-negative bacteria [107], fungi, and yeasts. It has been suggested that the antimicrobial activities of chitosan and its derivatives depend on several intrinsic factors such as the molecular structure, DDA, and molecular weight, and extrinsic factors such as species of microorganisms and environmental pH [108]. The exact mechanisms of the antimicrobial activities of chitosan are not clear; however, several possible modes of its antimicrobial action have been proposed. First, it is suggested that the electrostatic interaction between the polycationic structure of chitosan and the anionic cell surface proteins and lipopolysaccharide of microorganisms plays an important role. This interaction could change the permeability of the cell membrane of the microorganisms, resulting in leakage of intercellular components, and then cause cell death [109,110]. Therefore, the presence of amine groups in the chitosan molecular chain is critical to the antibacterial property. Fully acetylated chitosan, or chitin, loses its antibacterial activity due to the deprivation of the amine groups. Nevertheless, there does not seem to have a clear relationship between the number of amine groups or the DDA of chitosan and its antibacterial activity. In fact, seemingly conflicting results on the relationship between DDA and antibacterial activity of chitosan have been reported in the literature [111,112], which may partially be due to different types and sources of chitosan used in the studies and different bacteria being tested. Second, chitosan also acts as a chelating agent that selectively binds trace metals and thus inhibits the production of toxins and microbial growth [113], contributing to the antibacterial activity of chitosan especially when the environmental pH is above the pK_a. Other issues involved in the antibacterial activity of chitosan, such as the influence of the molecular weight and the effectiveness over Gram-positive and Gram-negative bacteria, are still debated [108,114,115]. However, it is believed that chitosan shows stronger antibacterial activity at lower pH [108,114].

5.5 CHITOSAN APPLICATION IN TISSUE ENGINEERING

5.5.1 Scaffold Fabrication Techniques

Tissue engineering is defined by Laurencin *et al.* as the application of biological, chemical, and engineering principles to the repair, restoration, or regeneration of living tissues by using biomaterials, cells, and factors alone or in combination [116]. Scaffold-based tissue engineering is one of the most well-studied approaches to regenerating different types of tissues, which involves seeding cells together with certain signaling molecules on a three-dimensional (3D) porous biodegradable matrix, culturing them *in vitro* and implanting them into *in vivo* defects. In this therapeutic approach, biomaterials, usually in the form of 3D porous scaffolds, play multiple significant roles to provide structural maintenance of the defect shape, void volume for vascularization, and serve as temporary extracellular matrix (ECM) for cell adhesion, proliferation, differentiation, and maturation. In addition, scaffolds can act as a delivery vehicle for bioactive molecules, growth factors, and cells to the defect site for tissue morphogenesis and defect healing [117–119].

A variety of 3D chitosan scaffolds have been fabricated using different techniques, attributing to the ease of processing of chitosan. Solvent casting, in combination with particulate leaching, was one of the first techniques developed to prepare porous scaffolds [120,121]. In this process, a polymer is first dissolved in an organic solvent. Particles (such as salt and sugar) with specific dimensions are then mixed with the polymer solution. The polymer/particles/solvent mixture is shaped into its final geometry by casting in a predefined mold and allowing for solvent evaporation. Following solvent evaporation, the polymer/particles composite material is then immersed in a suitable solvent to dissolve the particles, leaving behind a porous structure. For example, Lim *et al.* prepared porous chitosan scaffold from a mixture of acidic chitosan solution and sodium acetate particles as the porogen [122]. An increase in the ratio of sodium acetate resulted in an increase in the porosity and interconnectivity. For the scaffold prepared with 90% sodium acetate, many minute pores (7-30 μm) were present between the main pores (200-500 μm). Despite the simplicity and versatility, this technique is limited by the lack of control over the scaffold interconnectivity. Residual porogen particles are potentially associated with the resultant structures. In addition, the mechanical properties of the scaffolds prepared by this method are relatively low.

Another widely used method for the preparation of porous scaffolds is thermally induced phase separation [123–125]. In this process, the phase separation of the homogeneous polymer solution is achieved by cooling the solution below the solvent meting temperature, leading to solid-liquid demixing and resulting in the formation of frozen polymer-poor and polymer-rich phases. Subsequent removal of the solvent crystals in the demixed solution through vacuum sublimation leaves a porous polymer scaffold. The resultant structure porosity and the interconnectivity can be finely tuned by changing the various thermodynamic and kinetic parameters (such as types of polymers, polymer concentration, and thermal quenching strategy) [126–128]. Madihally *et al.*

were among the first to investigate the feasibility of fabricating highly porous 3D chitosan scaffolds with different architectures using this method [129]. By varying the freezing conditions, mean pore diameters of 1-250 μm could be obtained. This process generally produces highly interconnected porous scaffolds with porosities up to 95%. In addition to pure chitosan scaffolds, porous chitosan/ceramic composite scaffolds have also been fabricated using this method [130,131].

To eliminate the need for solid porogens and toxic organic solvents, a so-called gas-foaming technique is used. In this technique, a preshaped polymer substrate is placed in a chamber where the polymer substrate is exposed to high pressure CO_2 for several days to allow for CO_2 saturation in the polymer. The system is then gradually depressurized to the atmospheric level, resulting in the generation of gas bubbles that induce gas foaming and a sponge-like structure. For example, Ji *et al.* [132] fabricated cross-linked porous chitosan hydrogel scaffolds using dense CO_2 gas as a foaming agent. The chitosan hydrogel scaffolds had an average pore diameter of 30-40 μm. Temperature, reaction period, type of biopolymer, and cross-linker had a significant impact on the pore size and characteristics of the hydrogel produced by dense gas CO_2.

Many 3D porous chitosan scaffolds fabricated from the aforementioned methods lack sufficient mechanical properties necessary for load-bearing tissue engineering applications. Thus, a versatile scaffold development strategy based on the sintering of polymer microspheres has been developed by Laurencin *et al.* for the fabrication of mechanically competent 3D porous scaffolds and has showed great promise for regenerative applications [133–139]. In the classical heat sintering technique, 3D microsphere scaffolds are fabricated by heating orderly packed polymer microspheres above the glass transition temperature (T_g) of the polymer in a mold of desired size and shape. Polymer chains become flexible above the T_g and physically intertwined together between neighboring microspheres resulting in the formation of a 3D scaffold with an interconnected porous structure. The excellent scaffold mechanical strength originates from the bonding between neighboring microspheres and the mechanical properties can be controlled by varying the sintering temperature or sintering time. Furthermore, the pore size and pore volume of the scaffolds can be controlled by changing the duration/temperature of sintering and the microsphere diameter. Sintered chitosan microsphere scaffolds have been successfully prepared for bone tissue engineering by our laboratory using this technique [14,140]. In brief, chitosan microspheres were first prepared via the ionotropic gelation of chitosan solution droplets using tripolyphosphate ions as the cross-linking agent. The 3D chitosan scaffolds were then fabricated by packing chitosan microspheres in a stainless steel mold and sintering them together using a synergetic effect of both solvent and temperature. Jiang *et al.* have further developed 3D porous scaffolds fabricated from natural polymer chitosan and synthetic poly(lactide-*co*-glycolide) (PLGA) by heat sintering [14,35,141–143]. This hybrid scaffold system combines the mechanical excellence of PLGA with the functionality of chitosan through its reactive amine and hydroxyl groups.

One of the most intriguing features of chitosan is its ability to form hydrogels at mild conditions suitable to be administered in a minimally invasive manner. Various chitosan-based hydrogel systems that respond to stimuli such as light, pH, ionic concentration, chemical and physical reactions, and temperature have been developed as injectable systems for regenerative applications. For example, photo-cross-linkable chitosan systems could be obtained by introducing photosensitive moieties such as lactose and azide onto the chitosan backbone followed by photo-cross-linking [144]. Thermosensitive chitosan hydrogels are among the most extensively investigated *in situ* gelling systems. One approach to rendering thermogelling properties to chitosan is by grafting a synthetic polymer of low critical solution temperature to chitosan backbone. For example, *N*-isopropylacrylamide was graft polymerized onto chitosan molecular chain [74]. Due to the low critical solution temperature (32 °C) of poly(*N*-isopropylacrylamide) (PNIPAAm), a thermosensitive water-soluble chitosan-*g*-PNIPAAm gel was obtained. Another interesting type of thermogelling system is the neutral physiological temperature setting chitosan gel formed by treating aqueous acidic solution of chitosan with glycerophosphate salts or inorganic phosphate salts. The gelation has been attributed to a combination of favorable hydrogen bonding, electrostatic interactions, and hydrophobic interactions between polymer chains and the polyols bearing single anionic group such as glycerophosphate salt [145]. Our laboratory has shown that injectable thermosetting chitosan gels can also be developed by neutralizing chitosan solution using inorganic salts such as ammonium hydrogen phosphate (AHP) [146]. By adjusting the concentration of AHP, the gelling time can be varied from 1 min to several hours at 37 °C.

In addition to the microscale scaffolds described earlier, nanofiber-based nanoscale scaffolds, due to their structural similarity to natural ECM, are attractive matrices for tissue engineering applications. In recent years, nanofibrous scaffolds consisting of different types of polymer materials have been fabricated by electrospinning. In such a process, a polymer jet is ejected from the surface of a charged polymer solution when the applied electric potential overcomes the surface tension. The ejected jet under the influence of applied electrical field travels rapidly to the collector and collects in the form of nonwoven Web as the jet dries. Before reaching the collector, the jet undergoes a series of electrically driven bending instabilities that result in a series of looping and spiraling motions. In an effort to minimize the instability due to the repulsive electrostatic forces, the jet elongates to undergo a large amount of plastic stretching that consequently leads to a significant reduction in the fiber diameter and results in ultrathin fibers. By changing polymer concentration alone, it is possible to fabricate the fiber diameters in the range of a few nanometers to several micrometers while keeping other electrospinning parameters constant. In addition, the applied electrical potential also plays an important role in the fiber diameter obtained by electrospinning. Other factors that control the electrospinning process include polymer solution flow rate, distance between spinneret and collector, motion of the grounded target, and ambient parameters (temperature, humidity, and air velocity). For example, by altering the rotating targets and electrospinning conditions, electrospinning allows for the fabrication of complex structures such as aligned nanofibers, micro/nanofiber composites, and core-shell nanofibers. Due to their solubility in most acids, chitosan-based materials have been successfully electrospun to produce ECM-mimicking nanofibrous scaffolds for a variety of tissue engineering applications. Ohkawa *et al.* successfully prepared an electrospun nonwoven fabric of chitosan [147]. Optimal electrospinning conditions for solvent and concentration resulted in the formation of homogenous chitosan fibers with a mean diameter of 330nm. In order to combine the beneficial features of different components, chitosan-based blend or composite nanofibers have also been fabricated via electrospinning using chitosan and a second polymeric or ceramic component such as poly(lactic acid) [148,149], polycaprolactone [150], poly(L-lactic acid-*co*-ε-caprolactone) [151], poly(vinyl alcohol) [70,152–155], poly(ethylene oxide) (PEO) [156,157], poly(vinyl pyrrolidine) [107], poly(ethylene terephthalate) [158], or calcium based ceramics [159,160]. A collection of various miniaturized chitosan structures fabricated by different techniques is shown in Figure 5.4.

5.5.2 Chitosan-Based Scaffolds for Tissue Engineering Applications

Chitosan is a unique natural polymer that has potential in tissue engineering applications due to its controllable biodegradability, biocompatibility, antimicrobial activity, and functionalizability. A great number of studies have been performed to evaluate the cytocompatibility of chitosan

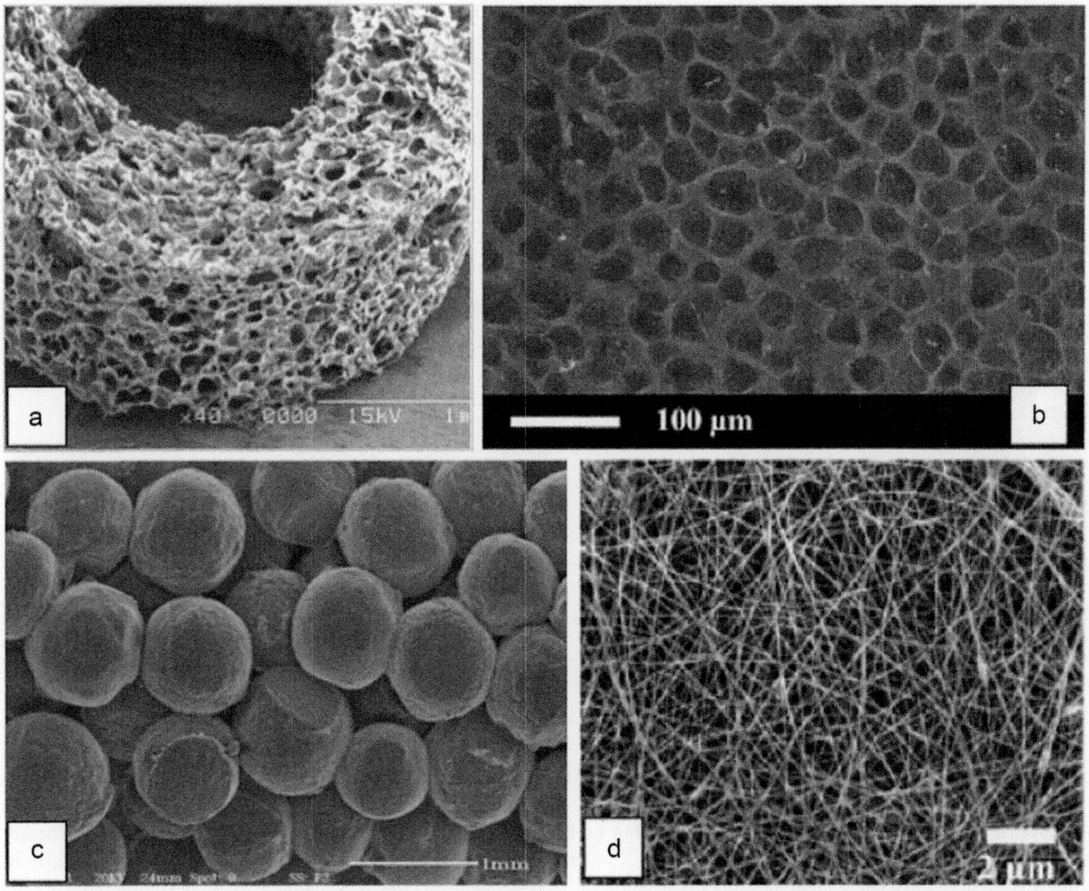

FIGURE 5.4 Different types of porous chitosan scaffolds fabricated by (a) thermally induced phase separation [129], (b) dense CO_2 gas foaming [132], (c) sintering chitosan microspheres, and (d) electrospinning [157]. Reprinted with permission. Copyright © 1999, 2005, 2011 Elsevier.

using a variety of cell types such as osteoblasts, chondrocytes, fibroblasts, nucleus pulposus cells, neural, and endothelial cells. Results have shown that chitosan is nontoxic and can support these types of cells to adhere and proliferate, suggesting that chitosan is biocompatible with these cell types and has the potential to be utilized for bone, cartilage, skin, intervertebral disc, ligament and tendon, and nerve and vascular tissue engineering.

5.5.2.1 Bone

A number of *in vitro* and *in vivo* studies have demonstrated the osteocompatibility and osteogenic potential of chitosan [161,162]. In the past decade, much research has focused on the development of chitosan-based systems for regenerating bone. Seol *et al.* [163] have shown the ability of freeze-dried chitosan sponges as effective scaffolding materials to support osteoblast growth and deposition of mineralized matrix. Although the feasibility and osteocompatibity of pure chitosan scaffolds have been demonstrated, pure chitosan scaffolds lack some key physical and biological properties, such as mechanical strength and osteoinductivity, to serve as successful artificial ECM for bone regeneration. Therefore, different strategies have been frequently used to overcome these barriers including the incorporation of additional components such as other polymeric materials, ceramics, growth factors, and/or cells into chitosan scaffolds. To improve the mechanical properties, materials such as alginate- and calcium-based ceramics have been used to make composites such as 3D composite chitosan sponges [164–166]. Incorporating calcium-based ceramics into chitosan scaffolds improves the osteoconductivity of the chitosan substrates [167]. Our laboratory has demonstrated the feasibility of developing both chitosan scaffold and chitosan/PLGA composite scaffold for bone tissue engineering [140,141]. The developed chitosan-/PLGA-sintered microsphere scaffolds have been shown to be biocompatible, biodegradable, and mechanically robust. For example, the compressive modulus and compressive strength of the composite scaffolds were found to be 200-450 MPa and 3.5-16 MPa, respectively, both of which were in the range of trabecular bone. Furthermore, the scaffolds possess various reactive groups (amine or hydroxyl groups) due to the presence of chitosan that allows for the chemical or ionic modifications to introduce desired functionalities to the system. Bioactivity was further introduced to the system by allowing negatively charged heparin to bind to protonated chitosan/PLGA scaffolds via ionic interaction. The fusion between adjacent microspheres contributes to the mechanical strength of the heparin immobilized chitosan/PLGA scaffold, while the void spaces among microspheres resulting from microsphere packing generate the total porosity. The uniform distribution of heparin was confirmed by confocal microscopy through the immobilization and visualization of fluorescence conjugated heparin molecules. The heparinized scaffold supported the adhesion, proliferation, and migration of MC3T3-E1 osteoblast-like cells (Figure 5.5a). In addition, MC3T3-E1 cell differentiation was enhanced as evidenced by the increased alkaline phosphatase activity and osteocalcin expression of cells on the heparinized scaffold. The ability of the heparinized scaffold to bind and release bone morphogenetic protein-2 and the *in vivo* bone regeneration capacity of the scaffold has been evaluated in a rabbit ulna critical-sized defect model (Figure 5.5b) [143].

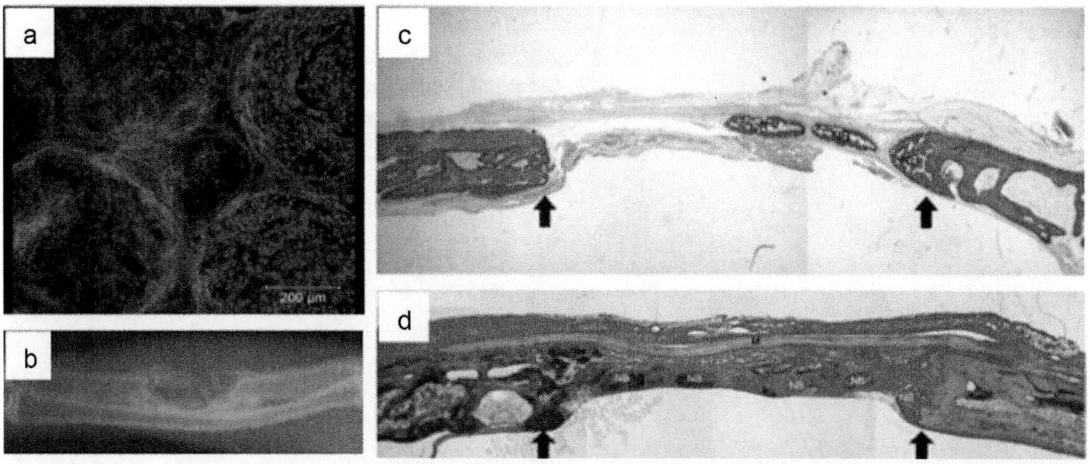

FIGURE 5.5 Porous chitosan scaffolds for bone tissue engineering. (a) Heparinized sintered chitosan/PLGA scaffold supported the adhesion, proliferation, and migration of MC3T3-E1 osteoblast-like cells [142]; (b) X-ray micrograph showing sintered chitosan/PLGA-based scaffold aided in the regeneration of critical-sized defect in a rabbit ulnar model [143]; histological view of rabbit calvarial defect after 4 weeks in control (no scaffold implanted) (c) and (d) covered by chitosan nanofiber membrane [168]. (For color version of this figure, the reader is referred to the online version of this chapter.) Reprinted with permission. Copyright © 2010 Elsevier, Wiley, and © 2005 American Academy of Periodontology.

The ECM of the native bone is a nanocomposite composed of hierarchically arranged collagen fibrils, hydroxyapatite, and proteoglycans in the nanoscale. To closely mimic the structure of bone, researchers have focused on the development of electrospun chitosan-based nanofibrous matrices for bone regeneration. Bhattarai *et al.* [157] fabricated electrospun chitosan/PEO nanofibers that promoted the adhesion of osteoblast cells and maintained characteristic cell morphology and cell phenotype during the *in vitro* culture. In an effort to develop biomimetic scaffolds with compositional and structural features similar to bone ECM, Zhang *et al.* [159] reported the synthesis of chitosan/hydroxyapatite nanocomposite nanofibers. During *in vitro* cell culture with human fetal osteoblast cells, the incorporation of hydroxyapatite has significantly stimulated the bone forming ability as evidenced by the cell proliferation, mineral deposition, and morphology observation, due to the excellent osteoconductivity of hydroxyapatite. After 15 days of culture, nanocomposite scaffolds produced much more mineral deposits with large mineral clumps than the plain chitosan scaffolds. Shin *et al.* [168] investigated the bone regenerative efficacy of chitosan nanofiber membranes using a 10 mm rabbit calvarial defect model. After 4 weeks, only a limited amount of bone (a few bony islands) was found in the defect alone group (Figure 5.5c). In contrast, a new bone bridge was observed when the calvarial defect was covered with chitosan nanofiber membranes (Figure 5.5d). Further histomorphometric evaluation demonstrated significantly higher amount of bone regenerated in the chitosan membrane group ($28.8\% \pm 6.9\%$) than in the control group ($10.6\% \pm 5.0\%$). Thus, chitosan-based nanofibers have demonstrated great potential as ECM-mimicking matrices for enhanced bone regeneration.

5.5.2.2 Cartilage

Articular cartilage damage is among the most encountered musculoskeletal diseases, eventually leading to total joint replacement if not treated properly. Due to the avascular structure of hyaline cartilage and its limited ability to self-regenerate, a number of surgical cartilage repair interventions have been used by orthopedic surgeons including allograft osteochondral transplantation, autologous chondrocyte transplantation, allogeneic juvenile cartilage transplantation (DeNovo NT®), mosaicplasty, and microfracture technique. In recent years, tissue engineering of cartilage has emerged as a promising approach to regenerating hyaline cartilage. Due to the structural similarity of chitosan to the proteoglycans that reside in cartilage matrix, chitosan hydrogels holds great potential as an artificial ECM to regenerate hyaline cartilage. Chitosan gels have shown to be able to maintain the round phenotype of chondrocytes. Hoemann *et al.* performed a series of studies using chitosan/glycerol phosphate/autologous blood hydrogel, in combination with the microfracture technique, for cartilage regeneration [169,170]. More hyaline cartilage was regenerated using a combination of chitosan-based hydrogel and microfracture than microfracture technique alone in rabbit and ovine models (Figure 5.6a and b). It has been demonstrated that the presence of chitosan gel

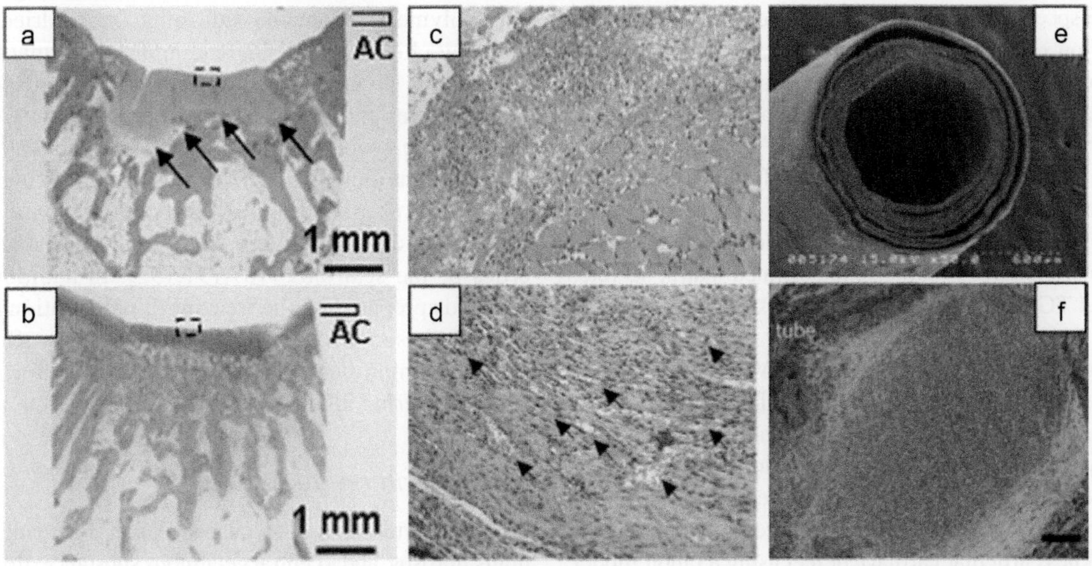

FIGURE 5.6 Porous chitosan scaffolds for the regeneration of soft tissues. (a) and (b): Chitosan hydrogel for the repair of cartilage in a rabbit model. More hyaline cartilage was regenerated using chitosan-based hydrogel together with microfracture (b) than microfracture alone (a) [170]. (c) and (d): Chitosan hydrogel for the repair of skin. More mature blood vessels formed in the group treated with chitosan hydrogel/FGF-2 (d) than in the nontreated control group (c) [171]. (e) and (f): Chitosan nanofibrous tubes (e) to repair a sciatic nerve defect in rat. Regenerated nerve tissue formed within the inner space of the tube and the sprouting of myelinated axons occurred (f) [172]. (For color version of this figure, the reader is referred to the online version of this chapter.) Reprinted with permission. Copyright © 2007 Elsevier, © 2005 & 2008 Wiley.

significantly increased inflammatory and marrow-derived stromal cell recruitment to the microdrill holes, increased vascularization of the provisional repair tissue, and also increased intramembranous bone formation and subchondral bone remodeling [173,174]. Moreover, Hoemann *et al.* have demonstrated that implantation of thermogelling chitosan gel mixed with blood in a microdrilled cartilage defect site led to recruitment of alternatively activated wound healing macrophages that may significantly promote subchondral repair [175]. Such biological mechanisms influencing the cellular responses to chitosan substrates have been proposed and confirmed by others. For example, studies have shown chitosan-mediated stimulation of macrophage function both *in vitro* and *in vivo* contributes to accelerated tissue regeneration [176–178]. A randomized, multicenter clinical trial has recently been completed to evaluate the chitosan/glycerol phosphate/autologous blood hydrogel for cartilage repair.

Signaling molecules such as growth factors and small peptide sequences and plasmid DNA encoding specific growth factors have frequently been incorporated into chitosan scaffolds for tissue engineering of cartilage. Lee *et al.* [179] encapsulated transforming growth factor-β1 (TGF-β1) into chitosan microspheres via an emulsion/ionic cross-linking method. The growth factor-containing microparticles were then loaded onto a porous chitosan/collagen/chondroitin sulfate scaffold to improve chondrocyte proliferation and glycosaminoglycan production. Other growth factors such as insulin- and NEL-like molecule-1 have been encapsulated into chitosan microspheres and investigated in other studies [180,181]. These growth factor-containing scaffolds could promote chondrocyte proliferation and deposition of cartilage specific ECM and stimulate the chondrogenic differentiation of stem cells. Furthermore, gene delivery has been used as a promising technique to achieve localized and sustained expression of growth factors. Guo *et al.* [182] incorporated plasmid DNA encoding TGF-β1 into porous chitosan-gelatin scaffolds and the scaffolds were seeded with chondrocytes. The plasmid DNA could be released in a sustainable manner up to 3 weeks and the chondrocytes with transferred gene on the gene-activated matrix secreted enhanced TGF-β1 as compared to cells seeded on scaffolds without DNA loading. In another study, Wang *et al.* [183] used a cationized chitosan derivative, *N,N,N*-trimethyl chitosan chloride, as a carrier for the plasmid DNA also encoding TGF-β1. The chitosan/DNA complex was then mixed with fibrin hydrogel and loaded together onto a PLGA sponge. The DNA-containing scaffold with seeded bone marrow mesenchymal stem cells (MSCs) successfully repaired a full-thickness articular cartilage defect using a rabbit model.

5.5.2.3 Skin

Attributing to the biodegradability, biocompatibility, and especially the antibacterial and antimicrobial activity, chitosan and its derivatives have been widely investigated as wound-dressing materials and as scaffolds for skin tissue engineering. Antibiotic-loaded chitosan-based scaffolds have been used to treat severe burn wounds. For example, artificial skin substitutes made of chitosan membranes or porous scaffolds could be used to release antibacterial agents such as nanotitanium dioxide [184] or silver sulfadiazine [185] and exhibit good antimicrobial properties. In addition to preventing infection, chitosan-based scaffolds can play more important roles as carriers for growth factors to accelerate skin wound healing. For example, Obara *et al.* [171,186] prepared a photo-cross-linkable chitosan consisting of chitosan molecules modified by lactose and photoreactive azide, which can be photo-cross-linked under brief UV irradiation to produce hydrogels. FGF-2 was dispersed into the aqueous azide-chitosan-lactose solution and photo-cross-linked to form a hydrogel, leading to accelerated wound closure (Figure 5.6a and d).

Chitosan-based scaffolds for skin tissue engineering were also fabricated to mimic the stratified microstructure of skin tissue and to provide a temporary ECM for cell infiltration and vascularization. Freeze-drying chitosan solutions has been frequently used to fabricate chitosan-based scaffolds for skin regeneration, which typically exhibit pore sizes of 100-200 μm and pore volume of more than 90% [187]. Asymmetric chitosan-based scaffolds have also been prepared by a controlled freezing-lyophilization process [188,189] or by applying an additional dense layer of biomaterials on top of the chitosan scaffolds, producing a bilayer structure to mimic that of the skin [190]. For example, a fibrin glue layer was polymerized on one side of a freeze-dried porous collagen-chitosan scaffold, leading to the formation of an asymmetric collagen-chitosan/fibrin glue scaffold [190]. Furthermore, chitosan-based porous scaffolds fabricated via freezing-lyophilization process are chemically cross-linked by various reagents such as glutaraldehyde, EDC and NHS, or dimethyl 3,3'-dithiobispropionimidate to modify the degradability, stability, and mechanical properties of the scaffolds [187,191,192]. In addition to the scaffolds prepared by the freezing-lyophilization method, chitosan-based hydrogels or nanofibrous scaffolds via electrospinning have also been investigated for skin tissue engineering applications [193,194].

5.5.2.4 Intervertebral Disc

The intervertebral disc is consisted of an outer ring, the annulus fibrosus, and a central gelatinous substance, the nucleus pulposus. This unique structure contributes to the normal disc functions such as giving the spine its flexibility, absorbing energy, and distributing loads applied to the body. The annulus fibrosus is a fibrocartilaginous tissue consisting of layers of lamellae rich in highly cross-linked collagen fibrils, whereas the nucleus pulposus is more amorphous with a small percentage of randomly oriented fibrils and a large content of proteoglycans [195]. Intervertebral disc degeneration occurs as people age, resulting in a decrease of water and proteoglycan contents and an increase of keratan sulfate and collagen contents.

While surgical interventions such as discectomy and spinal fusion may provide short-term pain relief, tissue engineers seek to regenerate intervertebral disc with biomaterials and cells. One of the major challenges is to select the proper scaffolding material that aids in the regeneration of the natural ECM of nucleus pulposus cells and mimics the mechanical properties of the native nucleus pulposus. Injectable hydrogels have been suggested to be ideal since hydrogels can be easily used as cell carriers and hydrogels can be administered in a minimally invasive manner [196]. Therefore, thermosensitive chitosan gels are promising in intervertebral disc tissue engineering. Alini *et al.* [197] showed that chitosan hydrogel was superior to collagen and hyaluronan hydrogels with regard to the synthesis and retention of proteoglycans. Cloyd *et al.* [198] showed that a hydrogel formulation composed of hyaluronic acid, PEG-*g*-chitosan, and gelatin approximated the mechanical properties of the nucleus pulposus serving as a suitable scaffold material. Roughley *et al.* [199] developed several temperature-sensitive chitosan-based hydrogels including chitosan/glycerophosphate gels, chitosan/glycerophosphate/hydroxyethyl cellulose gels, and chitosan/genipin gels. The majority of the proteoglycan produced by the encapsulated nucleus pulposus cells could be retained within the gels in an *in vitro* culture environment. Another challenge in intervertebral disc tissue engineering is the selection of appropriate cells. The use of autologous nucleus pulposus cells seems to be ideal; however, this approach is problematic as the density of the cells is sparse in nucleus pulposus and the removal of autologous tissue could induce or accelerate the degeneration of discs. Therefore, using stem cells to generate nucleus pulposus cells has become attractive. Different chitosan-based thermosensitive hydrogels have been tested for *in vitro* culture of stem cells potentially for the regeneration of intervertebral disc. For example, Richardson *et al.* [196] encapsulated human bone marrow-derived MSCs into a temperature-sensitive chitosan—glycerophosphate hydrogel. The encapsulated MSCs differentiate to a phenotype similar to nucleus pulposus cells and secrete proteoglycans and collagens in a ratio closely resembling that of nucleus pulposus.

5.5.2.5 Other Soft Tissues

Chitosan has also been investigated to regenerate other soft tissues such as ligament, tendon, nerve, and vascular tissue. To mimic the fibrous structure of native ligament and tendon, Funakoshi *et al.* [200] developed chitosan/hyaluronan hybrid fibers by a wet-spinning method, and the fibers were further fabricated into 3D scaffolds. The fibrous scaffolds had a tensile strength of approximately 200 MPa and were able to support the proliferation of tendon fibroblasts and the production of collagen matrix. Hayami *et al.* [201] fabricated a composite scaffold consisting of electrospun poly(ε-caprolactone-*co*-D,L-lactide) fibers and photo-cross-linked *N*-methacrylated glycol chitosan hydrogel. This composite scaffold was able to maintain ligament cell phenotype *in vitro*.

The regeneration of traumatized peripheral nerve has remained to be clinically challenging due to the various limitations of an autologous nerve graft. Chitosan has been studied as a candidate material for nerve regeneration particularly because of its neuroprotective properties [202,203]. Wang *et al.* [172] prepared aligned chitosan nanofiber meshes, which were further reeled on a bar to yield nanofibrous tubes to repair a sciatic nerve defect in rat. Results showed that the regenerated nerve tissue formed within the inner space of the tube and the sprouting of myelinated axons occurred (Figure 5.6e and f) [172]. To optimally guide axonal regeneration, Wang *et al.* [204] constructed a more sophisticated porous chitosan tubular scaffold, which had a knitted fibrous outer wall and an inner matrix with multiple axially oriented macrochannels and radically interconnected micropores. *In vitro* results showed that neuroblastoma cells grew along the oriented macrochannels, and the inner microporous structure was beneficial for cell ingrowth to the interior of the scaffold.

The application of chitosan-based hydrogels to regenerate vascular tissue has been partially driven by the cationic nature of chitosan. Specifically, heparin can be bound with chitosan through ionic interaction, aiding in the delivery of heparin-binding growth factors such as fibroblast growth factor-1 (FGF-1) [25], FGF-2 [205], vascular endothelial growth factor [206], epidermal growth factor [25], and TGF-β3 [207]. Ishihara *et al.* [25] incorporated heparin and FGF-1 or FGF-2 into a photo-cross-linkable chitosan hydrogel. The growth factors, which retained their biological activity and were released upon the hydrogel degradation, induced significant neovascularization. The prolonged effect of the incorporated growth factors was attributed to the ability of heparin to protect FGF-1 and FGF-2 from inactivation by acid, heat, and degradation by proteases. Several other studies also investigated other small molecules that would bind with heparin-binding growth factors. Nakamura *et al.* [208] fabricated a chitosan hydrogel by complexing chitosan with fucoidan that also binds with heparin-binding growth factors. The fucoidan-containing chitosan hydrogel protected FGF-2 from inactivation by heat and proteolysis, enhanced FGF-2 activity, and promoted significant neovascularization. Other studies investigated the use of a non-anticoagulant heparin derivative called periodate-oxidized heparin and other growth factors such as TGF-β3 to induce blood vessel formation, demonstrating the great potential of chitosan as a scaffolding material in vascular tissue engineering [205,207].

5.6 CHITOSAN APPLICATION IN DRUG DELIVERY

Biomaterials, cells, and growth factors are the three key components of tissue engineering. To achieve rapid tissue regeneration, biomaterials, in the form of tissue-specific scaffolds, should be utilized in combination with cells and/or growth factors to augment the regenerative potential of a tissue-engineered system. In such systems, growth factors play a pivotal role in tissue morphogenesis and regeneration. Direct injection of growth factors for tissue regeneration, although straightforward and may succeed to some extent, has significant drawbacks. The injected growth factors may quickly diffuse away from the injection site, and the short half-lives of growth factors require repeated and large concentrations to achieve a therapeutic effect [209]. In addition, the unlocalized delivery and distribution of these factors may lead to undesirable systemic effect and toxicity [210]. Therefore, controlled growth factor delivery from tissue engineering scaffolds has emerged as a more beneficial approach in tissue engineering and regenerative medicine, aiming to mimic the orchestrated endogenous growth factor production during natural tissue morphogenesis [211].

The natural polymer chitosan possesses the intrinsic advantages of being used as a delivery vehicle. Chitosan and its derivatives can be processed under mild conditions; therefore, growth factor inactivation due to otherwise harsh processing conditions can be avoided. In addition, chitosan can be fabricated into different forms and shapes to incorporate growth factors and other biomolecules for various *in vivo* applications.

As a straightforward approach, growth factors can be physically adsorbed on chitosan substrates. Chitosan films, porous sponges, or microgranules have been fabricated using a variety of techniques such as solvent evaporation, freeze-drying, and solution coagulation [212–214]. Growth factors were then loaded onto the scaffolds by soaking them in growth factor solutions. The factor-loaded bioactive scaffolds promoted cell growth and function, stimulating tissue regeneration as compared to chitosan scaffolds alone. Nevertheless, this type of factor-loading method generally results in an initial burst release followed by a short period of slow release for a few days, which is undesirable in many cases that require a long time-sustained factor release. Efforts were made to improve growth factor release profile by adding a negatively charged component into chitosan systems. Such an addition can potentially interact with the positively charged residues of growth factors and provide stability for the system and retard growth factor delivery [215]. Chitosan micro- or nanospheres were also developed as a carrier system for growth factor delivery [206,216,217]. The growth factors were more uniformly distributed within these particles, and the release was primarily controlled by diffusion. These growth factor-encapsulated micro- and nanosphere systems can be used directly for bone and cartilage tissue engineering and angiogenesis, or they can be loaded onto other forms of porous scaffolds for tissue regeneration.

Chitosan-based hydrogel carriers with specific growth factors show great potential in skin wound healing, angiogenesis, and vascularization. A number of chemically modified forms of chitosan were used to deliver bioactive agents to attain desired effects. Chitosan macromolecules with lactose moiety attached to its backbone can be prepared through the condensation reaction between carboxyl groups of lactose and amine groups of chitosan [24,205]. This injectable lactose-substituted chitosan hydrogel provides a noninvasive means to deliver growth factors. Several studies showed the feasibility of introducing photoreactive azide functionality to chitosan along with lactose moiety [25,171,186,218]. The modified azide-chitosan-lactose molecule can be photo-cross-linked under a brief UV irradiation to produce hydrogels. Such hydrogel systems are promising since growth factors can be dispersed into aqueous azide-chitosan-lactose solution, applied to an *in vivo* site, and readily cross-linked to gel to perform the functions. Additionally, chitosan-based hydrogels stabilize growth factors and preserve their biological activity. Growth factors then are released as the hydrogels degrade *in vivo*.

Physically entrapping and delivering growth factor from chitosan carriers has been successful only to a certain extent with several limitations. This is because sensitive growth factors are prone to undergo denaturation during the loading process, and their typically short half-lives in an *in vivo* milieu make them less effective. Thus, gene delivery has emerged as a promising technique to achieve localized, sustained expression of growth factors. Since plasmid DNA is more stable than recombinant growth factors, it seems to be a more favorable candidate to obtain sustained delivery systems. Chitosan in solution has a protonated structure and exhibits excellent affinity toward negatively charged plasmid DNA, thus favoring the formation of polyelectrolyte complex. These polyelectrolyte complexes formed by the condensation of DNA with chitosan are capable of delivering viable DNA to cells, which thereafter express the corresponding growth factors at the local site. Research has been performed to fabricate chitosan gene delivery matrices for a variety of tissue engineering applications and a detailed account of such systems is reviewed elsewhere [219].

5.7 CONCLUSIONS

Over the past several decades, chitosan has emerged as a promising biomaterial for biomedical applications due to its natural origin and structural similarity to glycosaminoglycans naturally presented in native tissues. In addition, the biocompatibility, biodegradability, antimicrobial activity, and functionalizability of chitosan make it an attractive biomaterial for tissue engineering and drug delivery. Moreover, the flexibility of the processing conditions of chitosan aids in the fabrication of versatile substrates as scaffolds for tissue regeneration or carriers for biological molecules. In order for chitosan to become successful in medical therapies, a few areas should be of particular importance in the future. First, it is critical to synthesize medical grade chitosan materials with controllable structure and properties. This will allow for the development of chitosan-based medical devices with controllable and predictable performance. Second, it is beneficial to chemically design chitosan derivatives with molecular and biological specificity through bulk material modification. Third, as tissue engineering evolves, it becomes imperative to combine advanced chitosan materials with modern micro-/nanofabrication techniques, stems cells, and developmental biology toward regenerating functional tissues. Despite all the challenges, chitosan holds great promise as a biomaterial for developing medical products and medical therapies.

ACKNOWLEDGMENTS

The authors gratefully acknowledge funding from the Raymond and Beverly Sackler Center for Biomedical, Biological, Physical and Engineering Sciences. Authors also acknowledge the funding from National Science Foundation (IIP-1311907 and EFRI-1332329).

REFERENCES

[1] C. Rouget, Des substances amylacées dans les tissus des animaux, spécialement des Articulés (chitine), C. R. Hebd. Séances Acad. Sci. 48 (1859) 792–795.

[2] G. Crini, P.-M. Badot, Application of chitosan, a natural aminopolysaccharide, for dye removal from aqueous solutions by adsorption processes using batch studies: a review of recent literature, Prog. Polym. Sci. 33 (2008) 399–447.

[3] W. Wan Ngah, L. Teong, M. Hanafiah, Adsorption of dyes and heavy metal ions by chitosan composites: a review, Carbohydr. Polym. 83 (2011) 1446–1456.

[4] G. Vanlerberghe, Chitosan derivative, method of making the same and cosmetic composition containing the same, US Patent 3,879,376, 1975.

[5] E.J. Ivani, Amino-polysaccharides and copolymers thereof for contact lenses and ophthalmic compositions, US Patent 4,447,562, 1984.

[6] H. Ueno, H. Yamada, I. Tanaka, N. Kaba, M. Matsuura, M. Okumura, et al., Accelerating effects of chitosan for healing at early phase of experimental open wound in dogs, Biomaterials 20 (1999) 1407–1414.

[7] H. Ueno, T. Mori, T. Fujinaga, Topical formulations and wound healing applications of chitosan, Adv. Drug Deliv. Rev. 52 (2001) 105–115.

[8] T.H. Fischer, R. Connolly, H.S. Thatte, S.S. Schwaitzberg, Comparison of structural and hemostatic properties of the poly-N-acetyl glucosamine Syvek Patch with products containing chitosan, Microsc. Res. Technol. 63 (2004) 168–174.

[9] R.V. Langer, P. Joseph, Tissue engineering, Science 260 (1993) 920–926.

[10] E. Khor, L.Y. Lim, Implantable applications of chitin and chitosan, Biomaterials 24 (2003) 2339–2349.

[11] S.A. Agnihotri, N.N. Mallikarjuna, T.M. Aminabhavi, Recent advances on chitosan-based micro-and nanoparticles in drug delivery, J. Control. Release 100 (2004) 5–28.

[12] I.A. Alsarra, S.S. Betigeri, H. Zhang, B.A. Evans, S.H. Neau, Molecular weight and degree of deacetylation effects on lipase-loaded chitosan bead characteristics, Biomaterials 23 (2002) 3637–3644.

[13] K.L.B. Chang, G. Tsai, J. Lee, W.-R. Fu, Heterogeneous N-deacetylation of chitin in alkaline solution, Carbohydr. Res. 303 (1997) 327–332.

[14] W.I. Abdel-Fattah, T. Jiang, G.E.-T. El-Bassyouni, C.T. Laurencin, Synthesis, characterization of chitosans and fabrication of sintered chitosan microsphere matrices for bone tissue engineering, Acta Biomater. 3 (2007) 503–514.

[15] H.K. No, Y.I. Cho, H.R. Kim, S.P. Meyers, Effective deacetylation of chitin under conditions of 15 psi/121 C, J. Agric. Food Chem. 48 (2000) 2625–2627.

[16] Z. Defang, Y. Gang, Z. Pengyi, F. Zhiwei, The modified process for preparing natural organic polymer flocculant chitosan, Chin. J. Environ. Sci. 22 (2001) 123–125.

[17] M. Zhang, A. Haga, H. Sekiguchi, S. Hirano, Structure of insect chitin isolated from beetle larva cuticle and silkworm (Bombyx mori) pupa exuvia, Int. J. Biol. Macromol. 27 (2000) 99–105.

[18] S. Mima, M. Miya, R. Iwamoto, S. Yoshikawa, Highly deacetylated chitosan and its properties, J. Appl. Polym. Sci. 28 (1983) 1909–1917.

[19] A. Domard, M. Rinaudo, Preparation and characterization of fully deacetylated chitosan, Int. J. Biol. Macromol. 5 (1983) 49–52.

[20] K. Kurita, T. Sannan, Y. Iwakura, Studies on chitin, 4. Evidence for formation of block and random copolymers of N-acetyl-D-glucosamine and D-glucosamine by hetero-and homogeneous hydrolyses, Die Makromolek. Chem. 178 (1977) 3197–3202.

[21] S.-i. Aiba, Studies on chitosan: 3. Evidence for the presence of random and block copolymer structures in partially N-acetylated chitosans, Int. J. Biol. Macromol. 13 (1991) 40–44.

[22] I. Aranaz, R. Harris, A. Heras, Chitosan amphiphilic derivatives. Chemistry and applications, Curr. Org. Chem. 14 (2010) 308.

[23] M. Kumar, R.A. Muzzarelli, C. Muzzarelli, H. Sashiwa, A. Domb, Chitosan chemistry and pharmaceutical perspectives, Chem. Rev. 104 (2004) 6017–6084.

[24] M. Fujita, M. Ishihara, M. Simizu, K. Obara, T. Ishizuka, Y. Saito, et al., Vascularization in vivo caused by the controlled release of fibroblast growth factor-2 from an injectable chitosan/non-anticoagulant heparin hydrogel, Biomaterials 25 (2004) 699–706.

[25] M. Ishihara, K. Obara, T. Ishizuka, M. Fujita, M. Sato, K. Masuoka, et al., Controlled release of fibroblast growth factors and heparin from photocrosslinked chitosan hydrogels and subsequent effect on in vivo vascularization, J. Biomed. Mater. Res. A 64 (2003) 551–559.

[26] M.-H. Ho, D.-M. Wang, H.-J. Hsieh, H.-C. Liu, T.-Y. Hsien, J.-Y. Lai, et al., Preparation and characterization of RGD-immobilized chitosan scaffolds, Biomaterials 26 (2005) 3197–3206.

[27] T. Masuko, N. Iwasaki, S. Yamane, T. Funakoshi, T. Majima, A. Minami, et al., Chitosan-RGDSGGC conjugate as a scaffold material for musculoskeletal tissue engineering, Biomaterials 26 (2005) 5339–5347.

[28] S. Itoh, A. Matsuda, H. Kobayashi, S. Ichinose, K. Shinomiya, J. Tanaka, Effects of a laminin peptide (YIGSR) immobilized on crab-tendon chitosan tubes on nerve regeneration, J. Biomed. Mater. Res. B Appl. Biomater. 73 (2005) 375–382.

[29] A. Matsuda, H. Kobayashi, S. Itoh, K. Kataoka, J. Tanaka, Immobilization of laminin peptide in molecularly aligned chitosan by covalent bonding, Biomaterials 26 (2005) 2273–2279.

[30] J. Li, H. Yun, Y. Gong, N. Zhao, X. Zhang, Investigation of MC3T3-E1 cell behavior on the surface of GRGDS-coupled chitosan, Biomacromolecules 7 (2006) 1112–1123.

[31] Y.J. Park, K.H. Kim, J.Y. Lee, Y. Ku, S.J. Lee, B.M. Min, et al., Immobilization of bone morphogenetic protein-2 on a nanofibrous chitosan membrane for enhanced guided bone regeneration, Biotechnol. Appl. Biochem. 43 (2006) 17–24.

[32] M. Suzuki, S. Itoh, I. Yamaguchi, K. Takakuda, H. Kobayashi, K. Shinomiya, et al., Tendon chitosan tubes covalently coupled with synthesized laminin peptides facilitate nerve regeneration in vivo, J. Neurosci. Res. 72 (2003) 646–659.

[33] T.W. Chung, Y.F. Lu, H.Y. Wang, W.P. Chen, S.S. Wang, Y.S. Lin, et al., Growth of human endothelial cells on different concentrations of Gly-Arg-Gly-Asp grafted chitosan surface, Artif. Organs 27 (2003) 155–161.

[34] T.-W. Chung, Y.-F. Lu, S.-S. Wang, Y.-S. Lin, S.-H. Chu, Growth of human endothelial cells on photochemically grafted Gly–Arg–Gly–Asp (GRGD) chitosans, Biomaterials 23 (2002) 4803–4809.

[35] C.T. Laurencin, T. Jiang, S.G. Kumbar, L.S. Nair, Biologically active chitosan systems for tissue engineering and regenerative medicine, Curr. Top. Med. Chem. 8 (2008) 354–364.

[36] Z. Dos Santos, A. Caroni, M. Pereira, D. da Silva, J. Fonseca, Determination of deacetylation degree of chitosan: a comparison between conductometric titration and CHN elemental analysis, Carbohydr. Res. 344 (2009) 2591–2595.

[37] X. Jiang, L. Chen, W. Zhong, A new linear potentiometric titration method for the determination of deacetylation degree of chitosan, Carbohydr. Polym. 54 (2003) 457–463.

[38] F. Nanjo, R. Katsumi, K. Sakai, Enzymatic method for determination of the degree of deacetylation of chitosan, Anal. Biochem. 193 (1991) 164–167.

[39] F. Niola, N. Basora, E. Chornet, P.F. Vidal, A rapid method for the determination of the degree of N-acetylation of chitin-chitosan samples by acid hydrolysis and HPLC, Carbohydr. Res. 238 (1993) 1–9.

[40] H. Sato, S.-I. Mizutani, S. Tsuge, H. Ohtani, K. Aoi, A. Takasu, et al., Determination of the degree of acetylation of chitin/chitosan by pyrolysis-gas chromatography in the presence of oxalic acid, Anal. Chem. 70 (1998) 7–12.

[41] M.R. Kasaai, Determination of the degree of N-acetylation for chitin and chitosan by various NMR spectroscopy techniques: a review, Carbohydr. Polym. 79 (2010) 801–810.

[42] M. Lavertu, Z. Xia, A. Serreqi, M. Berrada, A. Rodrigues, D. Wang, et al., A validated ^{1}H NMR method for the determination of the degree of deacetylation of chitosan, J. Pharm. Biomed. Anal. 32 (2003) 1149–1158.

[43] L. Heux, J. Brugnerotto, J. Desbrieres, M.-F. Versali, M. Rinaudo, Solid state NMR for determination of degree of acetylation of chitin and chitosan, Biomacromolecules 1 (2000) 746–751.

[44] M.R. Kasaai, A review of several reported procedures to determine the degree of N-acetylation for chitin and chitosan using infrared spectroscopy, Carbohydr. Polym. 71 (2008) 497–508.

[45] S. Sabnis, L.H. Block, Improved infrared spectroscopic method for the analysis of degree of N-deacetylation of chitosan, Polym. Bull. 39 (1997) 67–71.

[46] S.C. Tan, E. Khor, T.K. Tan, S.M. Wong, The degree of deacetylation of chitosan: advocating the first derivative UV-spectrophotometry method of determination, Talanta 45 (1998) 713–719.

[47] R.A. Muzzarelli, R. Rocchetti, Determination of the degree of acetylation of chitosans by first derivative ultraviolet spectrophotometry, Carbohydr. Polym. 5 (1985) 461–472.

[48] Y. Zhang, C. Xue, Y. Xue, R. Gao, X. Zhang, Determination of the degree of deacetylation of chitin and chitosan by X-ray powder diffraction, Carbohydr. Res. 340 (2005) 1914–1917.

[49] M.R. Kasaai, Various methods for determination of the degree of N-acetylation of chitin and chitosan: a review, J. Agric. Food Chem. 57 (2009) 1667–1676.

[50] Y. Shigemasa, H. Matsuura, H. Sashiwa, H. Saimoto, Evaluation of different absorbance ratios from infrared spectroscopy for analyzing the degree of deacetylation in chitin, Int. J. Biol. Macromol. 18 (1996) 237–242.

[51] M. Miya, R. Iwamoto, S. Yoshikawa, S. Mima, IR spectroscopic determination of CONH content in highly deacylated chitosan, Int. J. Biol. Macromol. 2 (1980) 323–324.

[52] A. Domard, Determination of N-acetyl content in chitosan samples by cd measurements, Int. J. Biol. Macromol. 9 (1987) 333–336.

[53] A. Baxter, M. Dillon, K. Anthony Taylor, G.A. Roberts, Improved method for ir determination of the degree of N-acetylation of chitosan, Int. J. Biol. Macromol. 14 (1992) 166–169.

[54] R.A. Muzzarelli, C. Lough, M. Emanuelli, The molecular weight of chitosans studied by laser light-scattering, Carbohydr. Res. 164 (1987) 433–442.

[55] M. Terbojevich, A. Cosani, B. Focher, E. Marsano, High-performance gel-permeation chromatography of chitosan samples, Carbohydr. Res. 250 (1993) 301–314.

[56] R.H. Chen, J.R. Chang, J.S. Shyur, Effects of ultrasonic conditions and storage in acidic solutions on changes in molecular weight and polydispersity of treated chitosan, Carbohydr. Res. 299 (1997) 287–294.

[57] R.G. Beri, J. Walker, E.T. Reese, J.E. Rollings, Characterization of chitosans via coupled size-exclusion chromatography and multiple-angle laser light-scattering technique, Carbohydr. Res. 238 (1993) 11–26.

[58] G.A. Roberts, J.G. Domszy, Determination of the viscometric constants for chitosan, Int. J. Biol. Macromol. 4 (1982) 374–377.

[59] V.F. Lee, Solution and shear properties of chitin and chitosan. Ph.D. Dissertation, University of Washington, University Microfilms, Ann Arbor, USA, 1974.

[60] M. Rinaudo, M. Milas, P.L. Dung, Characterization of chitosan. Influence of ionic strength and degree of acetylation on chain expansion, Int. J. Biol. Macromol. 15 (1993) 281–285.

[61] J. Brugnerotto, J. Desbrières, G. Roberts, M. Rinaudo, Characterization of chitosan by steric exclusion chromatography, Polymer 42 (2001) 09921–09927.

[62] G. Berth, H. Dautzenberg, The degree of acetylation of chitosans and its effect on the chain conformation in aqueous solution, Carbohydr. Polym. 47 (2002) 39–51.

[63] A. Gamzazade, V. Šlimak, A. Skljar, E. Štykova, S.S. Pavlova, S. Rogožin, Investigation of the hydrodynamic properties of chitosan solutions, Acta Polym. 36 (1985) 420–424.

[64] W. Wang, S. Bo, S. Li, W. Qin, Determination of the Mark-Houwink equation for chitosans with different degrees of deacetylation, Int. J. Biol. Macromol. 13 (1991) 281–285.
[65] I.A. Sogias, V.V. Khutoryanskiy, A.C. Williams, Exploring the factors affecting the solubility of chitosan in water, Macromol. Chem. Phys. 211 (2010) 426–433.
[66] K. Kurita, M. Kamiya, S.-I. Nishimura, Solubilization of a rigid polysaccharide: controlled partial N-acetylation of chitosan to develop solubility, Carbohydr. Polym. 16 (1991) 83–92.
[67] N. Kubota, Y. Eguchi, Facile preparation of water-soluble N-acetylated chitosan and molecular weight dependence of its water-solubility, Polym. J. 29 (1997) 123–127.
[68] H. Sashiwa, S.-I. Aiba, Chemically modified chitin and chitosan as biomaterials, Prog. Polym. Sci. 29 (2004) 887–908.
[69] D.-K. Kweon, S.-B. Song, Y.-Y. Park, Preparation of water-soluble chitosan/heparin complex and its application as wound healing accelerator, Biomaterials 24 (2003) 1595–1601.
[70] Y. Zhou, D. Yang, X. Chen, Q. Xu, F. Lu, J. Nie, Electrospun water-soluble carboxyethyl chitosan/poly (vinyl alcohol) nanofibrous membrane as potential wound dressing for skin regeneration, Biomacromolecules 9 (2007) 349–354.
[71] M. Prabaharan, Review paper: chitosan derivatives as promising materials for controlled drug delivery, J. Biomater. Appl. 23 (2008) 5–36.
[72] N. Bhattarai, J. Gunn, M. Zhang, Chitosan-based hydrogels for controlled, localized drug delivery, Adv. Drug Deliv. Rev. 62 (2010) 83–99.
[73] H.-l. Zhang, S.-H. Wu, Y. Tao, L.-Q. Zang, Z.-Q. Su, Preparation and characterization of water-soluble chitosan nanoparticles as protein delivery system, J. Nanomater. 2010 (2010) 1.
[74] J.H. Cho, S.-H. Kim, K.D. Park, M.C. Jung, W.I. Yang, S.W. Han, et al., Chondrogenic differentiation of human mesenchymal stem cells using a thermosensitive poly (N-isopropylacrylamide) and water-soluble chitosan copolymer, Biomaterials 25 (2004) 5743–5751.
[75] Y. Zhou, G. Ma, S. Shi, D. Yang, J. Nie, Photopolymerized water-soluble chitosan-based hydrogel as potential use in tissue engineering, Int. J. Biol. Macromol. 48 (2011) 408–413.
[76] I. Aranaz, M. Mengíbar, R. Harris, I. Paños, B. Miralles, N. Acosta, et al., Functional characterization of chitin and chitosan, Curr. Chem. Biol. 3 (2009) 203–230.
[77] T. Freier, H.S. Koh, K. Kazazian, M.S. Shoichet, Controlling cell adhesion and degradation of chitosan films by N-acetylation, Biomaterials 26 (2005) 5872–5878.
[78] T. Sannan, K. Kurita, Y. Iwakura, Studies on chitin, 2. Effect of deacetylation on solubility, Die Makromolek. Chem. 177 (1976) 3589–3600.
[79] K.T. Hwang, J.T. Kim, S.T. Jung, G.S. Cho, H.J. Park, Properties of chitosan-based biopolymer films with various degrees of deacetylation and molecular weights, J. Appl. Polym. Sci. 89 (2003) 3476–3484.
[80] C. Wenling, J. Duohui, L. Jiamou, G. Yandao, Z. Nanming, Z. Xiufang, Effects of the degree of deacetylation on the physicochemical properties and Schwann cell affinity of chitosan films, J. Biomater. Appl. 20 (2005) 157–177.
[81] C. Chatelet, O. Damour, A. Domard, Influence of the degree of acetylation on some biological properties of chitosan films, Biomaterials 22 (2001) 261–268.
[82] I. Amaral, A. Cordeiro, P. Sampaio, M. Barbosa, Attachment, spreading and short-term proliferation of human osteoblastic cells cultured on chitosan films with different degrees of acetylation, J. Biomater. Sci. Polym. Ed. 18 (2007) 469–485.
[83] Y. Xu, Y. Du, Effect of molecular structure of chitosan on protein delivery properties of chitosan nanoparticles, Int. J. Pharm. 250 (2003) 215–226.
[84] I. Amaral, M. Lamghari, S. Sousa, P. Sampaio, M. Barbosa, Rat bone marrow stromal cell osteogenic differentiation and fibronectin adsorption on chitosan membranes: the effect of the degree of acetylation, J. Biomed. Mater. Res. A 75 (2005) 387–397.
[85] J.D. Funkhouser, N.N. Aronson, Chitinase family GH18: evolutionary insights from the genomic history of a diverse protein family, BMC Evol. Biol. 7 (2007) 96.
[86] H. Zhang, S.H. Neau, In vitro degradation of chitosan by a commercial enzyme preparation: effect of molecular weight and degree of deacetylation, Biomaterials 22 (2001) 1653–1658.
[87] S.B. Rao, C.P. Sharma, Use of chitosan as a biomaterial: studies on its safety and hemostatic potential, J. Biomed. Mater. Res. 34 (1997) 21–28.
[88] K. Kurita, Y. Kaji, T. Mori, Y. Nishiyama, Enzymatic degradation of β-chitin: susceptibility and the influence of deacetylation, Carbohydr. Polym. 42 (2000) 19–21.
[89] H. Onishi, Y. Machida, Biodegradation and distribution of water-soluble chitosan in mice, Biomaterials 20 (1999) 175–182.
[90] L. Zeng, C. Qin, W. Wang, W. Chi, W. Li, Absorption and distribution of chitosan in mice after oral administration, Carbohydr. Polym. 71 (2008) 435–440.
[91] K. Tomihata, Y. Ikada, In vitro and in vivo degradation of films of chitin and its deacetylated derivatives, Biomaterials 18 (1997) 567–575.
[92] Y. Yang, W. Hu, X. Wang, X. Gu, The controlling biodegradation of chitosan fibers by N-acetylation in vitro and in vivo, J. Mater. Sci. Mater. Med. 18 (2007) 2117–2121.
[93] D.F. Williams, Definitions in biomaterials, in: Proceedings of a Consensus Conference of the European Society for Biomaterials, Chester, England, March 3-5, 1986, Elsevier, New York, 1987.
[94] D.F. Williams, On the mechanisms of biocompatibility, Biomaterials 29 (2008) 2941–2953.
[95] G.I. Howling, P.W. Dettmar, P.A. Goddard, F.C. Hampson, M. Dornish, E.J. Wood, The effect of chitin and chitosan on the proliferation of human skin fibroblasts and keratinocytes in vitro, Biomaterials 22 (2001) 2959–2966.
[96] F. Chellat, M. Tabrizian, S. Dumitriu, E. Chornet, P. Magny, C.H. Rivard, et al., In vitro and in vivo biocompatibility of chitosan-xanthan polyionic complex, J. Biomed. Mater. Res. 51 (2000) 107–116.
[97] A. Fakhry, G.B. Schneider, R. Zaharias, S. Şenel, Chitosan supports the initial attachment and spreading of osteoblasts preferentially over fibroblasts, Biomaterials 25 (2004) 2075–2079.
[98] A. Lahiji, A. Sohrabi, D.S. Hungerford, C.G. Frondoza, Chitosan supports the expression of extracellular matrix proteins in human osteoblasts and chondrocytes, J. Biomed. Mater. Res. 51 (2000) 586–595.
[99] Y. Yuan, P. Zhang, Y. Yang, X. Wang, X. Gu, The interaction of Schwann cells with chitosan membranes and fibers in vitro, Biomaterials 25 (2004) 4273–4278.
[100] J. Li, J. Pan, L. Zhang, X. Guo, Y. Yu, Culture of primary rat hepatocytes within porous chitosan scaffolds, J. Biomed. Mater. Res. A 67 (2003) 938–943.

[101] T.W. Chung, J. Yang, T. Akaike, K.Y. Cho, J.W. Nah, S.I. Kim, et al., Preparation of alginate/galactosylated chitosan scaffold for hepatocyte attachment, Biomaterials 23 (2002) 2827–2834.
[102] P.J. VandeVord, H.W. Matthew, S.P. DeSilva, L. Mayton, B. Wu, P.H. Wooley, Evaluation of the biocompatibility of a chitosan scaffold in mice, J. Biomed. Mater. Res. 59 (2002) 585–590.
[103] A. Wang, Q. Ao, Y. Wei, K. Gong, X. Liu, N. Zhao, et al., Physical properties and biocompatibility of a porous chitosan-based fiber-reinforced conduit for nerve regeneration, Biotechnol. Lett. 29 (2007) 1697–1702.
[104] J.D. Bumgardner, B.M. Chesnutt, Y. Yuan, Y. Yang, M. Appleford, S. Oh, et al., The integration of chitosan-coated titanium in bone: an in vivo study in rabbits, Implant Dent. 16 (2007) 66–79.
[105] M. Prasitsilp, R. Jenwithisuk, K. Kongsuwan, N. Damrongchai, P. Watts, Cellular responses to chitosan in vitro: the importance of deacetylation, J. Mater. Sci. Mater. Med. 11 (2000) 773–778.
[106] G. Molinaro, J.-C. Leroux, J. Damas, A. Adam, Biocompatibility of thermosensitive chitosan-based hydrogels: an in vivo experimental approach to injectable biomaterials, Biomaterials 23 (2002) 2717–2722.
[107] M. Ignatova, N. Manolova, I. Rashkov, Novel antibacterial fibers of quaternized chitosan and poly (vinyl pyrrolidone) prepared by electrospinning, Eur. Polym. J. 43 (2007) 1112–1122.
[108] M. Kong, X.G. Chen, K. Xing, H.J. Park, Antimicrobial properties of chitosan and mode of action: a state of the art review, Int. J. Food Microbiol. 144 (2010) 51–63.
[109] C.-S. Chen, W.-Y. Liau, G.-J. Tsai, Antibacterial effects of N-sulfonated and N-sulfobenzoyl chitosan and application to oyster preservation, J. Food Prot. 61 (1998) 1124–1128.
[110] B.O. Jung, C.H. Kim, K.S. Choi, Y.M. Lee, J.J. Kim, Preparation of amphiphilic chitosan and their antimicrobial activities, J. Appl. Polym. Sci. 72 (1999) 1713–1719.
[111] P.-J. Park, J.-Y. Je, H.-G. Byun, S.-H. Moon, S.-K. Kim, Antimicrobial activity of hetero-chitosans and their oligosaccharides with different molecular weights, J. Microbiol. Biotechnol. 14 (2004) 317–323.
[112] Y. Hu, Y. Du, J. Yang, Y. Tang, J. Li, X. Wang, Self-aggregation and antibacterial activity of N-acylated chitosan, Polymer 48 (2007) 3098–3106.
[113] R. Cuero, G. Osuji, A. Washington, N-Carboxymethylchitosan inhibition of aflatoxin production: role of zinc, Biotechnol. Lett. 13 (1991) 441–444.
[114] H.K. No, N. Young Park, S. Ho Lee, S.P. Meyers, Antibacterial activity of chitosans and chitosan oligomers with different molecular weights, Int. J. Food Microbiol. 74 (2002) 65–72.
[115] Y.-C. Chung, Y.-P. Su, C.-C. Chen, G. Jia, H.-l. Wang, J.G. Wu, et al., Relationship between antibacterial activity of chitosan and surface characteristics of cell wall, Acta Pharmacol. Sin. 25 (2004) 932–936.
[116] C.T. Laurencin, A. Ambrosio, M. Borden, J. Cooper Jr., Tissue engineering: orthopedic applications, Annu. Rev. Biomed. Eng. 1 (1999) 19–46.
[117] X.H. Liu, P.X. Ma, Polymeric scaffolds for bone tissue engineering, Ann. Biomed. Eng. 32 (2004) 477–486.
[118] G.E. Muschler, C. Nakamoto, L.G. Griffith, Engineering principles of clinical cell-based tissue engineering, J. Bone Joint Surg. Am. 86A (2004) 1541–1558.
[119] M. Spector, Biomaterials-based tissue engineering and regenerative medicine solutions to musculoskeletal problems, Swiss Med. Wkly. 136 (2006) 293–301.
[120] A.G. Mikos, A.J. Thorsen, L.A. Czerwonka, Y. Bao, R. Langer, D.N. Winslow, et al., Preparation and characterization of poly(l-lactic acid) foams, Polymer 35 (1994) 1068–1077.
[121] A.G. Mikos, G. Sarakinos, S.M. Leite, J.P. Vacant, R. Langer, Laminated three-dimensional biodegradable foams for use in tissue engineering, Biomaterials 14 (1993) 323–330.
[122] J. Lim, Y. Lee, J. Shin, K. Lim, Preparation of interconnected porous chitosan scaffolds by sodium acetate particulate leaching, J. Biomater. Sci. Polym. Ed. 22 (2011) 1319–1329.
[123] C. Schugens, V. Maquet, C. Grandfils, R. Jerome, P. Teyssie, Polylactide macroporous biodegradable implants for cell transplantation. II. Preparation of polylactide foams by liquid-liquid phase separation, J. Biomed. Mater. Res. 30 (1996) 449–461.
[124] R. Zhang, P.X. Ma, Poly(α-hydroxyl acids)/hydroxyapatite porous composites for bone-tissue engineering. I. Preparation and morphology, J. Biomed. Mater. Res. 44 (1999) 446–455.
[125] H.-W. Kang, Y. Tabata, Y. Ikada, Fabrication of porous gelatin scaffolds for tissue engineering, Biomaterials 20 (1999) 1339–1344.
[126] F.J. O'Brien, B.A. Harley, I.V. Yannas, L. Gibson, Influence of freezing rate on pore structure in freeze-dried collagen-GAG scaffolds, Biomaterials 25 (2004) 1077–1086.
[127] A. Hottot, S. Vessot, J. Andrieu, A direct characterization method of the ice morphology. Relationship between mean crystals size and primary drying times of freeze-drying processes, Dry Technol. 22 (2004) 2009–2021.
[128] A.I. Liapis, M.L. Pim, R. Bruttini, Research and development needs and opportunities in freeze drying, Dry Technol. 14 (1996) 1265–1300.
[129] S.V. Madihally, H.W. Matthew, Porous chitosan scaffolds for tissue engineering, Biomaterials 20 (1999) 1133–1142.
[130] Y. Zhang, M. Zhang, Synthesis and characterization of macroporous chitosan/calcium phosphate composite scaffolds for tissue engineering, J. Biomed. Mater. Res. 55 (2001) 304–312.
[131] F. Zhao, Y. Yin, W.W. Lu, J.C. Leong, W. Zhang, J. Zhang, et al., Preparation and histological evaluation of biomimetic three-dimensional hydroxyapatite/chitosan-gelatin network composite scaffolds, Biomaterials 23 (2002) 3227–3234.
[132] C. Ji, N. Annabi, A. Khademhosseini, F. Dehghani, Fabrication of porous chitosan scaffolds for soft tissue engineering using dense gas CO_2, Acta Biomater. 7 (2011) 1653–1664.
[133] M. Borden, M. Attawia, Y. Khan, C.T. Laurencin, Tissue engineered microsphere-based matrices for bone repair: design and evaluation, Biomaterials 23 (2002) 551–559.
[134] M. Borden, M. Attawia, Y. Khan, S. El-Amin, C. Laurencin, Tissue-engineered bone formation in vivo using a novel sintered polymeric microsphere matrix, J. Bone Joint Surg. Br. 86 (2004) 1200–1208.
[135] Y.M. Khan, D.S. Katti, C.T. Laurencin, Novel polymer-synthesized ceramic composite-based system for bone repair: an in vitro evaluation, J. Biomed. Mater. Res. A 69 (2004) 728–737.
[136] M. Kofron, J. Cooper, S. Kumbar, C. Laurencin, Novel tubular composite matrix for bone repair, J. Biomed. Mater. Res. A 82 (2007) 415–425.
[137] S.P. Nukavarapu, S.G. Kumbar, J.L. Brown, N.R. Krogman, A.L. Weikel, M.D. Hindenlang, et al., Polyphosphazene/nano-hydroxyapatite composite microsphere scaffolds for bone tissue engineering, Biomacromolecules 9 (2008) 1818–1825.

[138] E. Jabbarzadeh, T. Starnes, Y.M. Khan, T. Jiang, A.J. Wirtel, M. Deng, et al., Induction of angiogenesis in tissue-engineered scaffolds designed for bone repair: a combined gene therapy-cell transplantation approach, Proc. Natl. Acad. Sci. U.S.A. 105 (2008) 11099–11104.

[139] J.L. Brown, L.S. Nair, C.T. Laurencin, Solvent/non-solvent sintering: a novel route to create porous microsphere scaffolds for tissue regeneration, J. Biomed. Mater. Res. B Appl. Biomater. 86 (2008) 396–406.

[140] T. Jiang, C. Pilane, C.T. Laurencin, Fabrication of novel porous chitosan matrices as scaffolds for bone tissue engineering, in: C.T. Laurencin, E.A. Botchwey (Eds.), Nanoscale Materials Science in Biology and Medicine, Materials Research Society, Warrendale, Pennsylvania, 2005, pp. 187–192.

[141] T. Jiang, W.I. Abdel-Fattah, C.T. Laurencin, In vitro evaluation of chitosan/poly (lactic acid-glycolic acid) sintered microsphere scaffolds for bone tissue engineering, Biomaterials 27 (2006) 4894–4903.

[142] T. Jiang, Y. Khan, L.S. Nair, W.I. Abdel-Fattah, C.T. Laurencin, Functionalization of chitosan/poly (lactic acid-glycolic acid) sintered microsphere scaffolds via surface heparinization for bone tissue engineering, J. Biomed. Mater. Res. A 93 (2010) 1193–1208.

[143] T. Jiang, S.P. Nukavarapu, M. Deng, E. Jabbarzadeh, M.D. Kofron, S.B. Doty, et al., Chitosan-poly (lactide-co-glycolide) microsphere-based scaffolds for bone tissue engineering: in vitro degradation and in vivo bone regeneration studies, Acta Biomater. 6 (2010) 3457–3470.

[144] M. Ishihara, K. Nakanishi, K. Ono, M. Sato, M. Kikuchi, Y. Saito, et al., Photocrosslinkable chitosan as a dressing for wound occlusion and accelerator in healing process, Biomaterials 23 (2002) 833–840.

[145] C. Hoemann, J. Sun, A. Legare, M. McKee, M. Buschmann, Tissue engineering of cartilage using an injectable and adhesive chitosan-based cell-delivery vehicle, Osteoarthr. Cartil. 13 (2005) 318–329.

[146] L.S. Nair, T. Starnes, J.-W.K. Ko, C.T. Laurencin, Development of injectable thermogelling chitosan-inorganic phosphate solutions for biomedical applications, Biomacromolecules 8 (2007) 3779–3785.

[147] K. Ohkawa, D. Cha, H. Kim, A. Nishida, H. Yamamoto, Electrospinning of chitosan, Macromol. Rapid Commun. 25 (2004) 1600–1605.

[148] S. Torres-Giner, M.J. Ocio, J.M. Lagaron, Development of active antimicrobial fiber-based chitosan polysaccharide nanostructures using electrospinning, Eng. Life Sci. 8 (2008) 303–314.

[149] M. Peesan, R. Rujiravanit, P. Supaphol, Electrospinning of hexanoyl chitosan/polylactide blends, J. Biomater. Sci. Polym. Ed. 17 (2006) 547–565.

[150] N. Bhattarai, Z. Li, J. Gunn, M. Leung, A. Cooper, D. Edmondson, et al., Natural-synthetic polyblend nanofibers for biomedical applications, Adv. Mater. 21 (2009) 2792–2797.

[151] F. Chen, X. Li, X. Mo, C. He, H. Wang, Y. Ikada, Electrospun chitosan-P(LLA-CL) nanofibers for biomimetic extracellular matrix, J. Biomater. Sci. Polym. Ed. 19 (2008) 677–691.

[152] L. Li, Y.-L. Hsieh, Chitosan bicomponent nanofibers and nanoporous fibers, Carbohydr. Res. 341 (2006) 374–381.

[153] Y. Zhou, D. Yang, J. Nie, Electrospinning of chitosan/poly(vinyl alcohol)/acrylic acid aqueous solutions, J. Appl. Polym. Sci. 102 (2006) 5692–5697.

[154] Y.-T. Jia, J. Gong, X.-H. Gu, H.-Y. Kim, J. Dong, X.-Y. Shen, Fabrication and characterization of poly (vinyl alcohol)/chitosan blend nanofibers produced by electrospinning method, Carbohydr. Polym. 67 (2007) 403–409.

[155] Y. Zhang, X. Huang, B. Duan, L. Wu, S. Li, X. Yuan, Preparation of electrospun chitosan/poly(vinyl alcohol) membranes, Colloid Polym. Sci. 285 (2007) 855–863.

[156] B. Duan, C. Dong, X. Yuan, K. Yao, Electrospinning of chitosan solutions in acetic acid with poly(ethylene oxide), J. Biomater. Sci. Polym. Ed. 15 (2004) 797–811.

[157] N. Bhattarai, D. Edmondson, O. Veiseh, F.A. Matsen, M. Zhang, Electrospun chitosan-based nanofibers and their cellular compatibility, Biomaterials 26 (2005) 6176–6184.

[158] K.-H. Jung, M.-W. Huh, W. Meng, J. Yuan, S.H. Hyun, J.-S. Bae, et al., Preparation and antibacterial activity of PET/chitosan nanofibrous mats using an electrospinning technique, J. Appl. Polym. Sci. 105 (2007) 2816–2823.

[159] Y. Zhang, J.R. Venugopal, A. El-Turki, S. Ramakrishna, B. Su, C.T. Lim, Electrospun biomimetic nanocomposite nanofibers of hydroxyapatite/chitosan for bone tissue engineering, Biomaterials 29 (2008) 4314–4322.

[160] D. Yang, Y. Jin, Y. Zhou, G. Ma, X. Chen, F. Lu, et al., In situ mineralization of hydroxyapatite on electrospun chitosan-based nanofibrous scaffolds, Macromol. Biosci. 8 (2008) 239–246.

[161] P.R. Klokkevold, L. Vandemark, E.B. Kenney, G.W. Bernard, Osteogenesis enhanced by chitosan (poly-N-acetyl glucosaminoglycan) in vitro, J. Periodontol. 67 (1996) 1170–1175.

[162] R. Muzzarelli, M. Mattioli-Belmonte, C. Tietz, R. Biagini, G. Ferioli, M. Brunelli, et al., Stimulatory effect on bone formation exerted by a modified chitosan, Biomaterials 15 (1994) 1075–1081.

[163] Y.-J. Seol, J.-Y. Lee, Y.-J. Park, Y.-M. Lee, Y. Ku, I.-C. Rhyu, et al., Chitosan sponges as tissue engineering scaffolds for bone formation, Biotechnol. Lett. 26 (2004) 1037–1041.

[164] Z. Li, H.R. Ramay, K.D. Hauch, D. Xiao, M. Zhang, Chitosan-alginate hybrid scaffolds for bone tissue engineering, Biomaterials 26 (2005) 3919–3928.

[165] Y. Zhang, M. Zhang, Three-dimensional macroporous calcium phosphate bioceramics with nested chitosan sponges for load-bearing bone implants, J. Biomed. Mater. Res. 61 (2002) 1–8.

[166] Y. Zhang, M. Ni, M. Zhang, B. Ratner, Calcium phosphate-chitosan composite scaffolds for bone tissue engineering, Tissue Eng. 9 (2003) 337–345.

[167] T. Kawakami, M. Antoh, H. Hasegawa, T. Yamagishi, M. Ito, S. Eda, Experimental study on osteoconductive properties of a chitosan-bonded hydroxyapatite self-hardening paste, Biomaterials 13 (1992) 759–763.

[168] S.-Y. Shin, H.-N. Park, K.-H. Kim, M.-H. Lee, Y.S. Choi, Y.-J. Park, et al., Biological evaluation of chitosan nanofiber membrane for guided bone regeneration, J. Periodontol. 76 (2005) 1778–1784.

[169] C.D. Hoemann, M. Hurtig, E. Rossomacha, J. Sun, A. Chevrier, M.S. Shive, et al., Chitosan-glycerol phosphate/blood implants improve hyaline cartilage repair in ovine microfracture defects, J. Bone Joint Surg. Am. 87 (2005) 2671–2686.

[170] C. Hoemann, J. Sun, A. Chevrier, E. Rossomacha, G.-E. Rivard, M. Hurtig, et al., Chitosan-glycerol phosphate/blood implants elicit hyaline cartilage repair integrated with porous subchondral bone in microdrilled rabbit defects, Osteoarthr. Cartil. 15 (2007) 78–89.

[171] K. Obara, M. Ishihara, M. Fujita, Y. Kanatani, H. Hattori, T. Matsui, et al., Acceleration of wound healing in healing-impaired db/db mice with a photocrosslinkable chitosan hydrogel containing fibroblast growth factor-2, Wound Repair Regen. 13 (2005) 390–397.

[172] W. Wang, S. Itoh, K. Konno, T. Kikkawa, S. Ichinose, K. Sakai, et al., Effects of Schwann cell alignment along the oriented electrospun chitosan nanofibers on nerve regeneration, J. Biomed. Mater. Res. A 91 (2009) 994–1005.
[173] C. Marchand, G. Chen, N. Tran-Khanh, J. Sun, H. Chen, M.D. Buschmann, et al., Microdrilled cartilage defects treated with thrombin-solidified chitosan/blood implant regenerate a more hyaline, stable, and structurally integrated osteochondral unit compared to drilled controls, Tissue Eng. Part A 18 (2011) 508–519.
[174] A. Chevrier, C. Hoemann, J. Sun, M. Buschmann, Chitosan-glycerol phosphate/blood implants increase cell recruitment, transient vascularization and subchondral bone remodeling in drilled cartilage defects, Osteoarthr. Cartil. 15 (2007) 316–327.
[175] C.D. Hoemann, G. Chen, C. Marchand, N. Tran-Khanh, M. Thibault, A. Chevrier, et al., Scaffold-guided subchondral bone repair implication of neutrophils and alternatively activated arginase-1+ macrophages, Am. J. Sports Med. 38 (2010) 1845–1856.
[176] T. Mori, M. Murakami, M. Okumura, T. Kadosawa, T. Uede, T. Fujinaga, Mechanism of macrophage activation by chitin derivatives, J. Vet. Med. Sci. 67 (2005) 51.
[177] G. Peluso, O. Petillo, M. Ranieri, M. Santin, L. Ambrosic, D. Calabró, et al., Chitosan-mediated stimulation of macrophage function, Biomaterials 15 (1994) 1215–1220.
[178] T. Kosaka, Y. Kaneko, Y. Nakada, M. Matsuura, S. Tanaka, Effect of chitosan implantation on activation of canine macrophages and polymorphonuclear cells after surgical stress, J. Vet. Med. Sci. 58 (1996) 963.
[179] J.E. Lee, K.E. Kim, I.C. Kwon, H.J. Ahn, S.-H. Lee, H. Cho, et al., Effects of the controlled-released TGF-[beta]1 from chitosan microspheres on chondrocytes cultured in a collagen/chitosan/glycosaminoglycan scaffold, Biomaterials 25 (2004) 4163–4173.
[180] M. Lee, R.K. Siu, K. Ting, B.M. Wu, Effect of Nell-1 delivery on chondrocyte proliferation and cartilaginous extracellular matrix deposition, Tissue Eng. Part A 16 (2009) 1791–1800.
[181] P.B. Malafaya, J.T. Oliveira, R.L. Reis, The effect of insulin-loaded chitosan particle-aggregated scaffolds in chondrogenic differentiation, Tissue Eng. A 16 (2009) 735–747.
[182] T. Guo, J. Zhao, J. Chang, Z. Ding, H. Hong, J. Chen, et al., Porous chitosan-gelatin scaffold containing plasmid DNA encoding transforming growth factor-[beta]1 for chondrocytes proliferation, Biomaterials 27 (2006) 1095–1103.
[183] W. Wang, B. Li, Y. Li, Y. Jiang, H. Ouyang, C. Gao, In vivo restoration of full-thickness cartilage defects by poly(lactide-co-glycolide) sponges filled with fibrin gel, bone marrow mesenchymal stem cells and DNA complexes, Biomaterials 31 (2010) 5953–5965.
[184] C.-C. Peng, M.-H. Yang, W.-T. Chiu, C.-H. Chiu, C.-S. Yang, Y.-W. Chen, et al., Composite nano-titanium oxide-chitosan artificial skin exhibits strong wound-healing effect—an approach with anti-inflammatory and bactericidal kinetics, Macromol. Biosci. 8 (2008) 316–327.
[185] F.-L. Mi, Y.-B. Wu, S.-S. Shyu, J.-Y. Schoung, Y.-B. Huang, Y.-H. Tsai, et al., Control of wound infections using a bilayer chitosan wound dressing with sustainable antibiotic delivery, J. Biomed. Mater. Res. 59 (2002) 438–449.
[186] K. Obara, M. Ishihara, T. Ishizuka, M. Fujita, Y. Ozeki, T. Maehara, et al., Photocrosslinkable chitosan hydrogel containing fibroblast growth factor-2 stimulates wound healing in healing-impaired db/db mice, Biomaterials 24 (2003) 3437–3444.
[187] L. Ma, C. Gao, Z. Mao, J. Zhou, J. Shen, X. Hu, et al., Collagen/chitosan porous scaffolds with improved biostability for skin tissue engineering, Biomaterials 24 (2003) 4833–4841.
[188] J. Mao, L. Zhao, K. de Yao, Q. Shang, G. Yang, Y. Cao, Study of novel chitosan-gelatin artificial skin in vitro, J. Biomed. Mater. Res. A 64A (2003) 301–308.
[189] F.-L. Mi, S.-S. Shyu, Y.-B. Wu, S.-T. Lee, J.-Y. Shyong, R.-N. Huang, Fabrication and characterization of a sponge-like asymmetric chitosan membrane as a wound dressing, Biomaterials 22 (2001) 165–173.
[190] C.-M. Han, L.-P. Zhang, J.-Z. Sun, H.-F. Shi, J. Zhou, C.-Y. Gao, Application of collagen-chitosan/fibrin glue asymmetric scaffolds in skin tissue engineering, J. Zhejiang Univ. Sci. B 11 (2010) 524–530.
[191] H. Shi, C. Han, Z. Mao, L. Ma, C. Gao, Enhanced angiogenesis in porous collagen-chitosan scaffolds loaded with angiogenin, Tissue Eng. A 14 (2008) 1775–1785.
[192] I. Adekogbe, A. Ghanem, Fabrication and characterization of DTBP-crosslinked chitosan scaffolds for skin tissue engineering, Biomaterials 26 (2005) 7241–7250.
[193] N. Boucard, C. Viton, D. Agay, E. Mari, T. Roger, Y. Chancerelle, et al., The use of physical hydrogels of chitosan for skin regeneration following third-degree burns, Biomaterials 28 (2007) 3478–3488.
[194] L. Wu, H. Li, S. Li, X. Li, X. Yuan, X. Li, et al., Composite fibrous membranes of PLGA and chitosan prepared by coelectrospinning and coaxial electrospinning, J. Biomed. Mater. Res. A 92A (2010) 563–574.
[195] A.E. Park, S.D. Boden, Form and function of the intervertebral disk, in: T.A. Einhorn, R.J.O. Keefe, J.A. Buckwalter (Eds.), Orthopaedic Basic Science, AAOS, Rosemont, IL, 2007.
[196] S.M. Richardson, N. Hughes, J.A. Hunt, A.J. Freemont, J.A. Hoyland, Human mesenchymal stem cell differentiation to NP-like cells in chitosan-glycerophosphate hydrogels, Biomaterials 29 (2008) 85–93.
[197] M. Alini, P. Roughley, J. Antoniou, T. Stoll, M. Aebi, A biological approach to treating disc degeneration: not for today, but maybe for tomorrow, Eur. Spine J. 11 (2002) S215–S220.
[198] J.M. Cloyd, N.R. Malhotra, L. Weng, W. Chen, R.L. Mauck, D.M. Elliott, Material properties in unconfined compression of human nucleus pulposus, injectable hyaluronic acid-based hydrogels and tissue engineering scaffolds, Eur. Spine J. 16 (2007) 1892–1898.
[199] P. Roughley, C. Hoemann, E. DesRosiers, F. Mwale, J. Antoniou, M. Alini, The potential of chitosan-based gels containing intervertebral disc cells for nucleus pulposus supplementation, Biomaterials 27 (2006) 388–396.
[200] T. Funakoshi, T. Majima, N. Iwasaki, S. Yamane, T. Masuko, A. Minami, et al., Novel chitosan-based hyaluronan hybrid polymer fibers as a scaffold in ligament tissue engineering, J. Biomed. Mater. Res. A 74A (2005) 338–346.
[201] J.W.S. Hayami, D.C. Surrao, S.D. Waldman, B.G. Amsden, Design and characterization of a biodegradable composite scaffold for ligament tissue engineering, J. Biomed. Mater. Res. A 92A (2010) 1407–1420.
[202] R. Pangestuti, S.-K. Kim, Neuroprotective properties of chitosan and its derivatives, Mar. Drugs 8 (2010) 2117–2128.
[203] Y. Cho, R. Shi, R.B. Borgens, Chitosan produces potent neuroprotection and physiological recovery following traumatic spinal cord injury, J. Exp. Biol. 213 (2010) 1513–1520.

[204] A. Wang, Q. Ao, W. Cao, M. Yu, Q. He, L. Kong, et al., Porous chitosan tubular scaffolds with knitted outer wall and controllable inner structure for nerve tissue engineering, J. Biomed. Mater. Res. A 79 (2006) 36–46.
[205] M. Fujita, M. Ishihara, M. Shimizu, K. Obara, S. Nakamura, Y. Kanatani, et al., Therapeutic angiogenesis induced by controlled release of fibroblast growth factor-2 from injectable chitosan/non-anticoagulant heparin hydrogel in a rat hindlimb ischemia model, Wound Repair Regen. 15 (2007) 58–65.
[206] M. Huang, S.N. Vitharana, L.J. Peek, T. Coop, C. Berkland, Polyelectrolyte complexes stabilize and controllably release vascular endothelial growth factor, Biomacromolecules 8 (2007) 1607–1614.
[207] E.K. Yim, Liao I-c, K.W. Leong, Tissue compatibility of interfacial polyelectrolyte complexation fibrous scaffold: evaluation of blood compatibility and biocompatibility, Tissue Eng. 13 (2007) 423–433.
[208] S. Nakamura, M. Nambu, T. Ishizuka, H. Hattori, Y. Kanatani, B. Takase, et al., Effect of controlled release of fibroblast growth factor-2 from chitosan/fucoidan micro complex-hydrogel on in vitro and in vivo vascularization, J. Biomed. Mater. Res. A 85A (2008) 619–627.
[209] Y. Tabata, Significance of release technology in tissue engineering, Drug Discov. Today 10 (2005) 1639–1646.
[210] R.R. Chen, D.J. Mooney, Polymeric growth factor delivery strategies for tissue engineering, Pharm. Res. 20 (2003) 1103–1112.
[211] M.J. Whitaker, R.A. Quirk, S.M. Howdle, K.M. Shakesheff, Growth factor release from tissue engineering scaffolds, J. Pharm. Pharmacol. 53 (2001) 1427–1437.
[212] J.L. Lopez-Lacomba, J.M. Garcia-Cantalejo, J.V.S. Casado, A. Abarrategi, V.C. Magana, V. Ramos, Use of rhBMP-2 activated chitosan films to improve osseointegration, Biomacromolecules 7 (2006) 792–798.
[213] Y.J. Park, Y.M. Lee, S.N. Park, S.Y. Sheen, C.P. Chung, S.J. Lee, Platelet derived growth factor releasing chitosan sponge for periodontal bone regeneration, Biomaterials 21 (2000) 153–159.
[214] J.Y. Lee, K.H. Kim, S.Y. Shin, I.C. Rhyu, Y.M. Lee, Y.J. Park, et al., Enhanced bone formation by transforming growth factor-beta 1-releasing collagen/chitosan microgranules, J. Biomed. Mater. Res. A 76A (2006) 530–539.
[215] Y.J. Park, Y.M. Lee, J.Y. Lee, Y.J. Seol, C.P. Chung, S.J. Lee, Controlled release of platelet-derived growth factor-BB from chondroitin sulfate-chitosan sponge for guided bone regeneration, J. Control. Release 67 (2000) 385–394.
[216] J.E. Lee, S.E. Kim, I.C. Kwon, H.J. Ahn, H. Cho, S.H. Lee, et al., Effects of a chitosan scaffold containing TGF-β1 encapsulated chitosan microspheres on in vitro chondrocyte culture, Artif. Organs 28 (2004) 829–839.
[217] B.C. Cho, J.Y. Kim, J.H. Lee, H.Y. Chung, J.W. Park, K.H. Roh, et al., The bone regenerative effect of chitosan microsphere-encapsulated growth hormone on bony consolidation in mandibular distraction osteogenesis in a dog model, J. Craniofac. Surg. 15 (2004) 299–311.
[218] K. Ono, Y. Saito, H. Yura, K. Ishikawa, A. Kurita, T. Akaike, et al., Photocrosslinkable chitosan as a biological adhesive, J. Biomed. Mater. Res. 49 (2000) 289–295.
[219] T. Jiang, L.S. Nair, C.T. Laurencin, Chitosan composites for tissue engineering: bone tissue engineering scaffolds, Asian Chitin J. 2 (2006) 1–10.

Chapter 6

Poly(α-ester)s

Karen Burg
Department of Bioengineering and Institute for Biological Interfaces of Engineering, Clemson University, Clemson, SC, USA

Chapter Outline

6.1 ADVANTAGES OF ABSORBABLE POLY(α-ESTER)S

The poly(α-ester)s include polylactides, polyglycolides, poly(lactide-*co*-glycolides), polycaprolactones, and bacterial and other recombinant polyesters. The modularity of absorbable poly(α-ester)s provides distinct advantages. In theory, assuming one appreciates the material response to different synthesis and processing conditions, such a material may be custom-designed with the appropriate features and specifications to meet the needs of a given application, that is, to allow an absorbable material to remain in the body long enough for its intended use, but not long enough to cause detrimental effects. The material should degrade into degradants that do not remain in the body but which can readily be removed through normal processes. It is largely appreciated that there are no stealth biomaterials that escape the body's detection system; it is perhaps not so obvious that some level of a foreign body response is likely positive. That is, healing requires a stimulus for specific processes to occur (e.g., vascularization) and, without these stimuli, normal healing will not occur. Absorbable materials can be designed to provide stimuli that evolve with implantation time, as the needs evolve. These materials should be easily processed into a variety of shapes and forms, should be easy to sterilize, and should have reasonable shelf life. Because these materials are degradable, the long-term storage requirements are not standard; poly(α-ester)s are hydrolytically degraded and, as such, are very sensitive to even normal, everyday humidity. Many factors affect their bulk material properties, including chemical composition, molecular weight, initiators, and surface topography. Purposeful or unintended changes in one or more of these factors will cause a change in material characteristics, such as mechanical properties, water uptake, and degradation, which, in turn, will cause a change in cellular reaction to the material.

The stages of degradation for an absorbable polymer, as defined by Kronenthal in 1975, include hydration, strength loss, loss of mass integrity (i.e., cleavage of molecular chains), and mass loss (i.e., dissolution) [1]. Poly(α-ester)s are degraded by hydrolytic attack of the polymeric backbone *in vivo*; they are not influenced by enzymes. The poly(α-ester)s absorb relatively low amounts of water in the hydration stage as compared to other degradable systems such as collagen; however, crystallinity and opacity changes can occur within hours given external influences of water at 37 °C. This can be seen in amorphous systems produced following a quench cycle [2]. Often underappreciated is the cycling behavior that an absorbable poly(α-ester) can display following implantation. That is, upon immediate implantation, the low-molecular-weight species migrate from the device, resulting in a decrease in polydispersity index (or breadth of molecular weight distribution). However, with progressive loss of mass integrity, low-molecular-weight chains are produced, increasing the polydispersity index, and then lost, decreasing the polydispersity index. As these processes occur, the weight-average molecular weight slowly oscillates through local maxima and minima, while on a gradual decrease overall. Histologically, one also sees the cycling effect, almost taking the form of a chronic, foreign body response as new degradants are released, resulting in a response that is extended. If compared with the response to nondegradable materials, the extended response to absorbable poly(α-ester)s can be misinterpreted as a problematic situation.

6.2 POLYLACTIDES, POLYGLYCOLIDES, AND COPOLYMERS THEREOF

6.2.1 Structure and Characteristics

Polylactides are perhaps the most well known polyesters as they are clinically available. These intriguing materials are hydrolytically degraded, yielding acid by-products. Lactide ring-opening polymerization was reported in the 1930s [3]; however, at the time, the instability of the material was not appreciated as a positive feature, and it was not until the 1960s that the utility of polylactide to biomedicine was realized and reported [4]. Lactic acid can be derived from renewable resource-derived intermediates such as acetaldehyde or ethanol, or from chemicals such as acetylene or ethylene (i.e., coal or oil). Lactic acid can also be produced by bacteria, fungi, or yeasts, in the fermentation processes affected by agitation, pH, temperature, and atmosphere. Lactic acid and glycolic acid can be condensed [5] to form polylactic acid or polyglycolic acid, low-molecular-weight products, or to allow a condensation reaction and ringed product. A ring-opening reaction can then be performed to yield high-molecular-weight polylactide (Figure 6.1) or polyglycolide (Figure 6.2).

Polylactides have a very wide range of properties. There are two optically active stereoisomers; lactides rotate light either to the right (D, i.e., dextrorotatory) or to the left (L, i.e., levorotatory). Polylactide is hydrophobic due to the –CH3 side groups. The lactide methyl group provides a shielding effect (of the susceptible ester group) and renders polylactide more hydrophobic than polyglycolide; therefore, polylactide degrades slower than polyglycolide. Lactide can be formed by combining a D- and an L-lactic acid molecule, resulting in D,L-lactide (meso-lactide). Commonly, lactide monomers are melted in the presence of an initiator, which starts polymerization. Generally, a tin, zinc, or aluminum catalyst is added to enhance the reaction. Poly(α-ester)s are frequently mistermed in the literature. Polylactide or polyglycolide is the material produced by the ring-opening mechanism. Polylactic acid or polyglycolic acid is produced by the step-growth polymerization reaction. The latter are lower-molecular-weight species, from which complex 3D structures cannot be fashioned, while polylactide or polyglycolide indicates higher-molecular-weight materials that have the mechanical integrity to allow more complex 3D products to be fashioned. Polycondensation, or step growth polymerization, is known to yield low-molecular-weight material; however, mention has been made in the literature to the possibility of yielding higher-molecular-weight product [6].

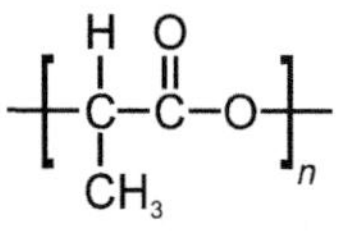

FIGURE 6.1 Polylactide structure. (For color version of this figure, the reader is referred to the online version of this chapter.)

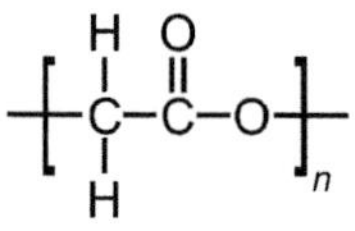

FIGURE 6.2 Polyglycolide structure. (For color version of this figure, the reader is referred to the online version of this chapter.)

Because of the less compact packing of enantiomers, poly (D,L-lactide) is amorphous, while poly (L-lactide) and polyglycolide are semicrystalline. Poly (L-lactide) is a brittle, transparent polymer with a melting point range of approximately 170-180 °C and a glass transition temperature of approximately 63 °C. Poly (D,L-lactide) is amorphous and therefore does not have a melting point; the glass transition temperature of poly (D,L-lactide) is approximately 55 °C. Polyglycolide is produced by ring-opening polymerization of glycolide. It has approximately 45-55% crystallinity and therefore is not soluble in many organic solvents. It has a high melting point (~225 °C) and a glass transition temperature of ~35 °C [7]. Polyglycolide degrades relatively quickly into acidic products. Accordingly, glycolide is often copolymerized with caprolactone, lactide, or trimethylene carbonate for biomedical application. Polycaprolactone has a melting temperature of approximately 60 °C and a very low (−60 °C) glass transition temperature and thus is flexible at room and body temperature. It has a "sticky" quality that can provide interesting benefits. Polyglycolide has no methyl groups and is highly crystalline, while polycaprolactone has many methyl groups, hence the fast degradation of polyglycolide in comparison (Figure 6.3). Degradation of polycaprolactone proceeds in two phases, beginning with hydrolytic degradation [8]. Polycaprolactone is considered semicrystalline and so has crystalline and amorphous domains. The crystalline

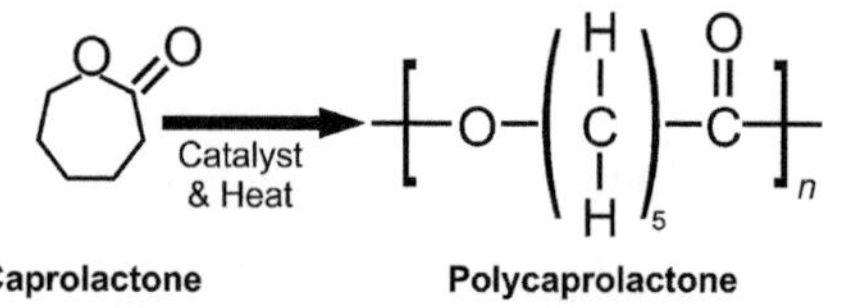

FIGURE 6.3 Ring-opening reaction to form polycaprolactone. (For color version of this figure, the reader is referred to the online version of this chapter.)

areas are areas of high regularity and high-density packing, while the amorphous areas are areas of low density and are therefore susceptibile to invasion of water.

The percent crystallinity may be assessed with differential scanning calorimetry, which will also provide the glass transition temperature and melting point. The glass transition temperature is the temperature below which a material behaves in a "glass-like" manner. Above this temperature, the material behaves in a rubbery manner as the molecular chains become increasingly mobile. Accordingly, as the temperature increases, water can more easily infiltrate the material. The melting temperature is really a temperature range over which the crystalline regions melt. The processing of an absorbable thermoplastic has enormous impact on the final characteristics of the material. For example, a polymer may be heated (causing molecular disorder) and then rapidly cooled to "lock" the disorder in place. This is likely to be more successful with a high molecular weight material, where the chains are not as mobile and likely to remain entangled. A quenched material can be annealed in order to increase crystallinity; specifically, a polymer can be maintained at a temperature between the glass transition temperature and the melting temperature in order to increase crystalline regions. A cloudy color often indicates a crystalline material. Processing a material in a solvent can cause solvent-induced crystallization, where the solvent acts as a plasticizer. Placing an object that has been quick quenched into 37 °C liquid can cause rapid crystallization (including a shift from clear to cloudy) and even distortion or buckling of the object.

A polymer is a collection of chains of varying length. Number-average molecular weight is the total weight divided by the number of molecules, while the weight-average molecular weight is the weighted average of molecules and is indicative of resistance to flow (viscosity) and of toughness. The polydispersity index gives an indication of the molecular weight range and is calculated by dividing the weight-average molecular weight by the number-average molecular weight. A value under 2 is considered appropriate for polylactides, polycaprolactones, or polyglycolides; the higher the number, the larger the range of chain lengths.

6.2.2 Processing

Hydrolytic degradation occurs because of the sensitive ester linkage. The combination of different monomers allows the degradation profile to be tuned (Figure 6.4).

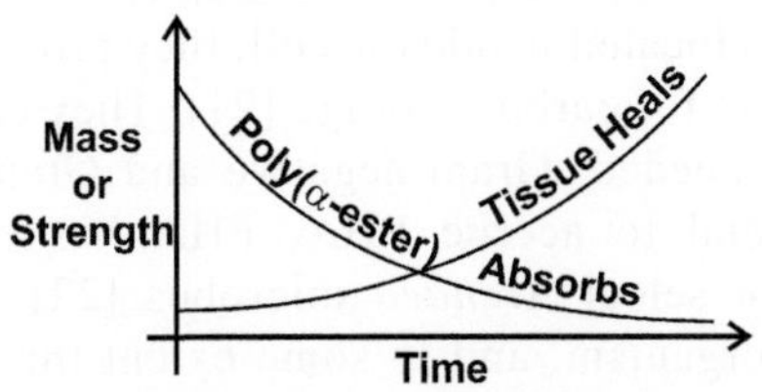

FIGURE 6.4 Theory of designed-to-degrade systems. (For color version of this figure, the reader is referred to the online version of this chapter.)

Molecular weight, polydispersity, crystallinity, and structure must be taken into consideration as one attempts to predict degradation rates; one cannot predict degradation based solely on chemical structure. That is, a high-molecular-weight poly (D,L-lactide) may have a degradation profile similar to that of a low-molecular-weight poly (L-lactide). Numerous studies have revealed the possibility of processing poly (D,L-lactide) to achieve a range of degradation profiles in *in vivo* studies [9]. The lactide materials can be modulated to achieve months to years degradation time. Lactide and glycolide content may be detected by proton nuclear magnetic resonance spectroscopy.

Monomers, water, and additives can all act as plasticizing agents. Vert and coworkers extensively studied the degradation of numerous polyesters and described the so-called surface-center effect [10]. It is difficult for water to move degradants out of the center of the material; hence, the center of the material becomes a location of low pH. In semicrystalline polymers, the water will first cleave the chains in the amorphous areas, while the crystals hold the material intact. When the water eventually reaches the crystalline areas, mass loss occurs. A higher percentage of low-length chains or plasticizers can depress the glass transition, resulting in a softer material. Once a chain reaches a specific length, it will solubilize, hence the importance of extracting materials prior to implantation to avoid relevant tissue response. Autocatalysis is caused by the production of new acid groups with hydrolysis, which lower the pH, creating enhanced degradation with increased acidity.

Indeed, polylactides, polycaprolactones, and polyglycolides are subject to a surface-center differential; that is, the processing of these materials (assuming thickness of a few millimeters) results in surface characteristics that differ from the central characteristics. Accordingly, a device with such characteristics will degrade differentially. Water can infiltrate the central portions of these devices, causing them to degrade; however, the degradant material will not be released until the exterior portion is breached. Once the exterior "shell" is breached, a bolus of acidic product is released into the surrounding area, potentially causing local tissue necrosis if the material is not rapidly cleared from the area. Key microenvironmental variables affecting degradation include pH, temperature, and rate of transport. Key polymeric variables affecting degradation include crystallinity, molecular weight, surface treatment, surface charge, and device volume.

In 1996, Park and Cima reported cell adhesion differences with respect to two different types of films [11]. The as-received material from which the films were made was the same, but one set of film was processed to have relatively low crystallinity or essentially amorphous qualities, and the other one was processed to semicrystallinity. Hepatocytes that were grown on the semicrystalline films formed spheroids very quickly, whereas the same cells on a relatively amorphous film spread on the films to form larger patches, eventually forming spheroids. A similar study was repeated later by Burg and coworkers [12] using films of molecular weight 45,000 and 270,000 Da. The results revealed a very interesting difference; over time, a higher number of cells attached to the semicrystalline films. The amorphous films swelled and crystallized once immersed in a 37 °C environment and the glass transition temperature was approached. That is, the purported amorphous films that one would have predicted to have high cell attachment, rather, had exactly the opposite. The environment changes the quality of the polymer or the characteristics of the polymer; there is an important interaction between the two and between the cells and the material, as well. Many seemingly simple processing variables have enormous impact on polymer characteristics.

The simple variables for double-emulsion fabrication of beads, for example, include the size of the beaker or the vessel, the size of the paddle, the speed, and the composition of the components. All of these factors result in a huge difference in the output; for example, cells respond differently to different radii of curvature. So, if one presents to the cells a bead that is oblong or perfectly round or very, very small or very large, the cell reaction to the respective surfaces will be very different. Additionally, the beads have changing characteristics *in vitro* or *in vivo* over time, particularly with respect to crystallinity, molecular weight, and polydispersity. Again, these differences will have a large effect on cellular behavior. In some instances, differences may be statistically significant, but not biologically significant, in other instances the differences may be both statistically and biologically significant. The challenge for the biomaterialist is to determine when there is a difference that is statistically and biologically significant and when the statistical difference really does not have biological impact. Those small, seemingly mundane characteristics and specifications are very important in determining how a material behaves in the body. Cells can react to topography; hence, one can sensibly design a material if one understands the "handshake" between the cell and the material. However, these materials change over time *in vitro* or *in vivo*, and that means the handshake changes over time.

6.3 BACTERIAL AND OTHER RECOMBINANT POLYESTERS

6.3.1 Structure and Characteristics

In 1958, an article was published detailing the development of a thermoplastic biopolymer material, through a fermentation process [13]. The article described how the polymer developed as submicron granules in and was later extracted from bacterial cells. Marchessault and coworkers published a paper in the late 1980s discussing the characteristics of

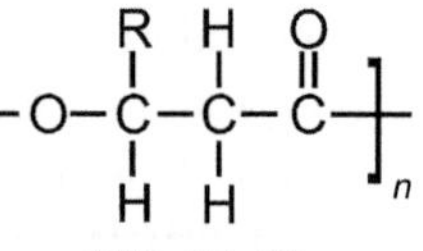

FIGURE 6.5 PHA structure. (For color version of this figure, the reader is referred to the online version of this chapter.)

poly(hydroxyalkanoate)s (PHAs) (Figure 6.5) and the many possible applications for PHA [14]. Indeed, the most well known PHA, poly(3-hydroxybutyric acid) (PHB), was identified in the 1920s [15] in the microorganism *Bacillus megaterium*. Low-molecular-weight PHB is found in the cell membrane of prokaryotes and eukaryotes [16].

PHA can be either biochemically or biologically produced [17,18]. PHAs are produced by microorganisms such as *B. megaterium*, while poly(ω-hydroxyalkanoate)s are produced via ring-opening polymerization. Biosynthesis of PHA leads to a higher molecular weight compared with that resulting from chemical means. However, biosynthesis of PHA limits control of the monomer structures in the PHA polymers; that is, the PHA polymerase specificity (or PHA synthase specificity) will influence the monomers incorporated into the polymers [19]. To generate PHA, microorganisms are grown in an aqueous solution containing starch, glucose, sucrose, fatty acids, and nutrients in wastewater at approximately 30-37 °C and atmospheric pressure; hence, it is considered more environmentally friendly than a chemical means of synthesis. PHAs synthesized by microorganisms are produced under limited nutrient and high carbon conditions. Because PHAs are of low molecular weight and have limited solubility, they do not increase cellular osmotic pressure; therefore, they may be readily stored. Although there has been an explosion of intellectual property based around PHAs, these materials have relatively limited clinical use and remain a high focus of research [20]. Unlike polyglycolides, polylactides, and polycaprolactones, the PHAs are degraded by surface erosion, that is, from the surface to the center. This characteristic provides a greater level of predictability and a tapered release of degradants. The PHAs may be fashioned with crystalline properties or may be fashioned with highly flexible properties. Generally, PHAs are biodegradable thermoplastics that have piezoelectric properties and can be manufactured in brittle to elastic form.

The PHAs are water-insoluble spherical inclusions or granules located inside the cell; they provide energy and a means for carbon storage [21]. They can be mobilized as needed. Gram-negative and Gram-positive bacteria tend to accrue PHA; PHA has even been detected in select *Archaea* microbes [22]. The synthesizing organism, and to some extent the substrate, determines the PHA monomer chain length distribution. The PHAs are termed short chain length (SCL) and medium chain length (MCL). The most common SCL PHA is PHB. The physicothermal properties of different PHAs are tunable; however, they are all considered biocompatible and biodegradable [23,24]. PHB is very rigid and very brittle, and has low elasticity [25]. Therefore, flexible 3HB-based copolymers such as P(3HB-*co*-3-hydroxyvalerate) [P(3HB-*co*-3HV)], P(3HB-*co*-3-hydroxyhexanoate) [P(3HB-*co*-3HHx)], P(3HB-*co*-3-hydroxy-4-methylvalerate) [P(3HB-*co*-3H4MV)], and P(3HB-*co*-MCL-3-hydroxyalkanoate) [P(3HB-*co*-mcl-3HA)] are recognized as more useful materials. PHA biosynthesis has been studied extensively over the past years; PHA biosynthesis can be summarized in eight pathways. Prokaryotic microorganisms, including bacteria and archaea, are responsible for most of the PHA production, although PHA has also been detected in select transgenic plants [26]. PHA oligomers were discovered in eukaryotes, specifically in mammalian tissue and blood [16]. In addition to carbon and energy storage, prokaryotic PHA also appears to be necessary for enhanced survival in the face of environmental stressors [27]. Prokaryotes grow quickly and allow mass production of PHA for biomedical applications.

A three-step PHB biosynthesis process, beginning with acetyl-coenzyme A (CoA), has been documented in the Gram-negative *Cupriavidus necator* (formerly termed *Ralstonia eutropha*) [28–30]. Hydroxyacyl-CoA monomers, produced by a condensation-reduction reaction, are polymerized by PHB synthase, which is the primary enzyme responsible for PHA biosynthesis. Different hydroxyacyl-CoA precursors, derived from alternate metabolic pathways, are involved for PHAs other than PHB. The PHA synthase is covalently bound to PHA at the surface of the granule, which is formed by assembly of PHA chains [31]. Noncovalent interactions allow additional proteins to interact with the granule [32]. PHAs can be synthesized by recombinant means, for example, in *Escherichia coli* [33]; additionally, investigators have studied PHA production in plants [34]. Biosynthesized PHA granules and PHA beads synthesized *in vitro* are a popular biomedical research focus, particularly for use as functionalized micro- or nanobeads [35,36]. Extracted PHA can be used to develop PHA micro- and nanoparticles *ex vivo*, as can PHA granules synthesized inside the cell. Whether accomplished *ex vivo* or not, a target protein can be integrated at the bead surface to a granule-associated protein. That is, proteins may be bound *in vitro* to prefabricated beads or protein may be naturally bound to beads as they form granules within the cell. In the case of natural binding, the PHA must be extracted under mild conditions in order to preserve granule shape, properties, and protein functionality. The production of PHA by this method is usually achieved using recombinant strains, not natural PHA producers. Generally, more complicated processes, such as drug encapsulation, involve *in vitro* bead preparation.

6.3.2 Processing

PHAs are being investigated for sutures or other implantables [37,38]. Intracellularly formed PHA granules can be stably maintained outside the cell as PHA beads [35,36]. PHA may be extracted from the cell in various ways because the focus is on isolating the material and not on preserving a particular morphology [39]. A common method of retrieving PHA is to disrupt the cells, collect the cellular matter, and perform the extraction, generally via solvent-based extraction. More specifically, the cells are usually lyophilized and resuspended in a solvent such as chloroform or methylene chloride in which PHA is soluble [40,41]. PHA is soluble in chloroform or methylene chloride; hence, after the cell debris is filtered and removed, methanol or ethanol is added to precipitate the PHA. Nonchlorinated solvents, such as acetone, only solubilize MCL PHA and will not solubilize PHB [42]. Solvent-based extraction of PHA results in a high purity level with minimal to no decrease in molecular weight, but large amounts of solvent are required, increasing the expense. Water can be used for the extraction but this approach would involve digestion and/or solubilization of non-PHA cellular material. Sodium hypochlorite is relatively inexpensive and may be used but results in significant degradation of PHA [43]. The combination of sodium hypochlorite and chloroform reduces degradation and allows a high-purity PHA [44]. Additional protocols have been developed, combining various detergents and/or enzymes, to solubilize cell components [45,46]. A very pure PHA with low levels of endotoxin may be obtained by extracting PHA with supercritical fluids, such as carbon dioxide. This process also removes lipid contaminants [47].

Only a few PHAs can be produced in the amounts needed for biomedical research [45,48,49]. The composition of PHA can be readily changed, thus allowing the selection of mechanical properties and degradation time to meet the needs of very specific clinical applications [20,50]. PHBs have been successfully used as films in cell culture studies [51] and as microparticles that promoted cell attachment.

With relatively recent approval of PHB sutures and mesh by the US Food and Drug Administration, it is anticipated that a wide range of PHA-based biomaterials will be examined in clinical trials. Homopolymers and copolymers of lactide and glycolide are widely used in drug delivery applications. However, lactide and glycolide copolymers undergo bulk degradation; therefore, the timing and amount of release cannot be fully controlled [52]. Due to their biodegradability, biocompatibility, and degradation by surface erosion, a large amount of research focus centered around determining the potential use of PHAs as drug carriers. It was found, for example, that the inclusion of fatty acids into PHB microspheres enhanced the rate of drug release. The ability to process a variety of PHAs will allow a larger range of controlled release properties.

Lee and coworkers [53] demonstrated that PHB could be efficiently produced *in vivo* by manipulating the environmental conditions. In studying *Alcaligenes latus*, for example, it was found that providing a more acidic environment, on the order of pH to 3-4, induced a high activity level of intracellular PHB depolymerase. Ren and coworkers [54] placed PHA-containing *P. putida* cells in phosphate buffer at different pHs and determined that a more basic pH, i.e., pH 11, was the most efficient.

PHA monomers provide an interesting starting point for the synthesis of new polymers [55,56]. Copolymerization of PHA monomers with commercially available monomers will generate an array of new copolymers. In the distant past, this line of inquiry did not generate much activity, possibly because of the high cost of PHA monomer production. In recent years, however, copolymers of lactide and PHB have been assessed for a wide range of applications [38,57,58]. Together, the poly(α-ester)s offer almost unlimited opportunities for the development of tunable biomedical implants.

REFERENCES

[1] R.L. Kronenthal, Biodegradable polymers in medicine and surgery, Springer, USA, 1975, pp. 119–137.

[2] K.J.L. Burg, S.W. Shalaby, Physicochemical changes in degrading polylactide films, J. Biomater. Sci. Polym. Ed. 9 (1) (1998) 15–29.

[3] W.H. Carothers, G.L. Dorough, F.V. Natta, Studies of polymerization and ring formation. X. The reversible polymerization of six-membered cyclic esters, J. Am. Chem. Soc. 54 (2) (1932) 761–772.

[4] R.K. Kulkarni, K.C. Pani, C. Neuman, F. Leonard, Polylactic acid for surgical implants, Arch. Surg. 93 (5) (1966) 839.

[5] J.C. Middleton, A.J. Tipton, Synthetic biodegradable polymers as orthopedic devices, Biomaterials 21 (23) (2000) 2335–2346.

[6] S.I. Moon, I. Taniguchi, M. Miyamoto, Y. Kimura, C.W. Lee, Synthesis and properties of high-molecular-weight poly (L-lactic acid) by melt/solid polycondensation under different reaction conditions, High Perform. Polym. 13 (2) (2001) S189–S196.

[7] D.E. Perrin, J.P. English, Polyglycolide and polylactide, in: A.J. Domb, J. Kost, D.M. Wiseman (Eds.), Handbook of biodegradable polymers, Harwood Academic Publishers, Amsterdam, The Netherlands, 1997, pp. 3–28.

[8] C.G. Pitt, F.I. Chasalow, Y.M. Hibionada, D.M. Klimas, A. Schindler, Aliphatic polyesters. I. The degradation of poly (ϵ-caprolactone) *in vivo*, J. Appl. Polym. Sci. 26 (11) (1981) 3779–3787.

[9] R.A. Kenley, M.O. Lee, T.R. Mahoney, L.M. Sanders, Poly (lactide-co-glycolide) decomposition kinetics *in vivo* and *in vitro*, Macromolecules 20 (10) (1987) 2398–2403.

[10] M. Vert, S. Li, H. Garreau, J. Mauduit, M. Boustta, G. Schwach, J. Coudane, Complexity of the hydrolytic degradation of aliphatic polyesters, Angew. Makromol. Chem. 247 (1) (1997) 239–253.

[11] A. Park, L.G. Cima, *In vitro* cell response to differences in poly-L-lactide crystallinity, J. Biomed. Mater. Res. 31 (1) (1996) 17–30.

[12] K.J.L. Burg, W.D. Holder, C.R. Culberson, R.J. Beiler, K.G. Greene, A.B. Loebsack, W.D. Roland, D.J. Mooney, C.R. Halberstadt, Parameters affecting cellular adhesion to polylactide films, J. Biomater. Sci. Polym. Ed. 10 (2) (1999) 147–161.

[13] R.M. Macrae, J.F. Wilkinson, Poly-beta-hydroxybutyrate metabolism in washed suspensions of *Bacillus cereus* and Bacillus megaterium, J. Gen. Microbiol. 19 (1) (1958) 210–222.

[14] R.H. Marchessault, T.L. Bluhm, Y. Deslandes, G.K. Hamer, W.J. Orts, P.R. Sundararajan, D.A. Holden, Poly (β-hydroxyalkanoates): biorefinery polymers in search of applications, Makromolekulare Chemie. Macromolecular Symposia 19 (1) (1988) 235–254.

[15] M. Lemoigne, Produits de dehydration et de polymerisation de l'acide betaoxobutyrique, Bull. Soc. Chim. Biol. 8 (1926) 770–782.

[16] R.N. Reusch, Poly-β-hydroxybutyrate/calcium polyphosphate complexes in eukaryotic membranes, Exp. Biol. Med. 191 (4) (1989) 377–381.

[17] J.E. Kemnitzer, S.P. McCarthy, R.A. Gross, Syndiospecific ring-opening polymerization of beta-butyrolactone to form predominantly syndiotactic poly (beta-hydroxybutyrate) using tin (IV) catalysts, Macromolecules 26 (23) (1993) 6143–6150.

[18] H.E. Valentin, D.L. Broyles, L.A. Casagrande, S.M. Colburn, W.L. Creely, P.A. DeLaquil, H.M. Felton, K.A. Gonzalez, K.L. Houmiel, K. Lutke, D.A. Mahadeo, T.A. Mitsky, S.R. Padgette, S.E. Reiser, S. Slater, D.M. Stark, R.T. Stock, D.A. Stone, N.B. Taylor, G.M. Thorne, M. Tran, K.J. Gruys, PHA production, from bacteria to plants, Int. J. Biol. Macromol. 25 (1) (1999) 303–306.

[19] G.Q. Chen, Q. Wu, The application of polyhydroxyalkanoates as tissue engineering materials, Biomaterials 26 (33) (2005) 6565–6578.

[20] S.P. Valappil, S.K. Misra, A.R. Boccaccini, I. Roy, Biomedical applications of polyhydroxyalkanoates, an overview of animal testing and *in vivo* responses, Expert Rev. Med. Devices 3 (6) (2006) 853–868.

[21] B.H. Rehm, Genetics and biochemistry of polyhydroxyalkanoate granule self-assembly: the key role of polyester synthases, Biotechnol. Lett. 28 (4) (2006) 207–213.

[22] A.J. Anderson, E.A. Dawes, Occurrence, metabolism, metabolic role, and industrial uses of bacterial polyhydroxyalkanoates, Microbiol. Rev. 54 (4) (1990) 450–472.

[23] G. Braunegg, G. Lefebvre, K.F. Genser, Polyhydroxyalkanoates, biopolyesters from renewable resources: physiological and engineering aspects, J. Biotechnol. 65 (2) (1998) 127–161.

[24] G.A. van der Walle, G.J. de Koning, R.A. Weusthuis, G. Eggink, Properties, modifications and applications of biopolyesters, Adv. Biochem. 618 (2001) 263–291.

[25] E.R. Olivera, M. Arcos, G. Naharro, J.M. Luengo, Unusual PHA biosynthesis, in: G.-Q. Chen (Ed.), Plastics from Bacteria: Natural Functions and Applications, Microbiology Monographs, 14, Springer-Verlag, Berlin, Heidelberg, 2010. pp. 133–186.

[26] Y. Poirier, S.M. Brumbley, Metabolic engineering of plants for the synthesis of polyhydroxyalkanaotes, in: G.-Q. Chen (Ed.), Plastics from Bacteria: Natural Functions and Applications, Microbiology Monographs, 14, Springer-Verlag, Berlin, Heidelberg, 2010, pp. 187–212.

[27] S. Castro-Sowinski, S. Burdman, O. Matan, Y. Okon, Natural functions of bacterial polyhydroxyalkanoates, in: G.-Q. Chen (Ed.), Plastics from Bacteria: Natural Functions and Applications, Microbiology Monographs, 14, Springer-Verlag, Berlin, Heidelberg, 2010, pp. 39–61.

[28] O.P. Peoples, A.J. Sinskey, Poly-beta-hydroxybutyrate biosynthesis in Alcaligenes eutrophus H16. Characterization of the genes encoding beta-ketothiolase and acetoacetyl-CoA reductase, J. Biol. Chem. 264 (26) (1989) 15293–15297.

[29] O.P. Peoples, A.J. Sinskey, Poly-beta-hydroxybutyrate (PHB) biosynthesis in Alcaligenes eutrophus H16. Identification and characterization of the PHB polymerase gene (phbC), J. Biol. Chem. 264 (26) (1989) 15298–15303.

[30] A. Steinbüchel, H.G. Schlegel, Physiology and molecular genetics of poly (β-hydroxyalkanoic acid) synthesis in Alcaligenes eutrophus, Mol. Microbiol. 5 (3) (1991) 535–542.
[31] R. Griebel, Z. Smith, J.M. Merrick, Metabolism of poly (β-hydroxybutyrate). I. Purification, composition, and properties of native poly (β-hydroxybutyrate) granules from Bacillus megaterium, Biochemistry 7 (10) (1968) 3676–3681.
[32] U. Pieper-Fürst, M.H. Madkour, F. Mayer, A. Steinbüchel, Identification of the region of a 14-kilodalton protein of Rhodococcus ruber that is responsible for the binding of this phasin to polyhydroxyalkanoic acid granules, J. Bacteriol. 177 (9) (1995) 2513–2523.
[33] S.Y. Lee, K.S. Kim, H.N. Chang, Y.K. Chang, Construction of plasmids, estimation of plasmid stability, and use of stable plasmids for the production of poly (3-hydroxy-butyric acid) by recombinant *Escherichia coli*, J. Biotechnol. 32 (2) (1994) 203–211.
[34] P. Suriyamongkol, R. Weselake, S. Narine, M. Moloney, S. Shah, Biotechnological approaches for the production of polyhydroxyalkanoates in microorganisms and plants—a review, Biotechnol. Adv. 25 (2) (2007) 148–175.
[35] K. Grage, A.C. Jahns, N. Parlane, R. Palanisamy, I.A. Rasiah, J.A. Atwood, B.H. Rehm, Bacterial polyhydroxyalkanoate granules: biogenesis, structure, and potential use as nano-/micro-beads in biotechnological and biomedical applications, Biomacromolecules 10 (4) (2009) 660–669.
[36] K. Grage, V. Peters, R. Palanisamy, B.H.A. Rehm, Polyhydroxyalkanoates: from bacterial storage compound via alternative plastic to bio-bead, in: B.H.A. Rehm (Ed.), Microbial Production of Biopolymers and Polymer Precursors: Applications and Perspectives, Caister Academic Press, Norfolk, UK, 2009, pp. 255–287.
[37] P. Furrer, S. Panke, M. Zinn, Efficient recovery of low endotoxin medium-chain-length poly ([R]-3-hydroxyalkanoate) from bacterial biomass, J. Microbiol. Methods 69 (1) (2007) 206–213.
[38] B. Hazer, A. Steinbüchel, Increased diversification of polyhydroxyalkanoates by modification reactions for industrial and medical applications, Appl. Microbiol. Biotechnol. 74 (1) (2007) 1–12.
[39] B. Kessler, R. Weusthuis, B. Witholt, G. Eggink, Production of microbial polyesters: fermentation and downstream processes, in: W. Babel, A. Steinbüchel (Eds.), Biopolyesters, Springer, Berlin, Heidelberg, 2001, pp. 159–182.
[40] H. Brandl, R.A. Gross, R.W. Lenz, R.C. Fuller, Pseudomonas oleovorans as a source of poly (β-hydroxyalkanoates) for potential applications as biodegradable polyesters, Appl. Environ. Microbiol. 54 (8) (1988) 1977–1982.
[41] R.G. Lageveen, G.W. Huisman, H. Preusting, P. Ketelaar, G. Eggink, B. Witholt, Formation of polyesters by Pseudomonas oleovorans: effect of substrates on formation and composition of poly-(R)-3-hydroxyalkanoates and poly-(R)-3-hydroxyalkenoates, Appl. Environ. Microbiol. 54 (12) (1988) 2924–2932.
[42] J. Yao, G. Zhang, Q. Wu, G.-Q. Chen, R. Zhang, Production of polyhydroxyalkanoates by Pseudomonas nitroreducens, Antonie Van Leeuwenhoek 75 (4) (1999) 345–349.
[43] E. Berger, B.A. Ramsay, J.A. Ramsay, C. Chavarie, G. Braunegg, PHB recovery by hypochlorite digestion of non-PHB biomass, Biotechnol. Tech. 3 (4) (1989) 227–232.
[44] S.K. Hahn, Y.K. Chang, S.Y. Lee, Recovery and characterization of poly (3-hydroxybutyric acid) synthesized in Alcaligenes eutrophus and recombinant Escherichia coli, Appl. Environ. Microbiol. 61 (1) (1995) 34–39.
[45] D. Byrom, Polymer synthesis by microorganisms: technology and economics, Trends Biotechnol. 5 (9) (1987) 246–250.
[46] G.J.M. De Koning, M. Kellerhals, C. Van Meurs, B. Witholt, A process for the recovery of poly (hydroxyalkanoates) from Pseudomonads Part 2: Process development and economic evaluation, Bioprocess Eng. 17 (1) (1997) 15–21.
[47] S.F. Williams, D.P. Martin, D.M. Horowitz, O.P. Peoples, PHA applications: addressing the price performance issue: I. Tissue engineering, Int. J. Biol. Macromol. 25 (1) (1999) 111–121.
[48] G. Chen, G. Zhang, S. Park, S. Lee, Industrial scale production of poly (3-hydroxybutyrate-co-3-hydroxyhexanoate), Appl. Microbiol. Biotechnol. 57 (1–2) (2001) 50–55.
[49] O. Hrabak, Industrial production of poly-β-hydroxybutyrate, FEMS Microbiol. Lett. 103 (2–4) (1992) 251–255.
[50] H. Abe, Y. Doi, H. Aoki, T. Akehata, Y. Hori, A. Yamaguchi, Physical properties and enzymic degradability of copolymers of (R)-3-hydroxybutyric and 6-hydroxyhexanoic acids, Macromolecules 28 (23) (1995) 7630–7637.
[51] Q. Zhao, G. Cheng, C. Song, Y. Zeng, J. Tao, L. Zhang, Crystallization behavior and biodegradation of poly (3-hydroxybutyrate) and poly (ethylene glycol) multiblock copolymers, Polym. Degrad. Stab. 91 (6) (2006) 1240–1246.
[52] C.W. Pouton, S. Akhtar, Biosynthetic polyhydroxyalkanoates and their potential in drug delivery, Adv. Drug Deliv. Rev. 18 (2) (1996) 133–162.
[53] S.Y. Lee, Y. Lee, F. Wang, Chiral compounds from bacterial polyesters: sugars to plastics to fine chemicals, Biotechnol. Bioeng. 65 (3) (1999) 363–368.
[54] D. Ren, R. Zuo, T.K. Wood, Quorum-sensing antagonist (5Z)-4-bromo-5-(bromomethylene)-3-butyl-2 (5H)-furanone influences siderophore biosynthesis in Pseudomonas putida and Pseudomonas aeruginosa, Appl. Microbiol. Biotechnol. 66 (6) (2005) 689–695.
[55] L.R. Rieth, D.R. Moore, E.B. Lobkovsky, G.W. Coates, Single-site β-diiminate zinc catalysts for the ring-opening polymerization of β-butyrolactone and β-valerolactone to poly (3-hydroxyalkanoates), J. Am. Chem. Soc. 124 (51) (2002) 15239–15248.
[56] S. Taguchi, M. Yamada, K.I. Matsumoto, K. Tajima, Y. Satoh, M. Munekata, K. Ohnob, K. Kohdab, T. Shimamurab, H. Kambec, S. Obata, A microbial factory for lactate-based polyesters using a lactate-polymerizing enzyme, Proc. Natl. Acad. Sci. U.S.A. 105 (45) (2008) 17323–17327.
[57] K.M. Nampoothiri, N.R. Nair, R.P. John, An overview of the recent developments in polylactide (PLA) research, Bioresour. Technol. 101 (22) (2010) 8493–8501.
[58] K.M. Schreck, M.A. Hillmyer, Block copolymers and melt blends of polylactide with Nodax™ microbial polyesters: preparation and mechanical properties, J. Biotechnol. 132 (3) (2007) 287–295.

Polyurethanes

Namdev B. Shelke*,†,‡,1, Rajaram K. Nagarale§,¶,1, Sangamesh G. Kumbar*,†,‡,§
*Institute for Regenerative Engineering, University of Connecticut Health Center, Farmington, Connecticut, USA
†Raymond and Beverly Sackler Center for Biological, Physical and Engineering Sciences, Yale University, New Haven, Connecticut, USA
‡Department of Orthopaedic Surgery, University of Connecticut Health Center, Farmington, Connecticut, USA;
§Department of Chemical, Materials and Biomedical Engineering, University of Connecticut, Farmington, Connecticut, USA
¶Department of Chemical Engineering, Indian Institute of Technology Kanpur, Uttar Pradesh, India
1Corresponding authors: e-mail: shelke@uchc.edu and rknagarale@gmail.com

Chapter Outline

7.1 INTRODUCTION

Polyurethanes (PURs) are versatile and are extensively known for their vastly differing mechanical and biological properties. They are considered as one of the most biocompatible materials; therefore, they are receiving increasing interest. They have been used in biomedical field for various applications due to the propensity to modify their unique properties and they are easily available materials. The properties such as process-ability, toughness, durability, surface functionality, flexibility, biocompatibility, and biostability makes them superior for the development of materials useful for drug delivery, tissue engineering and for the development of many medical devices [1]. Using these polymers, drug delivery systems such as matrix, reservoir, and stimuli-responsive devices have been developed [2–7], and in tissue engineering, they have been used widely for making scaffolds, vascular grafts, breast implants, wound healing materials, etc., successfully [8–11]. In addition, due to their unique elastic properties and durability, they are used in manufacturing of various blood bags; many extracorporeal or implantable medical devices such as various vascular catheters; the total artificial heart; bladders of the left ventricle assist device; small caliber vascular grafts for vascular access and arterial reconstruction or bypass (by-pass and replacement); artificial hearts, kidneys, and lungs, heart valves; ventricular assist devices; pacemaker leads; connector blocks; tissue replacement and augmentation; and sewing rings for heart valves as well as heart valve prostheses [12]. Polyurethane's referred to as a segmented polymers because of their soft and hard segments that provide them flexibility and strength, respectively. They are thermoplastic, linear, and elastomeric in nature where the elasticity as well as hardness of these polymers can be varied by varying the ratio of the soft and hard segments. Polyurethane's mostly consists of three building blocks, the soft segments usually diols of long-chain molecules of polyether, polyesters, polysiloxane, polycarbonate, etc., which impart flexibility;

the hard segments, which are usually the combination of diisocyanates; and the chain extender, in which chain extender also acts as a cross-linker [13–17]. By varying the ratio of these building blocks, PURs of required properties such as chemical, mechanical, compatibility with various hardness-softness, and hydrophilicity-hydrophobicity can be obtained [18,19]. Polyurethane's are also very useful for high-load-bearing applications such as bone scaffolds, and these properties can be achieved by choosing suitable hard segments and hard to soft segment composition [20–23]. In addition this suitable scaffold properties such as tendency to recover, durability, elastic character, and resistance to fatigue properties such as tensile strength, elasticity, compression, or shear can be achieved using PURs [12,24]. Various drug delivery and tissue engineering devices have been engineered by employing various techniques. The device engineering techniques such as hot melt extrusion, polymer casting with or without solvents, electrostatic and wet spinning of monofilaments, and dip coating or spraying of mandrels followed by coagulation and phase inversion have been widely used. These devices behave like natural tissues upon implantation due to the porous structures, smoothness, and biocompatibility [12,25–30].

Degradation characteristics of PURs are widely studied *in vitro* and *in vivo* using various compositions. The hydrolytically unstable PURs are synthesized using degradable monomers such as polyester diols, for engineering biodegradable micro-/nanoparticulate drug delivery systems and implants and for tissue engineering applications [13,31–35]. Whereas for long-term applications, hydrolytically stable PURs using polyether diols, polycarbonate diols, polysiloxane-diol, etc., were developed for engineering of various medical devices and tissue engineering scaffolds [36,37]. The effect of various monomer compositions as well as *in vitro* and *in vivo* exposures conditions has been studied on stability and mechanical properties of the PURs. Studies indicate that under physiological fluid conditions, the degradation of PURs is faster due to the effect of various cells, enzymes, chemicals, stress, and metal ions as compared to *in vitro*. This indicates that PURs have different interactions with physiological fluids than their interactions with *in vitro* buffers. Also stability studies using various PURs show that the significant changes in degradation profiles could be obtained by varying the monomer compositions. The poly(ester urethanes) degrade by magnitude faster than the poly(ether urethanes) (PEURs) or poly(carbonate urethanes) (PCURs) or poly(siloxane urethanes) (PSURs); this alteration in degradation gives excellent opportunity where they can be used successfully [38].

Polyurethane implant devices under physiological conditions undergo hydrolytic degradation, thrombosis, chain scission, calcification, and environmental stress cracking (ESC), and upon storage or during use, they undergo phase separation phenomenon. During the phase separation, the hard and soft segments tend to agglomerate due to chain mobilities and fall apart microscopically. This phenomenon could have positive effect when using as an implant device [12]. Polyurethanes have very well propensity of surface modification and could be tethered with suitable polymer ligand, which could help in enhancing the properties of the implant device. Therefore, surface-modified PURs are developed using poly(ethylene oxide-propylene oxide-ethylene oxide) (Pluronics®), polyethylene oxide (PEO), polyethylene glycol (PEG), etc., to enhance their biological properties [39,40]. Since the last several decades, researchers have published numerous patents [41–59] claiming their uses for various biomedical applications indicating their usefulness in biomedical field. The list of some patents related to PUR applications for drug delivery and tissue engineering is given in Table 7.1.

7.2 SYNTHESIS AND CHARACTERIZATION

7.2.1 Synthesis

Typically, segmented PURs are synthesized in two steps; in the first step, isocyanate-terminated prepolymer is obtained by reacting excess of diisocyanate with diol (2:1 ratio) [60–62]. Segmented PURs such as poly(ester urethanes), PEURs, PCURs, and PSURs are synthesized using diols of polyester, polyether, polycarbonate, and polysiloxane, respectively. In this step, polyols are reacted with excess amount of diisocyanates such as 4,4′-methylenebis (cyclohexyl isocyanate) ($H_{12}MDI$), isophorone diisocyanate (IPDI), or 1,6-hexamethylene diisocyanate (HDI), and as a result, prepolymer of polyol terminated with −NCO can be obtained. In second step, polyol terminated with isocyanate groups is reacted with chain extenders such as butane diol to get high-molecular-weight PURs. During the first step of PUR synthesis, urethane linkages are formed between two polyols, leaving one isocyanate group free at the end, and additional urethane linkages are formed between isocyanate-terminated prepolymers and chain extenders during the second step of synthesis. Two types of chain extenders are used most commonly, which are low-molecular-weight diol such as 1,4-butanediol and diamine such as ethylenediamine to obtain PURs [14,63–65].

Commercially available various biostable thermoplastic segmented PURs such as Tecoflex®, Pellethane®, Elasthane™, and Biomer are poly(ether urethanes) and are synthesized using polyether diol, which is usually polytetramethylene oxide (PTMO), whereas Bionate®, Myo Lynk™, Chronoflex®, and Carbothane® are PCURs, synthesized using polycarbonate diols. Some of these polymers also have plasticizers and antioxidants added for suitable mechanical properties and for longer shelf life purpose, and these PURs have excellent biostability [36,66]. Typically, the higher the soft segment content, the higher would be the flexibility and elasticity of the PURs, on other hand, the hydrophilicity

TABLE 7.1 List of Published Patents for Drug Delivery, Tissue Engineering, and Medical Devices Using Biomedical Polyurethanes

Polyurethane Form	Description	References
Implantable reservoir drug delivery device	Long-term delivery of various bioactive agents	[41–43]
Intravaginal drug delivery device (reservoir and matrix)	Anti-HIV, hormone, acrosin inhibitor delivery	[44–48]
Implantable drug-PUR matrix device	Delivery of bioactive agent	[49]
In vivo tissue engineering device	Cell delivery using scaffolds	[50]
A ventricular cuff	Method and apparatus for assisting a heart to pump blood	[51]
A tissue engineering scaffold	Porous scaffold for cell, tissue, or organ growth	[52]
Multiple coil spiral-wound vascular catheter	Surgical device suitable for accessing tissue target	[53]
Porous synthetic vascular grafts	Porosity allows connective tissue ingrowth into a wall of the prosthesis	[54]
Blood-contacting medical device	To improve biocompatibility of devices by heparin coating	[55]
Ventricular assist device	Receiving human heart in aligned position	[56]
Extrudable polyurethane for prosthetic devices	Useful for blood bags, transvenous cardiac pacemaker leads, tubing and catheters	[57]
Optically clear polyurethane	Cornea or lens	[58]
Polyurethane elastomer	Heart assist devices	[59]

will be depend on the hydrophilic nature of macrodiol. The degradation rates of PURs can be varied by varying the monomer compositions. For instance, hydrolytically unstable polyester diol-based poly(ester urethanes) have been developed as a resorbable material with the degradation rate higher than PEURs, PCURs, and PSURs. Degradable PURs can be synthesized using diols of PEG, adipate, polylactic acid (PLA), PEG-co-PLA, castor oil, vegetable oil, cellulose, starch natural rubber, etc. [61,67–69].

7.2.2 Characterization

A very useful technique such as Fourier transform infrared spectroscopy (FTIR) is regularly used for characterization and as a conformation tool to study PURs before and after degradation. The technique is also successfully used to follow PUR formation reactions. The characteristic peak associated with various groups in PURs is the absence of peak around 2260 cm^{-1} assigned to isocyanate (–N=C=O) group stretching that indicates utilization of all isocyanate groups during PUR synthesis. The poly(ester urethanes) have characteristic peak at 1740 cm^{-1}, which corresponds to -C=O st. of ester linkage; similarly, FTIR spectrum of poly(urethane ureas) (PUURs) shows peak at 1632 cm^{-1} due to the presence of urea generated by the reaction between isocyanate and amine groups [29,70–72]. In one of the PUR degradation studies, Fu *et al.* [73] studied degradation patterns of poly(ester urethane) composites made up of using nano-hydroxyapatite (n-HA). The relative intensities of characteristic stretching bands around 3420, 1734, 1627, and 1539 cm^{-1} attributed to –N–H st., –C=O st., and amides, respectively, decreased upon degradation. In another study, FTIR in combination with cryomicrotomy and scanning electron microscopy (SEM) was used by McCarthy *et al.* for *in vivo* degradation studies of commercially available PURs such as Pellethane™ 2363-80A, Tecoflex™ EG8OA, and Biomer using ovine implants. Thin sections of implanted PURs obtained by cryomicrotomy and FTIR attached to microscope show the chemical changes in PURs, and also SEM showed microstructural changes upon 18 months' implantation. Uniform pitting and fissuring (~2.0 μm depth) on the surface of Biomer implants whereas embrittlement with bulk superficial fissure infiltration up to 40 μm for Pellethane™ 23633 80A and Tecoflex™ EG 80A implants were seen upon degradation [74]. The nature of crystalline properties of PURs before and after degradation can be studied using X-ray diffraction technique (XRD); Fu *et al.* [73] noticed significant changes in the polymer crystallinity upon degradation using this technique. Thermal properties of PURs such as enthalpy content, glass transition temperature (T_g), and melt temperatures (T_m) are significantly affected by the extent of molecular weight, hard to soft segment ratios, heating and cooling history, and cross-linking. Differential scanning calorimetry (DSC) is widely used to reveal the enthalpy content (Δ_H cal/g) values corresponding to T_g and T_m peaks associated with soft or hard segments [75]. Bajsic and Rek [76] investigated morphological changes in PURs upon annealing as a function of temperature using DSC. Two different PURs were synthesized by varying hard segments of 4.4′-methylenebis(phenyl isocyanate)/1,4-butane diol (MDI/BD) and H_{12}MDI/BD with 35-57 wt% with respect to soft segments such as PTMO (M_W: 1000, 2000 Da) and polycaprolactone diol (PCL, M_W: 1250 Da). Annealing of these PURs at various temperatures such as 70, 130, 160, and 180 °C showed morphological changes upon annealing; also significant changes were seen in aromatic PURs as compared in aliphatic PURs. The morphological changes due to mixing of soft and hard segments were understood by comparing the changes in T_g of these segments before and after annealing. The values of T_g for soft domains increased with respect to increase in hard domains, which is attributed to mixing and interactions of soft domains with hard domains. These values increased when annealed at low temperatures and decreased when annealed at higher temperatures, which is attributed to mixing of soft and hard domain at low temperature and phase separation at higher temperature. On other hand, endotherms shifted to higher temperature with respect to increase in annealing temperature, which is attributed to well-arranged hard domains upon annealing at higher temperature. GPC technique is used widely to understand weight-average molecular weight (M_W) and number-average molecular weight (Mn) of the polymers; also this technique is used to determine molecular weights corresponding to the fractions generated upon degradation in various buffer solutions. Apart from these techniques, the techniques such as atomic force microscopy (AFM), nuclear magnetic resonance spectroscopy, and thermogravimetric analysis are also used for the characterization of PURs [32,70,77–79].

7.3 IMPACT OF COMPOSITION ON POLYURETHANE PROPERTIES

The excellent properties of PURs are useful for manufacturing of catheters, pacemakers, vascular grafts, and nanocarrier medical devices. However, the properties such as poor adherence to endothelial cells, fatigue nature, and uncontrolled biodegradation limit their applications in this field [80]. The presence of confluent endothelial cell layer can make device surface inert for excellent acceptability in physiological conditions upon implantation. This continuous cell monolayer on surface also plays important role by enhancing anticoagulant properties by expressing binding sites for anticoagulant factor without distracting blood flow [81,82]. Many efforts have been made to construct medical devices with various graft materials on the surface or inside the polymer to endothelize the biomaterials. Upon modification PURs functions like ECs where they help to protect the device from thrombosis and vasculogenesis.

7.3.1 Poly(Ether Urethanes)

Poly(ether urethanes) have excellent mechanical property, biocompatibility, and biostability; therefore, they have been used for various medical applications such as heart valves, artificial heart, catheters, vascular prostheses, pacemaker lead insulation, and drug delivery [7,25,83]. Polyether such as PTMO have been used for the synthesis of PEURs, and these polymers are stable to the hydrolytic degradation processes. However, PEURs undergo autoxidation (AO), metal ion oxidation (MIO), and ESC. The cardiac pacing device of Pellethane® 2363 80A (P80A), which is PEUR, undergoes autoxidation and ESC upon implantation. Prior to degradation, proteins form blood plasma adsorbs on the surface of PURs, and these proteins subsequently dictate the inflammatory responses. The initial step of protein adsorption is responsible for the degradation of PURs by releasing various enzymes and soluble factors by activating inflammatory cells [84,85]. Chan-Chan *et al.* [86] characterized biocompatibility and degradation of arginine-, glycine-, and aspartic acid-based PUURs. The studies showed decrease in molecular weights in pH 7.4 after 24 weeks, and under accelerated conditions such as acidic, alkaline, and oxidative media, PURs showed faster degradations. No adverse effects of PUURs synthesized using L-arginine on metabolism of human umbilical vein endothelial cells were noticed. On other hand, enhanced cell adhesion, spreading, and viability were noticed compared to the commercially available Tecoflex® PEUR. Grasl *et al.* [87] fabricated vascular grafts using commercially available PEUR such as Pellethane® in a small diameter. The grafts had similar mechanical properties as local vessels, and the bioactivity of the PEUR grafts was tested for mechanical homogeneity, cell proliferation, and expression of various molecules, for instance, expression of cytokines due to adhesion of endothelial cells (EC) to grafts *in vitro*. Endothelial cell attachment was noticed with interleukin-1b (IL-1b) and resulted with increase in molecules such as E-selectin, ICAM-1, and VCAM-1. The endothelial cell response was seen consistent, and no adverse effects were noticed. The SEM pictures of small-diameter vascular grafts fabricated using Pellethane® are as shown in Figure 7.1.

7.3.2 Poly(Carbonate Urethanes)

Failure of PUR devices upon implantation warranted the finding of new materials with acceptable and predictive *in vitro* and *in vivo* degradation profiles. Studies show that the presence of ether linkages in PURs used for device development results in autoxidation and ESC, which subsequently leads to failure of devices; therefore, the development of reduced ether content or ether-free PURs was proposed for the long-term applications *in vivo* [85,88,89]. Polyurethanes such as PCURs containing polycarbonate segments have been developed and tested for their *in vitro* and *in vivo* stability. The PCUR shows better *in vitro,*

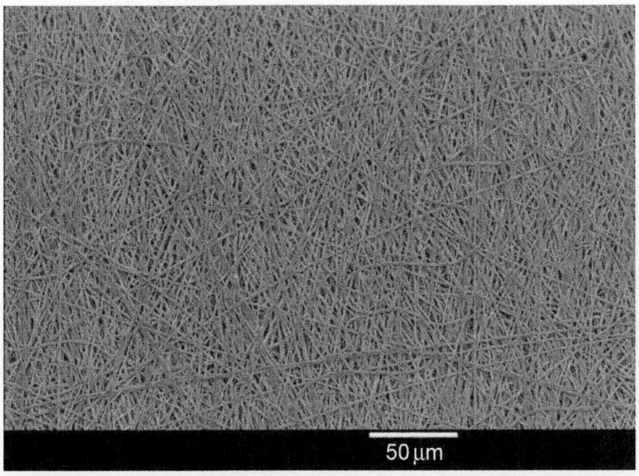

FIGURE 7.1 SEM image of the luminal side of an electrospun PUR prosthesis, X350. Reprinted with permission from Ref. [87]. © 2009 Wiley Periodicals, Inc.

in vivo, and chemical stability as compared to PEURs [32]. The presence of enzymes *in vivo* such as myeloperoxidase, cholesteryl esterase, K protease, B cathepsin, A2 phospholipase, α-2 macroglobulin, and monocyte-derived macrophages are believed to be responsible for the degradation of PURs [90,91]. Also the concentrations of these enzymes and the nature of hard segment present in PURs affect on the degradation rate of PURs. Poly(carbonate urethanes) synthesized using hard segments of H_{12}MDI underwent fast degradation when compared with PCURs synthesized using hard segments of HDI and MDI in the presence of cholesteryl esterase (CE). The concentration of CE also played important role in the degradation of PCURs, these polymers showed stability when enzyme concentration was 80 units/ml and significant change was seen in properties upon elevation in enzyme concentrations; also it was noticed that the PCUR surface saturated in the presence of low and high concentrations before achieving the saturation degradation by enzymes [92,93]. The degradation kinetics of poly(ester carbonate urethane ureas) (PECUUs) synthesized using mixture of PCL and poly(1,6-hexamethylene carbonate) as a soft segment and 1,4-diisocyanatobutane with putrescine as a hard segment have been studied. The increase of polycarbonate content in PURs yielded smooth films with increased tensile strength and breaking strains. The scaffolds of PECUU obtained by salt leaching method showed suitable *in vitro* muscle cell adhesion and growth when implanted subcutaneously in rats. These PECUU scaffold implants also degraded faster as compared to only PCURs, but degradation rates were slower than poly(ester urethanes). The comparative accelerated degradation using 20% hydrogen peroxide/0.1 M cobalt chloride solution at 37 °C of Bionate 80A (PCUR) showed overall lower degradation in comparison to Elastothane 80A (PEUR). Comparative degradation SEM images of PEUR and PCUR show pitting on surface of both polymers; in addition, PEUR had a wide range of pit sizes that were 1-30 μm upon 24-day treatment (Figure 7.2). When degradations were performed with 400% strain, extensive surface cracking perpendicular to the strain was

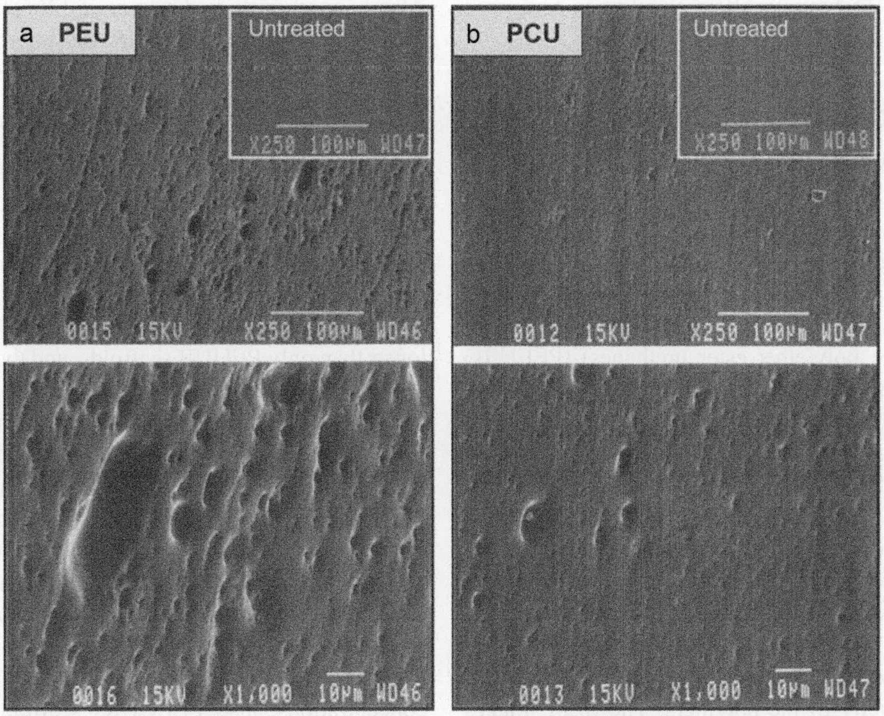

FIGURE 7.2 SEM image showing physical degradation after 24 days *in vitro* treatment: (a) PEU (PEUR) and (b) PCU (PCUR). Reprinted with permission from Ref. [94]. © 2009 Wiley Periodicals, Inc.

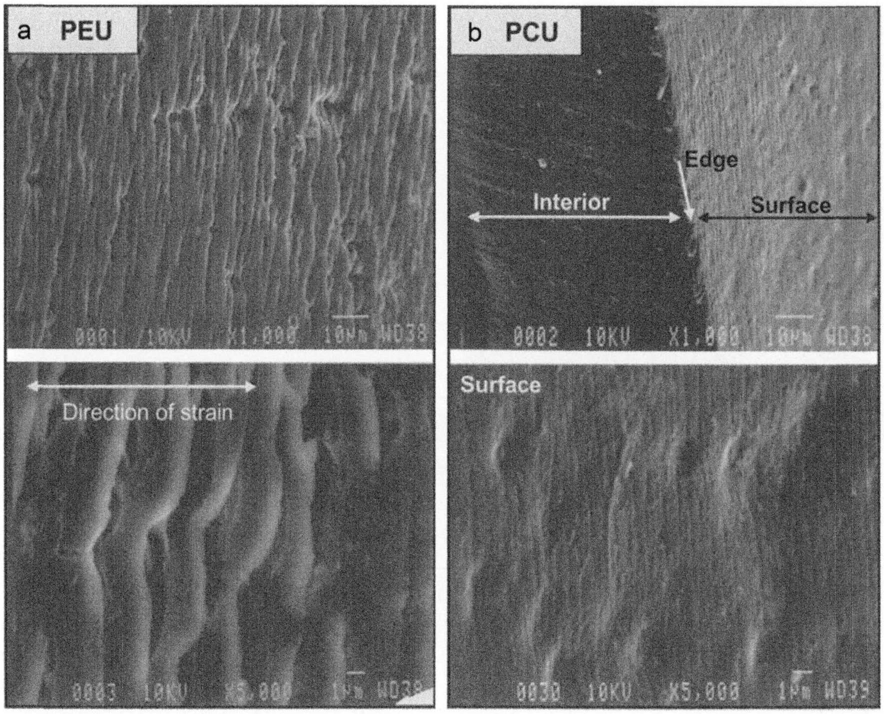

FIGURE 7.3 SEM image after 24 days *in vitro* treatment: (a) PEU and (b) PCU. Films were strained to 400% strain and allowed to recover. Reprinted with permission from Ref. [94]. © 2009 Wiley Periodicals, Inc.

noticed for PEUR, whereas for PCURs, no cracking was noticed despite having a surface. However, the surface of these polymers had wavelike appearance (Figure 7.3). In other studies, it was noticed that the solvent-casted sheets of PCUR such as Corethane® 80A, Bionate® 80A, and Chronoflex® AL80A have excellent stability as compared to PEURs such as Pellethane® 2363 80A [94,95].

7.3.3 Poly(Ether Ester Urethanes)

Polyurethanes with ester segments are generally hydrolytically unstable and they have significantly high degradation rate upon implantation in the human body as compared to degradation during storage or *in vitro*; therefore, they can be used successfully as a bioabsorbable materials. Gorna and Gogolewski [33] synthesized PURs using PEO, Pluronic®, and PCL diol. The poly(ether ester urethanes) (PEEURs) with various glass transition and melt temperature were highly degradable with 15-80% loss in molecular weight in 48 weeks upon implantation. The degree of PEEUR degradation was dependent on the ratio between PEO or Pluronic® and PCL, where higher degradation was noticed when PEO or Pluronic® content was higher with respect to PCL. Also the higher molecular weight loss was noticed for PEEURs, which was 96% as compared to only poly(ester urethanes), which was 4%, and this is attributed to presence of hydrophilic PEO segments into the PEEURs which imparts hydrophilicity to the polymers. The SEM images of PEEURs degradation patterns show decrease in surface smoothness up to 48 weeks, whereas minimal changes were seen for 72-week degradation sample (Figure 7.4a–c). Comparatively, higher molecular-weight PEO content films of PEEURs had high surface roughness at 48 weeks, and films were cracked at the end of 72 weeks (Figure 7.4d–f). Also in another study, Guan *et al.* [26] studied the effect of monomer composition on PURs mechanical and degradation properties using porous scaffolds. Poly(ester urethane urea) (PEUUR) and poly(ether ester urethane urea) (PEEUUR) were synthesized using PCL diol and PEO-PCL diol as soft segments, respectively, with 1,4-diisocyanatobutane (BDI) and putrescine as a hard segments. The degradation study of soft scaffolds developed using phase separation technique with pore size ranging between 1-159 μm and 80-97% porosity showed correlation between pore size and hydrophilicity. Also the breaking strains of PEEUUR scaffolds were lower than only PEUUR scaffolds, and the cell adhesion and growth on these scaffolds was similar. Hydrophilic nature of PEEUURs made scaffolds with smooth surface and enhanced rate of degradation. SEM images of PEUURs porous scaffolds developed using various concentrations showed different morphologies with respect to quenching temperatures (Figure 7.5). Also the PUR scaffolds developed using different polymer solution concentrations and same quenching temperature had different rates of degradation at the end of 8 weeks. For instance, the scaffolds of 5% PEUUR degraded faster than scaffolds developed using 8% and 10% solution concentrations; also PEEUURs scaffolds made from 5% polymer solutions degraded faster as compared to scaffolds

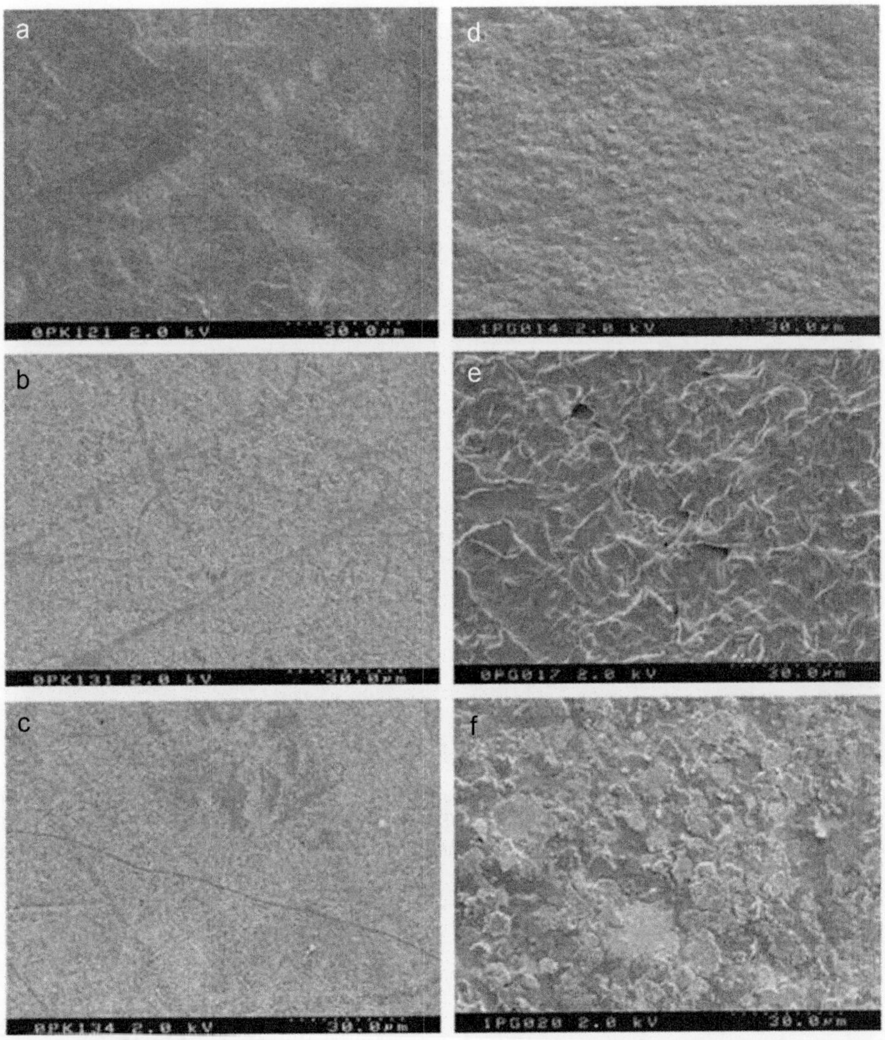

FIGURE 7.4 SEM images of selected PURs subjected to *in vitro* degradation: (a) PCL200-PEO600 (70:30) control, (b) PCL200-PEO600 (70:30) after 48 weeks, (c) PCL200-PEO600 (70:30) after 76 weeks, (d) PCL530-PEO2000 (50:50) control, (e) PCL530-PEO2000 (50:50) after 48 weeks, and (f) PCL530-PEO2000 (50:50) after 76 weeks. Reprinted with permission from Ref. [33]. © 2009 Wiley Periodicals, Inc.

made using 10% solution. This study also indicates that there is no difference in degradation rates when scaffolds were developed using same polymer concentrations but degradation rates varied by varying quenching temperatures (Figure 7.6a and b). The sterilization by gamma irradiation also has significant impact on PEURs; Gorna and Gogolewski [96] studied the impact of gamma radiation sterilization on PUR degradation as well as on mechanical properties using standard dose of 25 kGy. Poly(ester urethanes) based on PEO and PCL degraded to different level as revealed by change in molecular weight, mechanical properties, surface roughness, and contact angle. The only PCL-based polymers degraded to less extent, which is 10-12% decrease in molecular weight and 12% decrease in tensile strength as compared to 30-50% and 50% for polymers based on mixture of PCL and PEO. Properties such as surface roughness and contact angle also changed significantly with respect to decrease in molecular weight. These findings indicate that the rate of PEEURs degradation can be manipulated by varying factors such as hydrophilicity and porosity of the devices for their use as resorbable materials.

7.3.4 Poly(Siloxane Urethanes)

Polyurethanes with highly water-repellant properties can act as a useful barrier between implantable device and biological tissue by protecting device from water exchange resulting in biocompatibility. PURs with highly hydrophobic properties are synthesized using poly-dimethoxysilane (PDMS) macrodiol, and as a result, PSURs are obtained. Roohpour *et al.* have done comparative study for the impact of various compositions on water permeability and elasticity of PEURs and PSURs. An increase in microphase separation and elastic modulus upon increase of PDMS content was noticed. On other hand, the water permeability, water uptake, and surface energy of the membrane were decreased with respect to

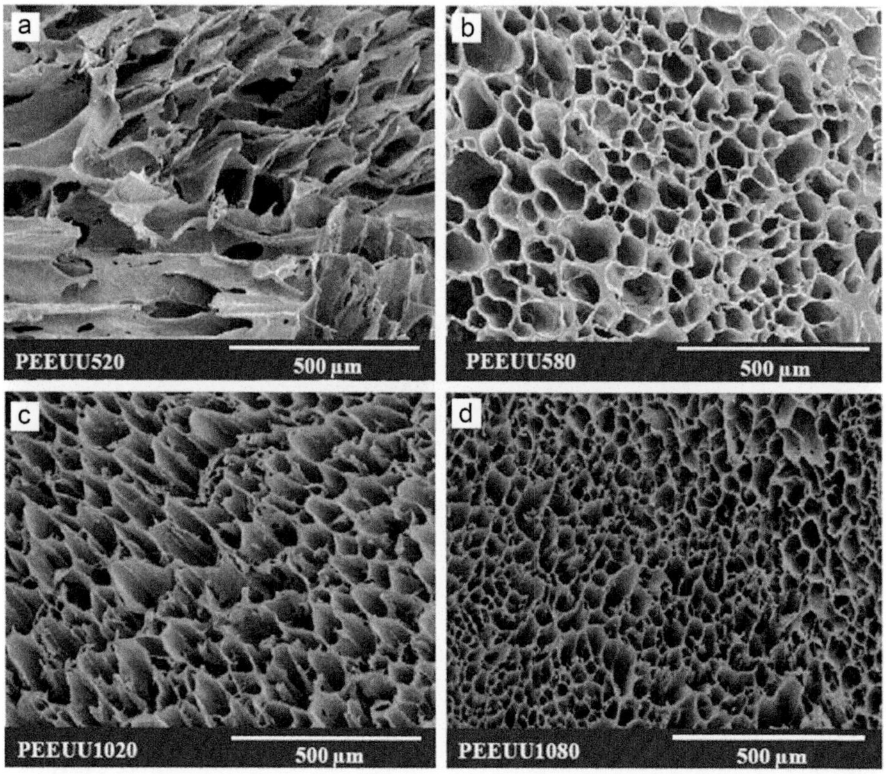

FIGURE 7.5 Electron micrographs of PEEUU (PEEUUR) scaffolds prepared from different PEEUU solution concentrations and quenching temperatures. (a) PEEUU520 (5%, −20 °C), longitudinal cross-section; (b) PEEUU580 (5%, −80 °C), longitudinal cross-section; (c) PEEUU1020 (10%, −20 °C), longitudinal cross-section; (d) PEEUU1080 (10%, −80 °C), longitudinal cross-section. Scale bar 500 μm. Reprinted with permission from Ref. [26]. © 2004 Elsevier Ltd.

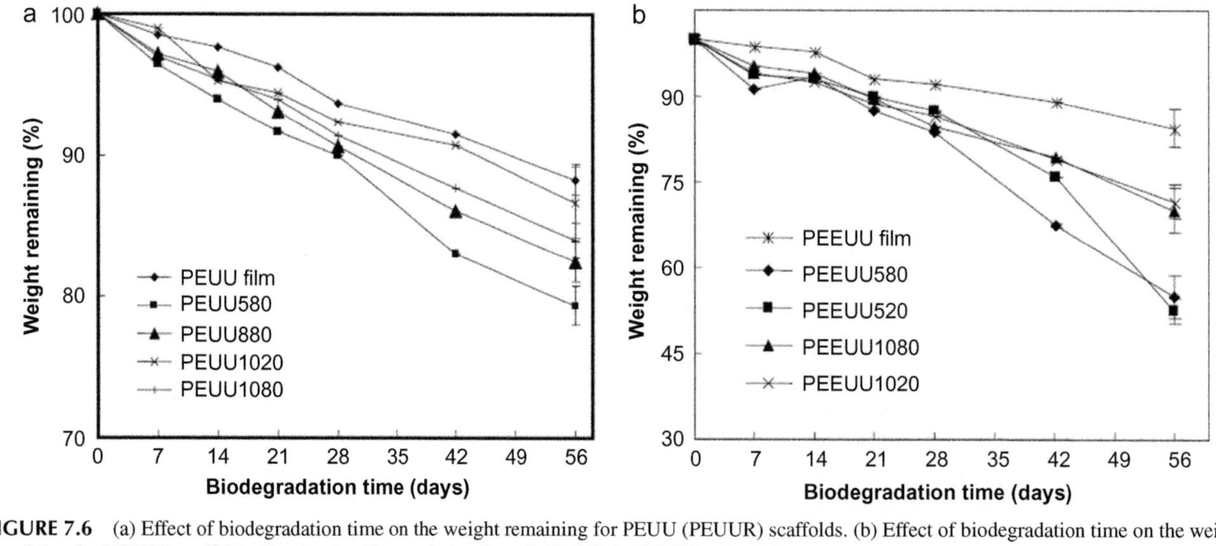

FIGURE 7.6 (a) Effect of biodegradation time on the weight remaining for PEUU (PEUUR) scaffolds. (b) Effect of biodegradation time on the weight remaining for PEEUU scaffolds. Reprinted with permission from Ref. [26]. © 2004 Elsevier Ltd.

increase of PDMS in PSURs as compared to PEUR. These studies indicate that the PSURs with high PDMS content could be effective for use as an implantable material due to the water resistance properties [97]. Martin *et al.* synthesized PSURs based on poly(hexamethylene oxide) (PHMO) and PDMS soft segments and MDI with BD as a hard segments. The biocompatibility of PSURs with various compositions of PHMO and PDMS was assessed by implantation into sheep for 3 months. The implants were examined microscopically and compared with the control implants of Pellethane® 2363-80A and Pellethane® 2363-55D. Polyurethanes based on PDMS had excellent flexibility, strength, and biostability,

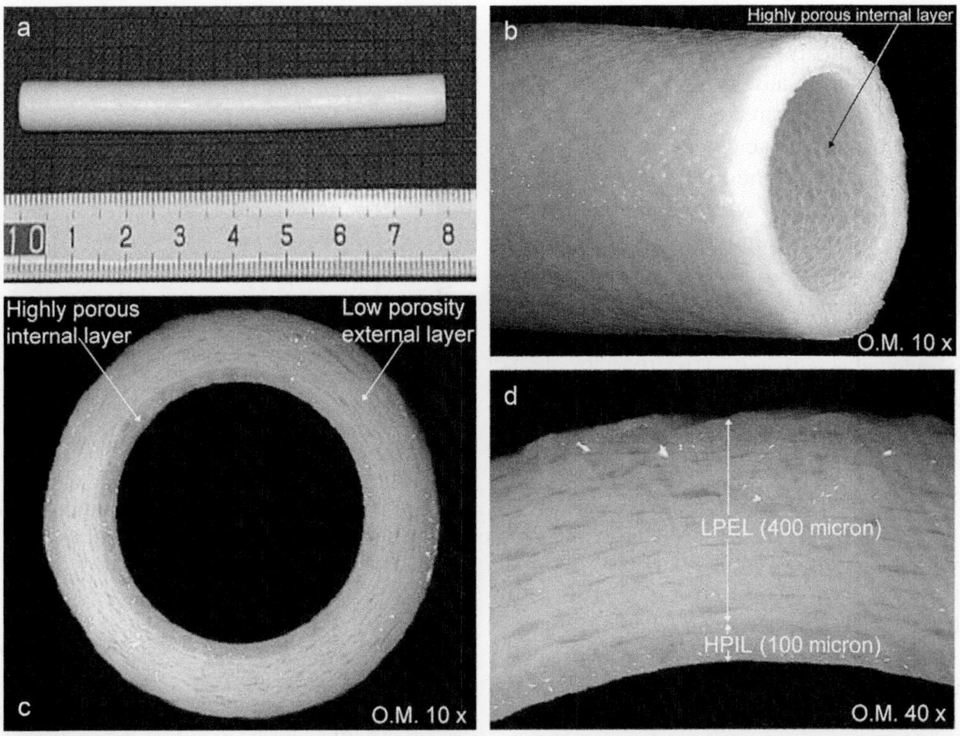

FIGURE 7.7 Representative PEU-PDMS graft used in the sheep implantation experiments. (a) longitudinal view of the graft; (b) tilted view of the graft showing the HPIL microfibrillar 3D structure; (c) cross-sectional view of the graft showing the sponge-like appearance of the graft wall composed of the LPEL fibrolamellar 3D structure and of the HPIL microfibrillar 3D structure (original magnification 10×); (d) higher magnification of the graft wall showing the extent of the HPIL (400 μm) and of the LPEL (100 μm)-500 μm total wall thickness (original magnification 40×). (For color version of this figure, the reader is referred to the online version of this chapter.) Reprinted with permission from Ref. [98]. © 2004 Elsevier Ltd.

and among various compositions, the PSUR containing 80% PDMS showed promising properties [13,98]. Soldani *et al.* developed small-diameter vascular grafts of PEU-PDMS semi-interpenetrating polymeric network (IPN), and results were compared with commercially available vascular grafts of ePTFE. PEU-PDMS grafts with 8 cm in length, 500 μm in thickness, and with 5 mm internal diameter were fabricated using spray, phase-inversion technique (Figure 7.7). For comparative study, grafts of PEU-PDMS and ePTFE were implanted using anastomotic technique in the common carotid artery of adult sheep, and biocompatibility studies were performed. PEU-PDMS grafts showed high patency rates, better handling, and compliance as compared to ePTFE grafts; also *in vivo*, these grafts have the ability of remodeling with no calcification sign and gradual degradation [98]. Same research group studied the effect of gamma irradiation sterilization on chemical, physical-mechanical, cytotoxic properties of microfibrillar vascular grafts of IPN of PEURs and PSURs. The infrared spectroscopy and light/scanning electron microscopy showed no significant changes due to gamma irradiation, and also the cytotoxicity study using L929 fibroblasts revealed no toxic effect. The results of mechanical properties and toxicity using PEU-PDMS polymers indicate that these polymers are stable [99]. Segmented PUR copolymers such as PEURs are used preferably for cardiovascular applications due to excellent blood-contacting properties and mechanical properties. However, *in vivo* upon long-term applications showed significant polymer degradation due to hydrolytic or oxidative mechanisms or both. These observations ascertain their degradation due to oxidation induced by metal ions. The comparative degradation between commercially available PDMS containing thermoplastic PUR elastomer such as Elast-Eon™ and polycarbonate containing PURs has been studied *in vitro*. Characterization techniques such as SEM, dynamic mechanical analyzer, and XRD indicate that PSURs such as Elast-Eon™ are stable as compared to PCURs. This could be due to phase separation phenomenon where intermixing and trapping of polycarbonate hard segments into soft segments render them for degradation; on other hand, phase-separated PDMS in Elast-Eon™ protects other segments from oxidation due to their hydrophobic nature [100].

7.3.5 Polyurethane and Natural Polymers

To study the hemocompatible of PURs in combination with natural polymers, Xu *et al.* [101] have synthesized copolymers of PUR with hyaluronic acid. The presence of hyaluronic acid could incorporate non thrombogenic nature to the PURs due to the presence of hyaluronic acid in biofluid. Hyaluronic acid when used as a chain extender for PUR synthesis created linear hydrophilic polymer structure and increased with increase in hyaluronic acid content. The significant decrease in platelet adhesion and red blood cell adhesion was observed as compared to only PUR that was 20 times higher than hyaluronic acid containing polymers. These studies also indicate that the hyaluronic acid-based PUR was highly cytocompatible, with endothelial cell adhesion and viability, and the hyaluronic acid-based PURs had similar surface smoothness as PURs. Some studies show that when PURs modified with derivative of natural polymer such as hydroxypropyl cellulose, the platelet adhesion could be reduced significantly [102,103]. To obtain biomaterials with excellent biological characteristics such as antithrombogenic properties while in contact with blood and tissues for long term, the modifications of PURs using extracellular matrix components such as collagen, glycosaminoglycans, and elastin are recommended. Also it was noticed that the biocompatibility of the PURs is correlated with phase separation phenomenon [104,105]. This indicates that synthesizing PUR devices using natural polymer is an excellent way of enhancing the biocompatibility of devices used in tissue engineering applications.

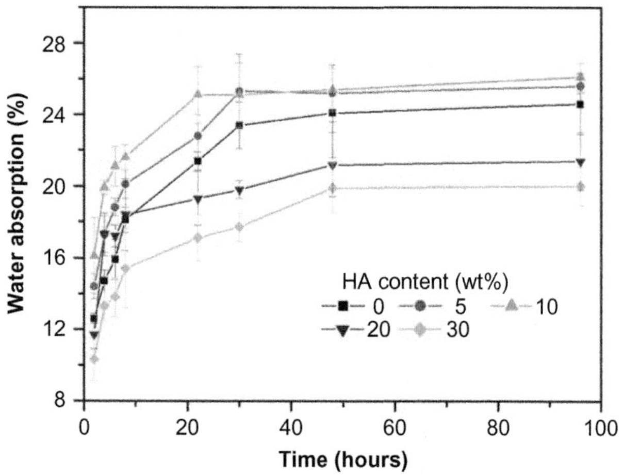

FIGURE 7.8 *In vitro* water uptake of n-HA/PCFC PUR composites at 37 °C. (For color version of this figure, the reader is referred to the online version of this chapter.) Reprinted with permission from Ref. [73]. © 2009 Wiley Periodicals, Inc.

7.3.6 Polyurethane Composites

The scaffolds used for tissue engineering applications such as bone implant need to have suitable mechanical properties and degradation behavior for suitable in growth of new bone upon scaffold implantation. Fu *et al.* [73] studied *in vitro* and *in vivo* degradation behavior of n-HA and poly(ester urethane) composites. This study indicates that n-HA content has significant effect on water absorption and degradation behavior of PURs *in vitro* and *in vivo*. The blending poly(ester urethanes) with 20% and 30% n-HA reduced the water uptake capacities of PURs whereas there was no difference noticed between water uptake capacities of PURs without n-HA and with 5% and 10% n-HA (Figure 7.8). *In vitro* degradation rates were higher when blended with 30% n-HA in basic pH as compared to acidic and neutral pH. There was no difference noticed between degradation profiles of samples with 5% and 10% n-HA under acidic conditions. Also under acidic condition the rates of degradations were higher without n-HA than 5% and 10% but lower than 20% and 30% n-HA content (Figure 7.9). The higher rates of PUR degradation are attributed to hydrophilic nature of composites in the presence of n-HA. *In vivo* studies also shows similar degradation patterns of composites with respect to increase in n-HA content. At the end of 12 weeks, upon subcutaneous implantation in rats, an average of 25%, 17%, and 5% weight loss was seen for PUR composites with 0%, 10%, and 30% n-HA content (Figure 7.10). This study indicates that the degradation profiles of PURs could be manipulated by varying n-HA contents for tissue engineering applications. Sawant *et al.* developed antimicrobial PUR surfaces composite films using silver (Ag) and gold (Au) nanoparticles of 20-27 nm size. Study indicates that Ag nanocomposites exhibited better antibiofilm properties than Au and casting method appeared to be better than swelling method in reducing the attachment (by a factor of 2). Composites reduced growth of organisms by six orders of magnitude, and protein and carbohydrate by two to five times. This study indicates that the nanocomposites of PUR with Ag prepared using casting method could be suitable for implant applications [78]. Yoshii *et al.* developed composite scaffolds using PURs synthesized using lysine triisocyanate and hydroxyapatite (HA) or β-TCP for the application of alternative graft material for bone defect treatments. The PUR/HA and PUR/TCP composite scaffolds were implanted in rats that had femoral defects, and the bioactivity of the composites was evaluated using X-rays, microcomputed tomography, and histology. *In vitro* cell cytotoxicity studies showed increased cell viability and proliferation, and also they supported the differentiation of osteoblastic cells relative to HA composites, and they had potential mechanical properties suitable for weight-bearing applications. From the histological images at week 4 showed growth of extensive bone matrix at HA/PUR and TCP/PUR composite surfaces. Images at higher magnification clearly revealed cellular infiltration into the matrix materials (Figure 7.11). When sections were stained using TRAP osteoclast-mediated composite, resorption was noticed at the interface of newly formed bone and implants (Figure 7.12). Hence, CaP/PUR composites have suitable osteoconductivity, biocompatibility, cellular infiltration, and appositional remodeling properties and could be potentially useful for excellent bioactive and weight-bearing orthopedic implants [106]. The initial mechanical property of scaffolds could diminish due to the high degree of interconnectivity that is useful for the effective cellular

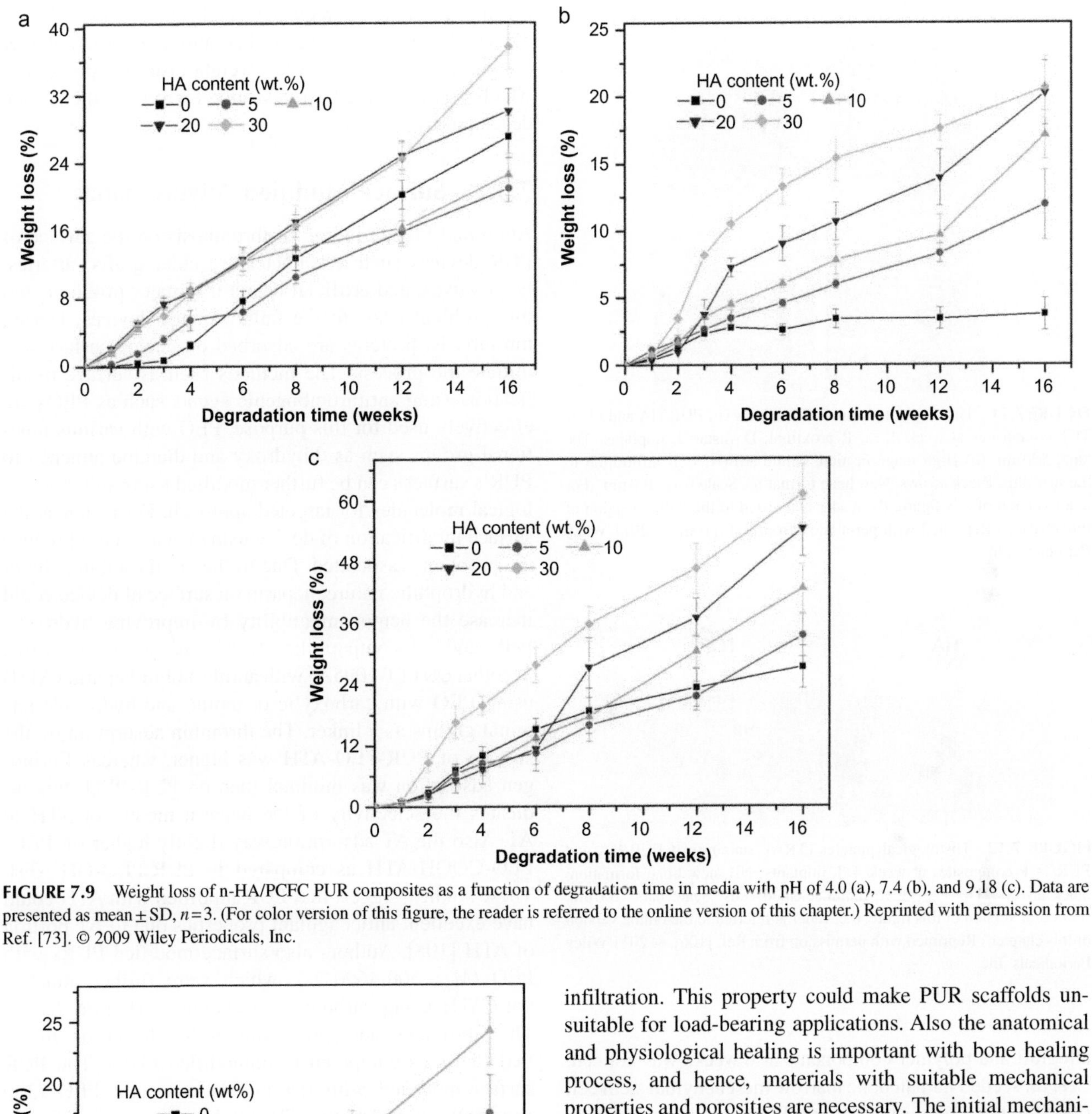

FIGURE 7.9 Weight loss of n-HA/PCFC PUR composites as a function of degradation time in media with pH of 4.0 (a), 7.4 (b), and 9.18 (c). Data are presented as mean ± SD, $n=3$. (For color version of this figure, the reader is referred to the online version of this chapter.) Reprinted with permission from Ref. [73]. © 2009 Wiley Periodicals, Inc.

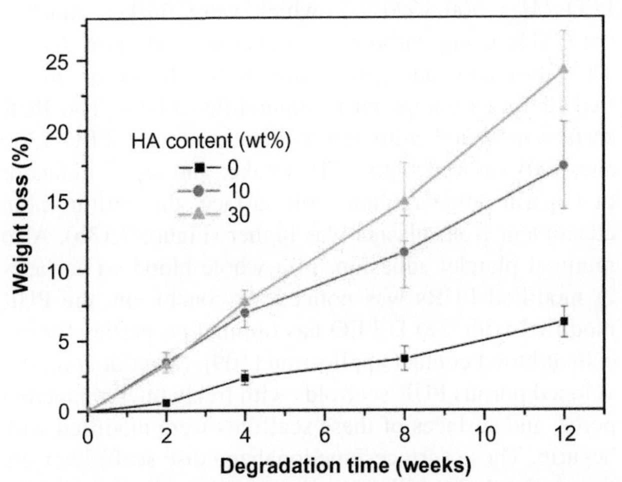

FIGURE 7.10 *In vivo* weight loss of n-HA/PCFC PUR composites with HA content of 0%, 10%, and 30%. Data are presented as mean ± SD, $n=3$. (For color version of this figure, the reader is referred to the online version of this chapter.) Reprinted with permission from Ref. [73]. © 2009 Wiley Periodicals, Inc.

infiltration. This property could make PUR scaffolds unsuitable for load-bearing applications. Also the anatomical and physiological healing is important with bone healing process, and hence, materials with suitable mechanical properties and porosities are necessary. The initial mechanical property of the PUR scaffold could be maintained intact by developing the composites by varying the compositions. Dumas *et al.* fabricated nonporous composite family using reactive compression molding of mineralized allograft bone particles (MBPs) with a biodegradable PUR binder. PUR binders are synthesized using polyester macro diols and a lysine-based polyisocyanate. Suitable compressive modulus and strength of 3-6 GPa and 107-172 MPa were achieved by varying the polyol molecular weights and surface chemistry of MBP via surface demineralization. The implantation of these MBP/PUR composites in New Zealand's white rabbits with bilateral femoral condyle plug defects showed extensive cellular infiltration deep into the interior of the implant. Also new bone formation and composite resorption of the

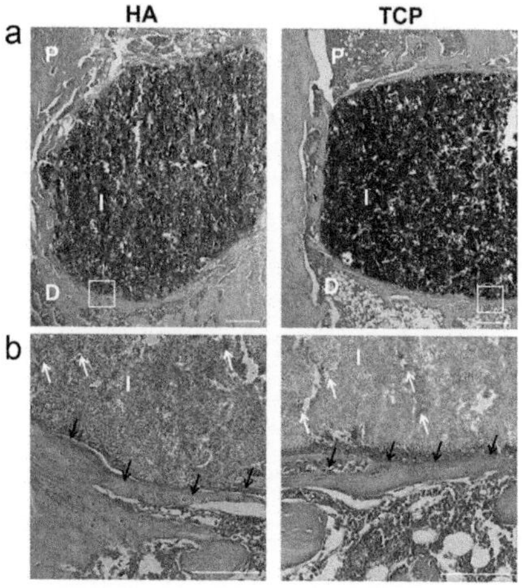

FIGURE 7.11 Histological pictures (HE staining) of PUR/HA and PUR/TCP composites at week 4. (a) P, proximal; D, distal; I, implants. The bars: 500 μm. (b) High magnification. White arrows: Cell infiltration to the scaffolds. Black arrows: New bone formation. Scale bars: 100 μm. (For color version of this figure, the reader is referred to the online version of this chapter.) Reprinted with permission from Ref. [106]. © 2011 Wiley Periodicals, Inc.

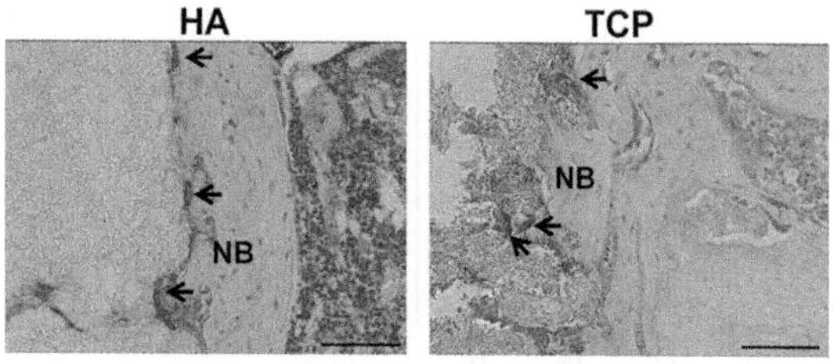

FIGURE 7.12 Histological pictures (TRAP staining) of PUR/HA and PUR/TCP composites at week 4. I, implants; NB, new bone formation. black arrows: TRAP positive multinucleated cells. Scale bars: 100 μm. (For color version of this figure, the reader is referred to the online version of this chapter.) Reprinted with permission from Ref. [106]. © 2011 Wiley Periodicals, Inc.

allograft and polymer components at 6 weeks was noticed. Li *et al.* seeded human bone marrow mesenchymal stem cell (hMSC) with composite of fibrin/PUR, and these composites were subjected to various mechanical stresses. The effect of compression, surface rotation frequency, and axial compression magnitude on the induction of chondrocyte-specific gene expression and protein synthesis was determined. The composites were precultured for 7 days followed by load application for 7 days. The dynamic compression and daily surface shear for 1 h induced chondrogenesis of hMSCs; also the mechanical stress significantly induced glycosaminoglycan synthesis and chondrocytic and transforming growth factor gene expressions. These studies indicate that the chondrogenesis of human MSCs stimulated by applying the mechanical stress can be further increased by using suitable frequency and compression [107]. These observations suggest that allograft bone/polymer composites have suitable weight-bearing properties, which could be used successfully for orthopedic device development.

7.3.7 Surface-Modified Polyurethanes

Abnormal blood clotting or thrombosis on the surface of PUR devices such as catheters vascular grafts, arteries, heart valves, and artificial hearts is a major problem, and this problem leads to the failure of the devices. During thrombosis, proteins are adsorbed on device surface and initiate the process. The methods such as surface modification using antithrombogenic agents such as PEOs are effectively used for this purpose. PEO with various functional groups such as dihydroxy and diamine attached to PUR's surfaces can be further modified using suitable biological molecules for targeted approach. For instance, the surface modification of device using heparin could reduce the problems associated. Due to the antithrombin activity and hydrophilic nature, heparin on surface of device could increase the hemocompatibility by improving hydrophilicity and cytocompatibility. Sask *et al.* surface-modified Tecothane® (TT-1095A) with antithrombin-heparin (ATH) using PEO with carboxylic or amino and hydroxide terminal groups as a linker. The thrombin adsorption on the surface of PUR-PEO-ATH was higher, whereas fibrinogen adsorption was minimal than on PUR-PEO; this indicates the selectivity of the heparin moiety of ATH to AT. Also the AT adsorption was slightly higher on PEU-PEO-COOH-ATH as compared to PUR-PEO-OH-ATH. These studies suggest that PUR modified with ATH could have excellent anticoagulant properties due to AT portion of ATH [108]. Authors also surface-modified PURs with PEO (M_w: 300-4600 Da), which were further attached with ATH using carbodiimide chemistry (Figure 7.13a). The fibrinogen adsorption was ~80% lower on modified PURs as compared to unmodified PURs. The PUR surface-modified with low-molecular-weight PEO (300 and 600) showed high ATH uptake, and as an indicator of heparin anticoagulation on surface, the antithrombin attachment from plasma was higher (Figure 7.13b). Also minimal platelet adhesion from whole blood on surfaces of modified PURs was noticed. In conclusion, the PUR modified with 600 D PEO has optimal properties for excellent blood contact application [109]. Schmidt *et al.* developed porous PUR scaffolds with freely interconnecting pores, and surfaces of these scaffolds were modified with heparin. These surface-functionalized disc scaffolds were tested for their ability to stimulate vessels and cellular growth under normoxia and chronic hypoxia for 10 days. Increased vascularization and cellularization were noticed

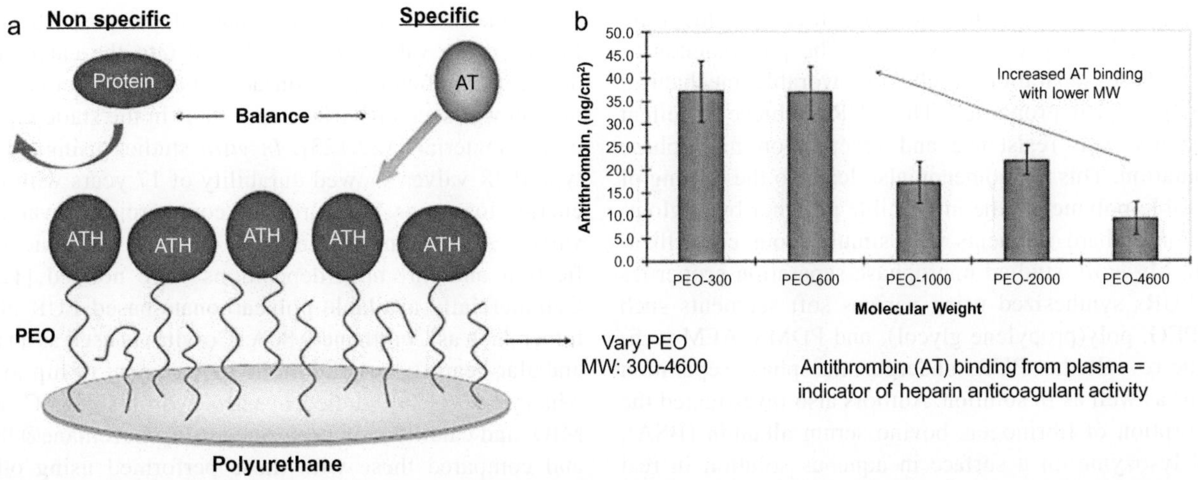

FIGURE 7.13 (a) Modification of PUR surface with an antithrombin-heparin complex for blood contact, (b) influence of molecular weight of PEO used as a linker/spacer. (For color version of this figure, the reader is referred to the online version of this chapter.) Reprinted with permission from Ref. [109]. © 2012 American Chemical Society.

when heparinized discs were used under normoxia. The addition of VEGF to the heparinized PUR discs resulted increase in vascularization and cellularization under normoxia and hypoxia. This indicates that under hypoxia the endogenous growth factor production was limited and an angiogenic response could occur when exogenous growth factors are delivered using PUR discs [110]. Poly(ester urethanes) are surface-modified using grafting technique by covalently bonding gelatin or silk fibroin using diamine aminolysis and glutaraldehyde cross-linking technique [111]. These surface-modified poly(ester urethane) films had low cell toxicity and good biocompatibility *in vitro;* also the subcutaneous implantation of these films showed better biocompatibility to peripheral tissues, and degradation rates were faster than nonsurface-modified PUR films. Williams *et al.* showed that Pellethane®, Tecoflex®, and Zytar® could act as potential substrates for subretinal pigment epithelial cell transplantation upon surface modification. Human retinal pigment epithelial cells (ARPE-19) were attached to Zytar® without air gas-plasma surface modification by 7 days, whereas only a few cells were attached to Pellethane® and Tecoflex®. The air gas-plasma treatment increased surface wettability of Pellethane® and Tecoflex®, and growth of a monolayer on surface of these PURs was well spread, and further cell growth was seen with a normal cobblestone' phenotype [112]. Lehle *et al.* have evaluated the biocompatibility of commercially available PURs used for the engineering of medical devices such as implantable pumps for ventricular assist devices (VAD). Human saphenous vein endothelial cells (HSVEC) and a mouse fibroblast cell line (L929) were cultivated on various PUR specimens, and *in vitro* cytotoxicity results were compared with tissue-cultured polystyrole (TCP) control. L929 cells covered surfaces of Tecothane®, Carbothane®, and Mecora specimens; however, incomplete layer for HSVEC monolayer was noticed onto the PUR. These studies also showed that there was less mitochondrial activity for all cell types attached to PUR surfaces and there was no cell proliferation. These results indicate the unfavorable results using commercially available PURs, indicating the need for surface modification [113]. Wu *et al.* developed a new fabrication processes for PUR-based transparent, thin, and flexible microfluidic devices. PUR microfluidic device with hydrophilic surface modification with PEO, high transparency (96%), and high bonding strength (326.4 kPa) showed low fibrinogen absorption, which is 80% lower than unmodified PDMS microfluidic PDMS device. This is attributed to microchannels, the integration of microfluidic interconnections and surface modification with hydrophilic PEO. These findings indicate that PEO-modified PUR could be effective for blood contacting micro fluidic applications [114].

7.4 PHASE SEPARATION BEHAVIOR

Thermodynamically, soft segments and hard segments in PURs are not compatible and therefore exhibit in a form of two-phase microstructure, and this phenomenon is called phase separation. The soft segments that provides connection between hard segments acts as an effective crosslinkers which are responsible for the maximum deformation of structure. In addition to this hard segments are also held together by physical interactions. Typically, hard segments are urethane linkages in which they have carbonyl and amine groups and result in agglomeration due to hydrogen bonding and strengthen the effect of hard segments over soft amorphous chains already present in polymer. This phenomenon is incomplete and accelerates under various conditions; therefore, these polymers are also called living polymers [115]. The phenomenon of microphase separation results in favorable mechanical and biological properties. The PURs achieve excellent elasticity, tear resistance and creep upon microphase separation. This phenomenon also leads to the turning of miscible polymer to the immiscible polymer by agglomeration of hard segments with simultaneous crystallization. Ma *et al.* studied microphase separation properties of PURs synthesized using various soft segments such as PEG, poly(propylene glycol), and PDMS. AFM technique reveals that PURs undergo microphase separation in air as well as in solution. Authors also investigated the adsorption of fibrinogen, bovine serum albumin (BSA), and lysozyme on a surface in aqueous solution in real time using quartz crystal microbalance with dissipation and surface plasmon resonance techniques. Decrease in protein adsorption on the surface of PURs upon phase separation was noticed, which is attributed to hydrated PEG segments [116]. Xu *et al.* did a similar study on the phase separation behavior of implanted PUUR using an array of AFM technique and correlated the surface structure with molecular interactions. AFM imaging results convey that hard segments undergo rearrangement and enrichment at the surface upon hydration and protein adhesion on polymer surface was less with respect to hydration time as revealed by protein modified AFM probe. Force measurement studies using BSA showed that the decrease in adhesion was due to polar hard domain region; these studies were also confirmed using nanogold-labeled protein conjugates. The nanogold-labeled protein conjugate study visualized protein adsorption at separate microstructures and indicated more protein adsorption on the more nonpolar hydrophobic soft segments as compared to hard segment domain. These studies indicate that the microphase separation influences on the protein and PUR surface interactions upon hydrations due to change in local surface microenvironment [117]. For instance the surface of blend PURs with suitable ratio of hard and soft domains indicated lowest fibrinogen-albumin adsorption ratio, which could be attributed to the similar scale of PUR hard domains on surface with respect to proteins [118,119].

7.5 CALCIFICATION

Mechanical and bioprosthetic heart valves are the two major types of valves that are in clinical use currently. These valves are durable but require anticoagulation therapy continuously upon implantation. Bioprostheses are less durable, causing tissue failure due to leaking in calcified valve, whereas mechanical valves require continuous anticoagulation therapy, which is catastrophic to the patient. Hence, they are not good solutions for the replacement of defective natural valves [120,121]. Polyurethane valve when implanted into the calves underwent calcification on surface whereas thrombus formation was seen on surface as well as in the static region of the material [122,123]. *In vitro* studies using prototype PUR valves showed durability of 17 years with low energy losses as compared to commercially available valves. And upon implantation of valve extrinsic calfication and thrombin depositions were noticed [124]. Commercially available polycarbonate-based PUR elastomer such as Corethane® 80A (Corvita) is used as an acetabular bearing material in the replacement of hip joint. Khan *et al.* [125] studied *in vitro* hydrolysis, ESC, and MIO, and calcification performed using Corethane® 80A and compared these with those performed using other commercially available PURs such as polyether-based Pellethane® 2363-80A (DOW chemicals), PHMO-PUR (CSIRO), and polycarbonate-based ChronoFlex® AL-80A (CardioTech). Among these polymers, Corethane® 80A showed good resistance to hydrolysis, ESC, MIO, and calcification followed by ChronoFlex® 80A, PHMO-PUR, and Pellethane® 80A. This study provides evidence for use of polycarbonate based Corethane® 80A as a resistance bearing layer in hip arthroplasty and for other joints possibly. The ideal bone graft substitute to be used for implantation should be porous with interconnected pores for ingrowth of capillaries and cell infiltration and degradable with respect to bone healing. In addition the bone material should have the ability to calcify *in vivo* and promote osteoblast differentiation by attracting MSCs [126,127]. Gorna and Gogolewski [127] synthesized cross-linked 3D biodegradable PURs scaffolds using PEG, PCL diols, and polydiamine-based diols with various ratios for bone graft substitute. The scaffolds were porous where size and geometry of the pores was related to chemical composition used while manufacturing the devices. These scaffolds underwent hydrolytic degradation *in vitro* and increased with respect to increase in PEG and the calcium-complexing moiety in the scaffolds. Interestingly, these PURs with various ratios of polydiol monomers and calcium-complexing moieties induced deposition of calcium phosphate crystals in to the scaffolds. Also atomic ratios of calcium-phosphorus were varied between 1.5 and 2.0 with respect to variation in chemical composition of PURs. In another study [33], authors studied *in vitro* calcification of PURs synthesized using PEO or Pluronic® F-68 and PCL diols. As shown in Figure 7.4, the *in vitro* degradation rates of these PEEURs increased with increase in hydrophilicity as well as increase in molecular weights of hydrophilic diol segments. On other hand, the *in vitro* calcification also increased with increase in hydrophilicity of the PEEURs and increase in molecular weights of hydrophilic diols as shown in Figure 7.14.

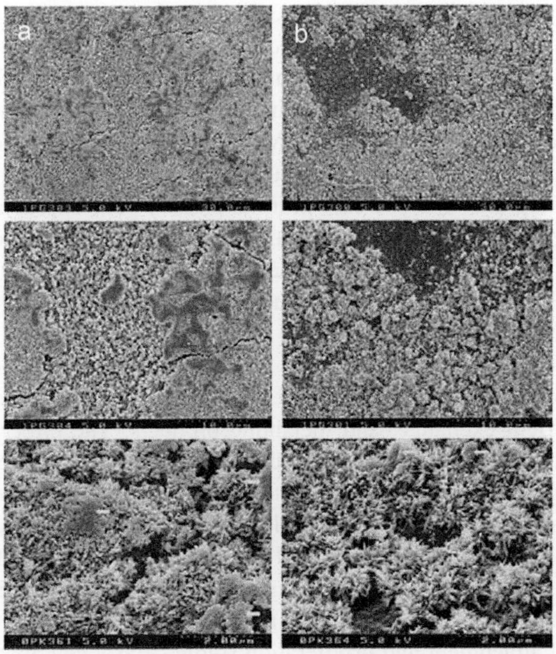

FIGURE 7.14 Calcium salt crystals formed on the surface of PUR films subjected to *in vitro* calcification: (a) PCL530-PEO2000 (70:30; white arrows show large plate like crystals within the plate-like aggregates) and (b) PCL2000-PEO600 (70:30; white arrows show smaller plate-like crystals sparsely distributed among the prevailing needlelike aggregates). Reprinted with permission from Ref. [33]. © 2009 Wiley Periodicals, Inc.

7.6 POLYURETHANE APPLICATIONS

7.6.1 Drug Delivery

The long-term drug release devices have various advantages such as patience compliance and excellent adherence rate. Efforts are made to modify PURs with suitable monomers and their compositions in an effort to develop zero-order drug release devices. The swelling, hydrophilicity/hydrophobicity, drug diffusion, degradation, and glass transition temperature of the PUR play important role in delivering the drug release. In general, higher swelling devices could release drug at faster rate as compared to less swelling devices. Also the swelling of PUR matrix devices could slow down the drug release rate due to the increased length of the passage through which drug diffusion occurs [128]. PUR's glass transition temperature (T_g) and degradation properties plays very important role in modulating drug release properties. Higher T_g, which means less polymer chain mobility, slows down the drug release rate due to the slower drug diffusion through the polymer chains. On other hand, the hydrophilicity and degradable nature of polymer enhances the drug release properties due to faster drug diffusion upon hydration and loss in molecular weight. Hence, modulations in properties such as glass transition temperature, hydrophilicity, and degradation of PUR could result in suitable drug release properties. PUR's chain mobility increases upon increase in temperature resulting in increased drug release rate through the mobile chains; also hydrogen-bonding or hydrophobic interactions between PURs and drugs are weakened [128]. Altering matrix porosity also plays very important role for the controlled drug release from the PUR-based devices. PUR matrix device loaded with cefadroxil and pore formers such as D-mannitol, BSA, and PEG 1450 increased release rate [129]. The solubility of pore former in solvent and in polymer impacts on the drug release rate behavior of matrix device. For instance, BSA is less soluble in solvent than D-mannitol and PEG 1450; therefore, better pores can be obtained using BSA as compared to D-mannitol and PEG. Therefore, enhanced cefadroxil release from PUR matrix loaded with BSA was obtained as compared to D-mannitol and PEG-loaded devices. Physical forms of drug such as acid, base, or salt also affect on release profiles due to the interactions with polymers. For instance, free base of vancomycin loaded in degradable PUR synthesized from lysine triisocyanate released 80% of the drug in 50 days with lower initial burst due to its high solubility in polymer and less solubility in release media, whereas salt of vancomycin loaded in the same matrix released 80% in 5 days; this indicates that vancomycin salt has less solubility in polymer and could generate pores with respect to vancomycin release [130]. The implant films of PUR based on IPDI, PCL, and PEG loaded with dexamethasone acetate could be significantly effective to decrease neovascularization after 7 days of subcutaneous implantation in mice. The PUR-dexamethasone acetate films efficiently modulate the key components of inflammation, angiogenesis, and fibrosis induced by sponge discs in an experimental animal model. The dexamethasone acetate-loaded PUR films were embedded in nonbiocompatible sponges and implanted subcutaneously in mice. Subcutaneous implants maintained linear release rate for at least 45 days with constant drug levels and for 120 days *in vitro* [131]. Similarly, for dexamethasone release, DaSilva *et al.* synthesized PURs using IPDI, PCL, and hydrazine, and in addition, devices were loaded with montmorillonite Na^+ clay nanoparticles. The presence of clay nanoparticles disturbed the phase separation behavior of the PUR, and also its presence helped to enhance drug release rate as well as PUR degradation as compared to PUR device loaded with dexamethasone without clay nanoparticles. These PURs released dexamethasone for a year with nearly constant release and with nearly 30% polymer weight loss. Similarly, controlled release of triamcinolone acetonide from implantable devices based on PUR synthesized using PCL has been studied. Sponge discs of PUR synthesized from IPDI and PCL were developed for the release of triamcinolone acetonide (TA) *in vitro* and *in vivo*. Study indicated that 64% of TA was released *in vitro* and about 81% of TA was released in 45 days *in vivo*. Upon TA-PUR sponge implantation, an inflammatory angiogenesis was inhibited efficiently; these studies indicate that PURs have the potential to control drug release and could be used for long-term application [132]. PUR scaffolds loaded with insulin-like growth factor-I and hepatocyte growth factor were tested for release properties and polymer degradation in phosphate-buffered saline with and without enzyme lipase to stimulate *in vivo* insulin release and polymer degradation [133]. Porous scaffolds developed using 5 and 8 wt% of PUR released about 80% insulin for more than a year in multiphase manner, and in presence of enzyme lipase, release was faster due to polymer degradation, which was about 90% in 7 weeks and then very slow up to about 5 months. The significant PUR weight loss (~95%) was seen in presence of lipase enzyme and negligible polymer weight loss was seen without lipase enzyme. This indicates that the polymer degradation plays an important role in releasing drug from the devices.

7.6.2 Tissue Engineering

Promising approaches such as tissue engineering and regenerative medicine have been in use for organ and tissue restoration purpose. This can be achieved by mimicking the properties of organs and tissues with the mechanism of physiological microenvironment regeneration. Polyurethane scaffolds are used widely for this application by the combination of surface modification, cell seeding, or the addition of growth factors so that they can mimic extracellular matrix properties. These scaffolds act as a support for the growing tissues and organs as well as navigator for cells involved in regeneration process. The outcome of scaffolds used for tissue constructs is significantly dependent on the effective vascularization upon implantation that helps for adequate cellular oxygen supply for long-term survival of implant. Various attempts have been made to improve the biocompatibility and maximum biological compliance of the scaffolds used for tissue regeneration applications using modified PURs. Mechanical and biological heart valve prostheses are currently in use, mechanical devices have long lifetime and biological devices have limited lifetime. Also when mechanical prostheses are used, oral anticoagulation treatment is needed for life time [134,135]. Thierfelder *et al.* [136] studied PUR scaffolds biocompatibility and compared with compatibility of biological prostheses, which is an established conventional aortic homograft valve used in modern heart surgery under similar conditions in a virgin bioreactor. PUR scaffold and homograft when seeded with human fibroblast and endothelial cells in a bioreactor showed adherence of endothelial cells and fibroblast on the surface and formation of extracellular matrix. PUR scaffolds and biological prostheses showed extracellular matrix formation, colonization, and inflammatory cell reaction. The study compares the behavior of the seeded human fibroblast and endothelial cells on PUR and homograft aortic valve scaffolds using a rotating seeding device. This study indicates that PUR scaffolds are biocompatible and could be a new option to replace an aortic valve. SEM and immunohistochemical analysis revealed the presence of confluent cell layer on the surface, presence of EC and FB before as well as after conditioning, and the formation of ECM during conditioning. This study also indicates higher expression of ECM on inner surface of scaffold, higher inflammatory response during the conditioning of homografts, and decrease in inflammatory gene expression upon endothelialization. Hence, the efficient colonization, the establishment of an ECM, and comparable inflammatory cell reaction to the scaffolds indicate the biocompatibility of PUR scaffolds.

Graft porosity and pore interconnectivity are important factors that can influence on host cell immigration, nutrient exchange, and metabolic products. The effect of Pellethane® electrospun vascular fabric porosity on cell migration has been investigated by [137]. The PUR electrospun grafts were fabricated into fine mesh, low porosity with 2 μm pore size and void fraction of 53%, and coarse mesh, high porosity with 4 μm pore size and void fraction of 80%, and implanted in rat model for up to 6 months for cell attachment and proliferation studies. Early cell attachment was noticed for fine pore mesh *in vitro* and *in vivo* significantly, and high cell population was noticed for coarse mesh grafts, and also rapid endothelial cell coverage on surfaces was noticed for both devices. The grafts indicated suitable degradation and mechanical properties before and after implantation over 6 months' study with reconstitution of host cells without affecting on any tissue or cell properties. Previously, same authors have investigated cell proliferation using small-diameter Pellethane® electrospun fabricated in vascular grafts with 2 μm pore size and void fraction of 70% into abdominal aorta in rat for up to 6 months [138]. Vascular grafts were well accepted and cell migration was started within a week with significant increase in attachment of CD34+ cells with consistent increase and unaltered cell proliferation over 6-month period. Hence it can be concluded that the nanostructured vascular grafts of PUR promoted cell immigration and continuing differentiation. Biodegradable PURs are highly biocompatible materials used widely as porous scaffolds in bone tissue scaffold engineering. Degradable PURs synthesized using PCL, P(CL-D,L-LA), or poly(γ-butyrolactone) and porous scaffolds were developed for soft tissue remodeling purpose and were highly haemocompatible [139]. Upon implantation for 21 days in rat model, a highly vascularized stroma formation was seen through interconnected pores with minimal number of macrophages on the device edges. This study indicates that the devices could be developed for cardiovascular regeneration using porous PUR scaffolds. Hill *et al.* studied osteogenesis and osteoblast behavior of PUR-HA scaffolds implanted in nude mice; this approach was used to create new bone. The comparative study between PUR and PUR-HA indicated that there was more bone formation in PUR-HA group [140]. The mechanical properties of PURs porous scaffolds were improved by blending PURs with bioactive fluor-hydroxyapatite (F-HA) and HA. Studies show that compressive strength and compressive modulus increased significantly with no effect on porosity of the bone scaffolds [141]. Herrmann *et al.* [142] tested PUR, collagen cell carriers, expanded PTFE, and expanded PTFE with spider silk protein1 and RGD peptide for cardiac patch construction. Cardiac patches were seeded using MSCs for 35 days and characterized using advanced techniques such as SEM, histological stains microscopy, fluorescence microscopy, and mitochondrial assay. Organized cell multilayers with seeding density of 0.75×10^6 cells/cm^2 were observed for PUR patches with increased mitochondrial activity. The cell seeding density for collagen cell carriers was low, and extended PTFE and extended PTFE with SSP1-RGG were not receptive for the cell seeding. This indicates that PURs crystallized cells most as compared to collagen cell carrier and PTPE patches; hence, they could be more appropriate for use as cardiac patches. Cerebral or brain aneurysm is a brain dysfunction related to blood vessels and results in localized dilation or ballooning of the vessels. Cerebral aneurysms have been treated by usual methods such as platinum coils; however, these devices are not stable because of chronic inflammation and coil compaction and due to the growth of aneurysm [143]. Rodriguez *et al.* have characterized PUR based on shape memory polymer foam for the treatment of intracranial aneurysms. In this study, SMP-PUR foam localized thrombogenicity, biocompatibility, and stability to serve as filler within aneurysm. These properties and healing of aneurysm were evaluated using low-vacuum SEM technique at various time points up to 90 days. It was observed that the healing was started within an hour of implantation with partial healing on day 30 and fully healing on day 90 in porcine. This indicates that SMP-PUR has the potential to heal aneurysm and could be used as an alternative for unstable traditional methods such as platinum coils [143].

7.6.3 Polyurethane Medical Devices

PUR biomedical devices such as heart valves, left ventricular assist devices, and blood pumps have short-term biocompatibility [1,12,144]. The failure of these devices for long-term application is due to the catastrophic failure of PURs used for the fabrication of these devices. The failure of heart valves fabricated using aromatic PURs such as Pellethane®, Biomer, Mitrathane®, and Cardiothane® is due to the calcification and stiffness that leads to catastrophic failure of organs [145]. The PEUUR (Biolon blood sacs) showed formation of micro pitting and surface erosion upon implantation in calves for 17 weeks [146]. The Tecothane® valve fabricated survived for flex life of 361 million cycles as shown by Butterfield *et al.* [147]. Aromatic PURs, for example, Pellethane®, Biomer, Cardiothane®, and Mitrathane® used for cardiac applications are withdrawn from the market because of various court issues. Thomas and Jayabalan [67,148] synthesized aliphatic PUURs using aliphatic unsaturated hydrocarbon polyol hydroxyl-terminated polybutadiene, H_{12}MDI and hexamethylene diamine. Authors report that the PUUR's HFL 18 PU synthesized in this study is fully aliphatic, ether-free physically cross-linked with suitable elastic modulus (6.84 ± 0.27 MPa). Also these PUURs have *in vivo* biostability, biocompatibility, and high flex life (721 ± 30 million cycles). Authors conclude that the physical cross-linking, which leads to three-dimensional structure in hard domain, could be the result of urea-urea hydrogen bonding. Microcrystallite was formed as a result of physical cross-linking, which also caused micro crystallite formation in hard segments. Also the unsaturated hydrocarbon polyol soft segment used for synthesis of HFL18-PU is stable against oxidative degradation as compared to polyether diol used for synthesis of previously reported PURs for fabrication of cardiac devices. The stability of HFL18-PU is attributed to three-dimensional physical cross-linking, which protected degradation of unsaturated groups. Apart from these results, authors mention that HFL18-PU has excellent long-term *in vivo* biostability in rat animal model, and three-dimensional cross-linking favors to high flex life, and biodurability; in addition to this microphase, separation favors to less calcification, platelet consumption, and optimum growth of endothelial cells. These studies indicate that HFL 18-PU is a potential candidate for the cardio vascular device development. The list of some PUR biomedical devices and polymers available in market [12,114,149–155] is as given in Table 7.2.

7.7 CONCLUSION

Polyurethanes are a versatile class of polymers with great control over their physicochemical properties based on the chemical composition. Segmented PURs are designed with well-defined degradation and mechanical properties combined with excellent biocompatibility that makes them attractive for the development of drug delivery, tissue engineering, and medical devices. Various PURs including PEURs, poly(ester urethanes), PCURs, PSURs, surface-modified PURs, and composite PURs have been developed for a variety of biomedical applications. Many research efforts are continued in the development of PURs for specific drug delivery and tissue regeneration application with a particular emphasis on biocompatibility and biodegradability.

ACKNOWLEDGMENTS

The authors gratefully acknowledge funding from the Raymond and Beverly Sackler Center for Biomedical, Biological, Physical, and Engineering Sciences. Authors also acknowledge the funding from National Science Foundation IIP-1311907 and EFRI-1332329.

TABLE 7.2 Biomedical Polyurethane Devices

Device name/Polyurethane	Applications	References
Actifit™	Medial and lateral meniscal scaffolds	[149]
BD Vialon™	IV Catheter	[12]
PELLETHANE2363™	Pacemaker lead insulation	[12]
Biomer™	Cardiac assist devices	[12]
Tri-leaflet valve	Prosthetic heart valve	[150]
Silver-coated polyurethane catheters	Long-term implants	[151]
Tecothane® TT-1095A	Microfluidic devices for blood-contacting applications	[114]
SOLOSITE®	Conformable wound gel dressing	[152]
Silverlon®	Wound dressing	[153]
Carbothane®	Venous central catheter	[154]
Dureflex®, Baymedix FR	Wound dressing films	[155]
Baymedix CL 100, Baymedix CD	Drug-releasing coatings	[155]

REFERENCES

[1] N.M.K. Lamba, K.A. Woodhouse, S.L. Cooper, Polyurethanes in Biomedical Applications *Boca Raton*, CRC Press, New York, 1998.

[2] Z. Li, Z. Zhang, K. Liu, X. Ni, J. Li, Biodegradable hyperbranched amphiphilic polyurethane multiblock copolymers consisting of poly(propylene glycol), poly(ethylene glycol), and polycaprolactone as in situ thermogels, *Biomacromolecules* 13 (12) (2012) 3977–3989.

[3] Z. Lijuan, Y. Lunquan, D. Mingming, L. Jiehua, T. Hong, W. Zhigao, F. Qiang, Synthesis and characterization of pH-sensitive biodegradable polyurethane for potential drug delivery applications, *Macromolecules* 44 (2011) 857–864.

[4] C. Mattu, R. Pabari, M. Boffito, S. Sartori, G. Ciardelli, Z. Ramtoola, Comparative evaluation of novel biodegradable nanoparticles for the drug targeting to breast cancer cells, *Eur. J. Pharm. Biopharm.* 85 (3) (2013) 463–472.

[5] K. Gupta, S. Pearce, A. Poursaid, H. Aliyar, P. Tresco, M. Mitchnik, P. Kiser, Polyurethane intravaginal ring for controlled delivery of dapivirine, a nonnucleoside reverse transcriptase inhibitor of HIV-1, *J. Pharm. Sci.* 97 (10) (2008) 4228–4239.

[6] S.H. Kuo, P. Kuzma, Long term drug delivery devices with polyurethane based polymers and their manufacture, US8343528 B2, US 12/907,717, 2013.

[7] N. Shelke, J. Clark, T. Johnson, J. Fabian, A. Tuitupou, J. Nebeker, M. Clark, D. Friend, P. Kiser, 90 Day vaginal delivery of poorly water soluble drug levonorgestrel from coaxially extruded polyurethane reservoirs, in: "Utah Biomedical Engineering Conference" Rice Eccles Stadium, Salt Lake City, Utah, USA, 2011.

[8] S. Grad, L. Kupcsik, K. Gorna, S. Gogolewski, M. Alini, The use of biodegradable polyurethane scaffolds for cartilage tissue engineering: potential and limitations, *Biomaterials* 24 (28) (2003) 5163–5171.

[9] M.R. Williamson, R. Black, C. Kielty, PCL-PU composite vascular scaffold production for vascular tissue engineering: attachment, proliferation and bioactivity of human vascular endothelial cells, *Biomaterials* 27 (19) (2006) 3608–3616.

[10] T. Shirota, H. He, H. Yasui, T. Matsuda, Human endothelial progenitor cell-seeded hybrid graft: proliferative and antithrombogenic potentials in vitro and fabrication processing, *Tissue Eng.* 9 (1) (2003) 127–136.

[11] J. Santerre, R. Labow, G. Adams, Re: the polyurethane foam covering the Même Breast Prosthesis: a biomedical breakthrough or a biomaterial tar baby? *Ann. Plast. Surg.* 29 (5) (1992) 477–478.

[12] R. Zdrahala, I. Zdrahala, Biomedical applications of polyurethanes: a review of past promises, present realities, and a vibrant future, *J. Biomater. Appl.* 14 (1) (1999) 67–90.

[13] D. Martin, L. Warren, P. Gunatillake, S. McCarthy, G. Meijs, K. Schindhelm, Polydimethylsiloxane/polyether-mixed macrodiol-based polyurethane elastomers: biostability, *Biomaterials* 21 (10) (2000) 1021–1029.

[14] G. Woo, M. Mittelman, J. Santerre, Synthesis and characterization of a novel biodegradable antimicrobial polymer, *Biomaterials* 21 (12) (2000) 1235–1246.

[15] Z. Ma, Y. Hong, D. Nelson, J. Pichamuthu, C. Leeson, W. Wagner, Biodegradable polyurethane ureas with variable polyester or polycarbonate soft segments: effects of crystallinity, molecular weight, and composition on mechanical properties, *Biomacromolecules* 12 (9) (2011) 3265–3274.

[16] G. Skarja, K. Woodhouse, Synthesis and characterization of degradable polyurethane elastomers containing and amino acid-based chain extender, *J. Biomater. Sci. Polym. Ed.* 9 (3) (1998) 271–295.

[17] J. Guan, M. Sacks, E. Beckman, W. Wagner, Biodegradable poly(ether ester urethane)urea elastomers based on poly(ether ester) triblock copolymers and putrescine: synthesis, characterization and cytocompatibility, *Biomaterials* 25 (1) (2004) 85–96.

[18] J. Park, K. Park, Y. Bae, PDMS-based polyurethanes with MPEG grafts: synthesis, characterization and platelet adhesion study, *Biomaterials* 20 (10) (1999) 943–953.

[19] Y.H. Lin, N.K. Chou, K.F. Chen, G.H. Ho, C.H. Chang, S.S. Wang, S.H. Chu, K.H. Hsieh, Effect of soft segment length on properties of hydrophilic/hydrophobic polyurethanes, *Polym. Int.* 56 (2007) 1415–1422.

[20] S. Guelcher, Biodegradable polyurethanes: synthesis and applications in regenerative medicine, *Tissue Eng. B Rev.* 14 (1) (2008) 3–17.
[21] I. Bonzani, R. Adhikari, S. Houshyar, R. Mayadunne, P. Gunatillake, M. Stevens, Synthesis of two-component injectable polyurethanes for bone tissue engineering, *Biomaterials* 28 (3) (2007) 423–433.
[22] E. David, M. Derek, A. Mauro, Degradation of synthetic polymeric scaffolds for bone and cartilage tissue repairs, *Soft Matter* 5 (2009) 938–947.
[23] S. Gogolewski, Biocompatible, biodegradable polyurethane materials with controlled hydrophobic to hydrophilic ratio, EP Patent App. 20, PCT/CH2004/000471, WO/2006/010278.
[24] J.P. Sheth, A. Aneja, G.L. Wilkes, E. Yilgor, G.E. Atilla, I. Yilgor, F.L. Beyer, Influence of system variables on the morphological and dynamic mechanical behavior of polydimethylsiloxane based segmented polyurethane and polyurea copolymers: a comparative perspective, *Polymer* 45 (20) (2004) 6919–6932.
[25] E. Ho, F. Damian, A. Tuitupou, N. Shelke, M. Clark, F. Friend, P. Kiser, A combination intravaginal ring for the protection against HIV and unwanted pregnancy, in: 37th Annual Meeting & Exposition of the Controlled Release Society, July 10–14, 2010, 2010.
[26] J. Guan, K. Fujimoto, M. Sacks, W. Wagner, Preparation and characterization of highly porous, biodegradable polyurethane scaffolds for soft tissue applications, *Biomaterials* 26 (18) (2005) 3961–3971.
[27] D. Sin, X. Miao, G. Liu, F. Wei, G. Chadwick, C. Yan, T. Friis, Polyurethane (PU) scaffolds prepared by solvent casting/particulate leaching (SCPL) combined with centrifugation, *Mater. Sci. Eng. C* 30 (1) (2010) 78–85.
[28] S. Agarwal, J. Wendorff, A. Greiner, Use of electrospinning technique for biomedical applications, *Polymer* 49 (26) (2008) 5603–5621.
[29] R. Nirmala, H.S. Kang, M. El-Newehy, R. Navamathavan, H.M. Park, H. Kim, Human osteoblast cytotoxicity study of electrospun polyurethane/calcium chloride ultrafine nanofibers, *J. Nanosci. Nanotechnol.* 11 (6) (2011) 4749–4756.
[30] H. Machado, R. Correia, J. Covas, Synthesis, extrusion and rheological behaviour of PU/HA composites for biomedical applications, *J. Mater. Sci. Mater. Med.* 21 (7) (2010) 2057–2066.
[31] J. Santerre, K. Woodhouse, G. Laroche, R. Labow, Understanding the biodegradation of polyurethanes: from classical implants to tissue engineering materials, *Biomaterials* 26 (35) (2005) 7457–7470.
[32] A. Mathur, T. Collier, W. Kao, M. Wiggins, M. Schubert, A. Hiltner, J. Anderson, In vivo biocompatibility and biostability of modified polyurethanes, *J. Biomed. Mater. Res.* 36 (2) (1997) 246–257.
[33] K. Gorna, S. Gogolewski, Biodegradable polyurethanes for implants. II. In vitro degradation and calcification of materials from poly(epsilon-caprolactone)-poly(ethylene oxide) diols and various chain extenders, *J. Biomed. Mater. Res.* 60 (4) (2002) 592–606.
[34] T. van Tienen, R. Heijkants, P. Buma, J. de Groot, A. Pennings, R. Veth, Tissue ingrowth and degradation of two biodegradable porous polymers with different porosities and pore sizes, *Biomaterials* 23 (8) (2002) 1731–1738.
[35] P.A. Gunatillake, D.J. Martin, G.F. Meijs, S.J. McCarthy, R. Adhikari, Designing biostable polyurethane elastomers for biomedical implants, *Aust. J. Chem.* 56 (6) (2003) 545–557.
[36] A. Simmons, J. Hyvarinen, R. Odell, D. Martin, P. Gunatillake, K. Noble, L. Poole-Warren, Long-term in vivo biostability of poly(dimethylsiloxane)/poly(hexamethylene oxide) mixed macrodiol-based polyurethane elastomers, *Biomaterials* 25 (20) (2004) 4887–4900.
[37] E. Christenson, M. Dadsetan, M. Wiggins, J. Anderson, A. Hiltner, Poly(carbonate urethane) and poly(ether urethane) biodegradation: in vivo studies, *J. Biomed. Mater. Res. A* 69 (3) (2004) 407–416.
[38] T. Nakajima-Kambe, Y. Shigeno-Akutsu, N. Nomura, F. Onuma, T. Nakahara, Microbial degradation of polyurethane, polyester polyurethanes and polyether polyurethanes, *Appl. Microbiol. Biotechnol.* 51 (2) (1999) 134–140.
[39] K. Fujimoto, H. Inoue, Y. Ikada, Protein adsorption and platelet adhesion onto polyurethane grafted with methoxy-poly(ethylene glycol) methacrylate by plasma technique, *J. Biomed. Mater. Res.* 27 (12) (1993) 1559–1567.
[40] K. Fujimoto, H. Tadokoro, Y. Ueda, Y. Ikada, Polyurethane surface modification by graft polymerization of acrylamide for reduced protein adsorption and platelet adhesion, *Biomaterials* 14 (6) (1993) 442–448.
[41] S. Kuo, P. Kuzma, Long term drug delivery devicew with polyurethane based polymers and their manufacture, US Patent 20130096496, 13/693659, Endo Pharmaceuticals Soulutions, Inc., 2013.
[42] P. Kuzma, H. Quandt, Long term drug delivery devices with polyurethane-based polymers and their manufacture, US Patent 8,460,274, 2013.
[43] S.H. Kuo, P. Kuzma, Long term drug delivery devices with polyurethane based polymers and their manufacture, Google Patents, 2013.
[44] P. Kiser, T. Johnson, J. Clark, N. Shelke, R. Rastogi, Intravaginal devices for drug delivery, in WO Patent App. PCT/US2012/047649, WO/2013/013172, University of Utah, Salt Lake City, Utah, 2012.
[45] G. Elliott, C. Gilligan, C. Lorimer, Intravaginal reservoir drug delivery devices, WO Patent App. PCT/EP2003/003, EP1494646A1, EP1494646B1, US20050042292, WO2003080018A1, Grant Elliott, Galen (Chemicals) Limited, Claire Gilligan, Colin Lorimer, 2003.
[46] S. Nabahi, Intravaginal drug delivery device, EP Patent App. 19, USP # 5,788,980, 1998.
[47] C. Gilligan, C. Passmore, Intravaginal drug delivery devices for the administration of testosterone and testosterone precursors, WO Patent App. PCT/IE1998/000, 1998.
[48] R. Zimmerman, P. Burck, R. Dunn, Contraceptive device, US4469671 A, US 06/468,436, Eli Lilly And Company,1984.
[49] K. Kamath, J. Barry, H. Sepideh, Polymeric coatings for controlled delivery of active agents, US 6,335,029 B1, Scimed Life Systems, Inc., Maple grove, MN (US), 2002.
[50] R.J. Zdrahala, I.J. Zdrahala, In vivo tissue engineering with biodegradable polymers, Google Patents, 2002.
[51] W. Easterbrook III, L. Gudis, M. Howansky, H. Levin, P. Michelman, R. Reinhardt, C. Sherman, J. Tsitlik, N. Ziselson, Method and apparatus for assisting a heart to pump blood, Google Patents, Cardio Technologies, Inc., 2001.
[52] E. Brady, A. Cannon, F. Farrell, G. Mccaffrey, A tissue engineering scaffold, WO Patent App. PCT/IE2000/000059, 2000.
[53] H. Nita, L.P. Jansen, P. Park, G. Samson, E. Engelson, J. Sarge, Optimized high performance multiple coil spiral-wound vascular catheter, Google Patents, Target Therpeutics, Inc., Fremont, Calif., USA, 1999.
[54] P. Zilla, D. Bezuidenhout, Porous synthetic vascular grafts with oriented ingrowth channels, WO Patent App. PCT/US1999/027, Medtronic, Inc., 1999.
[55] R.J. Tuch, Blood contacting medical device and method, Google Patents, Medtronic, Inc., 1999.
[56] R.V. Snyders, Ventricular assist device, Google Patents, 1987.
[57] M. Szycher, Extrudable polyurethane for prosthetic devices prepared from a diisocyanate, a polytetramethylene ether polyol, and 1, 4-butane diol, Google Patents, 1985.

[58] M. Szycher, Process for forming an optically clear polyurethane lens or cornea, Google Patents, 1983.

[59] M. Szycher, Polyurethane elastomer for heart assist devices, Google Patents, 1978.

[60] N.B. Shelke, T.M. Aminabhavi, Synthesis and characterization of methoxypolyethyleneglycol and lauric acid grafted novel polyurethanes for controlled release of nifedipine, *J. Appl. Polym. Sci.* 105 (2007) 2155–2163.

[61] N. Shelke, M. Sairam, S. Halligudi, T.M. Aminabhavi, Development of transdermal drug delivery films with castoroil based polyurethanes, *J. Appl. Polym. Sci.* 103 (2) (2007) 779–788.

[62] D. Cohn, T. Stern, M. González, J. Epstein, Biodegradable poly(ethylene oxide)/poly(epsilon-caprolactone) multiblock copolymers, *J. Biomed. Mater. Res.* 59 (2) (2002) 273–281.

[63] P. Sathiskumar, G. Madras, Synthesis, characterization, degradation of biodegradable castor oil based polyesters, *Polym. Degrad. Stab.* 96 (9) (2011) 1695–1704.

[64] P. Pissis, A. Kanapitsas, Y.V. Savelyev, E. Akhranovich, E. Privalko, V. Privalko, Influence of chain extenders and chain end groups on properties of segmented polyurethanes. II. Dielectric study, *Polymer* 39 (15) (1998) 3431–3435.

[65] R. Adhikari, P.A. Gunatillake, S.J. Mccarthy, G.F. Meijs, Effect of chain extender structure on the properties and morphology of polyurethanes based on H12MDI and mixed macrodiols (PDMS–PHMO), *J. Appl. Polym. Sci.* 74 (12) (1999) 2979–2989.

[66] A. Simmons, A. Padsalgikar, L. Ferris, L. Poole-Warren, Biostability and biological performance of a PDMS-based polyurethane for controlled drug release, *Biomaterials* 29 (20) (2008) 2987–2995.

[67] V. Thomas, M. Jayabalan, A new generation of high flex life polyurethane urea for polymer heart valve–studies on in vivo biocompatibility and biodurability, *J. Biomed. Mater. Res. A* 89 (1) (2009) 192–205.

[68] R. González-Paz, G. Lligadas, J. Ronda, M. Galià, A. Ferreira, F. Boccafoschi, G. Ciardelli, V. Cádiz, Study on the interaction between gelatin and polyurethanes derived from fatty acids, *J. Biomed. Mater. Res. A* 101 (4) (2013) 1036–1046.

[69] S. Obruca, I. Marova, L. Vojtova, Biodegradation of polyether-polyol-based polyurethane elastomeric films: influence of partial replacement of polyether polyol by biopolymers of renewable origin, *Environ. Technol.* 32 (9–10) (2011) 1043–1052.

[70] L. Chan-Chan, R. Solis-Correa, R. Vargas-Coronado, J. Cervantes-Uc, J. Cauich-Rodríguez, P. Quintana, P. Bartolo-Pérez, Degradation studies on segmented polyurethanes prepared with HMDI, PCL and different chain extenders, *Acta Biomater.* 6 (6) (2010) 2035–2044.

[71] G. Trovati, E.A. Sanches, S.C. Neto, Y.P. Mascarenhas, G.O. Chierice, Characterization of polyurethane resins by FTIR, TGA, and XRD, *J. Appl. Polym. Sci.* 115 (2010) 263–268.

[72] L. Widjaja, J. Kong, S. Chattopadhyay, V. Lipik, S. Liow, M. Abadie, S. Venkatraman, Triblock copolymers of ε-caprolactone, trimethylene carbonate, and L-lactide: effects of using random copolymer as hard-block, *J. Mech. Behav. Biomed. Mater.* 6 (2012) 80–88.

[73] S.Z. Fu, X.H. Meng, J. Fan, L.L. Yang, S. Lin, Q.L. Wen, B.Q. Wang, L.L. Chen, J.B. Wu, Y. Chen, In vitro and in vivo degradation behavior of n-HA/PCL-Pluronic-PCL polyurethane composites. *J. Biomed. Mater. Res. A* 102 (2) (2014) 479–486.

[74] S. McCarthy, G. Meijs, N. Mitchell, P. Gunatillake, G. Heath, A. Brandwood, K. Schindhelm, In-vivo degradation of polyurethanes: transmission-FTIR microscopic characterization of polyurethanes sectioned by cryomicrotomy, *Biomaterials* 18 (21) (1997) 1387–1409.

[75] G.A. Senich, W.J. MacKnight, Fourier transform Infrared thermal analysis of a segmented polyurethane, *Macromolecules* 13 (1) (1980) 106–110.

[76] E.G. Bajsic, V. Rek, DSC study of morphological changes in segmented polyurethane elastomers, *J. Elastom. Plast.* 32 (2) (2000) 162–182.

[77] M. Pergal, V. Antić, G. Tovilović, J. Nestorov, D. Vasiljević-Radović, J. Djonlagić, In vitro biocompatibility evaluation of novel urethane-siloxane co-polymers based on poly(ε-caprolactone)-block-poly(dimethylsiloxane)-block-poly(ε-caprolactone), *J. Biomater. Sci. Polym. Ed.* 23 (13) (2012) 1629–1657.

[78] S. Sawant, V. Selvaraj, V. Prabhawathi, M. Doble, Antibiofilm properties of silver and gold incorporated PU, PCLm, PC and PMMA nanocomposites under two shear conditions, *PLoS one* 8 (5) (2013) 1–9, e63311.

[79] D. Sarkar, J.C. Yang, A. Gupta, S. Lopina, Synthesis and characterization of L-tyrosine based polyurethanes for biomaterial applications, *J. Biomed. Mater. Res. A* 90 (1) (2009) 263–271.

[80] M. Frost, M. Meyerhoff, Synthesis, characterization, and controlled nitric oxide release from S-nitrosothiol-derivatized fumed silica polymer filler particles, *J. Biomed. Mater. Res. A* 72 (4) (2005) 409–419.

[81] O. Khan, M. Sefton, Endothelialized biomaterials for tissue engineering applications in vivo, *Trends Biotechnol.* 29 (8) (2011) 379–387.

[82] A. McGuigan, M. Sefton, The influence of biomaterials on endothelial cell thrombogenicity, *Biomaterials* 28 (16) (2007) 2547–2571.

[83] H. Planck, G. Egbers, I. Syre, Polyurethanes in Biomedical Engineering, Elsevier Science Publishers B.V., Amsterdam, 1984.

[84] K.B. Stokes, P.W. Urbanski, M.W. Davis, A.J. Coury, Pacemaker leads, in: A.E. Aubert, H. Ector (Eds.), Nine Years Experience with Polyurethane Leads, Elsevier Science Publishers, Amsterdam, The Netherlands, 1985, pp. 279–286.

[85] K. Stokes, R. McVenes, J.M. Anderson, Polyurethane elastomer biostability, *J. Biomater. Appl.* 9 (1995) 321–354.

[86] L. Chan-Chan, C. Tkaczyk, R. Vargas-Coronado, J. Cervantes-Uc, M. Tabrizian, J. Cauich-Rodriguez, Characterization and biocompatibility studies of new degradable poly(urea)urethanes prepared with arginine, glycine or aspartic acid as chain extenders, *J. Mater. Sci. Mater. Med.* 24 (7) (2013) 1733–1744.

[87] C. Grasl, H. Bergmeister, M. Stoiber, H. Schima, G. Weigel, Electrospun polyurethane vascular grafts: in vitro mechanical behavior and endothelial adhesion molecule expression, *J. Biomed. Mater. Res. A* 93 (2) (2010) 716–723.

[88] A. Brandwood, G.F. Meijs, P.A. Gunatillake, K.R. Noble, K. Shindhelm, E. Rizzardo, In vivo evaluation of polyurethanes based on novel macrodiols and MDI, *J. Biomater. Sci. Polym. Ed.* 6 (1) (1994) 41–54.

[89] A. Takahara, A.J. Coury, R.W. Hergenrother, S.L. Cooper, Effect of soft segment chemistry on the biostability of segmented polyurethanes. I. In vitro oxidation, *J. Biomed. Mater. Res.* 25 (1991) 341–356.

[90] Q. Zhao, A. McNally, K. Rubin, M. Renier, Y. Wu, V. Rose-Caprara, J. Anderson, A. Hiltner, P. Urbanski, K. Stokes, Human plasma alpha 2-macroglobulin promotes in vitro oxidative stress cracking of Pellethane 2363-80A: in vivo and in vitro correlations, *J. Biomed. Mater. Res.* 27 (3) (1993) 379–388.

[91] E. Christenson, S. Patel, J. Anderson, A. Hiltner, Enzymatic degradation of poly(ether urethane) and poly(carbonate urethane) by cholesterol esterase, *Biomaterials* 27 (21) (2006) 3920–3926.

[92] Y. Tang, R. Labow, J. Santerre, Enzyme induced biodegradation of polycarbonate-polyurethanes: dose dependence effect of cholesterol esterase, *Biomaterials* 24 (12) (2003) 2003–2011.

[93] H. Yi, G. Jianjun, L.F. Kazuro, H. Ryotaro, L.P. Anca, R.W. William, Tailoring the degradation kinetics of poly(ester carbonate urethane)urea thermoplastic elastomers for tissue engineering scaffolds, *Biomaterials* 31 (15) (2010) 4249–4258.

[94] E. Christenson, J. Anderson, A. Hiltner, Oxidative mechanisms of poly(carbonate urethane) and poly(ether urethane) biodegradation: in vivo and in vitro correlations, *J. Biomed. Mater. Res. A* 70 (2) (2004) 245–255.

[95] M. Tanzi, S. Farè, P. Petrini, In vitro stability of polyether and polycarbonate urethanes, *J. Biomater. Appl.* 14 (4) (2000) 325–348.

[96] K. Gorna, S. Gogolewski, In vitro degradation of novel medical biodegradable aliphatic polyurethanes based on -caprolactone and Pluronics® with various hydrophilicities, *Polym. Degrad. Stab.* 75 (1) (2002) 113–122.

[97] N. Roohpour, J. Wasikiewicz, D. Paul, P. Vadgama, I. Rehman, Synthesis and characterisation of enhanced barrier polyurethane for encapsulation of implantable medical devices, *J. Mater. Sci. Mater. Med.* 20 (9) (2009) 1803–1814.

[98] G. Soldani, P. Losi, M. Bernabei, S. Burchielli, D. Chiappino, S. Kull, E. Briganti, D. Spiller, Long term performance of small-diameter vascular grafts made of a poly(ether)urethane-polydimethylsiloxane semi-interpenetrating polymeric network, *Biomaterials* 31 (9) (2010) 2592–2605.

[99] E. Briganti, T. Al Kayal, S. Kull, P. Losi, D. Spiller, S. Tonlorenzi, D. Berti, G. Soldani, The effect of gamma irradiation on physical-mechanical properties and cytotoxicity of polyurethane-polydimethylsiloxane microfibrillar vascular grafts, *J. Mater. Sci. Mater. Med.* 21 (4) (2010) 1311–1319.

[100] R. Hernandez, J. Weksler, A. Padsalgikar, J. Runt, In vitro oxidation of high polydimethylsiloxane content biomedical polyurethanes: correlation with the microstructure, *J. Biomed. Mater. Res. A* 87 (2) (2008) 546–556.

[101] F. Xu, J. Nacker, W. Crone, K. Masters, The haemocompatibility of polyurethane-hyaluronic acid copolymers, *Biomaterials* 29 (2) (2008) 150–160.

[102] D. Macocinschi, M. Filip, C. Butnaru, C.D. Dimitriu, Surface characterization of biopolyurethanes based on cellulose derivatives, *J. Mater. Sci. Mater. Med.* 20 (2009) 775–783.

[103] D. Macocinschi, D. Filip, S. Vlad, M. Cristea, M. Butnaru, Segmented polyurethanes for medical applications, *J. Mater. Sci. Mater. Med.* 20 (2009) 1659–1668.

[104] D. Macocinschi, D. Filip, S. Vlad, Surface and mechanical properties of some new biopolyurethane composites, *Polym. Compos.* 31 (2010) 1956–1964.

[105] L. Moldovan, O. Craciunescu, O. Zarnescu, D. Macocinschi, D. Bojin, Preparation and characterization of new biocompatibilized polymeric materials for medical use, *J. Optoelectron. Adv. Mater.* 10 (4) (2008) 942.

[106] T. Yoshii, J. Dumas, A. Okawa, D. Spengler, S. Guelcher, Synthesis, characterization of calcium phosphates/polyurethane composites for weight-bearing implants, *J. Biomed. Mater. Res. B Appl. Biomater.* 100 (1) (2012) 32–40.

[107] Z. Li, S.J. Yao, M. Alini, M. Stoddart, Chondrogenesis of human bone marrow mesenchymal stem cells in fibrin-polyurethane composites is modulated by frequency and amplitude of dynamic compression and shear stress, *Tissue Eng. A* 16 (2) (2010) 575–584.

[108] K. Sask, L. Berry, A. Chan, J. Brash, Polyurethane modified with an antithrombin-heparin complex via polyethylene oxide linker/spacers: influence of PEO molecular weight and PEO-ATH bond on catalytic and direct anticoagulant functions, *J. Biomed. Mater. Res. A* 100 (10) (2012) 2821–2828.

[109] K. Sask, L. Berry, A. Chan, J. Brash, Modification of polyurethane surface with an antithrombin-heparin complex for blood contact: influence of molecular weight of polyethylene oxide used as a linker/spacer, *Langmuir* 28 (4) (2012) 2099–2106.

[110] C. Schmidt, D. Bezuidenhout, L. Higham, P. Zilla, N. Davies, Induced chronic hypoxia negates the pro-angiogenic effect of surface immobilized heparin in a polyurethane porous scaffold, *J. Biomed. Mater. Res. A* 98 (4) (2011) 621–628.

[111] Z. Shen, C. Kang, J. Chen, D. Ye, S. Qiu, S. Guo, Y. Zhu, Surface modification of polyurethane towards promoting the ex vivo cytocompatibility and in vivo biocompatibility for hypopharyngeal tissue engineering. *J. Biomater. Appl.* 28 (4) (2013) 607–616.

[112] R. Williams, Y. Krishna, S. Dixon, A. Haridas, I. Grierson, C. Sheridan, Polyurethanes as potential substrates for sub-retinal retinal pigment epithelial cell transplantation, *J. Mater. Sci. Mater. Med.* 16 (12) (2005) 1087–1092.

[113] K. Lehle, M. Stock, T. Schmid, S. Schopka, R. Straub, C. Schmid, Cell-type specific evaluation of biocompatibility of commercially available polyurethanes, *J. Biomed. Mater. Res. B Appl. Biomater.* 90 (1) (2009) 312–318.

[114] W. Wu, K. Sask, J. Brash, P. Selvaganapathy, Polyurethane-based microfluidic devices for blood contacting applications, *Lab Chip* 12 (5) (2012) 960–970.

[115] Z.S. Petrović, J. Ferguson, N. Hudson, I. Javni, M. Vraneš, The effect of hard segment structure on rheological properties of solutions of segmented polyurethanes, *Eur. Polym. J.* 28 (6) (1992) 637–642.

[116] C. Ma, Y. Hou, S. Liu, G. Zhang, Effect of microphase separation on the protein resistance of a polymeric surface, *Langmuir* 25 (16) (2009) 9467–9472.

[117] L.C. Xu, C. Siedlecki, Microphase separation structure influences protein interactions with poly(urethane urea) surfaces, *J. Biomed. Mater. Res. A* 92 (1) (2010) 126–136.

[118] L. Feng, J.D. Andrade, Structure and adsorption properties of fibrinogen, In: T.A. Horbett, J.L. Brash (Eds.), Proteins at Interfaces II, Fundamentals and Applications, ACS Sym. Ser.#602, American Chemical Society, Washington, DC, 1995, pp. 66–79.

[119] M.R. Brunstedt, N.P. Ziats, M. Schubert, P.A. Hiltner, J.M. Anderson, Protein adsorption onto poly(ether urethane ureas) containing methacrol 2138F: a surface-active amphiphilic additive, *J. Biomed. Mater. Res.* 27 (1993) 255–267.

[120] S.H. Chu, C.R. Hung, Y.J. Yang, Comparison of long term results of various card iac valvular prostheses, in: E. Bod nar, M. Yacoub (Eds.), Biologic and Bioprosthetic Valves, Yorke Medical Books, New York, 1984, pp. 805–816.

[121] J.B. Garcia-Bengochea, I. Casagrande, J.R. Gonzalez-Juanatey, L. Puig, J. Rubio, D. Duran, J. Sierra, The new Labcor-Santiago pericardial bioprosthesis, *J. Card. Surg.* 6 (1991) 613–619.

[122] M. Herold, H.B. Lo, H. Reul, et al., The Helmholtz-Institute tri-leaflet-polyurethane-heart valve prosthesis: design, manufacturing and first in-vitro and in-vivo results, in: H. Planck, I. Syre, M. Dauner, G. Egbers (Eds.), Polyurethanes in Biomedical Engineering, vol. II, Elsevier Science, Amsterdam, 1987, pp. 231–256.

[123] F. Nistal, V. Garcia-Martinez, E. Arbe, D. Fernandez, F. Mazzora, I. Gallo, In vivo experimental assessment of polytetrafluoroethylene trileaflet heart valve prosthesis, *J. Thorac. Cardiovasc. Surg.* 99 (1990) 1074–1081.

[124] J. Jansen, H. Reul, A synthetic three-leaflet valve, *J. Med. Eng. Technol.* 16 (1992) 27–33.

[125] I. Khan, N. Smith, E. Jones, D. Finch, R. Cameron, Analysis and evaluation of a biomedical polycarbonate urethane tested in an in vitro study and an ovine arthroplasty model. Part I: materials selection and evaluation, *Biomaterials* 26 (6) (2005) 621–631.

[126] S. Gogolewski, Nonmetallic materials for bone substitute, *Eur. Cells Mater.* 1 (2001) 54–55.

[127] K. Gorna, S. Gogolewski, Preparation, degradation, and calcification of biodegradable polyurethane foams for bone graft substitutes, *J. Biomed. Mater. Res. A* 67 (3) (2003) 813–827.

[128] Q. Guo, P.T. Knight, P.T. Mather, Tailored drug release from biodegradable stent coatings based on hybrid polyurethanes, *J. Control. Release* 137 (2009) 224–233.

[129] J. Kim, S. Kim, S. Lee, C. Lee, D. Kim, The effect of pore formers on the controlled release of cefadroxil from a polyurethane matrix, *Int. J. Pharm.* 201 (1) (2000) 29–36.

[130] B. Li, K. Brown, J. Wenke, S. Guelcher, Sustained release of vancomycin from polyurethane scaffolds inhibits infection of bone wounds in a rat femoral segmental defect model, *J. Control. Release* 145 (3) (2010) 221–230.

[131] S. Moura, L. Lima, S. Andrade, A. Da Silva-Cunha, R. Órefice, E. Ayres, G. Da Silva, Local drug delivery system: inhibition of inflammatory angiogenesis in a murine sponge model by dexamethasone-loaded polyurethane implants, *J. Pharm. Sci.* 100 (7) (2011) 2886–2895.

[132] F. Pinto, A. Da Silva-Cunha Junior, R. Oréfice, E. Ayres, S. Andrade, L. Lima, S. Moura, G. Da Silva, Controlled release of triamcinolone acetonide from polyurethane implantable devices: application for inhibition of inflammatory-angiogenesis, *J. Mater. Sci. Mater. Med.* 23 (6) (2012) 1431–1445.

[133] D. Nelson, P. Baraniak, Z. Ma, J. Guan, N. Mason, W. Wagner, Controlled release of IGF-1 and HGF from a biodegradable polyurethane scaffold, *Pharm. Res.* 28 (6) (2011) 1282–1293.

[134] N.L. De Campos, R.R. de Andrade, M.A. Silva, Oral anticoagulation in carriers of mechanical heart valve prostheses: experience of ten years, *Rev. Bras. Cir. Cardiovasc.* 25 (2010) 457–465.

[135] G. Hoffmann, G. Lutter, J. Cremer, Durability of bioprosthetic cardiac valves, *Dtsch. Arztebl. Int.* 105 (2008) 143–148.

[136] N. Thierfelder, F. Koenig, R. Bombien, C. Fano, B. Reichart, E. Wintermantel, C. Schmitz, B. Akra, In vitro comparison of novel polyurethane aortic valves and homografts after seeding and conditioning, *ASAIO J.* 59 (3) (2013) 309–316.

[137] H. Bergmeister, C. Schreiber, C. Grasl, I. Walter, R. Plasenzotti, M. Stoiber, D. Bernhard, H. Schima, Healing characteristics of electrospun polyurethane grafts with various porosities, *Acta Biomater.* 9 (4) (2013) 6032–6040.

[138] H. Bergmeister, C. Grasl, I. Walter, R. Plasenzotti, M. Stoiber, C. Schreiber, U. Losert, G. Weigel, H. Schima, Electrospun small-diameter polyurethane vascular grafts: ingrowth and differentiation of vascular-specific host cells, *Artif. Organs* 36 (1) (2012) 54–61.

[139] D. Jovanovic, G. Engels, J. Plantinga, M. Bruinsma, W. van Oeveren, A. Schouten, M. van Luyn, M. Harmsen, Novel polyurethanes with interconnected porous structure induce in vivo tissue remodeling and accompanied vascularization, *J. Biomed. Mater. Res. A* 95 (1) (2010) 198–208.

[140] C.M. Hill, Y.H. An, Q.K. Kang, L.A. Hartsock, S. Gogolewski, K. Gorna, Osteogenesis of osteoblast seeded polyurethane-hydroxyapatite scaffolds in nude mice, *Macromol. Symp.* 253 (1) (2007) 94–97.

[141] A. Asefnejad, A. Behnamghader, M. Khorasani, B. Farsadzadeh, Polyurethane/fluor-hydroxyapatite nanocomposite scaffolds for bone tissue engineering. Part I: morphological, physical, and mechanical characterization, *Int. J. Nanomedicine* 6 (2011) 93–100.

[142] F. Herrmann, A. Lehner, T. Hollweck, U. Haas, C. Fano, D. Fehrenbach, R. Kozlik-Feldmann, E. Wintermantel, G. Eissner, C. Hagl, B. Akra, In vitro biological and mechanical evaluation of various scaffold materials for myocardial tissue engineering. *J. Biomed. Mater. Res. A* (2013) doi:10.1002/jbm.a.34786.

[143] J. Rodriguez, F. Clubb, T. Wilson, M. Miller, T. Fossum, J. Hartman, E. Tuzun, P. Singhal, D. Maitland, In vivo response to an implanted shape memory polyurethane foam in a porcine aneurysm model. *J. Biomed. Mater. Res. A* (2013) doi:10.1002/jbm.a.34782.

[144] S.A. Korossis, J. Fisher, E. Ingham, Cardiac valve replacement: a bioengineering approach, *Biomed. Mater. Eng.* 10 (2000) 83–124.

[145] S.L. Hilbert, V.J. Ferrans, Y. Tomita, E.E. Eidbo, M. Jones, Evaluation of explanted polyurethane trileaflet cardiac valve prostheses, *J. Thorac. Cardiovasc. Surg.* 94 (1987) 419–429.

[146] L. Wu, D.M. Weisberg, J. Runt, G. Felder, A.J. Snyder, G. Rosenberg, An investigation of the in vivo stability of poly(ether urethaneurea) blood sacs, *J. Biomed. Mater. Res.* 11 (1999) 371–380.

[147] M. Butterfield, D.J. Wheatley, D.F. Williams, J. Fisher, A new design for polyurethane heart valves, *J. Heart Valve Dis.* 10 (1) (2001) 105–110.

[148] V. Thomas, M. Jayabalan, Studies on the effect of virtual crosslinking on the hydrolytic stability of novel aliphatic polyurethane ureas for blood contact applications, *J. Biomed. Mater. Res.* 56 (1) (2001) 144–157.

[149] K. Myers, N. Sgaglione, P. Kurzweil, A current update on meniscal scaffolds, *Oper. Tech. Sports Med.* 21 (2013) 75–81.

[150] T. Mackay, D. Wheatley, G. Bernacca, A. Fisher, C. Hindle, New polyurethane heart valve prosthesis: design, manufacture and evaluation, *Biomaterials* 17 (19) (1996) 1857–1863.

[151] A. Oloffs, C. Grosse-Siestrup, S. Bisson, M. Rinck, R. Rudolph, U. Gross, Biocompatibility of silver-coated polyurethane catheters and silver-coated Dacron material, *Biomaterials* 15 (10) (1994) 753–758.

[152] L.I.F. Moura, A.M.A. Dias, E. Carvalhoa, H.C. de Sousa, Recent advances on the development of wound dressings for diabetic foot ulcer treatment—a review, *Biomaterialia* 9 (2013) 7093–7114.

[153] N.E. Epstein, Do silver-impregnated dressings limit infections after lumbar laminectomy with instrumented fusion? *Surg. Neurol.* 68 (2007) 483–485.

[154] M.G. Knuttinen, S. Bobra, J. Hardman, R.C. Gaba, J.T. Bui, C.A. Owens, Review of evolving dialysis catheter technologies, *Semin. Internat. Rad.* 26 (2) (2009) 106–114.

[155] Bayer, "Medical device developments", "A hydroselective approach to healing", Baymedix A.

Chapter 8

Poly(Ester Amide)s: Recent Developments on Synthesis and Applications

Sara K. Murase, Jordi Puiggalí

Departament d'Enginyeria Química, Universitat Politècnica de Catalunya, Barcelona, Spain

Chapter Outline

8.1 INTRODUCTION

Aliphatic polyesters are the most important polymeric materials commercialized up to now for biomedical applications and in particular as absorbable biomaterials. However, they have several limitations (e.g., relatively high level of *in vivo* inflammation and lack of reactive sites on their backbone) that enhance the research focused on the development of new materials. Poly(ester amide)s (PEAs) have gained great attention since the integration of polyester and polyamide building blocks in a single molecule can provide a great opportunity of tuning final characteristics of materials. Specifically, these polymers can combine a degradable character, afforded by hydrolyzable ester groups (–COO–) placed in the backbone, with relatively good thermal and mechanical properties given by the strong intermolecular hydrogen bonding interactions that can be established between their amide groups (–NHCO–). In this way, the possibility to meet a good profile in terms of biodegradability, chemical, mechanical, physical, thermal, and biological properties is open.

PEAs can be synthesized from a great variety of monomers (e.g., α-amino acids, α,ω-amino alcohols, and carbohydrates), which can lead to different monomer distributions (e.g., ordered, blocky, or random) and molecular architectures (e.g., linear or hyperbranched chains). Polymers can also be obtained with variable ester/amide and aliphatic/aromatic ratios, hydrophilicity (e.g., by the incorporation of poly(ethylene oxide) blocks or by changing the length of polymethylene sequences), degree of functionalization, stereochemistry (e.g., by changing the D- and L-amino acid content), crystallinity, chain stiffness, and elastomeric reticulation. This high variability provides a set of materials that can be specifically designed to cover highly specialized tasks such as linking compounds with a pharmacological activity for drug delivery applications. PEAs have also been applied as nonviral delivery vectors, stimuli-sensitive implants, biodegradable elastomers, synthetic scaffolds, pH-sensitive polymer, and hydrogels.

PEAs were firstly obtained in 1979 from polyamides and polyesters by temperature-induced amide-ester interchange reactions. Since then, a considerable body of literature has been developed due to the noticeable characteristics and properties of the new polymers. For more detailed information on the synthesis and applications of

Natural and Synthetic Biomedical Polymers

PEAs, some specific reviews can be taken into account [1–3]. The present work is focused to the more recent developments performed with PEAs and highlights current activities on the biomedical field. The review is constituted by six sections: the first and the second ones dealing with generic issues such as the synthetic methods applied and the derived polymer microstructures, while the third and fourth sections summarize more specific subjects such as preparation of hyperbranched and stiff polymers. The two last sections concern to PEAs prepared from renewable resources, paying special attention to polymers derived from α-amino acids, and specific applications like scaffolds with multiple functionalities, light-responsive materials, or nanocomposites.

8.2 SYNTHESIS OF PEAs

PEAs have been synthesized by ring-opening polymerization and polycondensation methods. The first ones were mainly employed to get copolymers of α-hydroxy acids and α-amino acids (i.e., polydepsipeptides) and reported in the literature [4]. Recent works are focused to the use of enzymes (e.g., lipases) as new efficient catalysts for reaction of these morpholine-2,5-diones [5]. It has been demonstrated that the configuration of the α-amino acid moiety did not affect the enzyme-catalyzed polymerization, but in contrast, the configuration of the α-hydroxy acid moiety strongly influenced the polymerization behavior. Unfortunately, racemization of both units was demonstrated to take place during polymerization.

Polymerization of cyclic ester amides with large rings (e.g., 11-, 13-, and 14-membered rings) has also been studied [6–8]. In addition, it is well known that random copolymers can be easily prepared from mixtures of lactones and lactams, and even several works have recently been focused on the preparation of block copolymers from these macrocycles [9–11].

Similarly, PEAs having ether linkages in the main chain can be easily synthesized via ring-opening copolymerization of different ratios of a hydrophilic monomer (e.g., 3-morpholinone) and a hydrophobic monomer (e.g., ε-caprolactone) [12]. In comparison with the structure of traditional PEAs, these copolymers with additional ether linkages in the backbone had an enhanced hydrophilicity and flexibility. The water absorption and the *in vitro* degradation of these polymers were found to increase by increasing the morpholinone content. These new copolymers were assayed to get a controlled release of 5-fluorouracil, a drug with a strong antitumor activity but also with undesirable side effects. Interestingly, a sustained release that increased with the polymer hydrophilicity was determined.

Synthesis of PEAs by polycondensation has usually been performed by reacting diamide-diol, diester-diamine, ester-diamine, or diamide-diester monomers with dicarboxylic acid derivatives or diols (Figure 8.1). In addition, α,ω-amino alcohols can be also reacted with acid anhydrides or dicarboxylic acid derivatives.

Polycondensation can be performed in the melt state [13,14], which may be advantageous for industrial production since no posttreatment is necessary after reaction, by an interfacial polymerization [15,16], which can be performed at room temperature and therefore typical problems associated to melt condensation can be avoided, and solution polycondensation [17,18]. The latter can make use of condensing agents (e.g., carbodiimide, DCCI) to favor elimination of water molecules and even of activation groups (e.g., *p*-nitrophenol, pentachlorophenol, and *N*-hydroxysuccinimide) for the carboxylic acid in order to favor aminolysis reactions.

Some alternating PEAs can be synthesized via ring-opening polymerization of cyclic ester amides and via polycondensation of suitable linear AB monomers. Specifically, alternating PEAs have been easily obtained from succinic anhydride and α,ω-amino alcohols $H_2N–(CH_2)_xOH$ (x=2-6), avoiding tedious protection/activation reactions (Figure 8.2) [19]. Polymerization was only achieved in solution under mild conditions and using a carbodiimide as condensing agent. Bulk polycondensation at temperatures above the melting point of the AB monomers was not successful due to the formation of the thermodynamically stable *N*-(hydroxyalkyl) imides.

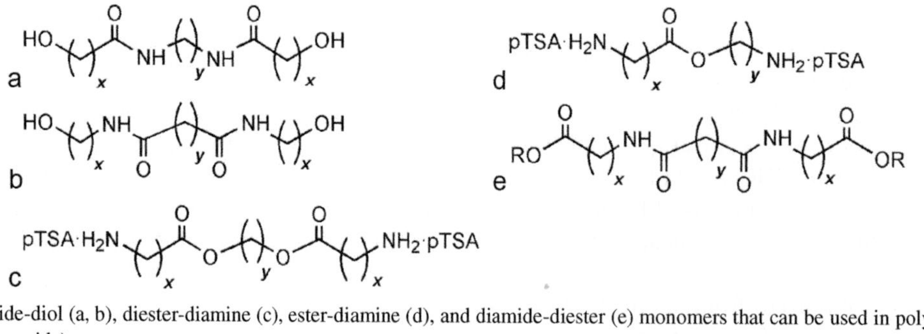

FIGURE 8.1 Diamide-diol (a, b), diester-diamine (c), ester-diamine (d), and diamide-diester (e) monomers that can be used in polycondensation reactions to get poly(ester amide)s.

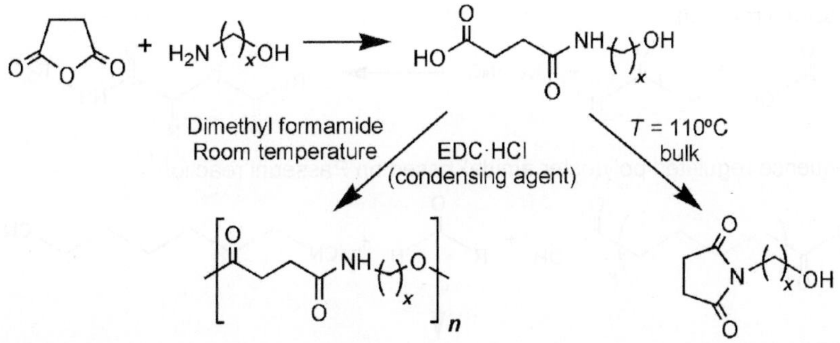

FIGURE 8.2 Synthesis of poly(ester amide)s from succinic anhydride and α,ω-amino alcohols by solution polycondensation. *N*-(hydroxyalkyl) imide by-products were on the contrary obtained by bulk polycondensation.

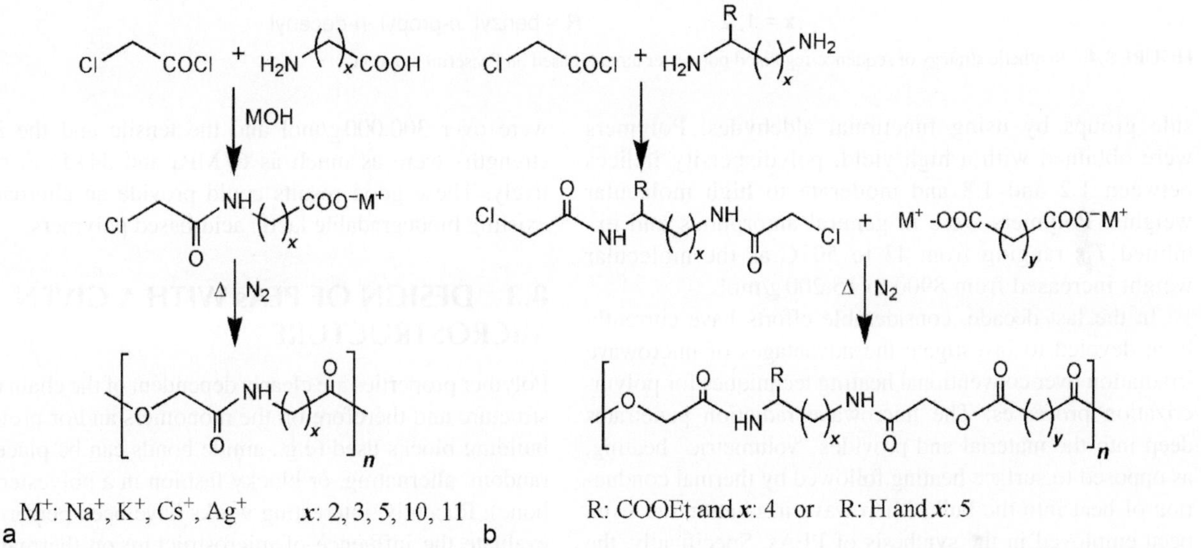

FIGURE 8.3 Synthesis of regular PEAs by a bulk polycondensation method based on the formation of metal halide salts. Derivatives of ω-amino acids and glycolic acid (a) and diamines, dicarboxylic acids, and glycolic acid (b).

Bulk polymerization at temperatures above the melting point of the AB monomers was on the contrary feasible when amino alcohols were reacted with adipic anhydride instead of succinic anhydride [6]. At temperatures of 170 °C, these PEAs could be converted into cyclic ester amides with an increasing yield as the ring size did. Cycles were also prone to ring-opening polymerization with nucleophilic initiators giving a method to get alternating PEAs with the typical advantages of this kind of polymerization (e.g., control of molecular weight, molecular weight distribution, and polymer topology).

It is well known that thermal polycondensation of halogenoacetates leads to polyglycolide through the elimination of a metal halide salt, which can be considered as the driving force of the polycondensation reaction. The method has recently applied to get PEAs constituted by an alternating sequence of glycolic acid and ω-amino acid units (Figure 8.3a) [20–24]. Monomers can be easily synthesized by the reaction of chloroacetyl chloride with the appropriated ω-amino acid and by a subsequent neutralization with a metal hydroxide. This process contrasts with the more usual and high time-consuming process based on selective protection and deprotection of reactive groups. In a similar way, PEAs derived from a diamine, a dicarboxylic acid, and glycolic acid were also effectively synthesized by metal halide elimination method (Figure 8.3b) [11,25].

The synthesis of sequence-regulated polymers from a set of monomers in a one-pot and efficient process is nowadays a great challenge since samples with unprecedented properties can be derived from a rational design of constitutive sequences [26,27]. Isocyanide-based multicomponent reactions are extremely powerful synthetic tools that can assemble three or more different components into one molecule in a one-pot process and specifically the ester-amide (EA) linkage can be formed from a dicarboxylic acid, an aldehyde, and a diisocyanide [28]. Therefore, Li *et al.* have proposed a method to prepare linear polymers with a sequence-regulated backbone repeating unit of ester-ester-amide-amide (Figure 8.4) [29]. The method allows also to incorporate functional

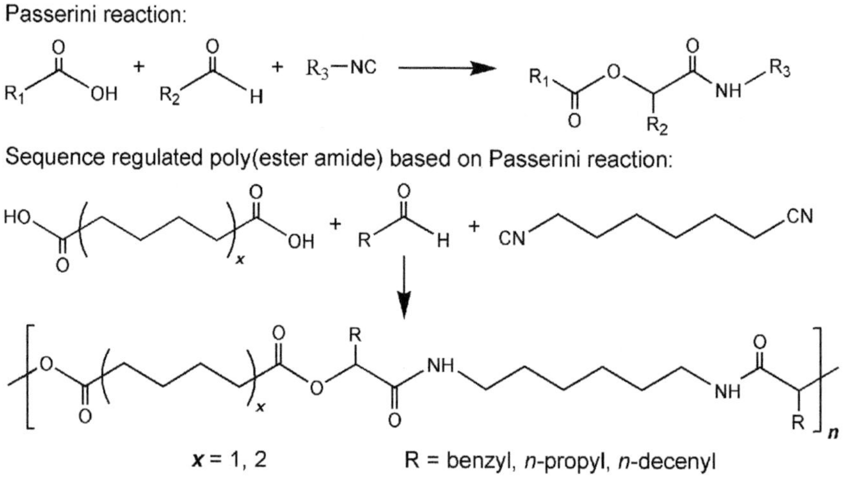

FIGURE 8.4 Synthetic strategy of sequence-regulated poly(ester amide) based on Passerini reaction.

side groups by using functional aldehydes. Polymers were obtained with a high yield, polydispersity indices between 1.2 and 1.8 and moderate to high molecular weights. Polymers were in general amorphous and exhibited T_gs ranging from 43 to 50 °C as the molecular weight increased from 8900 to 15,200 g/mol.

In the last decade, considerable efforts have currently been devoted to investigate the advantages of microwave irradiation over conventional heating techniques for polymerization processes. The microwave radiation penetrates deep into the material and provides "volumetric" heating, as opposed to surface heating followed by thermal conduction of heat into the bulk. Microwave irradiation has even been employed in the synthesis of PEAs. Specifically, the polycondensation of sebacic acid and α,ω-amino alcohols (3-aminopropanol, 2-aminoethanol, and 6-aminohexanol) was demonstrated to proceed at a much higher rate, to have higher energy efficiency and higher yield, and to render higher molecular weight upon microwave irradiation than conventional heating methods [30,31]. Microwave irradiation was also applied to synthesize poly(ε-caprolactam-*co*-ε-caprolactone)s directly from the anionic polymerization of mixtures of the two cyclic monomers [32]. The microwave-synthesized copolymers gave higher yield, higher amide composition, higher glass transition temperature, and equivalent molecular weight than the corresponding thermal products.

High molecular weights can be achieved by adding highly reactive coupling agents (chain extenders) during the last steps of melt polycondensation reactions. Bisoxazolines have been revealed as highly efficient coupling agents for polyesters (e.g., polylactide and polycaprolactone), giving rise to PEAs after coupling. For example, Tuominen and Sépala [33]. investigated the use of 2,2-bis(2-oxazoline) as a chain extender for a lactic acid-based carboxyl-terminated prepolymer (M_n 6200 g/mol). The highest molecular weights were over 300,000 g/mol and the tensile and the impact strengths were as much as 67 MPa and 34 kJ/m^2, respectively. These good results could provide an alternative to existing biodegradable lactic acid-based polymers.

8.3 DESIGN OF PEAs WITH A GIVEN MICROSTRUCTURE

Polymer properties are clearly dependent of the chain microstructure and therefore on the monomers and/or preformed building blocks used (e.g., amide bonds can be placed in a random, alternating, or blocky fashion in a polyester backbone). Recently, interesting works have been performed to evaluate the influence of microstructure on thermal properties [34]. Therefore, three different series of segmented PEAs were synthesized from preformed monomers such as α,ω-diols, α,ω-amino alcohols, or α,ω-diamines, which contained a built-in amide bond. These monomers were polymerized with dimethyl adipate and 1,4-butanediol and gave rise to PEAs containing either an isolated amide group, two adjacent amide groups, or three adjacent amide groups randomly distributed along the poly(butylene adipate) backbone. Furthermore, the amide concentration in each series was controlled by varying the ratio of the preformed monomer to 1,4-butanediol.

Calorimetric results clearly indicated that the length of the amide segment played a significant role in the polymer chain self-organization through hydrogen bonding interactions. Thus, crystallization behavior was found to be dependent on which functional groups participated in the crystalizable hard domains. Concretely, the amide groups mainly crystallized in the three adjacent amide series of PEAs, both the ester-rich domains and single EA sequences could crystallize in the isolated amide series, and single EA sequences and two or more EA sequences could crystallize in the two adjacent amide series (Figure 8.5). The

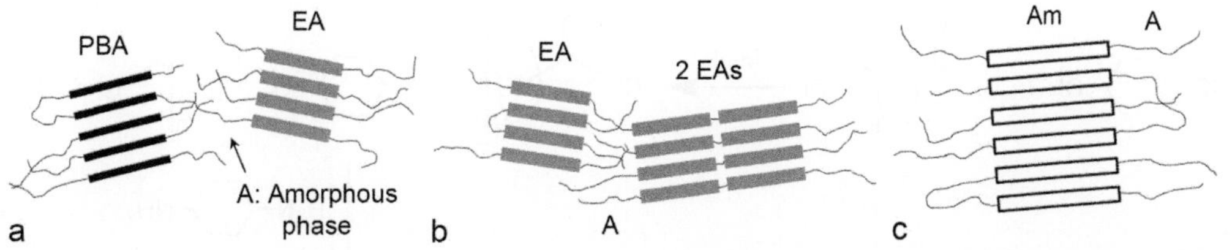

FIGURE 8.5 Representative scheme showing the crystalline phases present on PEAs based on isolated amide (a), two adjacent amide (b), and three amide groups (c) based on Ref. [34].

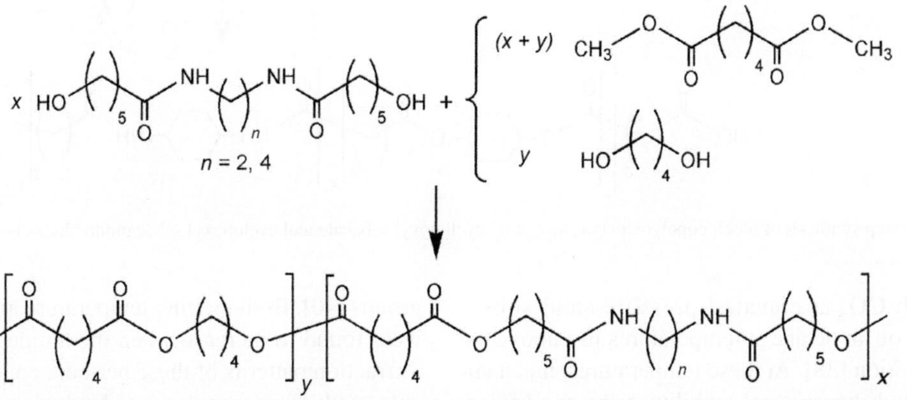

FIGURE 8.6 Synthesis of segmented poly(ester amide)s derived from bisamide-diols, dimethyl adipate, and 1,4-butanediol based on Ref. [35].

glass transition temperatures were logically dependent on the fraction of amide groups in the amorphous phase, which increased with increasing amide content. Even though all the studied polymers had the same amide content, thermal properties were found to be clearly determined by the type of crystalline phase and the length of the amide segment.

Segmented PEAs consisting of rigid amide segments and amorphous flexible ester segments can be used as thermoplastic elastomers. Amide-rich hard phase may contain crystalline lamellae and acts as a thermoreversible physical cross-linker for the ester-rich amorphous phase. This soft phase has a subambient glass transition temperature and contributes to the flexibility and extensibility of the polymer. Heating above the melting temperature of the hard domains usually results in phase mixing of hard and soft segments. Upon cooling, the hard and soft segments become incompatible, which leads to phase separation into microdomains and crystallization and thus to reformation of the physical cross-links. An effective way to improve phase separation corresponds to the use of short, symmetrical, and uniform amide blocks. The ratio of hard to soft segments affects the properties of segmented PEAs as the polymer changes from a more soft polyester to a hard polyamide.

The influence of the amount of hard segments in copoly(ester amide)s based on *N,N′*-1,4-butanediyl-bis[6-hydroxy-hexanamide], dimethyl adipate, and 1,4-butanediol (Figure 8.6) on the thermal, physical, and mechanical properties has been studied in detail [36]. Thus, by increasing the hard segment content from 10 to 85 mol%, the elastic modulus increased from 70 to 524 MPa, the stress at break increased from 8 to 28 MPa, and the strain at break decreased from 820% to 370%. Properties could be easily tuned for specific applications by varying the hard segment content. X-ray diffraction data demonstrated a semicrystalline character and pointed out a structure related to the typical α-form of polyamides. On the contrary, similar PEAs derived in this case from a bisamide-diol prepared from 1,2-diaminoethane and ε-caprolactone had a completely different crystalline structure with characteristics of the γ-crystalline phase present in odd-odd nylons [35]. The studied segmented PEAs usually showed two melt transitions, being more significant than appearing at higher temperature when the hard content increased. Melting peaks at low and high temperatures were ascribed to the fusion of crystals comprising single EA sequences and two or more EA sequences, respectively.

Degradation of these PEAs in PBS (pH 7.4) at 37 °C was very slow and took place preferentially through ester bond cleavage. The polymers with 25 and 50 mol% of amide content were noncytotoxic and sustained growth of fibroblasts onto polymer films. The *in vivo* degradation of these PEAs was slow (at least over a time period of 6 weeks) and consequently materials were proposed to be suitable as long-term biodegradable implants [37].

Interestingly, the related PEAs derived from 1,2-diaminoethane were able to be gas-foamed by

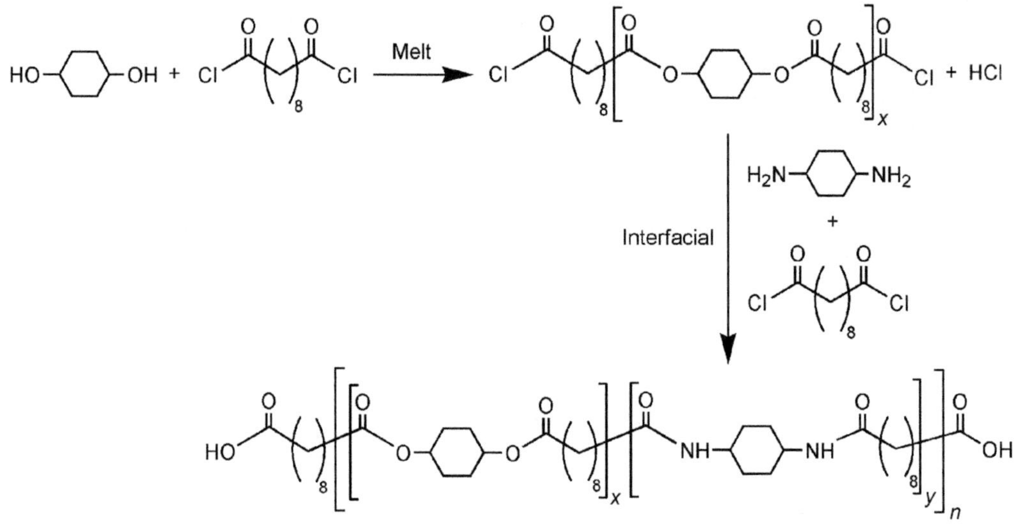

FIGURE 8.7 Two-step synthesis of block copolymers constituted by cyclohexyl sebacate and cyclohexyl sebacamide blocks based on Ref. [39].

saturation with CO_2 at elevated pressures and subsequent immersion in octane at temperatures just above the lower melt transition [38]. At these temperatures, mechanical properties and dimensional stability were guaranteed, while chain mobility was sufficient to nucleate and expand gas cells. Closed cell foams with a maximum porosity of ~90% and various pore sizes (2.5-100 μm) were obtained by changing the saturation pressure or the diamide to ester ratio of the PEA. Specifically, lower pore sizes were attained by increasing the pressure or the diamide content.

PEAs having a blocky microstructure were synthesized by Lecomte *et al.* [39]. following a two-step procedure, where acid chloride-terminated oligoesters of a predetermined length were first synthesized by melt polycondensation. Oligoamide moieties were subsequently grown from the oligoester chain ends by interfacial polycondensation (Figure 8.7). Block copolymers were specifically constituted by cyclohexyl sebacate and cyclohexyl sebacamide segments and differed on the oligoester segment length and the total amount of amide bonds incorporated. Microphase separation between amide moieties and ester moieties took place and led to thermal and mechanical properties within the limits set by each of the constitutive oligomeric blocks. New materials were shown to be hydrolyzable, noncytotoxic, and favorable for cell attachment. Hence, it was postulated their promising future in the area of medical devices or flexible tissue scaffolds. Furthermore, the mechanical and thermal properties of these PEAs, which can be tuned by varying their compositions, could be tailored to other dynamic industrial sectors, such as that of biodegradable packaging films.

Aliphatic copoly(ester amide)s constituted by a polyester block were also synthesized by two-step polycondensation reactions using adipate, 1,4-butanediamine and linear diols with chain lengths that ranged from 3 to 6 methylene groups [40]. Both melting temperature and thermal stability were found to increase when the amide content did. X-ray diffraction patterns of these periodic copolymers were variable, with some copolymers displaying patterns similar to homopolyesters, while samples with high amide content were characterized by molecular chain arrangements based on intermolecular hydrogen bond interactions. These were found to play a decisive role in the formation of a thermally stable crystalline region.

Poly(ether ester amide)s, based on poly(ethylene glycol), 1,4-butanediol, and a diamide-diester (Figure 8.8), were developed for tissue-engineering applications by Deschamps *et al.* [41]. Physical properties could be modulated by varying either the ester/amide ratio or the length of the hydrophilic ether blocks. Biocompatible and biodegradable polymers with varying hydrophilicity and adequate mechanical properties in the swollen state were therefore obtained. Melting and glass transition temperatures could be tuned from 110 to 150 °C and from −50 to −20 °C, respectively. Similarly, the modulus ranged from 200 to 400 MPa, the stress at break from 10 to 25 MPa, and the strain at break from 300% to 850%. Adhesion and growth of endothelial cells over these new materials was proved and was found to increase when the poly(ethylene glycol) content decreased. *In vivo* degradation of discs subcutaneously implanted in rats was also demonstrated.

Segmented PEAs having soft segments constituted by 1-3 oxyethylene units have been prepared by polycondensation of a series of monoaminooligo(oxyethylene) pentachlorophenyl succinates [42]. Monomers were obtained following a four-step procedure consisting on protection, activation and deprotection reactions. Solution and bulk polycondensation experiments were performed but only the last gave rise to high molecular weight polymers. Solubility and thermal properties of the materials were found to be

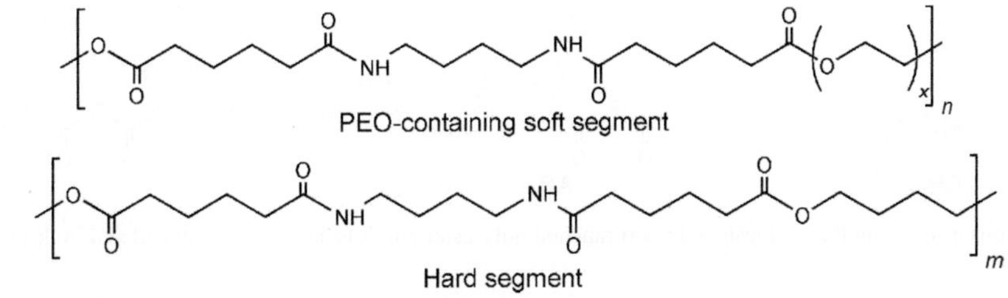

FIGURE 8.8 Chemical structure of soft and hard segments of PEAs based on poly(ethylene glycol), 1,4-butanediol, and a diamide-diester constituted by 1,4-butanediamine and adipic acid units.

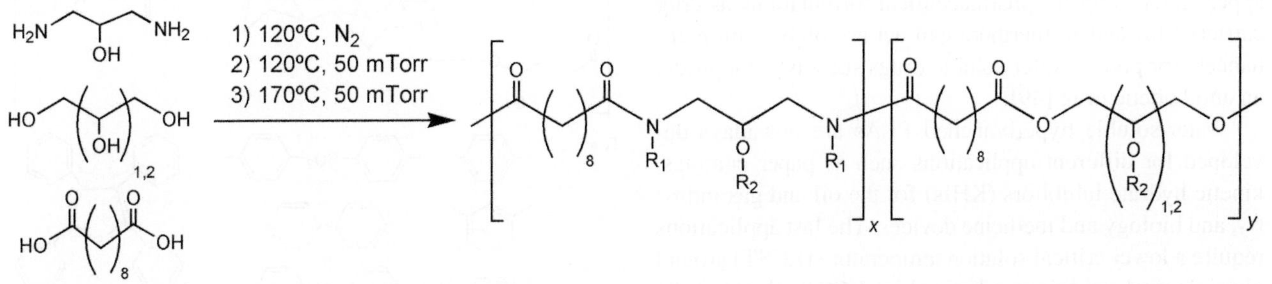

FIGURE 8.9 Synthesis of elastomeric poly(ester amide)s derived from 1,3-diamino-2-hydroxy-propane based on Ref. [44]. R_1 and R_2 represent either a single hydrogen or bond to either the *X*-segment or the *Y*-segment via amide bond or ester bond, respectively.

related to the length of the oligo(oxyethylene) segment and the presence of strongly interacting amide groups.

Development of biodegradable elastomers with chemical, physical, and mechanical properties of native extracellular matrix is currently necessary for tissue-engineering applications since implanted materials must support the *in vivo* dynamic mechanical environment. It is well known that natural biodegradable elastomers have processing difficulties and may lead to potentially dangerous immune responses upon implantation [43]. Synthetic biodegradable elastomers, typically composed of cross-linked networks of aliphatic molecules with polyester linkages, may have several problems associated to extremely high cross-linking densities, rapid biodegradation upon implantation, or limited functional groups. Therefore, Langer and coworkers [44] have proposed the synthesis of a new class of synthetic, elastomeric, and biodegradable PEAs based on the amino alcohol 1,3-diamino-2-hydroxy-propane (DAHP). This monomer has several advantages: nontoxicity, low cost, low melting point, disposal of primary amines susceptible of subsequent conjugation, and capability to form amide and ester bonds to enable tunable properties. The copolymerization of DAHP with a polyol (e.g., glycerol and D,L-threitol) and a diacid (e.g., sebacic acid) rendered new biodegradable elastomers (Figure 8.9). These cross-linked networks feature tensile Young's modulus on the order of 1 MPa and reversible elongations up to 92%. Polymers exhibit *in vitro* and *in vivo* biocompatibility and *in vivo* degradation half-lives up to 20 months. The favorable cellular responses suggested that the new class of elastomeric materials could be used as a biomedical device material platform.

8.3.1 Hyperbranched PEAs

Hyperbranched polymers are considered to be a good alternative to dendrimers for emerging applications in wide fields. Basically, they can be easily prepared in a low-cost and large-scale one-pot procedure, while typical properties of dendrimer structure can be maintained. Thus, a three-dimensional globular architecture, good solubility, low solution and bulk viscosity, and easy modification because of the abundance of terminal functional groups can be mentioned. Li *et al.* [45,46] have recently developed a strategy for synthesis of hyperbranched PEAs from commercially available dicarboxylic acids (A_2) and multihydroxyl secondary amine (CB_2) (Figure 8.10). Secondary amino groups were able to react with one carboxylic group and predominantly formed AB_2 dimers. These intermediates could be subjected to thermal polycondensation without any purification and in the absence of any catalyst to form aliphatic and semiaromatic hyperbranched PEAs bearing multihydroxyl end groups. Furthermore, the hyperbranched polymers possessed abundant hydroxyl end groups, which were potential reactive sites for subsequent modification. The alteration of these terminal units appears highly interesting to tune the solubility and thermal properties of the hyperbranched polymer.

Some hyperbranched PEAs have been commercialized under the trade name of Hybrane™ at a very competitive cost. These PEAs were obtained from diisopropanolamine and a monomer derived from a cyclic anhydride. The polycondensation was performed in bulk at relatively mild conditions in the absence of catalyst [47]. A large variety of structures with concomitant properties and industrial

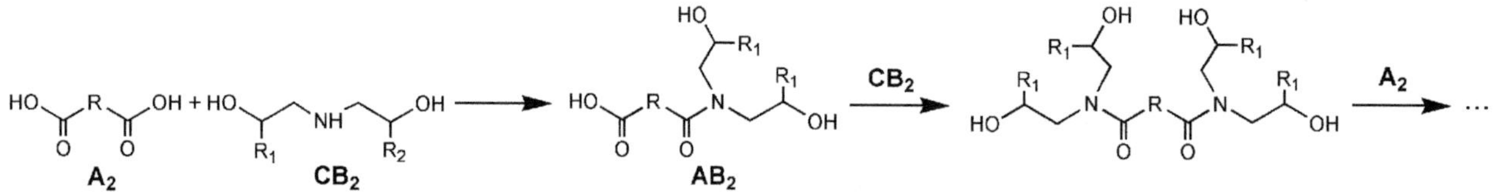

FIGURE 8.10 Main reactions involved in the synthesis of hyperbranched poly(ester amide)s derived from diacid and multihydroxyl secondary amines.

applications were obtained by choosing the appropriate anhydride and even modifying the end groups. Compounds appear interesting for pharmaceutical formulations as drug carriers [48] and furthermore can act as solubilization enhancers for poorly water-soluble drugs such as glimepiride, an antidiabetic drug [49].

Water-soluble hyperbranched PEAs are nowadays developed for different applications such as paper coatings, kinetic hydrate inhibitors (KHIs) for the oil and gas industry, and biology and medicine devices. The last applications require a lower critical solution temperature (LCST) around physiological conditions, whereas high LCST values may be useful as KHIs. Kelland [50] has recently demonstrated that PEAs based on a cyclic anhydride and diisopropanolamine can be effectively tuned by varying the hydrophobicity of the cyclic anhydride or by the addition of a less hydrophilic secondary amine (i.e., without hydroxyl groups).

Hyperbranched PEAs have also been prepared by thermal polycondensation of succinic anhydride and tris(hydroxymethyl)aminomethane followed by end-capping the terminal hydroxyl groups with different ratios of octadecyl alcohol and isophorone diisocyanate chains (IPDI) [51]. A high substitution degree of hydroxyl groups with IPDI resulted in higher degree of crystallization and a glass transition temperature up to 43 °C. The new resins had high photopolymerization rate and final unsaturation conversion and consequently appear as promising materials for UV-curable powder coating.

8.4 LIQUID CRYSTALS AND RIGID-CHAIN PEAs

The rigidity due to aromatic rings and the double bond character of carbonyl groups coupled with extensive hydrogen bonding between amide-amide and amide-ester moieties influences the ordering of aromatic-aliphatic PEAs, causing liquid crystalline (LC) behavior. Sudha and Pillai investigated the influence of various aromatic entities differing in size and structure on the intermolecular interactions and LC phase transitions of PEAs prepared by the polycondensation of diamide-diols with diacid chlorides [52,53]. Specifically, monomers were prepared by the aminolysis of γ-butyrolactone with hexamethylenediamine, *p*-phenylenediamine and *p*-xylylenediamine. The type of the structural units present in the polymer chain influenced

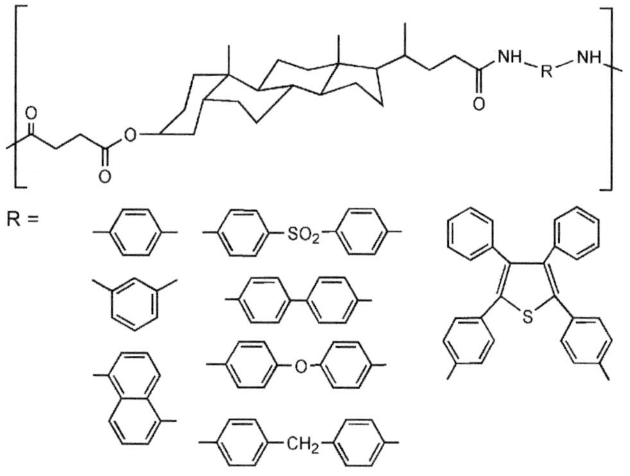

FIGURE 8.11 Chemical structure of liquid crystal PEAs having a steroid moiety.

the nature and extent of hydrogen bonding, which in turn affected the chain stiffness and the persistence length. It was morphologically observed that the formation of different forms of nematic/smectic/columnar/spherulitic phases in these PEAs was due to a balance of self-assembling through hetero-intermolecular hydrogen-bonded amide-amide and amide-ester networks and also through interplane mesogenic interactions. Both kinds of interactions were found to be greatly influenced by the structure and position of the aromatic moieties.

Main-chain LC PEAs have been synthesized using a dicarboxylic mesogen compound of biological origin and subsequent polycondensation with aromatic diamine monomers via the Yamazaki's phosphorylation reaction [54]. The dicarboxylic mesogen 3-(3-carboxypropionyl) lithocholic acid was selected due to the capability to act as both a mesogen and a bulky, flexible spacer, which results in improved solubility compared to polymers containing rigid mesogenic units. Moreover, the synthesized polymers were based on a steroid moiety (Figure 8.11) and consequently could be useful in biomedical applications. The new polymers were soluble in most of the organic solvents, had inherent viscosities in the range of 0.21-0.38 dL/g, and exhibited a nematic mesophase on shearing. Specifically, polymers derived from *p*-phenylenediamine showed a smectic G/nematic transition with dendritic texture, whereas other polymers exhibited a mesophase that suggested a nematic Schlieren texture.

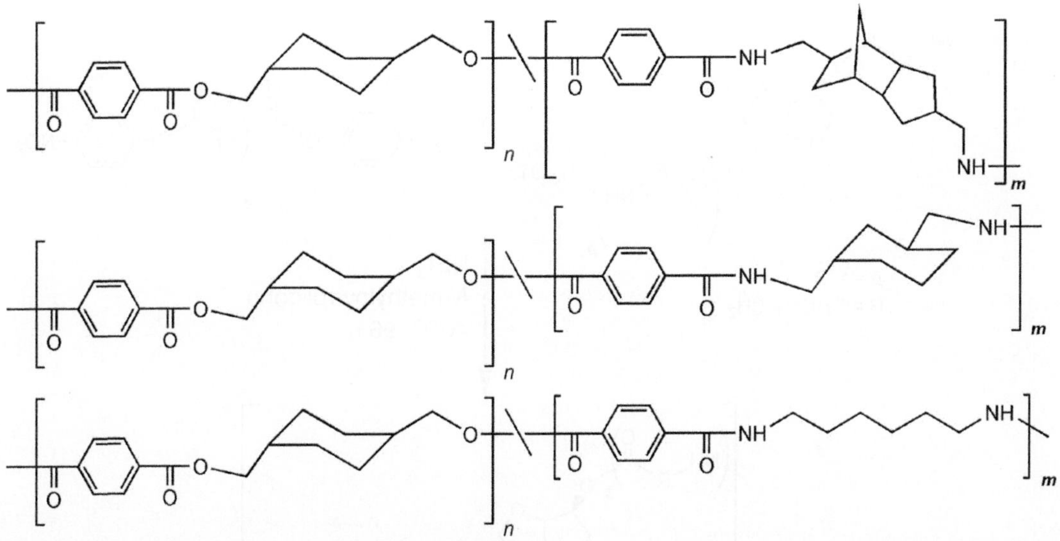

FIGURE 8.12 Chemical structure of cycloaliphatic liquid-crystalline PEAs.

Cycloaliphatic LC PEAs based on commercially poly(1,4-cyclohexanedimethylene terephthalate) (PCT) have also been synthesized with two cycloaliphatic diamines and a linear counterpart (1,6-hexamethylenediamine) (Figure 8.12) [55]. The compositions of the ester/amide units in the copolymers were varied up to 50% by the adjustment of the amounts of the diol and diamine in the feed. The introduction of amide linkages was found to induce nematic LC properties into the polyester backbone, which in turn increased the polymer chain alignment. Interestingly, the introduction of nematic LC phases into PCT was only possible when a low ratio of amide units (i.e., less than 25 mol%) was incorporated into the polymer backbone.

PEAs used as LC polymers have the advantage of having a reduced symmetry with respect to intractable fully aromatic homopolyamides and homopolyesters and thus may have a better melt processability. New hydrogen-bonded LC PEAs have been prepared from 1,4-terephthaloyl[bis-(3-nitro-*N*-anthranilic acid)] or 1,4-terephthaloyl[bis-(*N*-anthranilic acid)] in order to discern the influence of nitro groups on the phase behavior of LCPs [56]. Nitro groups play an important role in nonlinear optical (NLO) and electro-optical polymers as electron-acceptor groups. In addition, the orientational order of an LC system has importance in the alignment of NLO chromophores. Introduction of nitro groups was found to favor layer ordering, and consequently, the studied PEAs containing nitro groups exhibited polymorphism (smectic and nematic), whereas those without nitro groups showed only one phase transition (a nematic threaded texture).

Alternating PEAs consisting of a rigid diamide-diester derived from dimethyl terephthalate and 1,4-butanediamine and flexible diol spacer building blocks of differing lengths have been studied by solid-state MAS NMR relaxation in order to understand the impact of the spacer length on the molecular dynamics [57]. Results pointed out a more efficient crystallization when the spacer corresponded to even diols. Interestingly, copolymers derived from only one kind of diol spacer were built up of slightly thicker lamellae with respect to those derived from two diol spacers (i.e., 3.7 and 5.6 nm).

8.5 PEAs FROM RENEWABLE SOURCES

In concert with the depletion of oil resources, increasingly greater attention has been directed to effective utilization of plant biomass and carbohydrates as alternative renewable resources that can be steadily supplied and used for polymer syntheses. In addition, polymers derived from α-amino acids may offer many advantages since functional groups can be incorporated; biological properties like cell migration, adhesion, and biodegradability can be attained; and finally degradation products are expected to be nontoxic and readily metabolized/excreted from organisms.

8.5.1 Carbohydrate Derivatives

Synthetic polymers containing carbohydrates in the main chain have been considered as a new type of biodegradable and biocompatible polymeric materials. Furthermore and despite limitations derived from their multifunctionality, carbohydrates stand out as highly convenient raw materials for the synthesis of stereoregular polymers containing several stereocenters in the main chain, due to their easy availability and great stereochemical diversity. It has been demonstrated that physical properties and biological activity can be varied by controlling the tacticity and regicity of the samples. Different groups have performed a research

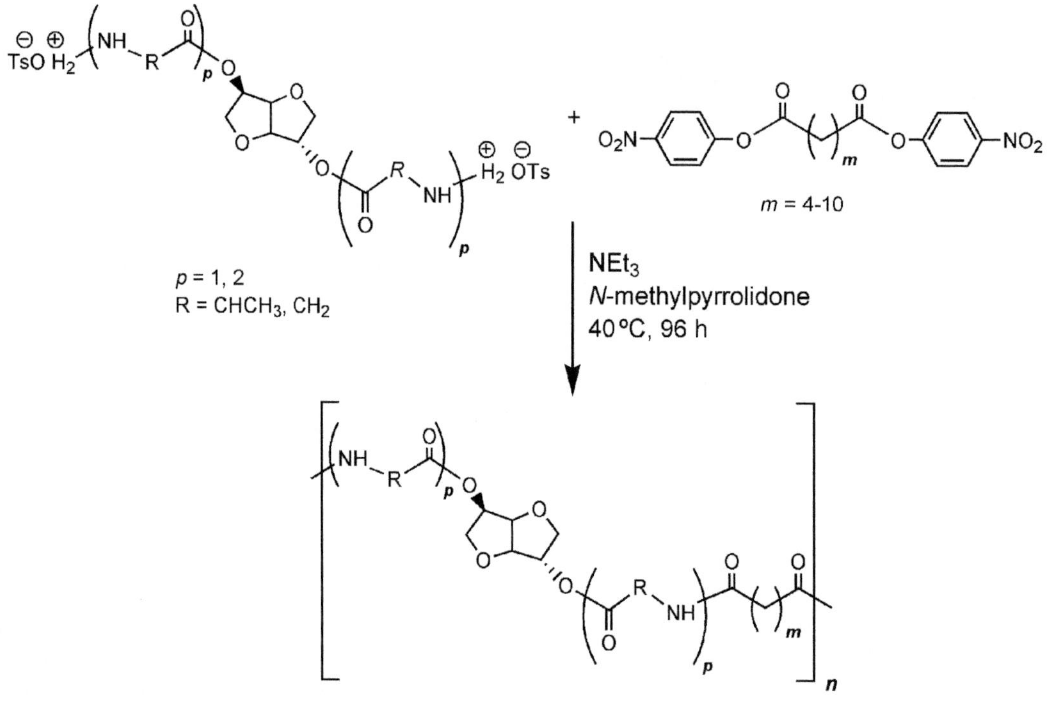

FIGURE 8.13 Synthesis of regular PEAs derived from 1,4:3,6-dianhydro-D-glucitol.

focused to the study of PEAs derived from tartaric acid, malic acid, arabinose, and dianhydrohexitoles.

Synthesis and biodegradation behavior of a series of PEAs composed of 1,4:3,6-dianhydro-D-glucitol, which is readily available from D-glucose, an α-amino acid (glycine and L-alanine), and an aliphatic dicarboxylic acid (from adipic to dodecanedioic acid) units (Figure 8.13), has been described by Okada and coworkers as new examples concerning the development of biodegradable polymers based on plant biomass resources [58]. The biodegradability of all these PEAs was found to depend on their molecular structure. Interestingly, most of them were degraded more rapidly than the corresponding polyesters having the same dicarboxylic acid unit when papain was employed, whereas this enhanced degradability was only found for PEAs containing dicarboxylic acid components with shorter methylene chain lengths when lipases (e.g., from *porcine pancreas*) were used. On the contrary, PEAs were less degradable than parent polyesters when studies were conducted on both composted soil and activated sludges.

Synthesis, structure, and properties of a set of alternating copoly(ester amide)s made from L-tartaric acid, succinic acid, and *n*-amino-1-alkanols (n=2, 3, and 6) (Figure 8.14) have been studied in order to understand the influence of the heterogeneity introduced in the polymer microstructure from the differences in constitution of succinic and tartaric units [59,60]. These PEAs contained equal amounts of amide and ester groups and were stereoregular and easily synthesized by the condensation of the *p*-toluene sulfonate salts of bis(aminoalkyl) succinates with active esters of tartaric acid. Polymers displayed

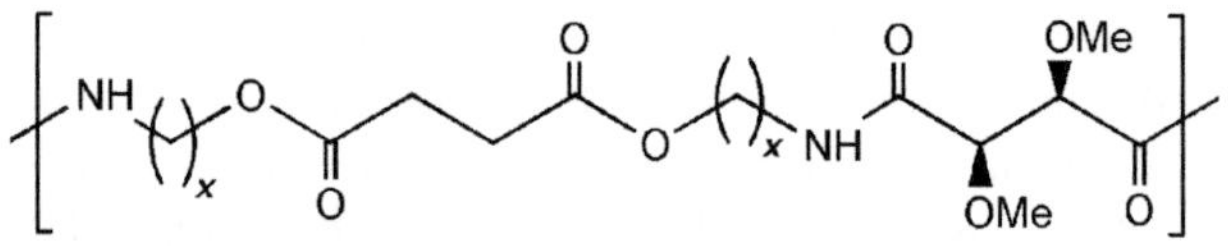

FIGURE 8.14 Chemical structure of regular PEAs made from L-tartaric acid, succinic acid, and *n*-amino-1-alkanols.

optical activity in solution, were crystalline in the solid state with melting points between 90 and 160 °C, and were relatively hygroscopic, although nonwater-soluble.

The hydrolytic degradation of a series of crystalline copoly(ester amide)s derived from L-tartaric and succinic acids, 1,6 hexanediamine, and 1,6-hexanediol with ester/amide groups ratios 3/97, 10/90, 15/85, and 20/80 has also been investigated [61]. Degradation proceeded with a notable increment in crystallinity and was clearly enhanced with the content in ester groups. Interestingly, degradation was accompanied by formation of cyclic succinimide units indicative of a particular scission mechanism based on the occurrence of intramolecular imidation reactions. The changes in structure and properties taking place upon hydrolytic degradation were also examined [62].

The synthesis, characterization, and properties of a series of PEAs obtained from L-malic acid, 6-aminohexanol, and 1,6-hexanediamine, in which the hydroxy side group of malic acid was protected as methyl ether, have been studied [63]. *Aregic* and *isoregic* polymers were obtained and, furthermore, the ester-amide ratio was varied (i.e., from 1:50 to 1:1). All studied PEAs were semicrystalline with T_m decreasing with the content in ester groups from 168 to 144 °C and T_g falling down in parallel from 60 to 10 °C. Degradation occurred through ester hydrolysis and cyclic imide formation and, logically, the rate increased with the ester content.

Biodegradable PEAs derived from isosorbide or isomannide have also been prepared following a classical two-step route where dianhydrohexitoles were firstly esterified with α-amino acids in the presence of *p*-toluenesulfonic acid and secondly the amino groups were liberated and polycondensed with *p*-nitrophenyl esters of aliphatic dicarboxylic acids [64]. The resulting PEAs had glass-transition temperatures in the range of 60-120 °C and were demonstrated to be biodegradable under the action of enzymes like chymotrypsin or lipases.

Carbohydrate units bearing methoxyl groups along the polymer backbone were found to increase the hydrophilicity and to enhance the hydrolytic degradability. These sugar-based polymers can be water-soluble, depending on the nature and/or the amount of the carbohydrate monomer in the polymer chain. In order to study the hydrophilic effect on degradation, a series of copoly(ester amide)s were prepared by random copolymerization of monomers derived from L-arabinitol and 5-aminopentanol [65]. Samples containing about 50% of the sugar-based monomer were water-soluble and their hydrolytic degradability experimented a noticeable increase. Spectroscopic investigations of the hydrolysis products provided evidence for succinimide ring formation, which is a characteristic degradation mechanism of PEAs containing four-carbon diacid units in their structure.

8.5.2 PEAs from Vegetable Oils and Fatty Diacids

Synthesis of polymers from vegetable oils is economically, scientifically, and environmentally significant because of their environmentally friendly nature, low cost, abundance, and possible biodegradability. Linseed oil, one of the most widely occurring vegetable oils, has been employed to prepare PEAs with superior characteristics over normal alkyds in terms of hardness, ease of drying, and water vapor resistance. A polymer with fatty lateral chains was easily obtained through condensation polymerization of phthalic anhydride and *N,N*-bis(2-hydroxyethyl) linseed oil [66]. The sample exhibited high thermal stability and good physicomechanical and chemical resistance properties to find application as a corrosion-protective coating.

Multiblock copolymers based on oligoamide (M_n, 2000 g/mol) and aliphatic oligoesters from a dimerized fatty acid and 1,4-butanediol were prepared by melt polycondensation under vacuum (~400 Pa) at 255-260 °C and using a Mg-Ti catalyst [67]. The copolymers showed a wide range of softness and processing flexibility. Calorimetric analysis indicated the existence of segregated amorphous phases corresponding to the oligoester and oligoamide blocks. The tensile properties confirmed a typical thermoplastic elastomeric behavior of the synthesized PEAs.

Bio-based PEAs have also been prepared from castor oil as a source of methyl 10-undecenoate and a bio-based methyl diester in the presence of a transesterification catalyst [68]. Aliphatic diols containing EA, monoamide, and diamide linkages were specifically synthesized from methyl 10-undecenoate through transesterification, amidation, and thiol-ene reactions (Figure 8.15). The incorporation of amide functions in the polyester backbone resulted in semicrystalline materials with melting points ranging from 22 to 127 °C and complex melting behaviors due to polymorphism and melting-crystallization processes. Young's modulus of these new PEAs could reach 363 MPa depending on the ratio of amide groups.

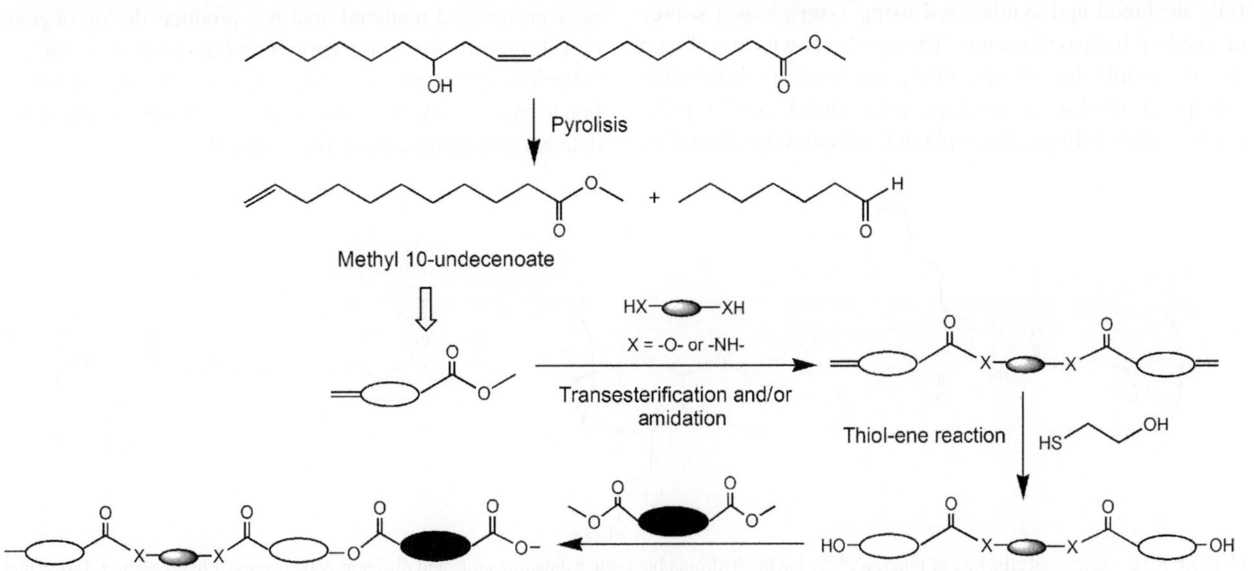

FIGURE 8.15 Synthesis of bio-based PEAs from methyl 10-undecenoate.

8.5.3 PEAs Derived from α-Amino Acids and Their Applications in the Biomedical Field

Amino acid-based poly(ester amide)s (AA-PEAs) constitute one of the most promising family of biodegradable synthetic polymers derived from natural amino acids. AA-PEAs have been fabricated into three-dimensional microporous hydrogels, micro/nanospheres, and electrospun fibrous membranes for many different biomedical applications, such as drug carriers, coatings for drug-eluting stents, transfection agents, and wound dressings [69–74].

Incorporation of amino acids with functional side chains in PEAs is a significant topic since conjugation of drugs or cell signaling molecules in tissue-engineering scaffolds may be possible and, consequently, potential applications of these materials can be increased. In addition, functional side chains may provide significant effects on polymer solubility, biocompatibility, and biodegradation. An easy procedure to synthesize this kind of polymers is based on the use of L-lysine to provide functionalizable pendant amine groups. An interesting example concerns to the incorporation of varying percentages of lysine into PEAs comprised of L-phenylalanine, 1,4-butanediol, and succinic acid by tuning the ratio of ε-protected L-lysine- and L-phenylalanine-derived monomers (Figure 8.16) [75]. Polymers prepared by this method had M_ns ranging from 26,900 to 56,600 g/mol and polydispersity indices from 1.19 to 1.38. The reactivity of the pendant amines in the final polymer was demonstrated by reaction with amino acid and tri(ethylene glycol) derivatives. It was also demonstrated that the hydrolytic and enzymatic (evaluated in media containing chymotrypsin or trypsin) degradability could be accelerated by increasing the content of lysine probably as a consequence of a greater hydrophilicity.

Functional amino AA-PEAs have also been successfully designed and synthesized using L-serine as a source of pendant hydroxyl groups. The synthesis was based on a *N*-carboxyanhydride ring-opening reaction that allowed the incorporation of serine residues and a typical solution polycondensation using a *p*-nitrophenyl-activated dicarboxylate [76]. Gels via photogelation were easily prepared after conversion of pendant hydroxyl groups into acrylate groups (Figure 8.17). The new family of functionalized AA-PEAs supported cell attachment and proliferation if samples reached a certain level of hydrophobicity.

A series of biodegradable functional amino acid-based PEAs were designed and synthesized by the solution *co*-polycondensation of amino acid (L-phenylalanine and DL-2-allylglycine)- and dicarboxylic acid-based monomers [77]. Polymers incorporated pendant carbon-carbon double bonds through the DL-2-allylglycine units and consequently the content on these functional groups could be adjusted by tuning the feed ratio of L-phenylalanine to DL-2-allylglycine monomers. The incorporation of the functional pendant carbon-carbon double bonds along the polymer chains could significantly expand the biomedical applications via their capability to either conjugate bioactive agents or prepare additional useful functional derivatives. Cross-linking of double bonds was successfully performed using poly(ethylene glycol) diacrylate and gave rise to hybrid hydrogels with a three-dimensional porous network structure [78]. The hydrophobicity, cross-linking density, and mechanical strength of these hydrogels increased with an increase in the allylglycine content, but the swelling and pore size decreased.

Another series of random PEAs with functional groups were synthesized by Jing and coworkers [79] with a two-step polycondensation from the benzyl ester of the dimethylolpropionic acid, a diamine derived from glycine, and sebacoyl dichloride (Figure 8.18). The benzyl ester group could be removed selectively by catalytic hydrogenolysis or converted into other active groups after polymerization. In this way, PEAs had the possibility of further connecting drugs or other bioactive molecules. The chosen diamine was derived from hexanediol and glycine and consequently had biodegradable ester groups and rendered nontoxic products during degradation. Therefore, these target polymers had good biodegradation behaviors and chemical functionality, i.e., an ideal combination to expect wide biomedical applications, especially in drug delivery and tissue-engineering scaffolds.

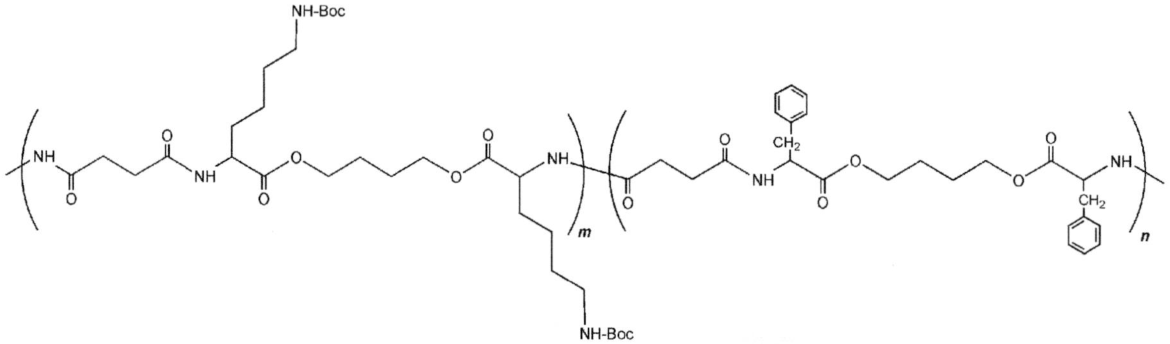

FIGURE 8.16 Chemical structure of random PEAs having hydrophobic L-phenylalanine units and different percentages of hydrophilic L-lysine units. After deprotection, PEAs are provided of functional amine pendant groups.

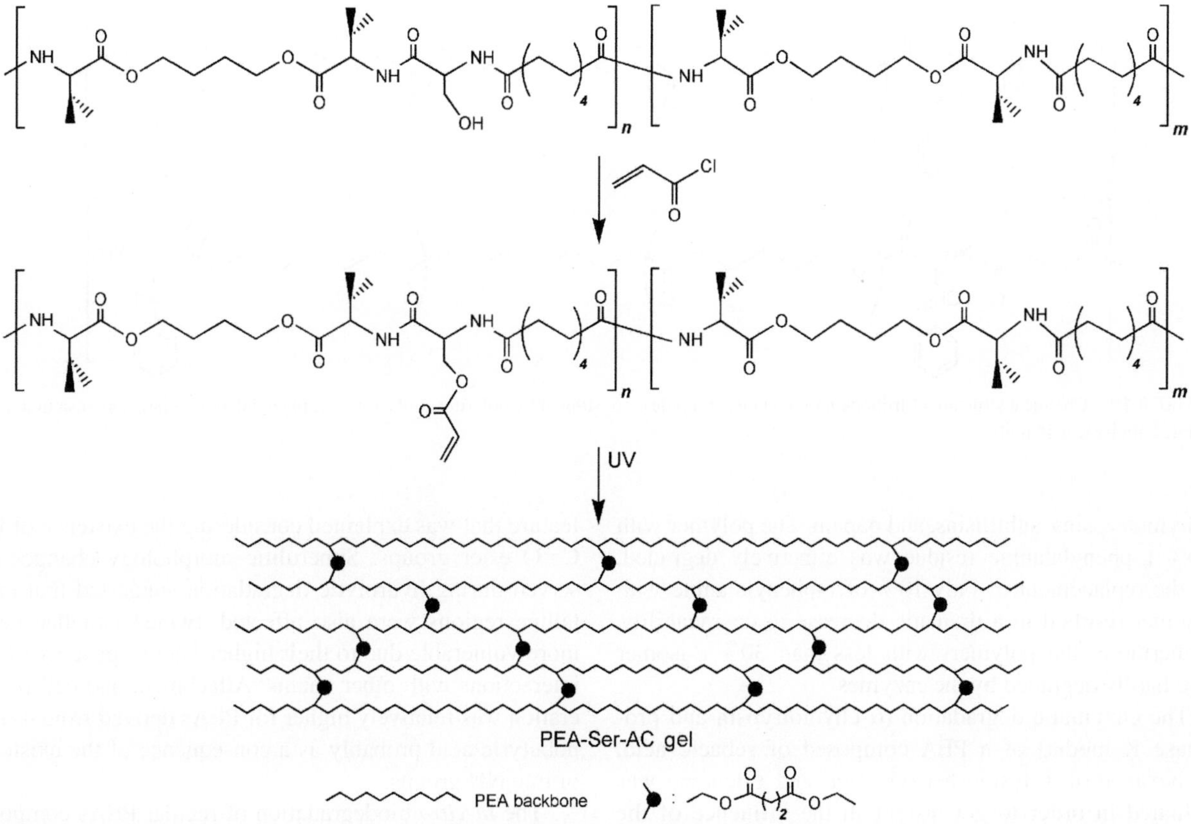

FIGURE 8.17 Functionalized poly(ester amide)s derived from L-serine with application as hydrogel.

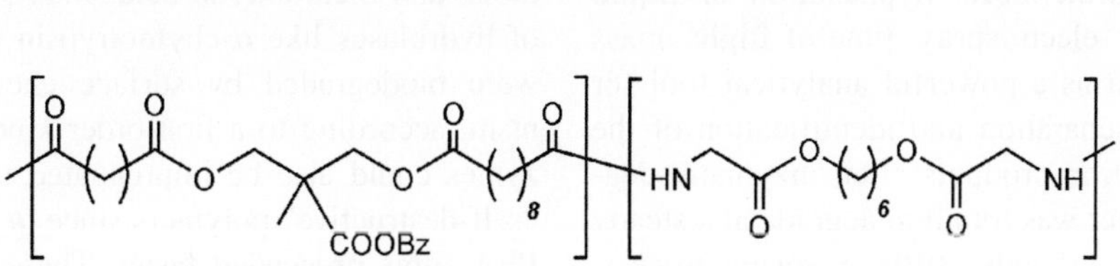

FIGURE 8.18 Chemical structure of PEAs having functionalizable benzyl ester groups by hydrogenolysis.

Recently, oligoethylene glycols (i.e., diethylene glycol and tetraethylene glycol) were also introduced into the PEA macromolecular backbone instead of aliphatic diols in order to improve hydrophilicity, solubility, and biodegradability [80,81]. In the same way, low molecular weight poly(ε-caprolactone) (PCL) has been used to replace the conventional aliphatic diols to provide more flexible backbone chains and improve hydrolytic degradation capability [82]. Thus, a series of biodegradable amino acid-based poly(ether ester amide)s consisting of three building blocks (i.e., PCL, L-phenylalanine, and unsaturated and saturated aliphatic acid units) were synthesized by a solution polycondensation (Figure 8.19). Polymers were obtained with very good yields and average molecular weights ranging from 6900 to 31,000 g/mol, depending on the original molecular weight of PCL. These PCL derivatives showed lower T_g than most of the related oligoethylene glycol-based PEAs and an improved solubility in common organic solvents, such as THF and chloroform. It was postulated that the presence of PCL segments in the molecular backbone should provide materials with highly promising biomedical applications in tissue engineering, drug/gene delivery, and wound healings.

Biodegradability of polymers containing α-amino acids can be easily controlled by changing their L/D ratio. Hydrolytic degradability and both chemical and physical properties should on the contrary be maintained. It is known that α-amino acids undergo racemization at high temperatures or under basic conditions, and therefore isomerization can occur during synthesis and even during spinning or molding processes. Detailed studies are interesting for better understanding the effects of the stereostructure on biodegradability. In this way, PEAs containing phenylalanine were synthesized by the polycondensation of di-*p*-nitrophenyl sebacate with diamines containing phenylalanine ester groups with different L/D ratios [83]. The PEAs were subjected to enzymatic degradation with proteases like

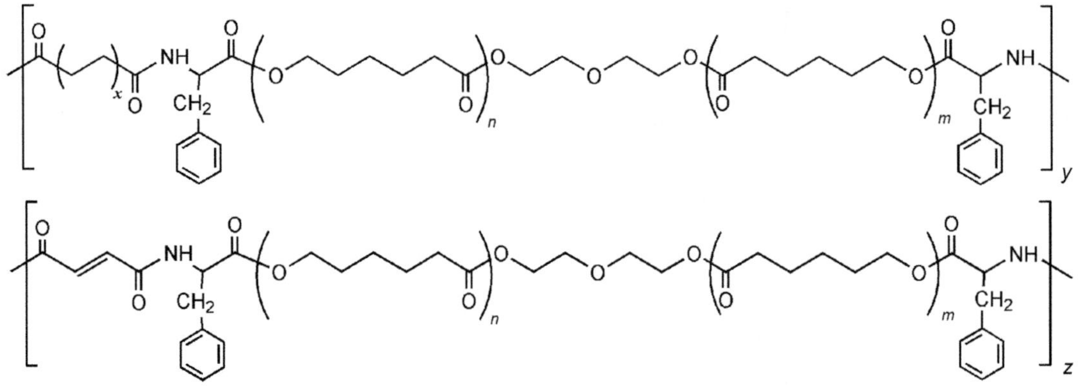

FIGURE 8.19 Chemical structure of triblock poly(ether ester amide)s constituted by poly(ε-caprolactone), L-phenylalanine (Phe), and unsaturated and saturated aliphatic acid units.

α-chymotrypsins, subtilisins, and papain. The polymer with 100% L-phenylalanine residue was effectively degraded, but the replacement of only 10% of L-phenylalanine with D-isomer resulted in a dramatic decrease in degradability. Furthermore, the polymers with less than 30% L-isomer were hardly degraded by the enzymes.

The enzymatic degradation (α-chymotrypsin and proteinase K media) of a PEA composed of sebacic acid, 1,6-hexanediol, L-lysine benzyl ester, and L-leucine was evaluated in order to get insight in the influence of the monomer sequence in multiblock copolymers on the degradation rate [84]. Furthermore, hyphenation of liquid chromatography with electrospray time-of-flight mass spectrometry was used as a powerful analytical tool for the chromatographic separation and identification of the water-soluble degradation products after enzymatic degradation. The copolymer was found to degrade at a steady rate with both enzymes through a surface erosion process. Interestingly, no accumulation of acidic byproducts was observed during the course of degradation and consequently the new PEA had a better behavior than conventional polyesters.

Hydrogen bonding interactions play also a fundamental role on crystallinity, degradability, and even on biocompatibility. Thus, Im and coworkers evaluated these interactions on PEAs having different positions of the amide and ester groups in the repeat unit [16]. Specifically, different amino acids (i.e., glycine and 4-amino butyric acid) and different diols (i.e., 1,6-hexanediol and 1,4-butanediol) were employed to get a diester monomer for the subsequent interfacial polymerization with sebacoyl dichloride. FTIR analysis indicated that for PEAs derived from glycine, only amide-amide hydrogen bonds between adjacent chains were established, whereas the PEAs derived from 4-amino butyric acid contained amide-amide and amide-ester hydrogen bonds, including NH groups and C=O ester groups in free state. Enzymatic degradation using papain demonstrated a greater degradation for samples derived from glycine, a feature that was explained considering the existence of free C=O ester groups. Spherulitic morphology changes observed during hydrolytic degradation suggested that crystalline regions were also affected, twined lamellae being more vulnerable due to their higher lateral space free from interactions with other chains. Attachment and cell proliferation was relatively higher for PEAs derived from 4-aminobutyric acid probably as a consequence of the existence of free NH groups.

The *in vitro* biodegradation of regular PEAs composed of naturally occurring hydrophobic α-amino acids, fatty diols, and dicarboxylic acids was studied in the presence of hydrolases like α-chymotrypsin and lipase [85]. PEAs were biodegraded by surface chemical erosion mechanism, according to a first-order kinetics. Interestingly, enzymes could also be impregnated into the PEAs to make "self-destructive" polymers since *in vivo* biodegradation of PEA films proceeded faster. These findings open a great potential for designing drug-sustained/controlled release devices and implantable surgical devices. Specifically, slow-release biodegradable PEAs have been impregnated with α-chymotrypsin (or trypsin), antibiotic (ciprofloxacin hydrochloride), and lytic bacteriophages to provide a new wound-healing preparation that appears ideal for the treatment of poorly vascularized and venous stasis ulcers [86]. This system appears highly interesting in the management of refractory wounds and could be an alternative to the antibiotic therapy when bacterial strains become resistant to common antimicrobial agents.

Incorporation of the amino acids L-2,4-diaminobutyric acid (DAB) and homocysteine (HCY) into a PEA backbone was investigated with the aim of imparting stimuli-responsive degradation properties. Studies on model compounds revealed that the pendant γ-amine or γ-thiol of DAB and HCY esters experimented intramolecular cyclizations at a much more rapid rate than background ester hydrolysis. Thus, stimuli-induced cyclization reactions (i.e., after deprotection of pendant groups by acids or reducing agents) were effective to cleave ester groups in the backbone of PEAs containing the indicated self-immolative spacers. In addition, a photochemically responsive PEA was prepared by simply changing the protecting group on the pendant amine [87].

The nonviral transfer method appears nowadays to be the most promising approach for gene delivery when clinical safety is considered. Synthetic polycations have attracted much attention as nonviral delivery vectors despite their high transfection efficiency and low toxicity could not simultaneously be achieved. Arginine-based poly(ester amide)s (Arg-PEAs) synthesized from diols (e.g., 1,4-butanediol and 1,6-hexanediol) have been demonstrated to exhibit excellent cell membrane-penetrating capability and low cytotoxicity [73]. However, these Arg-PEAs could not transfect the primary and stem cells with high efficiency probably as a consequence of the stiffness of the chain structure. Therefore, a new series of biodegradable L-Arg-PEAs was prepared having a controllable flexible structure by incorporating oligoethylene glycol units (Figure 8.20) [88]. Results clearly indicated that these PEAs were able to condense DNA and form a stable complex easily. Interestingly, transfection results obtained from a wide range of cell lines, primary cells, and stem cells showed that members of the series had comparable or better transfection efficiency than commercial reagents (e.g., Lipofectamine2000®) and even a much lower cytotoxicity. Furthermore, transfection efficiency was clearly improved when Arg-PEAs having more flexible chains were assayed.

A biodegradable pH-sensitive polymer system, namely, a dual amino acid (L-lysine and L-leucine)-based PEA with pendant –COOH groups (Figure 8.21), has been synthesized and characterized [89]. This polymer was developed to prepare microspheres for oral insulin delivery and to overcome drawbacks of previously studied polymers (e.g., methacrylate/methacrylic acid derivatives and poly(methacrylic acid)-graft-poly(ethylene glycol)), which were not biodegradable or even had an excessively hydrophilic nature. This new α-amino acid-based PEA had the advantage to be sensitive to the attack of specific enzymes like elastase and α-chymotrypsin, the major proteolytic enzymes in the small intestine where oral insulin is favored to be released and absorbed. Insulin could be encapsulated in the PEA microspheres by a solid-in-oil-in-oil technique with good sphericity and high efficiency. The pH-triggered *in vitro* release profiles indicated that the carrier could protect the loaded insulin in the acidic conditions of the stomach and provided a sustained release in the intestine. The leucine component in the PEA served to increase its hydrophobicity and improved the membrane permeability of protein drugs. A dose-dependent hypoglycemic effect was achieved when an oral dose of insulin-loaded PEA microspheres was given to streptozotocin-induced diabetic rats.

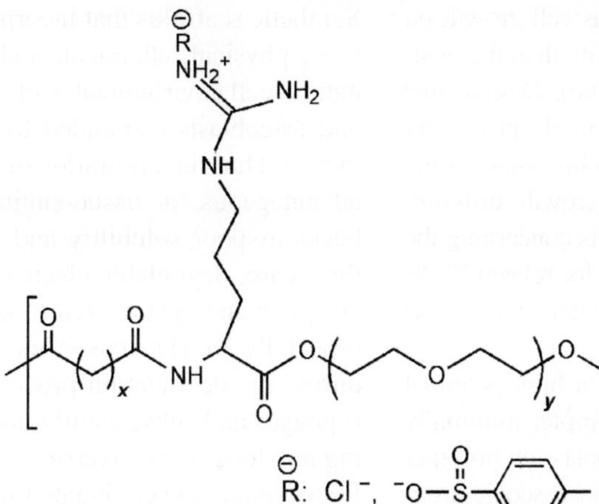

FIGURE 8.20 Chemical structure of PEAs containing L-arginine amino acids with interesting features as nonviral transfer vector.

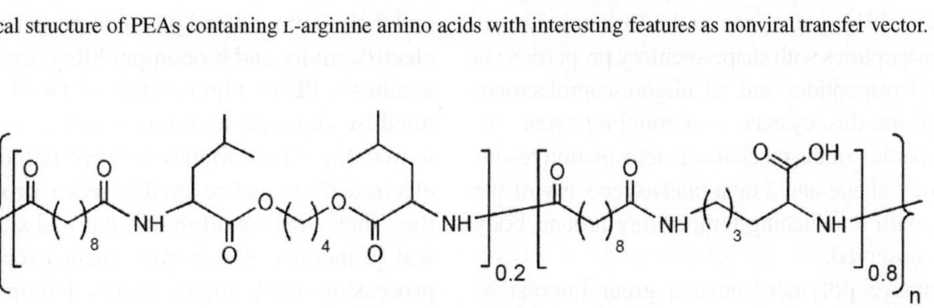

FIGURE 8.21 Chemical structure of the copoly(ester amide) based on L-lysine and L-leucine amino acids that has been developed to prepare microspheres for oral insulin delivery.

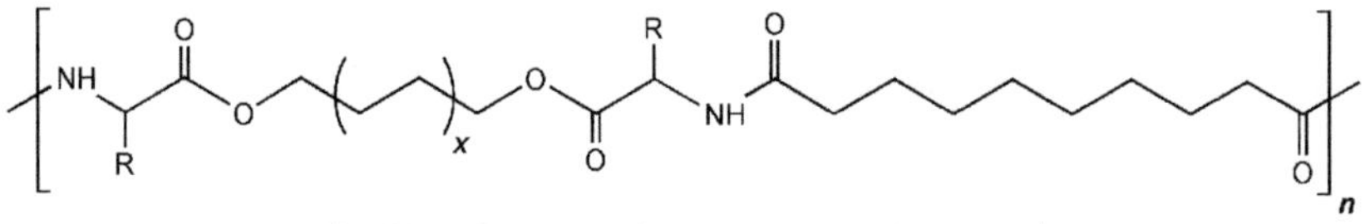

R: $-CH_3$, $-CH_2$-Ph, $-(CH_2)_4$NH-Boc, $-(CH_2)_4NH_3^+$ TFA^-
x: 1,3

FIGURE 8.22 Chemical structure of biodegradable poly(ester amide)s with potential interest as vascular constructs for therapeutic uses.

The design of biologically and mechanically responsive vascular constructs for therapeutic uses is currently a subject of great applied interest. Both naturally occurring (e.g., collagen) and synthetic biodegradable biomaterials (e.g., polyesters like poly(lactic acid) and poly(glycolic acid) or copolymers constituted by glycolic acid and lactic acid units) have been extensively studied. Synthetic materials have the advantage of a greater control over their final properties, but in general, the studied polyesters produced acidic products during degradation that could be toxic and induce undesired phenotype modulation in seeded vascular smooth muscle cells. In this way, biodegradable amino acid-based PEAs have been proposed as an alternative biomaterial to regulate human coronary artery smooth muscle cell (HCASMC) growth and phenotype. Specifically, a series of biodegradable PEAs (Figure 8.22) was synthesized from L-alanine, L-phenylalanine, and L-lysine using both solution and interfacial polymerizations [90]. The synthesized PEAs had glass-transition temperatures ranging from 10 to 40 °C, indicating their pliability under physiological conditions. HCASMC attachment and spreading was observed on all PEAs up to 7 days of culture, whereas cell growth on functionalized PEAs proceeded at a slower rate than the positive control (tissue-culture treated polystyrene). Despite this feature, lysine derivatives were considered highly promising materials because the pendant functional groups enabled the conjugation of molecules that regulate cell growth, differentiation, and signaling pathways. Similar results concerning the effect of lysine units have also been reported for related PEAs where succinyl chloride was employed instead of sebacoyl chloride [41].

Stimuli-sensitive implant materials have a high potential for applications in biomedicine as, for example, minimally invasive surgery. These implant materials consist on polymer systems that allow the variation of different macroscopic properties over a wide range by only small changes in the chemical structure. Feng *et al.* [91] synthesized multiblock copolymers based on oligodepsipeptides with shape-memory properties via coupling of oligodepsipeptides and an oligo(ε-caprolactone) diol using an aliphatic diisocyanate as a coupling agent. An almost complete fixation of the mechanical deformation resulting in the temporary shape and a quantitative recovery of the permanent shape with a switching temperature around body temperature were observed.

Stimuli-responsive polymers have a great interest as materials for advances in biotechnology and the biomedical field due to their wide applications as sensing, pollution control, drug delivery, biomimetic actuation, and catalysis. The thermosensitivity of any polymeric material can be regulated by controlling the hydrophobic-hydrophilic balance of the polymeric chains. Ohya *et al.* [92] induced thermosensitivity in the polydepsipeptide constituted by glycolic and aspartic acids by attaching the moderately hydrophobic isopropylamine group into its carboxylic side group. The new polymer was fully degradable *in vitro* at room temperature by cleavage of the ester bonds in the main chain. The polymer and its degradation products were nontoxic and biocompatible. Furthermore, the cloud point at 29 °C (between room and body temperature) makes the polymer attractive for implants and other biomedical applications.

8.6 MISCELLANEOUS APPLICATIONS OF PEAs

8.6.1 Scaffolds from Electroactive Samples and Electrospun Nanofibers

Synthetic scaffolds that incorporate multiple functionalities (e.g., physical, chemical, and biological) may mimic the natural cell environment. Cells such as fibroblasts, neurons, and osteoblasts responded to electrical fields *in vitro* and *in vivo*. Thus, incorporation of conducting polymers appears advantageous for tissue-engineering applications if drawbacks as poor solubility and processability are solved. In this sense, degradable electroactive polymers are currently designed and synthesized to meet biomedical applications, especially *in vivo* tissue engineering. Oligomers formed during the degradation process can be consumed by macrophages and subsequently undergo renal clearance, avoiding any long-term adverse effect *in vivo* [93]. Electroactive PEAs containing conjugated segments of amino-capped tetraaniline (PEA-*g*-TA) have been synthesized (Figure 8.23) and their chemical and physical properties, along with their electroactivity and biocompatibility, evaluated [94]. For the synthesis, PEAs with pendant –COOH groups (i.e., constituted by glutamic acid units) were considered for coupling tetraaniline (TA). Materials were demonstrated to possess electroactivity and reversible redox property. Furthermore, they showed good solubility, thermal stability, and mechanical properties, which gave them excellent properties for processing. Cell culture results demonstrated that electroactive PEA-*g*-TA copolymers doped with camphorsulfonic acid could be more effective to promote the differentiation

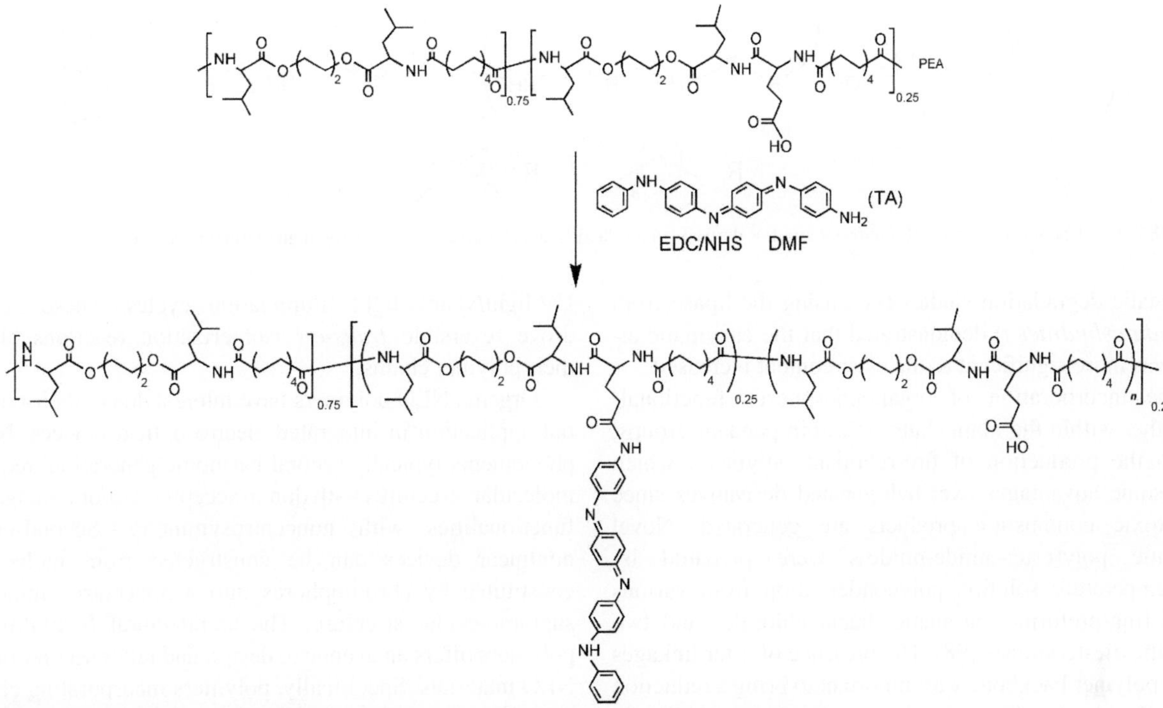

FIGURE 8.23 Chemical structure of electroactive poly(ester amide) having grafted tetraaniline units.

of mouse preosteoblastic MC3T3-E1 cells by pulsed electrical signal compared with the pure PEA and tissue-culture treated polystyrene used as control.

Electrospinning is a simple, versatile, reproducible, and economic technique that allows producing nonwoven fibrous mats that, for example, can be used for tissue-engineering applications. This electrostatic technique involves the use of a high-voltage field to charge the surface of a polymer solution droplet, held at the end of a capillary tube, and induce the ejection of a liquid jet towards a grounded target (collector). Continuous and homogeneous fibers with diameters varying from micro- to nano-dimensions can be produced by the careful selection of solution properties (e.g., viscosity, dielectric constant, volatility, and concentration) and operational parameters (e.g., strength of the applied electrical field, deposition distance, and flux). A considerable number of works have been focused on the electrospinning of polyesters and polyamides, but few studies deal with the use of PEAs. Probably the first one concerns to segmented aliphatic PEAs prepared by melt polycondensation of 1,4-butanediol, dimethyl adipate, and a preformed α,ω-amino alcohol based on ε-caprolactone and 1,4-diaminobutane [95]. In this case, the electrospinning optimal conditions for obtaining fibers in the nanometer range (180-450 nm) were found to be a voltage of 20 kV and the use of $CHCl_3$/HCOOH (9:1, v/v) as the solvent system. It was also observed that the increase on the amide content allowed getting smoother and more homogeneous ultrafine fibers. The process led to materials with a higher amorphous content than the pristine polymer.

Electrospinning conditions were also evaluated to prepare micro/nanofibers of a biodegradable PEA constituted by L-alanine, 1,12-dodecanediol, and sebacic acid [96]. 1,1,1,3,3,3-Hexafluroroisopropanol was the most appropriate solvent to obtain fibers in a wide range of electrospinning conditions that allowed tuning the final diameter size (1700-320 nm range). Fibers were also loaded with antimicrobial agents like silver and chlorhexidine, and the influence of agent concentration in the electrospinning solutions on the fiber diameter size was determined. New scaffolds supported cell adhesion and proliferation and showed a clear and well-differentiated antimicrobial effect against both Gram-positive and Gram-negative bacteria.

8.6.2 High-Performance Materials

Novel multiblock PEAs containing poly(L-lactide) and cycloaliphatic amide segments were synthesized from telechelic oligomer of α,ω-hydroxyl-terminated poly(L-lactide) (PLLA), 1,3-cyclohexylbis(methylamine), and sebacoyl chloride by a two-step interfacial polycondensation (Figure 8.24) [97]. The rigidity arising from the conformational restriction in the cyclic ring led to high thermal properties. The prepared polymers were expected to be biocompatible since the cycloaliphatic systems are nontoxic. PEAs having lower content of PLLA showed a crystallization pattern analogous to polyamides, whereas those having higher content of PLLA showed two crystalline phases characterized by polyester and polyamide segments.

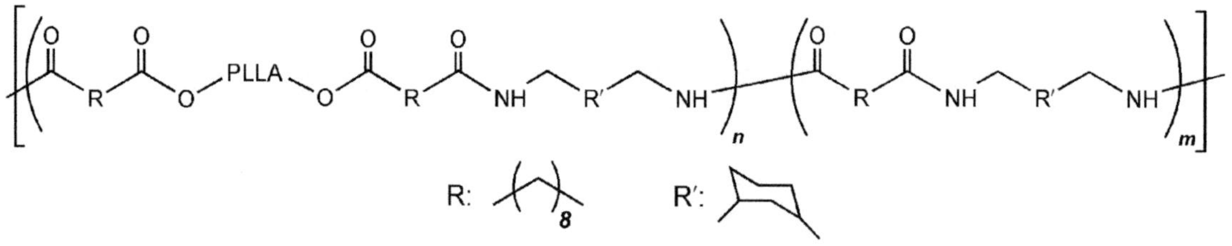

FIGURE 8.24 Chemical structure of poly(ester amide)s derived from L-lactide and characterized by having high thermal properties.

Enzymatic degradation studies (i.e., using the lipase from *Candida cylindracea*) demonstrated that the enzymatic attack was more significant as the ester content increased.

The incorporation of organophosphorus functionality, either within the main chain or within pendant groups, led to the production of fire-retardant polymers, which have some advantages over halogenated derivatives since less toxic combustion products are generated. Novel aromatic poly(ester-amide-imide)s were prepared by low-temperature solution polycondensation from various imide ring-preformed aromatic diacid chlorides and two aromatic diesteramines [98]. The presence of ester linkages in the polymer backbone was important to bring a reduction in the T_g values and a further increase in their solubility in organic solvents. Therefore, new copolymers had promising features to be employed as high-performance polymeric materials.

8.6.3 Optical Properties

Polymers endowed with light-responsive geometries are broadly relevant to a variety of emerging technologies, including novel sensory devices and optical recording media. These systems can also serve as biomimetic analogues that allow for an improved understanding of vital photoprocesses that are operative in many living organisms.

New azobenzene-modified polyesters and PEAs fitted with chiral, atropisomeric binaphthylene segments (Figure 8.25) were prepared by a series of polycondensation reactions carried out in polar solvent media [99]. The simple incorporation of ester groups in the polymer backbone provided significant improvements in organosolubility behaviors. The choice of both solvent and temperature allowed to modulate, over a relatively wide latitude, the photoregulated oscillations in optical rotatory power recorded for these new materials when stimulated with multiple UV-light/visible-light illumination cycles. These cycles drove reversible *trans-cis* isomerization reactions along their polymer chains.

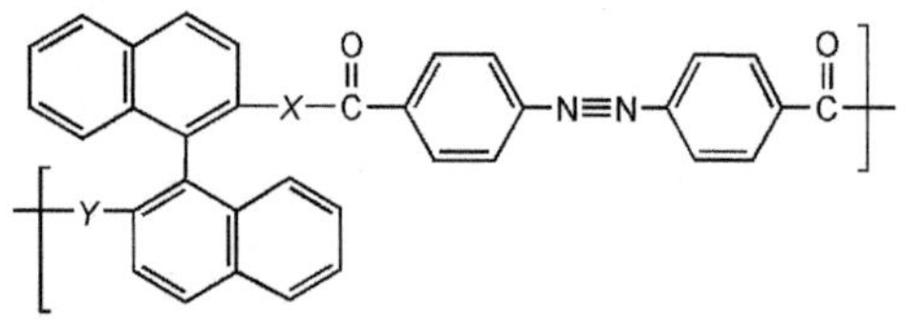

FIGURE 8.25 Chemical structure of light-responsive poly(ester amide)s.

Organic NLO polymers have interest due to their potential application in integrated electro-optical devices. NLO phenomena, typically second-harmonic generation, require molecular structures with donor-acceptor and/or conjugated functionalities with noncentrosymmetry. Second-order nonlinear devices can be constructed from molecules constituted by chromophores into a noncentrosymmetric supramolecular structure. The architectural flexibility of polymers offers an avenue to design and tailor second-order NLO materials. Specifically, polymers incorporating chiral units in order to prevent crystallization into centrosymmetric groups and donor-acceptor building blocks have been considered [100,101]. Thus, chiral PEAs prepared by the simple reaction of diacid chlorides with biphenolic azo chromophores and optically active isosorbide showed a positive solvatochromism in UV-visible absorption spectra. Glass transition temperatures and thermal stability of these PEAs were demonstrated to be high enough for NLO applications.

8.6.4 Composites and Nanocomposites Based on PEAs

Commercial success of biodegradable polymers used in commodity applications is limited due to their high cost relative to conventional thermoplastics. Starch has widely been investigated as filler because of its low cost, biodegradability, and renewability. Some studies have been performed with composites based on a PEA matrix, like the random BAK (i.e., a copolymer derived from 1,4-butanediol, adipic acid, and caprolactam) [102–104]. The use of agro-fibers as reinforcing components for thermoplastics is also a subject of research because they are renewable, biodegradable, and environmentally friendly. Some studies have been performed to evaluate the suitability of jute, one of the most common agro-fibers having high tensile modulus and low elongation at break, as reinforcement in BAK-based biocomposites [105].

Preparation of polymer nanocomposites is nowadays an important research subject since polymer properties can be enhanced (e.g., modulus, strength, thermal resistance, permeability, flammability resistance, and even biodegradability) and their range of applications extended by using molecular or nanoscale reinforcements rather than conventional fillers. Although very good future prospects exist, the present low level of production and the high costs associated to the biodegradable polymer matrix still restrict them for a wide range of applications. PEAs are recently receiving attention for use as biodegradable matrices in nanocomposite preparation [14–18]. Thus, barrier and mechanical properties of biodegradable melt-mixed PEA/octadecylamine-treated montmorillonite clay have been studied [106], the influence of nanoparticles (i.e., nano-$CaCO_3$ and nano-SiO_2) on the biodegradability of PEAs evaluated [107], and the effect of montmorillonite clay particles on the polymerization and crystallization kinetics [108,109].

8.7 CONCLUSIONS

Poly(ester amide)s are polymers with promising expectations since can combine degradability with good mechanical and thermal properties. The capability to form strong intermolecular hydrogen bond interactions and the presence of hydrolyzable ester groups bring a high potential for both commodity and specialty applications. PEAs can be synthesized by a wide diversity of methods (i.e., from ring-opening polymerization to standard polycondensation reactions) and from a wide variety of monomers. Therefore, these polymers have a high versatility concerning the disposition between ester and amide groups along the molecular chain and even the amide/ester ratio, aliphatic content, hydrophilicity, stereochemistry, crystallinity, chain stiffness, and reticulation. In fact, one of the major advantages of PEAs is the capability to tune easily their final properties by selecting the appropriate architecture and composition. It merits also great attention the possibility to prepare PEAs from renewable resources such as carbohydrates and even natural α-amino acids, which can provide biocompatibility.

It is clear from the present review that applications of PEAs in the biomedical field are continuously increasing. Thus, current developments cover a broad spectrum that mainly concerns to biodegradable and biocompatible elastomers, hyperbranched polymers as drug carriers, stiff-chain liquid crystals, functional polymers for conjugating drugs or cell signaling molecules, hydrogels, nonviral delivery vectors, biodegradable pH-sensitive systems, stimuli-sensitive implant materials, scaffolds with multiple functionalities, high-performance materials, and light-responsive polymers. Despite these advances, efforts appear still necessary to validate the promising properties and even to improve performance characteristics of new PEAs.

ACKNOWLEDGMENTS

Authors want to indicate the support by MINECO and AGAUR Grants (MAT2012-36205, 2009SGR-1208).

REFERENCES

[1] M. Okada, Chemical synthesis of biodegradable polymers, Prog. Polym. Sci. 27 (2000) 87–133.

[2] P.A.M. Lips, P.J. Dijkstra, Biodegradable polyesteramides, in: R. Smith (Ed.), Biodegradable Polymers for Industrial Applications, CRC Press, Boca Raton, 2005, pp. 109–139, Ch. 5.

[3] A. Rodríguez-Galán, L. Franco, J. Puiggalí, Degradable poly(ester amide)s for biomedical applications, Polymers 3 (2011) 65–99.

[4] Y. Feng, J. Guo, Biodegradable polydepsipeptides, Int. J. Mol. Sci. 10 (2009) 589–615.

[5] Y. Feng, D. Klee, H. Keul, H. Höcker, Lipase-catalyzed ring-opening polymerization of morpholine-2,5-dione derivatives: a novel route to the synthesis of poly(ester amide)s, Macromol. Chem. Phys. 201 (2000) 2670–2675.

[6] T. Fey, H. Keul, H. Höcker, Interconversion of alternating poly(ester amide)s and cyclic ester amides from adipic anhydride and α, ω-amino alcohols, Macromol. Chem. Phys. 204 (2003) 591–599.

[7] T. Fey, H. Keul, H. Höcker, Ring-opening polymerization of the cyclic ester amide derived from adipic anhydride and 1-amino-5-pentanol, Macromolecules 36 (2003) 3882–3889.

[8] T. Fey, H. Keul, H. Höcker, Ring-opening polymerization of the cyclic ester amide derived from adipic anhydride and 1-amino-5-hexanol in melt and in solution, Macromol. Symp. 215 (2004) 307–324.

[9] A.C. Draye, O. Persenaire, J. Brozek, J. Roda, T. Kosek, Ph. Dubois, Thermogravimetric analysis of poly(ε-caprolactam) and poly[(ε-caprolactam-*co*-(ε-caprolactone) polymers, Polymer 42 (2001) 8325–8332.

[10] B.J. Kim, J.L. White, Continuous polymerization of lactam–lactone block copolymers in a twin-screw extruder, J. Appl. Polym. Sci. 88 (2003) 1429–1437.

[11] B.J. Kim, J.L. White, Anionic copolymerization of lauryl lactam and polycaprolactone for the production of a poly(ester amide) triblock copolymer, J. Appl. Polym. Sci. 90 (2003) 3797–3805.

[12] M.X. Li, R.X. Zhuo, F.Q. Qu, Synthesis and characterization of novel biodegradable poly(ester amide) with ether linkage in the backbone chain, J. Polym. Sci. A Polym. Chem. 40 (2002) 4550–4555.

[13] J. Montané, E. Armelin, L. Asín, A. Rodríguez-Galán, J. Puiggalí, Comparative degradation data of polyesters and related poly(ester amide)s derived from 1,4-butanediol, sebacic acid, and α-amino acids, J. Appl. Polym. Sci. 85 (2002) 1815–1824.

[14] L. Asín, E. Armelin, J. Montané, A. Rodríguez-Galán, J. Puiggalí, Sequential poly(ester amide)s based on glycine, diols, and dicarboxylic acids: thermal polyesterification versus interfacial polyamidation. Characterization of polymers containing stiff units, J. Polym. Sci. A Polym. Chem. 39 (2001) 4283–4293.

[15] N. Paredes, A. Rodríguez-Galán, J. Puiggalí, Synthesis and characterization of a family of biodegradable poly(ester amide)s derived from glycine, J. Polym. Sci. A Polym. Chem. 36 (1998) 1271–1282.

[16] S.I. Han, B.S. Kim, S.W. Kang, H. Shirai, S.S. Im, Cellular interactions and degradation of aliphatic poly(ester amide)s derived from glycine and/or 4-amino butyric acid, Biomaterials 24 (2003) 3453–3462.

[17] R. Katsarava, Active polycondensation: from peptide chemistry to amino acid based biodegradable polymers, Macromol. Symp. 199 (2003) 419–429.

[18] Y. Fan, M. Kobayashi, H. Kise, Synthesis and biodegradation of poly(ester amide)s containing amino acid residues: the effect of the stereoisomeric composition of L- and D-phenylalanines on the enzymatic degradation of the polymers, J. Polym. Sci. A Polym. Chem. 40 (2003) 385–392.

[19] T. Fey, M. Hölscher, H. Keul, H. Höcke, Alternating poly(ester amide)s from succinic anhydride and α, ω-amino alcohols: synthesis and thermal characterization, Polym. Int. 52 (2003) 1625–1632.

[20] M. Vera, A. Rodríguez-Galán, J. Puiggalí, New method of synthesis of poly(ester amide)s derived from the incorporation of glycolic acid residues into aliphatic polyamide, Macromol. Rapid Commun. 25 (2004) 812–817.

[21] M. Vera, L. Franco, J. Puiggalí, Synthesis and characterization of poly(glycolic acid-*alt*-6-aminohexanoic acid) and poly(glycolic acid-*alt*-11-aminoundecanoic acid), Macromol. Chem. Phys. 205 (2004) 1782–1792.

[22] X. Ramis, J.M. Salla, J. Puiggalí, Kinetic studies on the thermal polymerization of *N*-chloroacetyl-11-aminoundecanoate potassium salt, J. Polym. Sci. A Polym. Chem. 43 (2005) 1166–1176.

[23] E. Botines, L. Franco, X. Ramis, J. Puiggalí, Synthesis of poly(glycolic acid-alt-12-aminododecanoic acid): the thermal polymerization kinetics of sodium *N*-chloroacetyl-12-aminododecanoate, J. Polym. Sci.A Poly. Chem. 44 (2006) 1199–1213.

[24] S.K. Murase, L. Franco, A. Rodríguez-Galán, J. Puiggalí, Copolymerization of potassium chloroacetate and potassium *N*-chloroacetyl-6-aminohexanoate, J. Appl. Polym. Sci. 126 (2012) 1425–1436.

[25] M. Vera, M. Admetlla, A. Rodríguez-Galán, J. Puiggalí, Synthesis, characterization and degradation studies of sequential poly(ester amide)s derived from glycolic acid, 1,6-hexanediamine and aliphatic dicarboxylic acids, Polym. Degrad. Stab. 89 (2005) 21–32.

[26] T. Terashima, T. Mes, T.F.A. De Greef, M.A.J. Gillissen, P. Besenius, A.R.A. Palmans, E.W. Meijer, Single-chain folding of polymers for catalytic systems in water, J. Am. Chem. Soc. 133 (2011) 4742–4745.

[27] K. Satoh, M. Matsuda, K. Nagai, M. Kamigaito, AAB-sequence living radical chain copolymerization of naturally occurring limonene with maleimide: an end-to-end sequence-regulated copolymer, J. Am. Chem. Soc. 132 (2010) 10003–10005.

[28] C.K.Z. Andrade, S.C.S. Takada, P.A.Z. Suarez, M.B. Alves, Revisiting the Passerini under eco-friendly reaction conditions, Synlett. 10 (2006) 1539–1541.

[29] X.-X. Deng, L. Li, Z.-L. Li, A. Lv, F.-S. Du, Z.-C. Li, Sequence regulated poly(ester-amide)s based on Passerini reaction, ACS Macro Lett. 1 (2012) 1300–1303.

[30] A. Borriello, L. Nicolais, X. Fang, S.J. Huang, D.A. Scola, Synthesis of poly(amide-ester)s by microwave methods, J. Appl. Polym. Sci. 103 (2007) 1952–1958.

[31] A. Borriello, L. Nicolais, S.J. Huang, Poly(amide-ester)s derived from dicarboxylic acid and aminoalcohol, J. Appl. Polym. Sci. 95 (2005) 362–368.

[32] X. Fang, R. Hutcheon, D.A. Scola, Microwave syntheses of poly(ε-caprolactam-*co*-ε-caprolactone), J. Polym. Sci. A Polym. Chem. 38 (2000) 1379–1390.

[33] J. Tuominen, J.V. Sépala, Synthesis and characterization of lactic acid based poly(ester-amide), Macromolecules 33 (2000) 3530–3535.

[34] P. Garg, H. Keul, D. Klee, M. Möller, Thermal properties of poly(ester amide)s with isolated, two adjacent and three adjacent amide groups within a polyester chain, Macromol. Chem. Phys. 210 (2009) 1754–1765.

[35] P.A.M. Lips, R. Broos, M.J.M. van Heeringen, P.J. Dijkstra, J. Feijen, Incorporation of different crystallizable amide blocks in segmented poly(ester amide)s, Polymer 46 (2005) 7834–7842.

[36] P.A.M. Lips, R. Broos, M.J.M. van Heeringen, P.J. Dijkstra, J. Feijen, Synthesis and characterization of poly(ester amide)s containing crystallizable amide segments, Polymer 46 (2005) 7823–7833.

[37] P.A.M. Lips, M.J.A. van Luyn, F. Chiellini, L.A. Brouwer, I.W. Velthoen, P.J. Dijkstra, J. Feijen, Biocompatibility and degradation of aliphatic segmented poly(ester amide)s: in vitro and in vivo evaluation, J. Biomed. Mater. Res. A 76A (2005) 699–710.

[38] P.A.M. Lips, I.W. Velthoen, P.J. Dijkstra, M. Wessling, J. Feijen, Gas foaming of segmented poly(ester amide) film, Polymer 46 (2005) 9396–9403.

[39] H.A. Lecomte, J.J. Liggat, A.S.G. Curtis, Synthesis and characterization of novel biodegradable aliphatic poly(ester amide)s containing cyclohexane units, J. Polym. Sci. A Polym. Chem. 44 (2006) 1785–1795.

[40] H. Tetsuka, Y. Doi, H. Abe, Synthesis and thermal properties of novel periodic poly(ester-amide)s derived from adipate, butane-1,4-diamine, and linear aliphatic diols, Macromolecules 39 (2006) 2875–2885.

[41] A.A. Deschamps, A.A. van Apeldoorn, J.D. de Bruijn, D.W. Grijpma, J. Feijen, Poly(ether ester amide)s for tissue engineering, Biomaterials 24 (2003) 2643–2652.

[42] R. Bizzarri, R. Solaro, P. Talamelli, E. Chiellini, Synthesis and characterization of new poly(ester-amide)s containing oligo(oxyethylene) segments, J. Bioactive Compatible Polym. 15 (2000) 43–59.

[43] S.H. Hsu, S.W. Whu, S.C. Hsieh, C.L. Tsai, D.C. Chen, T.S. Tan, Evaluation of chitosan–alginate–hyaluronate complexes modified by an RGD-containing protein as tissue-engineering scaffolds for cartilage regeneration, Artif. Organs 28 (2004) 693–703.

[44] C.J. Bettinger, J.P. Bruggeman, J.T. Borenstein, R.S. Langer, Amino alcohol-based degradable poly(ester amide) elastomers, Biomaterials 29 (2008) 2315–2325.

[45] X. Li, X. Lu, Y. Lin, J. Zhan, Y. Li, Z. Liu, X. Chen, S. Liu, Synthesis and characterization of hyperbranched poly(ester-amide)s from commercially available dicarboxylic acids and multihydroxyl primary amines, Macromolecules 39 (2006) 7889–7899.

[46] Y. Lin, Z. Dong, Y. Li, One-pot synthesis and characterization of hyperbranched poly(ester-amide)s from commercially available dicarboxylic acids and multihydroxyl secondary amines, J. Polym. Sci. A Polym. Chem. 46 (2008) 5077–5092.

[47] P. Froehling, Development of DSM's hybrane—hyperbranched polyesteramides, J. Polym. Sci.A Polym. Chem. 42 (2004) 3110–3115.

[48] S. Suttiriuengwong, J. Rolker, I. Smirnova, W. Arit, M. Seiler, L. Lüderitz, Y. Perez de Diego, P.J. Jansens, Hyperbranched polymers as drug carriers: microencapsulation and release, Pharm. Dev. Technol. 11 (2006) 55–70.

[49] S. Reven, J. Grdadolnik, J. Kristl, E. Zagar, Hyperbranched poly(esteramides) as solubility enhancers for poorly water-soluble drug glimepiride, Int. J. Pharm. 396 (2010) 119–126.

[50] M.A. Kelland, Tuning the thermoresponsive properties of hyperbranched poly(ester amide)s based on diisopropanolamine and cyclic dicarboxylic anhydrides, J. Appl. Polym. Sci. 121 (2011) 2282–2290.

[51] X. Cheng, Z. Huang, J. Liu, W. Shi, Synthesis and properties of semicrystalline hyperbranched poly(ester-amide) grafted with long alkyl chains used for UV-curable powder coatings, Prog. Org. Coat. 59 (2007) 284–290.

[52] J.D. Sudha, Synthesis and characterization of hydrogen-bonded thermotropic liquid crystalline aromatic-aliphatic poly(ester-amide)s from amido diol, J. Polym. Sci. A Polym. Chem. 38 (2000) 2469–2486.

[53] J.D. Sudha, C.K.S. Pillai, Synthesis and properties of amphotropic hydrogen bonded liquid crystalline (LC) poly(ester amide)s (PEA): effect of aromatic moieties on LC behavior, Polymer 46 (2005) 6986–6997.

[54] V.A.E. Shaikh, V.P. Ubale, N.N. Maldar, S.V. Lonikar, C.R. Rajan, S. Ponrathnam, Main-chain liquid crystalline poly(ester-amide)s containing lithocholic acid units, J. Appl. Polym. Sci. 100 (2006) 73–80.

[55] P. Deepa, C. Sona, M. Jayakannan, Synthesis and investigation of the effect of nematic phases on the glass-transition behavior of novel cycloaliphatic liquid-crystalline poly(ester amide)s, J. Polym. Sci. A Polym. Chem. 44 (2006) 5557–5571.

[56] K.Y. Sandhya, C.K.S. Pillai, K. Sree Kumar, New liquid-crystalline poly(ester amide)s: the role of nitro groups in the phase behavior, J. Polym. Sci. B Polym. Phys. 42 (2004) 1289–1298.

[57] P.J.M. Serrano, J.P.M. Van Duynhoven, R.J. Gaymans, R. Hulst, Morphology of alternating poly(ester amide)s based on 1,4-butylene established by ^{13}C solid-state NMR relaxation measurements, Macromolecules 35 (2002) 8013–8019.

[58] W.G. Kim, H.G. Yoon, J.Y. Lee, Biodegradable polymers based on renewable resources. V. Synthesis and biodegradation behavior of poly(ester amide)s composed of 1,4:3,6-dianhydro-D-glucitol, α-amino acid, and aliphatic dicarboxylic acid units, J. Appl. Polym. Sci. 81 (2001) 2721–2734.

[59] C. Regaño, A.M. De Ilarduya, J.I. Iribarren, S. Muñoz-Guerra, Poly(ester amide)s derived from L-tartaric acid and amino alcohols. II. Aregic polymers, J. Polym. Sci. A Polym. Chem. 38 (2000) 2687–2696.

[60] I. Villuendas, J.J. Bou, A. Rodríguez-Galán, S. Muñoz-Guerra, Alternating copoly(ester amide)s derived from amino alcohols and L-tartaric and succinic acids, Macromol. Chem. Phys. 202 (2001) 236–244.

[61] A. Alla, A. Rodríguez-Galán, S. Muñoz-Guerra, Hydrolytic and fungal degradation of polyamides derived from tartaric acid and hexamethylenediamine, Polymer 41 (2000) 2765–2772.

[62] A. Pérez-Rodríguez, A. Alla, J.M. Fernández-Santín, S. Muñoz-Guerra, Poly(ester amide)s derived from tartaric and succinic acids: changes in structure and properties upon hydrolytic degradation, J. Appl. Polym. Sci. 78 (2000) 486–494.

[63] C. Regaño, A. Alla, A. Martínez de Ilarduya, S. Muñoz-Guerra, Poly(ester amide)s derived from L-malic acid, Macromolecules 37 (2004) 2067–2075.

[64] Z. Gomurashvili, H.R. Kricheldorf, R. Katsarava, Amino acid based bioanalogous polymers. Synthesis and study of new poly(ester amide)s composed of hydrophobic α-amino acids and dianhydrohexitoles, J. Macromol. Sci. Pure Appl. Chem. 37A (2000) 215–227.

[65] I. Molina Pinilla, M. Bueno Martínez, J.A. Galbis, Carbohydrate-based copolymers. Hydrolysis of copoly(ester amide)s containing L-arabinose units, Macromolecules 35 (2002) 2985–2992.

[66] S. Ahmad, S.M. Ashraf, F. Zafar, Development of linseed oil based polyesteramide without organic solvent at lower temperature, J. Appl. Polym. Sci. 104 (2007) 1143–1148.

[67] A. Kozlowska, R. Ukielski, New type of thermoplastic multiblock elastomers—poly(ester-block-amide)s—based on oligoamide 12 and oligoester prepared from dimerized fatty acid, Eur. Polym. J. 40 (2004) 2767–2772.

[68] T. Lebarbé, L. Maisonneuve, T.H.N. Nguyen, B. Gadenne, C. Alfos, H. Cramail, Methyl 10-undecenoate as a raw material for the synthesis of renewable semi-crystalline polyesters and poly(ester-amide)s, Polym. Chem. 3 (2012) 2842–2851.

[69] K. Guo, C.C. Chu, Controlled release of paclitaxel from biodegradable unsaturated poly(ester amide)s/poly(ethylene glycol) diacrylate hydrogels, J. Biomater. Sci. Polym. Ed. 18 (2007) 489–504.

[70] L. Li, C.C. Chu, Nitroxyl radical incorporated electrospun biodegradable poly(ester amide) nanofiber membranes, J. Biomater. Sci. Polym. Ed. 20 (2009) 341–361.

[71] K. Guo, C.C. Chu, Biodegradable and injectable paclitaxel-loaded poly(ester amide)s, J. Biomed. Mater. Res. B Appl. Biomater. 89 (2009) 491–500.

[72] C.C. Chu, D.D. Sun, C.C. Chu, D.D. Sun, New electrospun synthetic biodegradable poly(ester amide) drug-eluting fibrous membranes for potential wound treatment, in: AATCC Symposium Proceedings on Medical, Nonwovens, and Technical Textiles, 2008, pp. 60–76.

[73] D. Yamanouchi, J. Wu, A.N. Lazar, K. Craig Kent, C.C. Chu, B. Liu, Biodegradable arginine-based poly(ester-amide)s as non-viral gene delivery reagents, Biomaterials 29 (2008) 3269–3277.

[74] S.H. Lee, I. Szinai, K. Carpenter, R. Katsarava, G. Jokhadze, C.C. Chu, Y. Huang, E. Verbeken, O. Bramwell, I. de Scheerder, M.K. Hong, In-vivo biocompatibility evaluation of stents coated with a new biodegradable elastomeric and functional polymer, Coron. Artery Dis. 13 (2002) 237–241.

[75] M.A. de Wit, Z. Wang, K.M. Atkins, K. Mequanint, E.R. Gillies, Syntheses, characterization, and functionalization of poly(ester amide)s with pendant amine functional groups, J. Polym. Sci. A Polym. Chem. 46 (2008) 6376–6392.

[76] M. Deng, J. Wu, C.A. Reinhart-King, C.C. Chu, Biodegradable functional poly(ester amide)s with pendant hydroxyl functional groups: synthesis, characterization, fabrication and in vitro cellular response, Acta Biomater. 7 (2011) 1504–1515.

[77] X. Pang, C.C. Chu, Synthesis, characterization and biodegradation of functionalized amino acid-based poly(ester amide)s, Biomaterials 31 (2010) 3745–3754.

[78] X. Pang, C.C. Chu, Synthesis, characterization and biodegradation of poly(ester amide)s based hydrogels, Polymer 51 (2010) 4200–4210.

[79] H.L.I. Guan, C. Deng, X.Y. Xu, Q.Z. Liang, X.S. Chen, X.B. Jing, Synthesis of biodegradable poly(ester amide)s containing functional groups, J. Polym. Sci. A Polym. Chem. Ed. 43 (2005) 1144–1149.

[80] K. Guo, C.C. Chu, Synthesis, characterization and biodegradation of novel poly(ether ester amide)s based on L-phenylalanine and oligoethylene glycol, Biomacromolecules 8 (2007) 2851–2861.

[81] K. Guo, C.C. Chu, Copolymers of unsaturated and saturated poly(ether ester amide)s: synthesis, characterization and biodegradation, J. Appl. Polym. Sci. 110 (2008) 1858–1869.

[82] K. Guo, C.C. Chu, Synthesis and characterization of poly(ε-caprolactone)-containing amino acid-based poly(ether ester amide) s, J. Appl. Polym. Sci. 125 (2012) 812–819.

[83] Y. Fan, M. Kobayashi, H. Kise, Synthesis and biodegradation of Poly (ester amide)s containing amino acid residues: the effect of the stereoisomeric composition of L- and D-phenylalanines on the enzymatic degradation of the polymers, J. Polym. Sci. A Polym. Chem. 40 (2002) 385–392.

[84] A. Ghaffar, G.J.J. Draaisma, G. Mihov, A.A. Dias, P.J. Schoenmakers, Sj van der Wal, Monitoring the in vitro enzyme-mediated degradation of degradable poly(ester amide) for controlled drug delivery by LC-ToF-MS, Biomacromolecules 12 (2011) 3243–3251.

[85] G. Tsitlanadze, M. Machaidze, T. Kviria, N. Djavakhishvili, C.C. Chu, R. Katsarava, Biodegradation of amino-acid-based poly(ester amide)s: in vitro weight loss and preliminary in vivo studies, J. Biomater. Sci. Polym. Ed. 15 (2004) 1–24.

[86] K. Markoishvili, G. Tsitlanadze, R. Katsarava, J.G. Morris Jr., A. Sulakvelidze, A novel sustained-release matrix based on biodegradable poly(ester amide)s and impregnated with bacteriophages and an antibiotic shows promise in management of infected venous stasis ulcers and other poorly healing wounds, Int. J. Dermatol. 41 (2002) 453–458.
[87] J.S. Mejia, E.R. Gillies, Triggered degradation of poly(ester amide) s via cyclization of pendant functional groups of amino acid monomers, Polym. Chem. 4 (2013) 1969–1982.
[88] J. Wu, D. Yamanouchi, B. Liu, C.C. Chu, Biodegradable arginine-based poly(ether ester amide)s as a non-viral DNA delivery vector and their structure–function study, J. Mater. Chem. 22 (2012) 18983–18991.
[89] P. He, Z. Tang, L. Lin, M. Deng, X. Pang, X. Zhuang, X. Chen, Novel biodegradable and pH-sensitive poly(ester amide) microspheres for oral insulin delivery, Macromol. Biosci. 12 (2012) 547–556.
[90] D.K. Knight, E.R. Gillies, K. Mequanint, Strategies in functional poly(ester amide) syntheses to study human coronary artery smooth muscle cell interactions, Biomacromolecules 12 (2011) 2475–2487.
[91] Y. Feng, M. Behl, S. Kelch, A. Lendlein, Biodegradable multiblock copolymers based on oligodepsipeptides having shape-memory properties, Macromol. Biosci. 9 (2009) 45–54.
[92] Y. Ohya, M. Toyohara, M. Sasakawa, H. Arimura, T. Ouchi, Thermosensitive biodegradable polydepsipeptide, Macromol. Biosci. 5 (2005) 273–276.
[93] T.J. Rivers, T.W. Hudson, C.E. Schmidt, Synthesis of a novel, biodegradable electrically conducting polymer for biomedical applications, Adv. Funct. Mater. 12 (2002) 33–37.
[94] H. Cui, Y. Liu, M. Deng, X. Pang, P. Zhang, X. Wang, X. Chen, Y. Wei, Synthesis of biodegradable and electroactive tetraaniline grafted poly(ester amide) copolymers for bone tissue engineering, Biomacromolecules 13 (2012) 2881–2889.
[95] P. Garg, D. Klee, H. Keul, M. Möller, Electrospinning of novel poly(ester amide)s, Macromol. Mater. Eng. 294 (2009) 679–690.
[96] L.J. del Valle, M. Roa, A. Díaz, M.T. Casas, J. Puiggalí, A. Rodríguez-Galán, Electrospun nanofibers of a degradable poly(ester amide). Scaffolds loaded with antimicrobial agents, J. Polym. Res. 19 (2012) 9792–9804.
[97] G.E. Luckachan, C.K.S. Pillai, Random multiblock poly(ester amide)s containing poly (L-lactide) and cycloaliphatic amide segments: synthesis and biodegradation studies, J. Polym. Sci. A Polym. Chem. 44 (2006) 3250–3260.
[98] G.S. Liou, S.H. Hsiao, Synthesis and properties of aromatic poly(ester amide)s with pendant phosphorus groups, J. Polym. Sci. A Polym. Chem. 40 (2002) 459–470.
[99] G.D. Jaycox, Stimuli-responsive polymers. VIII. Polyesters and poly(ester amides) containing azobenzene and chiral binaphthylene segments: highly adaptive materials endowed with light-, heat-, and solvent-regulated optical rotatory power, J. Polym. Sci. A Polym. Chem. 44 (2006) 207–218.
[100] B. Philip, K. Sreekumar, Optically active poly(ester-amide)s: synthesis and characterization, Polym. Int. 50 (2001) 1318–1323.
[101] B. Philip, K. Sreekumar, Thermal properties of some chiral polymers designed for NLO studies, J. Polym. Mater. 18 (2001) 365–370.
[102] T. Ferre, L. Franco, A. Rodriguez-Galan, J. Puiggali, Poly(ester amide)s derived from 1,4-butanediol, adipic acid and 6-aminohexanoic acid. Part II: composition changes and fillers, Polymer 44 (2003) 6139–6152.
[103] O. Martin, L. Averous, Comprehensive experimental study of a starch/polyesteramide coextrusion, J. Appl. Polym. Sci. 86 (2002) 2586–2600.
[104] L. Willett, F.C. Felker, Tensile yield properties of starch-filled poly(ester amide) materials, Polymer 46 (2005) 3035–3042.
[105] A.K. Mohanty, M.A. Khan, G. Hinrichsen, Influence of chemical surface modification on the properties of biodegradable jute fabrics—polyester amide composites, Composites A Appl. Sci. Manufacturing 31 (2000) 143–150.
[106] M. Krook, A.C. Albertsson, U.W. Gedde, M.S. Hedenqvist, Barrier and mechanical properties of montmorillonite/polyesteramide nanocomposites, Polym. Eng. Sci. 42 (2002) 1238–1246.
[107] X. Liu, Y. Zou, G. Cao, D. Luo, The preparation and properties of biodegradable polyesteramide composites reinforced with nano-$CaCO_3$ and nano-SiO_2, Mater. Lett. 61 (2007) 4216–4221.
[108] L. Morales, L. Franco, M.T. Casas, J. Puiggali, Poly(ester amide)/clay nanocomposites prepared by in situ polymerization of the sodium salt of N-chloroacetyl-6-aminohexanoic acid, J. Polym. Sci. A Polym. Chem. 47 (2009) 3616–3629.
[109] L. Morales, L. Franco, M.T. Casas, J. Puiggali, Crystallization studies on a clay nanocomposite prepared from a degradable poly(ester amide) constituted by glycolic acid and 6-aminohexanoic acid, Polym. Eng. Sci. 51 (2011) 1650–1661.

Chapter 9

Progress in Functionalized Biodegradable Polyesters

Markus Heiny[1], Jonathan Johannes Wurth[1,2], Venkatram Prasad Shastri[1,2]
[1]*Institute for Macromolecular Chemistry, University of Freiburg, Freiburg, Germany*
[2]*BIOSS - Centre for Biological Signalling Studies, University of Freiburg, Freiburg, Germany*

Chapter Outline

9.1 INTRODUCTION

9.1.1 Background

The past decade has witnessed a renewed interest in degradable polymers, especially those sourced from renewable resources [1]. This has occurred in part due to the not always uncontroversial emergence of a green economy and policies that promote green and eco-friendly technologies [2]. A family of polymers that has seen resurgence are the poly(α-hydroxy acids) (PHAs) especially poly(lactic acid) or polylactide (PLA) that is sourced from lactic acid, a by-product of the dairy industry. Ironically, degradable polyesters that have been a key element of the medical device and therapeutics industry are beginning to see new applications in packaging, waste management, daily consumer products, and recycling of plastics [3]. In this application space, PLA has seen use in coffee cups, garbage bags, and other household items. This new commercial opportunity provides much impetus for the development of new chemistries and strategies to improve processibility, performance, and reuse of polyesters.

Poly(glycolic acid) or polyglycolide (PGA), a polymer synthesized from the dilactone glycolide, has been in use in the medical arena as degradable sutures since the 1960s [4], and much of the development and application of degradable polyesters to date in the life sciences area is based on the experiences, lessons, and opportunities that have evolved from the *in vivo* use of PHAs. Over the past three to four decades, there has been a steady stream of applications involving copolymers of PGA, PLA, and other lactones in areas ranging from orthopedics [5], drug delivery [6], to tissue engineering [7,8]. For many *in vivo* medical applications, the degradation behavior (mass loss and kinetics) and degradation by-products are of primary concern as both of these parameters affect performance. Some of the earlier applications of degradable

Natural and Synthetic Biomedical Polymers

polyesters were in medical implants such as orthopedic pins, plates, and screws. For these medical devices, their function is dependent on mechanical properties and any premature loss of physical integrity can lead to devastating consequences [9]. Therefore, much of the earlier effort in functionalization of degradable polymers has focused on improving strength and degradation lifetime [10]. However, the current need for functionalized polyesters is driven by the emergence of new application domains and need for improvement in existing application spaces as discussed in the succeeding text.

9.1.2 Biocompatibility and Biodegradability Aspects

Biocompatibility can be loosely defined as the *tolerance of a foreign material in a biological ex vivo or in vivo environment* or may also be defined as *the extent to which a foreign material can change the bioenvironment*. For the purpose of this chapter, the latter definition is more relevant as the application space of polyesters has evolved beyond the medical realm to consumer products. So, biocompatibility can be envisioned as the sum total of toxicity, bioprocessibility, and function and this is highly subjective depending on the use criteria. Ideally speaking, a degradable polymer for *in vivo* use has to present low cellular and tissue toxicity, degradation products that elicit no adverse response and are metabolized by mammalian physiology, and performs its intended function without interfering with the functions of the site of implantation. Strategies to date have focused on altering monomer composition as discussed later. With regard to the new and emerging applications involving packaging and waste management such as garbage bags and renewable household products, the polymer has to survive exposure to moisture and pathogens (fungi and bacteria) and yet be compostable. Developing strategies to promote bioprocessing under anaerobic conditions can further enhance the attractiveness and utility of polyesters beyond the biomedical arena. A key step towards addressing many of the issues raised earlier is providing easy access to functional and chemical diversity in the polymer backbone.

The chemical makeup of a polymer dictates many of the performance criteria. These performance criteria vary for different application areas. In the case of *ex vivo* and oral applications, such as formulations for delivery of medication, selective degradation (selective stability) in the gut environment is important, and chemistries that are susceptible to the highly acidic and corrosive gastric environments are not desirable. While traditional PHAs are less likely to fare well in the gastric environment, altering backbone chemistry to increase hydrophobicity, thus diminishing wetting and water uptake, can significantly improve outcomes. Case in point, aliphatic polyanhydrides [11], while possessing a highly reactive anhydride linkage, have durability in hydrated *in vivo* environments, as the long carbon-chain backbone favors a surface erosion over the commonly observed bulk erosion in PHAs. This ensures that the hydrolytic events in a polyanhydride system are limited to the surface of the device. One can envisage the application of a similar paradigm in the design of the next generation of degradable polyesters for medical devices and packaging materials to increase their shelf life and durability in high humidity environments.

9.1.3 Need for Functionalization

Tissue engineering, the paradigm involving the synthesis of tissue equivalents *ex vivo* by combining cells with polymer scaffolds, has seen a sea of change with respect to achieving functional traits in engineered tissue. When it comes to scaffolds for tissue engineering, premature degradation of the scaffold is less of a concern as the cellular microenvironment alters the local variables such as pH and oxygen tension leading to some inherent variability from subject to subject. But more importantly, the deposition of extracellular matrix deposited by the cells can compensate for a loss in scaffold mass integrity [12]. For cell-contacting applications such as tissue engineering and targeted therapeutics, presentation of surface motifs that can promote specific and favorable cell behavior is desirable [13,14]. Therefore, chemical structures along the polymer backbone that can be postfunctionalized are desirable.

Degradable polymers are a cornerstone of nanomedicine [15]. Polymer-based targeted therapeutics has emerged as an important adjuvant to radiation and chemotherapy. With regard to the use of degradable polymers as micro- or nanoparticulate matrices for drug delivery, the bulk degradation behavior cannot be extrapolated to the smaller length scales and volumes, as many of the variables such as pH and water content do not have the same impact on the degradation behavior. Therefore, much of the structure-degradation behavior relationships need to be redefined. For example, while a phase separated structure in a macroscopic solid could lead to preferential degradation in one region, resulting in the induction of porosity and inhomogeneities in degradation, this effect can be dulled or eliminated when the size of the system (e.g., nanoparticles) approaches that of the phase domains. Nevertheless, in the case of degradable nanoparticulate carriers, changes to matrix integrity are seldom an issue as much of either the material is processed intracellularly or the matrix typically outlives the prescribed functional lifetime of the system. However, for drug-delivery systems, water uptake and swelling of the matrix are critical elements that can dramatically alter drug release kinetics. Additionally, targeting therapy requires biologically relevant moieties along the backbone. Therefore, in this case, developing strategies to control hygroscopicity of the polymer without significantly altering polymer crystallinity and thermal behavior is important in addition to biofunctionalization.

Degradable polymers also constitute an important component of vascular therapies [16]. For example, drug elution stents have fundamentally revolutionized interventional cardiology and improved clinical outcomes following endovascular stenting [17]. A polymer coating on the metal stent frame is necessary to impart the drug elution characteristics. Towards improving *in vivo* performance, metallic stents are currently being replaced by polymer stents, and this provides an opportunity to customize the stent surface to the recipient biochemistry. Such strategies require the development of highly functionalized polymers that can provide the necessary biological cues to foster functional repair and yet undergo degradation. PHAs, due to their chemistry, are well suited to fulfill this need. Polyesters, such as poly(ethylene terephthalate), have been the mainstay of synthetic large and medium diameter vascular grafts. However, for vessels smaller than 3 mm (small diameter vessels), at present, the only two options exist: autologous (or allogeneic) grafts and tissue-engineered grafts [18,19]. Therefore, ideally speaking, an *off-the-shelf* synthetic polymer graft is conceivable as biofunctionalization can provide an avenue for controlling and promoting favorable cellular processes. However, poor hemocompatibility (antithrombogenicity) of the synthetic polyesters remains their Achilles' heel. Surface functionalization provides an attractive and viable option for meeting this long-sought objective.

9.1.4 Concepts of Polymerization and Functionalization

Polyesters can be synthesized either by ring-opening polymerization (ROP) or polycondensation. Both of these approaches have merit in the manipulation of properties of degradable polymers. Commercially, ROP is the most widely used practice for the synthesis of PHAs for consumer applications due to the ease of scale up, acceptable purity, and cost considerations. However, for biomedical applications where cost pressures are low and purity and function are paramount, condensation polymerization can yield superior outcomes [20]. Although the class of degradable polymers is rather large and includes poly(butyrolactones), poly(dioxanone), aliphatic poly(carbonates), poly(anhydrides), and poly(hydroxyalkanoates), the focus of the subsequent sections will be on the PGA, PLA, and poly(caprolactone) (PCL) family of polymers, as these are the most widely used polymers in both medical and consumer products arena.

These degradable polyesters as discussed later are primarily synthesized from cyclic monomers. The chemistry of lactones presents several challenges with respect to making analogues that can be pre- or postfunctionalized. Notwithstanding, strategies to derivatize cyclic esters have been developed. One of the downsides of this approach is that it requires multistep synthesis with less than optimal yields and also a considerable knowledge of synthetic organic chemistry. Also, the presence of chiral centers in some monomers can add more complexity to the synthesis. So there is a need to develop simple approaches that are compatible with traditional polymerization strategies and yet can be implemented with basic synthetic chemistry skills. This is critical to ensure wide adaptability, as many of the end users of polymers in life sciences do not have formal training in polymer chemistry. Some of the earlier attempts in imparting functionality in degradable polyesters have focused on incorporation of segments (or blocks) that can improve processibility or affinity to aqueous environments, such as poly(ethylene glycol) (PEG)-PHAs block copolymers, and acrylate end-terminated PHA-PEG-PHA triblock copolymers for tissue-adhesive applications [21]. The presence of water-soluble PEG confers surfactant-like properties to the block polymer that has been leveraged in the fabrication of PHA nanoparticles that exhibit high circulation times due to the presence of a PEG-rich surface that inhibits opsonization [22]. However, as mentioned earlier, for applications such as targeted therapeutics, a plurality of functional moieties that are actually presented on the surface is desirable. In response to these new needs, a whole host of new chemistries and approaches have been developed, and a few key approaches are summarized in the following sections.

9.2 FUNCTIONALIZED POLYESTERS

Since there are several chapters in this book that are devoted to the medical applications of polyesters, which the readership is encouraged to read, the emphasis of this chapter therefore is on reviewing the strategies and chemistries that have been employed to derivatize polyester backbones. In order to ensure the relevance of these backbone modification approaches to current topics in medicine and the green technologies, the chapter focuses on paradigms that have generality to either a polymer family or a routine synthesis route.

The functionalization approaches presented are grouped under the chemical strategy and whenever possible placed in context of the polymerization method. It is important to note that while the content is discussed with a medical application in view, the functionalization strategies are also applicable to modifying performance characteristics of degradable polyesters for consumer applications as well. For example, a strategy for improving the wettability while improving mammalian cell interaction can also improve infestation and hosting of microbes necessary for degradation in composting environment. Likewise, strategies for introducing biofunctionality for targeted therapeutics can also be used to tether phase compatibilizers and antioxidants to the processibility of polymers such as PLA.

9.2.1 Polylactide and Polyglycolide

Since Carothers reported on the ROP of cyclic esters in 1932 [23] and the consequential patenting of high-molecular-weight PLA by DuPont in 1954, this polymer has evolved into one of the most prominent biodegradable polymers. It has been extensively explored and commercialized, not only as a packaging and textile material but also and especially for the medical field, encompassing such diverse applications as surgical sutures [24], drug-delivery systems [25], and orthopedic implants [5].

PGA in its pure form is applied to a lesser extent, due to its insolubility in most organic solvents, its high degradation rate, and the concomitant rapid loss of mechanical strength. Worth mentioning are its uses as degradable sutures [24] and as scaffold material in short-term tissue engineering, for example, of cartilage [26]. However, as a copolymer for lactide, glycolide plays an important role, and PLGA copolymers offer a wide range of tunable properties and find wide application as sutures, drug-delivery systems, and tissue engineering scaffolds.

9.2.1.1 General Aspects of Lactide Monomer and Lactide Polymerization

Lactic acid (2-hydroxypropionic acid) is a widely occurring natural α-hydroxy acid, either produced chemically or derived from renewable sources via fermentation mainly by lactic acid bacteria [27,28]. The presence of an asymmetric carbon atom results in the existence of the two stereoisomers D(−)- and L(+)-lactic acid whose optically pure forms can only be produced via the biotechnological pathway. The theoretically possible direct polycondensation of lactic acid has no practical significance for the production of PLA with sufficiently high molecular weight. Instead, the catalytic ROP of the lactide intermediate is most widely applied. Lactide is the six-membered cyclic dimer of lactic acid and therefore exists in three different stereoisomers, namely, L-lactide (or L,L-lactide), D-lactide (or D,D-lactide), and meso-lactide. It is obtained by catalytically depolymerizing low-molecular-weight polycondensates of the respective lactic acid. Of the conceivable lactide polymers (poly(L-lactide) (PLLA), poly(D-lactide), poly(D,L-lactide) (PDLLA; obtained by polymerizing a racemic mixture of D- and L-lactide), and meso-poly(lactide)), only PLLA and PDLLA have been extensively studied (Figure 9.1). The possibility to synthesize copolymers of varying stereochemistry allows for the tuning of mechanical properties and degradation behavior to a certain extent. However, ever more sophisticated applications of biomaterials demand functionalization beyond that method.

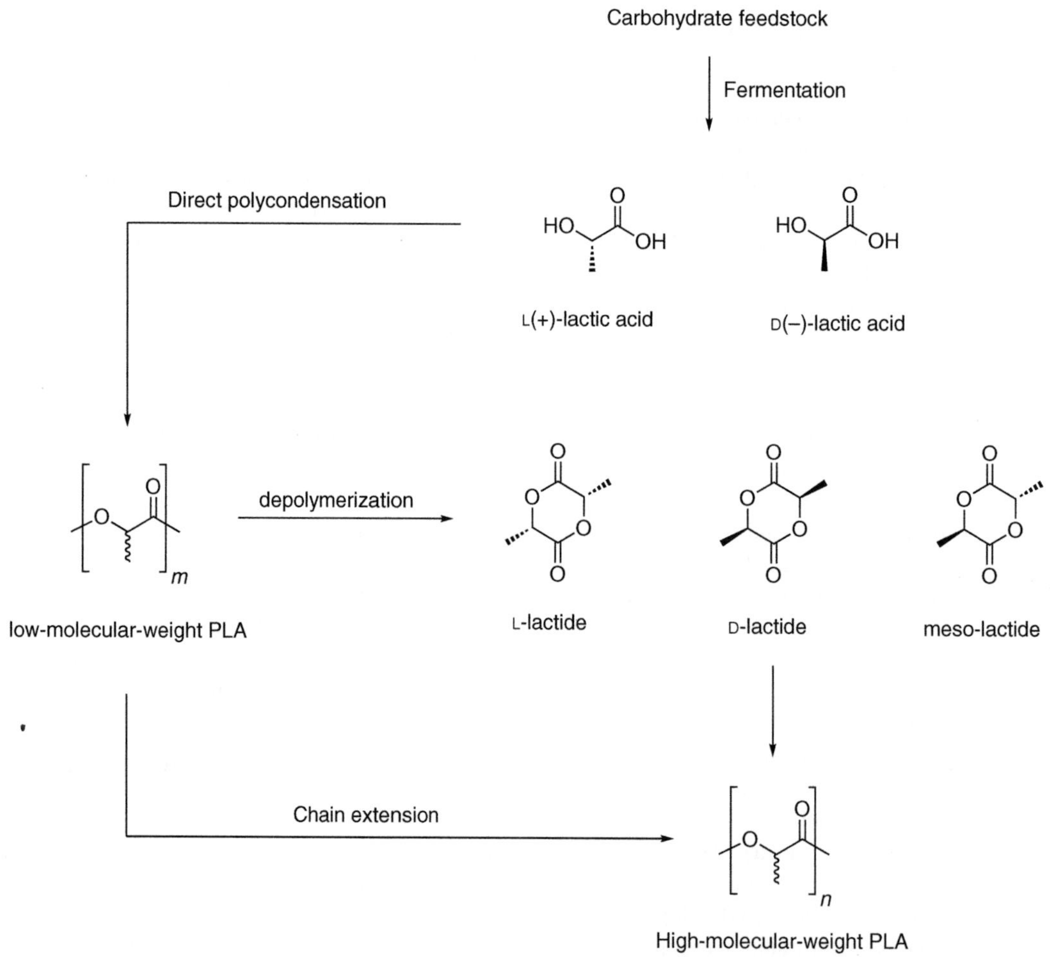

FIGURE 9.1 Overview of different routes towards PLA polymers.

Efforts to functionalize PLA are mainly based on the following three paradigms:

- PLA end-chain modification using functional initiating systems or functional terminating agents (endcapping)
- Modification along the PLA chain by copolymerization of lactide with modified lactide monomers or other functional comonomers
- Modification of the entire polymer backbone via post-polymerization functionalization (polymer-analogous reactions)

Another distinction should be made regarding the extent of functionalization. Whereas endcapping and copolymerization are performed during polymer synthesis and therefore involve all the polymer chains, the postpolymerization modifications do not necessarily affect the complete bulk of the material. Depending on the experimental procedure applied, it is also possible that only the PLA chains on the surface (of a manufactured device or likewise) are modified (surface functionalization).

9.2.1.2 Chain-End Functionalization of PLA

ROP of lactides is very well suited for end-chain functionalization of PLA. In the anionic initiation pathway, a functional anionic initiator (e.g. a functional metal alkoxide) can be used to start the polymerization reaction. Even better suited is the coordination-insertion pathway using an organometallic compound and a functional alcohol coinitiator that is easier to obtain than the corresponding anionic initiating agent. The anionic initiator and the alcohol coinitiator, respectively, are incorporated into and remain at the PLA chain end after polymerization (or in the middle of the chain if a divalent alcohol is applied) [29] (Figure 9.2). The impact of the single functional group introduced in this way naturally diminishes with increasing molecular weight of the polymer.

Introduction of such different moieties as double and triple bonds, aromatic rings, azides, maleimides, polyethers, and methacrylates is possible, many of which are susceptible to easy follow-up reactions (click reactions, Michael additions, etc.) [30]. The concept can easily be extended to hydroxyl group bearing biocompatible and biofunctional compounds like sugars [31,32] or boronic acids [33].

Extending the concept to mono- or α,ω-bifunctional macromolecular (co-)initiators (macroinitiators) leads to AB-diblock or ABA-triblock copolymers of PLA, respectively. A well-established method is the use of polyethylene glycol (PEG) as the coinitiating alcohol species in lactide ROP, resulting in adjacent hydrophilic polyether blocks. If the PEG initiator bears a functional group itself, further possibilities for functionalization, e.g., for the linkage of targeting moieties, are provided [34]. In an analogous manner, specifically synthesized macroinitiators can be used to obtain block copolymers of PLA with a certain profile of properties [35].

9.2.1.3 Functionalization of PLA via Copolymerization

Synthesis of functionalized PLA via copolymerization is mainly carried out applying ROP of lactide with functional

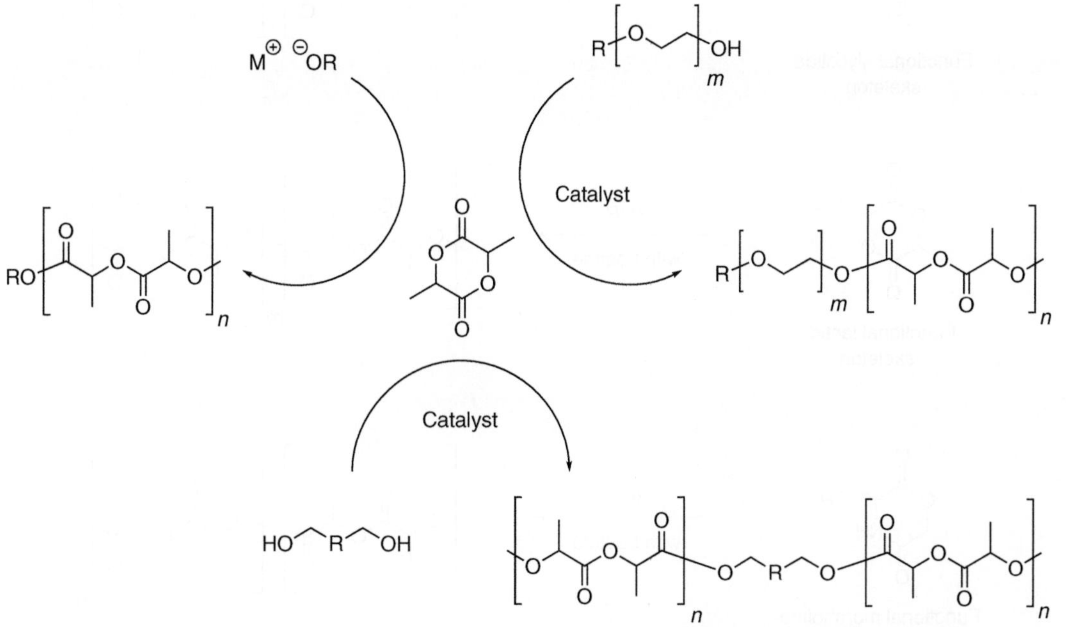

FIGURE 9.2 Examples of lactide ROP with anionic initiator and organometallic catalyst/alcohol initiator system for possible chain-end/midchain functionalization.

cyclic comonomers. Since the early work of Vert on the copolymerization of lactides with functional lactones [36], many studies have focused on this functionalization concept. It allows for the synthesis of functional high-molecular-weight PLA in a highly controlled way. The major shortcoming of this method is the often complex synthetic route to the functional cyclic comonomers. Most widely spread comonomers are functional lactides, functional glycolides, functional morpholine-2,5-dione derivatives, and functional cyclic carbonates. The functional group is typically linked via a methylene group at the 3-position of the six-membered ring (Figure 9.3). It is, furthermore, noteworthy that if the functional copolymer is not derived from the lactide skeleton, the resulting polymer chain does not necessarily consist only of ester linkages. In the case of the morpholine-2,5-dione copolymers, for example, poly(ester amides) are formed, and depending on the amount of comonomer incorporated into the copolymer chain, the material properties can be altered significantly.

One of the first works applying this approach was Kimura's synthesis of carboxylic acid-functionalized PHAs (mainly PGA) using the functional glycolide (benzyloxycarbonyl)methyl-1,4-dioxane-2,5-dione [37]. Since then, this regime has been successfully extended to functionalized lactides offering a way of introducing (protected) functional groups such as protected amine, alcohol, and carboxylic acid [38]. In a similar fashion, Feijen has reported on the synthesis of morpholine-2,5-dione derivatives bearing protected amine, carboxylic acid, and thiol groups and the copolymerization thereof with D,L-lactide to yield functional poly(ester amides) [39]. The morpholine diones are generally derived from (protected) amino acids and therefore present a way to introduce the biofunctional potential of peptides into PLA polymers [40,41].

Cyclic carbonates present another approach towards functional PLAs. Again, an increasing fraction of ester linkages along the polymer chain is substituted for carbonate groups as the comonomer incorporation goes up. An interesting example exploiting this approach is the copolymerization of lactide with a cyclic carbonate bearing an acryloyl group [42]. This group provides a versatile target for nucleophiles such as thiols and amines that can be readily linked to the polymer backbone via Michael addition reaction to further functionalize the copolymers.

If a PLA copolymer exclusively with ester linkages is required or desired but functional lactides are not available or only hard to obtain, copolymerization with functional lactones such as α-chloro-caprolactone [43] can be an alternative. The chlorine atom thus introduced into the copolymer chain can be substituted with an azide group amenable for click reactions with various alkynes. The strategies for the functionalization of PCL are discussed in detail later on in this chapter.

An extension of ROP towards functionalized PLA is presented by the possibility to copolymerize lactide with

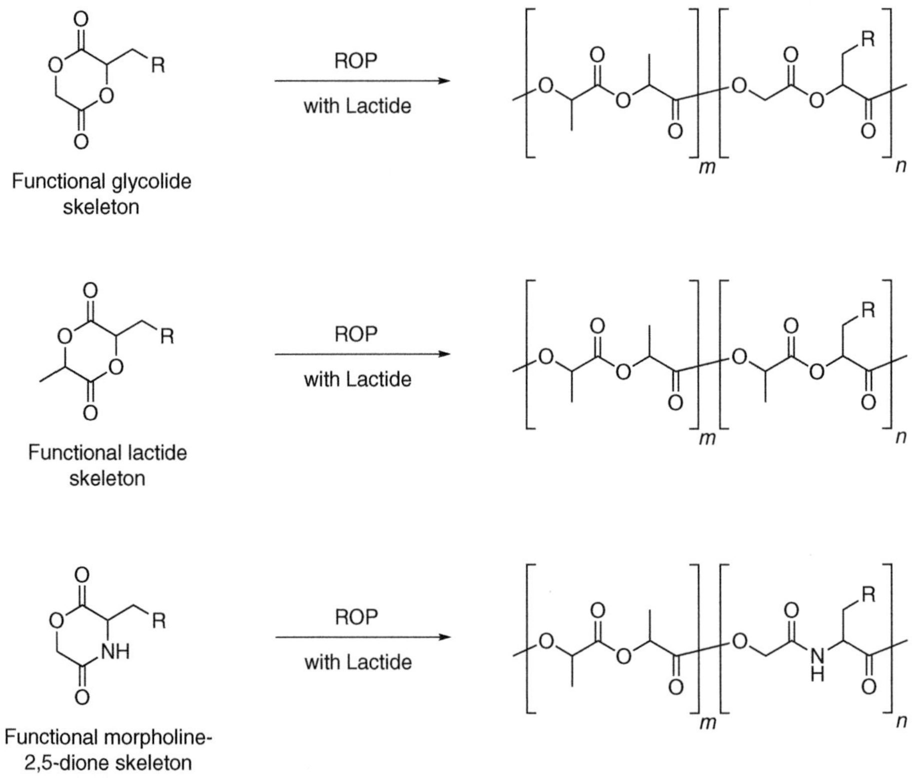

FIGURE 9.3 Functional glycolide, lactide, and morpholine-2,5-dione systems for the synthesis of functional lactide copolymers.

functional derivatized epoxides [44]. The epoxides can be copolymerized with lactide using an organotin compound and an alcohol as the initiating system. Key advantages of this method are the wide variety of introducible functional groups and the comparatively easy synthesis of the epoxide copolymers.

9.2.1.4 Postpolymerization Functionalization of PLA

Most of the aforementioned copolymers are subjected to further chemical reactions after copolymerization in order to extend or alter the functionality introduced. Apart from that a "classic" field of postpolymerization functionalization exists where PLA in its pure form is chemically or sometimes physically treated to introduce reactive sites into the polymer chains that can be used as starting points for functionalization reactions. These methods however are in most cases limited to the polymer surface since applying the procedures to the dissolved polymer would in many cases lead to severe chain degradation. In any case, polymer chain degradation is an issue that has to be kept in mind for most of the postpolymerization functionalization techniques.

Since PLA is a widely spread material for implants and tissue engineering scaffolds, surface functionalization is an important measure to improve the rather poor device-tissue interaction profile of this material. A few, more or less easy, methods for nonpermanent surface modification of PLA have been developed such as coating with proteins or entrapment of other polymers applying a swelling/entrapment/contraction cycle. However, for most applications, a permanent (that is covalent) linkage of functional molecules is desired. Various approaches have been developed to address this issue. One of the mildest forms of surface functionalization is probably the treatment of a PLA surface with benzophenone under UV irradiation. The benzophenone moiety is covalently linked to the PLA surface (photografting) and can readily be used as a starting point for radical grafting onto polymerization reactions [45] or linkage of other functional molecules [46].

Treatment of PLA devices with bases like alkali hydroxides leads to hydrolysis of ester bonds resulting in (reactive) carboxylic acid and hydroxyl groups on the surface of the material. The carboxylic acid groups can be reacted with hydroxyl or amine groups present in many (bioactive) compounds to form ester or amide bonds in order to covalently link the molecules to the material surface. By simply treating the PLA surface with a mixture of a low concentrated aqueous solution of sodium hydroxide and ethanol, a considerable increase in surface hydrophilicity can be achieved (without impacting the polymer bulk) resulting in an improved cell affinity of the material [47]. An alternative can be the application of the so-called direct aminolysis of PLA using an aliphatic diamine such as 1,6-hexamethylene diamine in order to produce amino groups on the surface [48,49]. Further modification towards bromine bearing surface moieties provides, for instance, an opportunity for grafting from polymerization via atom transfer radical polymerization [50] (Figure 9.4).

If a more thorough functionalization of the polymer bulk is desired, the strong base lithium *N*,*N*-diisopropylamide (LDA) can be used to abstract protons from the PLA

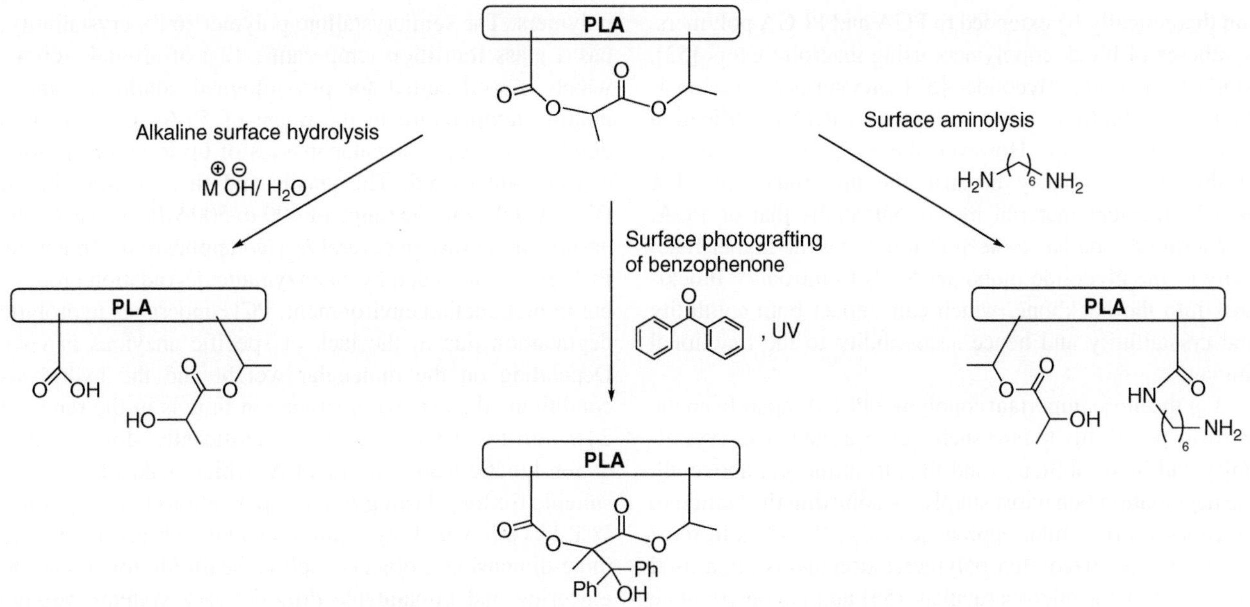

FIGURE 9.4 Examples for PLA surface functionalization.

chain creating carbanionic sites in the α-position to the carbonyl groups [51]. These sites can react with electrophilic compounds allowing for the introduction of various functionalities into the PLA chain. Since this method is carried out in solution (usually THF is used as a solvent), the functionalization encompasses not only the material surface but also the complete polymer bulk and therefore cannot be considered a classic surface functionalization procedure. As another consequence, chain degradation reactions lowering the molecular weight of the polymer are unavoidable. However, if the LDA concentration is kept sufficiently low, this effect can be kept to an acceptable minimum.

In an attempt to develop biodegradable polyesters with surface rather than the typical bulk erosion behavior, Xu reported on the synthesis of a wide range of PLA polymers with varying lipophilicities and chain lengths [20]. In the two-step approach presented, macromer diols with different alkanediol cores and PLA or PLGA chains of various lengths on either side of the initiator molecule have initially been synthesized. Those macromer diols are then linked with diacid dichlorides also varying in chain length to create polyesters of molecular weights ranging from 20,000 to 130,000 g mol^{-1}. The obtained polymers encompass a wide range of easily tunable physical properties such as glass transition temperature, crystallinity, and solubility. Furthermore, these polymers showed linear release profiles of Congo red as a model drug from the onset indicating good applicability of these materials in drug-delivery systems (Figure 9.5).

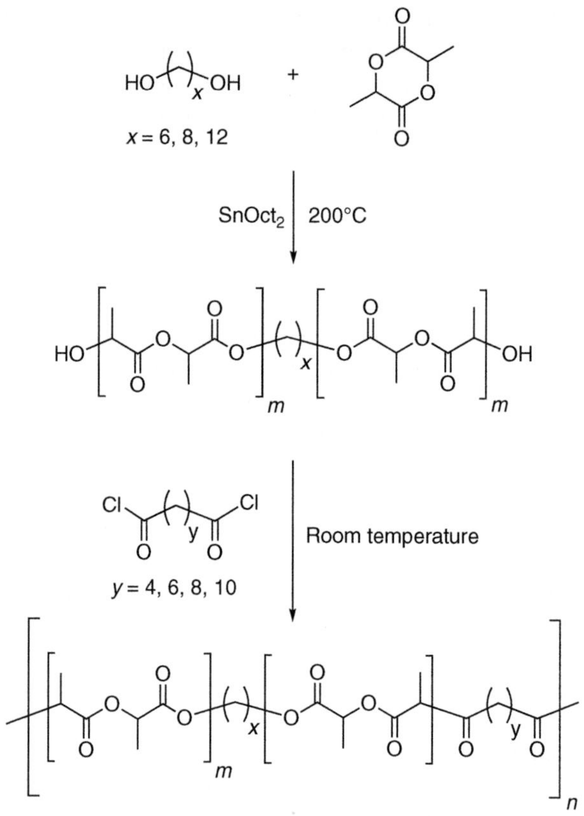

FIGURE 9.5 Synthetic route towards PLA polymers applying condensation reactions according to Xu [20].

9.2.1.5 PGA and PLGA Copolymers

The functionalization strategies for PLA introduced earlier can theoretically be extended to PGA and PLGA polymers. Syntheses of block copolymers using macroinitiators [52], ROP of functional glycolides [53], and surface functionalization with biotin [54] are examples for PGA modification reported in literature. However, there are only few reports in this field primarily because the importance of PLA as a biomedical material by far outweighs that of PGA. Furthermore, one has to keep in mind that the higher reactivity of the glycolide monomer tends to introduce blockiness into the backbone, which can impact both solubility and crystallinity and hence accessibility to the functional domains.

For the more important copolymer PLGA, apart from the intrinsic possibility to tune such factors as degree of crystallinity and hydrophilicity (and thus to influence, above all, the degradation behavior) simply by adjusting the lactide to glycolide ratio, similar approaches exist. PLGA is in most cases functionalized after polymerization and is often used as a material for microparticulate [55] and nanoparticulate drug-delivery systems [56].

9.2.2 Polycaprolactone

PCL is a widely used material in various applications due to its good physicochemical properties like biodegradability, mechanical properties, and good miscibility with other polymers. The semicrystalline polymer (60% crystallinity) has a glass transition temperature (T_g) of around −60 °C, which is well suited for physiological conditions and a melting temperature in the range of 59-64 °C. PCL with number average molecular masses of up to 80,000 g mol^{-1} can be synthesized. The tensile and flexural modulus of PCL, which is in the range of 400 to 500 MPa, respectively, ensures durability in several *in vivo* applications. In nature, PCL can be degraded by an enzymatic degradation process, but in mammalian environment, PCL undergoes hydrolytic degradation due to the lack of specific enzymes *in vivo*. Depending on the molecular weight and the hydrolysis conditions, the average degradation time is in the range of 24 month to 4 years. This is a significantly slower degradation kinetic than PGA or PLA, which makes PCL more suitable for long-lasting *in vivo* applications [57–59]. Since PCL has a low melting point, it can be melt-processed into three-dimensional objects such as scaffolds for tissue engineering and implantable drug-delivery systems through rapid prototyping technologies such as free-form fabrication and printing of polymer melts [60] or can be spun-cast from solution or deposited using electrospinning or electrospraying [57,58].

9.2.2.1 General Aspects of PCL Polymerization

Like PLA and PGA, PCL can be synthesized via polycondensation of 6-hydroxyhexanoic acid or via ROP of ε-caprolactone, although the latter method is preferred and is the most prevalent. Advantages of the ROP are the ability to achieve higher molecular weights, lower polydispersities, and higher monomer conversions. The polymerization can be initiated with a variety of initiators such as metal-based initiators, enzymatic catalysts based on lipases, or organic compounds and inorganic acids. Most commonly, the ROP of CL is initiated using a metal-based initiator such as stannous octoate or aluminum triflate in combination with an alcohol coinitiator [57].

Since PCL consists of ester units and inert ethylene units, methods to alter that structure demand the modification of the monomer or postfunctionalization of the polymer. Toward achieving a functional backbone in PCL, various concepts are described in literature.

- Functionalization at the homopolymer chain end by
  - initiator system [61]
  - endcapping [61]
- Co- and terpolymers via functionalization within the polymer chain by
  - copolymerization with another type of monomer [62,63]
  - copolymerization with a modified monomer [64]
- Polymeric analogous reaction at the polymer backbone via
  - an existing functional group in the backbone
  - creation of an accessible group

The functionalization of PCL by introducing a functional moiety can be done by two different strategies. In a direct approach, the desired functional moiety is introduced in the functional PCL to play the desired role. Thereby, this functional moiety has to be compatible with the chosen system to avoid side reactions. In an indirect approach, a connecting moiety is introduced that is later converted into the desired functional moiety. Thereby, the desired functionality can be introduced after polymerization; thus, more versatile moieties are allowed that are incompatible with the polymerization or the purification [65]. In the following sections, the different concepts are introduced and some example systems are presented.

9.2.2.2 Functionalization of PCL at the Chain End

A common method to functionalize PCL is the functionalization of the polymer at the chain end, either via a functional initiator or via a functional terminating agent. Examples are the chain-end functionalization of PCL with a thiol group via a thiol-containing initiator system [61]. The "functionalization from initiation" approach starts with a $Al(^{i}OPr)_3$: α-(2,4-dinitrophenylthio)ethanol initiating system and PCL is polymerized in living ROP. In a second step, the mercapto-protecting group is cleaved and thio-PCL is obtained. The chain-end functionalization of PCL is achieved using α-(2,4-dinitrophenylthio)acetic acid. Here, the acid moiety stops the polymerization and functionalizes the chain end. After a cleavage of the protecting group, the thiol-modified PCL is obtained. The thiol-containing polyester can be further functionalized with the desired moieties [61] (Figure 9.6).

9.2.2.3 Functionalization of PCL via Block Copolymerization

If a high-molecular-mass substance or a polymer is used as a functional component, AB block copolymers are obtained. Starting with a macro initiator or polymeric precursor/initiator, caprolactone is polymerized via a grafting from approach [66,67]. This concept allows the functionalization of PCL with chitosan [66] or with PEO-PPO block copolymers [67] via the OH moiety using $Sn(Oct)_2$ as a catalyst. This concept can be extended to bifunctional initiators or precursors that lead to triblock copolymers or to multifunctional systems that lead to star-shaped copolymers with functional core and PCL shell [68]. Alternatively, AB block copolymers can also be obtained inversely. PCL is synthesized first and the polymer is end functionalized by endcapping using a functional polymer, which is grafted on or grafted from PCL.

Another concept to obtain functional block copolymers of PCL is synthesizing the polymers using living ROP, whereas the monomers such as caprolactone, lactide, and glycolide are added subsequently [69]. Using this approach, defined block copolymers can be obtained with variable blocks and compositions.

9.2.2.4 Functionalization of PCL via Statistical Copolymerization

PCL is mainly synthesized via ROP using metal-based catalysts. This system is compatible with many other polymer syntheses and a variety of copolymers can be obtained depending on the components and the polymerization process [57].

One approach is the copolymerization of caprolactone with other cyclic monomers that undergo ROP, namely, lactide, glycolide [70], or functionalized epoxides [62,63]. In these systems, the second monomer is considered to bear the desired functionality or chemical moiety, and therefore, a wide range of novel functionalities and polymer properties are accessible. Statistical copolymers of caprolactone with lactides and glycolides are discussed earlier [70]. Beside these commonly used polyester monomers, a variety of other functional cyclic esters are described. Cyclic esters

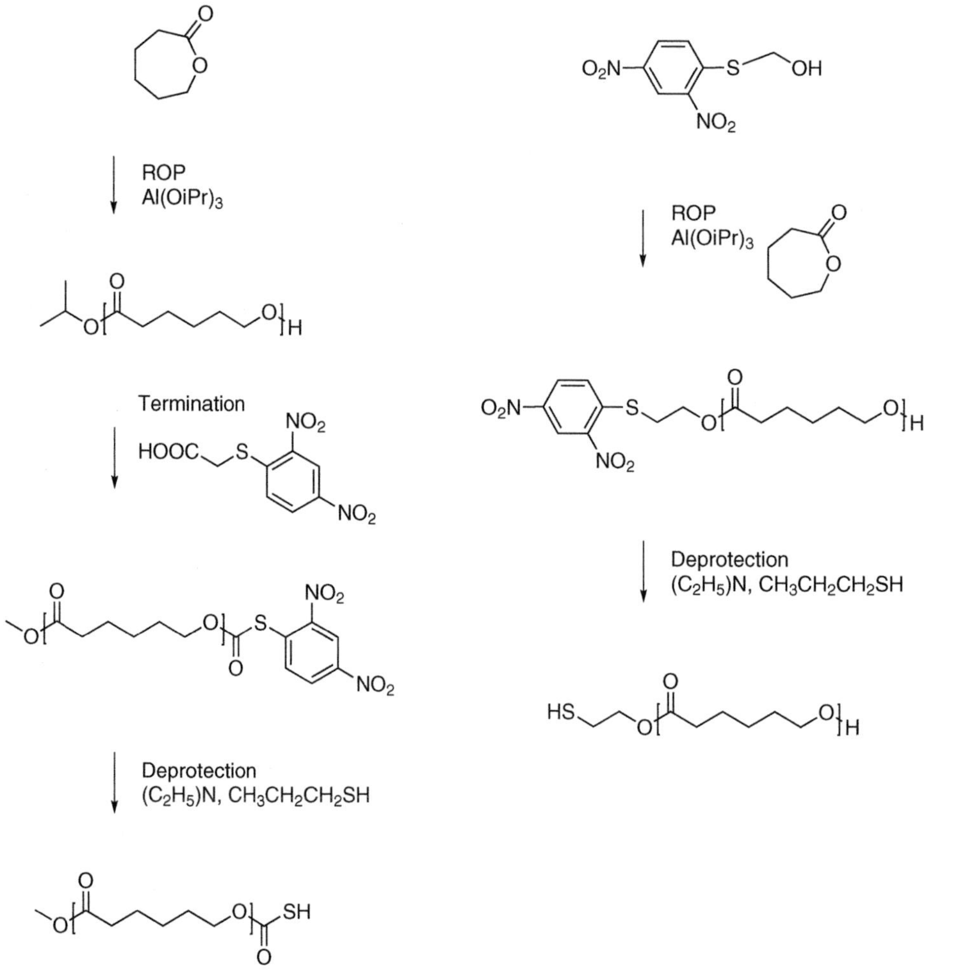

FIGURE 9.6 Chain-end functionalization according to Carrot [61].

such as β-propiolactone, γ-butyrolactone, δ-valerolactone, or pivalolactone are compatible with the ROP of caprolactone and the unsubstituted or substituted lactones can be copolymerized to introduce functionalities in PCL [71].

Copolymerization of caprolactone with epoxides opens a wide range of various functionalities that can be accessed. These functional epoxides can be based on substituted epichlorohydrin [63] or on other molecules. In our laboratory, we recently presented an epoxide system based on 2-methyl-4-pentenoic acid, which is esterified in a first step with a desired functional moiety and then epoxidized to obtain a functional epoxide that is compatible with caprolactone copolymerization [44,62]. This toolbox approach allows for the functionalization with different functional moieties on demand. One advantage of these statistical copolymers is that the copolymer composition can be controlled by the reactivity of the monomers, and therefore, defined copolymers can be obtained (Figure 9.7).

Another approach is the copolymerization of ε-caprolactone and substituted caprolactones that are functionalized in α- or γ-position with bromides, acrylates, olefins, ketones, or protected alcohols [71,72]. These systems allow the synthesis of a PCL polymer backbone with defined adjacent functional groups that can be further functionalized, whereas the copolymerization system is minimally influenced because of the similarity of the monomers.

The introduced adjacent functional moieties either can play their role in the copolymer or can be further functionalized. Pendant functional groups such as methyl acrylates can be further functionalized in click reactions without releasing by-products and therefore provide access to a larger variety of adjacent functional groups [73]. Furthermore, the functionalization of PCL with pendant acetylene groups using azide-terminated oligopeptides has also been reported [65] (Figure 9.8).

A limitation of both strategies is that the functional moieties that are introduced have to be compatible with the initiator system. If incompatible groups such as OH moieties or chelating groups are used, protecting groups are needed. On the other hand, the introduced adjacent moieties can be further functionalized postpolymerization that provides access to a large variety of functionalized PCLs.

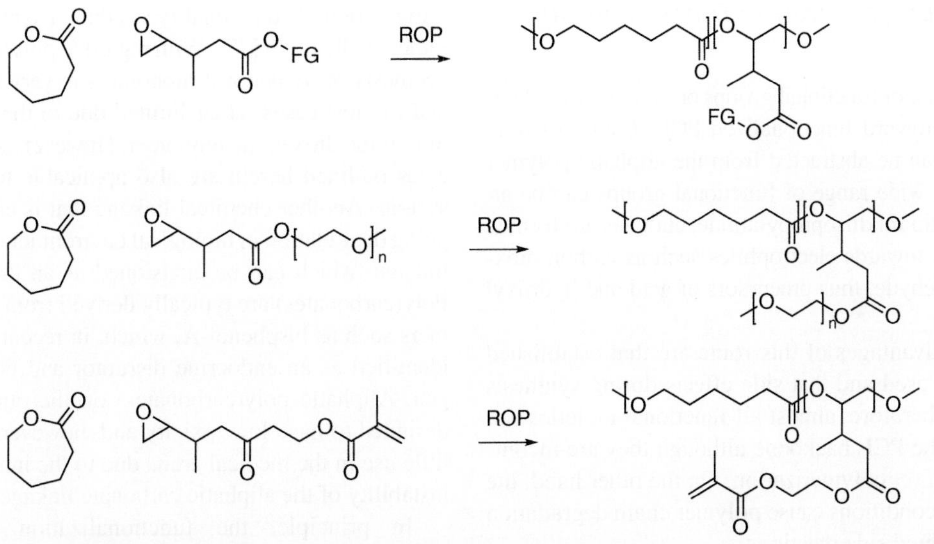

FIGURE 9.7 Functionalization via statistical copolymerization with functional epoxides [62].

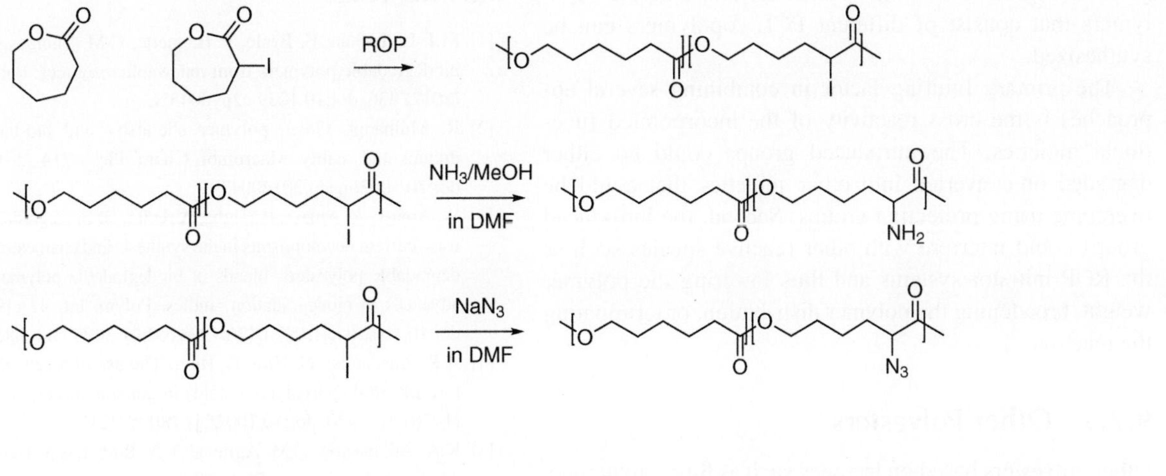

FIGURE 9.8 Statistical copolymers of caprolactone and caprolactone derivates that are further functionalized via postpolymerization reactions [64,65].

9.2.2.5 *Postpolymerization Functionalization of PCL*

The direct grafting of functional groups or moieties onto PCL is another path toward functionalized PCL. Using a strong base, a proton can be abstracted from the aliphatic polymer backbone and a wide range of functional groups can be attached. Using lithium diisopropylamide, enolates are formed that are reactive towards electrophiles such as carbon dioxide and benzaldehyde, thus precursors of acid and hydroxyl moieties [74].

The main advantages of this route are that established systems can be used and that side effects during synthesis are excluded. Therefore, almost all functional moieties can be coupled on the PCL backbone although they are incompatible with PCL copolymerization. On the other hand, the severe reaction conditions cause polymer chain degradation and other unwanted side reactions.

9.2.2.6 *Combinatorial Strategies*

Most of the introduced concepts are compatible with one another to some extent, and thus, a myriad of functional PCL (co)polymers are conceivable by combining the various strategies. The synthesized functional PCL copolymers can be further used as macroinitiator or can be used as a graft-on segment for other polymers. Also, block copolymers that consist of different PCL copolymers can be synthesized.

The primary limiting factor in combining several approaches is the cross reactivity of the incorporated functional moieties. The introduced groups could be either degraded or converted into other moieties that could be overcome using protecting groups. Second, the introduced groups could interfere with other reactive species such as the ROP initiator systems and thus lowering the polymer weight, broadening the polymer distribution, or terminating the reaction.

9.2.3 Other Polyesters

Other polyesters based on lactones such as β-propiolactone, γ-butyrolactone, δ-valerolactone, or pivalolactone play a minor role compared to PGA, PLA, PLGA, and PCL, and therefore, fewer concepts for functionalization are reported. Nevertheless, the introduced principles of functionalizations can in many cases be applied [71].

Poly(hydroxyalkanoates), which are produced through bacterial fermentation, are one of the oldest examples of polymers produced via the biosynthetic pathway [75,76]. Poly(hydroxyalkanoates) are essentially the energy reserves of anaerobic bacteria and are stored as granules in the cytoplasms of these organisms. In this system, hydrocarbons are converted into higher carbon-chain polyesters. Since monomers that can be processed by the bacteria are limited, introduction of functionality via the biosynthetic pathway is rather challenging [77]. With regard to poly(dioxanones), the synthesis of derivatized monomers is very time-consuming and in most cases rather limited due to the chemical structure of the dioxanone monomer. However, some of the strategies outlined herein are also applicable to the dioxanone system. Another chemical linkage that is capable of undergoing degradation in biological environment is the carbonate linkage, which can be envisioned as an ester-like linkage. Poly(carbonates) are typically derived from aromatic monomers such as bisphenol-A, which, in recent years, has been identified as an endocrine disruptor and potential carcinogen. Aliphatic poly(carbonates) on the other hand can be designed to have low toxicity and, however, have seen very little use in the medical arena due to the inherent hydrolytic instability of the aliphatic carbonate linkage.

In principle, the functionalization strategies presented may not be limited to functionalization of polyesters. One can, for instance, envisage the incorporation of polyester segments as chain extenders of soft segments in poly(urethanes), thereby providing access to functional groups in an important class of biomaterial especially for cardiovascular applications.

REFERENCES

[1] M.J.-L. Tschan, E. Brulé, P. Haquette, C.M. Thomas, Synthesis of biodegradable polymers from renewable resources. Polym. Chem. 3 (2012) 836, doi:10.1039/c2py00452f.

[2] R. Mülhaupt, Green polymer chemistry and bio-based plastics: dreams and reality. Macromol. Chem. Phys. 214 (2013) 159–174, doi:10.1002/macp.201200439.

[3] W. Amass, A. Amass, B. Tighe, A review of biodegradable polymers: uses, current developments in the synthesis and characterization of biodegradable polyesters, blends of biodegradable polymers and recent advances in biodegradation studies. Polym. Int. 47 (1998) 89–144, doi:10.1002/(sici)1097-0126(1998100)47:2<89::aid-pi86>3.0.co;2-f.

[4] A.R. Anscombe, N. Hira, B. Hunt, The use of a new absorbable suture material (polyglycolic acid) in general surgery. Br. J. Surg. 57 (1970) 917–920, doi:10.1002/bjs.1800571212.

[5] K.A. Athanasiou, C.M. Agrawal, F.A. Barber, S.S. Burkhart, Orthopaedic applications for PLA-PGA biodegradable polymers. Arthroscopy 14 (1998) 726–737, doi:10.1016/s0749-8063(98)70099-4.

[6] A. Kumari, S.K. Yadav, S.C. Yadav, Biodegradable polymeric nanoparticles based drug delivery systems. Colloids Surf. B: Biointerfaces 75 (2010) 1–18, doi:10.1016/j.colsurfb.2009.09.001.

[7] P.A. Gunatillake, R. Adhikari, Biodegradable synthetic polymers for tissue engineering, Eur. Cells Mater. 5 (2003) 1–16.

[8] I. Armentano, M. Dottori, E. Fortunati, S. Mattioli, J.M. Kenny, Biodegradable polymer matrix nanocomposites for tissue engineering: a review. Polym. Degrad. Stab. 95 (2010) 2126–2146, doi:10.1016/j.polymdegradstab.2010.06.007.

[9] M. Hiljanen-Vainio, P. Varpomaa, J. Seppälä, P. Törmälä, Modification of poly(L-lactides) by blending: mechanical and hydrolytic behavior. Macromol. Chem. Phys. 197 (1996) 1503–1523, doi:10.1002/macp.1996.021970427.

[10] R. Langer, K.S. Anseth, V.R. Shastri, Photopolymerizable degradable polyanhydrides with osteocompatibility. Nat. Biotechnol. 17 (1999) 156–159, doi:10.1038/6152.

[11] H.B. Rosen, J. Chang, G.E. Wnek, R.J. Linhardt, R. Langer, Bioerodible polyanhydrides for controlled drug delivery. Biomaterials 4 (1983) 131–133, doi:10.1016/0142-9612(83)90054-6.

[12] S. Sant, D. Iyer, A.K. Gaharwar, A. Patel, A. Khademhosseini, Effect of biodegradation and de novo matrix synthesis on the mechanical properties of valvular interstitial cell-seeded polyglycerol sebacate–polycaprolactone scaffolds. Acta Biomater. 9 (2013) 5963–5973, doi:10.1016/j.actbio.2012.11.014.

[13] X. Punet, et al., Enhanced cell-material interactions through the biofunctionalization of polymeric surfaces with engineered peptides. Biomacromolecules 14 (2013) 2690–2702, doi:10.1021/bm4005436.

[14] N. Patel, Spatially controlled cell engineering on biodegradable polymer surfaces, J. Federation Am. Soc. Exp. Biol. 1 (2) (1998) 1447–1454.

[15] A.Z. Wang, R. Langer, O.C. Farokhzad, Nanoparticle delivery of cancer drugs. Annu. Rev. Med. 63 (2012) 185–198, doi:10.1146/annurev-med-040210-16254.

[16] S. Ravi, E.L. Chaikof, Biomaterials for vascular tissue engineering. Regen. Med. 5 (2010) 107–120, doi:10.2217/rme.09.77.

[17] K. Kolandaivelu, et al., Stent thrombogenicity early in high-risk interventional settings is driven by stent design and deployment and protected by polymer-drug coatings. Circulation 123 (2011) 1400–1409, doi:10.1161/circulationaha.110.003210.

[18] L.E. Niklason, Functional arteries grown in vitro. Science 284 (1999) 489–493, doi:10.1126/science.284.5413.489.

[19] M. Poh, et al., Blood vessels engineered from human cells. Lancet 365 (2005) 2122–2124, doi:10.1016/s0140-6736(05)66735-9.

[20] X.-J. Xu, J.C. Sy, V.P. Shastri, Towards developing surface eroding poly(α-hydroxy acids). Biomaterials 27 (2006) 3021–3030, doi:10.1016/j.biomaterials.2005.12.006.

[21] A.S. Sawhney, C.P. Pathak, J.A. Hubbell, Bioerodible hydrogels based on photopolymerized poly(ethylene glycol)-co-poly(.alpha.-hydroxy acid) diacrylate macromers. Macromolecules 26 (1993) 581–587, doi:10.1021/ma00056a005.

[22] R. Gref, et al., Biodegradable long-circulating polymeric nanospheres. Science 263 (1994) 1600–1603, doi:10.1126/science.8128245.

[23] W.H. Carothers, G.L. Dorough, F.J. Van Natta, Studies of polymerization and ring formation. X. The reversible polymerization of six-membered cyclic esters, J. Am. Chem. Soc. 54 (1932) 761–772.

[24] C.K.S. Pillai, C.P. Sharma, Review paper: absorbable polymeric surgical sutures: chemistry, production, properties, biodegradability, and performance. J. Biomater. Appl. 25 (2010) 291–366, doi:10.1177/0885328210384890.

[25] S. Fredenberg, M. Wahlgren, M. Reslow, A. Axelsson, The mechanisms of drug release in poly(lactic-co-glycolic acid)-based drug delivery systems—a review. Int. J. Pharm. 415 (2011) 34–52, doi:10.1016/j.ijpharm.2011.05.049.

[26] N. Mahmoudifar, P.M. Doran, Chondrogenic differentiation of human adipose-derived stem cells in polyglycolic acid mesh scaffolds under dynamic culture conditions. Biomaterials 31 (2010) 3858–3867, doi:10.1016/j.biomaterials.2010.01.090.

[27] A.N. Vaidya, et al., Production and recovery of lactic acid for polylactide—an overview. Crit. Rev. Environ. Sci. Technol. 35 (2005) 429–467, doi:10.1080/10643380590966181.

[28] B.C. Saha, J. Woodward, ACS Symposium Series, American Chemical Society, Washington, DC, 1997.

[29] A.-C. Albertsson, I.K. Varma, Recent developments in ring opening polymerization of lactones for biomedical applications. Biomacromolecules 4 (2003) 1466–1486, doi:10.1021/bm034247a.

[30] K. Makiguchi, S. Kikuchi, T. Satoh, T. Kakuchi, Synthesis of block and end-functionalized polyesters by triflimide-catalyzed ring-opening polymerization of ε-caprolactone, 1,5-dioxepan-2-one, and rac -lactide. J. Polym. Sci. A Polym. Chem. 51 (2012) 2455–2463, doi:10.1002/pola.26631.

[31] K. Bernard, P. Degée, P. Dubois, Regioselective end-functionalization of polylactide oligomers with D-glucose and D-galactose. Polym. Int. 52 (2003) 406–411, doi:10.1002/pi.1050.

[32] S. Vuorinen, et al., Sugar end-capped poly-d, l-lactides as excipients in oral sustained release tablets. AAPS PharmSciTech 10 (2009) 566–573, doi:10.1208/s12249-009-9247-9.

[33] A.J. Cross, M.G. Davidson, D. García-Vivó, T.D. James, Well-controlled synthesis of boronic-acid functionalised poly(lactide)s: a versatile platform for biocompatible polymer conjugates and sensors. RSC Adv. 2 (2012) 5954, doi:10.1039/c2ra20373a.

[34] D.-H. Yu, Q. Lu, J. Xie, C. Fang, H.-Z. Chen, Peptide-conjugated biodegradable nanoparticles as a carrier to target paclitaxel to tumor neovasculature. Biomaterials 31 (2010) 2278–2292, doi:10.1016/j.biomaterials.2009.11.047.

[35] G. Dorff, M. Hahn, A. Laschewsky, A. Lieske, Optimization of the property profile of poly-L-lactide by synthesis of PLLA-polystyrene-block copolymers. J. Appl. Polym. Sci. 127 (2013) 120–126, doi:10.1002/app.37836.

[36] M. Vert, Biomedical polymers from chiral lactides and functional lactones: properties and applications. Makromol. Chem. Macromol. Symp. 6 (1986) 109–122, doi:10.1002/masy.19860060113.

[37] Y. Kimura, K. Shirotani, H. Yamane, T. Kitao, Ring-opening polymerization of 3(S)-[(benzyloxycarbonyl)methyl]-1,4-dioxane-2,5-dione: a new route to a poly(alpha-hydroxy acid) with pendant carboxyl groups. Macromolecules 21 (1988) 3338–3340, doi:10.1021/ma00189a037.

[38] W.W. Gerhardt, et al., Functional lactide monomers: methodology and polymerization. Biomacromolecules 7 (2006) 1735–1742, doi:10.1021/bm060024j.

[39] P.J.A. in't Veld, P.J. Dijkstra, J. Feijen, Synthesis of biodegradable polyesteramides with pendant functional groups. Makromol. Chem. 193 (1992) 2713–2730, doi:10.1002/macp.1992.021931101.

[40] D.A. Barrera, E. Zylstra, P.T. Lansbury, R. Langer, Synthesis and RGD peptide modification of a new biodegradable copolymer: poly(lactic acid-co-lysine). J. Am. Chem. Soc. 115 (1993) 11010–11011, doi:10.1021/ja00076a077.

[41] Z. Pang, et al., Controllable ring-opening copolymerization of L-lactide and (3S)-benzyloxymethyl-(6S)-methyl-morpholine-2,5-dione initiated by a biogenic compound creatinine acetate. J. Polym. Sci. A Polym. Chem. 50 (2012) 4004–4009, doi:10.1002/pola.26196.

[42] W. Chen, et al., Versatile synthesis of functional biodegradable polymers by combining ring-opening polymerization and post-polymerization modification via Michael-type addition reaction. Macromolecules 43 (2010) 201–207, doi:10.1021/ma901897y.

[43] F.-C. Chiu, S.-W. Wang, K.-Y. Peng, R.-S. Lee, Synthesis and characterization of amphiphilic PLA-(PαN3CL-g-PBA) copolymers by ring-opening polymerization and click reaction. Polymer 53 (2012) 3476–3484, doi:10.1016/j.polymer.2012.06.004.

[44] V.P. Shastri, Biodegradable Polymers from Derivatized Ring-Opened Epoxides. US 6730772, 2004.

[45] R.M. Rasal, B.G. Bohannon, D.E. Hirt, Effect of the photoreaction solvent on surface and bulk properties of poly(lactic acid) and poly(hydroxyalkanoate) films, http://onlinelibrary.wiley.com/store/10.1002/jbm.b.30980/asset/30980_ftp.pdf?v=1&t=hiypzu7h&s=c97c4765807e96c50836e74d93bfa1a4fc25c8bf, 2008.

[46] B. Guo, A. Finne-Wistrand, A.-C. Albertsson, Electroactive hydrophilic polylactide surface by covalent modification with tetraaniline. Macromolecules 45 (2012) 652–659, doi:10.1021/ma202508h.

[47] J. Yang, et al., Enhancing the cell affinity of macroporous poly(L-lactide) cell scaffold by a convenient surface modification method. Polym. Int. 52 (2003) 1892–1899, doi:10.1002/pi.1272.

[48] T.G. Kim, T.G. Park, Biodegradable polymer nanocylinders fabricated by transverse fragmentation of electrospun nanofibers through aminolysis, http://onlinelibrary.wiley.com/store/10.1002/marc.200800094/asset/1231_ftp.pdf?v=1&t=hizo7uah&s=72268303b8d3d5c4574c6f44b82c42b7889f7c2a, 2008.

[49] Y. Zhu, C. Gao, X. Liu, T. He, J. Shen, Immobilization of biomacromolecules onto aminolyzed poly(L-lactic acid) toward acceleration of endothelium regeneration. Tissue Eng. 10 (2004) 53–61, doi:10.1089/107632704322791691.

[50] F.J. Xu, X.C. Yang, C.Y. Li, W.T. Yang, Functionalized polylactide film surfaces via surface-initiated ATRP. Macromolecules 44 (2011) 2371–2377, doi:10.1021/ma200160h.

[51] B. Saulnier, S. Ponsart, J. Coudane, H. Garreau, M. Vert, Lactic acid-based functionalized polymers via copolymerization and chemical modification. Macromol. Biosci. 4 (2004) 232–237, doi:10.1002/mabi.200300087.

[52] I. Barakat, P. Dubois, C. Grandfils, R. Jérôme, Poly(e-caprolactone-b-glycolide) and poly(D, L-lactide-b-glycolide) diblock copolyesters: controlled synthesis, characterization, and colloidal dispersions. J. Polym. Sci. A Polym. Chem. 39 (2001) 294–306, doi:10.1002/1099-0518(20010115)39:2<294::aid-pola50>3.0.co;2-a.

[53] X. Jiang, E.B. Vogel, M.R. Smith, G.L. Baker, “Clickable” polyglycolides: tunable synthons for thermoresponsive, degradable polymers. Macromolecules 41 (2008) 1937–1944, doi:10.1021/ma7027962.

[54] K.-B. Lee, K.R. Yoon, S.I. Woo, I.S. Choi, Surface modification of poly(glycolic acid) (PGA) for biomedical applications. J. Pharm. Sci. 92 (2003) 933–937, doi:10.1002/jps.10556.

[55] Y. Wei, et al., A novel sustained-release formulation of recombinant human growth hormone and its pharmacokinetic. Pharmacodyn. Safety Profiles Mol. Pharm. 9 (2012) 2039–2048, doi:10.1021/mp300126t.

[56] O.C. Farokhzad, et al., Targeted nanoparticle-aptamer bioconjugates for cancer chemotherapy in vivo. Proc. Natl. Acad. Sci. U.S.A. 103 (2006) 6315–6320, doi:10.1073/pnas.0601755103.

[57] M. Labet, W. Thielemans, Synthesis of polycaprolactone: a review. Chem. Soc. Rev. 38 (2009) 3484–3504, doi:10.1039/b820162p.

[58] M.A. Woodruff, D.W. Hutmacher, The return of a forgotten polymer—polycaprolactone in the 21st century. Prog. Polym. Sci. 35 (2010) 1217–1256, doi:10.1016/j.progpolymsci.2010.04.002.

[59] J.E. Mark, Polymer Data Handbook, Oxford University Press, New York, 1999.

[60] M. Kammerer, et al., Valproate release from polycaprolactone implants prepared by 3D-bioplotting. Pharmazie 66 (2011) 511–516, doi:10.1691/ph.2011.0882.

[61] G. Carrot, J.G. Hilborn, M. Trollsås, J.L. Hedrick, Two general methods for the synthesis of thiol-functional polycaprolactones. Macromolecules 32 (1999) 5264–5269, doi:10.1021/ma990198b.

[62] J.J. Wurth, V.P. Shasti, Synthesis and characterization of functionalized poly(ε-caprolactone). J. Polym. Sci. 51 (2013) 3375–3382, doi:10.1002/pola.26734.

[63] K.Y. Cho, J.-K. Park, Synthesis and characterization of poly(ethylene glycol) grafted poly(ε-caprolactone). Polym. Bull. 57 (2006) 549–856, doi:10.1007/s00289-006-0658-4.

[64] S.E. Habnouni, S. Blanquer, V. Darcos, J. Coudane, S. El Habnouni, Aminated PCL-based copolymers by chemical modification of poly(α-iodo-ε-caprolactone- co -ε-caprolactone). Polymer 47 (2009) 6104–6115, doi:10.1002/pola.23652.

[65] B. Parrish, R.B. Breitenkamp, T. Emrick, PEG- and peptide-grafted aliphatic polyesters by click chemistry. J. Am. Chem. Soc. 127 (2005) 7404–7410, doi:10.1021/ja050310n.

[66] L. Xu, et al., Synthesis, characterization, and self-assembly of linear poly(ethylene oxide)-block-poly(propylene oxide)-block-poly(ε-caprolactone) (PEO-PPO-PCL) copolymers. J. Colloid Interface Sci. 393 (2013) 174–181, doi:10.1016/j.jcis.2012.10.051.

[67] L. Liu, L. Chen, Y.E. Fang, Self-catalysis of phthaloylchitosan for graft copolymerization of ε-caprolactone with chitosan. Macromol. Rapid Commun. 27 (2006) 1988–1994, doi:10.1002/marc.200600508.

[68] Q. Cai, J. Bei, S. Wang, Synthesis and properties of ABA-type triblock copolymers of poly(glycolide-co-caprolactone) (A) and poly(ethylene glycol) (B). Polymer 43 (2002) 3585–3591, doi:10.1016/S0032-3861(02)00197-0.

[69] M.-H. Huang, S. Li, J. Coudane, M. Vert, Synthesis and characterization of block copolymers of ε-caprolactone and DL-lactide initiated by ethylene glycol or poly(ethylene glycol). Macromol. Chem. Phys. 204 (2003) 1994–2001, doi:10.1002/macp.200350054.

[70] M. Srisa-ard, R. Molloy, N. Molloy, J. Siripitayananon, M. Sriyai, Synthesis and characterization of a random terpolymer of l-lactide, ε-caprolactone and glycolide. Polym. Int. 50 (2001) 891–896, doi:10.1002/pi.713.

[71] X. Lou, C. Detrembleur, R. Jérôme, Novel aliphatic polyesters based on functional cyclic (di)esters. Macromol. Rapid Commun. 24 (2003) 161–172, doi:10.1002/marc.200390029.

[72] P. Lecomte, et al., New prospects for the grafting of functional groups onto aliphatic polyesters. Ring-opening polymerization of α- or γ-substituted ε-caprolactone followed by chemical derivatization of the substituents. Macromol. Symp. 240 (2006) 157–165, doi:10.1002/masy.200650820.

[73] J. Rieger, et al., Versatile functionalization and grafting of poly(ε-caprolactone) by Michael-type addition. Chemical Commun. 41 (2005) 274–276, doi:10.1039/b411565a.

[74] M. Vert, Aliphatic polyesters: great degradable polymers that cannot do everything. Biomacromolecules 6 (2005) 538–546, doi:10.1021/bm0494702.

[75] R.W. Lenz, R.H. Marchessault, Bacterial polyesters: biosynthesis. Biodegradable Plastics Biotechnol. Biomacromol. 6 (2005) 1–8, doi:10.1021/bm049700c.

[76] Y. Poirier, C. Nawrath, C. Somerville, Production of polyhydroxyalkanoates, a family of biodegradable plastics and elastomers, in bacteria and plants. Bio/Technology 13 (1995) 142–150, doi:10.1038/nbt0295-142.

[77] Y.B. Kim, R.W. Lenz, R.C. Fuller, Poly(β-hydroxyalkanoate) copolymers containing brominated repeating units produced by Pseudomonas oleovorans. Macromolecules 25 (1992) 1852–1857, doi:10.1021/ma00033a002.

Chapter 10

Polyanhydrides

Muntimadugu Eameema*, Lakshmi Sailaja Duvvuri*, Wahid Khan*,[†], Abraham J. Domb[†]

*Department of Pharmaceutics, National Institute of Pharmaceutical Education and Research, Hyderabad, India

[†]School of Pharmacy, Faculty of Medicine, The Hebrew University of Jerusalem, Jerusalem, Israel

Chapter Outline

ABBREVIATIONS

FA fumaric acid
CPH 1,3 bis(*p*-carboxyphenoxyhexane)
P(CPP) poly(carboxyphenoxypropane)
PAA poly(adipic acid)
PBSAM *N,N´*-bis(L-alanine)-sebacoylamide
PCL poly(caprolactone)
PHB poly(hydroxybutyrate)
PLA poly(lactic acid)
PTMC poly(trimethylene carbonate)
RA ricinoleic acid

10.1 HISTORY OF POLYANHYDRIDES

The class of polyanhydrides is one of the most widely investigated biodegradable polymers, which have been investigated for more than three decades [1–3]. The historical development of polyanhydrides dates back to the 1900s when Bucher and Slade documented first synthesis of polyanhydrides upon heating isophthalic acid and terephthalic acid with acetic anhydride in 1909 [4]. This led to the investigation of polyanhydrides as textiles by Hill and Carothers in the 1930s [5]. However, due to high hydrolytic instability of anhydride linkages, they were not found to be suitable for textile applications.

Systematic development of polyanhydrides, as substitutes for polyesters in textile applications, was undertaken

Natural and Synthetic Biomedical Polymers

initially by Conix *et al.* [6]. They prepared and studied a number of aromatic and heterocyclic polyanhydrides that were found to be quite stable towards hydrolysis and possessed excellent film- and fiber-forming properties. A new class of heterocyclic crystalline polyanhydrides was introduced by Yoda [7,8] who synthesized a variety of five-membered heterocyclic dibasic acids and polymerized these compounds with acetic anhydride at 200-300 °C under vacuum and nitrogen atmosphere. These heterocyclic polymers have melting point in the range of 70-190°C and good film- and fiber-forming properties. Aliphatic polyanhydrides were considered immaterial due to their hydrolytic instability. It was not until the 1980s that a connection was made between polyanhydride degradation properties and biomedical applications. In 1980, Langer was the first to identify and accomplish the hydrolytic instability of these polymers for controlled drug delivery applications and used them as biodegradable carriers in various medical devices [9]. Since then, extensive research has been going on to explore the use of polyanhydrides for controlled delivery and has also landed up in two devices for clinical use, i.e., Glaidel® [10] and Septacin™ [11].

10.2 PROPERTIES OF POLYANHYDRIDES

Polyanhydrides have been considered to be useful biomaterials as carriers of drugs to various organs of the human body such as the brain, bone, blood vessels, and eyes. They can be prepared easily from available, low-cost resources and can be manipulated to meet desirable characteristics. Since last 25 years, intensive research has been conducted, which yielded hundreds of publications and patents describing new polymer with excellent film- and fiber-forming properties, structures, studies on chemical and physical characterization of these polymers, degradation and stability properties, toxicity studies, and applications of polymers for mainly controlled bioactive agents. However, due to their rapid degradation and limited mechanical properties, their main use has been limited to short-term controlled delivery of bioactive agents [1].

10.2.1 Distinctive Features and Limitations

The hydrolytic instability of polyanhydrides presents them with both advantages and disadvantages [12,13].

- Resources are easily available with low cost and are generally considered as safe dicarboxylic acid building blocks.
- One-step synthesis with no need for purification steps.
- A well-defined polymer structure with controlled molecular weight and that degrades hydrolytically at a predictable rate.
- Manipulation of polymer to release bioactive agents at a predictable rate for periods of weeks.
- They are processable by low-temperature injection molding or extrusion for mass product and have versatile properties, which can be varied by monomer selection, composition, surface area, and additives.
- They degrade to their respective diacids and completely eliminate from the body within a period of weeks to months.
- They can be sterilized by terminal gamma irradiation with minimal effect on polymer properties.
- They are highly instable to hydrolysis.
- They have poor mechanical strength.
- They possess poor fiber- and film-forming capacities.
- They are highly sensitive to heat and moisture (require a storage temperature of −20 °C or below).

10.2.2 Thermal Properties

The melting point, as determined by differential scanning calorimeter, of these aromatic polyanhydrides is much higher than aliphatic polyanhydrides. The melting point of aliphatic-aromatic copolyanhydrides is proportional to aromatic content. For this type of copolymers, there is characteristically a minimum T_m between 5 and 20 mol% of lower-melting component. The introduction of fatty acids in the copolymer chain lowers the melting point as compared to that of bulk polymer [14].

Properties of polymer matrix other than melting point are very important to obtain a proper drug release. Crystallinity is an important factor that controls a polymer's erosion rate and influences its solubility along with its melting point. Almost all polyanhydrides show some degree of crystallinity as manifested by its melting point. Homopolymers of aliphatic and aromatic diacids like sebacic acid (SA), bis(carboxyphenoxy)propane (CPP), bis(carboxyphenoxy) hexane (CPH), and fumaric acid (FA) are more crystalline (>50% crystallinity), whereas their copolymers were found to be less crystalline and degree of crystallinity increased with increasing mole ratio of either aliphatic or aromatic diacid content [15]. The decrease in crystallinity was because of random presence of other units in the polymer chain. A detailed analysis of copolymers of SA with aromatic and unsaturated monomers, CPP, CPH, and FA showed that copolymers with high ratios of SA and CPP or CPH were crystalline, while copolymers with equal ratios of SA and CPP or CPH were amorphous [16]. In contrast, P(FA: SA) series displayed high crystallinity despite comonomer ratio [15].

10.2.3 Solubility

The majority of polyanhydrides dissolve in solvents such as dichloromethane and chloroform. However, the aromatic polyanhydrides exhibit much lower solubility than aliphatic polyanhydrides. But the copolymers of two different aromatic monomers showed increased solubility with a decrease in T_m [15].

10.2.4 Mechanical Properties

Polyanhydrides show poor mechanical properties in comparison with other polymers such as polyesters. It was observed that increasing the CPP content in copolymer composition increases the tensile strength and elongation of various polyanhydrides tested [14]. Despite the low molecular weight (MW=6400) of poly(CPP-SA) (60:40), it has a higher tensile strength of 981 MPa (100 kgf/cm^2) than it has in the 20:80 composition (MW=18,900), 441 MPa (45 kgf/cm^2).

Transparent and flexible films of fatty acid polyanhydrides were reported with a tensile strength of 4-19 MPa and elongation at break in the range of 77-115%. Thus, introduction of nonlinear fatty acid structures in polyanhydrides provides hydrophobicity and flexibility to the polymers.

10.2.5 Stability

Polymers used in drug delivery applications should have not only good solubility required for film casting or microencapsulation but also stability in both solid and solution state. Polyanhydrides are considered as air-sensitive materials and are stored at a temperature of −20 °C under argon atmosphere [13]. A detailed analysis on stability of aromatic and aliphatic polymers in solid state and dry chloroform solution showed that poly(1,1-bis(*p*-carboxyphenoxy)methane) maintained their original molecular weight for at least 1 year in the solid state, whereas, aliphatic polyanhydrides, such as poly sebacic acid, showed decreased molecular weight over time. The decrease in molecular weight was explained by an internal anhydride interchange mechanism, as revealed from elemental and spectral analyses supported by the fact that this decrease in molecular weight was reversible and heating the depolymerized polymer at 180 °C for 20 min yielded the original high-molecular-weight polymers. However, under similar conditions, the hydrolyzed polymer did not increase in molecular weight [17]. In many cases, it was observed that the stability of polymers in the solid state or in organic solution did not correlate with its hydrolytic stability [17]. A similar decrease in molecular weight as function of time was also observed among the aliphatic-aromatic copolyanhydrides and imide-containing polyanhydrides [18,19].

10.3 SYNTHESIS OF POLYANHYDRIDES

Synthesis of polyanhydrides is carried out through generalized methods such as melt polycondensation, solution polymerization, use of coupling agents, and ring-opening polymerization.

10.3.1 Melt Condensation

Carothers and Hill first reported the synthesis of polyanhydrides through condensation. In this method, a prepolymer is formed by converting the carboxylic group to a mixed anhydride with acetic anhydride before subjecting the prepolymer to melt polycondensation and the schematic process is given in Scheme 10.1 [14].

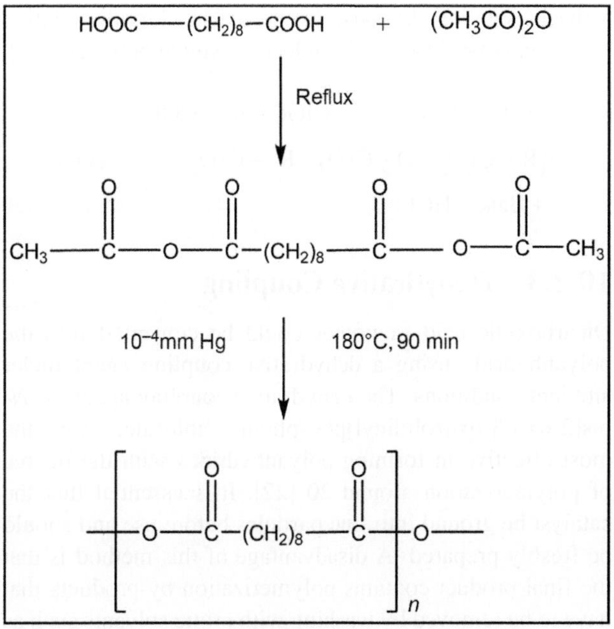

SCHEME 10.1 Synthesis of polyanhydrides through melt condensation.

The polyanhydride thus obtained was of low molecular weight. For most of the practical applications, high-molecular-weight polyanhydrides are desirable. Hence, a systematic study was undertaken to determine the factors that affected the polymer molecular weight. It was found that the critical factors were monomer purity, reaction time, and temperature, and an efficient system to remove the by-product, acetic anhydride is required. The highest-molecular-weight polymers were obtained using pure isolated prepolymers and heating them at 180 °C for 90 min with a vacuum of 10^{-4} mmHg, using a dry ice/acetone trap. Significantly higher molecular weights were obtained in shorter times by using coordination catalysts such as, cadmium acetate, earth metal oxides, and $ZnEt_2 \cdot H_2O$ [20].

10.3.2 Solution Polymerization

The solution polymerization is carried out by Schotten-Baumann technique. In this method, the solution of diacid chloride is added dropwise into an ice-cooled solution of a dicarboxylic acid. The reaction is facilitated by using an acid acceptor such as triethylamine. Polymerization takes place instantly on contact of the monomers and is essentially complete within 1 h. The solvents employed can be a single solvent or a mixture of solvents like dichloromethane, chloroform, benzene, and ethyl ether. It was found that the order of addition is very important in obtaining relatively high-molecular-weight polyanhydrides. Addition of a diacid solution dropwise to the diacid chloride solution consistently produced high-molecular-weight polymers [21].

$$\mathrm{HOOC-R-COOH+ClOC-R'-COCl} \xrightarrow{\text{Base}}$$
$$-\left(\mathrm{R-C(O)-O-C(O)-R'-C(O)-O-C(O)}\right)$$
$$-+\mathrm{Base-HCl}$$

10.3.3 Dehydrative Coupling

Dicarboxylic acid monomer could be converted into the polyanhydride using a dehydrative coupling agent under ambient conditions. The dehydrative coupling agent, *N′*,*N*-bis[2-oxo-3-oxazolidinyl]phosphonic chloride, was the most effective in forming polyanhydrides with the degree of polymerization around 20 [22]. It is essential that the catalyst be ground into fine particles before use and should be freshly prepared. A disadvantage of this method is that the final product contains polymerization by-products that have to be removed by washing with protic solvents such as methanol or cold dilute hydrochloric acid. The washing by protic solvents may evoke some hydrolysis of the polymer.

Coupling agents such as phosgene and diphosgene could also be used for the polyanhydride formation. Polymerization of SA using either phosgene or diphosgene as coupling agents with the amine-based heterogeneous acid acceptor, poly(4-vinyl pyridine), produced higher molecular weights in comparison to nonamine heterogeneous base K_2CO_3 [23].

10.3.4 Ring-Opening Polymerization

ROP offers an alternate approach to the synthesis of polyanhydrides used for medical applications. Albertsson and coworkers prepared adipic acid polyanhydride from cyclic adipic anhydride (oxepane-2,7-dione) using cationic (e.g., $AlCl_3$ and $BF_3{\cdot}(C_2H_5)_2O$), anionic (e.g., $CH_3COO^-K^+$ and NaH), and coordination-type inhibitors such as stannous-2-ethylhexanoate and dibutyltin oxide [24,25]. ROP takes place in two steps: (1) preparation of the cyclic monomer and (2) polymerization of the cyclic monomers [26].

10.4 CLASSES OF POLYANHYDRIDES

Since the introduction of polyanhydrides to the regime of polymers, hundreds of polyanhydride structures have been reported. These polymers are divided into conventional and advanced polyanhydrides, with various subclasses under each category. Different classes of polyanhydrides with their structures and examples are given in Table 10.1.

TABLE 10.1 Different Classes of Polyanhydrides [12,27]

S. No.	Class	Backbone	Examples	Drug Delivery Systems	References
1	Aliphatic polyanhydrides	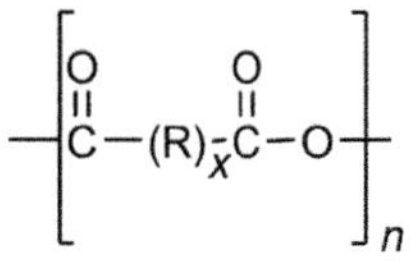 R = aliphatic moiety	PSA (x = 8)	PSA-co-Pluronic microspheres for the controlled release of nifedipine	[28]
			PAA (x = 4)	PAA microsphere gel for ocular delivery of timolol maleate	[29]
2	Unsaturated polyanhydrides	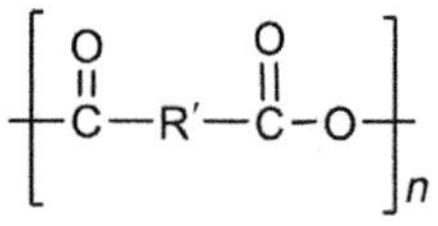 R′ = unsaturated moeity	PFA	Fumaric acid-based polyanhydrides as bioadhesive excipient for oral drug delivery	[30]
3	Aromatic polyanhydrides	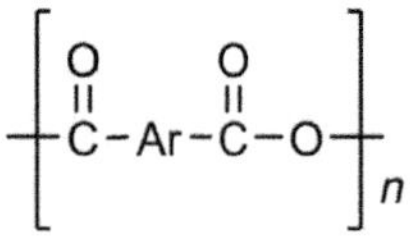 Ar = aromatic moiety	P(CPH)	Discs of zinc insulin-impregnated P(CPH) for controlled delivery	[31]

TABLE 10.1 Different Classes of Polyanhydrides—Cont'd

S. No.	Class	Backbone	Examples	Drug Delivery Systems	References
4	Aliphatic-aromatic polyanhydrides	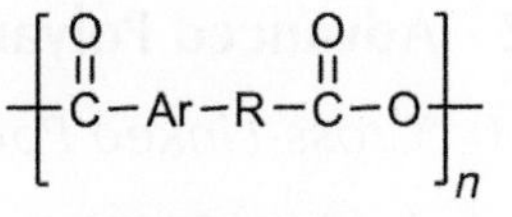 Ar = aromatic moiety	P(CPP-SA) copolymer	Human serum albumin-loaded P(CPP-SA) microspheres	[32]
5	Polyanhydride blends	Blends with polyanhydrides or other polyesters or polycarbonates	PSA blend PLA	Delivery of ofloxacin using PLA-PSA blend for the treatment of bone infection	[33]
6	Fatty acid-based polyanhydrides	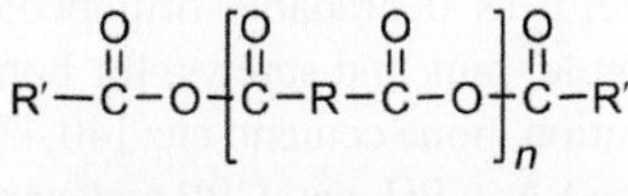 R = aliphatic, aromatic or heterocyclic moiety R′ = fatty acid residue	P(RA-SA)	P(RA-SA) biodegradable carrier for paclitaxel	[34]
7	Amino acid-based polyanhydrides	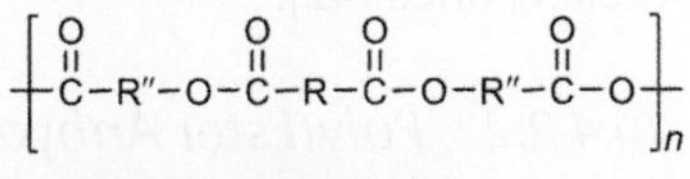 R = alkyl group or organic residue R″ = amino acid moiety	PSBAM	Nerve regeneration	[35]
8	Poly(anhydride-*co*-imide)	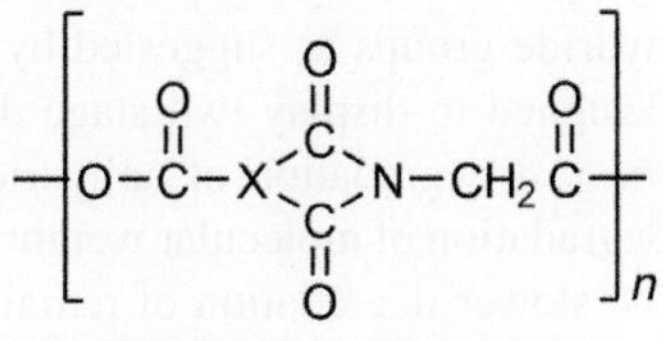 X = aliphatic or aromatic moiety	Poly(TMA-Tyr:SA:CPP)	Bovine serum albumin-loaded poly(TMA-Tyr:SA:CPP) microspheres	[36]
9	PEG containing polyanhydride	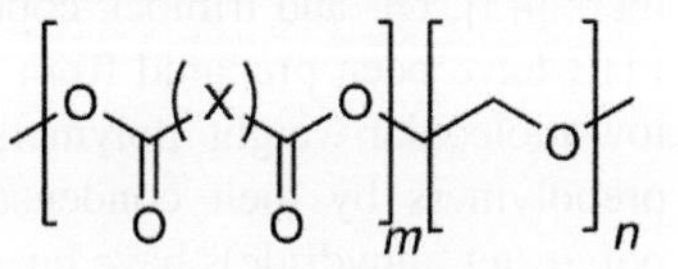 X = aliphatic or aromatic moiety	SA-CPP-PEG polyanhydride	Pulsatile delivery of parathyroid hormone	[37]
10	Photo-cross-linked polyanhydrides	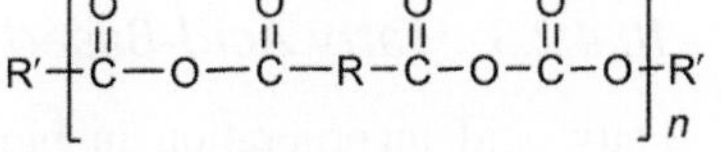 R = alkyl or aromatic chain R″ = vinyl, 2- propenyl or methacrylate group	MSA [R = $(CH_2)_8$] MCPH R = 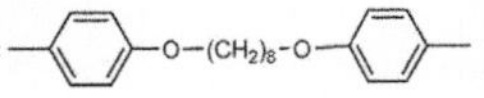	DNA delivery from photo-cross-linked MSA, MCPH, and their copolymers	[38]

n = number of monomer units.
Abbreviations—PSA, poly(sebacic acid); PAA, poly(adipic acid); P(CPH), 1,3 bis(*p*-carboxyphenoxyhexane); P(CPP-SA), poly(1,3 bis(*p*-carboxyphenoxypropane-sebacic acid) copolymer; PLA, poly(D,L-lactide); Poly(TMA-Tyr:SA:CPP), tyrosine-containing poly(anhydride-*co*-imide); P(RA-SA), poly(ricinoleic acid-sebacic acid) copolymer; PBSAM, *N*,*N*′-bis(L-alanine)-sebacoylamide; MSA, methacrylated sebacic acid; MCPH, methacrylated 1,3 bis(*p*-carboxyphenoxyhexane).

10.4.1 Conventional Polyanhydrides

10.4.1.1 Aliphatic Polyanhydrides

Aliphatic polyanhydrides synthesized from saturated diacid monomers are crystalline, melt at temperatures below 100 °C, and are soluble in chlorinated hydrocarbons. They are degraded and eliminated from the body within weeks [39].

10.4.1.2 Unsaturated Polyanhydrides

The unsaturated homopolymers are crystalline and insoluble in common organic solvents, whereas copolymers with aliphatic diacids are less crystalline and soluble in chlorinated hydrocarbons. They also provide a means for cross-linking through the double bonds that remain intact during polymerization process and thus their mechanical properties can be improved [40].

10.4.1.3 Aromatic Polyanhydrides

Aromatic homopolyanhydrides are insoluble in common organic solvents and melt at temperatures above 200 °C [15]. These properties limit the use of dictated polymers for films and microspheres using solvent or melt technique. Fully aromatic polymers that are soluble in chlorinated hydrocarbons and melt at temperatures below 100 °C were obtained copolymerization of aromatic diacids such as isophthalic acid (IPA), terephthalic acid (TA), CPP, or CPH. But due to their aromatic nature, they possess a slow degradation profile.

10.4.1.4 Aliphatic-Aromatic Polyanhydrides

These are the copolymers of aromatic and aliphatic diacid monomers. Polyanhydrides of diacid monomers containing poly(lactic acid) (PLA) and poly(hydroxybutyrate) (PHB) are synthesized by either melt or solution with molecular weights of up to 44,600 [17,18]. They were found to be less crystalline in nature and possessed slow degradation due to their aromatic content.

10.4.1.5 Polymer Blends

The physical and mechanical properties of polyanhydrides can be altered by minor modifications. Biodegradable polymer blends of polyanhydrides and polyesters have been investigated as drug carriers [41]. A polymeric blend of poly(trimethylene carbonate) (PTMC) with poly(adipic anhydride) and the matrix of PTMC-poly(adipic anhydride) blend was found to be biocompatible in *in vitro* and *in vivo* experiments and a promising candidate for controlled drug delivery erosion with tunable erosion rate achieved by varying the proportion of PTMC and poly(adipic anhydride) [42]. In general, polyanhydrides of different structures form uniform blends with a single melting temperature. Low-molecular-weight PLA, PHB, and poly(caprolactone) (PCL) are miscible with polyanhydrides, while high-molecular-weight polyesters (MW > 10,000) are not compatible with polyanhydrides.

10.4.2 Advanced Polyanhydrides

10.4.2.1 Cross-Linked Polyanhydrides

Cross-linked polyanhydrides are in the form of a three-dimensional network and have been developed for high mechanical strength and slow degradation. The photo-cross-linked polyanhydrides are prepared from the monomers having anhydride bonds and unsaturated endcaps, e.g., vinyl or 2-propenyl groups. These polyanhydrides are useful as degradable orthopedic fixation devices, for example, pins and screws, for bone augmentation and regeneration, bone cement, etc. [40,43,44]. Anhydride monomers like SA, CPH, and CPP endcapped with methacrylate functionalities with degradation rate varying from 2 days for SA to 1 year for CPH by varying network composition have been reported [42].

10.4.2.2 Poly(Ester Anhydride) Polymers

These polymers are one of the modifications of polyanhydrides whereby the polymer contains both ester and anhydride groups as suggested by the name itself. These are designed to display two-stage degradation profile wherein the fast degradation of anhydride linkages allows a rapid degradation of molecular weight of the polymer followed by the slower degradation of remaining oligomers, the rate of which is governed by the composition of polyester prepolymers [45]. Di- and triblock copolymers of PCL, PLA, and PHB have been prepared from carboxylic acid-terminated low-molecular-weight polymers copolymerized with SA prepolymers by melt condensation. Salicylic acid-based poly(ester anhydride)s have been reported to stimulate new bone formation [12].

10.4.2.3 Fatty Acid-Based Polyanhydrides

Fatty acid incorporation in biodegradable polymers provides flexibility, low melting temperature, hydrophobicity, and pliability. It degrades into naturally occurring compounds and thus is environment-friendly. Fatty acids have been incorporated in polymers as monomers using carboxylic acid functionality. Most fatty acids that are monofunctional in nature act as chain terminator in polymerization. Dimerization of unsaturated fatty acids via the unsaturation or creating a functional group on the fatty acid provides a bifunctional monomer suitable for polymerization. These copolymers are generally hydrophobic, soluble in common chlorinated organic solvents, and non- or semicrystalline; have low melting point (20-90 °C); and possess low mechanical strength. Ricinoleic acid (RA)-based polymers

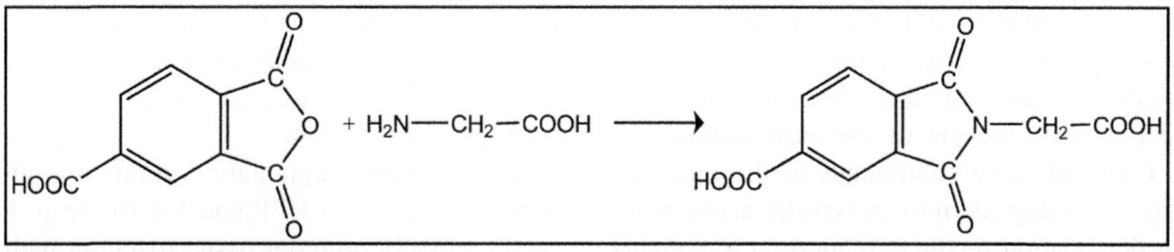

SCHEME 10.2 Synthesis of trimellitylimidoglycine.

are the newest addition to polyanhydride series. RA (cis-12-hydroxyoctadeca-9-eonoic acid) was found to be the most appropriate alternative for the synthesis of the fatty acid-based polyanhydrides. It is one of the few commercially available fatty acids that have the additional 12-hydroxy group. The advantage of RA is that it is a bifunctional fatty acid containing a hydroxyl group along the acid group and, therefore, can be incorporated into the polyanhydride backbone by the formation of an ester bond [46].

10.4.2.4 Amino Acid-Based Polyanhydrides

Linear amino acid-containing polyanhydrides were first developed in 1990 with improved physical and chemical properties [19]. They are synthesized by amidation of the amino group of an amino acid with a cyclic anhydride or by the amide coupling of two acids with a diacid chloride [47]. Introduction of amino acid in the polymer composition can elicit various functions such as functioning as carriers for controlled release of active agents or may provide a polymer whereby the polymer itself or the degradation products are the active agents. Various amino acids suitable for polymeric drug approach are natural α-amino acids like glycine, γ-amino butyric acid (brain transmitters), oligopeptides as peptide hormones, etc. [48].

10.4.2.5 Poly(Anhydride-co-Imide)s

These are also known as copolyimides synthesized via melt condensation polymerization and were found to exhibit good thermal resistance [47]. The imide segments in polymer backbone imparted high mechanical strength. Their insolubility in polar solvents was improved by using an imide diacid containing aliphatic-aromatic characteristic whereby the starting monomers were composed of aromatic acid anhydride and α-amino acid. Such polymers were found to undergo degradation via anhydride bond first followed by hydrolysis of imide bonds [49]. One such example of aliphatic-aromatic monomer is that obtained by trimellitic anhydride and glycine [50]. Polymers based on succinic acid, trimellitylimidoglycine, and trimellitylimidoalanine with compressive strengths in order of 50-60 MPa suitable for orthopedic applications have been reported [51]. An example of aliphatic-aromatic monomer obtained by reaction of trimellitic anhydride with glycine is given in Scheme 10.2 [50].

10.4.2.6 Polyanhydrides with Polyethylene Glycol Functionality

Polyanhydrides because of their surface erosion properties can be ideal materials for a constant rate release profile (a zero order). Furthermore, these polyanhydrides are very hydrophobic and their hydrolytic degradation may take relatively a long time, which is not suitable for pulsatile release. Hence, in order to achieve a tunable erosion kinetics, a two-component polyanhydride made of SA precursor and CPP precursor when copolymerized with polyethylene glycol (PEG) was found to retain the surface erosion of two-component polyanhydride while increasing the erosion rate due to increased hydrophilicity by PEG functionality. Relatively faster erosion rates can be achieved by adjusting PEG precursor content [37].

10.5 BIODEGRADABILITY

Biodegradation actually involves two complementary processes such as degradation and erosion. The process of degradation refers to the chain scission process during which polymer chains are cleaved to form oligomers and finally to form monomers. Degradation of polymers can take place passively by hydrolysis or actively by enzymatic reaction. Erosion designates the loss of material owing to monomers and oligomers leaving the polymer [52]. In ideal bulk erosion, material is lost from the entire polymer volume at the same time due to water penetrating the bulk. In this case, the erosion rate depends on the total amount of the material. In surface erosion, material is lost from the polymer matrix surface only. These are generally hydrophobic polymers wherein water cannot penetrate easily into the bulk. In ideal surface erosion, the erosion rate will be proportional to the surface area [53]. As polyanhydrides belong to the class of water-insoluble hydrophobic polymers, it is mandatory for these materials to degrade prior to erosion. These highly hydrophobic polyanhydrides exhibit ideal surface erosion because the rate of hydrolytic degradation at the surface will be much faster than the rate of water penetration into the bulk of the matrix [54].

Hydrolysis of the anhydride bond is base-catalyzed, and, thus, the rate of degradation of the polymer and the diffusion of oligomers and monomers formed by polymer degradation depends on the pH of the surrounding medium and solubility of these compounds in the medium. Since polyanhydrides degrade into carboxylic acids, solubility of these degradation products is more at higher pH [55]. Polyanhydrides in general degrade more rapidly in basic media than in acidic media [56]. At pH 7.4, pure poly(carboxyphenoxypropane) (P(CPP)) degrades in about 3 years and the rate of degradation increases with increase in pH, with the polymer degrading in just over 100 days at pH 10.0 [57].

Erosion of the polymer matrices depends on processes such as the rate of degradation, swelling, porosity, and ease of diffusion of oligomers and monomers from the matrices. Such erosion maintains constant surface area and hence leads to zero-order drug release [52]. Degradation of the polyanhydride, being a hydrolytically triggered process, depends on the rate of water uptake into the polymer matrix and pH of the surrounding medium. The rate of water uptake depends mainly on parameters like hydrophobicity and the crystallinity of the matrix, the porosity of the matrix, and the surface area/volume of the matrix. The higher the hydrophobicity, the lower is the water permeability of the matrix.

Type of monomers and their composition affect the mechanism and rate of degradation of polyanhydrides. All aliphatic polyanhydrides are rigid, crystalline material and their melting point increases with the monomer chain length. They usually erode fast and therefore are not much useful alone for pharmaceutical applications except some aliphatic polyanhydride such as P(FA:SA) having bioadhesive properties. Aromatic polyanhydrides are high-melting polymers and degrade slowly [15,58]. Combined properties of aliphatic and aromatic polyanhydride have been used to get the copolymer with improved mechanical characteristics and adjustable erosion times. It has been found that the initial molecular weight of polyanhydrides does not have a significant effect on the rate of degradation of the polymeric matrices [59]. Since there are no specific enzymes for anhydride bond cleavage, the degradation rate of polyanhydrides is unaffected by enzymes. Elevated temperatures can accelerate the erosion of polyanhydrides [54]. The rate of degradation of polyanhydrides can be modulated by carefully tuning one or more of the previously mentioned parameters.

10.6 BIOCOMPATIBILITY

Various biocompatibility studies reported on several polyanhydrides have shown them to be nonmutagenic and nontoxic. *In vitro* tests measuring teratogenic potential were also negative. Growth of two types of mammalian cells in tissue culture was also not affected by the polyanhydride polymers [60]; both the cellular doubling time and cellular morphology were unchanged when either bovine aorta endothelial cells or smooth muscle cells were grown directly on the polymeric substrate.

Subcutaneous implantation in rats of high doses of the 20:80 copolymer of CPP and SA for up to 8 weeks indicated relatively minimal tissue irritation with no evidence of local or systemic toxicity [61]. Since this polymer was designed to be used clinically to deliver an anticancer agent directly into the brain for the treatment of brain neoplasms, its biocompatibility in rat brain was also studied [62]. The tissue reaction of the polymer was compared to the reaction observed with two control materials used in surgery, oxidized cellulose absorbable hemostat (Surgicel®, Johnson and Johnson), and with absorbable gelatin sponge (Gelfoam®). The inflammatory reaction of the polymer was intermediate between the controls [62]. A closely related polyanhydride copolymer poly(CPP-SA) 50:50 was also implanted in rabbit brains and was found to be essentially equivalent to Gelfoam® in terms of biocompatibility evaluations [63]. In a similar study conducted in monkey brains, no abnormalities were noted in the CT scans and magnetic resonance image nor in the blood chemistry or hematology evaluations [64]. No systemic effects of the implants were observed on histological examinations of any of the tissues tested [65].

In the rabbit cornea bioassay, no evidence of inflammatory response was observed with any of the implants at any time. On an average, the bulk of the polymers disappeared completely between 7 and 14 days after the implantation [66]. In similar animal experiments in which polyanhydride matrices containing tumor angiogenic factor (TAF) were implanted in rabbit cornea, a significant vascularization response was observed without edema or white cells. Moreover, and most importantly from the biocompatibility standpoint, polymer matrices without incorporated TAF showed no adverse vascular response [67,68].

Based on the biocompatibility and safety preclinical studies carried out in rats [61,62], rabbits [63], and monkeys [64,65] reviewed here showing quite acceptability of the polyanhydrides for human use, a phase I/II clinical protocol was instituted [69]. In these clinical trials, a polyanhydride dosage form (Gliadel®) consisting of wafer polymer implants of poly(CPP-SA) 20:80 and containing the chemotherapeutic agent carmustine (BCNU) was used for the treatment of glioblastoma multiforme, a universally fatal form of brain cancer. In these studies, up to eight of these wafer implants were placed to line the surgical cavity created during the surgical debulking of the brain tumor in patients undergoing a second operation for surgical debulking of either a grade III or IV anaplastic astrocytoma. In keeping with the results of the earlier preclinical studies suggesting a lack of toxicity, no central or systemic toxicity of the treatment was observed during the course of treating

21 patients under this protocol. Phase III human clinical trials have demonstrated that site-specific delivery of BCNU from a P(CPP:SA) 20:80 wafer (Gliadel®) in patients with recurring brain cancer (glioblastoma multiforme) significantly prolongs patient survival [70]. Gliadel® finally got the approval from FDA as an adjunct therapy for the treatment of brain tumors.

10.7 APPLICATIONS

10.7.1 Drug Delivery

Polyanhydrides have been investigated as a candidate for controlled release devices for drugs treating eye disorders, chemotherapeutic agents, local anesthetics, anticoagulants, neuroactive drugs, and anticancer agents [12]. BCNU loaded in poly(CPP-SA) 20:80 wafer (Gliadel®) for treating brain tumors is approved for clinical use worldwide [70]. Septacin™ is a polyanhydride implant of a copolymer of EAD and sebacic acid in a 1:1 weight ratio that is being developed for osteomyelitis. It is a controlled release implant that contains gentamicin sulfate dispersed into a polyanhydride polymer matrix [11]. Both the marketed products are listed in Table 10.2.

10.7.2 Programmable Drug Release

Surface-eroding polymers have been used in the development of erodible polymer matrix that can release drugs or antigens in two phases with a first drug release period of 1-2 weeks and a second period lasting another week. Such systems could be beneficial for the local treatment of cancer because they allow switching from one drug to another or for vaccination to release antigens twice during a month. Loading different layers of laminated matrices with drugs, surface erosion releases the drug out of these layers one after another. Gopferich reported programmable drug release from polymeric implants that were manufactured using a combination of fast-eroding poly(1,3 bis[*p*-carboxyphenoxypropane]-*co*-sebacic acid) (p(CPP-SA) 20:80), a polyanhydride, and slow-eroding poly(D,L-lactic acid) [71].

10.7.3 Immunomodulation

Use of hydrophobic synthetic biodegradable polymeric biomaterials as immune modulators can eliminate the use of microbial-derived adjuvant that suffers from toxicity issues and other drawbacks. Moreover, these polymers can simultaneously serve as delivery devices for the antigens in the form of microspheres or nanospheres for enabling alternate routes of delivery and provide sustained release to facilitate single-dose vaccines and eliminate the need for booster shots [72]. Polyanhydride-based systems for antigen delivery have exhibited improved adjuvanticity, antigen stabilization, and enhanced immune responses [73,74]. A desirable feature of polyanhydrides as antigen carriers is the enhanced protein stability conferred by them [75,76]. Polyanhydrides are capable of stabilizing polypeptides and sustaining their release without the inclusion of potentially reactive excipients or stabilizers [37,72,77,78].

TABLE 10.2 Marketed Products of Polyanhydrides

	Gliadel®	Septacin™
		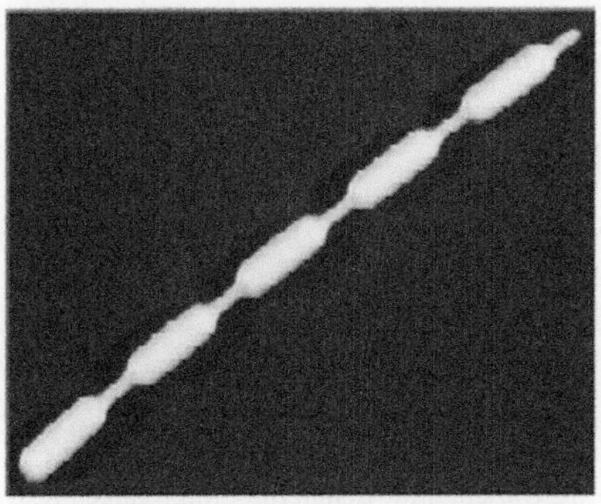
Drug	Carmustine	Gentamicin
Polymer composition	P(CPP:SA, 20:80)	P(EAD:SA, 50:50)
Therapeutic application	Glioblastoma multiforme	Osteomyelitis
Fabrication	Wafer	Beads

10.7.4 Protein Delivery

Biodegradable polymeric materials have been used successfully in protein delivery. Some of the characteristics of biodegradable carriers that can be manipulated to maintain protein stability include water swelling, hydrophobicity, and chemical nature of degradation products [79].

Since polyanhydrides are surface-eroding polymers, they have a distinct advantage for protein stabilization because they prohibit water penetration, thereby preventing proteins from denaturing due to physiological conditions. Unfortunately, increasing the aromatic content increases the hydrophobicity of the polymer, which may lead to protein aggregation along the hydrophobic domains inside the polymer matrix. One method of preventing protein destabilization is to increase the hydrophilicity of the polyanhydride. But aliphatic diacids do not reduce the hydrophobic interactions enough for stabilization or do not degrade in a useful time period [80]. A novel amphiphilic polyanhydride system has been developed based on copolymers of the anhydride monomers, CPH and 1,8-bis(*p*-carboxyphenoxy)-3,6-dioxaoctane (CPTEG), which contains oligomeric ethylene glycol. The incorporation of oligomeric ethylene glycol into the backbone of an aromatic polyanhydride creates the necessary hydrophilicity to create the amphiphilic environment needed for protein stabilization [81].

10.7.5 Tissue Engineering

Polyanhydrides have been developed into various systems with mainly bone tissue engineering applications in mind. These polymers have mechanical strength much lower than that of bone but have been combined with other polymers, such as poly(imide)s, to resolve this problem. Polyanhydrides have been developed into photo-cross-linkable systems, based on dimethacrylated anhydrides, and also injectable systems, but little interest into these polymers with regard to tissue engineering has been taken in the recent past [82].

REFERENCES

[1] M. Chasin, D. Lewis, R. Langer, Polyanhydrides for controlled drug delivery, Biopharm. Manufact. 1 (1988) 33–46.

[2] K. Mader, G. Bacic, A. Domb, H.M. Swartz, Characterization of microstructures in drug delivery systems by EPR spectroscopy, in: Proceedings of the International Symposium on Control Release of Bioactive Materials, 1995, p. 780.

[3] K.E. Uhrich, S.M. Cannizzaro, R.S. Langer, K.M. Shakesheff, Polymeric systems for controlled drug release, Chem. Rev. 99 (1999) 3181.

[4] J.E. Bucher, W.C. Slade, The anhydrides of isphthalic and terephthalic acids, J. Am. Chem. Soc. 31 (1909) 1319–1321.

[5] J.W. Hill, W.H. Carothers, Studies of polymerization and ring formation. XIV. A linear superpolyanhydride and a cyclic dimeric anhydride from sebacic acid, J. Am. Chem. Soc. 54 (1932) 1569–1579.

[6] A. Conix, Aromatic polyanhydrides, a new class of high melting fibre-forming polymers, J. Polymer Sci. 29 (1958) 343–353.

[7] N. Yoda, Synthesis of polyanhydrides. III. 1 Polyanhydrides of five-membered heterocyclic dibasic acids2, Die Makromol. Chem. 55 (1962) 174–190.

[8] N. Yoda, Syntheses of polyanhydrides. XII. Crystalline and high melting polyamidepolyanhydride of methylenebis (p-carboxyphenyl) amide, J. Polym. Sci. Part A Gen. Pap. 1 (1963) 1323–1338.

[9] H.B. Rosen, J. Chang, G.E. Wnek, R.J. Linhardt, R. Langer, Bioerodible polyanhydrides for controlled drug delivery, Biomaterials 4 (1983) 131–133.

[10] A. Olivi, Delivery of drugs to the brain by use at a sustained-release polyanhydride polymer system, New Technologies and Concepts for Reducing Drug Toxicities, CRC Press, Boca Raton, Florida, 1993, pp. 33.

[11] L. Chiu Li, J. Deng, D. Stephens, Polyanhydride implant for antibiotic delivery-from the bench to the clinic, Adv. Drug Deliv. Rev. 54 (2002) 963–986.

[12] N. Kumar, R.S. Langer, A.J. Domb, Polyanhydrides: an overview, Adv Drug Deliv Rev 54 (2002) 889–910.

[13] A.J. Domb, Polyanhydrides. EP Patent 1,339,776, 2010.

[14] A.J. Domb, R. Langer, Polyanhydrides. I. Preparation of high molecular weight polyanhydrides, J. Polym. Sci. Part A: Polym. Chem. 25 (1987) 3373–3386.

[15] A. Domb, Synthesis and characterization of biodegradable aromatic anhydride copolymers, Macromolecules 25 (1992) 12–17.

[16] A. Staubli, E. Mathiowitz, M. Lucarelli, R. Langer, Characterization of hydrolytically degradable amino acid containing poly (anhydride-co-imides), Macromolecules 24 (1991) 2283–2290.

[17] A.J. Domb, R. Langer, Solid-state and solution stability of poly (anhydrides) and poly (esters), Macromolecules 22 (1989) 2117–2122.

[18] A.J. Domb, C.F. Gallardo, R. Langer, Poly (anhydrides). 3. Poly (anhydrides) based on aliphatic-aromatic diacids, Macromolecules 22 (1989) 3200–3204.

[19] A. Staubli, E. Ron, R. Langer, Hydrolytically degradable amino acid-containing polymers, J. Am. Chem. Soc. 112 (1990) 4419–4424.

[20] A. Domb, R. Langer, Polyanhydrides: I. Preparation of high molecular weight polyanhydrides, J Polym Sci 25 (1987) 3373–3386.

[21] R. Subramanyam, A.G. Pinkus, Synthesis of poly(terephthalic anhydride) by hydrolysis of terephthaloyl chloride triethylamine intermediate adduct: characterization of intermediate adduct, J Macromol Sci Chem A22 (1985) 23.

[22] K. Leong, V. Simonte, R. Langer, Synthesis of polyanhydrides: melt-polycondensation, dehydrochlorination, and dehydrative coupling, Macromolecules 20 (1987) 705–712.

[23] A. Domb, E. Ron, R. Langer, Polyanhydrides. II. One step polymerization using phosgene or diphosgene as coupling agents, Macromolecules 21 (1988) 1925–1929.

[24] A.C. Albertsson, S. Lundmark, Synthesis of poly(adipic anhydride) by use of ketene, J. Macromol. Sci. A 25 (1988) 247–258.

[25] S. Lundmark, M. SjÖling, A.C. Albertsson, Polymerization of oxepan-2,7-dione in solution and synthesis of block copolymers of oxepan-2,7-dione and 2-oxepanone, J. Macromol. Sci. A 28 (1991) 15–29.

[26] N. Kumar, A.C. Albertsson, U. Edlund, D. Teomim, R. Aliza, A.J. Domb, Polyanhydrides, Wiley-VCH Verlag GmbH & Co. KGaA, New York, 2005.

[27] J.P. Jain, S. Modi, A.J. Domb, N. Kumar, Role of polyanhydrides as localized drug carriers, J. Control. Release 103 (2005) 541–563.

[28] N.B. Shelke, T.M. Aminabhavi, Synthesis and characterization of novel poly (sebacic anhydride-co-Pluronic F68/F127) biopolymeric

microspheres for the controlled release of nifedipine, Int. J. Pharm. 345 (2007) 51–58.
[29] A.C. Albertsson, J. Carlfors, C. Sturesson, Preparation and characterisation of poly (adipic anhydride) microspheres for ocular drug delivery, J. Appl. Polym. Sci. 62 (1996) 695–705.
[30] C.G. Thanos, Z. Liu, J. Reineke, E. Edwards, E. Mathiowitz, Improving relative bioavailability of dicumarol by reducing particle size and adding the adhesive poly (fumaric-co-sebacic) anhydride, Pharm. Res. 20 (2003) 1093–1100.
[31] E. Ron, T. Turek, E. Mathiowitz, M. Chasin, M. Hageman, R. Langer, Controlled release of polypeptides from polyanhydrides, Proc. Natl. Acad. Sci. U.S.A. 90 (1993) 4176–4180.
[32] L. Sun, S. Zhou, W. Wang, Q. Su, X. Li, J. Weng, Preparation and characterization of protein-loaded polyanhydride microspheres, J. Mater. Sci. Mater. Med. 20 (2009) 2035–2042.
[33] L. Chen, H. Wang, J. Wang, M. Chen, L. Shang, Ofloxacin-delivery system of a polyanhydride and polylactide blend used in the treatment of bone infection, J. Biomed. Mater. Res. B Appl. Biomater. 83 (2007) 589–595.
[34] A. Shikanov, B. Vaisman, M.Y. Krasko, A. Nyska, A.J. Domb, Poly (sebacic acid-co-ricinoleic acid) biodegradable carrier for paclitaxel: in vitro release and in vivo toxicity, J. Biomed. Mater. Res. A 69 (2004) 47–54.
[35] Z.Q. Zhang, X.M. Su, H.P. He, F.Q. Qu, Synthesis, characterization, and degradation of poly (anhydride-co-amide)s and their blends with polylactide, J. Polym. Sci. Part A Polym. Chem. 42 (2004) 4311–4317.
[36] M. Chiba, J. Hanes, R. Langer, Controlled protein delivery from biodegradable tyrosine-containing poly (anhydride-co-imide) microspheres, Biomaterials 18 (1997) 893–901.
[37] M.P. Torres, A.S. Determan, G.L. Anderson, S.K. Mallapragada, B. Narasimhan, Amphiphilic polyanhydrides for protein stabilization and release, Biomaterials 28 (2007) 108–116.
[38] D.J. Quick, K.K. Macdonald, K.S. Anseth, Delivering DNA from photocrosslinked, surface eroding polyanhydrides, J. Control. Release 97 (2004) 333–343.
[39] A.J. Domb, R. Nudelman, In vivo and in vitro elimination of aliphatic polyanhydrides, Biomaterials 16 (1995) 319–323.
[40] A.J. Domb, E. Mathiowitz, E. Ron, S. Giannos, R. Langer, Polyanhydrides. IV. Unsaturated and crosslinked polyanhydrides, J. Polym. Sci. Part A Polym. Chem. 29 (1991) 571–579.
[41] K.W. Leong, B.C. Brott, R. Langer, Bioerodible polyanhydrides as a drug carrier matrix, Polym. Prep. 25 (1984) 201–202.
[42] U. Edlund, A.C. Albertsson, Copolymerization and polymer blending of trimethylene carbonate and adipic anhydride for tailored drug delivery, J. Appl. Polymer Sci. 72 (1999) 227–239.
[43] K.S. Anseth, D.C. Svaldi, C.T. Laurencin, R. Langer, Photopolymerization of novel degradable networks for orthopedic applications, ACS Symposium Series, ACS Publications, 1997, pp. 189-202.
[44] K.S. Anseth, D.J. Quick, Polymerizations of multifunctional anhydride monomers to form highly crosslinked degradable networks, Macromol. Rapid Comm. 22 (2001) 564–572.
[45] R.F. Storey, D.Z. Deng, D.R. Peterson, T.P. Glancy, Poly (ester-anhydrides) and intermediates therefor. EP Patent 0,883,642, 2003.
[46] M. Sokolsky-Papkov, A. Shikanov, N. Kumar, B. Vaisman, A.J. Domb, Fatty acid based biodegradable polymers-synthesis and applications, Bull. Israel Chem. Soc. 23 (2008) 12–17.
[47] A.J. Domb, Biodegradable polymers derived from amino acids, Biomaterials 11 (1990) 686–689.
[48] E. Ron, A. Staubli, R.S. Langer, Poly (amide-and imide-co-anhydride) for biological application. Google Patents, 1991.
[49] K.E. Uhrich, T.T. Thomas, C.T. Laurencin, R. Langer, In vitro degradation characteristics of poly (anhydride-imides) containing trimellitylimidoglycine, J. Appl. Polymer Sci. 63 (1997) 1401–1411.
[50] A.J. Domb, N. Kumar, T. Sheskin, A. Bentolila, J. Slager, D. Teomim, Biodegradable polymers as drug carrier systems, in: S. Dumitriu (Ed.), second ed., Polymeric Biomaterials, Marcel Dekker, New York, 2002, pp. 91–121.
[51] K.E. Uhrich, A. Gupta, T.T. Thomas, C.T. Laurencin, R. Langer, Synthesis and characterization of degradable poly (anhydride-co-imides), Macromolecules 28 (1995) 2184–2193.
[52] J.A. Tamada, R. Langer, Erosion kinetics of hydrolytically degradable polymers, Proc. Natl. Acad. Sci. U.S.A. 90 (1993) 552–556.
[53] A. Gopferich, Mechanisms of polymer degradation and erosion, Biomaterials 17 (1996) 103–114.
[54] D.S. Katti, S. Lakshmi, R. Langer, C.T. Laurencin, Toxicity, biodegradation and elimination of polyanhydrides, Adv. Drug Deliv. Rev. 54 (2002) 933–961.
[55] L. Shieh, J. Tamada, I. Chen, J. Pang, A. Domb, R. Langer, Erosion of a new family of biodegradable polyanhydrides, J. Biomed. Mater. Res. 28 (1994) 1465–1475.
[56] K.W. Leong, B.C. Brott, R. Langer, Bioerodible polyanhydrides as drug-carrier matrices. I: characterization, degradation, and release characteristics, J. Biomed. Mater. Res. 19 (1985) 941–955.
[57] E.-S. Park, M. Maniar, J. Shah, Effects of model compounds with varying physicochemical properties on erosion of polyanhydride devices, J. Control. Release 40 (1996) 111–121.
[58] A. Domb, C. Gallardo, R. Langer, Poly(anhydrides). 3. Poly(anhydrides) based on aliphaticaromatic diacids, Macromolecules 22 (1989) 3200–3204.
[59] W. Dang, T. Daviau, P. Ying, Y. Zhao, D. Nowotnik, C.S. Clow, et al., Effects of GLIADEL® wafer initial molecular weight on the erosion of wafer and release of BCNU, J. Control. Release 42 (1996) 83–92.
[60] K.W. Leong, P.D. D'Amore, M. Marletta, R. Langer, Bioerodible polyanhydrides as drug-carrier matrices. II. Biocompatibility and chemical reactivity, J. Biomed. Mater. Res. 20 (1986) 51–64.
[61] C. Laurencin, A. Domb, C. Morris, V. Brown, M. Chasin, R. McConnell, et al., Poly(anhydride) administration in high doses in vivo: studies of biocompatibility and toxicology, J. Biomed. Mater. Res. 24 (1990) 1463–1481.
[62] R.J. Tamargo, J.I. Epstein, C.S. Reinhard, M. Chasin, H. Brem, Brain biocompatibility of a biodegradable, controlled-release polymer in rats, J. Biomed. Mater. Res. 23 (1989) 253–266.
[63] H. Brem, A. Kader, J.I. Epstein, R.J. Tamargo, A.J. Domb, R. Langer, et al., Biocompatibility of a biodegradable, controlled-release polymer in the rabbit brain, Sel. Cancer Ther. 5 (1989) 55–65.
[64] H. Brem, H. Ahn, R.J. Tamargo, M. Pinn, M. Chasin, A biodegradable polymer for intracraneal drug delivery: a radiological study in primates, J. Neurosurg. 68 (1988) 334–335.
[65] H. Brem, R.J. Tamargo, M. Pinn, M. Chasin, Biocompatibility of BCNU-loaded biodegradable polymer: a toxicity study in primates, J. Neurosurg. 68 (1988) 335–336.
[66] M. Rock, M. Green, C. Fait, R. Geil, J. Myer, M. Maniar, et al., Evaluation and comparison of biocompatibility of various classes of polyanhydrides, Polym. Preprints 32 (1991) 221–222.
[67] R. Langer, H. Brem, D. Tapper, Biocompatibility of polymeric delivery systems for macromolecules, J. Biomed. Mater. Res. 15 (1981) 267–277.

[68] R. Langer, D. Lund, K. Leong, J. Folkman, Controlled release of macromolecules: biological studies, J. Control. Release 2 (1985) 331–341.

[69] A. Domb, M. Maniar, S. Bogdansky, M. Chasin, Drug delivery to the brain using polymers, Crit. Rev. Ther. Drug Carrier Syst. 8 (1991) 1–17.

[70] H. Brem, S. Piantadosi, P.C. Burger, M. Walker, R. Selker, N.A. Vick, et al., Placebo-controlled trial of safety and efficacy of intraoperative controlled delivery by biodegradable polymers of chemotherapy for recurrent gliomas. The Polymer-brain Tumor Treatment Group, Lancet 345 (1995) 1008–1012.

[71] A. Gopferich, Bioerodible implants with programmable drug release, J. Control. Release 44 (1997) 271–281.

[72] S.K. Mallapragada, B. Narasimhan, Immunomodulatory biomaterials, Int. J. Pharm. 364 (2008) 265–271.

[73] K.W. Leong, J. Kost, E. Mathiowitz, R. Langer, Polyanhydrides for controlled release of bioactive agents, Biomaterials 7 (1986) 364–371.

[74] J. Tamada, R. Langer, The development of polyanhydrides for drug delivery applications, J. Biomater. Sci. Polym. Ed. 3 (1992) 315–353.

[75] J. Hanes, M. Chiba, R. Langer, Degradation of porous poly(anhydride-co-imide) microspheres and implications for controlled macromolecule delivery, Biomaterials 19 (1998) 163–172.

[76] M.J. Kipper, J.H. Wilson, M.J. Wannemuehler, B. Narasimhan, Single dose vaccine based on biodegradable polyanhydride microspheres can modulate immune response mechanism, J. Biomed. Mater. Res. A 76 (2006) 798–810.

[77] E. Ron, T. Turek, E. Mathiowitz, M. Chasin, M. Hageman, R. Langer, Controlled release of polypeptides from polyanhydrides, Proc. Natl. Acad. Sci. U.S.A. 90 (1993) 4176–4180.

[78] Y. Tabata, S. Gutta, R. Langer, Controlled delivery systems for proteins using polyanhydride microspheres, Pharm. Res. 10 (1993) 487–496.

[79] S.P. Schwendeman, H.R. Costantino, R.K. Gupta, R. Langer, Peptide, protein, and vaccine delivery from implantable polymeric systems: progress and challenges, in: K. Park (Ed.), Controlled Drug Delivery Challenges and Strategies, American Chemical Society, Washington, DC, 1997, pp. 229–267.

[80] B.M. Vogel, S.K. Mallapragada, Synthesis of novel biodegradable polyanhydrides containing aromatic and glycol functionality for tailoring of hydrophilicity in controlled drug delivery devices, Biomaterials 26 (2005) 721–728.

[81] M.P. Torres, B.M. Vogel, B. Narasimhan, S.K. Mallapragada, Synthesis and characterization of novel polyanhydrides with tailored erosion mechanisms, J. Biomed. Mater. Res. A 76 (2006) 102–110.

[82] M. Sokolsky-Papkov, K. Agashi, A. Olaye, K. Shakesheff, A.J. Domb, Polymer carriers for drug delivery in tissue engineering, Adv. Drug Deliv. Rev. 59 (2007) 187–206.

Chapter 11

Polyphosphazenes

Roshan James*,†, Meng Deng*,†, Sangamesh G. Kumbar*,†,‡,§, Cato T. Laurencin*,†,‡,§

*Institute for Regenerative Engineering, University of Connecticut Health Center, Farmington, Connecticut, USA

†Department of Orthopaedic Surgery, University of Connecticut Health Center, Farmington, Connecticut, USA

‡Raymond and Beverly Sackler Center for Biological, Physical and Engineering Sciences, University of Connecticut Health Center, Farmington, Connecticut, USA

§Department of Chemical, Materials and Biomolecular Engineering, University of Connecticut, Mansfield, Connecticut, USA

Chapter Outline

11.1 INTRODUCTION

Materials of both natural and synthetic origins are commonly used to interact with biological systems to achieve desired medical outcomes such as repair and regeneration of tissues and delivery of therapeutic factors. Implantable materials must be biocompatible and bioactive to ensure appropriate host response allowing it to function as expected. Among existing biomaterials, biodegradable polymers that present flexibility in chemistry and degrade into low-molecular-weight fragments have gained significant interest. The degradation products of biomaterials can be excreted or resorbed by the body [1].

Properties that need to be considered in the design of an ideal biodegradable biomaterial include (1) biocompatibility before, during, and after degradation; (2) controlled degradation that is complimentary to the regeneration process; (3) appropriate mechanical properties and maintain mechanical strength during the tissue regeneration process; (4) degradation products that is readily metabolized; and (5) processability to form constructs of complicated structures and shapes appropriate for the intended application. The incorporation of desired scaffold properties may be finely tuned by modulating parameters such as material chemistry, molecular weight, hydrophilicity/hydrophobicity, surface charge, water adsorption, degradation, and erosion mechanism.

Polyphosphazenes are inorganic-organic hybrid polymers comprising a unique synthetic polymer class. Nitrogen and phosphorus atoms alternate on the polymer backbone and are linked by alternating single and double bonds. Each phosphorus atom is substituted with two side groups R (organic or organometallic or a combination of different functional groups) [2–4]. The general structure for polyphosphazenes is shown as Figure 11.1 [5–7]. Polyphosphazenes are high-molecular-weight compounds, and their physicochemical and biological properties are largely dependent on the nature and composition of the substituted side groups [8–10]. More than 700 polyphosphazenes polymers have been reported for a variety of applications including drug delivery and tissue engineering.

This chapter on polyphosphazenes provides the reader an overview of the synthesis and side group chemistry in context to the degradation profile and biocompatibility. In addition, it reviews the medical applications developed using biodegradable polyphosphazenes specifically drug delivery matrices and tissue-engineering scaffolds.

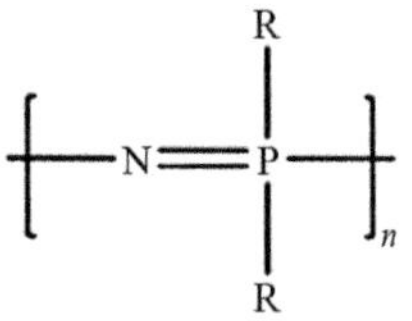

FIGURE 11.1 General structure of polyphosphazenes. R can be organic or organometallic or a combination of different functional groups.

11.2 SYNTHESIS OF POLYPHOSPHAZENES

Polyphosphazene synthesis was first attempted by H. N. Stokes in 1895 via thermal ring-opening polymerization of hexachlorocyclotriphosphazene ($[NPCl_2]_3$) (**a**; Figure 11.2), resulting in an insoluble cross-linked rubbery material susceptible to hydrolytic cleavage and thus decomposing into phosphates, ammonia, and hydrochloric acid. The instability and insolubility of this new polymer prevented its use for about seven decades [11]. The polymer (**b**, Figure 11.2) is hydrolytically unstable due to the highly reactive and polar phosphorus-chlorine bonds (P–Cl). In 1965, Allcock and Kugel reported the first successful synthesis of linear poly(dichlorophosphazene) (**b**, Figure 11.2), which was achieved by controlling the time and temperature for the thermal ring-opening polymerization of trimer **a** (Figure 11.2) [12]. This was followed up by Allcock in 1966, where they reported the synthesis of the first hydrolytically stable, soluble, and high-molecular-weight polyphosphazene by replacing the chlorine atoms with organic or organometallic nucleophiles, e.g., with alkoxide or aryloxide intermediates [13], primary or secondary amines [14], or organometallic reagents [15].

Polyphosphazenes are commonly synthesized via a two-step reaction scheme using the cyclic trimer hexachlorocyclotriphosphazene ($[NPCl_2]_3$). First step involves the synthesis of linear poly(dichlorophosphazene) (**b**, Figure 11.2) achieved via ring-opening polymerization of cyclic trimer at 250 °C [10]. An alternative approach for the synthesis of the macromolecular intermediate **b** is by living cationic polymerization of phosphoranimines at ambient temperature (Figure 11.2) [16–20]. This renders control of molecular weight and small polydispersities. The reaction scheme involves a catalyzed condensation of the monomer trichloro(trimethylsilyl)phosphoranimine, $(CH_3)_3Si{-}N{=}PCl_3$ (**c**, Figure 11.2) with loss of $(CH_3)_3SiCl$. In the second step, the reactive chlorine atoms on polymer **b** (Figure 11.2) are replaced via macromolecular substitution reaction with various organic nucleophiles as shown in Figure 11.3. This is in contrast to the direct synthesis approach used to polymerize various organic and inorganic macromolecules. Macromolecular substitution reactions allow the introduction of different side groups that govern the polymer properties in a controlled manner. In addition to single side group substitution, multiple substituents can be covalently linked to the polymer backbone. For example, substitution of two different types of side groups on the reactive intermediate polymer **b** can be achieved by sequential substitution, allowing the bulkier group to react first followed by the smaller reactive group in solution phase during substitution [10]. Secondary reactions can be used to incorporate desired functionality based on the incorporated side groups. In comparison to the conventional monomer-dependent polymerizations, macromolecular substitution is advantageous due to systematic control of polymer material properties via side group chemistry rather than monomer polymerizability. In the case of a mixed-substituent polymer having two side groups, one which is resistant to hydrolysis and the other being hydrolytically labile, the polymer erosion rate depends on the ratios of these two groups. Since many combinations of multiple side groups can be linked to the polymer backbone, it is possible to modulate the hydrolysis rate in a precisely controlled fashion. By macromolecular substitution, over 250 different side groups have been covalently linked to the backbone via replacement of the labile chlorine atoms [10,21].

11.3 BIODEGRADABLE POLYPHOSPHAZENES

11.3.1 Different Classes

Polyphosphazenes have an inherently flexible backbone due to the bonding nature of the phosphorus-nitrogen atoms in the backbone, and by varying the side group chemistry, numerous biodegradable polyphosphazenes have been synthesized. Substitution of chlorine atoms of the polymer **b** with hydrolytically labile groups such as amino acid esters (**d**, Figure 11.4) [6], imidazolyl (**e**, Figure 11.4) [22], glucosyl (**f**, Figure 11.4) [23], glyceryl (**g**, Figure 11.4) [24], and glycolate or lactate ester

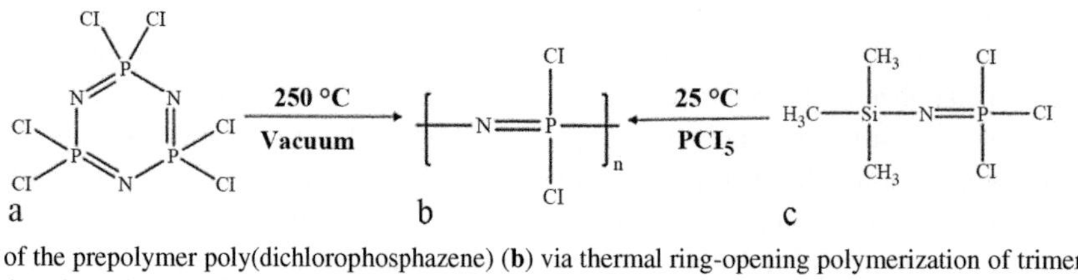

FIGURE 11.2 Synthesis of the prepolymer poly(dichlorophosphazene) (**b**) via thermal ring-opening polymerization of trimer hexachlorocyclotriphosphazene (**a**) or living cationic polymerization of a phosphoranamine monomer (**c**).

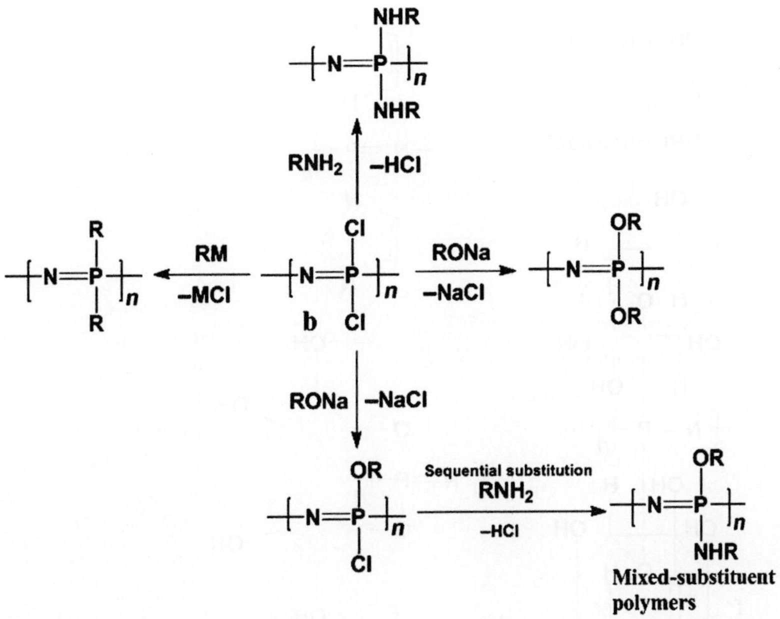

FIGURE 11.3 An illustration of the macromolecular substitution reactions that allow for a wide variety of organic side groups to be linked to the phosphorus atoms to form single-substituent polymers (simultaneous replacement of chlorine atoms) or mixed-substituent polymers (sequential substitution of chlorine atoms).

(**h**, Figure 11.4) [25] sensitizes the polymer to hydrolytic cleavage defined over a period of hours, days, months, or years at the physiological temperature depending on the molecular structure of the polymer. Table 11.1 summarizes the physicochemical properties, e.g., glass transition temperature (T_gs) of some of the representative polyphosphazenes used in biomedical applications [4–6,26–49]. Bulkier side groups result in a higher T_g and can thus be varied from −40 to 56 °C in the increasing order of PNEG < PNEA < PNEL < PNEPhA.

Aminated polyphosphazenes constitute the largest and most researched category of biodegradable polyphosphazenes and were first reported by Allcock and Kugel in 1966 [14]. Polyphosphazenes substituted with amino acid ester (**d**, Figure 11.4) and imidazole (**e**, Figure 11.4) side groups have great potential as biocompatible materials due to their sensitivity to hydrolysis leading to nontoxic degradation products. Allcock *et al.* synthesized a wide spectrum of amino acid ester-substituted polyphosphazenes as new biomaterials [5,6], and degradation rate of poly[(amino acid ester)phosphazenes] was accelerated by incorporating side groups with hydrolytically sensitive ester functions [38]. Furthermore, pH-sensitive polyphosphazenes were synthesized by incorporating aminoacethydroxamic acid as cosubstituents to poly[(amino acid ester)phosphazenes] [45]. More recently, polyphosphazenes substituted with glycylglycine dipeptides and those cosubstituted with the amino acid ester/purine or pyrimidine polyphosphazenes have been designed and synthesized to render excellent degradability and hydrogen bonding-forming ability [50,51]. The imidazole-substituted polymer (**e**, Figure 11.4) was found to be the most hydrolytically sensitive among the synthesized aminated phosphazenes. Furthermore, the hydrolysis rate of imidazole-substituted polyphosphazenes could be modulated by cosubstituting it with less hydrolytic sensitive groups such as *p*-methylphenoxy on the polymer backbone (**i** and **j**, Figure 11.4) [14]. Polyphosphazenes cosubstituted with glycylglycine dipeptide and hydrophobic phenylphenoxy group have been recently synthesized to render controlled degradability and hydrogen bonding-forming ability (**k**, Figure 11.4) [50,52].

In addition to the class of aminated phosphazenes, some alkoxy-substituted polyphosphazenes that are hydrolytically labile were synthesized such as (glyceryl-substituted polyphosphazene) (**g**, Figure 11.4) [4,23–25]. Hydrogels of this polymer were further synthesized by cross-linking with agents such as adipoyl chloride and hexamethylene diisocyanate. Other alkoxy-substituted polyphosphazene incorporating glycolic acid ester and lactic acid ester side groups (**h**, Figure 11.4) have been synthesized [25]. These polymers are not crystalline and hydrolyze much faster than the commonly used biodegradable polyesters such as polylactide and polyglycolide. Recently, Deng *et al.* developed a biodegradable etheric polyphosphazene, poly[(ethyl glycinato)$_1$ (methoxyethoxyethoxy)$_1$phosphazene] (PNEGMEEP), as a bioactive material for orthopedic applications [53].

11.3.2 Degradation Mechanisms

During polymer degradation, water molecules attack the organic side groups on the polyphosphazene. After the removal of the side groups from the polymer backbone, P-OH units

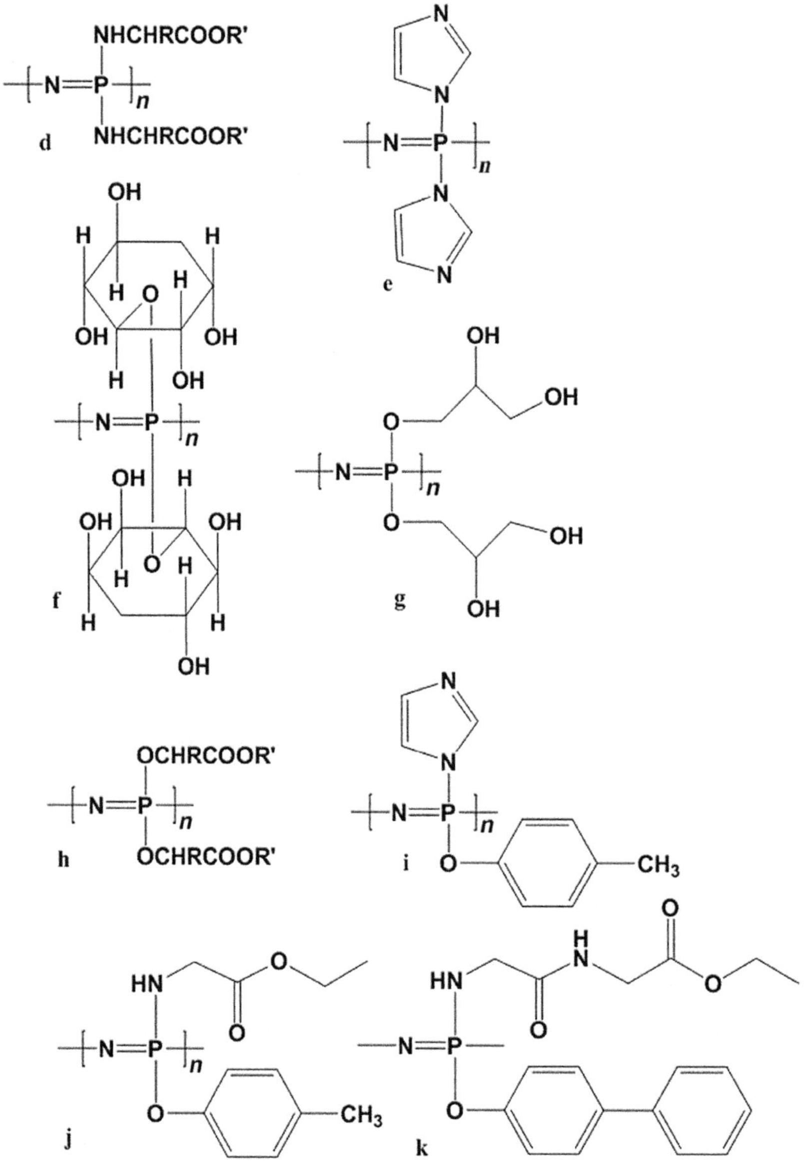

FIGURE 11.4 Structures of various biodegradable polyphosphazenes by substituting the chlorine atoms of poly(dichlorophosphazene) (**b**) with hydrolytically labile groups such as amino acid esters (**d**), imidazolyl (**e**), glucosyl (**f**), glyceryl (**g**), and glycolate or lactate esters (**h**), to name a few. Polymers (**i–k**) are representative biodegradable polyphosphazenes.

are formed followed by the migration of the proton from oxygen to nitrogen, which sensitizes the polymer backbone to hydrolysis. The polymer ultimately degrades into nontoxic degradation products comprising mainly of ammonia, phosphoric acid, and the corresponding side groups [33,48,54]. A generalized degradation scheme for degradable polyphosphazene is presented in Figure 11.5. In the case of a mixed-substituent polymer, the most hydrolytically sensitive group is cleaved first from the polyphosphazene backbone and is followed by bulkier side groups such as *p*-methylphenoxy, *p*-phenylphenoxy, tyrosine, or pyrrolidone [4–7,32–42,44,45,47–49]. The degradation products such as ammonium phosphate constitute a natural buffer system in the local microenvironment both *in vitro* and *in vivo* [33,48,54].

Allcock *et al.* studied the influence of ester groups and α-substituents on the hydrolytic sensitivity of amino acid ester-substituted polyphosphazenes [6]. Methyl, ethyl, *tert*-butyl, and benzyl esters of glycine-, alanine-, valine-, and phenylalanine-substituted polyphosphazenes were incubated in deionized water, and the molecular weight change of these polymers was followed by gel permeation chromatography analysis. The hydrolytic sensitivity of the polymers was found to increase as the size of the ester groups decreased. The hydrolytic sensitivity increased in the order

TABLE 11.1 Summary of glass transition temperatures (T_g's) of single-substituent polyphosphazenes illustrating the great tunability of physico-chemical properties by modulating side group chemistry. The selected polymers are poly[bis(ethyl glycinato)phosphazene] (PNEG), poly[bis(ethyl alanato)phosphazene] (PNEA), poly[bis(ethyl leucinato) phosphazene] (PNEL), poly[bis(ethyl phenylalaninato)phosphazene] (PNEPhA), and poly[bis(glycyl-ethylglycinato) phosphazene] (PNGEG).

Polyphosphazene Polymer	Side Group	T_g (°C)
PNEG	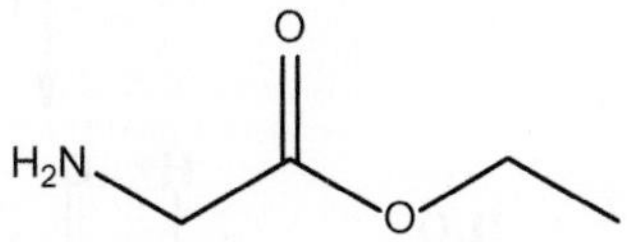	−40
PNEA	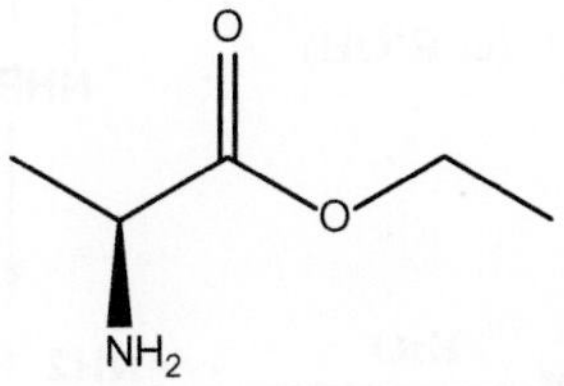	−10
PNEL	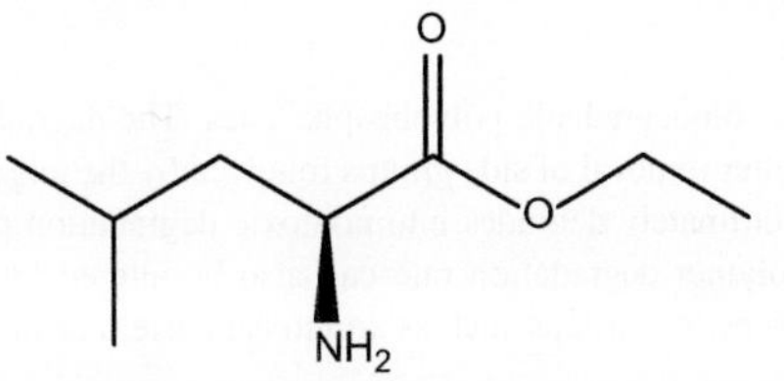	15
PNEPhA	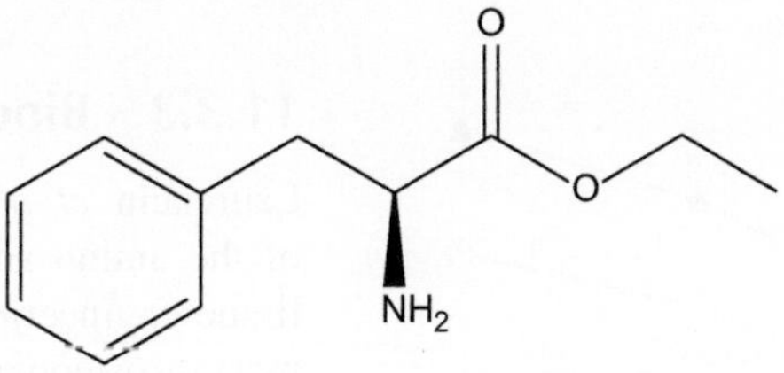	42
PNGEG	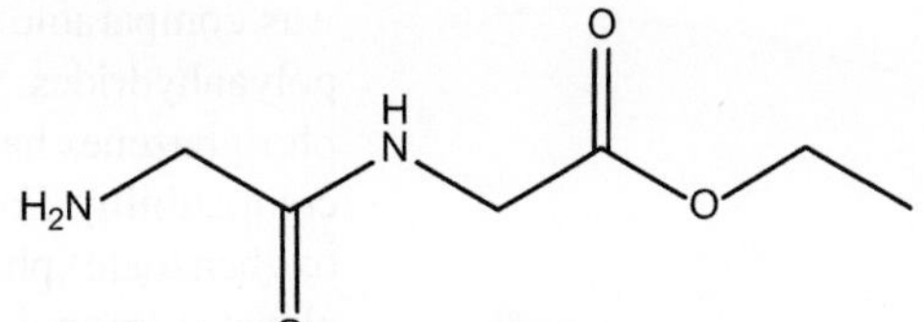	56

of benzyl esters < *tert*-butyl esters < ethyl esters < methyl esters. Similarly, the smaller the group linked to the α-carbon atom of the amino acid residue, the more hydrolytically sensitive the polymer was. Thus, phenylalanato units were least sensitive, and susceptibility to hydrolysis increased in order of phenylalanato units < valinato units < alaninato groups < glycinato side groups. The degradation products are phosphates, amino acid, alcohol, and ammonia. The physical properties of these polymers allow for facile processing and retention of dimensional stability at physiological temperatures, making them prospective biomaterials for short-term medical applications such as drug delivery matrices and tissue-engineering scaffolds. Laurencin *et al.* demonstrated that the degradation rate of ethyl glycinato-substituted polyphosphazene could be efficiently modulated by incorporating a hydrophobic cosubstituent, such as a methylphenoxy group (**j**, Figure 11.4) [42]. Figure 11.6 shows the percentage mass loss versus degradation time of poly[(ethyl glycinato)(*p*-methylphenoxy)phosphazene] of different compositions indicating increased degradation rate of the polymer with an increase in the percentage of the ethyl glycinato group. Various studies have demonstrated

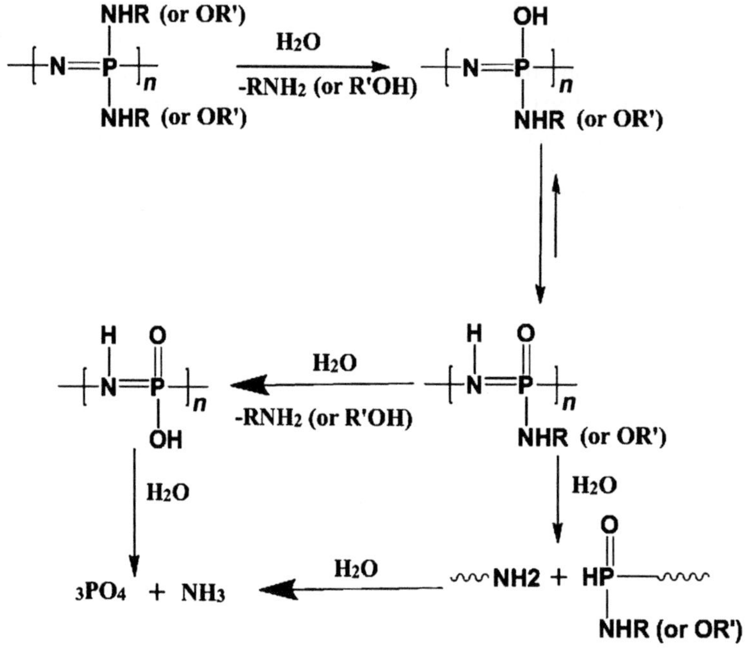

FIGURE 11.5 A general degradation pathway for biodegradable polyphosphazenes. The degradation is initiated by the water molecule attack on the organic side groups. The formation of P-OH units after removal of side groups followed by the migration of the proton from oxygen to nitrogen sensitizes the polymer skeleton to hydrolysis. The polymer ultimately degrades into nontoxic degradation products comprising mainly of ammonium phosphate buffer and the corresponding side groups. The polymer degradation rate can also be altered by modulating the side group chemistry. For example, polyphosphazenes substituted with hydrolytically sensitive groups such as amino acid esters or imidazolyl degraded faster than the bulkier hydrophobic groups-substituted polymers.

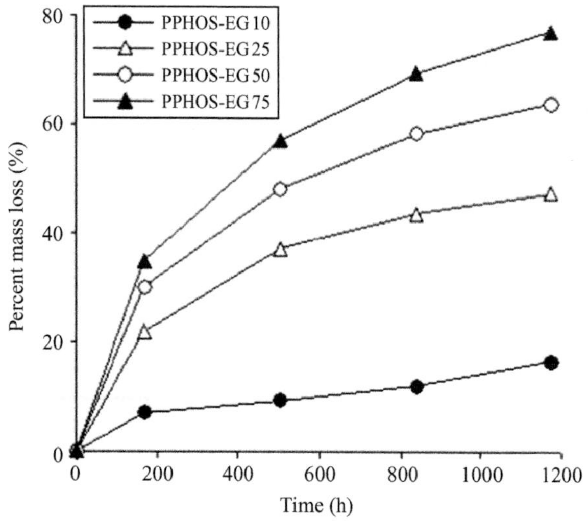

FIGURE 11.6 Percentage mass loss versus degradation time (hours) of ethyl glycinato-substituted polyphosphazenes of different compositions in 0.1 M sodium phosphate buffer at 37 °C and pH 7.4. The standard deviation (SD) of the mean is less than 3%. Reprinted from Ref. [42] with permission from John Wiley and Sons.

the versatility of the polyphosphazene synthesis process and the ability to have efficient control over polymer degradation by changing side group chemistry to suit the requirements of a specific biomedical application [42,48].

11.3.3 Biocompatibility

Laurencin *et al.* first characterized the biocompatibility of the amino acid ester-substituted polyphosphazenes for tissue-engineering applications [55]. The adhesion of primary rat osteoblasts (PRO) on biodegradable PNEG scaffolds was comparable to poly(lactide-*co*-glycolide) (PLGA) and polyanhydrides. Since then, several amino acid ester polyphosphazenes have been investigated for their *in vitro* cytocompatibility. For example, both poly[(ethyl alanato)$_1$(ethyl oxybenzoate)$_1$phosphazene] (PNEAEOB) and poly[(ethyl alanato)$_1$(propyl oxybenzoate)$_1$phosphazene] (PNEAPOB) supported the adhesion and proliferation of PRO cells and were found to be suitable as reinforcing materials for self-setting bone cements [56]. Cytotoxicity studies were conducted on Swiss 3T3 and HepG2 cell lines using the degradation solution of poly[bis(ethyl 4-aminobutyro)phosphazene] [40]. It did not affect the proliferation of the cell lines studied indicating the cytocompatibility of the novel polymer and its potential for *in vivo* applications [40]. Early studies focused on characterizing *in vivo* tissue compatibility of these amino acid ester-substituted polyphosphazenes. Alanine ethyl ester-substituted polyphosphazenes, namely, PNEA, PNEAmPh, and PNEAPhPh, were characterized using a rat subcutaneous implantation model [47]. These polymer formulations initially elicited a mild to moderate immune response that subsided over time. After 12 weeks, a minimal immune response was found for PNEAmPh and PNEAPhPh evidenced by the presence of few neutrophils, erythrocytes, and lymphocytes. Laurencin *et al.* showed that PNEGmPh-based bone grafts supported bone growth in New Zealand white rabbits when compared to PLGA grafts [57]. Histological evaluations revealed the presence of a thin layer of fibrous tissue at the implant interface and the adjacent bone at 1 week. After 12 weeks of implantation, the interface contained lamellar bone and showed the presence of fat and bone marrow cells adjacent to the implant, whereas PLGA implants were surrounded by a thin fibrous membrane that consisted of a few giant cells, occasional lymphocytes, and plasma cells. The superior biocompatibility of these novel polymers was further supported by the absence of implant fragmentation, fat necrosis, or granuloma formation [57]. Therefore, biodegradable polyphosphazenes are attractive candidate biomaterials for biomedical applications such as tissue engineering and drug delivery.

11.4 APPLICATIONS OF BIODEGRADABLE POLYPHOSPHAZENES IN TISSUE ENGINEERING

One of the most devastating and costly problems in health care is the loss of tissue and/or organ failure. The increasing demand for biologically compatible donor tissue and organ transplants (allografts) far outstrips the availability leading to an acute shortage. The available allografts have the potential to elicit an immune response, carry risk of disease transmission, and have reduced mechanical properties. Autografts are associated with issues such as donor-site morbidity and very limited availability. To overcome the limitations associated with the current graft options, tissue engineering has emerged as a promising strategy that provides *de novo* regenerated viable tissue substitutes. Tissue engineering can be defined as the application of biological, chemical, and engineering principles toward the repair, restoration, or regeneration of tissues using cells, factors, and biomaterials alone or in combination [58]. Biodegradable scaffolds play a crucial role in scaffold-based regenerative approach and their successful application is dependent on their inherent properties, composition, and scaffold properties. Biodegradable polyphosphazenes, due to their great biocompatibility, complete biodegradability, and synthesis versatility, have become increasingly popular in tissue-engineering application.

11.4.1 Bone Tissue Regeneration

Polyphosphazene films, prepared by solvent casting of either polyphosphazene polymers or polyphosphazene blends, have been actively studied for periodontal or bone tissue engineering. Duan *et al.* prepared poly[(ethyl alanato)$_{0.33}$(ethyl glycinato)$_{0.67}$phosphazene] (PGAP) films by solvent-casting polymer solutions under controlled humidity conditions (Figure 11.7a and b) [59]. It was shown that the high surface roughness of the patterned PGAP films increased protein adsorption and apatite deposition. The enhanced surface roughness of the honeycomb-patterned PGAP films also promoted osteogenic differentiation of MC3T3-E1 cells as evidenced by increased alkaline phosphatase activity, expression of type I collagen, and calcium deposition.

PLGA is a polyester known to have suitable compressive strength and satisfactory *in vitro* cell performance, while the material composition suffers from lack of bioactivity, uncontrolled bulk degradation, and acidic degradation products. Blending PLGA with polyphosphazene will combine the advantages of each parent polymer. Amino acid ester-substituted polyphosphazenes offer a variety of functional groups that allow for controlled degradation into phosphate and ammonia, thereby constituting a natural buffer. Deng *et al.* performed a series of studies to evaluate various polyphosphazene-PLGA blends in the form of thin disks for bone tissue engineering. In one study, the etheric biodegradable polyphosphazene PNEGMEEP was synthesized and blended with PLGA at weight ratios of 25:75 and 50:50 (PNEGMEEP to PLGA ratio) using a mutual solvent approach [53]. The blend films, although not completely miscible, were able to nucleate bone-like apatite, degrade into near-neutral pH products, support PRO growth and proliferation, and enhance osteoblast phenotype expression due to the presence of polyphosphazene. In order to improve miscibility, a mixed-substituent biodegradable polyphosphazene PNGEGPhPh (**k**, Figure 11.4) was synthesized and blended with PLGA (50:50 lactide to glycolide ratio) at weight ratios of 25:75 and 50:50 [52]. Completely miscible PNGEGPhPh-PLGA blends with enhanced mechanical properties were obtained as a result of strong hydrogen bonding between the glycylglycine dipeptide side groups and PLGA. *In vitro* studies show significantly higher osteoblast growth on the blend films than on the PLGA film (Figure 11.7c and d). Interestingly, the novel blend system was able to develop an *in situ* porous structure, which enabled cell infiltration and collagen tissue ingrowth throughout the void space between spheres as a result of polymer degradation [60]. The robust tissue ingrowth in the dynamic pore-forming scaffold attests to the versatility of this novel polymer platform leading to a new strategy in regenerative medicine, developing solid matrices that balance degradation with tissue formation (Figure 11.8). In addition, a wide range of miscible polyphosphazene-PLGA blend materials have been formulated by introducing extra hydrogen-bonding sites by the incorporation of dipeptide ester side groups to the polyphosphazene backbone [50]. These biocompatible and mechanically competent blend systems have shown a great promise for bone regeneration [61,62].

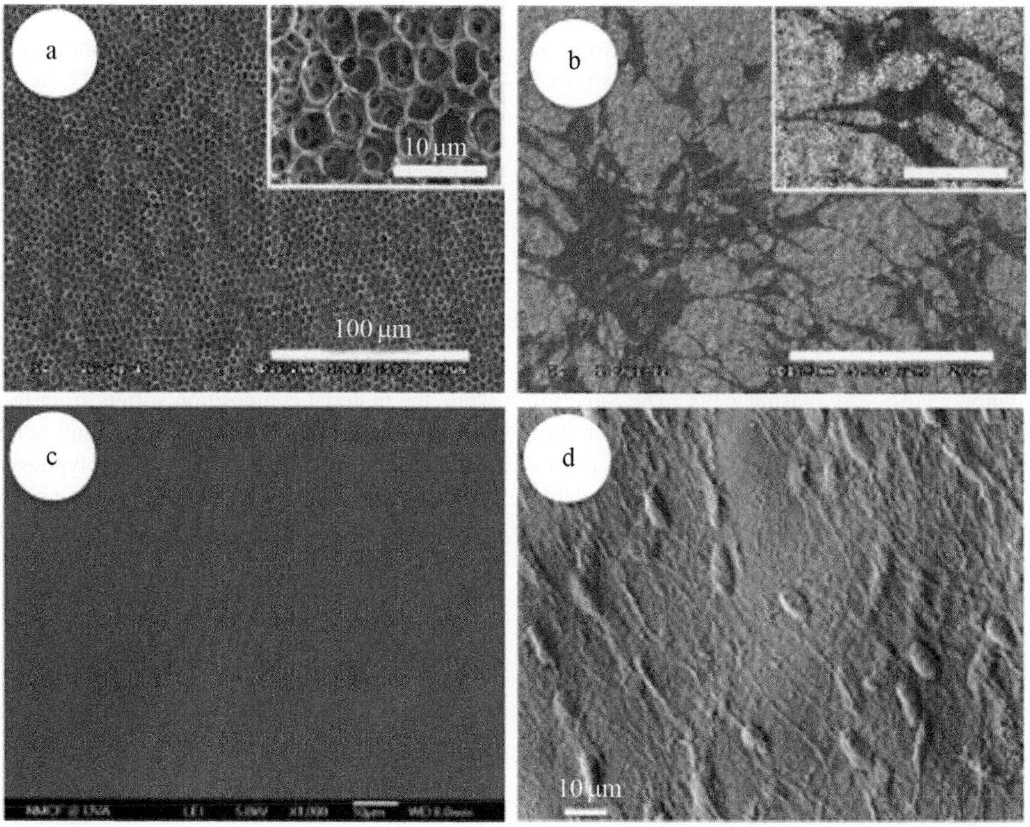

FIGURE 11.7 (a) Scanning electron micrograph of surface morphology of PGAP film with honeycomb-patterned topography; (b) SEM micrograph of MC3T3-E1 osteoblast cells on the patterned PGAP surface after 3 days of culture. (c) Surface morphology of miscible PNGEGPhPh-PLGA blends; (d) SEM micrograph of primary rat osteoblasts on PNGEGPhPh-PLGA surface after 21 days of culture. (b) Reprinted from Ref. [59] with permission from John Wiley and Sons and (d) Reprinted from Ref. [52] with permission from Elsevier.

Polymeric nanofibrous matrices due to their similarity to natural extracellular matrix (ECM) are attractive candidates as tissue-engineering scaffolds. Several polyphosphazene nanofibrous scaffolds have been successfully fabricated via an electrospinning process. Nair *et al.* fabricated poly[bis(*p*-methylphenoxy)phosphazene] nanofibrous matrices (Figure 11.9a), which supported the adhesion and proliferation of osteoblast-like MC3T3-E1 cells [44]. Bhattacharyya *et al.* fabricated poly[bis(ethyl alanato)phosphazene] (PNEA) and poly[bis(ethyl alanato) phosphazene]-nanohydroxyapatite PNEA-nHAp composite nanofibrous scaffolds for bone tissue-engineering applications [63,64]. Such polyphosphazene nanofiber structures closely mimic the ECM architecture and have shown improved cell performance over conventional scaffold architectures. Conconi *et al.* fabricated electrospun scaffolds from poly[(ethyl phenylalanato)$_{1.4}$(ethyl glycinato)$_{0.6}$phosphazene] and blends incorporating polylactic acid (PLA) and polycaprolactone (PCL) [65]. Osteoblast adhesion and growth were found to be less on electrospun polyphosphazene scaffolds than on electrospun PLA scaffolds. However, a synergic effect on cell proliferation was observed on polyphosphazene-PLA blends. Although the poor mechanical properties limit their applications to repair load-bearing defects, the inherited buffering capacity of polyphosphazenes to neutralize the acidic degradation of PLA makes this a unique system.

Inspired by the hierarchical structures that enable bone function, Deng *et al.* recently developed a mechanically competent 3D scaffold mimicking the bone marrow cavity and the lamellar structure of bone by orienting electrospun polyphosphazene-polyester blend nanofibers in a concentric manner with an open central cavity (Figure 11.9b and c) [66]. The 3D biomimetic scaffold exhibited mechanical characteristic similar to native bone. Compressive modulus of the scaffold was found to be within the range of human trabecular bone. When tuned to have desired properties, the concentric open macrostructures of nanofibers that structurally and mechanically mimic the native bone can be a potential scaffold design for accelerated bone healing.

Thermosensitive, injectable polyphosphazene hydrogels have attracted research interest in recent years. An injectable and thermosensitive poly(organophosphazene)-Arg-Gly-Asp (RGD) conjugate to enhance functionality was synthesized by covalently binding

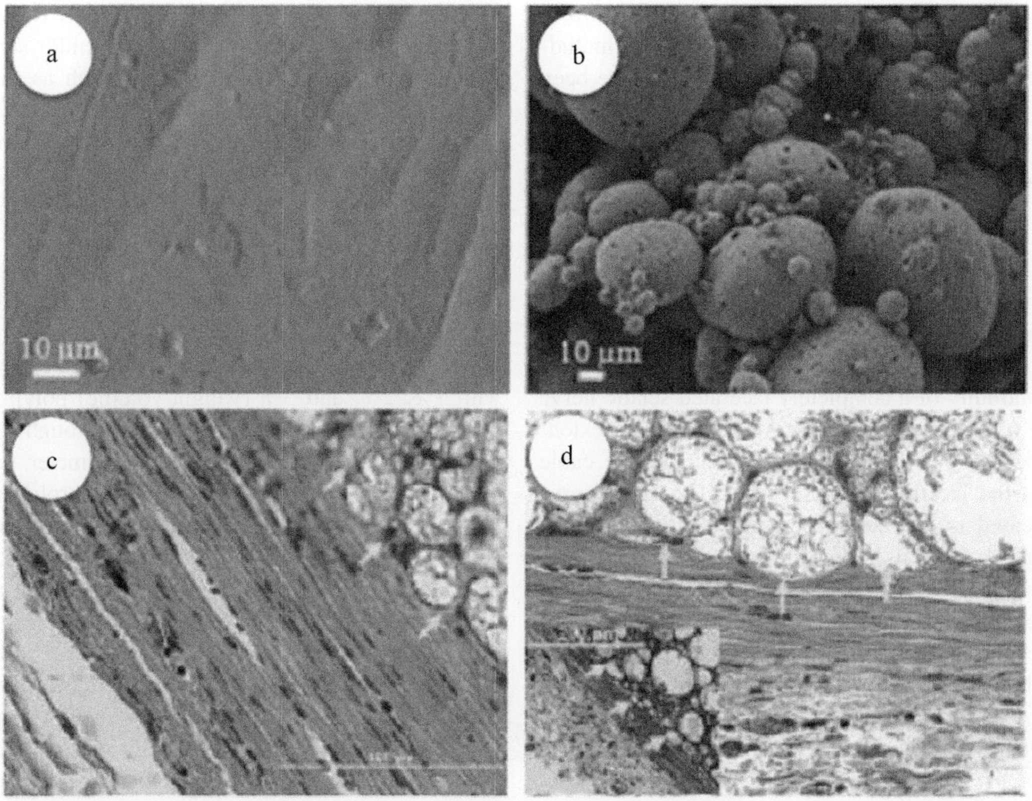

FIGURE 11.8 (a) Surface morphology of PNGEGPhPh-PLGA blend matrices; (b) surface morphology of PNGEGPhPh-PLGA blend matrices after 12 weeks of *in vitro* degradation, revealing the *in situ* formation of microspheres and interconnected porous structure; (c) and (d) histology images illustrating the formation of polymer spheres with pore system that is capable of accommodating cell infiltration and tissue ingrowth within the blend matrices at 12 weeks of implantation. (For color version of this figure, the reader is referred to the online version of this chapter.) Reprinted from Ref. [60] with permission from John Wiley and Sons.

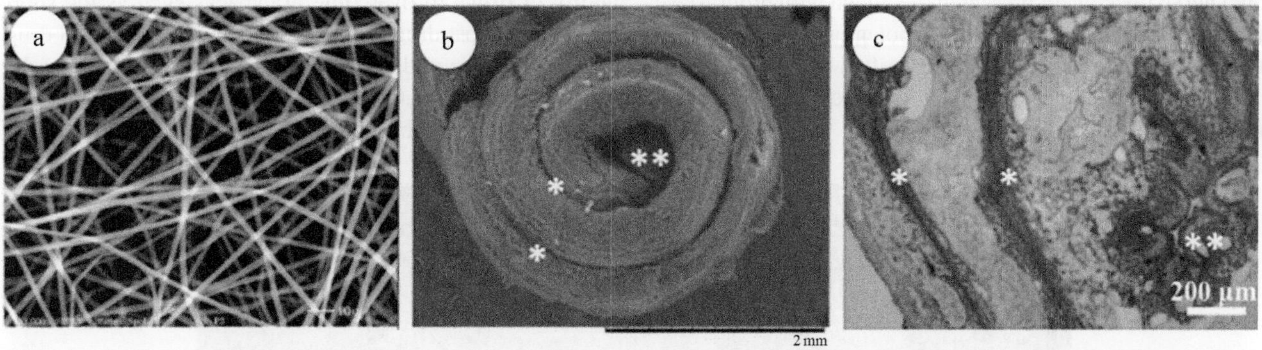

FIGURE 11.9 (a) Polyphosphazene nanofibrous scaffolds; (b) SEM image illustrating the morphologies of cell-seeded 3D biomimetic scaffolds after 28 days of culture. (c) Immunohistochemical staining for osteopontin (OPN), a prominent component of the mineralized ECM, illustrating a homogenous ECM distribution throughout the scaffold architecture at day 28. * indicates interlamellar space, whereas ** indicates central cavity. (For color version of this figure, the reader is referred to the online version of this chapter.) (b) Reprinted with permission from Ref. [44] American Chemical Society and (c) Reprinted from Ref. [66] with permission from John Wiley and Sons.

glycine-arginine-aspartic acid-glycine-serine to the carboxylic acid-terminated poly(organophosphazene) [17]. The synthesized poly(organophosphazene)-RGD conjugate was an injectable fluid at room temperature and immediately formed a hydrogel at 37 °C. Rabbit mesenchymal stem cells (rMSCs) encapsulated in the RGD-modified poly(organophosphazene) expressed significantly higher levels of osteocalcin, type I collagen, and calcium than those in poly(organophosphazene) without the RGD modification. The bioactive poly(organophosphazene)-RGD conjugate may be a suitable substrate material to induce osteogenic differentiation of MSCs for ectopic bone formation or delivering stem cells. In addition to the aforementioned matrices, several composite matrices including self-setting polyphosphazene bone cements have been developed [67–70].

11.4.2 Soft Tissue Regeneration

Besides bone repair, polyphosphazenes have shown tremendous potential for regenerating other types of tissue such as nerve [71–74], vessel [74,75], and tendon [74,76]. *In vivo* nerve repair studies were performed using tubes of poly[(ethyl alanato)$_{1.4}$(imidazolyl)$_{0.6}$phosphazene] (PEIP) to establish continuity of a completely lacerated sciatic nerve [71]. These scaffolds were found to have degraded extensively at 45 days postimplantation, leaving a tissue cable that bridged the lacerated nerve stumps. The PEIP nerve guides appeared to be more biocompatible than silicone tubes that are currently used, as significantly less scar tissue was induced and a thinner fibrotic capsule formed around these guides. In other studies, rat sciatic nerve stumps were transectioned and immediately sutured into the ends of 10 mm-long polymer tube composed of either poly[L-lactide-*co*-6-caprolactone] or PNEA. Histological analysis illustrated that the tissue cable was composed of Schwann cells and axons. Aldini *et al.* demonstrated that nerve regeneration in groups repaired with the polymer graft was comparable to autologous graft repair in Wistar rats [72]. Furthermore, the authors successfully stimulated and recorded the electrical muscle action potential in the regenerate nerve. Degradation of the polymer tube was evident at 90 and 180 days postsurgery, and myelinated and unmyelinated nerve fibers were present in the lumen with progressive increase in number. Muscle action potential between the autologous and the polymer conduit grafts was no different at 180 days, indicating that the conduits successfully permitted nerve regeneration [77].

Recent studies to augment rotator cuff repair explored the effect of fiber diameter and PNEAmPh surface functionalization on scaffold properties such as morphology, surface hydrophilicity, porosity, tensile properties, human mesenchymal stem cell (hMSC) adhesion, and proliferation. Peach *et al.* evaluated six PCL electrospun fiber matrices comprising average fiber diameters in the range of 400-500, 900-1000, 1400-1500, 1900-2000, 2900-3000, and 3900-4000 nm to augment and improve tendon tissue healing [78]. Matrices in the diameter range of 2900-3000 nm achieved the greatest proliferation of hMSCs while maintaining moderate tensile modulus and were thus dip-coated with 1%, 2%, and 3% (weight/volume) polyphosphazene solutions. Functionalization achieved through 2% polymer solutions did not affect average pore diameter, tensile modulus, suture retention strength, or cell proliferation compared to PCL controls. Functionalized PCL fiber matrices acquired a rough surface morphology and led to enhanced cell adhesion and superior cell-construct infiltration when compared to smooth PCL fiber matrices (Figure 11.10) [79]. Long-term *in vitro* hMSC cultures on both fiber matrices were able to produce clinically relevant moduli. The surface functionalization using PNEAmPh functionalization was instrumental in improving cell interactions with electrospun PCL matrices for the purpose of tendon repair.

11.4.3 Applications of Biodegradable Polyphosphazenes in Drug Delivery

Biodegradable polymer systems are widely investigated as controlled drug delivery matrices to enhance the bioavailability of drugs to overcome drawbacks such as the short half-life of most drug molecules and ensure localized drug delivery. Biodegradable polymers can circumvent the long-term biocompatibility concerns associated with many of the nondegradable polymer systems. In addition to

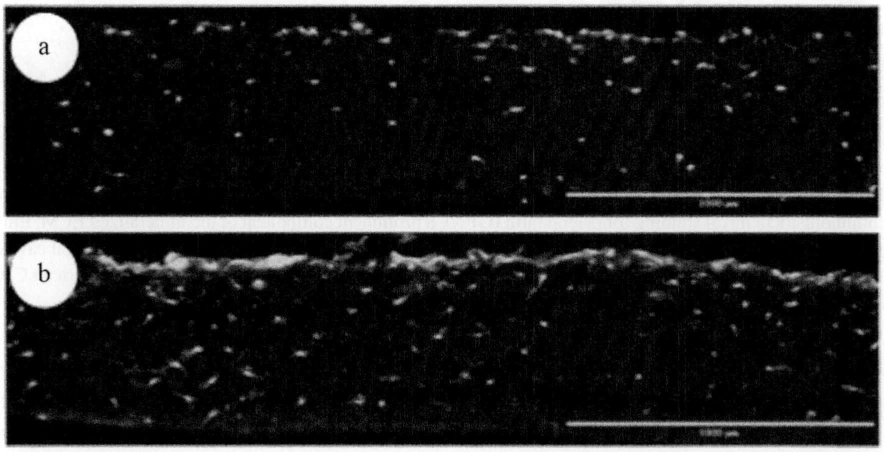

FIGURE 11.10 Permeation of hMSCs on (a) nonfunctionalized PCL and (b) PCL electrospun fibers functionalized by a 2% solution of PNEAmPh. Cells were allowed to culture for 4 weeks to observe cell infiltration. Functionalization promoted cell infiltration from the scaffold surface. Scaffold seeded at 30,000 cells per 1 × 1 cm scaffold. Scale bars = 1000 μm. (For color version of this figure, the reader is referred to the online version of this chapter.)

traditional polyesters, polyanhydrides, and poly(ortho esters), polyphosphazene polymers have recently received extensive interest as a biodegradable material platform for controlled drug release as evidenced by some of the studies discussed below.

Microspheres prepared by spray drying maintain their spherical geometry with a narrow size distribution with a mean diameter of 2-5 μm. Calceti *et al.* used suspension solvent evaporation, double emulsion-solvent evaporation, and suspension/double emulsion-solvent evaporation for the preparation of insulin-loaded polyphosphazene microspheres [80]. These preparation procedures produced spherical microparticles with a porous surface and a honeycomb internal structure (Figure 11.11).

Degradation studies using biodegradable polyphosphazenes clearly demonstrate that the polymer degradation rate can be efficiently tuned by incorporating different hydrophilic, hydrophobic, or bulky substituent groups. This

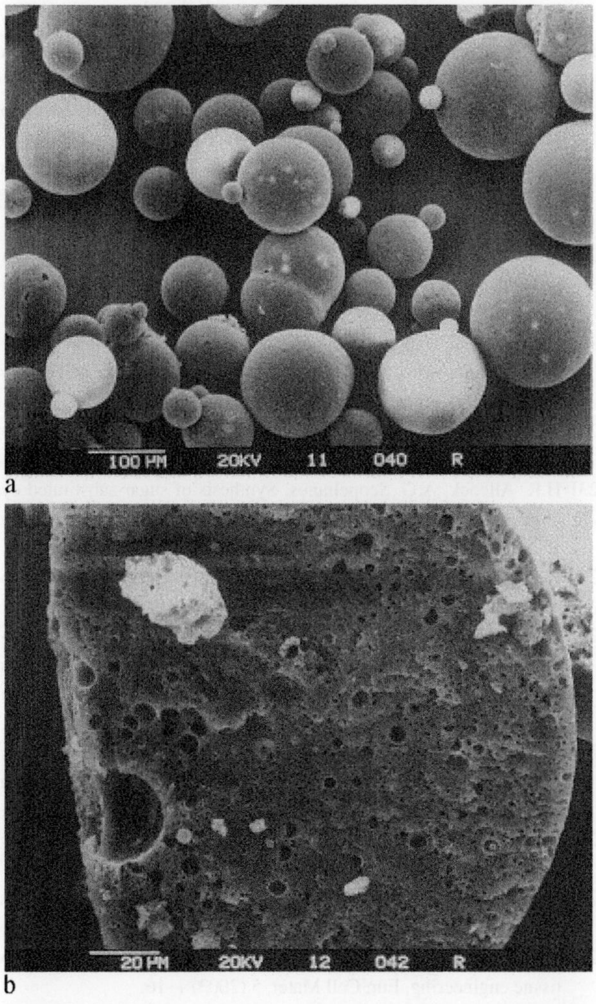

FIGURE 11.11 Scanning electron microscopy of intact (a) and (b) fractured microspheres obtained by double emulsion-solvent evaporation using 20% insulin/polymer. Reprinted from Ref. [80] with permission from Elsevier.

combined with the nontoxic degradation products of biodegradable polyphosphazenes distinguish them from other biodegradable polymers as attractive candidates for the controlled release of small and macromolecular drugs. The tunability in the degradation characteristics of these polymers is extremely versatile that one can formulate drug release systems based primarily on diffusion, erosion, or a combination of both. The first attempt in that direction was carried out by Allcock and coworkers by attaching various bioactive agents to a polyphosphazene backbone for subsequent hydrolytic release. Steroidal residues such as desoximetasone, estrone, 17β-estradiol, 17α-ethynylestradiol, estradiol 3-methyl ether, and 1,4-dihydroestradiol 3-methyl ether were attached to the polyphosphazene backbone by Allcock and Fuller via the sodium salt of the steroidal hydroxy function [32]. Hydrolytically stable polymers were obtained when the steroidal units were linked to phosphorus through an aryloxy residue, whereas linkage through an alkoxy residue led to hydrolytically labile polymers. The use of ethyl glycinato residues as cosubstituents yielded polymers that underwent chain cleavage in aqueous media. Similarly, local anesthetics such as procaine, benzocaine, chloroprocaine, butyl *p*-aminobenzoate, and 2-amino-4-picoline were attached to the polyphosphazene backbone by direct aminolysis in order to modulate the duration of biological activity of these compounds [81].

Laurencin *et al.* investigated the feasibility of imidazole-cosubstituted polyphosphazenes as a monolithic controlled drug delivery system [82]. Imidazole- and methylphenoxy-cosubstituted polyphosphazene matrices, polymer with 20%, 45%, and 80% imidazole content, were used for the *in vitro* release of small molecules such as progesterone, *p*-nitroaniline, and macromolecule such as ^{14}C-labeled bovine serum albumin. The prolonged release of low-molecular-weight molecules and macromolecules in combination with the inherent biodegradability and biocompatibility makes imidazole-substituted polymers (**i**, Figure 11.4) promising candidate matrices for developing drug delivery systems. The amino acid ester side groups confer biodegradability and biocompatibility to polyphosphazenes, while the imidazole group significantly increases the polymer degradation rate. Mixed-substituent polymers with both amino acid ester and imidazole groups have therefore been used for the controlled release of drugs to achieve desirable pharmaceutical dosages. Conforti *et al.* prepared a discoid matrix based on this mixed-substituent polymer with phenylalanine ethyl ester, imidazolyl group, and unsubstituted chlorine atoms in the side group ratio of 75:18:7 for the prolonged release of naproxen [83]. The feasibility of modulating the drug release rate was demonstrated *in vitro* by varying either the drug loading or the matrix thickness. In general, thinner disk matrices showed a faster release rate than thicker matrices. Ibim *et al.* developed a musculoskeletal delivery system for the controlled release of the anti-inflammatory agent colchicines to joints based on two types of polyphosphazenes, namely, poly[(imidazolyl)$_{0.4}$(*p*-methylphenoxy)$_{1.6}$phosphazene] (**i**, Figure 11.4, 20% imidazole-substituted) and PNEGmPh [84]. The degradation rate of PNEGmPh was much higher than that of polymer **i** (Figure 11.4) at physiological pH. Colchicine release was found to be 20% and 60% from polymer **i** (Figure 11.4) and PNEGmPh over the 21-day period of study, respectively.

11.5 CONCLUSIONS AND FUTURE TRENDS

Exploiting the synthetic flexibility of polyphosphazene, a library of biodegradable materials has been synthesized with significant prospect for tissue engineering and drug delivery. For instance, the polyphosphazene polymers have been explored as a promising matrix platform capable of controlled delivery of both small molecules and macromolecules through the mechanisms of diffusion, erosion, alone, or in combination. Successful translation of biodegradable polymer developments requires a further understanding of *in vivo* host responses to biomaterials and close collaboration among chemists, biologists, chemical engineers, material scientists, and clinicians.

ACKNOWLEDGMENTS

The authors gratefully acknowledge funding from the Raymond and Beverly Sackler Center for Biomedical, Biological, Physical and Engineering Sciences. Authors also acknowledge the funding from National Science Foundation (IIP-1311907 and EFRI-1332329).

REFERENCES

[1] L.S. Nair, C.T. Laurencin, Biodegradable polymers as biomaterials, Prog. Polym. Sci. 32 (8–9) (2007) 762–798.

[2] H.R. Allcock, Recent advances in phosphazene (phosphonitrilic) chemistry, Chem. Rev. 72 (4) (1972) 315–356.

[3] H.R. Allcock, W.J. Cook, D.P. Mack, Phosphonitrilic compounds. XV. High molecular weight poly[bis(amino)phosphazenes] and mixed-substituent poly(aminophosphazenes), Inorg. Chem. 11 (11) (1972) 2584–2590.

[4] H.R. Allcock, S.R. Pucher, Polyphosphazenes with glucosyl and methylamino, trifluoroethoxy, phenoxy, or (methoxyethoxy)ethoxy side groups, Macromolecules 24 (1) (1991) 23–34.

[5] H.R. Allcock, et al., Synthesis of poly[(amino acid alkyl ester)phosphazenes], Macromolecules 10 (4) (1977) 824–830.

[6] H.R. Allcock, S.R. Pucher, A.G. Scopelianos, Poly[(amino acid ester)phosphazenes]: synthesis, crystallinity, and hydrolytic sensitivity in solution and the solid state, Macromolecules 27 (5) (1994) 1071–1075.

[7] L.S. Nair, D.S. Katti, C.T. Laurencin, Biodegradable polyphosphazenes for drug delivery applications, Adv. Drug Deliv. Rev. 55 (4) (2003) 467–482.

[8] H.R. Allcock, Heteroatom Ring Systems and Polymers, Academic Press, New York, 1967.

[9] H.R. Allcock, Phosphorus-Nitrogen Compounds. Cyclic, Linear and High Polymeric Systems, Academic Press, New York, 1972.

[10] H.R. Allcock, Chemistry and Applications of Polyphosphazenes, Wiley Interscience, Hoboken, NJ, 2003.

[11] H.R. Allcock, Developments at the interface of inorganic, organic, and polymer chemistry, Chem. Eng. News. 63 (11) (1985) 22–36.

[12] H.R. Allcock, R.L. Kugel, Synthesis of high polymeric alkoxy- and aryloxyphosphonitriles, J. Am. Chem. Soc. 87 (18) (1965) 4216–4217.

[13] H.R. Allcock, R.L. Kugel, K.J. Valan, Phosphonitrilic compounds. VI. High molecular weight poly(alkoxy- and aryloxyphosphazenes), Inorg. Chem. 5 (10) (1966) 1709–1715.

[14] H.R. Allcock, R.L. Kugel, Phosphonitrilic compounds. VII. High molecular weight poly(diaminophosphazenes), Inorg. Chem. 5 (10) (1966) 1716–1718.

[15] H.R. Allcock, C.T.-W. Chu, Reaction of phenyllithium with poly(dichlorophosphazene), Macromolecules 12 (4) (1979) 551–555.

[16] C.H. Honeyman, et al., Ambient temperature synthesis of poly(dichlorophosphazene) with molecular weight control, J. Am. Chem. Soc. 117 (26) (1995) 7035–7036.

[17] H.R. Allcock, et al., Living cationic polymerization of phosphoranimines as an ambient temperature route to polyphosphazenes with controlled molecular weights, Macromolecules 29 (24) (1996) 7740–7747.

[18] J.M. Nelson, H.R. Allcock, Synthesis of triarmed-star polyphosphazenes via the "living" cationic polymerization of phosphoranimines at ambient temperatures, Macromolecules 30 (6) (1997) 1854–1856.

[19] H.R. Allcock, et al., Ambient-temperature direct synthesis of poly(organophosphazenes) via the living cationic polymerization of organo-substituted phosphoranimines, Macromolecules 30 (1) (1997) 50–56.

[20] H.R. Allcock, et al., Polyphosphazene block copolymers via the controlled cationic, ambient temperature polymerization of phosphoranimines, Macromolecules 30 (7) (1997) 2213–2215.

[21] H.R. Allcock, Hybrids of hybrids: nano-scale combinations of polyphosphazenes with other materials, Appl. Organometal. Chem. 24 (8) (2010) 600–607.

[22] H.R. Allcock, T.J. Fuller, K. Matsumura, Hydrolysis pathways for aminophosphazenes, Inorg. Chem. 21 (2) (1982) 515–521.

[23] H.R. Allcock, A.G. Scopelianos, Synthesis of sugar-substituted cyclic and polymeric phosphazenes and their oxidation, reduction, and acetylation reactions, Macromolecules 16 (5) (1983) 715–719.

[24] H.R. Allcock, S. Kwon, Glyceryl polyphosphazenes: synthesis, properties, and hydrolysis, Macromolecules 21 (7) (1988) 1980–1985.

[25] H.R. Allcock, S.R. Pucher, A.G. Scopelianos, Synthesis of poly(orgnaophosphazenes) with glycolic acid ester and lactic acid ester side groups: prototypes for new bioerodible polymers, Macromolecules 27 (1) (1994) 1–4.

[26] M.V. Chaubal, et al., Polyphosphates and other phosphorus-containing polymers for drug delivery applications, Crit. Rev. Ther. Drug Carrier Syst. 20 (4) (2003) 295–315.

[27] S. Cohen, et al., Design of synthetic polymeric structures for cell transplantation and tissue engineering, Clin. Mater. 13 (1–4) (1993) 3–10.

[28] M. Gleria, R. De Jaeger, Phosphazenes: A Worldwide Insight, Nova Science Publishers, New York, 2004.

[29] P.A. Gunatillake, R. Adhikari, Biodegradable synthetic polymers for tissue engineering, Eur. Cell Mater. 5 (2003) 1–16.

[30] D. Katti, C.T. Laurencin, Synthetic biomedical polymers for tissue engineering and drug delivery, in: G.O. Shonaike, S.G. Advani

(Eds.), Advanced Polymeric Materials: Structure Property Relationships, CRC Press, Boca Raton, FL, 2003, pp. 479–525.
[31] H.R. Kricheldorf, O. Nuyken, G. Swift, Handbook of Polymer Synthesis, second ed., CRC Press, Boca Raton, FL, 2004.
[32] H.R. Allcock, T.J. Fuller, Phosphazene high polymers with steroidal side groups, Macromolecules 13 (6) (1980) 1338–1345.
[33] A.K. Andrianov, A. Marin, Degradation of polyaminophosphazenes: effects of hydrolytic environment and polymer processing, Biomacromolecules 7 (5) (2006) 1581–1586.
[34] E.W. Barrett, et al., Patterning poly(organophosphazenes) for selective cell adhesion applications, Biomacromolecules 6 (3) (2005) 1689–1697.
[35] M.T. Conconi, et al., *In vitro* culture of rat neuromicrovascular endothelial cells on polymeric scaffolds, J. Biomed. Mater. Res. A 71A (4) (2004) 669–674.
[36] J. Crommen, J. Vandorpe, E. Schacht, Degradable polyphosphazenes for biomedical applications, J. Control. Release 24 (1–3) (1993) 167–180.
[37] J.H. Crommen, E.H. Schacht, E.H. Mense, Biodegradable polymers. I. Synthesis of hydrolysis-sensitive poly[(organo)phosphazenes], Biomaterials 13 (8) (1992) 511–520.
[38] J.H. Crommen, E.H. Schacht, E.H. Mense, Biodegradable polymers. II. Degradation characteristics of hydrolysis-sensitive poly[(organo) phosphazenes], Biomaterials 13 (9) (1992) 601–611.
[39] Y. Cui, et al., Novel micro-crosslinked poly(organophosphazenes) with improved mechanical properties and controllable degradation rate as potential biodegradable matrix, Biomaterials 25 (3) (2004) 451–457.
[40] M. Gümüsdereliolu, A. Gür, Synthesis, characterization, in vitro degradation and cytotoxicity of poly[bis(ethyl 4-aminobutyro)phosphazene], React. Funct. Polym. 52 (2) (2002) 71–80.
[41] C.T. Laurencin, et al., A highly porous 3-dimensional polyphosphazene polymer matrix for skeletal tissue regeneration, J. Biomed. Mater. Res. 30 (2) (1996) 133–138.
[42] C.T. Laurencin, et al., Use of polyphosphazenes for skeletal tissue regeneration, J. Biomed. Mater. Res. 27 (7) (1993) 963–973.
[43] K.Y. Lee, D.J. Mooney, Hydrogels for tissue engineering, Chem. Rev. 101 (7) (2001) 1869–1880.
[44] L.S. Nair, et al., Fabrication and optimization of methylphenoxy substituted polyphosphazene nanofibers for biomedical applications, Biomacromolecules 5 (6) (2004) 2212–2220.
[45] L.Y. Qiu, K.J. Zhu, Novel biodegradable polyphosphazenes containing glycine ethyl ester and benzyl ester of amino acethydroxamic acid as cosubstituents, J. Appl. Polym. Sci. 77 (13) (2000) 2987–2995.
[46] E. Schacht, et al., Biomedical applications of degradable polyphosphazenes, Biotechnol. Bioeng. 52 (1) (1996) 102–108.
[47] S. Sethuraman, et al., In vivo biodegradability and biocompatibility evaluation of novel alanine ester based polyphosphazenes in a rat model, J. Biomed. Mater. Res. A 77 (4) (2006) 679–687.
[48] A. Singh, et al., Effect of side group chemistry on the properties of biodegradable L-alanine cosubstituted polyphosphazenes, Biomacromolecules 7 (3) (2006) 914–918.
[49] W. Yuan, et al., Asymmetric penta-armed poly(e-caprolactone)s with short-chain phosphazene core: synthesis, characterization, and *in vitro* degradation, Polym. Int. 54 (9) (2005) 1262–1267.
[50] N.R. Krogman, et al., Miscibility of bioerodible polyphosphazene/poly(lactide-co-glycolide) blends, Biomacromolecules 8 (4) (2007) 1306–1312.
[51] N.R. Krogman, et al., Synthesis of purine- and pyrimidine-containing polyphosphazenes: physical properties and hydrolytic behavior, Macromolecules 41 (22) (2008) 8467–8472.
[52] M. Deng, et al., Dipeptide-based polyphosphazene and polyester blends for bone tissue engineering, Biomaterials 31 (18) (2010) 4898–4908.
[53] M. Deng, et al., Biomimetic, bioactive etheric polyphosphazene-poly(lactide-co-glycolide) blends for bone tissue engineering, J. Biomed. Mater. Res. A 92A (1) (2010) 114–125.
[54] S.G. Kumbar, et al., In vitro and in vivo characterization of biodegradable poly(organophosphazenes) for biomedical applications, J. Inorg. Organomet. Polym. Mater. 16 (4) (2006) 365–385.
[55] C.T. Laurencin, et al., Osteoblast culture on bioerodible polymers: studies of initial cell adhesion and spread, Polym. Adv. Tech. 3 (6) (1992) 359–364.
[56] L.S. Nair, et al., Synthesis, characterization, and osteocompatibility evaluation of novel alanine-based polyphosphazenes, J. Biomed. Mater. Res. A 76 (1) (2006) 206–213.
[57] C.T. Laurencin, et al., The biocompatibility of polyphosphazenes. Evaluation in Bone, in: 24th Annual Meeting in Conjunction with 30th International Symposium, San Diego, United States, Society for Biomaterials, 1998.
[58] R. Langer, J. Vacanti, Tissue engineering, Science 260 (5110) (1993) 920–926.
[59] S. Duan, et al., Osteocompatibility evaluation of poly(glycine ethyl ester-co-alanine ethyl ester)phosphazene with honeycomb-patterned surface topography. J. Biomed. Mater. Res. A 101 (2) (2013) 307–317, doi:10.1002/jbm.a.34282.
[60] M. Deng, et al., In situ porous structures: a unique polymer erosion mechanism in biodegradable dipeptide-based polyphosphazene and polyester blends producing matrices for regenerative engineering, Adv. Funct. Mater. 20 (17) (2010) 2794–2806.
[61] M. Deng, et al., Miscibility and in vitro osteocompatibility of biodegradable blends of poly[(ethyl alanato) (p-phenyl phenoxy) phosphazene] and poly(lactic acid-glycolic acid), Biomaterials 29 (3) (2008) 337–349.
[62] A.L. Weikel, et al., Miscibility of choline-substituted polyphosphazenes with PLGA and osteoblast activity on resulting blends, Biomaterials 31 (33) (2010) 8507–8515.
[63] S. Bhattacharyya, et al., Biodegradable polyphosphazene-nanohydroxyapatite composite nanofibers: scaffolds for bone tissue engineering, J. Biomed. Nanotechnol. 5 (1) (2009) 69–75.
[64] S. Bhattacharyya, et al., Electrospinning of poly[bis(ethyl alanato) phosphazene] nanofibers, J. Biomed. Nanotechnol. 2 (1) (2006) 36–45.
[65] M.T. Conconi, et al., Electrospun polyphosphazene nanofibers for in vitro osteoblast culture, in: A. Andrianov (Ed.), Polyphosphazenes for Biomedical Applications, Wiley-Interscience, Hoboken, NJ, 2009.
[66] M. Deng, et al., Biomimetic structures: biological implications of dipeptide-substituted polyphosphazene-polyester blend nanofiber matrices for load-bearing bone regeneration, Adv. Funct. Mater. 21 (14) (2011) 2641–2651.
[67] Y.E. Greish, et al., Low temperature formation of hydroxyapatite-poly(alkyl oxybenzoate)phosphazene composites for biomedical applications, Biomaterials 26 (1) (2005) 1–9.
[68] Y.E. Greish, et al., Composite formation from hydroxyapatite with sodium and potassium salts of polyphosphazene, J. Mater. Sci. Mater. Med. 16 (7) (2005) 613–620.

[69] Y.E. Greish, et al., Formation of hydroxyapatite-polyphosphazene polymer composites at physiologic temperature, J. Biomed. Mater. Res. A 77 (2) (2006) 416–425.

[70] Y.E. Greish, et al., Formation and properties of composites comprised of calcium-deficient hydroxyapatites and ethyl alanate polyphosphazenes, J. Mater. Sci. Mater. Med. 19 (9) (2008) 3153–3160.

[71] F. Langone, et al., Peripheral nerve repair using a poly(organo)phosphazene tubular prosthesis, Biomaterials 16 (5) (1995) 347–353.

[72] N.N. Aldini, et al., Peripheral nerve reconstruction with bioabsorbable polyphosphazene conduits, J. Bioact. Compat. Pol. 12 (1) (1997) 3–13.

[73] Q.-S. Zhang, et al., Synthesis of a novel biodegradable and electroactive polyphosphazene for biomedical application, Biomed. Mater. 4 (3) (2009) 035008.

[74] M. Deng, et al., Polyphosphazene polymers for tissue engineering: an analysis of material synthesis, characterization and applications, Soft Matter 6 (2010) 3119–3132.

[75] P. Carampin, et al., Electrospun polyphosphazene nanofibers for *in vitro* rat endothelial cells proliferation, J. Biomed. Mater. Res. A 80A (3) (2007) 661–668.

[76] D. Huang, G. Balian, A.B. Chhabra, Tendon tissue engineering and gene transfer: the future of surgical treatment, J. Hand Surg. 31 (5) (2006) 693–704.

[77] N. Nicoli Aldini, et al., Guided regeneration with resorbable conduits in experimental peripheral nerve injuries, Int. Orthop. 24 (3) (2000) 121–125.

[78] M.S. Peach, et al., Design and optimization of polyphosphazene functionalized fiber matrices for soft tissue regeneration, J. Biomed. Nanotechnol. 8 (1) (2012) 107–124.

[79] M.S. Peach, et al., Polyphosphazene functionalized polyester fiber matrices for tendon tissue engineering: in vitro evaluation with human mesenchymal stem cells, Biomed. Mater. 7 (4) (2012) 045016.

[80] P. Caliceti, F.M. Veronese, S. Lora, Polyphosphazene microspheres for insulin delivery, Int. J. Pharm. 211 (1–2) (2000) 57–65.

[81] H.R. Allcock, P.E. Austin, T.X. Neenan, Phosphazene high polymers with bioactive substituent groups: prospective anesthetic aminophosphazenes, Macromolecules 15 (3) (1982) 689–693.

[82] C.T. Laurencin, et al., Controlled release using a new bioerodible polyphosphazene matrix system, J. Biomed. Mater. Res. 21 (10) (1987) 1231–1246.

[83] A. Conforti, et al., Anti-inflammatory activity of polyphosphazene-based naproxen slow-release systems, J. Pharm. Pharmacol. 48 (5) (1996) 468–473.

[84] S.M. Ibim, et al., In vitro release of colchicine using poly(phosphazenes): the development of delivery systems for musculoskeletal use, Pharm. Dev. Technol. 3 (1) (1998) 55–62.

Chapter 12

Pseudo Poly(Amino Acids) Composed of Amino Acids Linked by Nonamide Bonds such as Esters, Imino Carbonates, and Carbonates

Kush N. Shah[1], Walid P. Qaqish[1], Yang H. Yun
Department of Biomedical Engineering, The University of Akron, Akron, Ohio, USA

Chapter Outline

12.1 INTRODUCTION

The principal rationale for developing biomaterials in the past 60 years has changed from the concept of inertness, where the materials have minimal interactions with the body, to bioactivity, in which the materials are designed to resorb, degrade, release drugs, and/or control the cellular microenvironment [1,2]. With the advancement of tissue engineering, regenerative medicine, and nanotechnology, a new set of biomaterials is required to support the challenging requirements of engineering tissues and biomedical devices that function on the atomic and molecular scales. A family of biomaterials that could meet the demanding specifications of these applications are "pseudo" poly(amino acids), which have been originally developed by Kohn and Langer [3–6]. This class of polymers is formulated by modifying amino acid(s) with chain extender(s) incorporated into their backbone by peptide and/or nonpeptide bonds (Figure 12.1). Numerous linkers are available to tune chemical and mechanical properties of the amino acid to match the desired applications of the resulting polymer, and by carefully choosing these linkers, the "pseudo" poly(amino acids) and their degradation products can be noncytotoxic [7–11].

Cells efficiently link together sequences of amino acids using peptide bonds, and the resulting proteins—upon assembling into their final tertiary or quaternary structures—have vast biological functions. The monomers of poly(amino acids) are naturally occurring, are nonimmunogenic, and can be metabolized by the host. Since proteins are the building blocks of cells and organs, the advantages of applying

[1]Made equal contributions.

Natural and Synthetic Biomedical Polymers

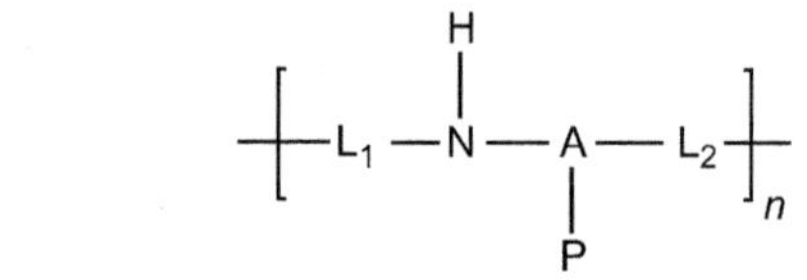

FIGURE 12.1 General scheme of "pseudo" poly(amino acids) where H–N–A is the amino acid, L_1 and L_2 are linking groups, and P is the protecting group.

poly(amino acids) to biomedical device development are readily apparent. For example, Spira and coworkers examined poly(amino acids) for sutures and skin grafts [12]. They have demonstrated that side chains of poly(amino acids) could be functionalized for controlling the physicochemical properties of the material to promoting cell adhesion and growth on scaffolds and to conjugate targeting moieties and drugs on delivery devices [2,13]. One such example is poly(L-lysine)-drug combinations of methotrexate and pepstatin use to improve therapeutic effects [13,14].

Despite their potential as biomaterials, poly(amino acids) are difficult to process and develop into biomedical devices. As an example of these challenges, the hydrogen bonds formed within the polymer result in poor mechanical properties and unpredictable swelling characteristics in a physiological setting [15]. In order to circumvent these issues, sequences of amino acids can be modified using alternating peptide and nonpeptide bonds. Use of non-peptide bonds, such as esters, urethanes, and carbonates, results in the generation of "pseudo" poly(amino acids). These polymers can be adapted for existing polymer fabrication methods such as emulsions, electrospinning, solvent casting, and injection molding. As of 2013, the Food and Drug Administration (FDA) has approved two tyrosine-derived "pseudo" poly(amino acids) as components of medical devices. TYRX, Inc. has developed AIGISRx as an antibacterial envelope for holding pacemakers or implantable cardioverter-defibrillators and PIVIT A/B™ ST for hernias [2,16]. AREVA Medical, Inc. has advanced ReZolve®2, a resorbable drug-eluting stent that is currently in clinical trials, by utilizing a tyrosine-based polycarbonate [2]. As is demonstrated by these two devices, the backbone chemistry of "pseudo" poly(amino acids) is ideal for attaching functional groups for applications in drug delivery, tissue engineering, and nanotechnology. Thus, this class of polymers could be ideal for developing solutions for the needs of the next generation of biomaterials.

12.2 SYNTHESIS OF "PSEUDO" POLY(AMINO ACID)

The amino acids that have been used for the synthesis of "pseudo" poly(amino acids), to date, are tyrosine, lysine, glycine, phenylalanine, leucine, valine, isoleucine, and methionine (Figure 12.2). The choice of amino acid depends on the desired physical properties such as hydrophobicity, tacticity, and thermal characteristics and the availability of side chains for cross-linking or functionalization, as necessitated by the desired application of the final polymer. Figure 12.3 shows the general reaction schemes for modifying and cross-linking the monomers. Amino acids such as phenylalanine, lysine, and tyrosine are primarily used for their hydrophobic natures, which are passed on to the synthesized polymer. The phenolic group on tyrosine is an excellent site for esterification including carbonate esters. Lysine, on the other hand, is primarily employed for urethane synthesis because of its widely reported cellular attachment properties. Additionally, the presence of two terminal amine groups can be used for forming diisocyanate. Thus, the choice of amino acid is crucial to tailor appropriate and desirable physicochemical properties of the final "pseudo" poly(amino acid).

12.3 ESTER-BASED "PSEUDO" POLY(AMINO ACIDS)

Esters are the most common functional groups used to synthesize "pseudo" poly(amino acids) because of their well-characterized reaction schemes for condensation reactions and the availability of the multiple reactants and catalysts. Table 12.1 summarizes ester-based "pseudo" poly(amino acids). The presence of hydroxyl (on serine, threonine,

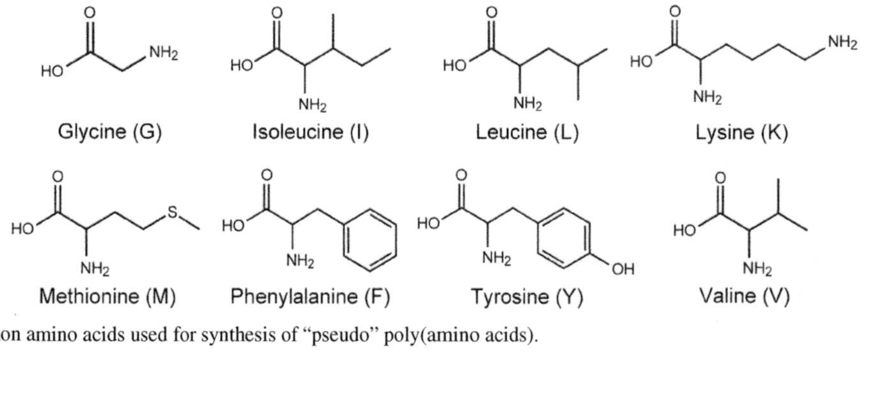

FIGURE 12.2 Common amino acids used for synthesis of "pseudo" poly(amino acids).

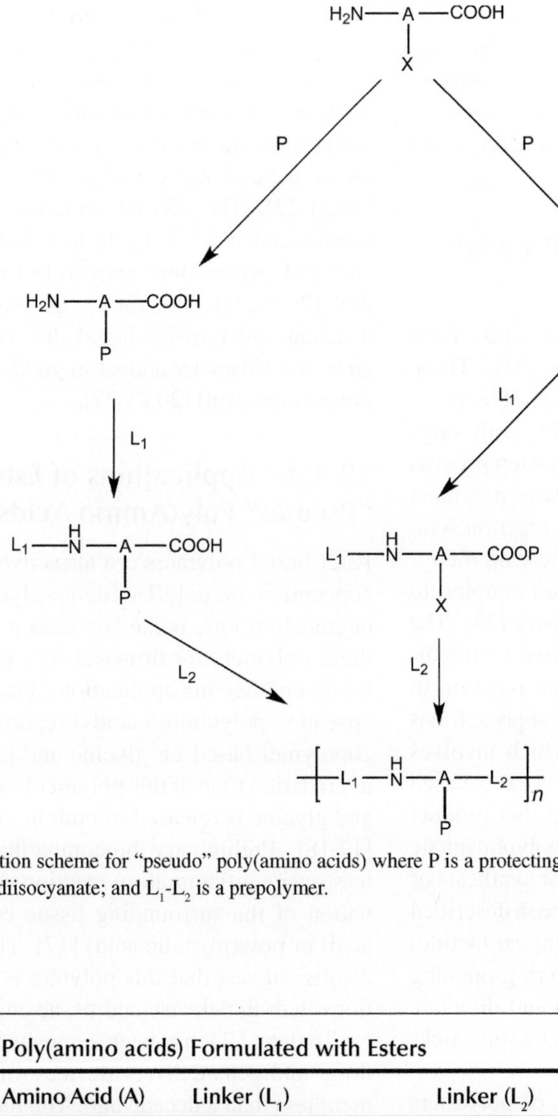

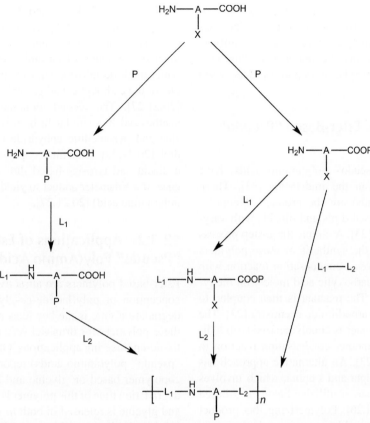

FIGURE 12.3 General reaction scheme for "pseudo" poly(amino acids) where P is a protecting group; L_1 is either diisocyanate, polyol, or diamine; L_2 is either dicarboxylic acid or diisocyanate; and L_1-L_2 is a prepolymer.

TABLE 12.1 "Pseudo" Poly(amino acids) Formulated with Esters

Investigator	Amino Acid (A)	Linker (L_1)	Linker (L_2)	Protecting Group (P)
Schakenraad *et al.* [17,18].	Glycine	Lactic acid	–	–
Won *et al.* [19]	Phenylalanine, leucine, valine, isoleucine, or methionine	1,3-Propanediol, 1,4-butanediol, or 1,6-hexanediol	Adipoyl chloride or sebacoyl chloride	–
Mallakpour and Zeraatpisheh [20,21]	Valine, leucine, tyrosine, or phenylalanine	Trimellitic anhydride	*N,N'*-(pyromellitoyl)-bis-dimethyl ester tyrosine	Methanol
Chu and Guo [22]	Phenylalanine	Succinate, adipate, or sebacate	Fumarate	–
Kohn [23]	Tyrosine	Desamino tyrosine	Succinic acid, glutaric acid, diglycolic acid, adipic acid, 3-methyl adipic acid, suberic acid, dioxaoctanedioic acid, or sebacic acid	Methanol, ethanol, isopropanol, butanol, isobutanol, sec-butanol, hexanol, octanol, dodecyl, or phenol
Sengupta and Lopina [24,25]	Tyrosine	Desamino tyrosine	Phosphate	Hexanol (on tyrosine) and ethanol (on phosphate)

and tyrosine) and carboxylic acid is ideal for these types of reactions. Common chemistry for esterification includes carbodiimide and acid-catalyzed reactions with sulfuric acid or thionyl chloride, and the majority of "pseudo" poly(amino acids) are carboxyl ester or phosphoester-based polymers.

12.3.1 Synthesis of Ester-Based "Pseudo" Poly(Amino Acids)

The first ester-based "pseudo" poly(amino acids) have been reported by Kohn in the mid-1990s [23]. These "pseudo" poly(amino acids) are the product of copolymerization of tyrosine-based diols and diacids with varying carbon chain lengths [23]. A simple three-step process has been established for the synthesis of these polymers [23]. The first step involves a condensation reaction with an alcohol to protect the carboxylic acid moiety on the tyrosine (Table 12.1) [23]. The reactant is then coupled to desamino tyrosine using carbodiimide chemistry [23]. The resulting biphenolic monomer is copolymerized with a diacid (Table 12.1) using another condensation reaction to yield the final polymer [23]. An alternative approach has been developed by Sengupta and Lopina, which involves the synthesis of a phosphoester utilizing hexyl ester-based biphenolic monomers [24,26]. Polymerizing this product by a condensation reaction yields L-tyrosine polyphosphate (LTP) (Table 12.1). The simplest approach for synthesis of ester-based "pseudo" poly(amino acids) has been described by Schakenraad *et al.* [17,18]. Poly(glycine-*co*-lactide) (Table 12.1) is synthesized with a single-step ring-opening polymerization of cyclo(glycine-lactic acid) and dilactide monomers to yield a polymer with ester and amide linkages [17,18].

Won *et al.* [19]. have reported synthesis of polyesters with valine, leucine, isoleucine, methionine, and phenylalanine (Table 12.1). This three-step process involves synthesis of a diester and a dinitro compound that are copolymerized [19]. An amino acid is first coupled with a diol (with 3, 4, or 6 methylene groups) in the presence of tosyl to yield a diester with acid salts of diamine at the terminal ends. The second monomer, di-*p*-nitrophenyl ester of carboxylic acids, is synthesized by a condensation reaction of adipoyl or sebacoyl chloride with *p*-nitro phenol. The final polymerization step involves an arduous condensation reaction in the presence of a strong proton abstractor between acid salt of bis(amino acid-alkyne diester) and di-*p*-nitrophenyl ester of dicarboxylic acids. Following along the same lines, Chu and Guo [22] have copolymerized a mixture of nitro phenyl ester of succinate, adipate, or sebacate and nitrophenyl fumarate with toluenesulfonic acid salt of phenylalanine butane-1,4-diester. The addition of fumarate derivative to the monomer mixture provides an unsaturated double bond in the polymer backbone that can be functionalized for specific biomedical applications. Mallakpour and Zeraatpisheh [20,21,27] have reported synthesis of "pseudo" poly(amino acids) using tyrosine, valine, leucine, and phenylalanine. The synthesis involves formation of an amino acid-based tertiary amide with trimellitic anhydride to yield a diacid compound, where one of the carboxylic acid groups belongs to the amino acid [20,21,27]. The second monomer for polymerization is synthesized in a similar fashion with methyl ester of tyrosine and pyromellitic anhydride to yield a tyrosine-based diol [20,21,27]. The final copolymerization step between a diacid and tyrosine-based diol is achieved in the presence of a Vilsmeier adduct to yield an ester-based "pseudo" poly(amino acid) [20,21,27].

12.3.2 Applications of Ester-Based "Pseudo" Poly(Amino Acids)

Ester-based polymers are attractive biomaterials since the conception of poly(lactide-*co*-glycolide). Control of the degradation time is the key reason for the development of these polymers for drug-delivery devices and scaffolds in tissue-engineering applications. One of the first ester-based "pseudo" poly(amino acids) reported in the literature is a copolymer based on glycine and lactic acid [17,18]. This degradation time of this polymer is approximately 10 weeks and glycine is released in both *in vitro* and *in vivo* settings [17,18]. Preliminary biocompatibility studies indicate a less severe inflammatory reaction upon histological examination of the surrounding tissue compared to poly(lactic acid) or poly(glycolic acid) [17]. These characteristics and results suggest that this polymer is promising for applications in drug delivery and tissue engineering.

To date, LTP has been extensively applied for preparing drug- and gene-delivery devices. Shortly after its development less than a decade ago, Yun has successfully prepared nanoparticles (Figure 12.4) [9,28,29] and microspheres (Figure 12.5) with this polymer and applied them in a wide

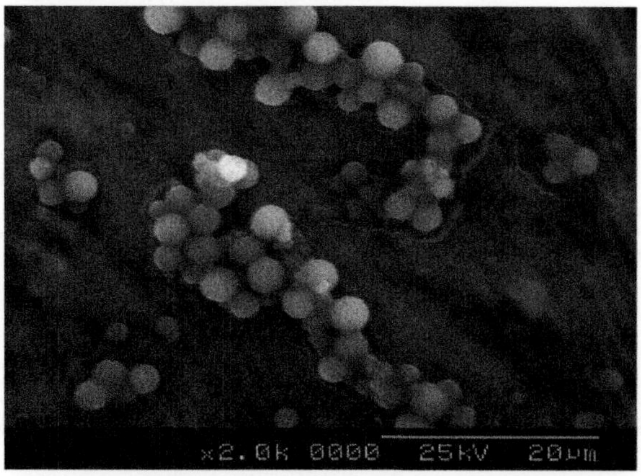

FIGURE 12.4 SEM of LTP nanoparticles.

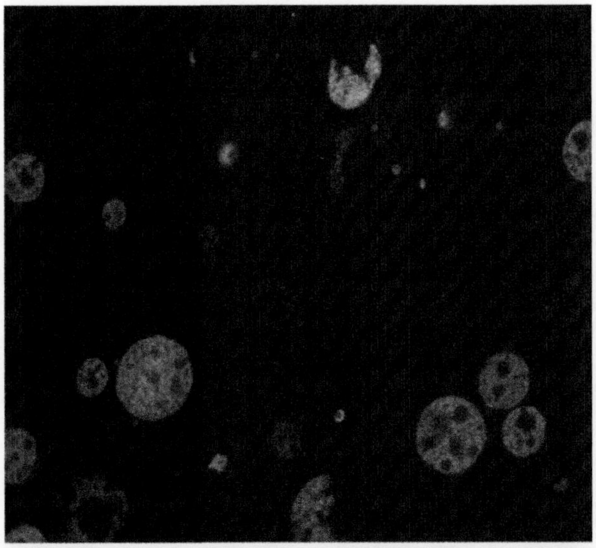

FIGURE 12.5 Confocal image of LTP microspheres. The fabrication method of water-in-oil-in-water emulsion shows internal structure that could be used to load aqueous-based drugs. (For color version of this figure, the reader is referred to the online version of this chapter.)

range of therapies because of LTP's short degradation period of 7 days [9]. Another property that makes LTP ideal for nanoparticles and other implantable applications is the nontoxic degradation products of L-tyrosine, desaminotyrosine, phosphates, and alcohols [9,29]. Furthermore, LTP nanoparticles, 9 days after injection into the rat foot pad, do not have signs of necrosis or inflammation within the tissue [7]. Only minimal leukocyte activity is observed, where these cells have normal nuclei and morphology [7]. The lack of nitric oxide production by bone marrow cells of animals injected with LTP nanoparticles also suggests a lack of innate immune response [7].

Biomedical applications of LTP nanoparticles include targeted therapy for cancers [30], bacterial infections [28], and gene therapy [7]. Folic acid-targeted LTP nanoparticles loaded with silver carbene complex 22 have shown significantly higher toxicity to HeLa cells after a 30-min exposure time when compared to cells exposed to the drug alone [30]. The LD_{50} value of the targeted nanoparticles is 12 μM, while the free drug, cisplatin, and untargeted nanoparticles show no change in toxicity [30]. Similarly, mice inoculated with *Pseudomonas aeruginosa*, a model of cystic fibrosis, and treated with silver carbine complex 10 loaded nanoparticles show a 20% survival advantage over mice treated with blank LTP nanoparticles [28]. When used in gene therapy, LTP nanoparticles loaded with plasmid DNA encoding for β-galactosidase (β-gal) successfully transfected the uterine horn in a rat model, while no transfection was observed for the sham or plasmid DNA loaded PLGA nanoparticles (Figure 12.6) [7]. This successful transfection *in vivo* opens new possibilities for applying nonviral gene delivery as therapies.

The library of poly(esters) synthesized by the Kohn uses a combinatorial approach to provide 112 polymers with varying degrees of hydrophobicity, tensile strength, flexibility, and cell proliferation efficacy [23]. This library can be used to pick and choose polymers with specific properties for specific applications. If an application has a specific requirement for hydrophobicity, this library can offer multiple polymers with varying degrees of tensile strength, flexibility, and cell-adhesion properties to suit specific needs. Although not extensively characterized for biocompatibility, these polymers follow a general trend where the absence of oxygen in the backbone shows an inverse correlation between hydrophobicity and fibroblast proliferation [23]. In contrast, all hydrophobic polymers with oxygen in the backbone supported cellular proliferation [23]. Additionally, a direct correlation can be observed between the structure and properties for these polymers, where the hydrophobicity, the contact angle, and the glass transition temperature are directly related to the length of the polymer and the degree of side-chain branching [23]. The tensile strength of the polymers is directly related to the number of methylene groups present in the backbone and/or the extent of branching [23]. Investigating a family of polymers with various chemical and physical properties can be useful for predicting cellular attachment and mechanical functions while designing scaffolds for tissue engineering.

12.4 AMIDE-BASED "PSEUDO" POLY(AMINO ACIDS)

An exception to the definition for "pseudo" poly(amino acids) as previously established is a polymer synthesized by Chen *et al.* [31,32], because its backbone is comprised of lysine alternating with an isophthaloyl group and lacks nonamide linkages (Table 12.2). Hydrophobicity of the polymer is increased by conjugating ethanol or hydrophobic amino acids, such as valine, leucine, and phenylalanine, as protecting groups on the carboxyl terminal of lysine via carbodiimide chemistry. This protected lysine is then copolymerized with isophthaloyl chloride via single-phase polymerization in the presence of a base (Table 12.2) [32].

12.4.1 Applications Poly(Amides) as "Pseudo" Poly(Amino Acids)

Poly(lysine-*co*-phthalamide) has been characterized with side chains of poly(ethylene glycol) (PEG) or Jeffamine M-1000®, a hydrophilic PEG analogue [33–35]. These polymers show minimal cytotoxicity when tested against HeLa and Chinese hamster ovarian cells, where viability is at least 60% at polymer concentrations of 2 mg/ml [34]. Poly(lysine-*co*-phthalamide) is generally nonhemolytic at physiological pH, but the incorporation of

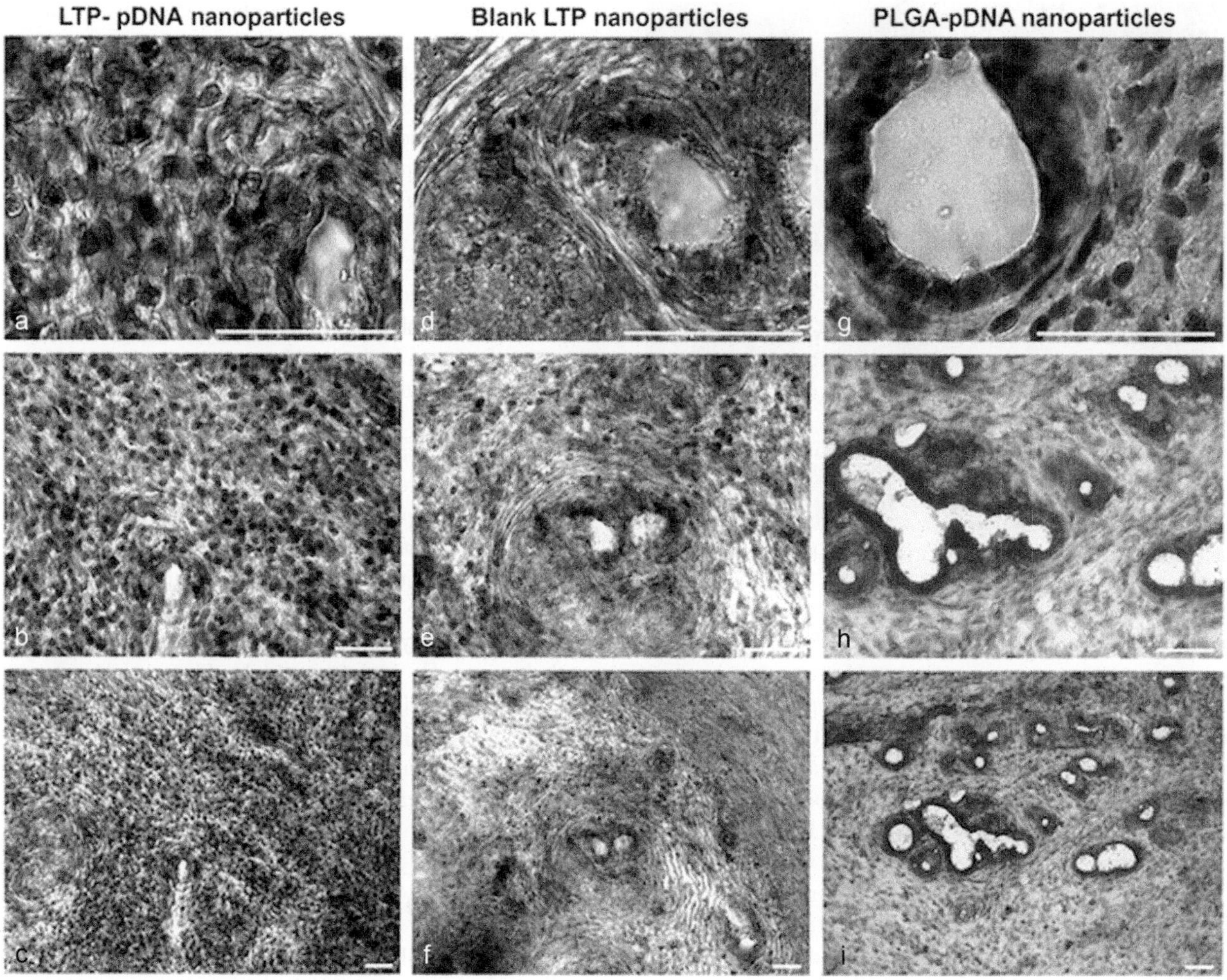

FIGURE 12.6 Representative images of rat uterus injected with (a–c) LTP-pDNA nanoparticles with 10×, 20×, and 63× objectives showing the presence of cells stained with X-gal, which is cleaved by the enzyme β-gal and is converted into a blue-colored product. The tissue injected with LTP-pDNA nanoparticles shows expression of β-gal. Rat uterus injected with (d–f) blank LTP and (g–i) PLGA-pDNA nanoparticles lack the expression of β-gal. The scale bars represent 50 μm for all images. Reproduced with permission from Ref. [7]. (For color version of this figure, the reader is referred to the online version of this chapter.)

TABLE 12.2 "Pseudo" Poly(amino acids) Formulated with Amides

Investigator	Amino Acid (A)	Linker (L_1)	Linker (L_2)	Protecting Group (P)
Chen *et al.* [31,33].	Lysine	Isophthaloyl chloride	–	Valine, leucine, phenylalanine, or ethanol

Jeffamine M-1000® increases the pH and hemolytic activities [34]. Modifying the side chains with amino acids valine, leucine, and phenylalanine imparts pH-responsive properties onto the polymer, which could be useful for many biomedical applications [36]. Polymers with PEG side chains have been developed into micelles and show enhanced uptake in spheroids formed with HeLa cells [37]. These polymers have also been conjugated to fluorescent dyes, which could be used for drug-tracking applications [38,39].

12.5 CARBONATE-BASED "PSEUDO" POLY(AMINO ACIDS)

Carbonates (ester of carbonic acid) are functional groups in which the carbon atom is surrounded by three oxygen atoms. This structure consists of two single and one double bonds. Carbonate-based "pseudo" poly(amino acids) are the first bioactive polymers to be synthesized and have been reported initially by Kohn [40,41]. The molecular weights of these polymers range from 120 to 450 kDa [40,41]. Their main degradation products include alcohols, the amino acid tyrosine, and the tyrosine analogue, desamino tyrosine [40]. Furthermore, *in vivo* studies suggest that the degradation products from poly(DTE carbonate) are less acidic than poly(glycolic acid) PGA and poly(L-lactic acid) PLLA [40]. In addition, tailor-made polymers can be synthesized to suit specific applications by varying amounts of monomers, incorporating side chains, or copolymerizing with other polymers such as PEG [40,41].

12.5.1 Synthesis of Carbonate-Based "Pseudo" Poly(Amino Acids)

The synthesis of carbonate-based "pseudo" poly(amino acids) requires three steps, in which the first step involves protection of the carboxylic acid group on tyrosine with a protecting group (Table 12.3) [40]. The variety of protecting groups available for this reaction allow for the ability to tailor the physicochemical properties of the final polymer [40]. The products of these reactions are then conjugated to desamino tyrosine using carbodiimide chemistry. The resulting biphenolic monomers are then subjected to a phosgenation reaction that produces a tyrosine-based poly(carbonate) homopolymers (Table 12.3) [40]. A variant of this homopolymer has also been reported by the Kohn group, where the biphenolic monomers are copolymerized with PEG via a similar reaction (Table 12.3) [41].

12.5.2 Applications of Carbonate-Based "Pseudo" Poly(Amino Acids)

"Pseudo" poly(amino acids) formulated with carbonates or poly(DTE carbonates) are hydrophobic, amorphous polymers that have been developed primarily for tissue engineering. The FDA master file for poly(DTE carbonates) highlights the excellent biocompatibility and limited toxicity of these polymers [42]. Electrospun scaffolds of poly(DTE carbonate) have been developed for reconstructive surgeries in anterior cruciate ligament injuries [43,44]. Initial mechanical characterizations demonstrate the scaffolds' abilities to retain approximately 87% of the original tensile strength and Young's modulus after 30 weeks in phosphate-buffered saline [44]. These properties are far superior when compared to PLLA, which only retains 7% of its initial strength under similar testing conditions [43]. Poly(DTE carbonates) display better overall biocompatibility with improved physicochemical and cellular compatibility properties when compared to PLLA [44]. In addition, an *in vivo* study using a rabbit transcortical bone pin model shows no inflammatory responses within the new bone formation [45]. Poly(DTE carbonate) films of 200-300 μm thickness also have been used to treat mandibular bone defects [45,46]. These films exhibited greater bone growth and a modest foreign-body response in a rabbit model when compared to bioactive glass [46].

Poly(carbonates) also have been developed for drug-delivery therapies [47,48]. Poly(DTH carbonate) microspheres have been utilized for intracranial controlled release of dopamine [40,49]. These microspheres do not exhibit a burst release phase and release dopamine at a controlled rate, where 15% of the total drug is released over a period of 180 days with an average dose of 1-2 μg/day [49].

12.6 URETHANE-BASED "PSEUDO" POLY(AMINO ACIDS)

Urethanes, or carbamates, are combinations of an amine and a carboxylic ester (NH–COO) and typically formulated using an addition reaction between an isocyanate (–N=C=O) and an alcohol. Polymers made with urethanes are an important class of biomaterials because of their excellent mechanical properties, specifically their elasticity [23,26]. Polyurethanes also have high tensile strength; excellent resistances to fatigue, wear and tear; and exceptional elasticity [50,51]. Although the majority of commercial polyurethanes are designed for lasting biostability, exposure to harsh biological conditions postimplantation has resulted in toxic, carcinogenic, and/or immunogenic by-products [52]. These physical and chemical breakdowns have caused several companies to reevaluate their formulations and/or to discontinue their polymers; thus, development of "pseudo" poly(amino acids) with urethane functional groups is a radical departure from biostable to biodegradable polymers that could positively impact implantable devices that require elastomeric materials [11,53].

TABLE 12.3 "Pseudo" Poly(amino acids) Formulated with Carbonates

Investigator	Amino Acid (A)	Linker 1 (L_1)	Linker 2 (L_2)	Protecting Group (P)
Kohn [40]	Tyrosine	Desamino tyrosine	Carbonate	Methanol, ethanol, isopropanol, butanol, isobutanol, sec-butanol, hexanol, octanol, dodecyl, or phenol
Kohn [41]	Tyrosine	Desamino tyrosine and poly(ethylene glycol)	Carbonate	Methanol, ethanol, isopropanol, butanol, isobutanol, sec-butanol, hexanol, octanol, dodecyl, or phenol

12.6.1 Synthesis of Urethane-Based "Pseudo" Poly(Amino Acids)

The basic reaction scheme for formulating urethane-based "pseudo" poly(amino acids) requires a reaction of a diisocyanate with a diol in the presence of stannous octoate as a catalyst. Depending on their properties, either the diol or the diisocyanate could serve as the hard segment (h) or soft segment (s) as reported in Table 12.4. Several poly(urethanes) also include a diol- or diamine-based chain extender to yield a higher molecular weight product.

To minimize the cytotoxicity of the polymer's degradation products and possibly to support cellular attachment and growth, Woodhouse [50,54–62] has formulated lysine diisocyanate as a hard segment and either glycine-leucine diamine diester of cyclohexanedimethanol (CDM) or phenylalanine diamine diester of CDM as the chain extender (Table 12.4). These chain extenders are conjugated with Fischer esterification between the amino acid(s) and CDM [56,57]. Another variation developed by Woodhouse [55] is a lysine-based polyurethane with a commercially available CDM chain. In addition, they also have developed "pseudo" poly(amino acids) with poly(caprolactone diol) (PCL-diol) and PEG as soft segments (Table 12.4) [55–57]. Alternative lysine-based polyurethanes have been developed by Guan [31], where PCL-diol is the soft segment and 1,4-diisocyanatohexane is the hard segment, and Zhang [32], who uses lysine diisocyanate coupled with the hydroxyl groups on a glucose linker to yield a highly branched copolymer. For tyrosine-based polyurethanes, Sengupta and Lopina [24–26] have synthesized polymers using a combination of PCL-diol or PEG serving as the soft segment and 1,6 diisocyanatohexane or methylenebis (dicycloisocyanate) serving as the hard segment. The chain extender employed for synthesis of the final polymer is a tyrosine-based biphenolic monomer (with a hexyl protecting group on tyrosine) [33,34].

12.6.2 Applications of Urethane-Based "Pseudo" Poly(Amino Acids)

Urethane-based "pseudo" poly(amino acids) have been investigated for tissue-engineering scaffolds and drug-delivery applications due to their physicochemical properties and biodegradation pathways [11,67]. By reducing the toxicity associated with polyurethanes, these polymers have desirable properties for biomedical applications including use in blood-contacting devices such as artificial heart valves and catheters. These polymers are also attractive candidates for tissue engineering and can be easily fabricated into films and fibrous scaffolds [10,11,67].

The elasticity of polyurethanes is ideal for high-stress environments such as those found in the cardiovascular system [10,11,54,67–69]. The "pseudo" poly(amino acids) developed by the Woodhouse have been applied for cardiac tissue engineering because of their favorable tensile strength (~8 MPa) and elongation (~900%) [50,54–59]. These polymers also show less than 10% loss in viability of cardiomyocytes that are derived from mouse embryonic stem cells [54,69]. The L-tyrosine-based polyurethanes (LTU), developed by Sengupta and Sarkar, can be tuned for a specific application by blending polymers synthesized with either PEG or PCL-diols [10,64,65]. These polymers exhibit a wide range of mechanical properties, such as tensile strengths ranging from 2 to 18 MPa and elongations ranging from 214% to 1513%, and have mechanical characteristics needed for tissue engineering of human arteries and skin [10,11,67]. Furthermore, cytotoxicity experiments using LTU electrospun scaffolds and films have shown biocompatibility with primary cell lines. Shah *et al.* reported approximately 97% viability for the different blends of LTUs on a human dermal fibroblast (HDF) cell line [11]. Cells exposed to 500 μg/mL degradation products from L-tyrosine-based LTUs and their blends maintained approximately 94% cell viability (Figure 12.7), while the degradation products from PLGA, used as a control, has a significantly lower viability, compared to the media only controls [11].

TABLE 12.4 "Pseudo" Poly(amino acids) Formulated with Urethanes

Investigator	Amino Acid (A)	Linker (L_1)	Linker (L_2)	Protecting Group (P)
Woodhouse [50,54–62]	Lysine diisocyanate (h)	PCL-diol (s) or PEG (s)	Phenylalanine diester (c), 1,4-cyclohexanedimethanol (c), or glycine-leucine diamine (c)	Methanol (lysine)
Guan [63]	Lysine (c)	PCL-diol (s)	1,4-Diisocyanato butane (h)	Ethanol
Sengupta and Sarkar [64,65]	Tyrosine (c)	Desamino tyrosine (c)	PCL-diol (s) or PEG-diol (s) and hexamethylene diisocyanate (h) or methylenebis (dicycloisocyanate) (h)	Hexanol
Zhang [66]	Lysine diisocyanate (h)	Glucose	–	–

(s) is the soft segment, (h) is the hard segment, and (c) is the chain extender.

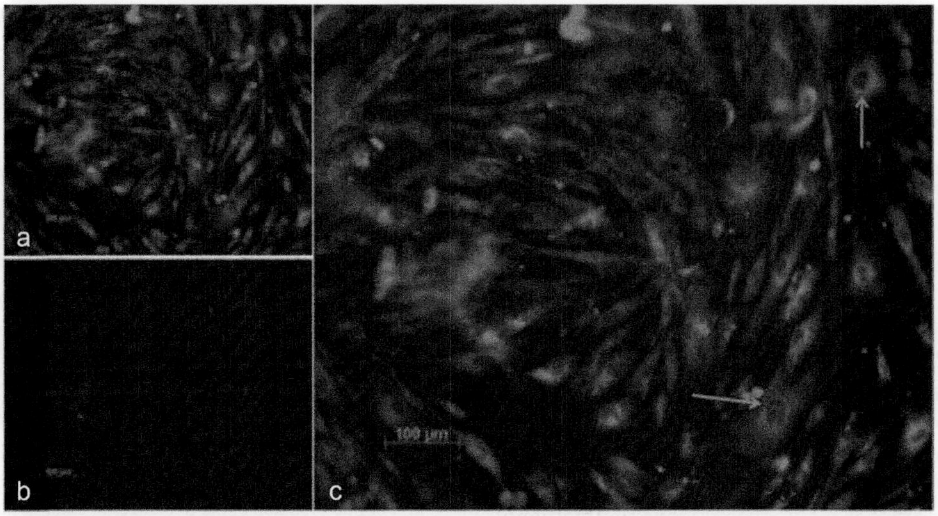

FIGURE 12.7 Representative images of human dermal fibroblasts exposed to 800 µg/ml of PCL-C-DTH degradation products using LIVE/DEAD® Cell Vitality assay obtained at day 42 for (a) live cells, (b) dead cells, and (c) combined image. Representative enlarged nuclei have been pointed out by arrows. Reproduced with permission from Ref. [11]. (For color version of this figure, the reader is referred to the online version of this chapter.)

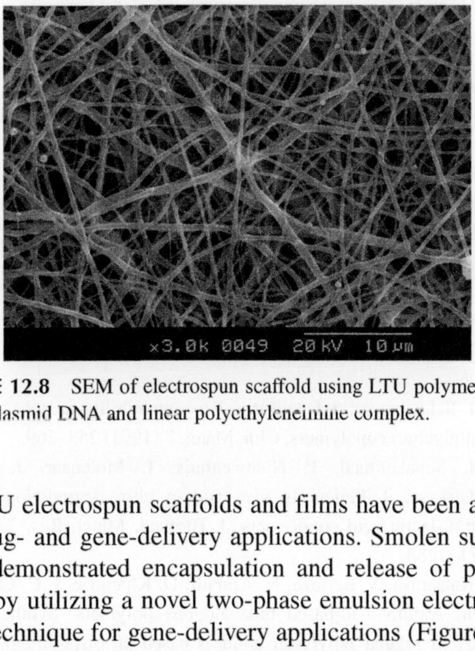

FIGURE 12.8 SEM of electrospun scaffold using LTU polymer loaded with a plasmid DNA and linear polyethyleneimine complex.

LTU electrospun scaffolds and films have been applied for drug- and gene-delivery applications. Smolen successfully demonstrated encapsulation and release of plasmid DNA by utilizing a novel two-phase emulsion electrospinning technique for gene-delivery applications (Figure 12.8) [70]. This emulsion process yielded scaffolds with unique secondary dome structures embedded throughout the scaffold. Unlike "bead on a string" structures, the sizes of these domes range in size from 20 to 200 µm, and they are hollow [70]. Furthermore, *in vitro* experiments show controlled release of plasmid DNA over a 5-week period (approximately 130 ng) while maintaining the DNA's functional and structural integrity [70]. LTUs also have shown the ability to support cellular growth and attachment via focal adhesion (Figure 12.9a and b) [71]. The doubling times for primary HDFs seeded on LTUs range from 37 to 90 h and is dependent upon the hydrophobic or hydrophilic characteristics of the soft segment [7–11]. These polymers are promising for applications in tissue engineering and regenerative medicine because electrospun fibers of "pseudo" poly(amino acids) mimics the architecture of the extracellular matrix, provides support specific cellular adhesion and growth, and can be incorporated with drugs.

12.7 CONCLUSIONS

To better accommodate the demands of the aging population, continual evolution of tunable, bioactive, and functional polymers is paramount for the advancement of medical devices. The development of "pseudo" poly(amino acids) over the past two decades has shown the potential of this family of polymers to be versatile and adaptable for treating specific diseases. The formulations of these polymers cover a wide range of functional groups imparting unique and desirable properties and limiting the toxicity of these polymer and their degradation products. Ester-based "pseudo" poly(amino acids), such as LTP nanoparticles, have shown excellent potential as drug-delivery devices for chemotherapies, antibacterial treatments, and gene therapies, while a library of esters developed by Kohn can be applied for developing scaffolds for a variety of tissue-engineering applications. The slow degradation of carbonate-based "pseudo" poly(amino acids) is ideal for applications that require mechanical loading or long-term drug releases. As a result, the FDA has approved the L-tyrosine-based PIVIT A/B™ and AIGISRx as nondegradable meshes for hernia and cardiovascular devices. Polyurethane-based "pseudo" poly(amino acids) such as LTU have high tensile strength, excellent resistance against wear and tear, and high elasticity. These polymers are ideal candidates for cardiovascular applications and tissue engineering of soft tissues.

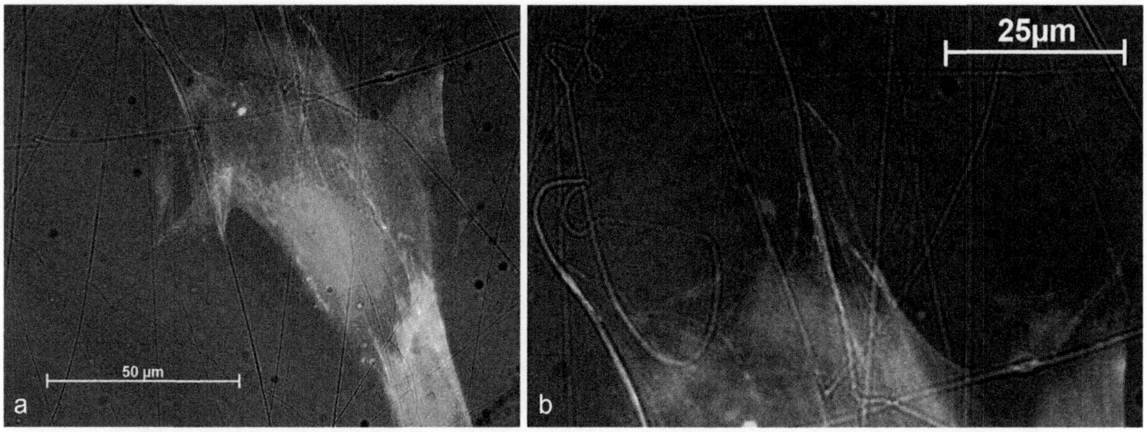

FIGURE 12.9 Immunofluorescence staining of human dermal fibroblast attachment to electrospun LTU. The green color shows the staining of actin using Alexa Fluor 488 phalloidin, the blue color shows Hoechst nuclear staining, and red colors show antivinculin antibody conjugated with TRITC. (a) Image of a single cell and (b) close up of focal adhesion sites. Reproduced with permission. (For interpretation of the references to color in this figure legend, the reader is referred to the online version of this chapter.)

When electrospun, the resulting scaffolds mimic the structure of extracellular matrix. Thus, further development of these polymers is needed to meet the demands of the next generation of biomaterials and support the advancement of medical devices, tissue engineering, regenerative medicine, and nanotechnology.

REFERENCES

[1] E. Rabkin, F.J. Schoen, Cardiovascular tissue engineering, Cardiovasc. Pathol. 11 (2002) 305–317.

[2] B.D. Ratner, A.S. Hoffman, F.J. Schoen, J.E. Lemons, Biomaterials science: an introduction to materials in medicine, Elsevier Academic Press, San Diego, CA, 2013.

[3] J. Kohn, R. Langer, A new approach to the development of bioerodible polymers for controlled release applications employing naturally occurring amino acids, Proc. ACS Division Polym. Mat. Sci. Eng. 51 (1984) 119–121.

[4] J. Kohn, R. Langer, Poly(iminocarbonates) as potential biomaterials, Biomaterials 7 (1986) 176–182.

[5] J. Kohn, R. Langer, Polymerization reactions involving the side chains of alpha.-l-amino acids, J. Am. Chem. Soc. 109 (1987) 817–820.

[6] J. Kohn, R. Langer, A New Approach to the Development of Bioerodible Polymers for Controlled Release Applications Employing Naturally Occurring Amino Acids, American Chemical Society, Washington, DC, 1984.

[7] A.J. Ditto, J.J. Reho, K.N. Shah, J.A. Smolen, J.H. Holda, R.J. Ramirez, Y.H. Yun, In vivo gene delivery with l-tyrosine polyphosphate nanoparticles, Mol. Pharm. 10 (2013) 1836–1844.

[8] A.J. Ditto, P.N. Shah, L.R. Gump, Y.H. Yun, Nanospheres formulated from l-tyrosine polyphosphate exhibiting sustained release of polyplexes and in vitro controlled transfection properties, Mol. Pharm. 6 (2009) 986–995.

[9] A.J. Ditto, P.N. Shah, S.T. Lopina, Y.H. Yun, Nanospheres formulated from l-tyrosine polyphosphate as a potential intracellular delivery device, Int. J. Pharm. 368 (2009) 199–206.

[10] P.N. Shah, S.T. Lopina, Y.H. Yun, Blends of novel l-tyrosine-based polyurethanes and polyphosphate for potential biomedical applications, J. Appl. Polym. Sci. 114 (2009) 3235–3247.

[11] P.N. Shah, Y.H. Yun, Cellular interactions with biodegradable polyurethanes formulated from l-tyrosine, J. Biomater. Appl. 27 (2013) 1017–1031.

[12] M. Spira, J. Fissette, W. Hall, S.B. Hardy, F.J. Gerow, Evaluation of synthetic fabrics as artificial skin grafts to experimental burn wounds, J. Biomed. Mater. Res. 3 (1969) 213–234.

[13] R.P. Lanza, R.S. Langer, J. Vacanti, Principles of Tissue Engineering, Elsevier/Academic Press, Amsterdam, Boston, 2007.

[14] H.J. Ryser, W.C. Shen, Conjugation of methotrexate to poly(l-lysine) increases drug transport and overcomes drug resistance in cultured cells, Proc. Natl. Acad. Sci. U.S.A. 75 (1978) 3867–3870.

[15] R.L. Reis, D. Cohn, Polymer based systems on tissue engineering, replacement and regeneration, Kluwer Academic Publishers, Dordrecht, The Netherlands, 2002.

[16] M.N. Melkerson, Tyrx antimicrobial mesh—premarket notification, 2006.

[17] J.M. Schakenraad, P.J. Dijkstra, Biocompatibility of poly (dl-lactic acid/glycine) copolymers, Clin. Mater. 7 (1991) 253–269.

[18] J.M. Schakenraad, P. Nieuwenhuis, I. Molenaar, J. Helder, P. Dijkstra, J. Feijen, In vivo and in vitro degradation of glycine/dl-lactic acid copolymers, J. Biomed. Mater. Res. 23 (1989) 1271–1288.

[19] R. Katsarava, V. Beridze, N. Arabuli, D. Kharadze, C.C. Chu, C.Y. Won, Amino acid-based bioanalogous polymers. Synthesis, and study of regular poly(ester amide)s based on bis(a-amino acid) a, w-alkylene diesters, and aliphatic dicarboxylic acids, J. Polym. Sci. Part A Polym. Chem. 37 (1999) 391–407.

[20] S. Mallakpour, F. Zeraatpisheh, Pseudo-poly(amino acid)s: study on construction and characterization of novel chiral and thermally stable nanostructured poly(ester-imide)s containing different trimellitylimido-amino acid-based diacids and pyromellitoyl-tyrosine-based diol, Colloid Polym. Sci. 289 (2011) 1055–1064.

[21] S. Mallakpour, F. Zeraatpisheh, The nanocomposites of zinc oxide/l-amino acid-based chiral poly (ester-imide) via an ultrasonic route: synthesis, characterization, and thermal properties, J. Appl. Polym. Sci. 289 (2012) 1055–1064.

[22] K. Guo, C. Chu, Copolymers of unsaturated and saturated poly (ether ester amide)s: synthesis, characterization, and biodegradation, J. Appl. Polym. Sci. 110 (2008) 1858–1869.

[23] S. Brocchini, K. James, V. Tangpasuthadol, J. Kohn, Structure–property correlations in a combinatorial library of degradable biomaterials, J. Biomed. Mater. Res. 42 (1998) 66–75.
[24] A. Sengupta, S.T. Lopina, Properties of l-tyrosine based polyphosphates pertinent to biomedical applications, Polymer 46 (2005) 2133–2140.
[25] A. Sengupta, S.T. Lopina, Synthesis and characterization of l-tyrosine based novel polyphosphates for potential biomaterial applications, Polymer 45 (2004) 4653–4662.
[26] A. Sengupta, S.T. Lopina, L-tyrosine-based backbone-modified poly(amino acids), J. Biomater. Sci. Polym. Ed. 13 (2002) 1093–1104.
[27] S. Mallakpour, F. Zeraatpisheh, M.R. Sabzalian, Construction, characterization and biological activity of chiral and thermally stable nanostructured poly(ester-imide)s as tyrosine-containing pseudo-poly(amino acid)s, J. Polym. Environ. 20 (2012) 117–123.
[28] K.M. Hindi, A.J. Ditto, M.J. Panzner, D.A. Medvetz, D.S. Han, C.E. Hovis, J.K. Hilliard, J.B. Taylor, Y.H. Yun, C.L. Cannon, The antimicrobial efficacy of sustained release silver-carbene complex-loaded l-tyrosine polyphosphate nanoparticles: characterization, in vitro and in vivo studies, Biomaterials 30 (2009) 3771–3779.
[29] P.N. Shah, A.A. Puntel, S.T. Lopina, Y.H. Yun, Development and in vitro cytotoxicity of microparticle drug delivery system for proteins using l-tyrosine polyphosphate, Colloid Polym. Sci. 287 (2009) 1195–1205.
[30] A.J. Ditto, K.N. Shah, N.K. Robishaw, M.J. Panzner, W.J. Youngs, Y.H. Yun, The interactions between l-tyrosine based nanoparticles decorated with folic acid and cervical cancer cells under physiological flow, Mol. Pharm. 9 (2012) 3089–3098.
[31] R. Chen, M.E. Eccleston, Z. Yue, N.K.H. Slater, Synthesis and ph-responsive properties of pseudo-peptides containing hydrophobic amino acid grafts, J. Mater. 19 (2009) 4217–4224.
[32] R. Chen, M.E. Eccleston, Z. Yue, N.K.H. Slater, Synthesis and ph-responsive properties of pseudo-peptides containing hydrophobic amino acid grafts, J. Mater. Chem. 19 (2009) 4217.
[33] R. Chen, Z. Yue, M.E. Eccleston, N.K.H. Slater, Aqueous solution behaviour and membrane disruptive activity of ph-responsive pegylated pseudo-peptides and their intracellular distribution, Biomaterials 29 (2008) 4333–4340.
[34] R. Chen, Z. Yue, M.E. Eccleston, S. Williams, N.K.H. Slater, Modulation of cell membrane disruption by ph-responsive pseudo-peptides through grafting with hydrophilic side chains, J. Control. Release 108 (2005) 63–72.
[35] Z. Yue, M.E. Eccleston, N.K.H. Slater, Pegylation and aqueous solution behaviour of ph responsive poly(l-lysine iso phthalamide), Polymer 46 (2005) 2497–2505.
[36] R. Chen, S. Khormaee, M.E. Eccleston, N.K.H. Slater, The role of hydrophobic amino acid grafts in the enhancement of membrane-disruptive activity of ph-responsive pseudo-peptides, Biomaterials 30 (2009) 1954–1961.
[37] V.H. Ho, N.K.H. Slater, R. Chen, Ph-responsive endosomolytic pseudo-peptides for drug delivery to multicellular spheroids tumour models, Biomaterials 32 (2011) 2953–2958.
[38] M.E. Eccleston, C.F. Kaminski, N.K.H. Slater, M.S.J. Briggs, Optical characteristics of responsive biopolymers; co-polycondensation of tri-functional amino acids and cy-3 bis-amine with diacylchlorides, Polymer 45 (2004) 25–32.
[39] X. Dai, M.E. Eccleston, Z. Yue, N.K.H. Slater, C.F. Kaminski, A spectroscopic study of the self-association and inter-molecular aggregation behaviour of ph-responsive poly(l-lysine isophthalamide), Polymer 47 (2006) 2689–2698.
[40] S.L. Bourke, J. Kohn, Polymers derived from the amino acid l-tyrosine: polycarbonates, polyarylates and copolymers with poly(ethylene glycol), Adv. Drug Deliv. Rev. 55 (2003) 447–466.
[41] C. Yu, J. Kohn, Tyrosine–peg-derived poly(ether carbonate)s as new biomaterials, Biomaterials 20 (1999) 253–264.
[42] M.H.C. Resurreccion-Magno, Optimization of tyrosine-derived polycarbonate terpolymers for bone regeneration scaffolds, Dissertation, Rutgers University, New Brunswick, NJ, 2012.
[43] S.L. Bourke, J. Kohn, M.G. Dunn, Preliminary development of a novel resorbable synthetic polymer fiber scaffold for anterior cruciate ligament reconstruction, Tissue Eng. 10 (2004) 43–52.
[44] N.B.S. Tovar, M. Jaffe, N.S. Murthy, J. Kohn, C. Gatt, M.G. Dunn, A comparison of degradable synthetic polymer fibers for anterior cruciate ligament reconstruction, J. Biomed. Mater. Res. A 93 (2010) 738–747.
[45] J. Kohn, W.J. Welsh, D. Knight, A new approach to the rationale discovery of polymeric biomaterials, Biomaterials 28 (2007) 4171–4177.
[46] A.J. Asikainen, J. Noponen, C. Lindqvist, M. Pelto, M. Kellomaki, H. Juuti, H. Pihlajamaki, R. Suuronen, Tyrosine-derived polycarbonate membrane in treating mandibular bone defects. An experimental study, J. R. Soc. Interface 3 (2006) 629–635.
[47] L. Sheihet, R.A. Dubin, D. Devore, J. Kohn, Hydrophobic drug delivery by self-assembling triblock copolymer-derived nanospheres, Biomacromolecules 8 (2005) 2726–2731.
[48] L. Sheihet, K. Piotrowska, R.A. Dubin, J. Kohn, D. Devore, Effect of tyrosine-derived triblock copolymer compositions on nanosphere self-assembly and drug delivery, Biomacromolecules 8 (2007) 998–1003.
[49] Z. Dong, Synthesis of Four Structurally Related Tyrosine-Derived Polycarbonates and In Vitro Study of Dopamine Release from Poly(Desaminotyrosyl-Tyrosine Hexyl Ester Carbonate), Master's Thesis, Rutgers University, New Brunswick, NJ, 1993.
[50] N.M.K. Lamba, K.A. Woodhouse, S.L. Cooper, Polyurethanes in Biomedical Applications, CRC Press, Boca Raton, 1998.
[51] R. Yoda, Elastomers for biomedical applications, J. Biomater. Sci. Polym. Ed. 9 (1998) 561–626.
[52] J. Lannutti, D. Reneker, T. Ma, D. Tomasko, D. Farson, Electrospinning for tissue engineering scaffolds, Mater. Sci. Eng. C 27 (2007) 504–509.
[53] A. Marcos-Fernandez, G.A. Abraham, J.L. Valentin, J.S. Roman, Synthesis and characterization of biodegradable non-toxic poly(ester-urethane-urea)s based on poly(e-caprolactone) and amino acid derivatives, Polymer 47 (2006) 785–798.
[54] I.C. Parrag, P.W. Zandstra, K.A. Woodhouse, Fiber alignment and coculture with fibroblasts improves the differentiated phenotype of murine embryonic stem cell-derived cardiomyocytes for cardiac tissue engineering, Biotechnol. Bioeng. 109 (2012) 813–822.
[55] G.A. Skarja, K.A. Woodhouse, Synthesis and characterization of degradable polyurethane elastomers containing an amino acid-based chain extender, J. Biomater. Sci. Polym. Ed. 9 (1998) 271–295.
[56] G.A. Skarja, K.A. Woodhouse, Structure-property relationships of degradable polyurethane elastomers containing an amino acid-based chain extender, J. Appl. Polym. Sci. 75 (2000) 1522–1534.
[57] J.D. Fromstein, K.A. Woodhouse, Elastomeric biodegradable polyurethane blends for soft tissue applications, J. Biomater. Sci. Polym. Ed. 13 (2002) 391–406.

[58] J.P. Santerre, K. Woodhouse, G. Laroche, R.S. Labow, Understanding the biodegradation of polyurethanes: from classical implants to tissue engineering materials, Biomaterials 26 (2005) 7457–7470.
[59] D.N. Rockwood, K.A. Woodhouse, J.D. Fromstein, D.B. Chase, J.F. Rabolt, Characterization of biodegradable polyurethane microfibers for tissue engineering, J. Biomater. Sci. Polym. Ed. 18 (2007) 734–758.
[60] G.A. Skarja, K.A. Woodhouse, In vitro degradation and erosion of degradable, segmented polyurethanes containing an amino-acid based chain extender, J. Biomater. Sci. Polym. Ed. 12 (2001) 851–873.
[61] S.L. Elliott, J.D. Fromstein, J.P. Santerre, K.A. Woodhouse, Identification of biodegradation products formed by l-phenylalanine based segmented polyurethanes, J. Appl. Polym. Sci. 13 (2002) 693–711.
[62] P.H. Blit, K.G. Battiston, K.A. Woodhouse, J.P. Santerre, Surface immobilization of elastin-like polypeptides using fluorinated surface modifying additives, J. Biomed. Mater. Res. 96 (2011) 648–662.
[63] J. Guan, M.S. Sacks, E.J. Beckman, W.R. Wagner, Synthesis, characterization, and cytocompatibility of elastomeric, biodegradable poly(ester-urethane)ureas based on poly(caprolactone) and putrescine, J. Biomed. Mater. Res. 61 (2002) 493–503.
[64] A. Sengupta, Synthesis and characterization of l-tyrosine based novel biodegradable polyphosphates and polyurethanes for biomaterial applications, Dissertation, The University of Akron, Akron, OH, 2003.
[65] D. Sarkar, Development and characterization of l-tyrosine based polyurethanes for tissue engineering applications, Dissertation, The University of Akron, Akron, OH, 2007.
[66] J.Y. Zhang, E.J. Beckman, J. Hu, G.G. Yang, S. Agarwal, J.O. Hollinger, Synthesis, biodegradability, and biocompatibility of lysine diisocyanate-glucose polymers, Tissue Eng. 8 (2002) 771–785.
[67] P.N. Shah, R.L. Manthe, S.T. Lopina, Y.H. Yun, Electrospinning of l-tyrosine polyurethanes for potential biomedical applications, Polymer 50 (2009) 2281–2289.
[68] P.N. Shah, Biocompatibility Analysis and Biomedical Device Development Using Novel L-Tyrosine Based Polymers, Dissertation, The University of Akron, Akron, OH, 2009.
[69] I.C. Parrag, The Development of Elastomeric Biodegradable Polyurethane Scaffolds for Cardiac Tissue Engineering, Dissertation, The University of Toronoto, Toronto, ON, 2010.
[70] J.A. Smolen, Emulsion electrospinning for producing dome-shaped structures within l-tyrosine polyurethane scaffolds for gene delivery, Master's Thesis, The University of Akron, Akron, OH, 2010.
[71] K. Popat, Nanotechnology in Tissue Engineering and Regenerative Medicine, CRC Press, Boca Raton, 2011.

Chapter 13

Polyacetals

Sheiliza Carmali, Steve Brocchini
UCL School of Pharmacy, University College London, London, United Kingdom

Chapter Contents

13.1 INTRODUCTION

Biomedical polymers that undergo hydrolytic degradation at mild acidic pH values may have some advantage for use in regions of low pH within the body (e.g., gastrointestinal tract) or where there are acidic pH gradients, e.g., endocytic pathway, within malignant tissue, or sites of infection. Acid-labile polymers have the potential to be used where hydrolytic degradation allows more efficient polymer clearance from the body or to release a biologically active molecule. There can be significant pH differences when a molecule moves from the blood compartment (pH 7.2-7.4) to malignant tissue (often 0.5-1.0 pH units lower than in normal tissue [1]) or to intracellular compartments (pH 4.0-6.5 [2–4]). Stimuli-responsive polymers have been thoroughly reviewed [5–7] and the use of a degradable element that is susceptible to acid hydrolysis has been examined as the responsive component within hydrogels or colloids. Such systems are often explored as a means to alter favorably the pharmacokinetics and biodistribution of a biologically active molecule (e.g., drug and siRNA) at an acidic pH value [8].

Several excellent reviews have recently been published that describe the broad field of degradable biomedical polymers [9,10]. Well-known hydrolytically degradable polymers developed or being developed for biomedical used include homo- and copolymers of polyamides (usually derived from amino acids), polyesters, polyanhydrides, poly(ortho ester)s, poly(amido amines), and poly(β-amino esters). This chapter is focused on polyacetals, which undergo hydrolysis at acidic pH values. The term polyacetal as used here also includes polyketals. Poly(ortho ester)s [11] and polymers with other degradable elements such as imine [12–14], hydrazone [15,16], and aconityl acid [17] also undergo faster hydrolytic degradation rates at acidic pH values, however, these polymers will not be described here.

Efforts to develop polyacetals for potential biomedical applications have increased for several reasons. Polyacetals can be prepared relatively easily without the requirement of overly stringent drying conditions, which are often necessary for poly(ortho esters), for example. The degradation by-products from acetal hydrolysis do not include an acid as is the case for polyanhydrides, polycarbonates, or polyesters, so there is no acid-driven autocatalysis during polyacetal degradation. One disadvantage of polyacetals is that 1 mol of aldehyde is generated for each acetal moiety that degrades, which raises a potential toxicity issue. Polyketals are often considered because a ketone is produced as a degradation product for every ketal unit rather than an aldehyde for each acetal moiety. A strategy to avoid potential toxicity due to the excessive generation of aldehyde by-products includes using a macromonomer with a single acetal function, which when polymerized will give a polymer with a single degradable acetal element in the main chain. In this way, the amount of the aldehyde formed is very small. Another strategy is to use a cross-linker with embedded acetal functionality that is used to make network polymers. Such cross-linkers are used in relatively low proportion compared to

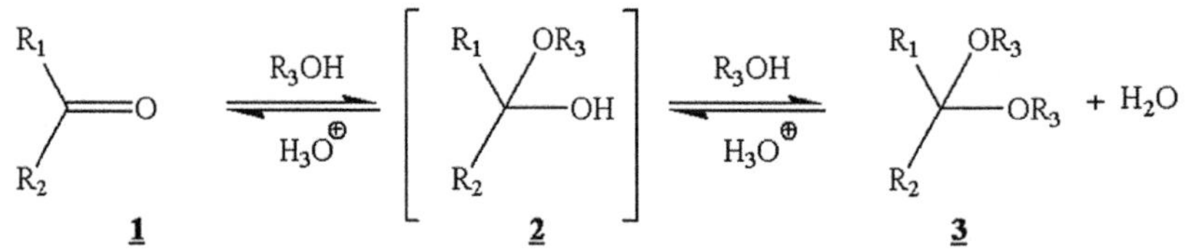

R_1, R_2 = alkyl group or H
R_3 = alkyl group

FIGURE 13.1 Formation of an acetal/ketal 3 starting from an aldehyde/ketone 1 in the presence of an alcohol and acid. Mechanistically, acetal/ketal formation in these conditions yields a hemiacetal intermediate 2 and a mole of water. Removal of the water formed during reaction can be difficult to achieve during a polymerization reaction in an effort to obtain high-molecular-weight polyacetal.

the monomers within the polymer main chain. The resulting cross-linked polymer would be acid-labile and this strategy has been used, for example, to fabricate colloid or network systems that are designed to release a drug-active substance [18–23].

An acetal (e.g., 3; Figure 13.1) can be prepared by the equilibrium reaction of two equivalents of an alcohol and one equivalent of a compound possessing an aldehyde (or ketone to generate a ketal). The hemiacetal intermediate 2 is hydrolytically labile to both base and acid, but the resulting acetal product 3 is only labile to hydrolysis at acidic pH values. When polyacetals are prepared by acid catalysis, it is important to remove or neutralize any residual acid to ensure the polymer is stable enough to isolate and for storage.

Historically, polyacetals have been long known. Commodity polyacetals are often produced by addition polymerization through a carbonyl double bond (e.g., formaldehyde) and the terminal hydroxyls must be endcapped [24], often using an anhydride to inhibit depolymerization of the final polymer. Homopolymers derived from formaldehyde and copolymers have been produced (Mn = 20,000-100,000) [25] with the uncapped homopolymer first being prepared by Staudinger in the 1920s. Capping the formaldehyde-derived polyacetal (known as polyformaldehyde or polyoxymethylene) with acetic anhydride gives a thermally stable, melt-processible plastic [26], which was commercialized (Delrin®). Acetal copolymers have also been developed including Celcon® derived from trioxane and ethylene oxide and Hostaform®, which is derived from trioxane and cyclic ethers. Since the development of these first polyacetals, other commodity polyacetals have been developed including Ultraform®, a trioxane copolymer; Tenac®, a formaldehyde homopolymer; Tarnoform®, a trioxane-dioxolane copolymer; and Jupital®, a trioxane copolymer.

Polyacetals that have been examined for biomedical applications are often prepared by step or condensation polymerizations. Utilizing a diol monomer and an aldehyde to prepare a polymer requires removal of 1 equivalent of water per acetal (Figure 13.1). Acetal exchange reactions can be used where the small molecule is an alcohol with a lower boiling point than water (e.g., methanol) is generated by reaction of an acetal with a diol monomer. This is done in an effort to more efficiently eliminate the small molecule by-product during polymerization to obtain polyacetals reproducibly with sufficiently high molecular weight to be useful. However, protocols that rely on elimination of water continue to be used to prepare especially, when release of the aldehyde is required [27]. Small amounts of water or alcohol, in the presence of residual acid, will result in polymer degradation, so it can be difficult to obtain the desired molecular weight characteristics reproducibly. However, the distillation of methanol is viable as evidenced by the use of 2,2-dimethoxypropane as a surrogate for acetone to make polyketals [28].

To avoid the need to remove a small, protic molecule, Heller [29] in 1980 showed that polyacetals could be easily prepared from divinyl ethers 4 and diols 5 (Figure 13.2a). Utilizing two monomers such as a diol and a divinyl ether to make polyacetal has subsequently been followed by others [31–33], but the inherent limitation of this approach to achieve high-molecular-weight polymer is the need to ensure both monomers are highly pure and used accurately at 1:1 stoichiometry. Divinyl ethers are labile to acid hydrolysis to give an alcohol, so some care is required to ensure that pure divinyl ethers are carefully stored, preferably in the presence of a base. Such hydrolysis can lead to oligomerization of the divinyl ether. The diol monomer, especially those derived from poly(ethylene glycol) (PEG), can retain small amounts of water. Clearly, the presence of vinyl ether hydrolysis products or residual water in the diol must be avoided and monitored to achieve reproducible polymerizations.

Some aspects of these inherent characteristics were addressed a few years later in the early 1980s when the preparation of polyacetals was described using a single (A-B) monomer 7 comprised of a vinyl ether and a hydroxyl moiety that could be polymerized to give a polyacetal [30] (Figure 13.2b). However, the vinyl ether moiety in monomers such as 7 must still be protected from hydrolysis, and as will be seen, the approach in Figure 13.2a is more flexible and has been followed to prepare different polyacetals for potential biomedical use [23,31,34].

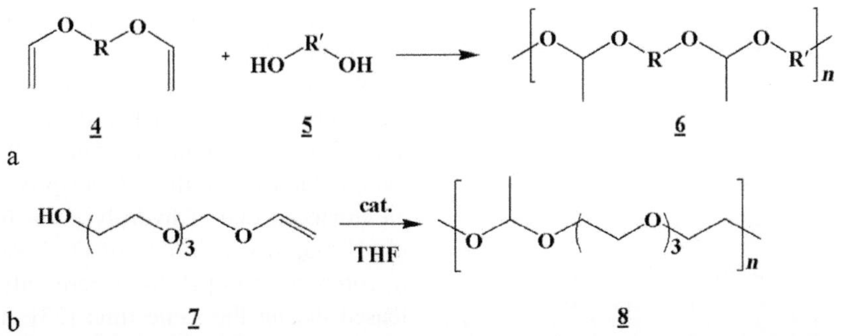

FIGURE 13.2 Preparation of acetal using vinyl ethers. (a) Polyacetals 6 are prepared using two monomers, a divinyl ether 4 and a diol 5 [29]. This synthetic route allows copolymerization with two or more diols (or divinyl ethers) as a means to vary polymer properties. (b) Polyacetals 8 can be prepared directly using a single a-b monomer 7 [30]. This more simplified approach precludes the need to use an exact stoichiometric equivalence of a diol(s) that is of sufficient purity required to obtain polymers with sufficiently high molecular weights (e.g., greater than ~10,000 g/mol).

13.2 BIOMEDICAL APPLICATIONS

13.2.1 Polymer-Drug Conjugates

Conjugation of water-soluble polymers to bioactive agents, such as drugs and proteins, has resulted in clinical products, especially for the PEGylation of proteins [35–39]. Many polymers have been examined for drug conjugation including PEG and copolymers of *N*-(2-hydroxypropyl)-methacrylamide (HPMA), which are not hydrolytically degradable [40,41]. To avoid potential accumulation of these and other nonhydrolytically degradable polymers, they and their conjugates are used where their solution structure is below the renal threshold. For random coiled, water-soluble polymers such as HPMA and PEG, the renal threshold is generally at a polymer molecular weight in the range of about 30-40 kDa, preferably with a narrow molecular weight distribution. This molecular weight can vary depending on the nature of the conjugate, e.g., hydrophobicity of the drug or amount of drug conjugated to the polymer, for example, because different conjugates will have different solution characteristics. The renal threshold is based on the solution structure of the polymer conjugate.

Examples of degradable, water-soluble polymers used for conjugation that have been clinically evaluated include poly(glutamic acid) [42–45] and co-blocked polymers of aspartic acid and PEG [46–48]. These are poly(amino acid)-derived polymers so they have amide bonds. While these polymers are susceptible to enzymatic degradation [49], they are not as hydrolytically acid-labile upon parenteral administration as polyacetals would be.

Polymer-drug conjugates have been extensively studied since the mid-1980s as a means to favorably alter the bioavailability of cytotoxic drugs to achieve higher concentrations within malignant tissue [50]. Some tumors at certain stages of development will be more permeable than healthy tissue [51]. This tissue permeability gradient, which is often called the enhanced permeability and retention (EPR) effect [52], has inspired much research to develop polymer cytotoxic drug conjugates and, in some cases, to develop conjugates to treat infection [53].

Because of their large size, polymer-drug conjugates will display prolonged circulation times in the blood compartment, allowing the conjugate to extravasate into more permeable, diseased tissue. Some retention of the conjugate in diseased tissue is possible because there may not be suitable lymphatic drainage. Since the polymer-drug conjugate is primarily confined to the blood compartment and not systemically distributed, a polymer-drug conjugate can provide an excellent means to mask the toxicity of the conjugated drug. Several polymer-drug conjugates have been clinically evaluated [54–58] and are continuing to be evaluated [44,59–62]; however, in the case of cancer, the ultimate criterion for clinical success will be decreased mortality. Clinical trials need to be carefully designed to determine efficacy.

Once a polymer-drug conjugate is localized within the tumor environment, there is a need to release the drug. Much work has been done utilizing peptides to link the drug to a polymer [63,64] or an antibody [65,66]. Such peptides have been shown to be degradable within the tumor interstitial spaces or intracellularly to release the drug. There are examples of clinical products for antibody-drug conjugates [67,68].

Following the lead of Heller [29], *co*-polyacetals **10** were examined with pendent chains [31,34] for drug conjugation (Figure 13.3). Diol drugs [23] have also been used as a comonomer so that drug can be incorporated within the main chain of the polyacetal, e.g., **9** (Figure 13.3). Malignant tissue can often display a lower pH than healthy tissues due to its enhanced metabolic processes. The polyacetal conjugates that have been described have been designed to degrade to release a cytotoxic drug at a tumor site.

We reported the preparation of polyacetal-doxorubicin conjugates that allow a pH-dependent release [31,34]. These amino-pendent polyacetals (APEGs) **10**

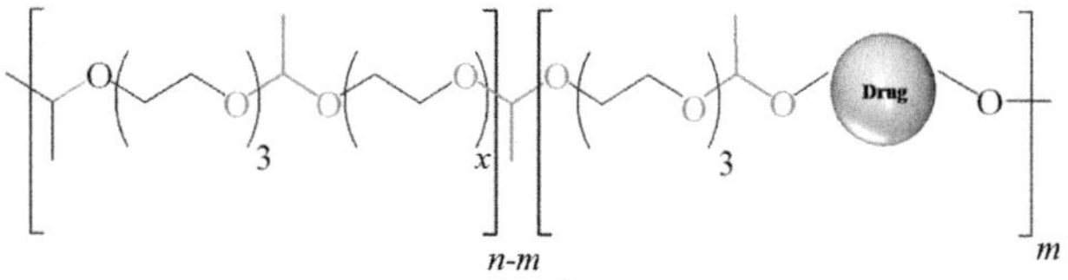

9

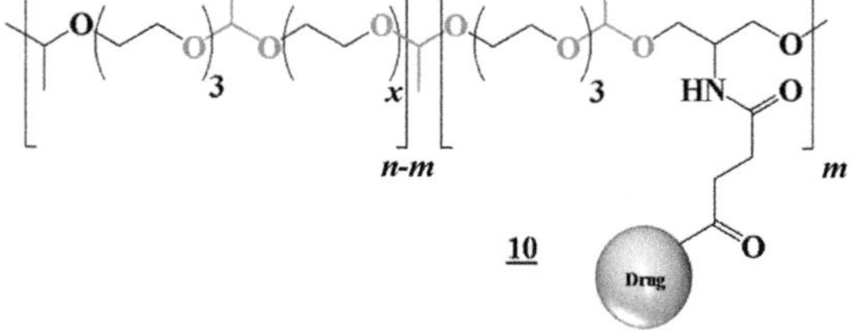

10

FIGURE 13.3 Copolyacetals capable of drug conjugation [23,31,34]. Both types shown are derived from polymerizations utilizing two diol monomers and one divinyl ether monomer. Drug incorporation in the main chain (e.g., polymer **9**) requires the drug to be the monomer or a part of a monomer. If the entire drug molecule is the monomer, then polyacetal degradation will directly yield the drug. Pendent chain conjugates **10** require drug release from monomer unit associated with the pendent chain. (For color version of this figure, the reader is referred to the online version of this chapter.)

displayed fast rates of hydrolysis at acidic pH, which would make them potentially suitable as biodegradable carriers for improved tumor targeting of anticancer drugs. *In vivo* studies in which APEG-doxorubicin conjugates were compared to HPMA copolymer-doxorubicin showed a more prolonged plasma circulation half-life for the polyacetal conjugates (approximately twofold), which ultimately led to a 1.4-fold increase in doxorubicin tumor targeting [34].

We reported another *ter*-polymerization with different monomers to evaluate another class of polyacetal copolymers with a pendent chain for potential conjugation [69]. Again, a divinyl ether monomer was used along with PEG diol (3400 g/mol) and a diol derived from des-aminotyrosine-tyrosine. Polymerization gave polyacetals with weight-average molecular weights ranging from 24,000 to 71,000 g/mol (PDI 1.6-2.9). Interestingly, the dipeptide diphenol incorporation into the polymer main chain was less than that of the aliphatic PEG incorporation. Polymer degradation was again pH-dependent and faster overall for polyacetals derived from the divinyl ether and PEG diol only. Air-water contact measurements were lower for the *ter*-polyacetals than for the polyacetals derived from PEG alone, indicating that surface properties could be imparted from the more hydrophobic diphenol monomer. Thermal and some solution properties of the *ter*-polyacetals were more influenced by the characteristics of PEG in the polymer main chain than the structure of the pendent chains that were examined from the peptide diol. The dipeptide diol monomers used in these polyacetals are the same ones used in the preparation of pseudo-poly(amino acids), which have been extensively developed for clinical use by Kohn and coworkers [70].

We also reported the *ter*-polymerization strategy with the intent to combine DES (diethylstilboestrol), which is a diol, into the polymer backbone [23]. Hydrolytic degradation studies with these bioresponsive polymers showed an enhanced rate of hydrolysis of the polymer backbone at pH 5.5, in which 65% of DES was released over 96 h, in comparison to pH 7.4, where only 4% of DES was released during the same time [23]. This DES containing *ter*-polyacetal has continued to be developed by Vicent *et al.* [32,71].

More recently, camptothecin (CPT) was conjugated to poly(1-hydroxymethylethylene hydroxymethylformyl) (PHF or Fleximer®) and the conjugate **11** (XMT-1001) is currently in clinical development [72,73]. The polyacetal is derived from the exhaustive oxidation of dextran [74]. CPT is poorly soluble and prone to rapid inactivation through lactone ring hydrolysis *in vivo* [75]. As a cytotoxic compound, CPT has been used in many efforts to develop other CPT-polymer conjugates [45,76,77]. The sodium salt of CPT has also been examined in phase I trials, but only modest responses were observed while severe toxicities remain.

CPT is conjugated to XMT-1001 **11** through a glycine-succinate tether, which undergoes intramolecular trans-acetylation to release CPT from the polymer in the form of camptothecin-20-*O*-(*N*-succinimidoglycinate) (CPT-SI). The remaining succinimide ester tether on CPT then undergoes a second hydrolysis reaction to form the intermediate camptothecin-20-*O*-(*N*-succinamidoyl-glycinate) (CPT-SA). Both CPT intermediates can then hydrolyze to give free CPT [72]. This two-step release mechanism can in principle enable a slow sustained systemic release of CPT to potentially improve the distribution of CPT [78] to the tumor and also to stabilize CPT in the lactone form [79]. Moreover, the release of CPT-SI and CPT-SA results in lower renal toxicity than has been observed for other CPT conjugates [74].

13.2.2 Particulate-Associated Systems

Noncovalently associating an active agent within a nano- or microsized particulate such as a micelle, liposome, or microparticle has been another strategy to optimize pharmacokinetics and to alter the biodistribution of an active pharmaceutical agent with a narrow therapeutic window. Analogous to covalent conjugates, the large size of the particulate is exploited to achieve extended circulation times and to exploit permeability gradients that may exist at diseased tissue sites. In the case of polyacetal-derived particulates, degradation of acetal (or ketal) moieties is a key mechanism used in an effort to release the biological agent at the site of action where often there can be a slight

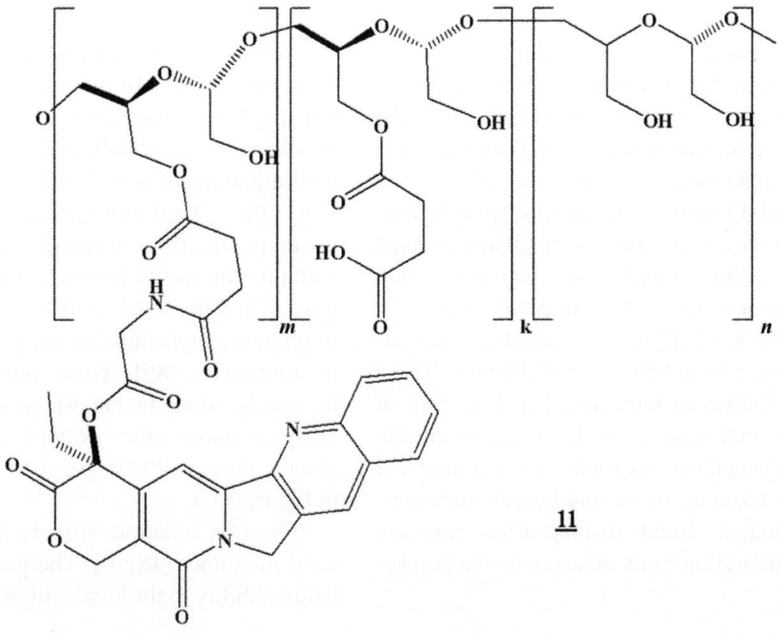

11

pH drop compared to the physiological pH in the blood compartment.

Murthy and colleagues have developed different polyketals that have been used to form particulates with a range of biologically active agents [28,80,81]. Poly(1,4-phenylene acetone dimethyleneketal) (PPADK) **14** (Figure 13.4a) is a hydrophobic polyketal derived from 1,4-benznendimethanol **12** and 2,2-dimethoxypropane **13**, which is the dimethyl ketal of acetone. The PPADK polyketal **14** is prepared by ketal exchange and removal of methanol by distillation. The steroid, dexamethasone, which is poorly soluble, was shown to be encapsulated into PPADK **14** particulates that ranged in size from about 200 to 600 nm [28]. PPADK **14** alone [28] displayed a hydrolysis half-life of 102 h at pH 7.4 and 35 h at pH 5.0, which is often assumed to be the pH encountered within the lysosome during endocytic uptake of particulates into a cell.

Aliphatic polyketals **17** (Figure 13.4b) were then prepared using varying proportions of two diols such as cyclohexanedimethanol **16** and acyclic diols **5** such as 1,4-butanediol or 1,5-pentanediol [82]. The hydrolysis half-life of the polyketals **17** could be modulated with different proportions of diol monomers. Preparing a polyketal **17** with repeat units derived from approximately 87% diol **16** and 13% pentanediol **5** displayed a hydrolysis half-life of about 2 days at pH 4.5. These polyketals were stable for weeks to hydrolysis at pH 7.4. The

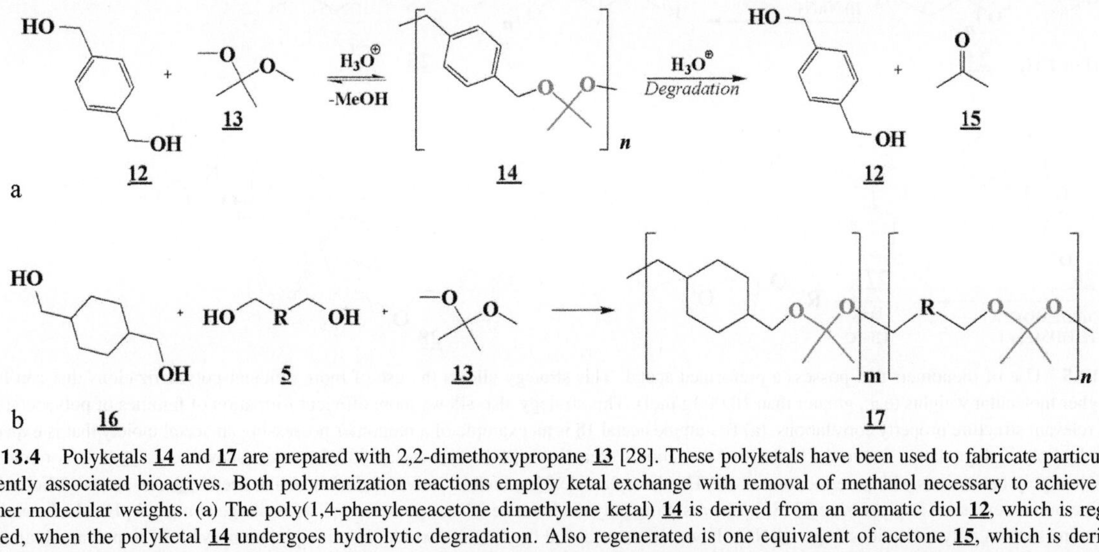

FIGURE 13.4 Polyketals **14** and **17** are prepared with 2,2-dimethoxypropane **13** [28]. These polyketals have been used to fabricate particulates with noncovalently associated bioactives. Both polymerization reactions employ ketal exchange with removal of methanol necessary to achieve appropriate polymer molecular weights. (a) The poly(1,4-phenyleneacetone dimethylene ketal) **14** is derived from an aromatic diol **12**, which is regenerated, as expected, when the polyketal **14** undergoes hydrolytic degradation. Also regenerated is one equivalent of acetone **15**, which is derived from 2,2-dimethoxypropane **13**. (b) Polyketal **17** is a copolymer derived from two diols **15** and **5**. Using various proportions of these diols allows polyketals **17** with varying hydrophobicities to be prepared. Such families of polyketals can display different degradation properties and presumably a range of particle-related properties such as size and encapsulation efficiency. (For color version of this figure, the reader is referred to the online version of this chapter.)

potential for these polyketals **17** to deliver different types of biologically active agents has been described for imatinib to treat inflammation [82] and siRNA to inhibit Nox2-NADPH expression to inhibit production of reactive oxygen species to improve cardiac function following a heart attack [83].

Polyketals **14** and **17** were designed to display hydrolysis half-lives that are faster than particulates derived from polyesters (e.g., PLGA) and slower than poly(ortho esters) or poly(β-amino esters) [28]. More recently [84], microparticles of polyketal **17** were fabricated with encapsulated apoptosis p38 inhibitor (SB239063). These particles were also fabricated with an alkyl derivative of *N*-acetylglucosamine that was noncovalently anchored into the surface of the polyketal microparticles. It was found that these particles were taken up by cultured cardiomyocytes. In a rat study, the treated animals displayed less coronary tissue damage after infarction when treated with the polyketal particles.

Another strategy to prepare polyacetal particulates involves using acetal monomers that have orthogonal functionality for polymerization. Such functionality can be used to achieve efficient polymerizations that can avoid some of the limitations of acetal exchange reactions and the use of vinyl ethers. This strategy is conducive to preparing libraries of polymers in a practical manner without too much variation in the polymerization conditions. Murthy employed this approach with acyclic diene ketal monomers to prepare polyketals that were used to prepare acid-labile microparticles [85]. These monomers were polymerized by acyclic diene metathesis polymerization. Although not comprehensive, other examples utilizing this approach of placing the acetal/ketal function into a monomer are shown in Figure 13.5.

The first example (Figure 13.5a) uses the bis-amino ketal monomer [18] **18**. The presence of the basic amines lends stability to the ketal within the monomer **18**, while the

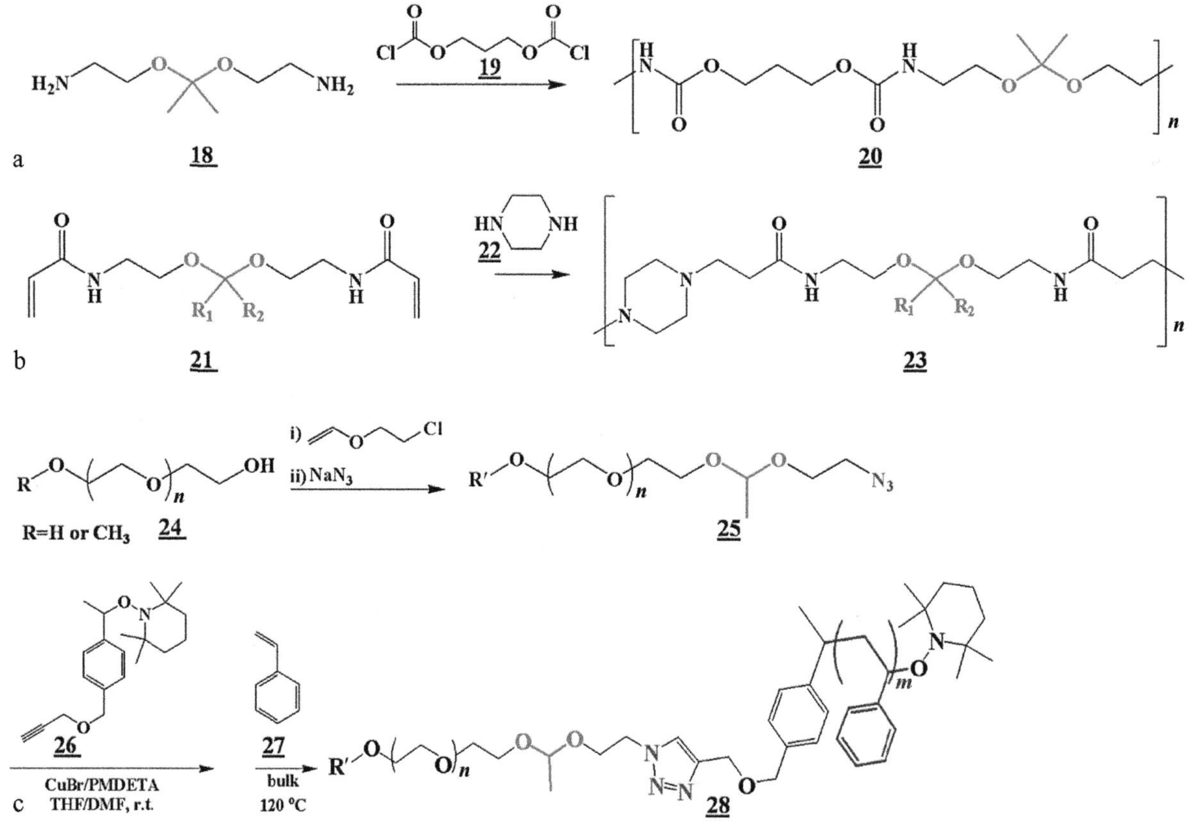

FIGURE 13.5 Use of monomers that possess a preformed acetal. This strategy allows the use of more efficient polymerizations that can be used to achieve higher molecular weights (e.g., greater than 10,000 g/mol). This strategy also allows more efficient formation of families of polyacetals/ketals to determine relevant structure property correlations. (a) Bis-amine acetal **18** is an example of a monomer possessing an acetal moiety that is expected to be resistant to hydrolytic degradation due to the basicity of the amines. Reaction with isocyanates, chloroformates (shown, e.g., **19** [18]), or acid chlorides can be efficient polymerizations with pure monomers. (b) The diacrylamide **21**, which is derived from bis-amine acetal **18**, can act as a Michael acceptor for bis-amine monomers, such as piperazine **22** [86]. The resulting polyamidoamine **23** can possess a positive charge at neutral pH, which can be used to complex anions such as polynucleotides (e.g., siRNA); however, some acetal degradation would, in principle, be possible. The polyamidoamine **23** can undergo degradation by retro-Michael reaction and acetal degradation. (c) An example is a precursor macromonomer **25**, which possesses a single acetal moiety that upon derivatization with **26** is utilized in a nitroxide-mediated polymerization with styrene **27** to give the block copolymer **28** with an acetal linking to the two blocks of the polymer [87]. (For color version of this figure, the reader is referred to the online version of this chapter.)

amines can also undergo efficient polymerization with diisocyanates and dichloroformates (such as **19**) to form the acid-labile polyketals **20**. This approach using bis-amino ketal monomer **18** gave polymers with number-average molecular weights of about 20,000 g/mol in most cases and sometimes higher (e.g., 42,000 g/mol) with polydispersities less than 2. Frechet and colleagues used the polyketal copolymers **20** to fabricate microparticles [18]. Ketal hydrolysis at different rates in mildly acidic conditions as a function of the polymer hydrophobicity could lead to optimal release of the bioactive (e.g., therapeutic protein or vaccine) [18,88]. The ability to tune acetal/ketal degradation rates as a function of hydrophobicity is a key strategy used to exploit acetal/ketal hydrolysis as a stimuli-responsive switch. This approach is also being utilized in particulate-associated strategies for drug delivery and in tissue engineering efforts [89].

A related strategy described by Frechet (Figure 13.5b) utilizes a bis-diene amide monomer **21** that possesses an internal acetal/ketal moiety [86]. Monomer **21** is a bis-acrylamide capable of conjugate or Michael addition at unsubstituted double bonds. A diamine such as piperazine **22** can be used to prepare acetal/ketal polyamidoamines **23**. One characteristic of polyacetal polyamidoamines **23** is they will be cationic and have been used in studies to complex oligonucleotide analogues (e.g., siRNA and DNA).

The Michael accepting ability of the bis-acrylamide **21** is somewhat low due to the low electron withdrawing capacity of the acrylamides; however, it is sufficient to allow polymerization to occur so that polyamidoamines **23** with number-average molecular weights ranging from 3000 to 10,000 g/mol with polydispersities of about 2 can be obtained. However, using a nondegradable bis-acrylamide and piperazine **22**, it was possible to isolate the corresponding polyamidoamine with a molecular weight of 52,000 g/mol, which illustrates how the lability often encountered in acetal/ketal-containing molecules can limit molecular weight. The hydrolysis half-life at pH 5 for the dimethylketal polyamidoamine **23** was very short (0.03 day), whereas the methyl acetal-polyamidoamine **23** displayed a half-life at pH 5 of 81 days. Although number-average molecular weights of the two polymers were 3.3 and 9.8 g/mol, respectively, this does not account for the large difference in hydrolysis half-life as a benzyl ethylene glycol with 4 (CH_2-CH_2-O) repeat units in the benzyl *para* position of the acetal-polyamidoamine **23** displayed a half-life of 3 days at pH 5.

Use of acetal/ketal monomers with reactive polymerizable functionality allows easier access to families of related polyacetals. This facilitates determination of structure-activity correlations that influence many properties including hydrolysis half-life. Hydrolytic stability is often a function of whether the degradable unit is a ketal or acetal, along with other features of the polymer structure (e.g., hydrophilicity or neighboring group effects). The use of a polymerizable bis-diene Michael acceptor as an A-A monomer, but without the internal acetal or ketal moiety, is also utilized to make Ferruti's degradable polyamidoamines [90] and Langer's poly(β-amino acids) [91,92].

Other monomers such as diacrylates **29** and **30** can also be used to prepare polyacetal/ketal polyamidoamines [19,22] or they can be polymerized directly to form particlates [22]. Frechet and colleagues have pioneered the use of the benzaldehyde-derived cross-linker **29** to prepare hydrogel microparticles encapsulating different active agents [19]. One advantage of this cross-linker is it allowed the introduction of substituents in the *para* position of the benzaldehyde benzene ring to modulate the acid lability of the acetal. The bis-acrylamide acetal cross-linker with a *p*-methoxy substituent produced microgels that hydrolyzed rapidly within the pH 5.0 environment that is encountered in phagolysosomes. Hydrolysis was proven to be extremely rapid at pH 5.0, with a half-life of 5.5 min, whereas the system remained stable at pH 7.4, with a half-life of 24 h [80].

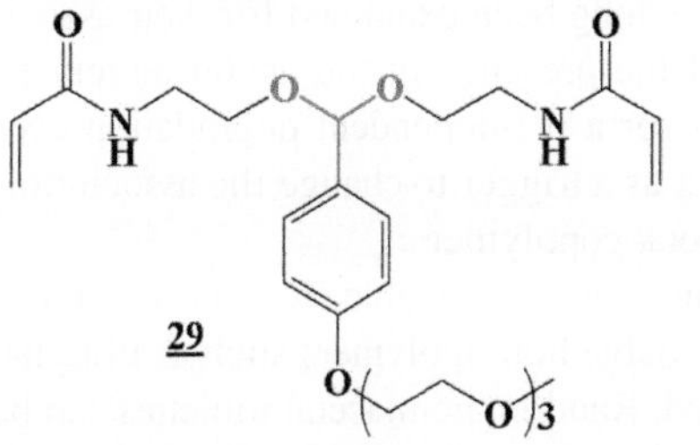

29

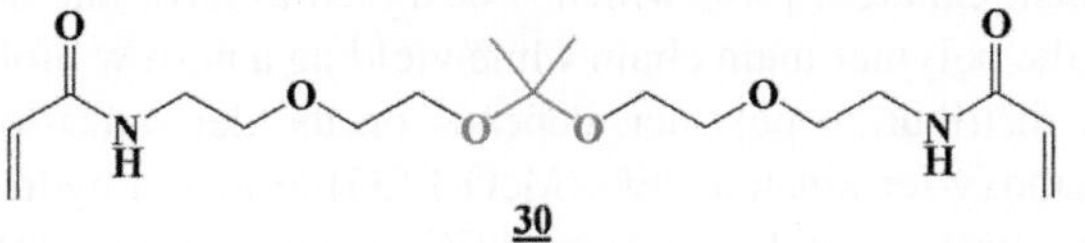

30

Particulates derived from cross-linkers such as **29** were also used to encapsulate plasmid DNA [19]. Retention of the DNA at physiological pH was achieved while being completely released in acidic conditions expected in the lysosome. Trapping the plasmid DNA within the cross-linked microparticle prevented enzymatic degradation of the DNA when exposed to serum nucleases. This system was then evaluated *in vivo*; the ability of protein-loaded particles to provide immunity against tumors in mice was investigated, using ovalbumin as the model [93]. The corresponding cross-linker with the triglyme moiety was also used to prepare acid-degradable cationic nanoparticles for protein-based vaccine development [19].

The aliphatic monomers and cross-linkers such as **30** were also used to examine the properties of the resulting particulates for lysosomotropic delivery [22]. Particulates derived from **30** and related aliphatic monomers displayed good biocompatibility and fast hydrolysis rates at acidic pH values. This aliphatic cross-linker undergoes degradation to generate acetone as the carbonyl-related degradation product due to ketal hydrolysis, which is a relatively nontoxic metabolic intermediate of fatty acid oxidization. A related acetal amine-functionalized acrylate monomer was also synthesized [94]. Due to its ability to generate small molecules upon hydrolysis, pH-sensitive polyacrylamide particles functionalized with the cell-penetrating peptide, poly(arginine), were also prepared [94] and these particles led to an efficient cellular uptake while displaying no cytotoxic effects at the concentrations that were evaluated.

An acetal-containing macromonomer can also be utilized (Figure 13.5c) [87]. This strategy again allows the synthesis of higher molecular-weight polymers by efficient means. A narrow molecular weight macromonomer **25** that is based on PEG **24**, which was then derivatized with **26** for nitroxide-mediated polymerization with styrene **27**, was used [87]. This strategy also allows for narrower molecular weight-distributed polymers to be prepared. Clearly, a polymer such as **28** only has a single acetal moiety within its main chain [87]. Hydrophilic-hydrophobic block copolymers have long been examined for their association properties, and the presence of the acetal moiety provides the possibility for a pH-dependent degradation event to occur that can act as a trigger to change the association properties of such block copolymers.

Placement of a degradable element within nonhydrolytically degradable homopolymers such as PEG has also been investigated. Random hemiacetal moieties can be formed in PEG substrates by oxidation with Fenton's reagent [95]. A recent example [96], which is designed to have one acetal in the polymer main chain while yielding a narrow molecular distribution polymer, depends on the derivatization of methoxy-terminated PEG (MeO-PEG) to give a hydroxyl-terminated acetal-containing PEG macromonomer (MeO-PEG-acetal-OH). The hydroxyl-acetal-derivatized PEG can be used as a macromonomer in anionic ring opening polymerization with ethylene oxide to give a methoxy-PEG-OH with a single internal acetal group.

Acetals are widely used as protection groups in synthetic chemistry since they are stable to base. This allows chemical transformations to be conducted in the presence of carbonyls and diols that could undergo competitive reaction. Likewise, a carbonyl function can be masked during a radical or anionic polymerization. Once the polymer is made, exposure to mild acid will reregenerate the carbonyl, which can be further functionalized. Maynard and coworkers describe the polymerization of acetal methacrylates (e.g., 3,3′-diethoxypropyl methacrylate) by controlled radical polymerization to yield a narrow molecular weight distribution polymer with a masked pendent acetal [97]. Acid hydrolysis generates a pendent aldehyde that can undergo reaction with amino-oxy-derivatized molecules of interest [97]. A related example describes how a pendent acetal polymer could be used to pattern surfaces. Following acetal hydrolysis, the regenerated aldehyde can be used to derivatize the surface [98].

We used a related strategy to prepare block copolymers **34** (Figure 13.6) [99] from the reactive precursor polymer **32** [100]. Active ester precursor polymers such as **32** were developed to be used as common intermediates to produce families of polymers for study [101]. In the example shown in Figure 13.6, the coblocked pendent ketal polymer **34** is a latent hydrophilic-hydrophobic polymer that is capable of solution aggregation. Acid hydrolysis of the pendent ketal leads to the more hydrophilic polymer **35** and deaggregation. Figure 13.7 shows acetalated dextran **38**, which was prepared directly from dextran **36** using the vinyl ether **37** [107]. Reaction on a polymer to achieve acetalization (or ketalization) is a different process than using an acetal monomer or cross-linker that is then polymerized. The ketal functionalized dextran **38** was used to form biocompatible microparticles that upon exposure to intracellular pH values (e.g., pH 5) undergo hydrolysis of the ketal moieties. Another example of a ketalization polymer-analogous reaction is the ketalization of poly(vinyl alcohol) with cyclohexanone [108].

13.2.3 Polymer-Oligonucleotide Complexes

Nonviral carriers have been examined in attempts to protect and try to achieve intracellular delivery of oligonucleotides such as plasmid DNA and siRNA [109]. Polyacetals would be expected to undergo enhanced rates of degradation upon endocytic uptake. In addition to the aforementioned acetal/ketal polyamidoamines **23** (Figure 13.5b),

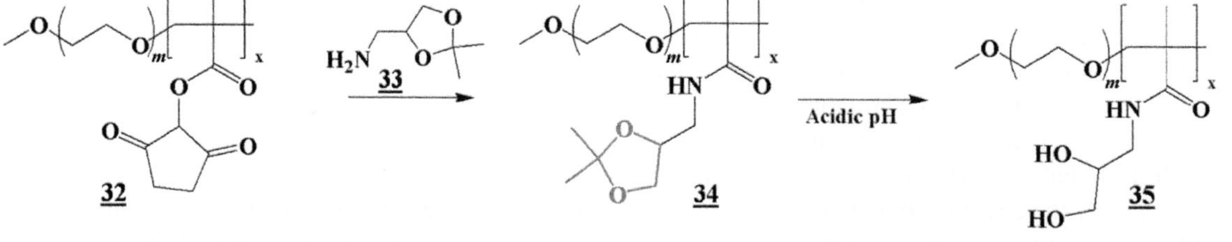

FIGURE 13.6 Preparation of hydrophilic-hydrophobic block copolymer **34** through the coupling reaction between precursor **32** and the amino ketal **33**. Block copolymer **34** is designed to associate and then dissociate upon hydrolysis of the ketal to give the more hydrophilic block copolymer **35** [99]. (For color version of this figure, the reader is referred to the online version of this chapter.)

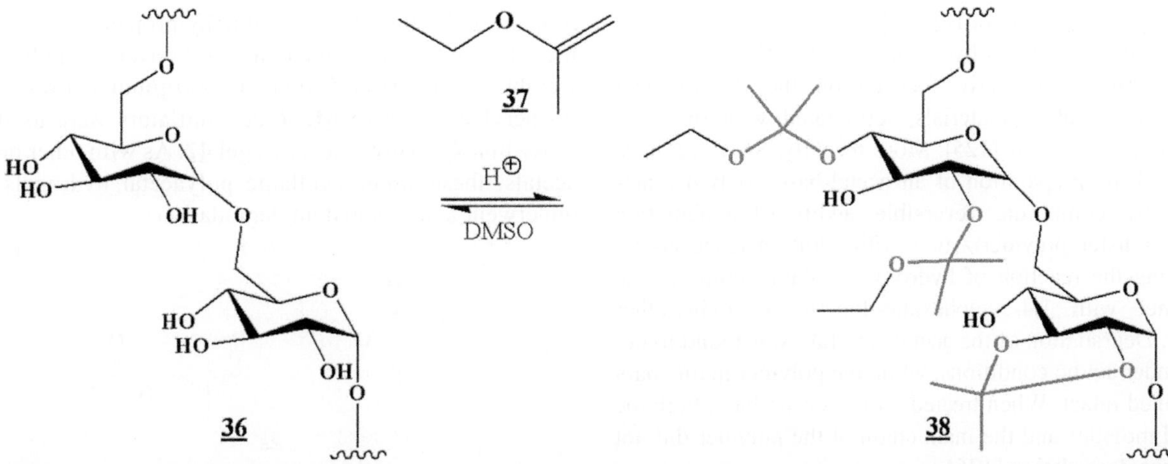

FIGURE 13.7 The polymer-analogous reaction to give the hydrophobized ketal dextran 38 [102]. The acetal modification is achieved by reaction of dextran 36 with 2-ethoxypropene 37 to give the poorly water-soluble dextran ketal 38. This change in dextran solubility enables 38 to be fabricated into particulates using nonaqueous media [103–106]. (For color version of this figure, the reader is referred to the online version of this chapter.)

other work has been described to determine the potential for polyacetals to be used for endocytic uptake of complexed oligonucleotides.

Knorr and coworkers prepared two types of acid-degradable nonviral gene carriers, by polymerizing oligoethylenimine with different acetal or ketal monomers [14]. Oligo-polyacetal complexes displayed an improved toxicity profile compared to those made with acid-stable polymer analogues. The kinetics of hydrolysis were measured and confirmed the pH-dependent degradation profile of the acetal functions [14]. Knorr's group also copolymerized oligoethylenimine with acetal PEG derivatives, forming a dynamic PEG-polycation copolymer [13]. Polyplexes formed with these degradable polymers displayed high transfection efficiency and improved toxicity profiles when compared to their acid-stable counterparts. However, these cationic particles displayed unspecific interactions with blood components [110]. PEG and other hydrophilic neutral polymers have been examined in efforts to reduce surface charge exposure of cationic copolymers in attempts to reduce the toxicity of the polyplexes. However, incorporating PEG has also been shown to diminish cellular uptake [12,13,111].

The synthesis of graft *ter*-polymers that are composed of a hydrophobic, membrane-disruptive backbone onto which hydrophilic PEG chains have been grafted through acid-degradable acetal linkers has also been reported [112]. These polymers were shown to also have the capacity to deliver oligonucleotides and macromolecules into the cytoplasm of hepatocytes. Improved endosomal escape and more efficient release of siRNA to the cytoplasm was reported with low cytotoxicity [113] for some of the polyketal derivatives. In a similar fashion, a branched PEI was also ketalized and showed enhanced transfection and RNA interference and reduced cytotoxicity [114,115].

Acid-degradable cationic nanoparticles were synthesized [116] using an aliphatic monomer with a pendant primary amine group and a cleavable acetal linkage [22]. These nanoparticles were designed to cause swelling and osmotic destabilization of the endosome, while the cationic branches were cleaved from the polymeric backbone to allow DNA dissociation from the polymer. This approach resulted in higher transfection efficiency for the degradable nanoparticles than PEI polyplexes at low concentrations [116].

Micelles prepared by self-assembling DNA with PEG-conjugated poly(ketalized serine) underwent structural changes at mildly acidic pH to release oligonucleotide into the cytoplasm. Results demonstrated that upon acid hydrolysis of the ketal linkages, poly(ketalized serine) was converted into neutral poly(serine), which destabilized the micelles [117]. Another approach was based on a host-guest vector relying on the self-assembly of cationic β-cyclodextrin derivatives with a poly(vinyl alcohol) backbone bearing PEG. Adamantane or cholesterol was conjugated to the polymer through an acid-sensitive benzylidene-acetal linkage [118,119]. These systems were investigated for their ability to promote supramolecular complex formation with pDNA- and siRNA-cyclodextrin derivatives. *In vitro* studies showed that they were less toxic than 25 kDa bPEI while maintaining higher transfection efficiencies [118].

13.2.4 Hydrogels

Hydrogels are generally prepared from a polymer chain network, creating a colloidal gel that can have a water content from 70% to almost 100% [120]. Hydrogels can be highly absorbent and permeable [120]. Medical and pharmaceutical uses of hydrogels range from wound dressing [121], skin grafts [122], oxygen-permeable contact lenses [123], to delivery of drugs [124].

Several studies have focused on the development of cyclic-acetal biomaterials, terminated with diol and carbonyl end groups [125]. More recently, Sui *et al.* have reported the preparation of an acetal-based polymer network by combining reversible addition-fragmentation chain-transfer polymerization with addition reactions by allowing the reaction of hydroxyl pendant groups of the polymer with 1,4-cyclohexane-dimethanol divinylether [126]. Degradation of the acetal structure was found to occur under acidic conditions, while the polymer main chain remained intact. When treated with a strong base, both the acetal moieties and the main chain of the polymer did not undergo degradation [126].

The synthesis of hyperbranched polyacetals via a melt transacetalization polymerization process of an AB_2 monomer bearing a single hydroxyl group and a dialkyl acetal has also been described [89]. The bulk degradation rates of these hyperbranched polyacetals at pH 4 revealed a strong dependence on the hydrophobicity of the peripheral alkyl substituents.

Thermogels that take advantage of the hydrolytic properties of polyacetals have also been explored. Schacht *et al.* have reported the synthesis of graft copolymers comprised of a polyacetal backbone with pendant PEG chains. Incorporation of FITC-BSA at 1 wt.% into the thermogels resulted in a sustained release over about 100 days at pH 7.4 and 40 days at pH 5.5. Release of FITC-BSA occurred by a process that appears to be erosion-controlled [8].

13.2.5 Polyacetals in Tissue Engineering

While much recent work has been described to develop polyacetals for drug delivery applications, and historically they have been used as implant materials, more recently, they have been examined as potential scaffold materials in tissue engineering. Implants of Delrin (polyoxymethylene) to repair heart valves were examined, but there was too much swelling *in vivo* [127]. However, this polyacetal has been used as an orthopedic implant [128] and as an orthopedic implant-coating material [129,130] to interface with bone tissue as this polyacetal has a similar modulus to bone. Ultrasound is used in the diagnosis of osteoporosis and porous polyacetal blocks were found useful to gain insights into bone porosity and ultrasonic properties [131].

Cross-linked cyclic polyacetals have been examined in recent years for use in regenerative medicine [125] in an effort to minimize the inflammatory response often caused by degradation side products (e.g., acids from polyesters). One example is the copolymer hydrogel poly[poly(ethylene glycol)-*co*-cyclic-acetal] **42** (Figure 13.8). The diol **39** is prepared from trimethanol propane and 3-hydroxy-2,2-dimethylpropanal [132]. The prepolymer **40** was isolated at weight-average molecular weights ranging from 5000 to 44,000 g/mol and was then bis-acrylated to give **41**. Redox initiators were used to cross-link **41** to give the hydrogel **42**. As with other polyacetals, these water-swellable polyacetal hydrogels **42** underwent acid-dependent degradation.

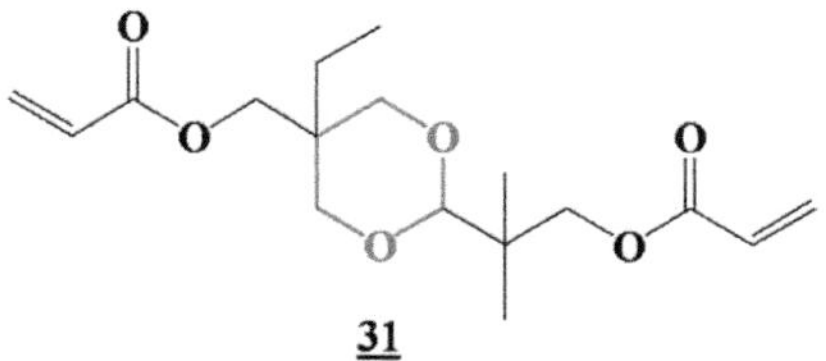

31

Other cross-linked polyacetals based on diol **39** have also been described [125]. The corresponding diacrylate **31**, which is derived from diol **39**, was prepared and then directly exposed to redox initiators to form the corresponding cross-linked polyacetal. The diacrylate **31** can also be mixed in different proportions with a PEG-diacrylate (PEGDA) to form analogous hydrogel networks. These networks are related to that shown for hydrogel **42** and together have allowed a wide range of materials of similar composition to be evaluated. These cyclic-acetal hydrogels are being evaluated for potential craniofacial tissue engineering applications [133]. Composites of these cyclic polyacetals and uniformly distributed hydroxyapatite particles have been shown to facilitate osteogenic differentiation of encapsulated bone marrow stromal cells by potentially promoting osteogenic signal expression. It is also thought that lack of acidic by-products from acetal degradation of these osteogenic materials contributes to their biocompatibility [133,134]. The diol **39** has also been allylated to be used with four-arm thiols in the presence of initiators to undergo a thiol-ene reaction to give hydrogel networks for study in tissue engineering [135].

13.3 CONCLUSIONS

There has been a recent resurgence in the development of polyacetals for a wide range of biomedical applications ranging from drug delivery to tissue engineering. While new acid-labile polymers such as **43** [136] (Figure 13.9) are also being developed and which do not generate acidic by-products, it does seem that in recent years, a considerable amount of research has been conducted to move the polyacetal field forward. Efforts have also focused on biocompatibility to ensure the carbonyl degradation products are not toxic. The strategies to achieve reproducible polymerizations with suitable molecular weight characteristics bode well to exploit the unique acid-labile characteristics of polyacetals in biomedicine.

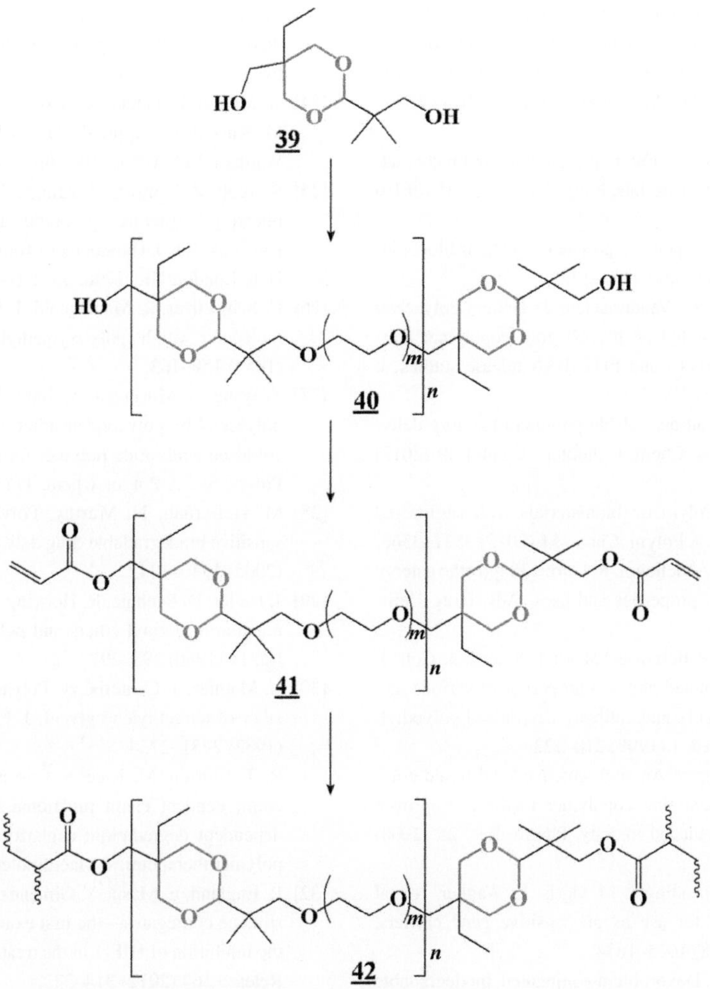

FIGURE 13.8 Monomer 39 is used to prepare cyclic polyacetal hydrogels such as 42 for potential use in tissue regeneration applications [125,132]. (For color version of this figure, the reader is referred to the online version of this chapter.)

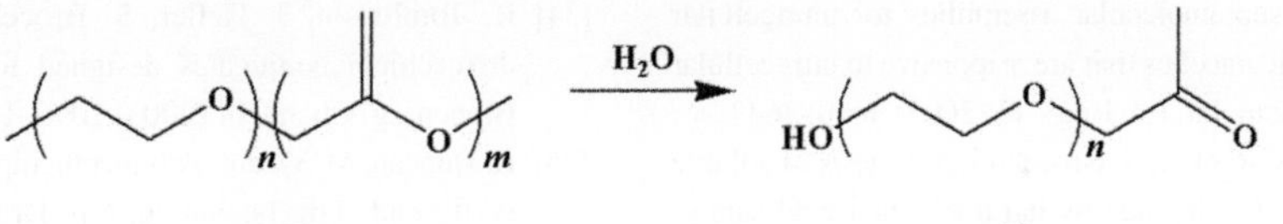

FIGURE 13.9 Polymer 43 displays enhanced hydrolysis rates at acidic pH values [136].

ACKNOWLEDGMENTS

S. C. gratefully acknowledges funding from the UCL School of Pharmacy and PolyTherics Ltd. S. B. is grateful for funding from NIHR Biomedical Research Centre at Moorfields Hospital and the UCL Institute of Ophthalmology, Moorfields Special Trustees, the Helen Hamlyn Trust (in memory of Paul Hamlyn), Fight for Sight, and The Freemasons Grand Charity. S. B. is also grateful for funding from the UK Engineering and Physical Sciences Research Council (EPSRC) for the EPSRC Centre for Innovative Manufacturing in Emergent Macromolecular Therapies. Financial support from the consortium of industrial and governmental users for the EPSRC Centre is also acknowledged.

REFERENCES

[1] I. Tannock, D. Rotin, Acid pH in tumors and its potential for therapeutic exploitation perspectives, Cancer Res. 49 (1989) 4373–4384.

[2] A. Akinc, R. Langer, Measuring the pH environment of DNA delivered using nonviral vectors: implications for lysosomal trafficking, Biotechnol. Bioeng. 78 (2002) 503–508.

[3] A. Asokan, M. Cho, Exploitation of intracellular pH gradients in the cellular delivery of macromolecules, J. Pharm. Sci. 91 (2002) 903–913.

[4] B. Ulery, L. Nair, C. Laurencin, Biomedical applications of biodegradable polymers, J. Polym. Sci. B Polym. Phys. 49 (2011) 832–864.

[5] A. Esser-Kahn, S. Odom, N. Sottos, S. White, J. Moore, Triggered release from polymer capsules, Macromolecules 44 (2011) 5539–5553.

[6] D. Roy, J. Cambre, B. Sumerlin, Future perspectives and recent advances in stimuli-responsive materials, Prog. Polym. Sci. 35 (2010) 278–301.

[7] E. Gil, S. Hudson, Stimuli-responsive polymers and their bioconjugates, Prog. Polym. Sci. 29 (2004) 1173–1222.

[8] E. Schacht, V. Toncheva, K. Vandertaelen, J. Heller, Polyacetal and poly(ortho ester)-poly(ethylene glycol) graft copolymer thermogels: preparation, hydrolysis and FITC-BSA release studies, J. Control. Release 116 (2006) 219–225.

[9] S. Binauld, M. Stenzel, Acid-degradable polymers for drug delivery: a decade of innovation, Chem. Commun. (Camb.) 49 (2013) 2082–2102.

[10] S. Deshayes, A. Kasko, Polymeric biomaterials with engineered degradation, J. Polym. Sci. A Polym. Chem. 51 (2013) 3531–3566.

[11] J. Heller, J. Barr, S. Ng, K. Abdellauoi, R. Gurny, Poly(ortho esters): synthesis, characterization, properties and uses, Adv. Drug Deliv. Rev. 54 (2002) 1015–1039.

[12] P. Erbacher, T. Bettinger, P. Belguise-Valladier, S. Zou, J. Coll, J. Behr, J. Remy, Transfection and physical properties of various saccharide, poly(ethylene glycol), and antibody-derivatized polyethylenimines (PEI), J. Gene Med. 1 (1999) 210–222.

[13] V. Knorr, M. Ogris, E. Wagner, An acid sensitive ketal-based polyethylene glycol-oligoethylenimine copolymer mediates improved transfection efficiency at reduced toxicity, Pharm. Res. 25 (2008) 2937–2945.

[14] V. Knorr, V. Russ, L. Allmendinger, M. Ogris, E. Wagner, Acetal linked oligoethylenimines for use as pH-sensitive gene carriers, Bioconjug. Chem. 19 (2008) 1625–1634.

[15] H. Yoo, E. Lee, T. Park, Doxorubicin-conjugated biodegradable polymeric micelles having acid-cleavable linkages, J. Control. Release 82 (2002) 17–27.

[16] Y. Bae, S. Fukushima, A. Harada, K. Kataoka, Design of environment-sensitive supramolecular assemblies for intracellular drug delivery: polymeric micelles that are responsive to intracellular pH change, Angew. Chem. Int. Ed. Engl. 42 (2003) 4640–4643.

[17] M. DuBois Clochard, S. Rankin, S. Brocchini, Synthesis of soluble polymers for medicine that degrade by intramolecular acid catalysis, Macromol. Rapid Commun. 21 (2000) 853–859.

[18] S. Paramonov, E. Bachelder, T. Beaudette, S. Standley, C. Lee, J. Dashe, J. Fréchet, Fully acid-degradable biocompatible polyacetal microparticles for drug delivery, Bioconjug. Chem. 19 (2008) 911–919.

[19] S. Goh, N. Murthy, M. Xu, J. Fréchet, Cross-linked microparticles as carriers for the delivery of plasmid DNA for vaccine development, Bioconjug. Chem. 15 (2004) 467–474.

[20] K. Broaders, S. Pastine, S. Grandhe, J. Fréchet, Acid-degradable solid-walled microcapsules for pH-responsive burst-release drug delivery, Chem. Commun. 47 (2011) 665–667.

[21] E. Gillies, J. Fréchet, pH-Responsive copolymer assemblies for controlled release of doxorubicin, Bioconjug. Chem. 16 (2005) 361–368.

[22] Y. Kwon, S. Standley, A. Goodwin, E. Gillies, J. Fréchet, Directed antigen presentation using polymeric microparticulate carriers degradable at lysosomal pH for controlled immune responses, Mol. Pharm. 2 (2005) 83–91.

[23] M. Vicent, R. Tomlinson, S. Brocchini, R. Duncan, Polyacetal-diethylstilboestrol: a polymeric drug designed for pH-triggered activation, J. Drug Target. 12 (2004) 491–501.

[24] G. Melton, E. Peters, R. Arisman, Engineering Thermoplastics, in: M. Kutz (Ed.), Applied Plastics Engineering Handbook, Elsevier, Waltham MA, USA, 2011, pp. 9–10.

[25] S. Teoh, Z. Tang, G. Hastings, Thermoplastics in biomedical applications: structures, properties and processing, in: J. Black, G.W. Hastings (Eds.), Handbook of Biomaterial Properties, Chapman and Hall, London/UK, 1998, pp. 270–302.

[26] C. Schweitzer, R. MacDonald, J. Punderson, Thermally stable high molecular weight polyoxymethylenes, J. Appl. Polym. Sci. 1 (2) (1959) 158–163.

[27] Y. Wang, H. Morinaga, A. Sudo, T. Endo, Synthesis of amphiphilic polyacetal by polycondensation of aldehyde and polyethylene glycol as an acid-labile polymer for controlled release of aldehyde, J. Polym. Sci. A Polym. Chem. 49 (2011) 596–602.

[28] M. Heffernan, N. Murthy, Polyketal nanoparticles: a new pH-sensitive biodegradable drug delivery vehicle, Bioconjug. Chem. 16 (2005) 1340–1342.

[29] J. Heller, D. Penhale, R. Helwing, Preparation of polyacetals by the reaction of divinyl ethers and polyols, J. Polym. Sci. Polym. Lett. Ed. 18 (1980) 293–297.

[30] L. Mathias, J. Canterberry, Polyacetal formation of the monovinyl ether of tetraethylene glycol, J. Polym. Sci. Polym. Chem. Ed. 20 (1982) 2731–2734.

[31] R. Tomlinson, M. Klee, S. Garrett, J. Heller, R. Duncan, S. Brocchini, Pendent chain functionalized polyacetals that display pH-dependent degradation: a platform for the development of novel polymer therapeutics, Macromolecules 35 (2002) 473–480.

[32] R. England, E. Masiá, V. Giménez, R. Lucas, M. Vicent, Polyacetal-stilbene conjugates—the first examples of polymer therapeutics for the inhibition of HIF-1 in the treatment of solid tumours, J. Control. Release 164 (2012) 314–322.

[33] E. Ruckenstein, H. Zhang, Novel copolymer networks via the combination of polyaddition and anionic polymerization, J. Polym. Sci. A Polym. Chem. 39 (2001) 117–126.

[34] R. Tomlinson, J. Heller, S. Brocchini, R. Duncan, Polyacetal-doxorubicin conjugates designed for pH-dependent degradation, Bioconjug. Chem. 14 (2003) 1096–1106.

[35] R. Duncan, M. Vicent, Polymer therapeutics-prospects for 21st century: the end of the beginning, Adv. Drug Deliv. Rev. 65 (2013) 60–70.

[36] R. Duncan, R. Gaspar, Nanomedicine(s) under the microscope, Mol. Pharm. 8 (2011) 2101–2141.

[37] R. Duncan, Drug-polymer conjugates: potential for improved chemotherapy, Anticancer Drugs 3 (1992) 175–210.

[38] M. Nucci, R. Shorr, The therapeutic value of poly(ethylene glycol)-modified proteins, Adv. Drug Deliv. Rev. 6 (1991) 133–151.

[39] G. Pasut, F. Veronese, State of the art in PEGylation: the great versatility achieved after forty years of research, J. Control. Release 161 (2011) 461–472.

[40] P. Goddard, I. Williamson, J. Brown, L. Hutchinson, J. Nicholls, K. Petrak, Soluble polymeric carriers for drug delivery-Part 4: tissue autoradiography and whole-body tissue distribution in mice, of N-(2-hydroxypropyl)methacrylamide copolymers following intravenous administration, J. Bioact. Compat. Polym. 6 (1991) 4–24.

[41] L. Seymour, R. Duncan, Effect of molecular weight of N-(2-hydroxypropyl)methacrylamide copolymers on body distribution and rate of excretion after subcutaneous, intraperitoneal and

intravenous administration to rats, J. Biomed. Mater. Res. 21 (1987) 1341–1358.

[42] M. Barz, R. Luxenhofer, R. Zentel, M. Vicent, Overcoming the PEG-addiction: well-defined alternatives to PEG, from structure-property relationships to better defined therapeutics, Polym. Chem. 2 (2011) 1900–1918.

[43] L. Paz-Ares, et al., Phase III trial comparing paclitaxel poliglumex vs docetaxel in the second-line treatment of non-small-cell lung cancer, Br. J. Cancer 98 (2008) 1608–1613.

[44] K.S. Albain, C.P. Belani, P. Bonomi, K.J. O'Byrne, J.H. Schiller, M. Socinski, PIONEER: a phase III randomized trial of paclitaxel poliglumex versus paclitaxel in chemotherapy-naive women with advanced-stage non-small-cell lung cancer and performance status of 2, Clin. Lung Cancer 7 (2006) 417–419.

[45] J. Singer, R. Bhatt, J. Tulinsky, K. Buhler, E. Heasley, P. Klein, P. de Vries, Water-soluble poly-(L-glutamic acid)-gly-camptothecin conjugates enhance camptothecin stability and efficacy in vivo, J. Control. Release 74 (2001) 243–247.

[46] M. Yokoyama, S. Inoue, Preparation of adriamycin-conjugated poly(ethy1ene glycol)-poly(aspartic acid) block copolymer. A new type of polymeric anticancer agent, Makromol. Chem. Rapid Commun. 435 (1987) 431–435.

[47] M. Yokoyama, M. Miyauchi, N. Yamada, T. Okano, Polymer micelles as novel drug carrier: adriamycin-conjugated poly (ethylene glycol)-poly (aspartic acid) block copolymer, J. Control. Release 11 (1990) 269–278.

[48] M. Yokoyama, M. Miyauchi, N. Yamada, T. Okano, Y. Sakurai, K. Kataoka, S. Inoue, Characterization and anticancer activity of the micelle-forming polymeric anticancer drug adriamycin-conjugated poly(ethylene glycol)-poly(aspartic acid) block copolymer, Cancer Res. 50 (1990) 1693–1700.

[49] T. Sedlačík, H. Studenovská, F. Rypáček, Enzymatic degradation of the hydrogels based on synthetic poly(α-amino acid)s, J. Mater. Sci. Mater. Med. 22 (2011) 781–788.

[50] R. Petros, J. DeSimone, Strategies in the design of nanoparticles for therapeutic applications, Nat. Rev. Drug Discov. 9 (2010) 615–627.

[51] H. Maeda, J. Wu, T. Sawa, Y. Matsumura, K. Hori, Tumor vascular permeability and the EPR effect in macromolecular therapeutics: a review, J. Control. Release 65 (2000) 271–284.

[52] Y. Matsumura, H. Maeda, A new concept for macromolecular therapeutics in cancer chemotherapy: mechanism of tumoritropic accumulation of proteins and the antitumor agent smancs, Cancer Res. 46 (1986) 6387–6392.

[53] R. Sinha, G. Kim, S. Nie, D. Shin, Nanotechnology in cancer therapeutics: bioconjugated nanoparticles for drug delivery, Mol. Cancer Ther. 5 (2006) 1909–1917.

[54] M.L. Graham, PEGaspargase: a review of clinical studies, Adv. Drug Deliv. Rev. 55 (2003) 1293–1302.

[55] R. Reddy, M. Modi, S. Pedder, Use of PEGinterferon alfa-2a (40 KD) (PEGasys) for the treatment of hepatitis C, Adv. Drug Deliv. Rev. 54 (2002) 571–586.

[56] G. Molineux, PEGfilgrastim: using PEGylation technology to improve neutropenia support in cancer patients, Anticancer Drugs 14 (2003) 259–264.

[57] J. Fang, T. Sawa, T. Akaike, H. Maeda, Tumor-targeted delivery of polyethylene glycol-conjugated d-amino acid Oxidase for antitumor therapy via enzymatic generation of hydrogen peroxide, Cancer Res. 62 (2002) 3138–3143.

[58] Y.-S. Wang, S. Youngster, M. Grace, J. Bausch, R. Bordens, D. Wyss, Structural and biological characterization of PEGylated recombinant interferon alpha-2b and its therapeutic implications, Adv. Drug Deliv. Rev. 54 (2002) 547–570.

[59] D. Bissett, J. Cassidy, J. de Bono, F. Muirhead, M. Main, L. Robson, D. Fraier, M. Magnè, C. Pellizzoni, M. Porro, R. Spinelli, W. Speed, C. Twelves, Phase I and pharmacokinetic (PK) study of MAG-CPT (PNU 166148): a polymeric derivative of camptothecin (CPT), Br. J. Cancer 91 (2004) 50–55.

[60] L. Seymour, D. Ferry, D. Anderson, S. Hesslewood, P. Julyan, R. Poyner, J. Doran, A. Young, S. Burtles, D. Kerr, Hepatic drug targeting: phase I evaluation of polymer-bound doxorubicin, J. Clin. Oncol. 20 (2002) 1668–1676.

[61] P. Vasey, S. Kaye, R. Morrison, C. Twelves, P. Wilson, R. Duncan, A. Thomson, L. Murray, T. Hilditch, T. Murray, S. Burtles, D. Fraier, E. Frigerio, Phase I clinical and pharmacokinetic study of PK1 N -(2-hydroxypropyl) methacrylamide copolymer doxorubicin: first member of a new class of chemotherapeutic agents—drug-polymer conjugates, Clin. Cancer Res. 5 (1999) 83–94.

[62] C. Li, S. Wallace, Polymer-drug conjugates: recent development in clinical oncology, Adv. Drug Deliv. Rev. 60 (2008) 886–898.

[63] G. Pasut, F. Veronese, Polymer–drug conjugation, recent achievements and general strategies, Prog. Polym. Sci. 32 (2007) 933–961.

[64] J. Khandare, T. Minko, Polymer–drug conjugates: progress in polymeric prodrugs, Prog. Polym. Sci. 31 (2006) 359–397.

[65] L. Ducry, B. Stump, Antibody-drug conjugates: linking cytotoxic payloads to monoclonal antibodies, Bioconjug. Chem. 21 (2010) 5–13.

[66] S. Doronina, B. Mendelsohn, T. Bovee, C. Cerveny, S. Alley, D. Meyer, E. Oflazoglu, B. Toki, R. Sanderson, R. Zabinski, A. Wahl, P. Senter, Enhanced activity of monomethylauristatin F through monoclonal antibody delivery: effects of linker technology on efficacy and toxicity, Bioconjug. Chem. 17 (2006) 114–124.

[67] J. Flygare, T. Pillow, P. Aristoff, Antibody-drug conjugates for the treatment of cancer, Chem. Biol. Drug Des. 81 (2013) 113–121.

[68] S. Alley, N. Okeley, P. Senter, Antibody-drug conjugates: targeted drug delivery for cancer, Curr. Opin. Chem. Biol. 14 (2010) 529–537.

[69] J. Rickerby, R. Prabhakar, M. Ali, J. Knowles, S. Brocchini, Water-soluble polyacetals derived from diphenols, J. Mater. Chem. 15 (2005) 1849–1856.

[70] S. Bourke, J. Kohn, Polymer derived from the amino acid L-tyrosine: polycarbonates, polyarylates and copolymers with poly(ethylene glycol), Adv. Drug Deliv. Rev. 55 (2003) 447–466.

[71] V. Gimenez, C. James, A. Arminan, R. Schweins, A. Paul, M. Vicent, Demonstrating the importance of polymer-conjugate conformation in solution on its therapeutic output: diethylstilbestrol (DES)-polyacetals as prostate cancer treatment, J. Control. Release 159 (2012) 290–301.

[72] A. Yurkovetskiy, R. Fram, XMT-1001, a novel polymeric camptothecin pro-drug in clinical development for patients with advanced cancer, Adv. Drug Deliv. Rev. 61 (2009) 1193–1202.

[73] M. Papisov, A. Hiller, A. Yurkovetskiy, M. Yin, M. Barzana, S. Hillier, A. Fischman, Semisynthetic hydrophilic polyals, Biomacromolecules 6 (2005) 2659–2670.

[74] V. Yurkovetskiy, A. Hiller, S. Syed, M. Yin, X. Lu, J. Fischman, M. Papisov, Synthesis of a macromolecular camptothecin conjugate with dual phase drug release, Mol. Pharm. 1 (2004) 375–382.

[75] Q.-Y. Li, Y.-G. Zu, R.-Z. Shi, L.-P. Yao, Review camptothecin: current perspectives, Curr. Med. Chem. 13 (2006) 2021–2039.

[76] D. Yu, P. Peng, S. Dharap, Y. Wang, M. Mehlig, P. Chandna, H. Zhao, D. Filpula, K. Yang, V. Borowski, G. Borchard, Z. Zhang, T. Minko, Antitumor activity of poly(ethylene glycol)-camptothecin conjugate: the inhibition of tumor growth in vivo, J. Control. Release 110 (2005) 90–102.

[77] J. Cheng, K. Khin, M. Davis, Antitumor activity of beta-cyclodextrin polymer-camptothecin conjugates, Mol. Pharm. 1 (2004) 1831–1893.

[78] M. Walsh, S. Hanna, J. Sen, S. Rawal, C. Cabral, A. Yurkovetskiy, R. Fram, T. Lowinger, W. Zamboni, Pharmacokinetics and antitumor efficacy of XMT-1001, a novel, polymeric topoisomerase I inhibitor, in mice bearing HT-29 human colon carcinoma xenografts, Clin. Cancer Res. 18 (2012) 2591–2602.

[79] C. Conover, R. Greenwald, A. Pendri, C. Gilbert, K. Shum, Camptothecin delivery systems: enhanced efficacy and tumor accumulation of camptothecin following its conjugation to polyethylene glycol via a glycine linker, Cancer Chemother. Pharmacol. 42 (1998) 407–414.

[80] N. Murthy, Y. Thng, S. Schuck, M. Xu, J. Fréchet, A novel strategy for encapsulation and release of proteins: hydrogels and microgels with acid-labile acetal cross-linkers, J. Am. Chem. Soc. 124 (2002) 12398–12399.

[81] N. Murthy, M. Xu, S. Schuck, J. Kunisawa, N. Shastri, J. Fréchet, A macromolecular delivery vehicle for protein-based vaccines: acid-degradable protein-loaded microgels, Proc. Natl. Acad. Sci. U. S. A. 100 (2003) 4995–5000.

[82] S. Yang, M. Bhide, I. Crispe, R. Pierce, N. Murphy, Polyketal copolymers: a new acid sensitive delivery vehicle for treating acute inflammatory diseases, Bioconjug. Chem. 19 (2009) 1164–1169.

[83] J. Sy, G. Seshadri, S. Yang, M. Brown, T. Oh, S. Dikalov, N. Murthy, M. Davis, Sustained release of a p38 inhibitor from non-inflammatory microspheres inhibits cardiac dysfunction, Nat. Mater. 7 (2008) 863–868.

[84] W. Gray, P. Che, M. Brown, X. Ning, N. Murthy, M. Davis, N-acetylglucosamine conjugated to nanoparticles enhances myocyte uptake and improves delivery of a small molecule p38 inhibitor for post-infarct healing, J. Cardiovasc. Transl. Res. 4 (2011) 631–643.

[85] S. Khaja, S. Lee, N. Murthy, Acid-degradable protein delivery vehicles based on metathesis chemistry, Biomacromolecules 8 (2007) 1391–1395.

[86] R. Jain, S. Standley, J. Frechet, Synthesis and degradation of pH-sensitive linear poly (amidoamine)s, Macromolecules 40 (2007) 452–457.

[87] K. Satoh, J. Poelma, L. Campos, B. Stahl, C. Hawker, A facile synthesis of clickable and acid-cleavable PEO for acid-degradable block copolymers, Polym. Chem. 3 (2012) 1890.

[88] E. Bachelder, T. Beaudette, K. Broaders, S. Paramonov, J. Dashe, J. Frechet, Acid-degradable polyurethane particles for protein-based vaccines: biological evaluation and in vitro analysis of particle degradation products, Mol. Pharm. 25 (2008) 876–884.

[89] S. Chatterjee, S. Ramakrishnan, Hyperbranched polyacetals with tunable degradation rates, Macromolecules 44 (2011) 4658–4664.

[90] P. Ferruti, M. Marchisio, R. Duncan, Poly(amido-amine)s: biomedical applications, Macromol. Rapid Commun. 23 (2002) 332–355.

[91] A. Akinc, D. Lynn, D. Anderson, R. Langer, Parallel synthesis and biophysical characterization of a degradable polymer library for gene delivery, J. Am. Chem. Soc. 125 (2003) 5316–5323.

[92] A. Akinc, D. Anderson, D. Lynn, R. Langer, Synthesis of poly(beta-amino ester)s optimized for highly effective gene delivery, Bioconjug. Chem. 14 (2003) 979–988.

[93] S. Standley, Y. Kwon, N. Murthy, J. Kunisawa, N. Shastri, S. Guillaudeu, L. Lau, J. Fréchet, Acid-degradable particles for protein-based vaccines: enhanced survival rate for tumor-challenged mice using ovalbumin model, Bioconjug. Chem. 15 (2004) 1281–1288.

[94] J. Cohen, A. Almutairi, J. Cohen, M. Bernstein, S. Brody, D. Schuster, J. Fréchet, Enhanced cell penetration of acid-degradable particles functionalized with cell-penetrating peptides, Bioconjug. Chem. 19 (2008) 876–881.

[95] B. Reid, S. Tzeng, A. Warren, K. Kozielski, J. Elisseeff, Development of a PEG derivative containing hydrolytically degradable hemiacetals, Macromolecules 43 (2010) 9588–9590.

[96] C. Dingels, S. Müller, T. Steinbach, C. Tonhauser, H. Frey, Universal concept for the implementation of a single cleavable unit at tunable position in functional poly(ethylene glycol)s, Biomacromolecules 14 (2013) 448–459.

[97] R. Li, R. Broyer, H. Maynard, Well-defined polymers with acetal side chains as reactive scaffolds synthesized by atom transfer radical polymerization, J. Polym. Sci. A Polym. Chem. 44 (2006) 5004–5013.

[98] K. Christman, H. Maynard, Protein micropatterns using a pH-responsive polymer and light, Langmuir 21 (2005) 8389–8393.

[99] J. Fletcher, A. Godwin, E. Pedone, B. Jahangeer, G. Buckton, S. Brocchini, The use of precursor polymers to prepare new excipients, J. Drug Deliv. Sci. Technol. 15 (2005) 295–299.

[100] A. Godwin, M. Hartenstein, A. Muller, S. Brocchini, Narrow molecular weight distribution precursor polymers, Angew. Chem. Int. Ed. Engl. 40 (2001) 594–597.

[101] E. Pedone, X. Li, N. Koseva, O. Alpar, S. Brocchini, An information rich biomedical polymer library, J. Mater. Chem. 13 (2003) 2825–2837.

[102] P. Wich, J. Fréchet, Degradable dextran particles for gene delivery applications, Aust. J. Chem. 65 (2012) 15.

[103] E. Bachelder, T. Beaudette, K. Broaders, J. Dashe, J. Fréchet, Acetal-derivatized dextran: an acid-responsive biodegradable material for therapeutic applications, J. Am. Chem. Soc. 130 (2008) 10494–10495.

[104] J. Cohen, T. Beaudette, J. Cohen, K. Broaders, E. Bachelder, J. Fréchet, Acetal-modified dextran microparticles with controlled degradation kinetics and surface functionality for gene delivery in phagocytic and non-phagocytic cells, Adv. Mater. 22 (2010) 3593–3597.

[105] W. Chen, F. Meng, R. Cheng, Z. Zhong, pH-Sensitive degradable polymersomes for triggered release of anticancer drugs: a comparative study with micelles, J. Control. Release 142 (2010) 40–46.

[106] C. Ornelas-Megiatto, P. Shah, P. Wich, J. Cohen, J. Tagaev, J. Smolen, B. Wright, M. Panzner, W. Youngs, J. Fréchet, C. Cannon, Aerosolized antimicrobial agents based on degradable dextran nanoparticles loaded with silver carbene complexes, Mol. Pharm. 9 (2012) 3012–3022.

[107] K. Kauffman, C. Do, S. Sharma, M. Gallovic, E. Bachelder, K. Ainslie, Synthesis and characterization of acetalated dextran polymer and microparticles with ethanol as a degradation product, Appl. Mater. Interface 4 (2012) 4149–4155.

[108] I. Chikhacheva, V. Zubov, M. Puolokainen, Polymer-analog reactions of polyvinyl alcohol under the action of microwave radiation, Russ. J. Gen. Chem. 81 (2011) 545–549.

[109] C. Scholz, E. Wagner, Therapeutic plasmid DNA versus siRNA delivery: common and different tasks for synthetic carriers, J. Control. Release 161 (2012) 554–565.
[110] P. Chollet, M.C. Favrot, A. Hurbin, J.-L. Coll, Side-effects of a systemic injection of linear polyethylenimine-DNA complexes, J. Gene Med. 4 (2002) 84–91.
[111] M. Ogris, S. Brunner, S. Schüller, R. Kircheis, E. Wagner, PEGylated DNA/transferrin-PEI complexes: reduced interaction with blood components, extended circulation in blood and potential for systemic gene delivery, Gene Ther. 6 (1999) 595–605.
[112] N. Murthy, J. Campbell, N. Fausto, A. Hoffman, P. Stayton, Design and synthesis of pH-responsive polymeric carriers that target uptake and enhance the intracellular delivery of oligonucleotides, J. Control. Release 89 (2003) 365–374.
[113] M. Shim, Y. Kwon, Acid-responsive linear polyethylenimine for efficient, specific, and biocompatible siRNA delivery, Bioconjug. Chem. 20 (2009) 488–499.
[114] M. Shim, Y. Kwon, Controlled delivery of plasmid DNA and siRNA to intracellular targets using ketalized polyethylenimine, Biomacromolecules 9 (2008) 444–455.
[115] Y. Kwon, Before and after endosomal escapes: roles of stimuli-converting siRNA/polymer interactions in determining gene silencing efficiency, Acc. Chem. Res. 45 (2012) 1077–1088.
[116] I. Ko, A. Ziady, S. Lu, Y. Kwon, Acid-degradable cationic methacrylamide polymerized in the presence of plasmid DNA as tunable non-viral gene carrier, Biomaterials 29 (2008) 3872–3881.
[117] M. Shim, Y. Kwon, Acid-transforming polypeptide micelles for targeted nonviral gene delivery, Biomaterials 31 (2010) 3404–3413.
[118] A. Kulkarni, K. DeFrees, S.-H. Hyun, D. Thompson, Pendant polymer:amino-β-cyclodextrin: siRNA guest:host nanoparticles as efficient vectors for gene silencing, J. Am. Chem. Soc. 134 (2012) 7596–7599.
[119] A. Kulkarni, W. Deng, S.-H. Hyun, D. Thompson, Development of a low toxicity, effective pDNA vector based on noncovalent assembly of bioresponsive amino-β-cyclodextrin:adamantane-poly(vinyl alcohol)-poly(ethylene glycol) transfection complexes, Bioconjug. Chem. 23 (2012) 933–940.
[120] A. Hoffman, Hydrogels for biomedical applications, Adv. Drug Deliv. Rev. 64 (2012) 18–23.
[121] G. Eccleston, Wound dressings in: M.E. Aulton (Ed.), Pharmaceutics: The Design and Manufacture of Medicines,, Elsevier, Churchill Livingstone, 2007, pp. 598–605.
[122] K. Lee, D. Mooney, Hydrogels for tissue engineering, Chem. Rev. 101 (2001) 1869–1880.
[123] V. Compañ, A. Andrio, A. López-Alemany, E. Riande, M. Refojo, Oxygen permeability of hydrogel contact lenses with organosilicon moieties, Biomaterials 23 (2002) 2767–2772.
[124] N. Peppas, P. Bures, W. Leobandung, H. Ichikawa, Hydrogels in pharmaceutical formulations, Eur. J. Pharm. Biopharm. 50 (2000) 27–46.
[125] E. Falco, M. Patel, J. Fisher, Recent developments in cyclic acetal biomaterials for tissue engineering applications, Pharm. Res. 25 (2008) 2348–2356.
[126] X.-C. Sui, Y. Shi, Z.-F. Fu, Novel degradable polymer networks containing acetal components and well-defined backbones, Aust. J. Chem. 63 (2010) 1497.
[127] T. Larmi, P. Karkola, Shrinkage and degradation of the Delrin occluder in tilting-disc valve prosthesis, J. Thorac. Cardiovasc. Surg. 68 (1974) 66–69.
[128] M. Thompson, M. Northmore-Ball, K. Tanner, Tensile mechanical properties of polyacetal after one and six months' immersion in Ringer ' s solution, J. Mater. Sci. Mater. Med. 12 (2001) 883–887.
[129] K. Strazar, A. Cör, V. Antolic, Biological impact of polyacetal particles on loosening of isoelastic stems, Biomacromolecules 7 (2006) 2507–2511.
[130] S. Kurtz, C. Muhlstein, A. Edidin, Surface morphology and wear mechanisms of four clinically relevant biomaterials after hip simulator testing, J. Biomed. Mater. Res. 52 (2000) 447–459.
[131] K. Lee, M. Choi, Phase velocity and normalized broadband ultrasonic attenuation in polyacetal cuboid bone-mimicking phantoms, J. Acoust. Soc. Am. 121 (2007) EL263–EL269.
[132] S. Kaihara, S. Matsumura, J. Fisher, Synthesis and properties of poly [poly(ethylene glycol)-co-cyclic acetal] based hydrogels, Macromolecules 40 (2007) 7625–7632.
[133] M. Patel, K. Patel, J. Caccamese, D. Coletti, J. Sauk, J. Fisher, Characterization of cyclic acetal hydroxyapatite nanocomposites for craniofacial tissue engineering, J. Biomed. Mater. Res. A 94 (2010) 408–418.
[134] M. Betz, J. Caccamese, D. Coletti, J. Sauk, J. Fisher, Tissue response and orbital floor regeneration using cyclic acetal hydrogels, J. Biomed. Mater. Res. A 90 (2009) 819–829.
[135] K. Wang, J. Lu, R. Yin, L. Chen, S. Du, Y. Jiang, Q. Yu, Preparation and properties of cyclic acetal based biodegradable gel by thiol-ene photopolymerization, Mater. Sci. Eng. C33 (2013) 1261–1266.
[136] P. Lundberg, B. Lee, S. van den Berg, E. Pressly, A. Lee, C. Hawker, N. Lynd, Poly[(ethylene oxide)-co-(methylene ethylene oxide)]: a hydrolytically-degradable poly(ethylene oxide) platform, MacroLetters 1 (2012) 1240–1243.

Chapter 14

Biomaterials and Tissue Engineering for Soft Tissue Reconstruction

Iwen Wu*,‡, Jennifer Elisseeff*,†,‡

*Translational Tissue Engineering Center, Johns Hopkins University, Baltimore, Maryland, USA

†Wilmer Eye Institute, Johns Hopkins University, Baltimore, Maryland, USA

‡Department of Biomedical Engineering, Johns Hopkins University, Baltimore, Maryland, USA

Chapter Outline

14.1 INTRODUCTION

Soft connective tissues function to provide structural support and establish contour. Soft tissue defects present a challenging problem as the aesthetic function is intricately linked with patient psychological well-being. As such, the benefits of improved aesthetic outcomes must be weighed against the risks from the complicated reconstructive surgeries often required. Soft tissue defects can arise from trauma, medical conditions, or after ablative procedures such as tumor resection. In addition to the aesthetic function, there are also many negative biological sequelae of scar formation involved in soft tissue loss and reconstruction [1].

Traumatic injuries such as those sustained from explosive devices and automotive accidents often result in extensive soft tissue loss. Laceration and dog bite repair account for over 320,000 reconstructive cases performed in 2012, making it the second most common condition requiring reconstructive surgery [2]. Despite advances in reconstructive procedures, regeneration of damaged soft connective tissues has yet to be achieved, and autologous flap reconstructions are often the only option for replacing areas of soft tissue loss. Significant contour defects and scarring often persist, resulting in poor aesthetic outcomes for individuals who have sustained soft tissue trauma.

In addition to traumatic injury, soft tissue loss can also arise secondary to medical conditions or as a result of surgical management of cancer. Lipodystrophy is a condition characterized by dysfunction or degenerative loss of adipose tissue and can lead to severe insulin resistance and dyslipidemia. There are both acquired and congenital forms and it is often associated with antiretroviral therapy for HIV patients [3]. Patients with this condition have a unique need for soft tissue replacements as they typically do not have adequate autologous adipose tissue available for grafting. Tumor resection is the most common reconstructive surgery carried out, with over 4.2 million operations performed annually for tumor removal [2]. Without a suitable soft tissue replacement, patients typically opt for contralateral reduction, prosthetic implants, or autologous flap reconstructions to restore symmetry after lumpectomy or mastectomy procedures [4]. The limited options available highlight the need for an alternative that avoids the risk of complications for prosthetic implants and negates the prerequisite of donor site tissue for flap reconstruction.

14.1.1 Current Clinical Strategies

Soft tissue defects are managed using different approaches depending on the defect size, location, and availability of donor sites tissue. For small volume corrections, dermal fillers or fat grafting can be used to fill the defect space and restore contour. Dermal fillers range from naturally occurring components of the extracellular matrix such as collagen, cross-linked forms of hyaluronic acid (HA), and calcium hydroxyapatite, to synthetic polymers including polylactic acid and polymethylmethacrylate [5]. However, most dermal fillers are not permanent and require patients to return regularly for repeat injections to maintain a proper volume correction, and there are a number of known complications that can occur with application of specific fillers [6].

Autologous fat grafting is another option and involves harvesting tissue by liposuction at one site of the body and reinjecting the lipoaspirate to fill a soft tissue defect. The procedure suffers from inconsistent outcomes due to highly variable rates of resorption. To rectify the unpredictability of graft take, several groups have established protocols to standardize and improve outcomes through harvest, preparation, and grafting techniques [7,8]. However, there are still complications that can arise from central necrosis of the transplanted lipoaspirate as poor viability results from rupture of adipocytes and lack of oxygen and vascular supply post transplantation [9]. Cyst formation and calcification often occur secondary to the necrosis and can interfere with radiological detection of breast cancer [10].

One of the most common reconstructive procedures to correct loss of soft tissue volume is autologous flap reconstruction, where a tissue flap from a donor site in the body is used for the reconstruction. Advances in reconstructive surgery have decreased the associated morbidity by attempting to keep the donor site muscle intact when possible [11]. However, the basic requirement of a donor site to repair a soft tissue defect at another location in the body inevitably results in donor site morbidity [12,13]. Availability of donor tissue can also be a limiting factor for patients who are thin or those who have multiple injury sites. Patients who undergo mastectomy have the additional option of prosthetic implants. However, complications can arise including capsular contracture, implant rupture, and infection, often requiring a second surgery to resolve the problem [14].

14.2 BIOMATERIALS

14.2.1 Synthetic Polymers

Synthetic polymers commonly used in numerous biomedical devices offer the distinct advantage of high level of control over the chemical properties of the polymer. As a scaffold for adipose tissue engineering, polylactic acid, polyglycolic acid, and copolymers incorporating both have been widely investigated due to their ability to degrade over time. These polymers degrade by bulk hydrolytic degradation of the ester bonds and the rate of degradation can be modulated by changing the molar mass and molar ratios of the monomers.

Cells can be combined with synthetic polymers to create a cell-scaffold construct for adipose tissue engineering. Several groups have seeded 3T3-L1 preadipocytes [15–17] or adipose-derived stem cells [18,19] on polyglycolic acid meshes and observed adipogenesis over time both *in vitro* and *in vivo*. Electrospun polylactic acid nanofibers were shown to support adipogenic differentiation of bone marrow stem cells *in vitro*. The copolymer of poly(lactic-*co*-glycolic acid) (PLGA) is more commonly used since the degradation rate of the polymer can be modulated by altering the molar ratios of the two monomers. Injectable PLGA spheres were used to deliver stem cells and it was found that predifferentiating stem cells attached to the PLGA spheres improved the adipose tissue formation *in vivo* [20–23]. In a different model of adipose tissue formation, PLGA was loaded in closed chambers incorporating a vascular pedicle for angiogenesis but researchers mainly observed granulation and fibrous tissue formation [24,25]. PLGA hollow fibers were also used for formation of cell aggregates and adipogenic differentiation of stem cells showing cells could persist in the hollow fibers *in vivo* [26]. Other groups have also used solid forms of PLGA scaffolds where cells are seeded directly on the scaffold and differentiated towards adipogenesis before *in vivo* application [27–31].

In addition to the solid polymer scaffolds previously described, polymers can also take the form of a hydrogel. Hydrogels are characterized as water-soluble polymers that can be cross-linked to form a hydrogel scaffold that shows high swelling properties in aqueous solutions. Polyethylene glycol is a commonly used hydrogel in which cells are typically encapsulated within the hydrogel prior to cross-linking. Photopolymerized polyethylene glycol hydrogels supported adipogenic differentiation of embryonic [32] and adult mesenchymal stem cells [33], although it was found that the addition of bFGF to hydrogel scaffolds greatly improved vascularization and cell infiltration *in vivo* [34].

14.2.2 Naturally Derived Scaffolds

Scaffolds for tissue engineering can also be derived from natural sources and generally have high biocompatibility. Some examples of naturally derived biomaterials are alginate, silk, and chitosan. Alginate is a natural polysaccharide derived from seaweed. Adipogenic differentiated stem cells were encapsulated in alginate and implanted subcutaneously [35]. To render alginate susceptible to degradation, an oxidized form of alginate was also investigated [36] as well as combined collagen alginate microspheres to better mimic the native extracellular matrix environment [37]. Similar to alginate, chitosan is a natural polysaccharide that is isolated from chitin. So far, researchers have used it in combination with HA to form an injectable hydrogel [38] or with PLGA to form a porous scaffold [31] for adipose tissue engineering. Silk is another naturally derived scaffold that is fibrous in nature and has high mechanical strength. In adipose tissue engineering, it has been applied as a scaffold for adipogenic differentiation of stem cells and subsequently implanted in rodent models [39–42]. The silk scaffolds alone produced a fibrous tissue but, when seeded with adipogenic differentiated cells, maintained their adipogenic state with lipid-laden cells observed over time.

14.2.3 Extracellular Matrix-Based Materials

A subset of naturally derived scaffolds is extracellular matrix-based biomaterials. This includes extracted components of the extracellular matrix such as collagen, dextran, fibrin, Matrigel, and HA, as well as bulk extracellular matrix obtained through decellularization procedures. These natural ECM-based scaffolds typically have high biocompatibility, and since they originate from the matrix, they contain the cell-binding domains required for the survival and growth of anchorage-dependent cells.

14.2.3.1 Collagen

Collagen is perhaps the most studied, as it forms the structural basis for much of the extracellular matrix of our tissues. Collagen can be used as a scaffold in which cells can be seeded directly on the collagen matrix and differentiated with adipogenic media [43,44]. Collagen sheets with an elastin component have also been tested for adipose engineering, based on a scaffold designed to be a skin substitute [45]. Collagen sponges are a porous form of the collagen-based scaffold and provide increased surface area and permeability [46,47]. A few groups have also incorporated gelatin microspheres in the collagen scaffolds for controlled release of bFGF in a collagen sponge [48] and collagen gel [49] to improve adipogenesis. Collagen microparticles or beads have been fabricated and used as microcarriers for cell delivery [50,51]. In a modified design based on the collagen gel, short embedded collagen fibers were incorporated in a collagen gel to improve structural support and limit cell-mediated contraction of the constructs [52].

14.2.3.2 Gelatin

Gelatin is derived from collagen, typically from bovine sources, and can be used in a number of forms similar to collagen. For application in adipose tissue engineering, it has primarily been used as a gelatin sponge onto which cells are seeded and differentiated before *in vivo* implantation [19,43,53,54]. Gelatin has also been used as microparticles for controlled release of bFGF in several systems to aid adipogenesis [48,49,55,56].

14.2.3.3 Fibrin

Fibrin is a blood component involved in the clotting process and can be used as a matrix for tissue engineering applications. Polymerization of fibrin is initiated with thrombin, and the fibrin gel can be naturally degraded *in vivo*. Fibrin matrices have been used for culture of cells *in vitro* as well as *in vivo* delivery for stem cells and in conjunction with endothelial cells [50,57–63]. One of the groups using fibrin matrices has also observed that providing a mechanical support for the developing construct greatly enhances generation of new adipose tissues [59,60].

14.2.3.4 Hyaluronic Acid

HA is a linear polysaccharide and is a natural component of the extracellular matrix. HA can be chemically modified to modulate its material and biological properties, making it a popular choice as a biomaterial for tissue engineering. An HA sponge fabricated from an ester derivative of HA allowed evaluation of the adipogenic potential for sponges with varying pore sizes and degrees of esterification [64–66]. In another formulation, an HA gel was formed with varying degrees of amidation of carboxyl groups and studied for application in adipose tissue engineering [67]. The two were compared in a study that investigated the *in vivo* response to the scaffolds alone and found less inflammation in the HA gels with dodecyl amidation and improved adipogenesis compared to the esterified HA sponges [68]. Additionally, cross-linked HA can be used as a hydrogel for adipogenesis [69] and can be combined with extracts from adipose tissue to improve bioactivity [70]. A polyethylene glycol-HA composite with encapsulated VEGF in heparin nanoparticles was investigated by one group to see if the added growth factor would enhance adipogenesis [71]. In a different approach to develop a material that can be modulated by physiological signals, a chitosan-HA hydrogel was developed for insulin delivery in a glucose responsive hydrogel upon swelling [38]. Through this design, the scaffold could deliver insulin, an adipogenic inductive factor, locally at the site of delivery after swelling.

14.2.3.5 Matrigel

Some of the most successful cell-free approaches to adipose tissue regeneration use Matrigel, a basement membrane extract derived from the EHS sarcoma in a mouse. Researchers found that subcutaneous injections of Matrigel with bFGF resulted in *de novo* adipose tissue formation, regardless of the site of injection [72,73]. Gelatin microspheres have also been used for controlled release of bFGF in Matrigel and were found to be more effective at lower concentrations of bFGF compared to doping the bFGF directly in Matrigel [55,56]. Other groups have also established a model in which Matrigel and bFGF are enclosed in a chamber in which a vascular pedicle is inserted to observe adipogenesis in a closed environment isolated from the surrounding tissue such that only circulating cells from the vasculature would contribute to new tissue formation [24,74,75].

14.2.3.6 Tissue-Based Protein Extracts

Extracellular matrix proteins are a promising class of biocompatible scaffold materials since they comprise the natural environment for cells in the body. Similar to the approach of extracting basement membrane proteins from the EHS sarcoma to produce Matrigel, investigators have also explored using different tissues as the starting material to form tissue extracts composed of extracellular matrix proteins. Groups have isolated proteins from the adipose matrix in a similar preparation to Matrigel, producing a protein extract enriched for adipose matrix components that can form a hydrogel at physiological temperature [76,77]. Additional tissue-derived protein extracts have been produced using dermis [78], skeletal muscle [79], and cardiac muscle tissue [80]. Adipose stem cells grown on these tissue-derived hydrogels show spontaneous adipogenic differentiation, suggesting that bioactive factors from the tissues partition with these extracted components. Another group has also incorporated adipose tissue extracts into electrospun nanofibers, showing that the nanofibers support stem cell attachment and survival [81].

14.2.3.7 Bulk Decellularized Extracellular Matrices

Another class of extracellular matrix-based biomaterials is derived through decellularization of bulk tissues to obtain an acellular matrix. Typically, tissues are processed with various solvents such as detergents, acids, and enzymes to remove the cellular component and prevent antigenicity. Lipoaspirate taken from patients undergoing liposuction procedures can be decellularized and milled to a powder to produce an injectable adipose ECM powder [82–85]. Adipose ECM powders can be formed into a thermoresponsive hydrogel by acid solubilization so that it undergoes gelation upon injection [86]. Intact adipose tissue from patients who undergo procedures where bulk tissue is excised, such as abdominoplasty or brachioplasty procedures, can also be decellularized to produce an extracellular matrix that remains intact [87,88]. This adipose extracellular matrix can also be solubilized using pepsin and formulated into microcarrier beads [89] or foams [90] for adipose tissue engineering. Placental tissue has also been decellularized and explored as a tissue source for use as a scaffold for adipose tissue regeneration [69,91,92].

14.3 CELL SOURCES

14.3.1 Cells of Adipose Tissue

The parenchymal cells of adipose tissue are adipocytes, terminally differentiated lipid-filled cells providing a key energy reserve for the rest of the body. Although they are the primary cells transplanted in fat grafting procedures, their inability to replicate, high metabolic demands, and fragile nature limit their utility in adipose tissue regeneration. As an alternative, many researchers have looked to progenitor cells that can be readily expanded and are capable of differentiating into adipocytes. Another prominent cell type in adipose tissue is endothelial cells due to the extensive vascularization of adipose tissue. Each adipocyte in the body has immediate access to at least one blood vessel, and the processes of adipogenesis and vascularization are tightly coupled during development. Endothelial cells have often been used in conjunction with adipose stem cells to encourage vascularization of constructs for nutrient diffusion to cells and to encourage adipogenesis [37,41,57,62,93].

14.3.2 Adult Stem Cells

One of the early sources of adult stem cells to be investigated was bone marrow. A subset of multipotent progenitor cells was found to reside in the bone marrow niche and could be induced to differentiate down specific mesenchymal lineages [94]. The bone marrow aspirate is collected from the iliac crest, followed by selection for plastic-adherent cells.

Adipose-derived stem cells present an intriguing cell population for adipose tissue engineering due to the relative ease of access for cell harvesting by liposuction. A resident population of multipotent cells has been identified in the stromal vascular fraction of adipose tissue and is likely to be the compartment in which adipocyte precursor cells reside. These adipose-derived cells have also been shown to differentiate down multiple mesenchymal lineages including adipose, bone, cartilage, neurons, and myocytes. In comparison to mature adipocytes, ASCs are easily expanded in culture and have a fibroblastic morphology. When exposed to adipogenic induction media, the cells begin to undergo adipogenic differentiation and start accumulating lipid.

The adipose-derived stem cells are typically isolated from adipose tissue by enzymatic digestion, generating a heterogeneous cell population of stromal vascular cells. Adipose-derived stem cells express many of the mesenchymal stem cell surface markers, including CD 13, CD 29, CD 44, CD 71, CD 73, CD 90, CD 105, and STRO-1 while being negative for the hematopoietic lineage markers CD14, CD16, CD45, CD56, CD61, CD62E, CD104, CD106, and endothelial markers CD31 and von Willebrand factor [95]. Initially a heterogeneous population of stromal vascular cells, selection of the plastic-adherent population depletes many cells of hematopoietic origin. After *in vitro* expansion over several passages, cells tend to show decreased expression of CD11, CD14, CD34, CD45, CD86, and HLA-DR while becoming uniformly positive for the other stromal cell surface markers with successive passages, suggesting that a more homogeneous population of cells is established over time [96,97]. Continuing development of point of care systems to automate stromal vascular cell isolation can potentially allow the use of autologous cells for treatment to be widely adopted in the clinic.

14.3.3 Embryonic Stem Cells

Embryonic stem cells are pluripotent cells that have high proliferative capacity in culture and can be expanded through far more passages compared to adult stem cells without reaching senescence. The cells are isolated from the inner cell mass of blastocysts and can be differentiated towards numerous somatic cells with varying phenotypes. The use of embryonic stem cells in research has been fraught with controversy and is subject to political policies that can limit access to their use. Unfortunately, this vulnerability to ethical and political pressures hinders research using embryonic stem cells, as they are subject to the political environment of changing government administrations.

14.3.4 Induced Pluripotent Stem Cells

In 2006, researchers discovered that it was possible to induce adult somatic cells to revert back to an embryonic-like state by genetically reprogramming the cells to express the transcription factors Oct3/4, Sox 2, and Klf4 [98–100]. Cells expressing these factors exhibited key characteristics of embryonic stem cell pluripotency, including embryoid body and teratoma formation, and proved capable of differentiating into cells of all three germ layers. As a result, any adult somatic cell may be used as a cell source, thereby circumventing the ethical issues and political influences controlling the use of embryonic stem cells in research.

iPS cells have been successfully differentiated down the adipogenic lineage showing similar efficacy as embryonic stem cells for adipogenic differentiation [101]. Thus far, the use of iPS cells in adipose tissue engineering has been limited, and consistent directed differentiation of cell populations still remains a challenge.

14.4 DISCUSSION

Biomaterials-based approaches to engineering soft connective tissue can potentially offer an alternative to current tissue transfer techniques that require donor site tissue. The advantages of an off-the-shelf adipose tissue replacement include availability and avoiding donor site morbidity. Advances in the field of regenerative medicine contribute to the development of more biomimetic materials that provide the foundation for tissue repair and regeneration. Researchers have uncovered the increasingly complex role of adipose tissue in maintaining the metabolic homeostasis of the body, and hopefully, approaches to regenerate adipose tissue can eventually restore functionality of the repaired tissue as well.

REFERENCES

[1] A.E. Brissett, D.A. Sherris, Scar contractures, hypertrophic scars, and keloids, Facial Plast Surg 17 (2001) 263–272.
[2] American Society of Plastic Surgeons, 2012 Plastic Surgery Statistics Report, 2012.
[3] A. Carr, et al., A syndrome of peripheral lipodystrophy, hyperlipidaemia and insulin resistance in patients receiving HIV protease inhibitors, AIDS 12 (1998) F51–F58.
[4] P.G. Cordeiro, Breast reconstruction after surgery for breast cancer, N. Engl. J. Med. 359 (2008) 1590–1601.
[5] S.S. Johl, R.A. Burgett, Dermal filler agents: a practical review, Curr. Opin. Ophthalmol. 17 (2006) 471–479.
[6] N.J. Lowe, C.A. Maxwell, R. Patnaik, Adverse reactions to dermal fillers: review, Dermatol. Surg. 31 (2005) 1616–1625.
[7] S.R. Coleman, Structural fat grafting: more than a permanent filler, Plast. Reconstr. Surg. 118 (2006) 108S–120S.
[8] K.J. Butterwick, P.K. Nootheti, J.W. Hsu, M.P. Goldman, Autologous fat transfer: an in-depth look at varying concepts and techniques, Facial Plast Surg Clin North Am 15 (2007) 99–111, viii.
[9] S.L. Spear, H.B. Wilson, M.D. Lockwood, Fat injection to correct contour deformities in the reconstructed breast, Plast. Reconstr. Surg. 116 (2005) 1300–1305.
[10] J. Carvajal, J.H. Patino, Mammographic findings after breast augmentation with autologous fat injection, Aesthet. Surg. J. 28 (2008) 153–162.
[11] R.J. Allen, P. Treece, Deep inferior epigastric perforator flap for breast reconstruction, Ann. Plast. Surg. 32 (1994) 32–38.
[12] A. Cornejo, S. Ivatury, C.N. Crane, J.G. Myers, H.T. Wang, Analysis of free flap complications and utilization of intensive care unit monitoring, J. Reconstr. Microsurg. 29 (2013) 473–479.
[13] A.K. Alderman, E.G. Wilkins, H.M. Kim, J.C. Lowery, Complications in postmastectomy breast reconstruction: two-year results of the Michigan Breast Reconstruction Outcome Study, Plast. Reconstr. Surg. 109 (2002) 2265–2274.
[14] S. Agha-Mohammadi, C. De La Cruz, D.J. Hurwitz, Breast reconstruction with alloplastic implants, J. Surg. Oncol. 94 (2006) 471–478.
[15] C. Fischbach, et al., Three-dimensional in vitro model of adipogenesis: comparison of culture conditions, Tissue Eng. 10 (2004) 215–229.
[16] C. Fischbach, et al., Generation of mature fat pads in vitro and in vivo utilizing 3-D long-term culture of 3T3-L1 preadipocytes, Exp. Cell Res. 300 (2004) 54–64.
[17] B. Weiser, et al., In vivo development and long-term survival of engineered adipose tissue depend on in vitro precultivation strategy, Tissue Eng Part A 14 (2008) 275–284.
[18] J.A. Lee, et al., Biological alchemy: engineering bone and fat from fat-derived stem cells, Ann. Plast. Surg. 50 (2003) 610–617.
[19] S.D. Lin, K.H. Wang, A.P. Kao, Engineered adipose tissue of predefined shape and dimensions from human adipose-derived mesenchymal stem cells, Tissue Eng Part A 14 (2008) 571–581.
[20] Y.S. Choi, et al., Adipogenic differentiation of adipose tissue derived adult stem cells in nude mouse, Biochem. Biophys. Res. Commun. 345 (2006) 631–637.
[21] Y.S. Choi, S.N. Park, H. Suh, Adipose tissue engineering using mesenchymal stem cells attached to injectable PLGA spheres, Biomaterials 26 (2005) 5855–5863.
[22] H.J. Chung, T.G. Park, Injectable cellular aggregates prepared from biodegradable porous microspheres for adipose tissue engineering, Tissue Eng Part A 15 (2009) 1391–1400.

[23] S.W. Kang, S.W. Seo, C.Y. Choi, B.S. Kim, Porous poly(lactic-co-glycolic acid) microsphere as cell culture substrate and cell transplantation vehicle for adipose tissue engineering, Tissue Eng Part C Methods 14 (2008) 25–34.

[24] K.J. Cronin, et al., New murine model of spontaneous autologous tissue engineering, combining an arteriovenous pedicle with matrix materials, Plast. Reconstr. Surg. 113 (2004) 260–269.

[25] J.H. Dolderer, et al., Spontaneous large volume adipose tissue generation from a vascularized pedicled fat flap inside a chamber space, Tissue Eng. 13 (2007) 673–681.

[26] S.M. Morgan, et al., Formation of a human-derived fat tissue layer in P(DL)LGA hollow fibre scaffolds for adipocyte tissue engineering, Biomaterials 30 (2009) 1910–1917.

[27] M. Neubauer, et al., Adipose tissue engineering based on mesenchymal stem cells and basic fibroblast growth factor in vitro, Tissue Eng. 11 (2005) 1840–1851.

[28] C.W. Patrick Jr., P.B. Chauvin, J. Hobley, G.P. Reece, Preadipocyte seeded PLGA scaffolds for adipose tissue engineering, Tissue Eng. 5 (1999) 139–151.

[29] C.W. Patrick Jr., B. Zheng, C. Johnston, G.P. Reece, Long-term implantation of preadipocyte-seeded PLGA scaffolds, Tissue Eng. 8 (2002) 283–293.

[30] J. Xu, Y. Chen, Y. Yue, J. Sun, L. Cui, Reconstruction of epidural fat with engineered adipose tissue from adipose derived stem cells and PLGA in the rabbit dorsal laminectomy model, Biomaterials 33 (2012) 6965–6973.

[31] W. Wang, B. Cao, L. Cui, J. Cai, J. Yin, Adipose tissue engineering with human adipose tissue-derived adult stem cells and a novel porous scaffold, J. Biomed. Mater. Res. B Appl. Biomater. 101 (2013) 68–75.

[32] A.T. Hillel, S. Varghese, J. Petsche, M.J. Shamblott, J.H. Elisseeff, Embryonic germ cells are capable of adipogenic differentiation in vitro and in vivo, Tissue Eng Part A 15 (2009) 479–486.

[33] A. Alhadlaq, M. Tang, J.J. Mao, Engineered adipose tissue from human mesenchymal stem cells maintains predefined shape and dimension: implications in soft tissue augmentation and reconstruction, Tissue Eng. 11 (2005) 556–566.

[34] M.S. Stosich, et al., Vascularized adipose tissue grafts from human mesenchymal stem cells with bioactive cues and microchannel conduits, Tissue Eng. 13 (2007) 2881–2890.

[35] W. Jing, et al., Ectopic adipogenesis of preconditioned adipose-derived stromal cells in an alginate system, Cell Tissue Res. 330 (2007) 567–572.

[36] W.S. Kim, et al., Adipose tissue engineering using injectable, oxidized alginate hydrogels, Tissue Eng Part A 18 (2012) 737–743.

[37] R. Yao, R. Zhang, F. Lin, J. Luan, Biomimetic injectable HUVEC-adipocytes/collagen/alginate microsphere co-cultures for adipose tissue engineering, Biotechnol. Bioeng. 110 (2013) 1430–1443.

[38] H. Tan, J.P. Rubin, K.G. Marra, Injectable in situ forming biodegradable chitosan-hyaluronic acid based hydrogels for adipose tissue regeneration, Organogenesis 6 (2010) 173–180.

[39] E. Bellas, K.G. Marra, D.L. Kaplan, Sustainable three-dimensional tissue model of human adipose tissue, Tissue Eng Part C Methods 19 (2013) 745–754.

[40] E. Bellas, et al., Sustained volume retention in vivo with adipocyte and lipoaspirate seeded silk scaffolds, Biomaterials 34 (2013) 2960–2968.

[41] J.H. Kang, J.M. Gimble, D.L. Kaplan, In vitro 3D model for human vascularized adipose tissue, Tissue Eng Part A 15 (2009) 2227–2236.

[42] J.R. Mauney, et al., Engineering adipose-like tissue in vitro and in vivo utilizing human bone marrow and adipose-derived mesenchymal stem cells with silk fibroin 3D scaffolds, Biomaterials 28 (2007) 5280–5290.

[43] S.D. Lin, et al., Engineering adipose tissue from uncultured human adipose stromal vascular fraction on collagen matrix and gelatin sponge scaffolds, Tissue Eng Part A 17 (2011) 1489–1498.

[44] S. Neuss, et al., Long-term survival and bipotent terminal differentiation of human mesenchymal stem cells (hMSC) in combination with a commercially available three-dimensional collagen scaffold, Cell Transplant. 17 (2008) 977–986.

[45] M. Keck, et al., Adipose tissue engineering: three different approaches to seed preadipocytes on a collagen-elastin matrix, Ann. Plast. Surg. 67 (2011) 484–488.

[46] W. Tsuji, et al., Simple and longstanding adipose tissue engineering in rabbits, J Artif Organs 16 (2013) 110–114.

[47] D. von Heimburg, et al., Preadipocyte-loaded collagen scaffolds with enlarged pore size for improved soft tissue engineering, Int. J. Artif. Organs 26 (2003) 1064–1076.

[48] Y. Hiraoka, et al., In situ regeneration of adipose tissue in rat fat pad by combining a collagen scaffold with gelatin microspheres containing basic fibroblast growth factor, Tissue Eng. 12 (2006) 1475–1487.

[49] A.V. Vashi, et al., Adipose tissue engineering based on the controlled release of fibroblast growth factor-2 in a collagen matrix, Tissue Eng. 12 (2006) 3035–3043.

[50] B. Frerich, K. Winter, K. Scheller, U.D. Braumann, Comparison of different fabrication techniques for human adipose tissue engineering in severe combined immunodeficient mice, Artif. Organs 36 (2012) 227–237.

[51] J.P. Rubin, J.M. Bennett, J.S. Doctor, B.M. Tebbets, K.G. Marra, Collagenous microbeads as a scaffold for tissue engineering with adipose-derived stem cells, Plast. Reconstr. Surg. 120 (2007) 414–424.

[52] E. Gentleman, E.A. Nauman, G.A. Livesay, K.C. Dee, Collagen composite biomaterials resist contraction while allowing development of adipocytic soft tissue in vitro, Tissue Eng. 12 (2006) 1639–1649.

[53] L. Hong, I. Peptan, P. Clark, J.J. Mao, Ex vivo adipose tissue engineering by human marrow stromal cell seeded gelatin sponge, Ann. Biomed. Eng. 33 (2005) 511–517.

[54] L. Hong, I.A. Peptan, A. Colpan, J.L. Daw, Adipose tissue engineering by human adipose-derived stromal cells, Cells Tissues Organs 183 (2006) 133–140.

[55] Y. Kimura, M. Ozeki, T. Inamoto, Y. Tabata, Time course of de novo adipogenesis in matrigel by gelatin microspheres incorporating basic fibroblast growth factor, Tissue Eng. 8 (2002) 603–613.

[56] Y. Tabata, et al., De novo formation of adipose tissue by controlled release of basic fibroblast growth factor, Tissue Eng. 6 (2000) 279–289.

[57] J. Borges, et al., Engineered adipose tissue supplied by functional microvessels, Tissue Eng. 9 (2003) 1263–1270.

[58] S.W. Cho, et al., Enhancement of adipose tissue formation by implantation of adipogenic-differentiated preadipocytes, Biochem. Biophys. Res. Commun. 345 (2006) 588–594.

[59] S.W. Cho, et al., Engineering of volume-stable adipose tissues, Biomaterials 26 (2005) 3577–3585.

[60] S.W. Cho, et al., Engineered adipose tissue formation enhanced by basic fibroblast growth factor and a mechanically stable environment, Cell Transplant. 16 (2007) 421–434.

[61] N. Torio-Padron, N. Baerlecken, A. Momeni, G.B. Stark, J. Borges, Engineering of adipose tissue by injection of human preadipocytes in fibrin, Aesthetic Plast Surg 31 (2007) 285–293.

[62] F. Verseijden, et al., Vascularization of prevascularized and non-prevascularized fibrin-based human adipose tissue constructs after implantation in nude mice, J Tissue Eng Regen Med 6 (2012) 169–178.
[63] F. Verseijden, et al., Comparing scaffold-free and fibrin-based adipose-derived stromal cell constructs for adipose tissue engineering: an in vitro and in vivo study, Cell Transplant. 21 (2012) 2283–2297.
[64] A. Borzacchiello, et al., Structural and rheological characterization of hyaluronic acid-based scaffolds for adipose tissue engineering, Biomaterials 28 (2007) 4399–4408.
[65] M. Halbleib, T. Skurk, C. de Luca, D. von Heimburg, H. Hauner, Tissue engineering of white adipose tissue using hyaluronic acid-based scaffolds. I: in vitro differentiation of human adipocyte precursor cells on scaffolds, Biomaterials 24 (2003) 3125–3132.
[66] K. Hemmrich, et al., Implantation of preadipocyte-loaded hyaluronic acid-based scaffolds into nude mice to evaluate potential for soft tissue engineering, Biomaterials 26 (2005) 7025–7037.
[67] K. Hemmrich, et al., Autologous in vivo adipose tissue engineering in hyaluronan-based gels—a pilot study, J. Surg. Res. 144 (2008) 82–88.
[68] N.P. Rhodes, Inflammatory signals in the development of tissue-engineered soft tissue, Biomaterials 28 (2007) 5131–5136.
[69] L.E. Flynn, G.D. Prestwich, J.L. Semple, K.A. Woodhouse, Proliferation and differentiation of adipose-derived stem cells on naturally derived scaffolds, Biomaterials 29 (2008) 1862–1871.
[70] J.R. Sarkanen, P. Ruusuvuori, H. Kuokkanen, T. Paavonen, T. Ylikomi, Bioactive acellular implant induces angiogenesis and adipogenesis and sustained soft tissue restoration in vivo, Tissue Eng Part A 18 (2012) 2568–2580.
[71] H. Tan, Q. Shen, X. Jia, Z. Yuan, D. Xiong, Injectable nanohybrid scaffold for biopharmaceuticals delivery and soft tissue engineering, Macromol. Rapid Commun. 33 (2012) 2015–2022.
[72] N. Kawaguchi, et al., De novo adipogenesis in mice at the site of injection of basement membrane and basic fibroblast growth factor, Proc. Natl. Acad. Sci. U.S.A. 95 (1998) 1062–1066.
[73] K. Toriyama, et al., Endogenous adipocyte precursor cells for regenerative soft-tissue engineering, Tissue Eng. 8 (2002) 157–165.
[74] G.P. Thomas, et al., Zymosan-induced inflammation stimulates neo-adipogenesis, Int J Obes (Lond) 32 (2008) 239–248.
[75] F. Stillaert, et al., Host rather than graft origin of Matrigel-induced adipose tissue in the murine tissue-engineering chamber, Tissue Eng. 13 (2007) 2291–2300.
[76] S. Uriel, et al., The role of adipose protein derived hydrogels in adipogenesis, Biomaterials 29 (2008) 3712–3719.
[77] C.J. Poon, et al., Preparation of an adipogenic hydrogel from subcutaneous adipose tissue, Acta Biomater. 9 (2013) 5609–5620.
[78] M.H. Cheng, et al., Dermis-derived hydrogels support adipogenesis in vivo, J Biomed Mater Res A 92 (2010) 852–858.
[79] K.M. Abberton, et al., Myogel, a novel, basement membrane-rich, extracellular matrix derived from skeletal muscle, is highly adipogenic in vivo and in vitro, Cells Tissues Organs 188 (2008) 347–358.
[80] K. Matsuda, et al., Adipose-derived stem cells promote angiogenesis and tissue formation for in vivo tissue engineering, Tissue Eng Part A 19 (2013) 1327–1335.
[81] M.P. Francis, et al., Electrospinning adipose tissue-derived extracellular matrix for adipose stem cell culture, J Biomed Mater Res A 100 (2012) 1716–1724.
[82] J.S. Choi, et al., Decellularized extracellular matrix derived from human adipose tissue as a potential scaffold for allograft tissue engineering, J Biomed Mater Res A 97 (2011) 292–299.
[83] J.S. Choi, et al., Human extracellular matrix (ECM) powders for injectable cell delivery and adipose tissue engineering, J. Control. Release 139 (2009) 2–7.
[84] Y.C. Choi, et al., Decellularized extracellular matrix derived from porcine adipose tissue as a xenogeneic biomaterial for tissue engineering, Tissue Eng Part C Methods 18 (2012) 866–876.
[85] B.S. Kim, J.S. Choi, J.D. Kim, Y.C. Choi, Y.W. Cho, Recellularization of decellularized human adipose-tissue-derived extracellular matrix sheets with other human cell types, Cell Tissue Res. 348 (2012) 559–567.
[86] D.A. Young, D.O. Ibrahim, D. Hu, K.L. Christman, Injectable hydrogel scaffold from decellularized human lipoaspirate, Acta Biomater. 7 (2011) 1040–1049.
[87] L.E. Flynn, The use of decellularized adipose tissue to provide an inductive microenvironment for the adipogenic differentiation of human adipose-derived stem cells, Biomaterials 31 (2010) 4715–4724.
[88] I. Wu, Z. Nahas, K.A. Kimmerling, G.D. Rosson, J.H. Elisseeff, An injectable adipose matrix for soft-tissue reconstruction, Plast. Reconstr. Surg. 129 (2012) 1247–1257.
[89] A.E. Turner, C. Yu, J. Bianco, J.F. Watkins, L.E. Flynn, The performance of decellularized adipose tissue microcarriers as an inductive substrate for human adipose-derived stem cells, Biomaterials 33 (2012) 4490–4499.
[90] C. Yu, et al., Porous decellularized adipose tissue foams for soft tissue regeneration, Biomaterials 34 (2013) 3290–3302.
[91] L. Flynn, G.D. Prestwich, J.L. Semple, K.A. Woodhouse, Adipose tissue engineering in vivo with adipose-derived stem cells on naturally derived scaffolds, J Biomed Mater Res A 89 (2009) 929–941.
[92] L. Flynn, J.L. Semple, K.A. Woodhouse, Decellularized placental matrices for adipose tissue engineering, J Biomed Mater Res A 79 (2006) 359–369.
[93] F. Verseijden, et al., Prevascular structures promote vascularization in engineered human adipose tissue constructs upon implantation, Cell Transplant. 19 (2010) 1007–1020.
[94] M.F. Pittenger, et al., Multilineage potential of adult human mesenchymal stem cells, Science 284 (1999) 143–147.
[95] P.A. Zuk, et al., Human adipose tissue is a source of multipotent stem cells, Mol. Biol. Cell 13 (2002) 4279–4295.
[96] J.B. Mitchell, et al., Immunophenotype of human adipose-derived cells: temporal changes in stromal-associated and stem cell-associated markers, Stem Cells 24 (2006) 376–385.
[97] K. McIntosh, et al., The immunogenicity of human adipose-derived cells: temporal changes in vitro, Stem Cells 24 (2006) 1246–1253.
[98] K. Takahashi, et al., Induction of pluripotent stem cells from adult human fibroblasts by defined factors, Cell 131 (2007) 861–872.
[99] K. Takahashi, S. Yamanaka, Induction of pluripotent stem cells from mouse embryonic and adult fibroblast cultures by defined factors, Cell 126 (2006) 663–676.
[100] J. Yu, et al., Induced pluripotent stem cell lines derived from human somatic cells, Science 318 (2007) 1917–1920.
[101] D. Taura, et al., Adipogenic differentiation of human induced pluripotent stem cells: comparison with that of human embryonic stem cells, FEBS Lett. 583 (2009) 1029–1033.

Chapter 15

Dendrimers and Its Biomedical Applications

Umesh Gupta, Omathanu Perumal[1]

Department of Pharmaceutical Sciences, College of Pharmacy, South Dakota State University, Brookings, South Dakota, USA

[1]Corresponding author: e-mail: omathanu.perumal@sdstate.edu

Chapter Outline

15.1 INTRODUCTION

Dendrimers are highly branched spherical polymers with core-shell architecture. This unique class of highly symmetrical branched polymers is characterized by (i) a multifunctional initiator core, (ii) repeating branched units attached to the core, and (iii) a high density of terminal functional groups attached to the outermost shell (Figure 15.1). Although hyperbranched polymers were first reported by Vogtle and coworkers [1], dendrimers were first reported by Donald Tomalia [2]. During the same period, Newkome *et al.* independently reported the synthesis of similar macromolecules known as "arborols" [3]. Different types of bifunctional molecules have been used to form the core in dendrimers, e.g., ammonia, ethylenediamine (EDA), diaminobutane (DAB), triazine, polyethylene glycol (PEG), bis(4-fluorophenyl) sulfone, and carbosilane. Dendrimers are synthesized using a suitable core and the synthetic sequence follows the pattern like the branch of a tree. Although different types of dendrimers have been reported (Table 15.1), the most commonly used dendrimers for biomedical applications are polyamidoamine (PAMAM) and polypropylene imine (PPI) dendrimers (Figure 15.2).

Dendrimers have significant advantages over linear polymers that includes the following: (i) The dendrimers are characterized by a high degree of symmetry and are formed by controlled synthesis with discrete number of functionalities [4]; (ii) dendrimers are monodisperse polymers unlike linear polymers; (iii) dendrimers have a high density of functional groups and are water soluble; (iv) the functional groups are placed in close proximity, which enables multivalent binding; (v) the chemical reactivity and water solubility of dendrimers are higher than their linear polymers; (vi) the dendrimers have lower viscosity than linear polymers with similar molecular weight [5–7]; (vii) the multifunctionality and spherical structure

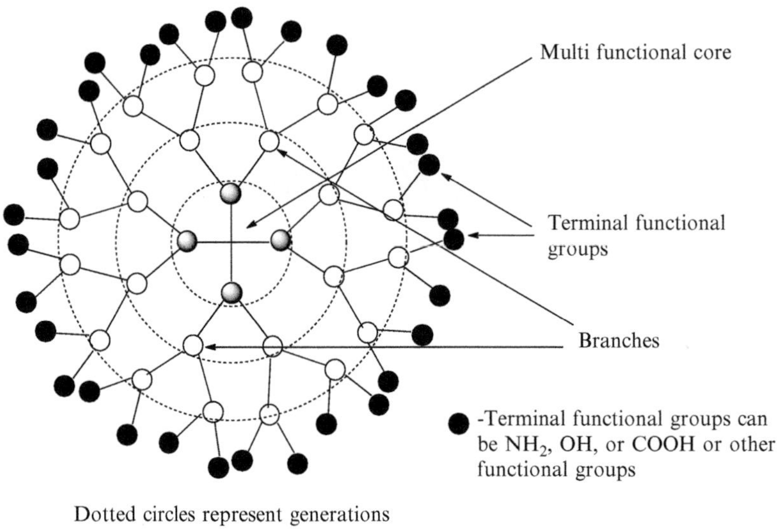

FIGURE 15.1 Schematic structure of dendrimer.

TABLE 15.1 Different Types of Dendrimers

Polyamidoamine (PAMAM) dendrimers
Polypropylene imine (PPI) dendrimers
Polyester dendrimers
Polylysine dendrimers
Poly (2,2-bis(hydroxymethyl) propionic acid dendrimer
Polypropyleneetherimine dendrimer
Peptide dendrimers
Carbohydrate dendrimers
Triazine dendrimers
Melamine dendrimer
Phosphorous dendrimer
Tecto dendrimers

of the dendrimers can result in cooperative binding with drugs and biological membranes; (viii) the core-shell architecture of dendrimers facilitates host-guest entrapment and affords better stability to the encapsulated molecules; (ix) molecules can be loaded by encapsulation, noncovalent interactions or covalent binding to the surface functional groups; and (x) the number, size, and type of surface functional groups can be easily tailored for various applications. Dendrimers are characterized by different generations (G) based on the number of branching units and the number of functional groups doubles with each generation. The size of the dendrimers varies from 10 to 130 Å in diameter depending on the dendrimer generation.

Given its multifunctionality, dendrimers have been widely explored for catalysis, for chiral recognition, as unimolecular micelles for host-guest chemistry and light harvesting, and as liquid crystalline materials [8,9]. Dendrimers have been extensively investigated for biomedical applications that include multifunctional delivery vehicles to carry multiple drug molecules, diagnostic agents, and/or targeting ligands [10]. Dendrimer architecture has also been used to design therapeutic molecules. A polyanionic-based dendrimer (VivaGel® by Starpharma) was approved in 2006 by the US Food and Drug Administration (FDA) as a vaginal antiviral gel for the prevention of HIV. Dendrimers have also been used in commercial immunoassay kits (Stratus® CS) to immobilize the antibody onto the solid phase and to increase the assay sensitivity as well as reduce the analysis time. The chapter discusses the general properties of dendrimers and its various biomedical applications.

15.2 SYNTHESIS AND CHARACTERIZATION

Unlike the traditional polymers that often have poorly defined structure, the dendrimer synthesis can be precisely controlled at each step. Dendrimers can be synthesized using "bottom-up" (divergent) or "top-down" (convergent) synthetic approaches. The basic difference between convergent and divergent strategies is in the growth and propagation of dendrimers [11]. In the divergent method, the dendrimer is synthesized from the core through outward propagation of the branches. In contrast, in convergent method, the growth starts from the end-surface groups and the branching units are coupled together using a suitable core unit. Other synthetic approaches include double exponential growth, click chemistry approach, and lego chemistry; however, the convergent and divergent

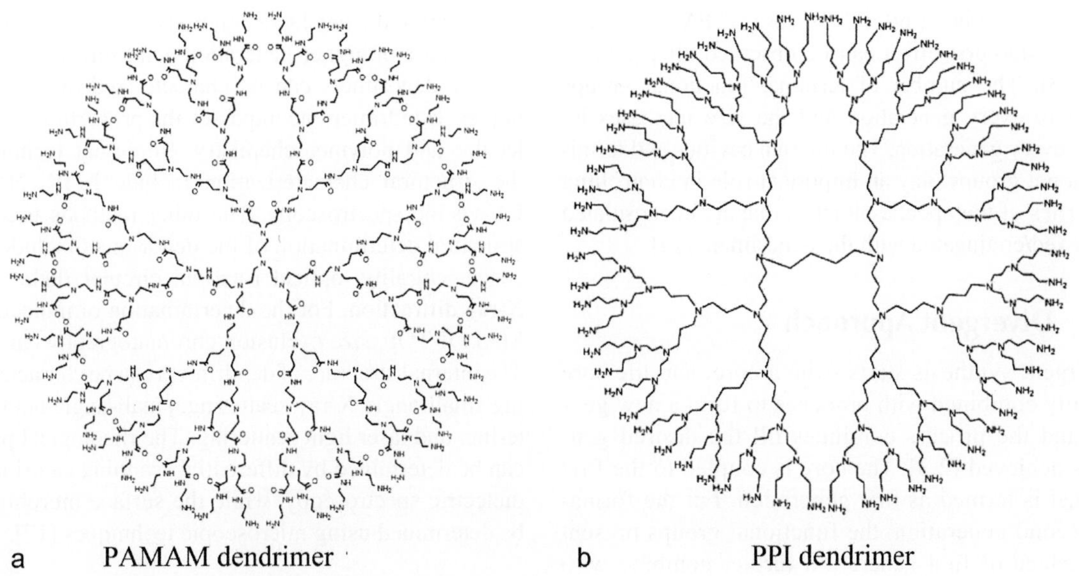

FIGURE 15.2 Chemical structures of (a) PAMAM and (b) PPI dendrimers.

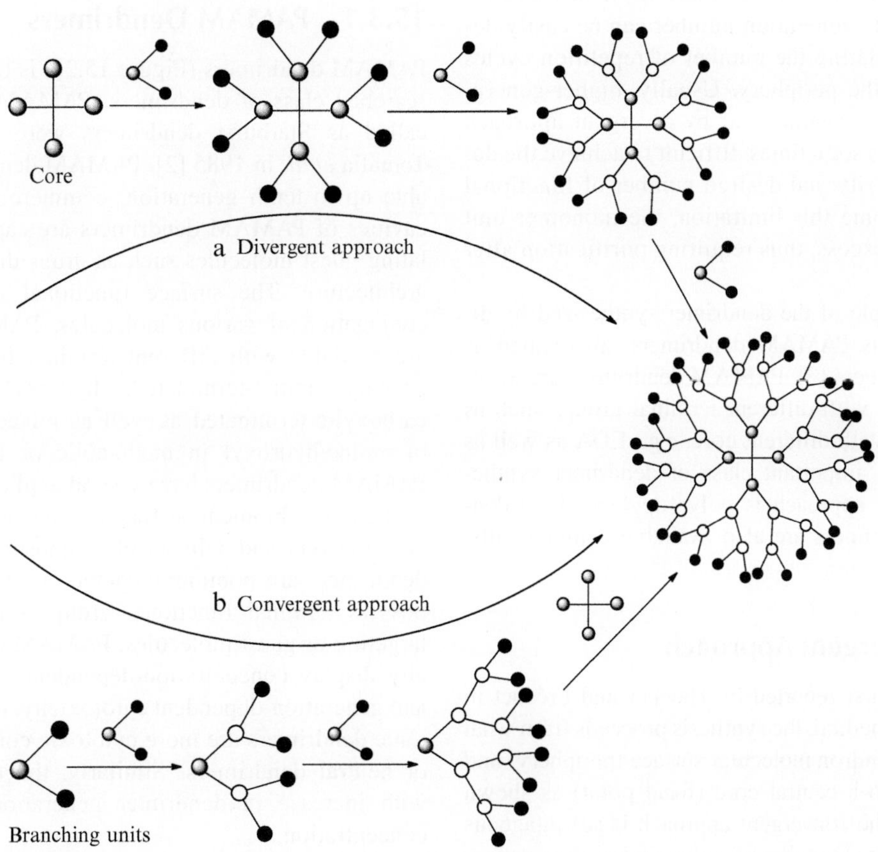

FIGURE 15.3 Two most common synthetic strategies for dendrimers. (a) Divergent synthetic approach. (b) Convergent synthetic approach.

approaches are the most commonly used approaches for the synthesis of dendrimers (Figure 15.3).

In general, the lower-generation dendrimers (G1–G3) have open structure and are highly asymmetric in shape compared to higher generations (>G4) in which the structure becomes more globular and compact. At higher generation, dendrimers become densely packed as they extend out to the periphery and form a closed membrane-like structure [12]. When a critical stage is reached, the dendrimers cannot grow because of steric hindrance. This is sometimes referred to as "starburst effect." In case of PAMAM dendrimers, the starburst effect can be observed after generation 10 [13]. The number of terminal functional groups doubles with every generation and the size increases by 1 nm with every generation. The internal cavities and terminal functional groups play an important role in controlling the properties of therapeutic moieties that are encapsulated or complexed/conjugated with the dendrimers [14].

15.2.1 Divergent Approach

The divergent synthesis starts from a core, and the core is gradually combined with branches to form a new generation, and the process continues till the desired generation is achieved [2,3]. The core is coupled to the first branch that is termed as first generation. For the formation of second generation, the functional groups present on the surface of first generation further combine with the branching units (Figure 15.3a). Each repetition cycle leads to the addition of one more layer of branches ("generation"). The generation number can be easily determined by calculating the number of repetition cycles from the core to the periphery. Usually, higher-generation dendrimers are synthesized by divergent approach [15]. However, it is sometimes difficult to achieve the desired degree of purity and desired number of functional groups. To overcome this limitation, the monomer unit is often added in excess, thus requiring purification after each step.

The best example of the dendrimer synthesized by divergent approach is PAMAM dendrimers, also called as "Starburst" dendrimers [2]. PAMAM dendrimers are available commercially with different terminal groups such as]OH and]NH_2 or with different core, e.g., EDA as well as ammonia. Another important class of dendrimer synthesized by divergent approach is polypropylene (PPI) dendrimers. PPI dendrimers are also available commercially, with DAB or EDA core.

15.2.2 Convergent Approach

The method was first reported by Hawker and Fréchet in 1990 [16]. In this method, the synthesis proceeds from what will become the dendron molecular surface (periphery) and proceeds inward to a central core (focal point) as shown in Figure 15.3b. The convergent approach is advantageous because only a limited number of active sites are present per reaction, thus reducing structural defects in the end product. This method offers advantage over divergent approach in synthesizing more homogenous and pure dendrimer. Further, the mass difference between the by-product and desired product is so large that it is easy to separate and purify. However, convergent strategy is often limited to the synthesis of lower-generation dendrimers because steric hindrance is encountered when large dendrons are reacted with a small core in case of higher-generation dendrimers [16].

The dendrimers can be characterized by several techniques. Dendrimer encompasses the properties of both molecular and polymer chemistry. Important techniques for the structural characterization include FTIR, NMR, and UV-visible spectroscopy. The other methods used for the structural determination of the dendrimers include fluorescence, chirality, optical rotation, circular dichroism, and X-ray diffraction. For the determination of molecular mass MALDI-TOF, size exclusion chromatography can be used. The internal structure of dendrimers can be characterized using small-angle X-ray scattering, small-angle neutron scattering, and laser light scattering. The rheological properties can be determined by differential scanning calorimetry and dielectric spectroscopy, while the surface morphology can be determined using microscopic techniques [17].

15.3 DENDRIMER TYPES

15.3.1 PAMAM Dendrimers

PAMAM dendrimers (Figure 15.2a) is by far the most researched class of dendrimers. PAMAM dendrimers also called as Starburst dendrimers were first reported by Tomalia *et al.* in 1985 [2]. PAMAM dendrimers are available up to tenth generation, commercially. The internal cavities of PAMAM dendrimers are capable of encapsulating guest molecules such as drugs due to their unique architecture. The surface functional groups allow the conjugation of various molecules. PAMAM dendrimers are available with different terminal functionalities like primary amine-terminated, hydroxyl-terminated, and carboxylic-terminated as well as mixed terminal groups of amine/hydroxyl in methanolic or aqueous solution. PAMAM dendrimers have several applications in pharmaceutical and biomedical fields. It has also been used for gene delivery and delivery of diagnostic agents. PAMAM dendrimers are nonimmunogenic, are water soluble, and possess terminal functional groups for binding various targeting or guest molecules. PAMAM dendrimers generally display concentration-dependent, charge-dependent, and generation-dependent cytotoxicity. In general, the cationic dendrimers are more cytotoxic compared to anionic or neutral dendrimers. Similarly, the toxicity increases with increase in dendrimer generation and dendrimer concentration.

15.3.2 PPI Dendrimers

In this class of dendrimers, ethylenediamine or diaminobutane is used as the core, while acrylonitrile molecules are used as the branching units. The chemical architecture of PPI dendrimers closely resembles PAMAM dendrimers. However, the internal architecture in PPI dendrimers is less polar compared to the PAMAM and possess alkyl chains with amide groups as repeating units (Figure 15.2b). PPI dendrimers were first reported by Vogtle [1], while a practically feasible synthetic strategy for PPI dendrimer was reported by Brabander-van den Berg and Meijer [18]. PPI dendrimers are commercially available as Astramol™ supplied and manufactured by DSM Netherlands up to generation 5 [18].

15.3.3 Carbohydrate Dendrimers

Carbohydrate dendrimers also called as glycodendrimers are either carbohydrate-coated, carbohydrate-centered or fully carbohydrate-based dendrimers. These dendrimers are useful in variety of biological applications such as studying protein-carbohydrate interactions that are important in many intercellular processes [19]. Sugar coating of dendrimers has been studied for the delivery of number of drugs and specially targeting drugs to the liver. Mannose and galactose have been used as ligands to target liver for the delivery of antimalarial drugs like primaquine phosphate [20].

15.3.4 Triazine Dendrimers

In this class of dendrimers, triazine trichloride and other triazine-based compounds are used as the core. Triazine dendrimers have been reported for a variety of applications including its use in cancer, in nonviral DNA and RNA delivery systems, in sensing applications, and as bioactive materials. Triazine trichloride offers several advantages including the ability to diversify the chemical functionality without the need for protecting groups. The nucleophilic aromatic substitution on its ring occurs sequentially in temperature-dependent manner, thus obviating the need for functional group manipulation [21].

15.3.5 Peptide Dendrimers

These dendrimers have peptide linkages on the surface or in dendritic architecture consisting of amino acids as branching or core units. Due to the therapeutic and biological relevance, these dendrimers have been used in various therapeutic areas including cancer, antimicrobials, antiviral, central nervous system, analgesia, and allergy. This class of dendrimers possesses excellent ability for drug delivery applications. Another interesting application of peptide dendrimers is that they can be used as contrast agents for magnetic resonance imaging (MRI), magnetic resonance angiography, fluorogenic imaging, and serodiagnosis [22,23].

15.3.6 Miscellaneous

Given that the dendrimers are a versatile polymer and can be tailor designed for various applications, many different dendrimer chemistries have been reported (Table 15.1). These include tecto dendrimers, chiral dendrimers, amphiphilic, hybrid, poly(amidoamine-organosilicon) (PAMAMOS), and liquid crystalline dendrimers. Tecto dendrimers are a composite dendrimer in which the dendrimer itself acts as core and is surrounded by other dendrimers. Liquid crystalline dendrimers are composed of mesogenic (liquid crystalline) monomers, e.g., mesogen-functionalized carbosilane dendrimers, whereas PAMAMOS are inverted unimolecular micelles that consist of hydrophilic, nucleophilic PAMAM interiors, and hydrophobic organosilicon exteriors [24]. Hybrid dendrimers are graft copolymers in which the dendrimer is composed of both linear and dendritic polymers.

15.4 DRUG LOADING IN DENDRIMERS

The unique nanoarchitecture, container properties, compact globular shape, monodispersity, and controllable surface functional groups make dendrimers as effective drug delivery carrier [25]. Dendrimers act as drug carrier mainly by two ways: by physical entrapment of drug molecules inside the dendritic structure (noncovalent interaction) and/or covalent attachment of drug molecules to surface functional groups in the dendrimers (Figure 15.4 and Table 15.2). Noncovalent interactions include hydrophobic, Van der Waal's, and electrostatic interactions [8,26–29]. Covalent conjugation of drugs to dendritic surface can be easily achieved due to the high density of surface functional groups. Covalent conjugation of the drugs to dendrimers is widely used for achieving higher drug payload, while the noncovalent interactions have been mainly used to improve the water solubility of insoluble drugs [30–32].

15.4.1 Noncovalent Interactions

The early reports that dendrimers can be used to encapsulate compounds were based on the studies on the encapsulation of pyrene dye [33]. Later, Meijer and coworkers (1994) reported the concept of "dendritic boxes" [34]. They demonstrated this by trapping small molecules such as rose bengal and *p*-nitrobenzoic acid inside G4 PPI dendrimers. Then, a shell was formed on the surface of the dendrimer by conjugating the terminal amines to an amino acid (L-phenylalanine), and guest molecules were stably encapsulated inside this "box." The guest molecules can be liberated from the dendritic box by hydrolyzing the outer shell of the dendrimer. The shape of the guest and the architecture of the box and its cavities determine the number of guest molecules that can be entrapped in the dendrimers [34,35].

Dendrimer offers excellent host-guest property, which leads to the noncovalent interactions for the encapsulation of drug molecules (Figure 15.4). The core is hydrophobic, while the shell is hydrophilic in dendrimers. This analogous micelle-like behavior makes it a promising carrier for drugs and

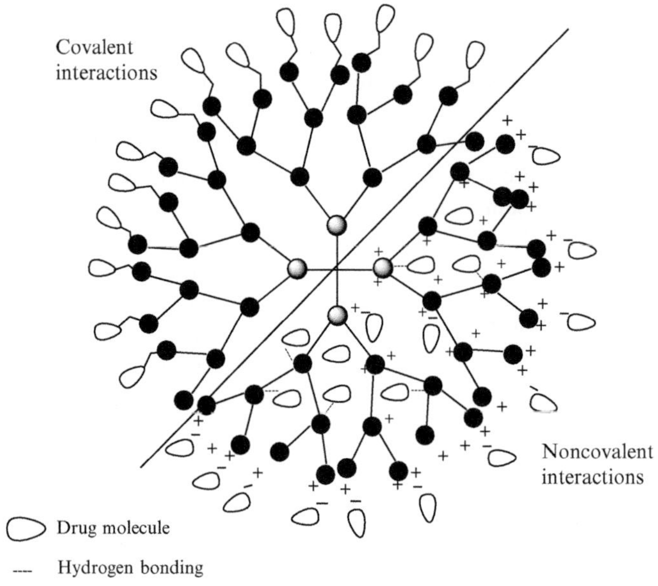

FIGURE 15.4 Possible drug loading strategies in dendrimer. Physical encapsulation of drugs results due to noncovalent interactions such as hydrogen bonding, Van der Waal's interactions, and ionic interactions, while the conjugation is usually with the surface functional groups.

TABLE 15.2 Representative List of Drugs Loaded in Dendrimers

Dendrimer	Non-covalent interactions	Covalent interactions
PAMAM	Indomethacin [27], Methotrexate [32], Doxorubicin [39], Ibuprofen [40,41], SN-38 [64], 5-Fluorouracil [26,68] Nadilfloxacin [69], Tropicamide, Pilocarpine nitrate [72], Flurbiprofen [88], Naproxen, Diflunisal, Ketoprofen [89], Piroxicam [90], 5-Amino Salicylic Acid [91], Tamsulosin [92]	Ibuprofen [31], Propanolol [63], Methotrexate [64], Dexamethasone [93] Paclitaxel [94], Cisplatin [95], Penicillin [96], Venlafaxine [97]
PPI	Cyclosporine A [98] Indomethacin, Amphotericin B, Famotidine [99]; Rifampicin [100], Lamivudine [101], Betamethasone [102]	
Polylysine	Artemether [20], Chloroquine Phosphate [103]	
Polyester	Camptothecin [104]	Doxorubicin [110]
Polyether	Methotrexate [105]	
Polyglycerol	Paclitaxel [106]	
Triazine dendrimers	Methotrexate and 6-mercaptopurine [107], Camptothecin [108], Paclitaxel [109]	Paclitaxel [111]

PAMAM-polyamidoamine dendrimer; PPI-polypropylene dendrimer.

hydrophobic guest molecules to form inclusion complexes in the dendrimer [15,33]. Drugs form noncovalent inclusion complex with dendrimers with the internal tertiary amine and surface amine groups through electrostatic and hydrophobic interactions. The terminal hydrophilic groups in the dendrimer help in drug solubilization [36].

Many drugs have been successfully encapsulated and solubilized using dendrimers (Table 15.2). Dendrimer-mediated solubilization depends on several factors such as generation size, dendrimer concentration, pH, core, temperature, and terminal functionality [37,38]. The encapsulation efficiency increases as the generation number and concentration of dendrimer increases [38]. Ke *et al.* [39] reported 95% encapsulation of doxorubicin in PAMAM dendrimer. Similarly, Chauhan *et al.* [27] reported high drug encapsulation of indomethacin in G4-NH_2, G4-OH, and G4.5 PAMAM dendrimers through electrostatic interactions. In another study, ~40 ibuprofen molecules were loaded per molecule of G4 PAMAM dendrimers at pH 10.5 resulting in high water solubility of ibuprofen [40]. Kolhe *et al.* [41]

reported the incorporation of 78 and 32 ibuprofen molecules per dendrimer, respectively, in G4 and G3 PAMAM dendrimers. The drug loading in dendrimers can be further enhanced by modifying the surface functional groups with hydrophilic moieties such as PEG. PEGylated dendrimers have been reported to have higher capacity to encapsulate water-insoluble drugs. The PEGylated dendrimers also have lesser toxicity compared to the cationic dendrimers [26].

15.4.2 Covalent Interactions

The surface functional groups in the dendrimers can be utilized for the conjugation of drugs, diagnostic agents, targeting ligands, for surface modification for various biomedical applications (Figure 15.3). Various synthetic strategies have been used to conjugate drug molecules, genetic materials, targeting ligands, dyes and imaging agents, etc., to dendrimers. Usually, the dendritic surface with hydroxyl, carboxyl, or amine termination can be used to form ester or amide, or other linkage with drug depending on the chemical functional groups in the drug. This covalent bond can be hydrolyzed inside the cell by endosomal or lysosomal enzymes to release the drug. Covalent conjugation leads to higher drug loading and controlled release of drugs [36]. Additionally, the drug-dendrimer conjugates can be used to passively target water-insoluble drugs to the tumor by utilizing the enhanced permeation and retention effect. A higher drug payload was achieved by Kolhe *et al* [31]. when they covalently conjugated approximately 58 ibuprofen molecules per G4 PAMAM-OH dendrimer molecule. The conjugate increased the cellular uptake and anti-inflammatory activity of ibuprofen in human lung epithelial carcinoma A549 cells [31].

15.5 BIOMEDICAL APPLICATIONS

15.5.1 Drug Delivery Applications

The dendrimer properties such as multifunctional surface, polyvalency, hyperbranched architecture, monodispersity, and small size make it a versatile carrier for various drug delivery applications. The internal core and/or the surface functional groups can be utilized to load drugs in the dendrimer. The hydrophilic surface functional groups can be used to improve the water solubility of insoluble drugs. Further, the dendrimers can be used as penetration enhancers to increase the transcellular and paracellular transport of drugs through various biological membranes. The polyvalency of the dendrimers can be utilized to develop multifunctional drug delivery systems for the conjugation of drugs, diagnostic agents, and targeting ligands [37]. In addition, the dendrimer surface can be modified to enhance drug loading, drug release, and drug solubility; modify biodistribution; and improve the biocompatibility of dendrimers.

15.5.1.1 Controlled and Targeted Drug Delivery

Drug targeting using nanocarriers has gained increased attention in the last few years. In particular, the drug targeting is important in cancer to maximize the therapeutic potential at the tumor site and at the same time minimize the side effects of the cytotoxic drug in the normal tissue. To this end, the altered physiology in the diseased tissue can be effectively utilized for targeted drug delivery. The enhanced permeability of tumor blood vessels and the poor lymphatic drainage in tumor can be exploited for the passive targeting of anticancer drugs using nanocarriers. This is referred to as the enhanced permeability and retention (EPR) effect [42]. Further, the change in the tumor microenvironment such as pH, enzyme, temperature, and redox potential can be used to further control the drug release at the target site [43]. Active targeting approaches use ligands that can bind to overexpressed receptors on the tumors to achieve specific targeting and enhanced cell uptake of nanocarriers. Dendrimers offer unique advantages as a targeted drug delivery carrier. The passive targeting of anticancer drugs can be achieved by encapsulation or conjugation of drugs to the dendrimer to increase the circulation half-life and avoid clearance by the reticulo-endothelial system. This can be achieved by altering the dendrimer generation, the surface functional groups, and the size. Further the dendrimer-drug conjugate/complex can significantly improve the water solubility to enable the delivery of chemotherapeutic agents [44,45]. Dendrimers in addition to enhancing the cell uptake of the drug by endocytosis can also bypass the drug efflux pumps to overcome drug resistance [45]. Patri *et al.* compared the efficacy of covalently attached MTX with a noncovalent complex with PAMAM dendrimer (G5). The MTX-conjugated dendrimers were found to be more stable compared to noncovalent MTX/dendrimer complexes [32].

Various targeting ligands such as folic acid [46,47], dextran [48] and hyaluronic acid [49], and antibodies [50,51] have been conjugated to dendrimers for active targeting of anticancer agents [52,53]. The ligand-specific receptors are usually overexpressed on tumor cells, and the ligand receptor binding leads to internalization of drug-dendrimer conjugate followed by release of the drug inside the cancer cells (Figure 15.5). Among the different ligands for cancer targeting, folic acid has been widely investigated. Quintana *et al.* [46] conjugated folic acid, fluorescein, and methotrexate covalently to PAMAM dendrimers to target, image, and enable intracellular drug delivery. Surface modification with folic acid improved the targeting of methotrexate and a 100-fold increase in cytotoxicity compared to the free drug in KB cells. The attachment of folate molecules to G5 PAMAM dendrimer dramatically enhanced the binding to folate receptors by 2500- to

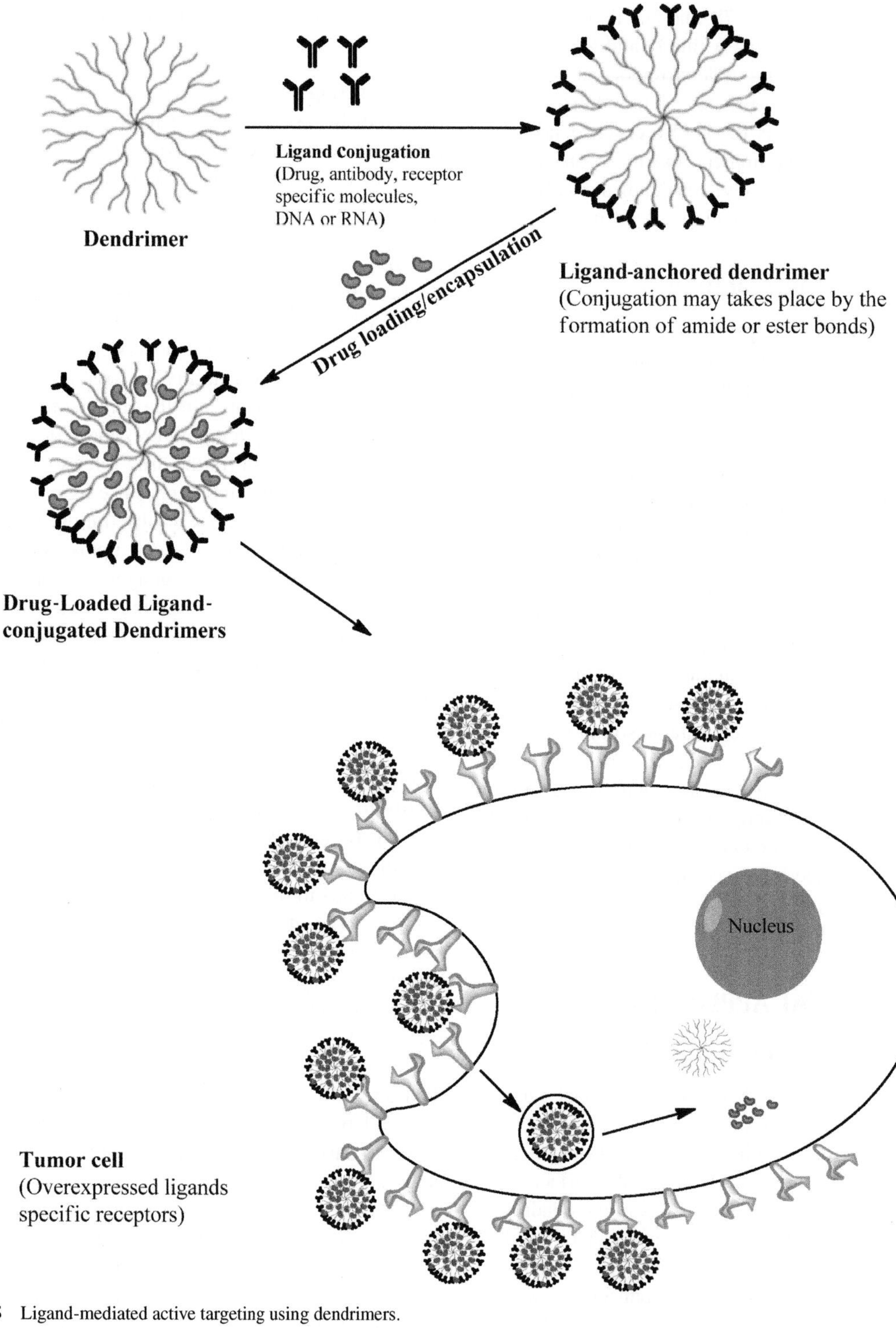

FIGURE 15.5 Ligand-mediated active targeting using dendrimers.

170,000-fold due to the multivalent folate presentation on the dendrimer surface [54]. The same research group also reported enhanced binding of methotrexate to folate receptor by 4300-fold higher when delivered using dendrimer, resulting in enhanced cytotoxicity of the conjugate in KB cells [55]. Monoclonal antibody has been conjugated to PAMAM dendrimer to target prostate-specific antigen (PSA) in prostate cancer. The antibody-conjugated dendrimer has been found to bind specifically to PSA positive cells (LNCaP; human prostate cancer cell lines) but not with PSA negative cells (PC-3; human prostate cancer cell lines from bone) [50].

The drug release from the dendrimers can be controlled by using cleavable linkers such as amides, esters, and hydrazones. The influence of spacer chemistry on drug release from drug-dendrimer conjugates was demonstrated in a study by Perumal *et al.* [56]. Methylprednisolone, an anti-inflammatory drug, was conjugated to PAMAM G4-hydroxyl dendrimer using glutaric acid and succinic acid as a spacer. Dendrimer conjugate prepared with succinic acid spacer showed higher anti-inflammatory activity in A549 lung epithelial cells compared to the dendrimer-drug conjugate formed using the glutaric acid spacer due to the higher rate of intracellular drug release from the former conjugate. In another study, doxorubicin was covalently attached through hydrazone linkage to PAMAM dendrimer. The dendrimer-doxorubicin conjugate reduced the drug distribution in the liver and heart compared to the free drug. Further, the hydrazone linkage provides pH-dependent release of doxorubicin inside the tumor cells [57].

Dendrimers have also been used for radiotherapy such as boron neutron capture therapy. This consists of two components. The first component is a stable isotope of boron (Boron-10) that can be concentrated in tumor cells by attaching targeting ligands. The second is a beam of low-energy neutrons. Boron-10 in or adjacent to the tumor cells disintegrates after capturing a neutron, and the high-energy heavy-charged particles produced destroy only the cancer cells that are in close proximity while sparing the adjacent normal cells [58]. In another study, boronated PAMAM dendrimers were designed to target the epidermal growth factor receptor, a cell surface receptor that is frequently overexpressed in brain tumor. *In vivo* studies showed that, after intratumoral injection, the conjugates were present in tenfold greater concentration in brain tumor than in normal brain tissues [59].

15.5.1.2 Oral Drug Delivery

Oral drug delivery is the most preferred drug administration route due to convenience, cost-effectiveness, and high patient compliance. The challenges in oral drug delivery include aqueous solubility, membrane permeability, and chemical and enzymatic stability of drugs. Dendrimers have been explored as potential oral drug delivery vehicles to address these challenges. The effect of dendrimers on drug solubility and transepithelial drug transport is highly dependent on the surface charge, size, generation, concentration, and treatment time [60]. Cationic amine-terminated PAMAM dendrimers show higher tissue uptake and lower drug transport across oral epithelial membrane. On the other hand, the negatively charged carboxyl-terminated PAMAM dendrimers show greater membrane transport and lower tissue uptake [60,61]. However, the higher-generation carboxyl-terminated PAMAM dendrimers have shown higher tissue uptake than lower-generation dendrimers. The dendrimers were found to be transported through both transcellular and paracellular transport pathways [60]. Cationic dendrimers decreased the transepithelial electrical resistance and modulated the tight junction. An increase in generation increased the effects on tight junction. On the other hand, the carboxyl-terminated PAMAM dendrimer showed generation-dependent effects on tight junction. The lower-generation carboxyl-terminated dendrimers (G0.5-G1.5 COOH) did not alter the tight junction, while the higher-generation dendrimers (G2.5 and G3.5 COOH) modulated the tight junction. On the other hand, hydroxyl-terminated PAMAM dendrimers did not have any effect on the tight junction. The dendrimers can be transported transcellularly by nonspecific endocytosis [60]. The toxicity of dendrimers was highly dependent on the surface charge with cationic dendrimers being most cytotoxic followed by anionic and neutral dendrimers [60]. The higher-generation dendrimers is more cytotoxic than lower-generation dendrimers. An increase in dendrimer concentration also causes a proportional increase in cytotoxicity. A higher maximum-tolerated dose is reported for anionic and neutral dendrimers (300 mg/kg) compared to cationic dendrimers (50 mg/kg) [60]. The cytotoxicity can be reduced by surface modification using fatty acids, acetylation, and PEGylation. The lipophilic groups (lauroyl and acetyl) in addition to reducing the cytotoxicity also can increase the permeability of dendrimers [62]. On the other hand, PEGylation can decrease the permeability of dendrimers.

Dendrimers have been used to improve the oral bioavailability of hydrophobic drugs such as propanolol, camptothecin analogues, silbylin, doxorubicin, and naproxen [60,63]. The dendrimers increased the oral bioavailability by increasing water solubility, drug penetration, and bypassing drug efflux pump. SN-38 which is a potent topoisomerase inhibitor suffers from poor aqueous solubility. Complexation of SN-38 with amine-terminated G4 PAMAM dendrimer increased the drug permeability by 10-fold and 100-fold higher cell uptake than the free drug. However, the complex is unstable under physiological conditions in the gastrointestinal tract leading to premature drug release. A conjugate of carboxyl-terminated G3.5 PAMAM dendrimer and SN-38 was used to address the premature drug release and stability of SN-38 [64].

15.5.1.3 Skin Drug Delivery

Skin, the largest organ in the body, is a potential site for drug delivery for both systemic and localized therapies. It is one of the most successful nonoral routes for systemic drug delivery with a handful of approved transdermal products in the market. Transdermal drug delivery has several advantages including the ease and flexibility of drug administration, sustained drug delivery for prolonged periods, overcoming first-pass metabolism in the liver and gastrointestinal stability as well as other gastrointestinal side effects of drugs, and more importantly a high patient compliance. However, drug delivery through the skin is limited by the highly impermeable and lipophilic stratum corneum layer. To this end, dendrimers can be a potential delivery vehicle to improve drug penetration across the skin by increasing drug solubility, drug partitioning, and interaction with skin lipids to increase skin penetration [65].

The structure-skin permeability relationship of dendrimer has been studied by Venuganti et al [66] using fluorescent-labeled PAMAM dendrimers and confocal laser scanning microscopy. The skin penetration of dendrimer is inversely related to the molecular weight of the dendrimer where the lower-generation dendrimers show higher skin penetration than higher-generation dendrimers. Cationic dendrimers show higher skin penetration than neutral and carboxyl terminated PAMAM dendrimers. The skin penetration increases with increase in treatment time. Surface modification of PAMAM dendrimers with acetylation and carboxylation further increases the drug transport through the skin, while conjugation of oleic acid increases dendrimer retention in the skin [67]. Iontophoresis is a physical method that uses a small electric current to increase drug transport through the skin. Cationic dendrimers were found to be suitable drug carriers for iontophoresis-mediated drug delivery through the skin [66]. The dendrimers were mainly found to be transported through intercellular lipids and hair follicles.

Dendrimers have been used to deliver drugs through the skin such as anticancer, anti-inflammatory drugs, and photosensitizers. The vehicle has a significant influence on skin penetration enhancing the effect of dendrimers. Venuganti and Perumal [68]) found that dendrimers in lipophilic vehicle increased the drug penetration of 5-fluorouracil, a hydrophilic anticancer drug by increasing partitioning into the skin. The cationic lower generation PAMAM dendrimer showed higher penetration enhancement than other dendrimers. The skin penetration of nonsteroidal anti-inflammatory drugs such as indomethacin, diflunisal, and ketoprofen was increased by dendrimers by increasing the drug's water solubility [27,69,70]. Dendrimers increased the *in vivo* skin penetration of the anti-inflammatory drugs in animals by two- to threefolds compared to the free drug. The dendrimers have also been reported to increase the skin penetration of 8-methoxypsoralen by increasing the drug aqueous solubility [71].

15.5.1.4 Delivery Through Other Routes of Drug Administration

Apart from oral, parenteral, and transdermal routes, dendrimer-mediated drug delivery has also been studied through other routes such as ocular, pulmonary, and nasal routes [43]. In case of ocular delivery, dendrimers have been used to increase the drug penetration and drug residence time in the eye. The topical bioavailability of pilocarpine nitrate and tropicamide was enhanced using dendrimers [72]. PAMAM dendrimers improved the bioavailability of pilocarpine nitrate and also increased precorneal residence time in animals. In another study, the ocular absorption of the puerarin and gatifloxacin was enhanced using dendrimers [73,74].

Dendrimers have been used for localized drug delivery to the lungs by pulmonary administration. Methylprednisolone, an anti-inflammatory agent, was delivered *in vivo* in an asthma mouse model by conjugating with hydroxyl-terminated PAMAM dendrimer [75]. The conjugate significantly improved the residence time of methylprednisolone and reduced the allergen-induced inflammation in the lungs. Bai *et al.* [76] reported 40% increase in bioavailability of enoxaparin in rats after complexing with amine-terminated PAMAM dendrimers. This drug is a low-molecular-weight heparin used to treat deep vein thrombosis. PAMAM dendrimers have also been used to increase the *in vivo* nasal absorption of macromolecules such as insulin and calcitonin [77].

15.5.2 Gene Delivery

Gene therapy includes the transfection of DNA to express a target protein or using an antisense oligonucleotide or small interfering RNA to silencing the target gene. However, the major challenges in gene therapy are the poor cell uptake and enzymatic stability of the nucleic acids. The high negative charge of nucleic acids and the high molecular weight limit the cellular uptake of DNA, siRNA, and antisense oligonucleotides. Further, the susceptibility to nuclease degradation also limits the stability of the nucleic acids. Compared to the viral vectors, the nonviral vectors (polymers and lipids) are safer and offer wider flexibility for gene delivery. Cationic polymers and lipids have been widely used as gene transfection agents. The negatively charged nucleic acids can be electrostatically complexed with the cationic delivery vehicles. The excess positive charge in the complex can enable cell uptake by binding to the negatively charged cell membrane and endocytosis-mediated uptake of the complex. Further, the complex can stabilize the nucleic acid against nuclease degradation. The high density of surface functional groups in the dendrimers makes it a potential gene delivery vehicle. Unlike most of the linear polymers, the spherical dendritic polymers can form a compact complex known as dendriplexes [78]. The dendriplex once endocytosed into the cell is transported by the endo-lysosomal pathway before reaching the cytoplasm and ultimately into the nucleus (Figure 15.6). The high density of surface amine groups in dendrimers can act as

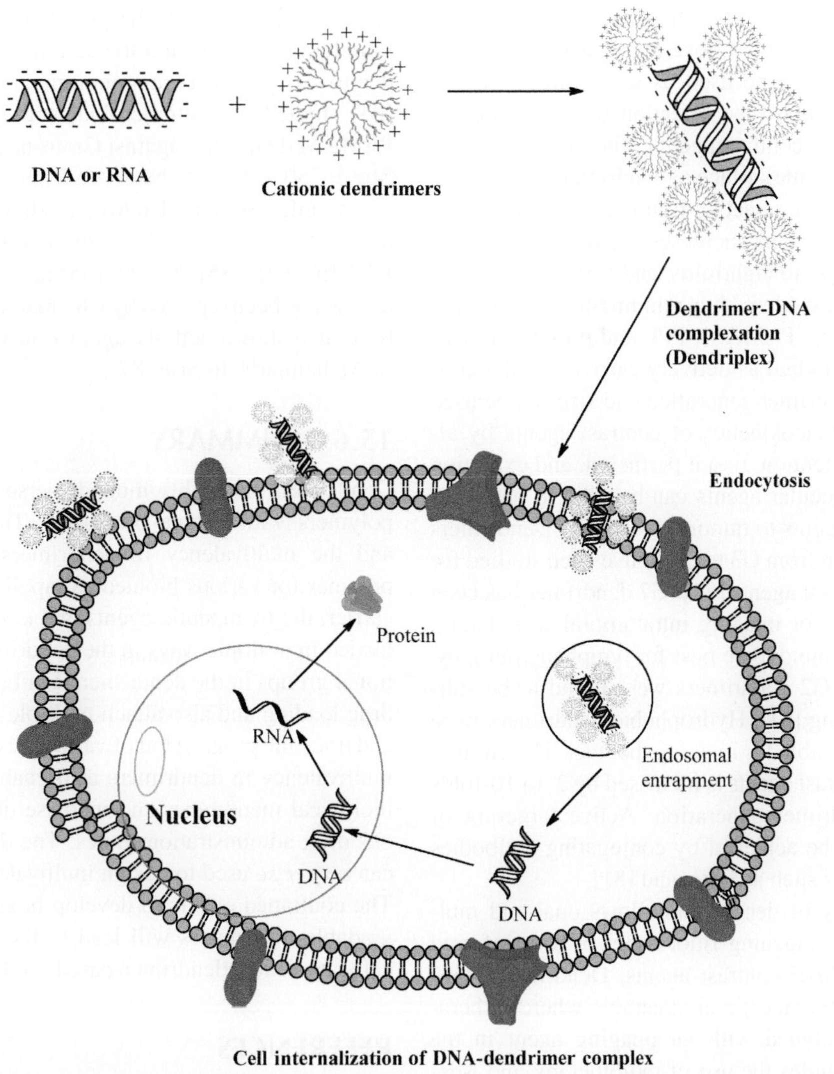

FIGURE 15.6 Dendriplex and its intracellular uptake.

a proton sponge in the endosomal compartment, leading to osmotic swelling and release of the complex from the endosomes into the cytoplasm. It has been suggested that the dendriplex can also be translocated into the nucleus [79].

PAMAM and PPI dendrimers are widely used as gene transfection agents. The transfection efficiency is dependent on the dendrimer/nucleic acid (N/P) ratio, the dendrimer type, the dendrimer generation, and the cell type [37,79]. An increase in N/P ratio generally results in a more compact and smaller complex. Although G3-G1-G0 PAMAM dendrimers have been used to form stable complex with nucleic acids, an optimum nucleic acid binding has been reported with G3 and G4 dendrimers. Solvolytic thermal degradation of dendrimers results in fractured or "imperfect" dendrimers and has been shown to improve transfection by over 50-fold [80]. The cytotoxicity and transfection efficiency of dendrimers have been improved by encapsulating in liposomes (dendrosomes) or conjugating with other polymers such as cyclodextrin or PEG [79].

15.5.3 Diagnostic Applications

MRI is a diagnostic technique that provides anatomical images of organs and blood vessels using a defined magnetic field. The image is the nuclear resonance signal of water, which is assigned to its place of origin. Addition of contrast agents (paramagnetic metal cations) improves sensitivity and specificity of the method. Low-molecular-weight gadolinium chelates such as GD(III) diethylenetriaminepentaacetic acid (GD(III)-DTPA) and Gd(III) *N*, *N*′, *N*″, *N*-tetracarboxymethyl-1,4,7,10-tetraazacyclododecane (Gd(III)-DOTA) are widely used MR contrast agents. However, these low-molecular-weight agents are rapidly cleared from the blood limiting their application for time-dependent imaging. Further, the low molar relaxivity necessitates the use of higher dose that can lead to metal toxicity. Dendrimers can be used as macromolecular delivery vehicles to increase the retention time and present multiple copies of the contrast agents. The presentation of the Gd(III) chelates on the surface of dendrimers enables the contrast agent to freely interact with surrounding water. The multivalency of dendrimers increases the relaxivity of the contrast agents by six- to eightfolds, and further, it reduces the dose of the contrast agent, thus minimizing the concerns of gadolinium toxicity. PAMAM, PPI, and polylysine dendrimers have been studied as delivery carriers for the contrast agents. The dendrimer generation and size can be used to modify the pharmacokinetics of contrast agents by altering their blood retention, tissue perfusion, and excretion rate. The macromolecular agents can be used to passively target the contrast agents to tumor vasculature. Dendrimers varying in generation from G3 to G10 have been studied for the delivery of contrast agents [81]. G7 dendrimer has been found to be optimal for imaging intratumoral vasculature, while G8 has been found to be best for lymphangiography. On the other hand, G2 dendrimers were found to be suitable for renal imaging [81]. Hydrophobic dendrimers were found to be more suitable for imaging the liver. The circulation half-life of contrast agents is increased by 2- to 10-folds depending on dendrimer generation. Active targeting of contrast agents can be achieved by conjugating antibodies and targeting ligands such as folic acid [81].

The multivalency of dendrimers allows dual and multimodal imaging by attaching fluorescent or near-infrared probes along with MRI contrast agents. Dendrimers have also been used for theranostic applications where a therapeutic agent is combined with an imaging agent in the dendrimer. This includes the use of radiotherapy and MRI contrast agent or magnetic nanoparticle with a therapeutic agent [81].

15.5.4 Therapeutic Applications

Apart from serving as delivery carriers, the dendrimer architecture has been used to design therapeutic agents. VivaGel™ (SPL7013, Starpharma, Australia) is a dendrimer-based vaginal microbicide approved by US FDA for the prevention of HIV infection. SPL7013 is a polyanionic dendrimer with a benzylhydramine-amide-lysine core and 32 naphthalene disulfonate surface functional groups [82]. This multivalent microbicide attaches to the gp120 protein on the surface of the virus to block virus attachment to cells and thus prevent infection. It is available as a Carbopol gel for vaginal application.

Cationic dendrimers have been reported to show antimicrobial [83–85], antiviral, and anti-inflammatory [86] properties. The dendrimer's antimicrobial activity is attributed to cationic charge of the dendrimers. Chen *et al.* [84] reported the synthesis of PPI dendrimers with quaternary ammonium groups on the periphery that also showed antimicrobial activity, and the activity was influenced by the length of quaternary ammonium group. Calabretta *et al.* [83] studied the antimicrobial activity of PEGylated G5 PAMAM dendrimer against Gram-negative bacteria strains. The EC50 value of these dendrimers ranged from 0.9 to 1.5 mg/mL. Anionic dendrimers showed 36-fold more cytotoxicity against *Bacillus subtilis* compared to eukaryotic HUVEC cells [85]. The anti-inflammatory activity of dendrimer has been reported by Chauhan et al. [86]. Dendrimers have also shown activity against amyloid protein involved in Alzheimer's disease [87].

15.6 SUMMARY

Dendrimers are highly monodisperse multivalent branched polymers with unique properties. The structural diversity and the multivalency in dendrimers make it a versatile polymer for various biomedical applications. As a delivery carrier, the therapeutic agents and diagnostic agents can be loaded in multiple ways in the dendrimer. The surface functional groups in the dendrimers can be used to achieve high drug loading and also attach multiple agents to target, treat, and track the progression of various diseases. In addition, the multivalency in dendrimers also enables it to interact with biological membranes and increase drug transport by various drug administration routes. The dendrimer architecture can be *per se* used to design multivalent therapeutic agents. The continued efforts to develop biocompatible and biodegradable dendrimers will lead to the introduction of clinically applicable dendrimer-based products in the near future.

REFERENCES

[1] E. Buhleier, W. Wehner, F. Vogtle, Cascade and nonskid–chain–like syntheses of molecular cavity topologies, Synthesis 2 (1978) 155–158.

[2] D.A. Tomalia, H. Baker, J.R. Dewald, M. Hall, G. Kallos, S. Martin, J. Roeck, J. Ryder, P. Smith, A new class of polymers: starburst-dendritic macromolecules, Polym. J. 17 (1985) 117–132.

[3] G.R. Newkome, Z. Yao, G.R. Baker, V.K. Gupta, Cascade molecules: a new approach to micelles, arborol, J. Org. Chem. 50 (1985) 2003–2004.

[4] A.M. Naylor, W.A. Goddard, G.E. Kiefer, D.A. Tomalia, Starburst dendrimers: molecular shape control, J. Am. Chem. Soc. 111 (1989) 2339–2341.

[5] M.W.P.L. Baars, E.W. Meijer, Host-guest chemistry of dendritic molecules, Top. Curr. Chem. 210 (2000) 131–182.

[6] B. Klajnert, M. Bryszewska, Dendrimers: properties and applications, Acta Biochim. Pol. 48 (2001) 199–208.

[7] T.H. Mourey, S.R. Turner, M. Rubenstein, J.M.J. Fréchet, C.J. Hawker, K.L. Wooley, Unique behavior of dendritic macromolecules: intrinsic viscosity of polyether dendrimers, Macromolecules 25 (1992) 2401–2406.

[8] F. Aulenta, W. Hayes, S. Rannard, Dendrimers: a new class of nanoscopic containers and delivery devices, Eur. Polym. J. 39 (2003) 1741–1771.

[9] F. Zeng, S.C. Zimmerman, Dendrimers in supramolecular chemistry: from molecular recognition to self-assembly, Chem. Rev. 97 (1997) 1681–1712.
[10] C.C. Lee, J.A. MacKay, J.M.J. Fréchet, F.C. Szoka, Designing dendrimers for biological applications, Nat. Biotechnol. 23 (2005) 1517–1526.
[11] P. Hodge, Polymer science branches out, Nature 362 (1993) 18–19.
[12] G. Caminati, N.J. Turro, D.A. Tomalia, Photophysical investigation of starburst dendrimers and their interactions with anionic and cationic surfactants, J. Am. Chem. Soc. 112 (1990) 8515–8522.
[13] M. Fischer, F. Vögtle, Dendrimers: from design to applications—a progress report, Angew. Chem. Int. Ed. 38 (1999) 884–905.
[14] D.A. Tomalia, A.M. Naylor, W.A. Goddard III., Starburst dendrimers: molecular-level control of size, shape, surface chemistry, topology, and flexibility from atoms to macroscopic matter, Angew. Chem. Int. Ed. Engl. 29 (1990) 138–175.
[15] J.M.J. Fréchet, Designing dendrimers for drug delivery, Pharm. Sci. Technol. Today 2 (2000) 393–401.
[16] C.J. Hawker, J.M.J. Fréchet, Preparation of polymers with controlled molecular architecture. A new convergent approach to dendritic macromolecules, J. Am. Chem. Soc. 112 (1990) 7638–7647.
[17] A.M. Caminade, R. Laurent, J.P. Majoral, Characterization of dendrimers, Adv. Drug Deliv. Rev. 57 (15) (2005) 2130–2146.
[18] E.M.M. Brabander-van den Berg, E.W. Meijer, Poly (propylene imine) dendrimers: large scale synthesis by heterogeneously catalyzed hydrogenation, Angew. Chem. Int. Ed. Engl. 32 (1993) 1308–1311.
[19] W.B. Turnbull, J.F. Stoddart, Design and synthesis of glycodendrimers, Rev. Mol. Biotechnol. 90 (2002) 231–255.
[20] D. Bhadra, S. Bhadra, N.K. Jain, PEGylated lysine based copolymeric dendritic micelles for solubilization and delivery of artemether, J. Pharm. Pharm. Sci. 8 (2005) 467–482.
[21] J. Lim, E.E. Simanek, Triazine dendrimers as drug delivery systems: from synthesis to therapy, Adv. Drug Deliv. Rev. 64 (9) (2012) 826–835.
[22] T. Bruckdorfer, O. Marder, F. Albericio, From production of peptides in milligram amounts for research to multi-tons quantities for drugs of the future, Curr. Pharm. Biotechnol. 5 (2004) 29–43.
[23] L. Crespo, G. Sanclimens, M. Pons, E. Giralt, M. Royo, F. Albericio, Peptide and amide bond-containing dendrimers, Chem. Rev. 105 (2005) 1663–1681.
[24] P.R. Dvornic, A.M. de Leuze-Jallouli, M.J. Owen, S.V. Perz, Radially layered poly(amidoamine-organosilicon) dendrimers, Macromolecules 33 (2000) 53–66.
[25] S. Stevelmans, J.C.M.V. Hest, J.F.G.A. Jansen, D.A.F.G. Van Boxtel, E.M.M. De Berg, E.W. Meijer, Synthesis, characterization and guest–host properties of inverted unimolecular dendritic micelles, J. Am. Chem. Soc. 118 (1996) 7398–7399.
[26] D. Bhadra, S. Bhadra, S. Jain, N.K. Jain, A PEGylated dendritic nanoparticulate carrier of fluorouracil, Int. J. Pharm. 257 (2003) 111–124.
[27] A. Chauhan, P. Diwan, N.K. Jain, K. Raghavan, Composition and complexes containing a macromolecular polymer as potential anti-inflammatory agents, US Patent 20030180250 A1, 2003.
[28] B. Devarakonda, R.A. Hill, M.M. De Villiers, The effect of PAMAM dendrimer generation size and surface functional group on the aqueous solubility of nifedipine, Int. J. Pharm. 284 (2004) 133–140.
[29] S. Svenson, D.A. Tomalia, Dendrimers in biomedical applications-reflections on the field, Adv. Drug Deliv. Rev. 57 (15) (2005) 2106–2129.
[30] H.L. Crampton, E.E. Simanek, Dendrimers as drug delivery vehicles: non-covalent interactions of bioactive compounds with dendrimers, Polym. Int. 56 (2007) 489–496.
[31] P. Kolhe, J. Khandare, O. Pillai, S. Kannan, M. Lieh-Lai, R.M. Kannan, Preparation, cellular transport, and activity of polyamidoamine-based dendritic nanodevices with a high drug payload, Biomaterials 27 (2006) 660–669.
[32] A.K. Patri, J.F. Kukowska-Latallo, J.R. Baker Jr., Targeted drug delivery with dendrimers: comparison of the release kinetics of covalently conjugated drug and non-covalent drug inclusion complex, Adv. Drug Deliv. Rev. 57 (2005) 2203–2214.
[33] J. Hawker, K.L. Wooley, J.M.J. Fréchet, Unimolecular micelles and globular amphiphiles: dendritic macromolecules as novel recyclable solubilization agents, J. Chem. Soc. Perkin Trans. I 12 (1993) 1287–1297.
[34] J.F.G.A. Jansen, E.M.M. De Berg, E.W. Meijer, Encapsulation of guest molecules into a dendritic box, Science 266 (1994) 1226–1229.
[35] J.F.G.A. Jansen, E.W. Meijer, E.M.M. De Berg, The dendritic box: shape-selective liberation of encapsulated guests, J. Am. Chem. Soc. 117 (1995) 4417–4418.
[36] A. D'Emanuele, D. Attwood, Dendrimer–drug interactions, Adv. Drug Deliv. Rev. 57 (2005) 2147–2162.
[37] U. Boas, P.M.H. Heegaard, Dendrimers in drug research, Chem. Soc. Rev. 33 (2004) 43–63.
[38] U. Gupta, H.B. Agashe, A. Asthana, N.K. Jain, Dendrimers: novel polymeric nano-architectures for solubility enhancement, Biomacromolecules 7 (2006) 649–658.
[39] W. Ke, Y. Zhao, R. Huang, C. Jiang, Y. Pei, Enhanced oral bioavailability of doxorubicin in a dendrimer drug delivery system, J. Pharm. Sci. 97 (2008) 2208–2216.
[40] O.M. Milhem, C. Myles, N.B. McKeown, D. Attwood, A. D'Emanuele, Polyamidoamine starburst dendrimers as solubility enhancers, Int. J. Pharm. 197 (2000) 239–241.
[41] P. Kolhe, E. Misra, R.M. Kannan, S. Kannan, M. Lieh-Lai, Drug complexation, in vitro release and cellular entry of dendrimers and hyperbranched polymers, Int. J. Pharm. 259 (2003) 143–160.
[42] R. Duncan, Polymer conjugates for tumor targeting and intracytoplasmic delivery. The EPR effect as a common gateway? Pharm. Sci. Technol. Today 2 (1999) 441–449.
[43] S. Mignani, S. El Kazoouli, M. Bousmina, J.P. Majora, Expand classical drug administration ways by emerging routes using dendrimer drug delivery systems: a concise review, Adv. Drug Deliv. Rev. 66 (2013) 1316–1330.
[44] J.R. Baker Jr., Dendrimer-based nanoparticles for cancer therapy, Hematology 1 (2009) 708–719.
[45] J.B. Wolinsky, M.W. Grinstaff, Therapeutic and diagnostic applications of dendrimers for cancer treatment, Adv. Drug Deliv. Rev. 60 (2008) 1037–1055.
[46] A. Quintana, E. Raczka, L. Piehler, I. Lee, A. Myc, I. Majoros, A.K. Patri, T. Thomas, J. Mule, J.R. Baker Jr., Design and function of a dendrimer-based therapeutic nanodevice targeted to tumor cells through the folate receptor, Pharm. Res. 19 (2002) 1310–1316.
[47] P. Singh, U. Gupta, A. Asthana, N.K. Jain, Folate and folate-PEG-PAMAM dendrimers: synthesis, characterization and targeted anticancer drug delivery potential in tumor induced mice, Bioconjug. Chem. 19 (11) (2008) 2239–2252.
[48] A. Agarwal, U. Gupta, A. Asthana, N.K. Jain, Dextran conjugated dendrtitic nanoconstructs as potential vectors for anti-cancer agent, Biomaterials 30 (21) (2009) 3588–3596.

[49] M. Han, Q. Lv, X.J. Tang, Y.L. Hu, D.H. Xu, F.Z. Li, W.Q. Liang, J.Q. Gao, Overcoming drug resistance of MCF-7/ADR cells by altering intracellular distribution of doxorubicin via MVP knockdown with a novel siRNA polyamidoamine-hyaluronic acid complex, J. Control. Release 163 (2) (2012) 136–144.

[50] A.K. Patri, A. Myc, J. Beals, T.P. Thomas, N.H. Bander, J. Baker Jr., Synthesis and in vitro testing of J591 antibody–dendrimer conjugates for targeted prostate cancer therapy, Bioconjug. Chem. 15 (2004) 1174–1181.

[51] R. Shukla, T.P. Thomas, J.L. Peters, A.M. Desai, J. Kukowska-Latallo, A.K. Patri, A. Kotlyar, J.R. Baker Jr., HER2 specific tumor targeting with dendrimer conjugated anti-HER2 mAb, Bioconjug. Chem. 17 (2006) 1109–1115.

[52] M. Najlah, A. D'Emanuele, Synthesis of dendrimers and drug-dendrimer conjugates for drug delivery, Curr. Opin. Drug Discov. Dev. 10 (6) (2007) 756–767.

[53] D.A. Tomalia, L.A. Reyna, S. Svenson, Dendrimers as multi-purpose nanodevices for oncology drug delivery and diagnostic imaging, Biochem. Soc. Trans. 35 (1) (2007) 61–67.

[54] S. Hong, P.R. Leroueil, I.J. Majoros, B.G. Orr, J.R. Baker Jr., The binding avidity of a nanoparticle-based multivalent targeted drug delivery platform, Chem. Biol. 14 (2007) 107–115.

[55] T.P. Thomas, B. Huang, S.K. Choi, J.E. Silpe, A. Kotlyar, A.M. Desai, H. Zong, J. Gam, M. Joice, J.R. Baker Jr., Polyvalent dendrimer-methotrexate as a folate receptor targeted cancer therapeutic, Mol. Pharm. 9 (2012) 2669–2676.

[56] O. Perumal, J. Khandare, P. Kolhe, S. Kannan, M. Lieh-Lai, R.M. Kannan, Effects of branching architecture and linker on the activity of hyperbranched polymer-drug conjugates, Bioconjug. Chem. 20 (2009) 842–846.

[57] H.R. Ihre, O.L.P. De Jesus, F.C. Szoka, J.M.J. Frechet, Polyester dendritic systems for drug delivery applications: design, synthesis, and characterization, Bioconjug. Chem. 13 (2002) 443–452.

[58] R.F. Barth, D.M. Adams, A.H. Soloway, F. Alam, M.V. Darby, Boronated starburst dendrimer-monoclonal antibody immunoconjugates: evaluation as a potential delivery system for neutron capture therapy, Bioconjug. Chem. 5 (1994) 58–66.

[59] G. Wu, R.F. Barth, W. Yang, M. Chatterjee, W. Tjarks, M.J. Ciesielski, R.A. Fenstermaker, Site-specific conjugation of boron-containing dendrimers to anti-EGF receptor monoclonal antibody cetuximab (IMC-C225) and its evaluation as a potential delivery agent for neutron capture therapy, Bioconjug. Chem. 15 (2004) 185–194.

[60] S. Sadekar, H. Ghandehari, Transepithelial transport and toxicity of PAMAM dendrimers: implications for oral drug delivery, Adv. Drug Deliv. Rev. 64 (6) (2012) 571–588.

[61] R. Wiwattanapatapee, B.C. Gomez, N. Malik, R. Duncan, Anionic PAMAM dendrimers rapidly cross adult rat intestine in vitro: a potential oral delivery system? Pharm. Res. 17 (2000) 991–998.

[62] R. Jevprasesphant, J. Penny, D. Attwood, N.B. McKeown, A. D'Emanuele, Engineering of dendrimer surfaces to enhance transepithelial transport and reduce cytotoxicity, Pharm. Res. 20 (2003) 1543–1550.

[63] A. D'Emanuele, J.P. Jevprasesphant, J. Penny, D. Attwood, The use of a dendrimer-propranolol prodrug to bypass efflux transporters and enhance oral bioavailability, J. Control. Release 95 (2004) 447–453.

[64] R.B. Kolhatkar, P. Swaan, H. Ghandehari, Potential oral delivery of 7-ethyl-10-hydroxy-camptothecin (SN-38) using poly (amidoamine) dendrimers, Pharm. Res. 25 (7) (2008) 1723–1729.

[65] M. Sun, A. Fan, Z. Wang, Y. Zhao, Dendrimer-mediated drug delivery to the skin, Soft Matter 8 (2012) 4301–4305.

[66] V.V.K. Venuganti, P. Sahdev, M. Hildreth, X. Guan, O.P. Perumal, Structure-skin permeability relationship of dendrimers, Pharm. Res. 28 (2011) 2246–2260.

[67] Y. Yang, S. Sunogrot, C. Stowell, J. Ji, C.W. Lee, J.W. Kim, S.A. Khan, S. Hong, Effect of size, surface charge, and hydrophobicity of poly(amidoamine) dendrimers on their skin penetration, Biomacromolecules 13 (2012) 2154–2162.

[68] V.V.K. Venuganti, O.P. Perumal, Poly(amidoamine) dendrimers as skin penetration enhancers: influence of charge, generation and concentration, J. Pharm. Sci. 98 (2009) 2345–2356.

[69] Y. Cheng, H. Qu, M. Ma, Polyamidoamine (PAMAM) dendrimers as biocompatible carriers of quinolone antimicrobial study: an in vitro study, Eur. J. Med. Chem. 42 (2007) 1032–1038.

[70] Y. Cheng, N. Man, T. Xu, R. Fu, X. Wang, X. Wang, L. Wen, Transdermal delivery of nonsteroidal anti-inflammatory drugs mediated by polyamidoamine (PAMAM) dendrimers, J. Pharm. Sci. 96 (2007) 595–602.

[71] K. Borowska, B. Laskowska, A. Magon, M. Mysliwiec Pyda, S. Wolowiec, PAMAM dendrimers as solubilizers and hosts for 8-methoxypsoralene enabling transdermal diffusion guest, Int. J. Pharm. 398 (2010) 185–189.

[72] T.F. Vandamme, L. Brobeck, Polyamido amine dendrimers as ophthalmic vehicles for ocular delivery of pilocarpine nitrate and tropicamide, J. Control. Release 102 (2005) 23–38.

[73] C. Yao, W. Wang, X. Zhou, T. Qu, H. Mu, R. Liang, A. Wang, K. Sun, Effects of poly(amidoamine) dendrimers on ocular absorption of puerarin using microdialysis, J. Ocul. Pharmacol. Ther. 27 (6) (2011) 565–569.

[74] C. Durairaj, R.S. Kadam, J.W. Chandler, S.L. Hutcherson, U.B. Kompella, Nanosized dendritic polyguanidilyated translocators for enhanced solubility, permeability, and delivery of gatifloxacin, Invest. Ophthalmol. Vis. Sci. 51 (11) (2010) 5804–5816.

[75] R. Inapagolla, B.R. Guru, Y.E. Kurtoglu, X. Gao, M. Lieh-Lai, D.J.P. Bassett, R.M. Kannan, In vivo efficacy of dendrimer-methylprednisolone conjugate formation for the treatment of lung inflammation, Int. J. Pharm. 339 (2010) 140–147.

[76] S. Bai, C. Thomas, F. Ahsan, Dendrimers as a carrier for pulmonary delivery of enoxaparin, a low molecular weight heparin, J. Pharm. Sci. 96 (2007) 2090–2106.

[77] Z. Dong, H. Katsumi, T. Sakane, A. Yamamoto, Effects of polyamidoamine (PAMAM) dendrimers on the nasal absorption of poorly absorbable drugs in rats, Int. J. Pharm. 30 (2010) 244–252.

[78] J.D. Eichman, A.U. Bielinska, J.F. Kukowska-Latallo, J.R.J. Baker, The use of PAMAM dendrimers in the efficient transfer of genetic material into cells, Pharm. Sci. Technol. Today 3 (2000) 232–245.

[79] C. Dufes, I.F. Uchegbu, A.G. Schatzlein, Dendrimers in gene delivery, Adv. Drug Deliv. Rev. 57 (2005) 2177–2202.

[80] M.X. Tang, C.T. Redemann, F.C. Szoka Jr., In vitro gene delivery by degraded polyamidoamine dendrimers, Bioconjug. Chem. 7 (6) (1996) 703–714.

[81] M. Longmire, P.L. Choyke, H. Kobayashi, Dendrimer-based contrast agents for molecular imaging, Curr. Top. Med. Chem. 8 (2008) 1180–1186.

[82] R. Rupp, S.L. Rosentha, L.R. Stanberry, Vivagel (SPL7013 Gel): a candidate dendrimer microbicide for the prevention of HIV and HSV infection, Int. J. Nanomedicine 5 (2007) 581–588.

[83] M.K. Calabretta, A. Kumar, A.M. McDermott, C. Cai, Antibacterial activities of poly(amidoamine) dendrimers terminated with amino and poly(ethylene glycol) groups, Biomacromolecules 8 (2007) 1807–1811.

[84] C.Z. Chen, N.C. Beck-Tan, P. Dhurjati, T.K. van Dyk, R.A. LaRossa, S.L. Cooper, Quaternary ammonium functionalized poly(propylene imine) dendrimers as effective antimicrobials: structure–activity studies, Biomacromolecules 1 (2000) 473–480.

[85] S.R. Meyers, F.S. Juhn, A.P. Griset, N.R. Luman, M.W. Grinstaff, Anionic amphiphilic dendrimers as antibacterial agents, J. Am. Chem. Soc. 130 (2008) 14444–14445.

[86] A.S. Chauhan, S. Sridevi, K.B. Chalasani, Dendrimer-mediated transdermal delivery: enhanced bioavailability of indomethacin, J. Control. Release 90 (2003) 335–343.

[87] B. Klajnert, M. Cortijo-Arellano, J. Cladera, M. Bryszewska, Influence of dendrimer's structure on its activity against amyloid fibril formation, Biochem. Biophys. Res. Commun. 345 (2006) 21–28.

[88] A. Asthana, A.S. Chauhan, P.V. Diwan, N.K. Jain, Poly(amidoamine) (PAMAM) dendritic nanostructures for controlled site-specific delivery of acidic anti-inflammatory active ingredient, AAPS PharmSciTech. 6 (2005) E536–E542.

[89] C. Yiyun, X. Tongwen, Dendrimers as potential drug carriers. Part I. Solubilization of non-steroidal anti-inflammatory drugs in the presence of polyamidoamine dendrimers, Eur. J. Med. Chem. 40 (2005) 1188–1192.

[90] R.N. Prajapati, R.K. Tekade, U. Gupta, V. Gajbhiye, N.K. Jain, Dendimer-mediated solubilization, formulation development and in vitro-in vivo assessment of piroxicam, Mol. Pharm. 6 (3) (2009) 940–950.

[91] H. Namazi, M. Adeli, Dendrimers of citric acid and poly (ethylene glycol) as the new drug-delivery agents, Biomaterials 26 (10) (2005) 1175–1183.

[92] Z. Wang, Y. Itoh, Y. Hosaka, I. Kobayashi, Y. Nakano, I. Maeda, F. Umeda, J. Yamakawa, M. Kawase, K. Yag, Novel transdermal drug delivery system with polyhydroxyalkanoate and starburst polyamidoaminedendrimer, J. Biosci. Bioeng. 95 (5) (2003) 541–543.

[93] A. Choksi, K.V. Sarojini, P. Vadnal, C. Dias, P.K. Suresh, J. Khandare, Comparative anti-inflammatory activity of poly(amidoamine) (PAMAM) dendrimer-dexamethasone conjugates with dexamethasone-liposomes, Int. J. Pharm. 449 (1–2) (2013) 28–36.

[94] J.J. Khandare, S. Jayant, A. Singh, Dendrimer versus linear conjugate: influence of polymeric architecture on the delivery and anticancer effect of paclitaxel, Bioconjug. Chem. 17 (6) (2006) 1464–1472.

[95] N. Malik, E.G. Evagorou, R. Duncan, Dendrimer-platinate: a novel approach to cancer chemotherapy, Anticancer Drugs 10 (8) (1999) 767–776.

[96] H. Yang, S.T. Lopina, Penicillin V-conjugated PEG-PAMAM star polymers, J. Biomater. Sci. 10 (2003) 1043–1056, Polymer Edition 14.

[97] H. Yang, S.T. Lopina, Extended release of a novel antidepressant, venlafaxine, based on anionic polyamidoamine dendrimers and poly(ethylene glycol) containing semi-interpenetrating networks, J. Biomed. Mater. Res. A 72 (2005) 107–114.

[98] K.W. Chooi, A.I. Gray, L. Tetley, Y. Fan, I.F. Uchegbu, The molecular shape of poly(propylenimine) dendrimer amphiphiles has a profound effect on their self assembly, Langmuir 26 (4) (2010) 2301–2316.

[99] U. Gupta, H.B. Agashe, N.K. Jain, Polypropylene imine dendrimer mediated solubility enhancement: effect of pH and functional groups of hydrophobes, J Pharm. Pharmaceut. Sci. 10 (3) (2007) 358–367.

[100] P.V. Kumar, H. Agashe, T. Dutta, N.K. Jain, PEGylated dendritic architecture for development of a prolonged drug delivery system for an antitubercular drug, Curr. Drug Deliv. 4 (2007) 11–19.

[101] T. Dutta, M. Garg, N.K. Jain, Targeting of efavirenz loaded tuftsin conjugated poly(propyleneimine) dendrimers to HIV infected macrophages in vitro, Eur. J. Pharm. Sci. 34 (2–3) (2008) 181–189.

[102] Z. Sideratou, D. Tsiourvas, C.M. Paleos, Solubilization and release properties of PEGylated diaminobutane poly(propylene imine) dendrimers, J. Colloid Interface Sci. 242 (2001) 272–276.

[103] P. Agrawal, U. Gupta, N.K. Jain, Glycoconjugated peptide dendrimers-based nanoparticulate system for the delivery of chloroquine phosphate, Biomaterials 28 (22) (2007) 3349–3359.

[104] M.T. Morgan, Y. Nakanishi, D.J. Kroll, A.P. Griset, M.A. Carnahan, M. Wathier, N.H. Oberlies, G. Manikumar, M.C. Wani, M.W. Grinstaff, Dendrimer-encapsulated camptothecins: increased solubility, cellular uptake, and cellular retention affords enhanced anticancer activity in vitro, Cancer Res. 66 (24) (2006) 11913–11921.

[105] K. Kono, M. Liu, J.M.J. Fréchet, Design of dendritic macromolecules containing folate or methotrexate residues, Bioconjug. Chem. 10 (6) (1999) 1115–1121.

[106] T. Ooya, J. Lee, K. Park, Effects of ethylene glycol based grafts, star-shaped and dendritic polymers on solubilization and controlled release of paclitaxel, J. Control. Release 93 (2003) 121–127.

[107] M.F. Neerman, H.T. Chen, A.R. Parrish, E.E. Simanek, Reduction of drug toxicity using dendrimers based on melamine, Mol. Pharm. 1 (5) (2004) 390–393.

[108] V.J. Venditto, K. Allred, C.D. Allred, E.E. Simanek, Intercepting the synthesis of triazine dendrimers with nucleophilic pharmacophores: a general strategy toward drug delivery vehicles, Chem. Commun. 37 (2009) 5541–5542.

[109] K.K. Bansal, D. Kakde, U. Gupta, N.K. Jain, Development and characterization of triazine based dendrimers for delivery of antitumor agent, J. Nanosci. Nanotechnol. 10 (12) (2010) 8395–8404.

[110] O.L. Padilla De Jesus, H.R. Ihre, L. Gagne, et al., Polyester dendritic systems for drug delivery applications: in vitro and in vivo evaluation, Bioconjug. Chem. 13 (2002) 453–461.

[111] S.T. Lo, S. Stern, J.D. Clogston, J. Zheng, P.P. Adiseshaiah, M. Dobrovolskaia, J. Lim, A.K. Patri, X. Sun, E.E. Simanek, Biological assessment of triazine dendrimer: toxicological profiles, solution behavior, biodistribution, drug release and efficacy in a PEGylated, paclitaxel construct, Mol. Pharm. 7 (4) (2010) 993–1006.

Chapter 16

Design Strategies and Applications of Citrate-Based Biodegradable Elastomeric Polymers

Jinshan Guo[1], Dianna Y. Nguyen[1], Richard T. Tran[1], Zhiwei Xie[1], Xiaochun Bai[2,3], Jian Yang[1,2,3]

[1]*Department of Bioengineering, Materials Research Institute, The Huck Institute of The Life sciences, The Pennsylvania State University, University Park, Pennsylvania, USA*

[2]*Department of Orthopedics, The third affiliated hospital, Southern Medical University, Guangzhou, China*

[3]*Academy of Orthopedics of Guangdong Province, Guangzhou, China*

Chapter Outline

16.1 INTRODUCTION

Biodegradable polymers are superior to traditional nondegradable polymers since they do not need subsequent surgical removal after being implanted in bodies. Thus, they have gained widespread application in biomedical areas in recent years, such as tissue engineering, drug delivery, gene delivery, and bioimaging [1–4]. Among biodegradable polymers, biodegradable elastomeric polymers have received increasing attention because their compliance under force can closely resemble the elastic nature of many soft tissues such as heart valves, blood vessels, tendons, cartilage, and bladder [5–7]. Elastomers are usually amorphous polymers with relatively low glass transition temperature (T_g). When used as an implantable material, the presence of at least one segment with a glass transition temperature (T_g) lower than room temperature or at least body temperature (37 °C) is necessary to make sure the polymer is in an elastic state and considerable segmental motion is possible in the temperature range used. Polyesters like poly(ε-caprolactone) (PCL), poly(dioxanone) (PDO), and poly(δ-valerolactone); some poly(carbonate)s; and their copolymers synthesized by ring-opening polymerization (ROP) have low T_gs and can be used as elastomers. This method is restricted by the stability of cyclic monomers used, which are usually 5-, 6-, or 7-membered rings. Thus, other ROP-derived polymers such as poly(L-lactide) (PLLA), poly(glycolide) (PGA), and their copolymers (PLGA) have higher T_gs and are in glassy state, brittle, and stiff at the intended use (room or body) temperature [2,3,8,9]. Compared with ROP, polycondensation provides more options. Diacid, diol, or hydroxyalkanoate monomers used for polyester synthesis may have long aliphatic chains to ensure the softness and elasticity of the obtained polyesters via polycondensation [1,7,10,11]. Among various polymerization methods used in polycondensation, such as thermal polymerization, microbial synthesis, and enzymatic polymerization, catalyst-free and solvent-free thermal polymerization are the most commonly used [1,7,10,11]. Multifunctional monomers including unsaturated monomers can be used in polycondensation, conferring the (pre) polymers with thermal or photo-cross-linkable (or curable) properties [1,2,10–12]. Among the cross-linkable elastomeric polymers, poly(polyol sebacate) [10,12–15] and citric acid-based poly(diol citrate) [2,16–27] are the two most widely researched. In this chapter, we will focus on the design strategies and applications of citrate-based biodegradable elastomeric polymers (CABEs).

Citric acid is a natural, weak organic acid that abundantly exists in many vegetables and fruits, especially citrus fruits like lemon and lime, where the citrate concentration can reach up to 8% after drying. As an intermediate in the tricarboxylic acid cycle (TCA cycle, also known as the citric acid cycle or Krebs cycle), which occurs in the metabolism of all aerobic organisms, citric acid is nontoxic and multifunctional. Thus, it is an excellent choice of a starting material for biodegradable polymer synthesis. Possessing three carboxyl groups and one hydroxyl group, citric acid has been widely used as a chelating or binding agent of various metal ions and metal oxide nanoparticles [28,29]. A majority of the body's citrate content is located in skeletal tissues and plays a large role in metabolism, calcium chelation, hydroxyapatite (HA) formation, and regulation of the thickness of bone apatite structure [30–33]. Citrate not only functions as a calcium-solubilizing agent but is also a strong bound and integral part of the bone nanocomposite [33]. Citrate also has a unique and innate ability to induce HA formation in simulated body fluid (SBF) [34]. The rich pendent –COOH groups in CABEs, like poly(1,8-octanediol citrate) (POC) and poly(poly(ethylene glycol) maleate citrate) (PEGMC), can also chelate calcium ions. The polymers and their composites with HA have proven to be promising orthopedic biomaterials that can promote the biomineralization process and increase osteoblast adhesion and mineralization, thus enhancing osteointegration [2,35–39]. In all CABE designs, citric acid participates in prepolymer formation through polycondensation with diols, and it also preserves pendent functionalities for postpolymerization through esterification to produce a cross-linked polyester network. Cross-linking confers elasticity and mechanical stability to the polymers similar to extracellular matrix (ECM), in which collagen and elastin are also all cross-linked polymers [40]. In addition to the multifunctionality and biocompatibility of citric acid, the sodium form of citric acid, sodium citrate, is an anticoagulant often used in hospitals [17]. This implies that CABEs may also possess suitable hemocompatibility when in contact with blood, which will benefit the application of CABEs in blood-contacting applications. Because of all the advantages stated earlier, the research on CABEs is just unfolding. In this book chapter, we will discuss the design strategies and applications of previous CABEs and the outlook of their future development in biomedical engineering.

16.2 DESIGN STRATEGIES OF CABEs

Material design strategies are always driven by application demands. As the first polymer series of CABE family, poly(diol citrate)s were initially designed for soft tissue engineering applications like blood vessels, so the chain length of diols used to react with citric acid should be long enough to guarantee the obtained polymers to be elastomeric, and 1,8-octanediol (OD) was proved to be the right diol, so was the born of POC. After that, a lot of modifications were made on POC by introducing additional diols or chain extension, to adjust the elasticity, mechanical strength, and degradation profile of the obtained polymers; bring new properties; and introduce new cross-linking mechanisms to the obtained elastomers (Figure 16.1). To obtain hydrophilic CABEs, water-soluble diol such as poly(ethyl glycol) (PEG) was also used. On the other hand, novel design strategies can also bring new application directions. The creations of biodegradable photoluminescent polymer (BPLP) [2,20,21] and injectable citrate-based mussel-inspired bioadhesives (iCMBA) [22,23] brought bioimaging and bioadhesive applications to CABE family, respectively.

The design strategies of CABEs also adapted the characteristics of the monomers used. Citric acid is a multifunctional monomer containing three carboxyl groups and one hydroxyl group; in CABEs, it not only participates in prepolymer formation by reacting with diols but also preserves pendent functionalities for postpolymerization or concurrent/postmodification with other molecules such as amine-containing molecules. In addition to citric acid, other acids were also used. By adjusting the feeding ratio of diol to citric acid, hydroxyl group-terminated CABEs were synthesized, which can be used for further chain extension by using hydroxyl groups reaction with diisocyanate or initiating lactones ROP. Additional diols except OD, such as double bond-containing diols, were also introduced into CABE systems to adjust the polymer properties (Figure 16.1).

16.2.1 Poly(Diol Citrate) Synthesis

As mentioned earlier, CABE prepolymers are synthesized via a simple thermal polycondensation process by reacting citric acid with diol monomers (Figure 16.2 and Table 16.1). Citric acid confers CABEs with pendent functionality for postmodification or concurrent modification and cross-linking. While the elasticity of CABEs mostly depends on the chain length of the diol used, if

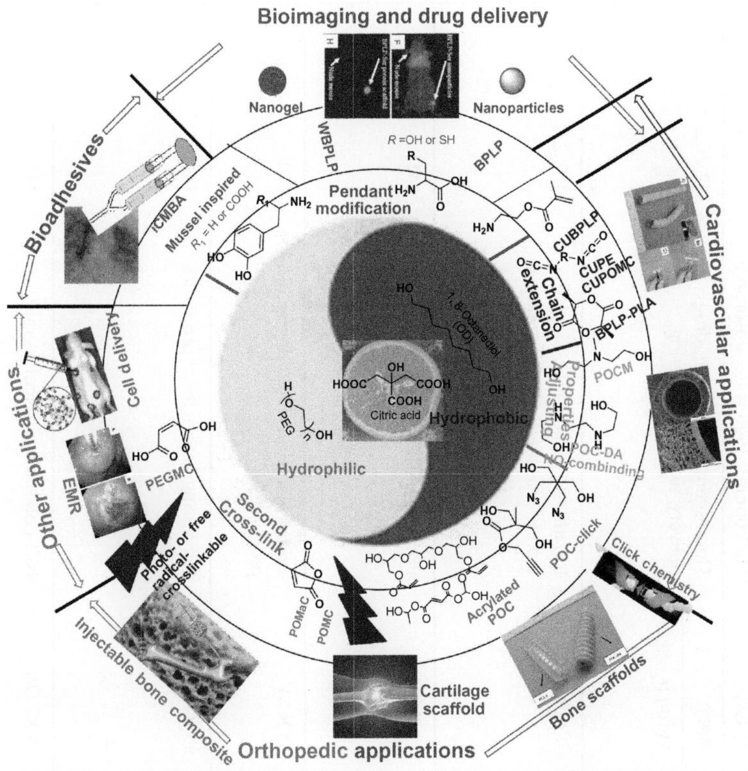

FIGURE 16.1 Design strategies and applications of citrate-based biodegradable elastomeric polymers (CABEs). (For color version of this figure, the reader is referred to the online version of this chapter.)

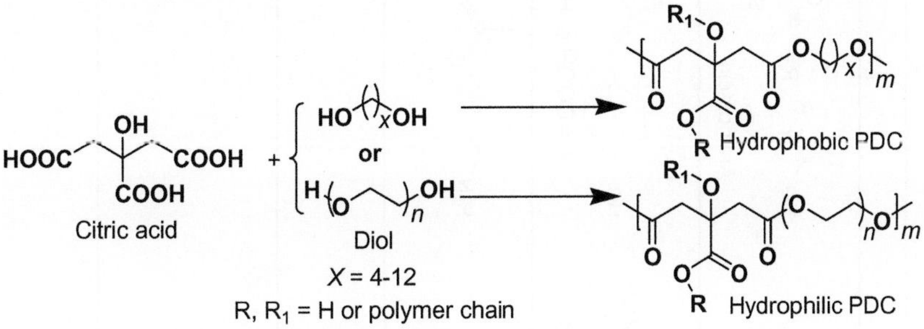

FIGURE 16.2 Synthesis schematic of poly(diol citrate) prepolymer family.

the chain length of the diol used is too short, like ethylene glycol or 1,4-butanediol, the obtained polymers are likely to be brittle and stiff. Conversely, long aliphatic chains may increase elasticity but may also slow down material degradation. Although polymers like poly(1,12-didecanediol-*co*-citrate) (PDDC), synthesized by the polycondensation of citric acid and 1,12-dodecanediol, have shape-memory properties [49], the most used hydrophobic diol is 1,8-octanediol (OD) as it is the longest diol that can dissolve in water or PBS.

By adjusting postpolymerization temperature and cross-linking (curing) time, the mechanical properties, degradation profiles, and surface energies of the cross-linked polyester networks, POC can be tuned to fit a wide range of tissue engineering applications (see Table 16.2) [2,16–18,35,50–56]. An increase in postpolymerization temperature and cross-linking time and the application of vacuum resulted in a network with increased mechanical strength due to the increased cross-linking densities. As shown in Table 16.2, the range of mechanical properties of POC potentially meets the needs for the engineering of various soft tissues including blood vessels, nerve, cartilage, and the bladders. The preliminary biocompatibility evaluation showed that POC supported the attachment and proliferation of human aortic smooth muscle cells (HASMC), endothelial cells (ECs), and 3T3 fibroblasts cells without any surface modifications [16,18]. Histological analysis of POC films subcutaneously implanted in Sprague-Dawley (SD) rats further confirmed that POC elicited minimal inflammatory responses. After 4-month implantation, the thickness of fibrous capsule was smaller than the reported values of the widely used commercial biodegradable polymer, PLGA [16,18].

TABLE 16.1 Citrate-Based Biodegradable Elastomers: Composition, Name, Properties, and Their Applications

Monomers[a]					
Diol or Polyol	**Citric Acid and Diacid**	**Pendent Modification[b]**	**Polymer[c]**	**Properties and Applications**	**Refs.**
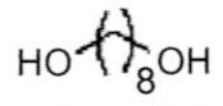 1,8-octanediol (OD)			POC	Hydrophobic, suitable elasticity; vascular graft, bone composite, drug delivery	[2,16–18,29,35]
PEG			PEGC	Hydrophilic, high water adsorption	[25–27]
OD		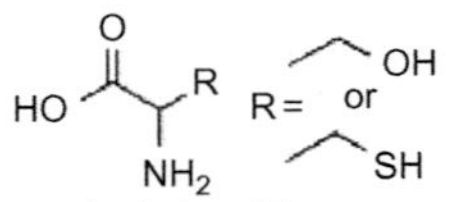 Amino acid	BPLP	Fluorescent; bioimaging, drug delivery, tissue engineering	[2,20,21]
PEG			WBPLP		
PEG	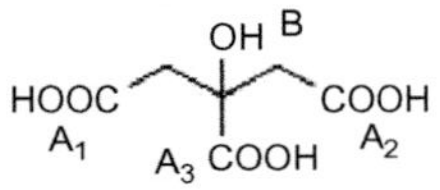 Citric acid (CA)	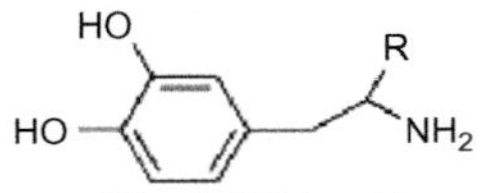 R′ = COOH or H	iCMBA	Bioadhesive, injectable; bioglue, wound closure, tissue bioadhesive	[22,23]
OD		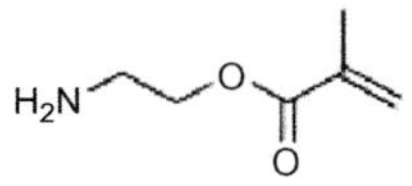	cMA-POC	Photo-cross-linkable	[41]
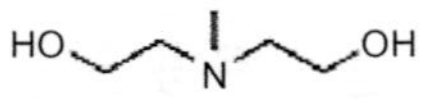 MDEA			POCM	Adjust mechanical and degradation properties	[18]
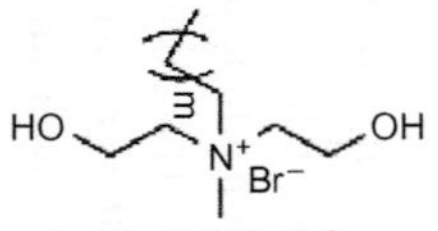 $m = 2,4,6$ or 8			POC-Q(m+2)	Conferring antibacterial properties to PDC	[42]
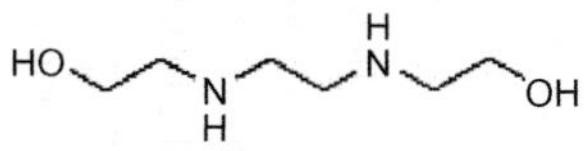			POC-DA	NO-combining; vascular graft	[43–45]
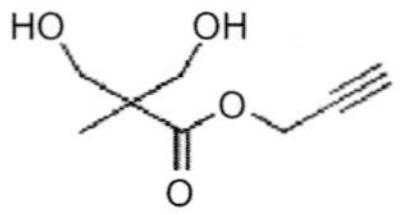			POC-click[d]	Enhanced mechanical properties, surface conjugate—easy—vascular graft, bone composites	Unpublished

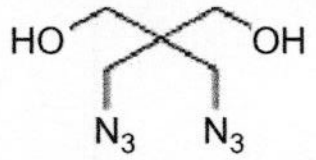				
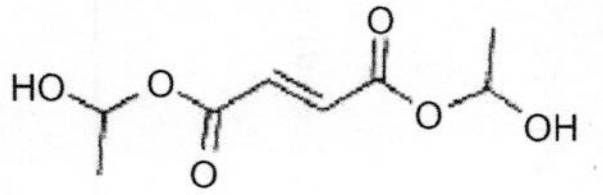 or 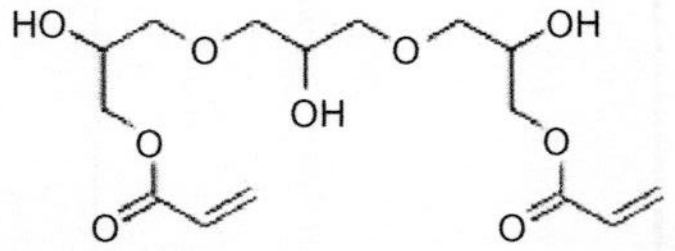		Acrylated POC	Photo-cross-linkable	[46]
OD	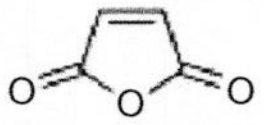	POMaC[e]	Photo-cross-linkable	[24]
PEG or OD	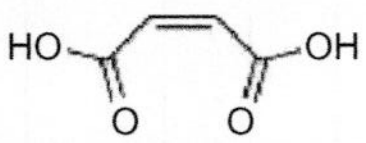	PEGMC or POMC[f]	Photo- or free radical cross-linkable; cell delivery, injectable bone composite	[27,35–37,47]

Pre-polymer	Chain Extension	Polymer[c]	Properties and Applications	Refs.
Pre-POC	HDI[g]	CUPE	Enhanced mechanical properties; vascular graft, bone composites	[19]
BPLP	HDI[g]	CUBPLP	Fluorescent, enhanced mechanical properties; vascular graft, bioimaging	[48]
Pre-POMC[g]	HDI[g]	CUPOMC	Photo-cross-linkable	[47]
BPLP	ROP of lactide	BPLP-LA	Fluorescent PLA	Unpublished

[a]*For all the hydrophobic polymers, like POC, BPLP, POCM, POC-DA, POMaC, POC-click, 1,8-octanediol, and citric acid, that composed the main chain backbone of the polymers, functional diols or polyols were used as additional monomers.*

[b]*Pendent modification refers to the modification of the pendent functional groups on citrate-based prepolymers. The modification can be conducted concurrently with (BPLP, WBPLP, and iCMBA) or after (cMA-POC) the polycondensation process of diol and citric acid.*

[c]*Here, polymers are the final cross-linked polymers, except for BPLP, WBPLP, and BPLP-LA. The full names of polymer abbreviations are listed as follows: POC, poly(1, 8-octanediol citrate); PEGMC, poly(poly(ethylene glycol) maleate citrate); BPLP, biodegradable photoluminescent polymer; WBPLP, water-soluble BPLP; iCMBA, injectable citrate-based mussel-inspired bioadhesive; CUPE, cross-linked urethane-doped polyester; CUBPLP, cross-linked urethane-doped BPLP.*

[d]*POC-click is formed by thermo-cross-linking the mixture of pre-POC-N_3 (azide-containing POC prepolymer) and pre-POC-Al (alkyne-containing POC prepolymer); the process applies synchronous binary cross-link mechanism, esterification, and thermal click reaction, and the residual azide groups on the surface of POC-click film or scaffold paved the way of surface bioconjugation through strain-promoted alkyne-azide cycloaddition (SPAAC), another copper-free click reaction.*

[e]*POMaC also applies dual cross-link mechanism, thermal esterification, and photo-cross-linking of double bonds.*

[f]*PEGMC was used as hydrogel at room or body temperature; thus, only double bonds serve as cross-linking functionality.*

[g]*HDI: 1,6-hexamethylene diisocyanate.*

TABLE 16.2 Mechanical and Degradation Properties of Citrate-Based Biodegradable Elastomers and Some Selected Soft Tissues

Polymer	Tensile Strength (MPa)	Modulus (MPa)	Elongation (%)	Degradation rate	Tested cells and Animals	Refs.
POC	2.93-11.15	1.85-13.98	117-502	100% in 26 weeks[a]	HASMCs/HAECs Sprague-Dawley rats	[18]
POMaC	2.45-9.94	0.05-1.52	51-441	15-76% in 10 weeks[b]	3T3 fibroblasts Sprague-Dawley rats	[24]
CBPLP[c]	2.2-7.6	2.4-8.9	140-272	100% in 31 weeks[d]	3T3 fibroblasts, nude mice	[20]
POC-DA	1.49-10.71	5.91-32.64	201-290	20% loss in 6 weeks[e]	PASMC/HASMC/HUVEC Sprague-Dawley rats	[43–45]
Acrylated POC	2.8-15.7	7.4-75.9	86.1-260	27-35% in 2 months[f]	–	[46]
POC-click	18.3-41.32	16.6-275.9	78-323.9	100% in 34 weeks[g]	3T3 fibroblasts, HUVEC Sprague-Dawley rats	Unpublished
CUPE	14-37	2.2-32	217-309	15% in 8 weeks[h]	3T3 fibroblasts Sprague-Dawley rats	[19]
CUPOMC	1-10.5	0.5-5.8	175-220	-	3T3 fibroblasts	[47]
CUBPLP	1.2-49.41	0.2-52	240-450	13-22 h in 0.05 M NaOH	3T3 fibroblasts	[48]
Porcine aortic heart valve (radial)	2.4	6.4	134.8	–	–	[50]
Porcine aortic heart valve (circumferential)	8.3	44.7	48.7	–	–	[50]
Ulnar peripheral nerve	9.8-21.6	–	8-21	–	–	[51]
Human coronary artery	1.4-11.14	–	–	–	–	[52]
Bovine elastin	–	1.1	–	–	–	[53]
Human ACL	24-112	–	–	–	–	[54]
Human cartilage	3.7-10.5	–	–	–	–	[52]
Smooth muscle relaxed	–	0.006	300	–	–	[55]
Porcine lung	–	0.005	–	–	–	[56]

[a]*For POC (80 °C, 2 days), incubated in PBS (pH 7.4, 37 °C).*

[b]*For POMaC with different maleic anhydride ratios and thermo- or photocured under different conditions, incubated in PBS (pH 7.4, 37 °C).*

[c]*CBPLP refers cross-linked BPLP.*

[d]*For CBPLP-Cys0.8 (80 °C, 2 days), incubated in PBS (pH 7.4, 37 °C).*

[e]*For POC-DA with different DA ratios (80 °C, 4 days), incubated in PBS (pH 7.4, 37 °C).*

[f]*Samples with different compositions were cross-linked at 80 °C for 0.5 days, followed by 120°C, 1 days and 120 °C, vacuum, 1 day. Degradation was conducted in PBS (pH 7.4, 37 °C).*

[g]*For POC-click3 (100 °C, 3 days), incubated in PBS (pH 7.4, 37 °C).*

[h]*For CUPE1.2 (80 °C, 2 days), incubated in PBS (pH 7.4, 37 °C).*

Although POC, as a representative poly(diol citrate), is soft and elastic, it is still considered relatively weak, with a tensile strength typically no more than 10 MPa at dry state (Table 16.2). This is already much lower than that of human anterior cruciate ligament (38 MPa) and may become even lower when fabricated into porous scaffolds and/or used *in vivo* at wet state. To modify the material properties, functionalities, and processability of poly(diol citrates), various diols, diacids, and/or diamines were introduced to CABEs, either to adjust the material properties or to introduce a second cross-linking mechanism. Pendent group (carboxyl or hydroxyl) modification and chain extension of CABEs were also conducted to improve the mechanical properties and functionalities of CABEs [22,23,27,35–37]. This will be discussed in details in the following sections.

16.2.2 Molecular Design of CABEs

Based on poly(diol citrate), several molecular design strategies were adopted for CABE syntheses. First, the introduction of additional functionalities or additional cross-link mechanisms using various additional diols or diacids were studied. By applying this strategy, a number of CABEs have recently been developed, including poly(1,8-octanediol-*co*-citrate-*co*-MDEA) (POCM) [18], POC with quaternary ammonium salt (POC-Q) (antibacterial POC) [42], diazeniumdiolated poly(1,8-octanediol citrate) (POC-DA) [43–45], poly(1,8-octamethylene maleate (anhydride) citrate) (POMaC) [24], poly(octamethylene maleate citrate) (POMC) [39], POC-click (unpublished), acrylated POC [46], and PEGMC [27,35,36] (Table 16.1).

Second, the modification of the pendent functionality of CABEs through reactions with amines or amino acids was developed, which can proceed concurrently or after the polycondensation of various CABE polymers, such as BPLP [2,20,21], iCMBA [22,23], and cross-linked methacrylated POC (cMA-POC) and cross-linked methacrylated poly(1,12-dodecanediol citrate) (cMA-PDDC) [41] (Table 16.2).

Third the chain extension through reacting with chain extenders such as 1,6-hexamethylene diisocyanate (HDI) was also developed. Such strategy has resulted in cross-linked urethane-doped polyester elastomers (CUPEs) based on POC [19], cross-linked urethane-doped BPLP (CUBPLP) [44] and cross-linked urethane-doped POMC (CUPOMC) [45]. This can also happen through initiating lactones' ROP based on BPLP polymers to form fluorescent polylactone, BPLP-polylactone (BPLPL) (unpublished) (Table 16.2).

16.2.2.1 Additional Diols

As shown in Figure 16.3, various diol or diacid monomers have been used in CABE systems to adjust material properties or to introduce additional functionalities. After introducing an amine-containing diol (*N*-methyldiethanolamine, MDEA) into POC, the resulting POCM showed enhanced mechanical strength and faster degradation rates (Tables 16.1 and 16.2). As reported, POCM10% and POCM5% showed mass losses of 72% and 48%, respectively, after degrading in PBS at 37 °C for 4 weeks. This is much higher than that of POC degraded in the same period (about 20%) [18]. Higher degradation rates of POCM polymers benefit from the positive charges and high water solubility of MDEA because the positive charges can neutralize the negative charges of degradation products, thus promoting the reaction balance of hydrolysis degradation to move forward.

By introducing quaternary ammonium salt diol into POC, Wynne *et al.* conferred antibacterial properties to the resulting POC-Q through a convenient and cost-effective thermal polycondensation process [42]. These materials have a tailored surface and strong antibacterial properties that make them good candidates as biodegradable packaging materials.

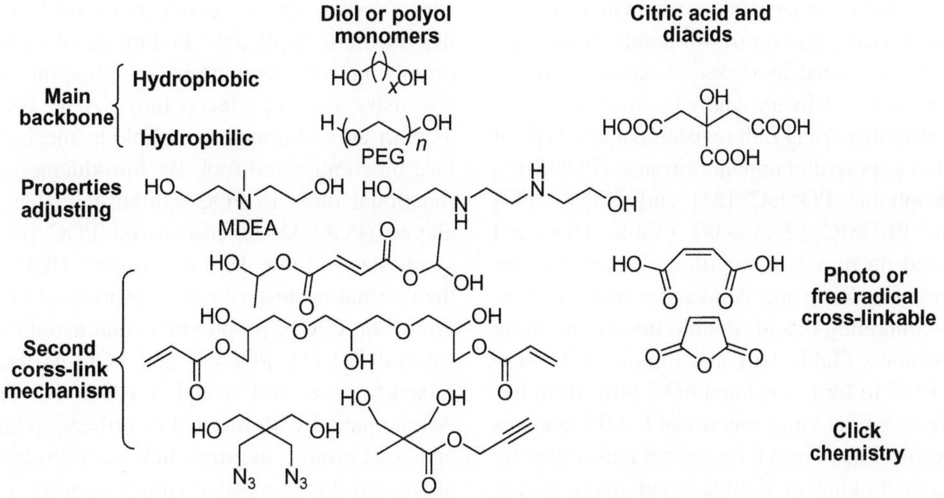

FIGURE 16.3 Diol and diacid monomers used in citrate-based biodegradable elastomers (CABEs) system.

In cardiovascular applications, nitric oxide (NO) released by vascular ECs has been shown to be a potent antithrombotic and antineointimal hyperplasia (NIH) agent by inhibiting platelet adhesion/activation and leukocyte chemotaxis, as well as smooth muscle cell (SMC) proliferation and migration. To increase the potential of CABEs in cardiovascular applications a NO-binding diol, *N*, *N*-bis(2-hydroxyethyl) ethylenediamine was introduced into POC polymer by Ameer's Lab to form a NO-releasing diazeniumdiolated POC, called POC-DA (Table 16.1) [43–45]. POC-DA possessed similar tensile strength (1.49-10.71 MPa) to POC, a higher Young's modulus (5.91-32.64 MPa), and an elongation at break in the range of 201-290% (Table 16.2). Cross-linked POC-DA polymer can deliver different doses of NO for 3 days by varying the amine content and the exposure time to pressurize NO gas without a significant impact on copolymer degradation rate [43]. Suitable NO-releasing dose was shown to be beneficial to the proliferation of human umbilical vein endothelial cells (HUVEC), while the proliferation of HASMC was significantly inhibited [44].

16.2.2.2 Additional Cross-Link Mechanism

The introduction of functional diols or diacids, such as double bond-containing diols or diacids and click moiety-containing diols, into CABE system, can confer the system with a second cross-link mechanism, such as free radical cross-linking of double bonds or click reaction between alkyne and azide groups (Figure 16.3). The introduction of a secondary cross-linking mechanism can adjust the mechanical properties and degradation properties of CABEs and can also confer CABEs with an additional and effective surface functionality for further bioactive molecule surface conjugation. The second cross-link mechanism is important, especially in the case of PEG-based hydrogel systems used for cell delivery or tissue bioadhesives at room or body temperature, where thermo-cross-link mechanism is not appropriate. In recent years, photo-cross-linkable biodegradable materials have attracted increased attention in tissue engineering, drug or cell delivery, and wound repair applications [57,58]. Recently, Yang Lab registered a new type of CABEs referred to as poly(diol maleate citrates) (PDOMC), containing hydrophobic POMaC [24] and POMC [47] and hydrophilic PEGMC [27,36–38] (Table 16.1 and Figure 16.3) based on previous poly(diol citrate). Similar work was also done by Zhao and Ameer, but instead of introducing vinyl-containing diacid, double bond-containing diol or triol monomers (Table 16.1 and Figure 16.3) were introduced into POC to form acrylated POC [46]. Both hydrophobic and hydrophilic vinyl functional CABE systems can be quickly cross-linked into a thermoset elastomers by either thermo-cross-linking or double bond photo-/redox cross-linking, or both of them, namely, dual cross-linking mechanism (DCM) [24,27,36–38,47]. The DCM allows the polymer to be quickly cross-linked by redox initiators or ultraviolet (UV) light to preserve valuable pendent carboxyl and hydroxyl groups for potential bioconjugation. PDOMC networks cross-linked by this route also show a pH-dependent swelling capability, which is useful in pH-sensitive drug delivery applications. The free radical (either photo- or redox) cross-linking method can also be combined with a thermo-cross-linking mechanism to further cross-link the network to fine-tune the mechanical and degradation properties in order to meet a variety of soft tissue engineering applications. As shown in Table 16.2, POMaC, POMC, and acrylated POC elastomer families have a wide range of mechanical properties (Young's modulus of 0.05-75.9 MPa, tensile strength of 2.45-15.7 MPa, and elasticity of 51-441%) that can be modulated through adjusting monomer ratios, photoinitiator or redox initiator concentrations, and the use of DCM. Cells seeded onto the surface of POMC and POMaC films, or encapsulated in PEGMC hydrogels, exhibited normal spread morphologies. *In vivo* host response studies show a decline in inflammatory response and reduction in capsule thickness over a 4-week period, and no tissue necrosis was found throughout the animal studies.

The ideal bioelastomer-based implant materials not only should be soft and elastic, possess suitable mechanical properties to match with the target tissue or organ, and be biocompatible to minimize adverse biological responses but also should be amenable to surface modification with bioactive molecules such as growth factors, cell-binding peptides, or signaling molecules to positively and selectively recognize, interact with, support, and promote the appropriate cellular responses, thus accelerating the regeneration of the target tissue or organ [59–63]. Although most CABEs possess some –COOH and –OH groups on their surface [16–18], these groups are not effective enough, especially for surface bioconjugation. As one of the most effective surface/interface reactions, click chemistry [64–69], especially copper-free click chemistry [70–73] that is more applicable in biorelated systems, endows a promising way for surface conjugation. Therefore, click chemistry was introduced into CABE system and served as both an additional cross-linking mechanism and a surface bioconjugation tool. By introducing azide and alkyne functional diols in POC syntheses, azide (POC-N_3)- and alkyne (POC-Al)-functionalized POC prepolymers were synthesized (Table 16.1 and Figure 16.3). To fully utilize the thermal postesterification process of –COOH and –OH groups on CABE prepolymers and avoid the use of copper catalyst [74,75], POC-N_3 and POC-Al prepolymers were mixed together and heated at 100 °C for designated times. Accompanied with thermal esterification between –COOH and –OH groups, thermal click reactions between azide and alkyne groups proceeded simultaneously; thus, dual cross-linked (esterification and thermal click reaction) POC-click elastomers were formed in a one-step postpolymerization process rather than a two-step process [24]. The DCM and the rigid property of triazole rings resulting from the click reaction [67] confer POC-click with significantly enhanced mechanical strength. The tensile strength of POC-click can reach as high as 40 MPa (Table 16.2), which is comparable to or even higher than that of CUPE (see Table 16.2, will be discussed in the following paragraphs), while the degradation time of POC-click has no significant increase as CUPE does when comparing with POC (Table 16.2). The degradation profile of POC-click performed a "first slow then fast" pattern, which is sometimes more favorable in bioengineering applications for the good maintenance of mechanical properties before the fulfillment of bioregeneration process and fast degradation after. Furthermore, the residual azide groups on the surface of POC-click bioelastomers paved the way for convenient and high-yield surface conjugation of bioactive molecules through strain-promoted alkyne-azide cycloaddition (SPAAC), a copper-free click reaction.

16.2.2.3 Pendent Group Modification

As a multifunctional monomer, citric acid contains three carboxyl groups and one hydroxyl group. Even after being synthesized into prepolymers with different diols, there are still some –COOH and –OH groups preserved for pendent group modifications (concurrent with or after polycondensation process) with either $-NH_2$ or –OH-containing molecules (react with –COOH groups) or –COOH, –COCl, or –NCO-containing molecules (react with –OH groups). When an excess amount of diol was used to react with citric acid, hydroxyl-terminated POC or other CABEs' prepolymers were obtained that can react with diisocyanate molecules such as HDI to extend the polymer chains (such as CUPE [19], CUBPLP [47], and CUPOMC [57]). This is referred to as chain extension reaction and will be discussed in Section 16.2.2.4. In this section, we will discuss the pendent modification of poly(diol citrates) with amino acids, L-3,4-dihydroxyphenylalanine (L-DOPA) or dopamine, and other amine-containing molecules (Figure 16.3). Pendent group modification of poly(diol citrates) brought some remarkable and intriguing properties such as photoluminescent (BPLP) and bioadhesive (iCMBA) properties to the resulting polymers.

16.2.2.3.1 Development of BPLPs

Biodegradable fluorescent polymers have attracted a lot of attention in targeting drug delivery, bioimaging, and tissue engineering. The most reported fluorescent biodegradable polymers are made by either conjugating or encapsulating organic dyes or quantum dots (QDs) with biodegradable polymers [20,21,66,75–80]. However, the low photobleaching resistance of organic dyes or the unacceptable toxicity of inorganic QDs largely limited the applications of these fluorescent biodegradable polymers. Therefore, developing fully biodegradable and biocompatible fluorescent polymer is urgently needed.

Recently, based on pure natural citric acid and all 20 essential α-amino acids as well as biocompatible aliphatic diols, Yang *et al.* developed a family of novel aliphatic BPLPs, referred to as BPLPs [20,21]. Unlike nondegradable aromatic fluorescent polymers or organic dyes commonly used in the lighting industry and bioimaging applications, BPLPs are aliphatic biodegradable oligomers synthesized from biocompatible and biodegradable monomers through a convenient and cost-effective thermal polycondensation process. Although whether (L-) α-amino acids contribute to the formation of BPLP backbones or not is still unclear, we conjecture that the synthesis of BPLPs is a kind of pendent modification of poly(diol citrate) concurrent with the polycondensation process (Figure 16.4). Briefly, BPLPs were synthesized by reacting one of the 20 natural (L-) α-amino acids with citric acid and diols (aliphatic diols such as OD or macrodiols such as PEG) at 140 °C for a certain time, which depends on the diol used and the feeding ratios of amino acid over other monomers.

Among BPLPs, BPLP-cysteine (BPLP-Cys, using L-cysteine) and BPLP-serine (BPLP-Ser, using L-serine) display the best fluorescent properties in terms of fluorescence intensity and quantum yield. The quantum yield of BPLP-Cys can be as high as 62.3% (Figure 16.5) [20].

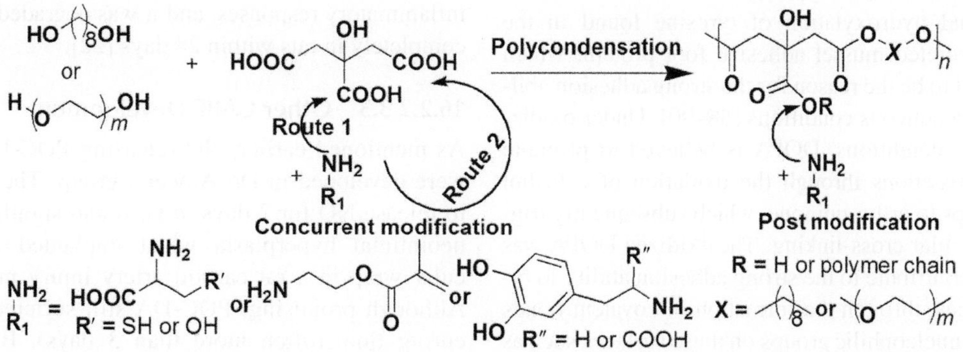

FIGURE 16.4 Pendent (concurrent with or after polycondensation process) modification of citrate-based biodegradable elastomers (CABEs).

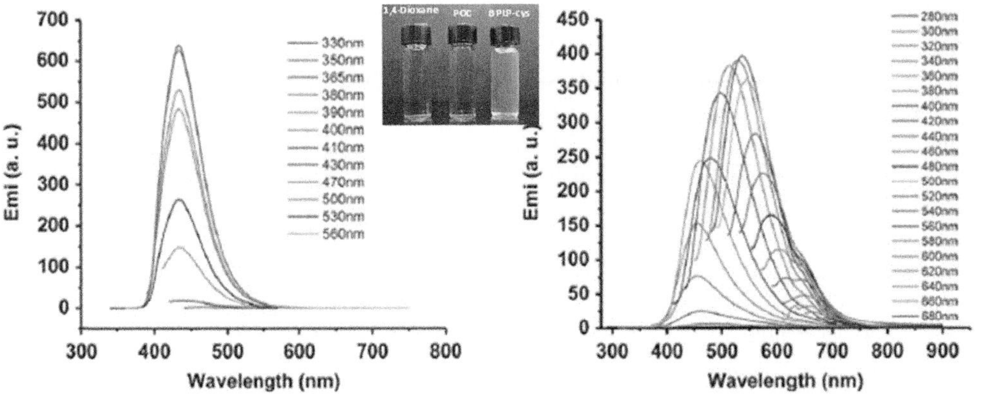

FIGURE 16.5 Emission spectra of BPLP-Cys (left) and BPLP-Ser (right) solution (in 1, 4-dioxane) at different excitation wavelengths [21]; photograph of 1, 4-dioxane, BPLP-Cys, and POC in the presence of an ultraviolet light source (middle). (For color version of this figure, the reader is referred to the online version of this chapter.)

The fluorescence emission wavelength can be tuned from blue to red by using different amino acids in BPLP syntheses. BPLPs in different modalities (polymer solution, film, scaffold, and nanoparticles) have all shown strong fluorescence. Due to pendent –COOH and –OH groups mostly provided by citric acid, BPLPs can be further polymerized into elastomeric cross-linked BPLP (CBPLP). After cross-linking, CBPLPs exhibited improved mechanical strengths compared to POC (Table 16.2) [20].

16.2.2.3.2 Development of iCMBA

In the past two decades, bioadhesives, tissue sealants, and hemostatic agents have been widely used in clinical surgical practices for blood loss control and wound healing [81]. Although the existing tissue bioadhesives like fibrin glues, cyanoacrylate tissue adhesives, gelatin glues, and polyurethane adhesives are commercially available and have been used in many clinical applications [22,23,82–87], these bioadhesives were largely limited by their toxicity or poor mechanical and adhesive strength, especially in wet conditions. In recent years, inspired by the high underwater adhesive strength of some maritime creatures, such as blue mussel *Mytilus edulis* [88], researchers developed a new family of adhesives based on the catechol-containing amino acid called L-3,4-dihydroxyphenylalanine (L-DOPA), a posttranslational hydroxylation of tyrosine found in the structure of secreted mussel adhesive foot protein, which was discovered to be the reason for the strong adhesion ability of mussel in aqueous conditions [88–90]. Under oxidizing or alkaline conditions, DOPA is believed to promote cross-linking reactions through the oxidation of catechol hydroxyl groups to ortho-quinone, which subsequently triggers intermolecular cross-linking. The oxidized DOPA was found to also contribute to the strong adhesion ability to biological surfaces, through the formation of covalent bonds with available nucleophilic groups on these surfaces such as $-NH_2$, –SH, –OH, and –COOH groups [89,91–94].

By introducing L-DOPA or its analog dopamine into poly[poly(ethylene glycol) citrate] (PEGC), Yang *et al.* developed a novel family of biodegradable and strong wet-tissue adhesives, referred to as injectable citrate-based mussel-inspired tissue bioadhesives (iCMBAs) [22,23]. iCMBAs were synthesized using a facile and cost-effective polycondensation reaction of FDA-approved and inexpensive monomers including citric acid and PEG, concurrent with the pendent modification of dopamine/L-DOPA in a one-pot synthesis process (Table 16.1 and Figure 16.4) [22]. The introduction of catechol group into the structure of iCMBA prepolymers conferred them with strong adhesion to wet-tissue surfaces as well as cross-linking capacity for bulk cohesive strength. The existence of hydrolytically degradable ester bonds formed by polycondensation in the backbone of iCMBA prepolymers made this family of adhesives readily biodegradable, which makes iCMBA significantly superior over other mussel-inspired bioadhesives, such as multiarmed PEG-based ones, which are essentially nondegradable [95]. In addition, the properties of iCMBAs, such as mechanical properties as well as degradation rate, could be tuned by adjusting the molecular weight of PEG and the feeding ratio of dopamine/L-DOPA [22]. iCMBAs exhibited excellent *in vitro* and *in vivo* cytocompatibility. *In vivo* studies of iCMBA did not induce any significant inflammatory responses, and it was degraded and absorbed completely in rats within 28 days [22].

16.2.2.3.3 Other CABE Development

As mentioned earlier, NO-releasing POC-DA elastomers were developed in Dr. Ameer's group. They were shown to release NO for 3 days *in vitro* and significantly reduce neointimal hyperplasia when implanted as a perivascular wrap in a rat carotid artery injury model [43–45]. Although promising, POC-DA still suffered from a long curing time (often more than 3 days). By introducing double bonds through post-pendent group modification of poly(diol citrate) prepolymers with 2-aminoethyl methacrylate, Wang *et al.* developed a family of photo-cross-linkable poly(diol citrate) (see Table 16.1 and Figure 16.3). After blending with miscible diazeniumdiolated NO donors followed by *in situ* and fast UV cross-linking, a long-lasting NO-releasing elastomer was formed [41]. The NO-containing polymer network could be cured within 3 min under UV light and could release NO for at least 2 weeks, which is much longer than that of previous POC-DA elastomers (released NO for only 3 days). This may be attributed to the *in situ* encapsulation of miscible diazeniumdiolated NO donors in the fast UV-curing process of the elastomers rather than NO-adsorbing after elastomer formation in the case of POC-DA. These materials may be very useful in cardiovascular application.

Polyol monomer like xylitol was also used to compose biodegradable polymer poly(xylitol citrate) (PXC) with citric acid by a simple thermal polycondensation. The abundant pendent hydroxyl groups on PXC, mostly introduced by xylitol, were used to react with methacrylic anhydride to obtain double bond functional PXC (PXCma), which can be formed into a bioelastomer network by photo-cross-linking [96].

16.2.2.4 *Chain Extension of CABEs*

Chain extension often refers to the postmodification of one polymer by using a chain extender (such as diisocyanate) or initiating a second polymerization by the polymer itself. In CABEs, the syntheses of CUPE and BPLPL copolymers such as BPLP-PLA are classified as chain extension reactions and will be discussed in the following sections.

16.2.2.4.1 Cross-Linked Urethane-Doped Polyester

Diisocyanate is often used in the chain extension reactions of biopolymers such as PLA, PCL, and their copolymers [97–101]. The combination of hard segment and soft segment may confer the resulting polyurethanes with shape-memory property [100,101]. Polyurethane is an important type of elastomeric polymer for biomedical applications [1,9,11]. Chain extension or cross-linking by diisocyanate can be adapted to many −OH-terminated or −OH-containing polymers or prepolymers [102]. The convenience of the urethane chemistry has made it into a very popular way of polymer chain extension method in biomaterial designs.

Although a lot of CABEs, such as POC, BPLP, and POMC, have shown great potential for tissue engineering, however, they are weak in mechanical strength especially when they were molded into porous scaffolds and used *in vivo* at a wet state. For example, POC underwent a significant loss in peak stress from 2.93 ± 0.09 MPa (film) to 0.3 ± 0.1 MPa (scaffold) when molded into porous scaffolds [103]. To obtain stronger elastomer, Dey *et al.* took advantage of excellent elasticity of cross-linked polyester

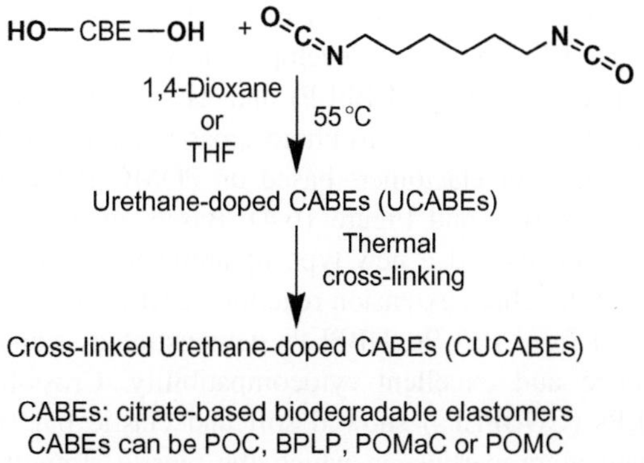

FIGURE 16.6 Synthesis and cross-linking process of urethane-doped citrate-based biodegradable elastomers (UCABEs).

network and the strong mechanical strength of polyurethanes and developed a new family of CUPE based on POC [19]. CUPE prepolymers were synthesized from diluted pre-POC solution in 1,4-dioxane (3 wt%), which was reacted with various molar ratios of HDI to pre-POC (0.9, 1.2, and 1.5) at 55 °C with continuous stirring for several days, until the characteristic absorbent peak at 2267 $cm^{-1}$of isocyanate (NCO) group in FTIR spectra disappeared (Figure 16.6). Similar to POC, some pendent carboxyl and hydroxyl groups originally from citric acid were preserved on CUPE prepolymers. This made them still cross-linkable by conducting a postpolymerization process. The cross-linking density between polymer chains can be adjusted by controlling the postpolymerization temperature and duration. The doped urethane bonds in the polyester served as a chain extender and enhanced hydrogen bonding within the network to produce elastomers with tensile strength as high as 41.07 ± 6.85 MPa while the elongation at break was still maintained at over 200% [19]. Amazingly, a simple urethane-doping chemical modification on POC resulted in an elastomer with almost 30 times higher tensile strength from 1.54 Ma of POC (80 °C, 4 days) to 44.98 MPa of CUPE1.2 (80 °C, 3 days).

CUPE polymers could be tuned to meet a variety of needs by varying the length of diols used in pre-POC synthesis, the feeding ratios of HDI over pre-POC, and postpolymerization conditions [104]. Preliminary cytocompatibility results showed that 3T3 fibroblast and SMCs were able to adhere and proliferate on a CUPE surface with a growth rate comparable to that on a PLLA control. Unlike previous POC, the higher molecular weights and nonsticky nature of CUPE prepolymers allow the use of other scaffold fabrication techniques such as thermally induced phase separation technique (TIPS) and electrospinning [19] in addition to salt-leaching method. The soft and elastic three-dimensional porous scaffold made by TIPS technology showed a highly porous structure, and the thin scaffold sheets allowed for even seeding, growth, and distribution of 3T3 fibroblasts.

Stimulated by the success of CUPE, other urethane-doped CABEs were also developed in our Lab, including urethane-doped BPLP (UBPLP) and its cross-linked form (CUBPLP) [48], as well as photo-cross-linkable urethane-doped polyester elastomers based on POMC (CUPOMC) [47] (Table 16.1 and Figure 16.4). Based on the BPLPs, Yi *et al.* developed a new type of urethane-doped BPLP (UBPLP) by chain extension reaction of BPLP using HDI. Inherited from BPLPs, UBPLPs demonstrated strong fluorescence and excellent cytocompatibility. Cross-linked UBPLPs (CUBPLPs) showed soft and elastic but strong mechanical properties, in which the tensile strength can reach 49.41 ± 6.17 MPa with a corresponding elongation at break of $334.87\% \pm 26.31\%$. Even after being molded into porous triphasic vascular scaffolds, CUBPLP showed strong mechanical properties with a burst pressure of 769.33 ± 70.88 mmHg and suture retention strength of 1.79 ± 0.11 N. Without cross-linking, UBPLP can be fabricated into stable and photoluminescent nanoparticles by a facile nanoprecipitation method. With a quantum yield as high as 38.65%, both CUBPLP triphasic scaffold and UBPLP nanoparticles could be noninvasively detected *in vivo*. UBPLPs represent another innovation in fluorescent biomaterial design and may offer great potential in advancing the field of tissue engineering and drug delivery where bioimaging has gained increasing interest. Besides UBPLPs and their cross-linked form CUBPLPs, another urethane-doped CABE based on photo-cross-linkable POMC (CUPOMC) was also developed in our lab. CUPOMC possesses tunable mechanical properties and degradation profiles. CUPOMCs could be either thermo-cross-linked or UV cross-linked providing fabrication flexibility for these polymers and fostering more convenient applications to various biomedical areas than previously developed thermal-curable biodegradable elastomers. Preliminary cell culture studies *in vitro* demonstrate that CUPOMCs could be good candidate materials for cell delivery carriers. The development of CUPOMCs expanded the choices of available biodegradable elastomers for broad biomedical applications like soft tissue engineering.

16.2.2.4.2 Biodegradable Photoluminescent Polylactones

The development of BPLP not only brought new applications for CABEs, especially in bioimaging and targeting drug delivery areas, but also sparked the innovation on developing biodegradable photoluminescent aliphatic polylactone biomaterials. Fluorescence imaging has gained increasing attention in drug delivery and tissue engineering where biodegradable polymers are usually conjugated with photobleaching organic dyes or toxic QDs. Given that BPLP exhibited excellent photostability and biocompatibility, the authors' group has started using BPLP to initiate the ROP of lactones for biodegradable photoluminescent polylactone (BPLPL) syntheses such as BPLPL-PLA. Developing biodegradable polylactone biomaterials represents new innovation on already widely used polylactone materials.

16.3 APPLICATIONS OF CABEs

16.3.1 Cardiovascular Applications

Cardiovascular disease remains the leading cause of morbidity and mortality in the world with more than 54% of the deaths in the United States [105]. For many patients, suitable vein autografts are not always available [106] necessitating the use of synthetic grafts. Although synthetic grafts such as polyethylene terephthalate (PET) grafts and expanded polytetrafluoroethylene (ePTFE) grafts have demonstrated adequate performance when replacing large blood vessels (diameter > 6 mm) [107], they also reduced long-term patency compared to autografts because of thrombosis, restenosis, and calcium deposition, especially when used in small-diameter blood vessels [108,109]. Additionally, PET and ePTFE are inert materials and nondegradable. PET and ePTFE are rigid and display mismatched compliance with the native arteries, which increases thrombosis and neointimal hyperplasia, the main causes of graft failure. Thus, biodegradable elastomers have been developed to match the soft and elastic properties of blood vessels, provide suitable biocompatibility, and allow functionalization to mediate the biological responses of native tissues [1,2,5,7–12]. Among biodegradable elastomers, CABEs emerge as an important type of materials for biomedical applications. Herein, we will discuss the applications of CABEs in cardiovascular tissue engineering.

16.3.1.1 Vascular Scaffold Designs

POC has been studied for a wide range of tissue engineering applications, especially soft tissue engineering applications such as blood vessels because of the soft, elastic, and tunable mechanical properties. To address the mechanical, compartmental, and microarchitectural requirement of small blood vessels, Yang *et al.* developed an implantable tubular biphasic POC scaffold composed of concentric nonporous and porous layers to mimic the intimal and medial vessel layers (Figure 16.7) [18,103,110]. The inside nonporous phase provides a continuous surface for EC adhesion, proliferation, and differentiation, as well as mechanical strength and elasticity. The outside porous phase serves as a three-dimensional layer to facilitate the expansion and maturation of SMCs and the establishment of an appropriate ECM to constitute the media. The mechanical properties of the whole construct were comparable to that of native arteries and veins. Cell culture experiment results using human aortic ECs and SMCs, along with the minimal

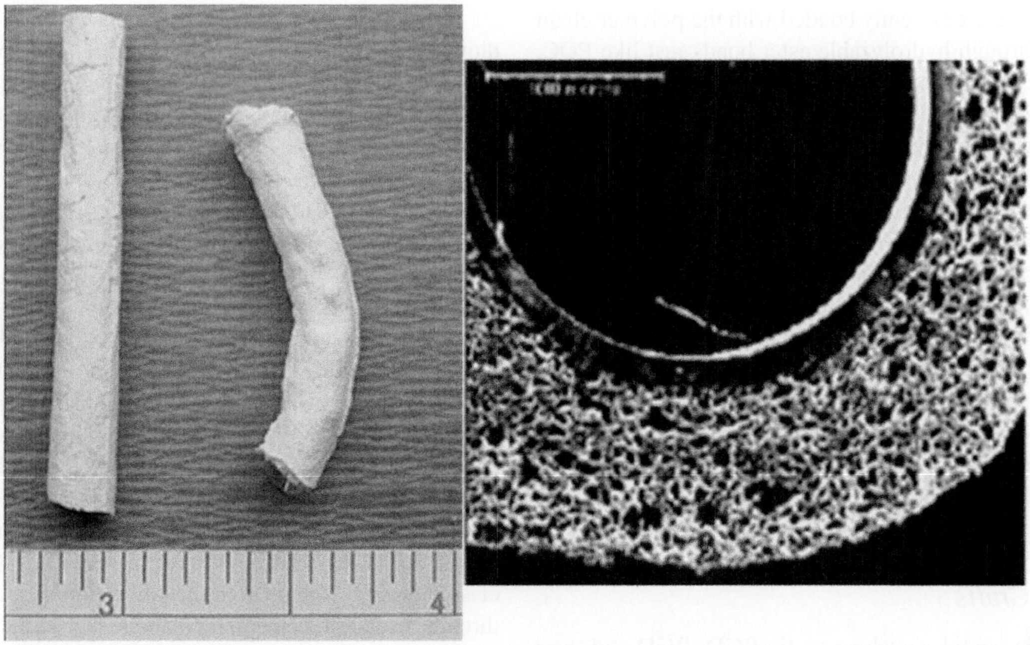

FIGURE 16.7 Photograph (left) and SEM (right) image of soft POC biphasic scaffold consisting of an inner nonporous lumen surrounded by an outer concentric porous layer [18,103]. (For color version of this figure, the reader is referred to the online version of this chapter.)

foreign body response found when subcutaneously implanted in rats, supported that POC might serve as a candidate material for small-diameter blood vessel tissue engineering [18,103,110].

Although POC is biocompatible, soft, and elastic, the mechanical properties of POC are relatively weak. After chain extension using HDI, CUPE possesses much better mechanical properties with a tensile strength up to 40 MPa (for CUPE1.8 film) [19]. Even after being molded into similar tubular biphasic scaffolds as POC, the tensile strength of CUPE biphasic scaffold was still 5 MPa with an elongation still higher than 150% [111]. The burst pressure of CUPE nonporous tube were much higher than corresponding POC scaffolds polymerized under similar conditions. The burst pressure of CUPE biphasic scaffold was found to increase from about 1600 to 2600 mmHg with the thickness of nonporous CUPE phase increasing from 160 to 384 μm. This result suggested that by simply varying the thickness of the inner nonporous layer, the burst pressure of CUPE scaffold could be adjusted to accurately match the burst pressure of the vessels being replaced [111]. Thin (~200 μm), strong, elastic, and porous CUPE scaffold sheets that are bonded together using a layer-by-layer approach were also developed by Tran *et al.* [112]. CUPE thin sheets allowed for even cell distribution and can be bonded together through ECM secreted by cells in each sheet layers. The layer-by-layer technology was considered as an alternative way to construct complex tissue scaffold such as blood vessel scaffolds.

Similar to CUPE, fluorescent CUBPLPs were also used to fabricate triphasic small-diameter vascular grafts by Zhang and Yang to replicate the stratified architecture of native vessels [48]. Different to biphasic scaffolds made by POC and CUPE [18,103,111], triphasic scaffolds were composed of a rough inner lumen surface, middle layer of porous scaffold with pore size of 1-20 μm, and outer layer of porous scaffold with pore size of 150-250 μm [48]. A rough surface is more favorable for ECs [113], and pore size of 1-20 μm is preferable for the compartmentalization of ECs and SMCs simulating the elastic lamina in native vessels [103]. Pore sizes of 150-250 μm are ideal for the growth of fibroblast and the formation of ECM [114]. The burst pressure and suture retention of CUBPLP triphasic scaffold could reach 800 mmHg and 1.79 N, respectively, which meet the requirements for off-the-shelf surgical implantation. By using CUBPLP, the scaffolds also possess *in vivo* detectable fluorescent properties, which will be very useful for the applications where fluorescence imaging may play an important role.

Although CUPE and other urethane-doped CABEs possess much higher mechanical strengths compared to POC, the applications of them were limited by the time-consuming synthesis process, low polymer concentration due to their low solubility in some common solvents, and prolonged degradation time. To address these problems, POC-click elastomer was developed by introducing click moieties containing diols and the usage of thermal click reaction (Table 16.1 and Figure 16.3). Although the structure of POC-click elastomer contains triazole rings, which are difficult to degrade, these triazole rings are covalently bonded with the polymer chain backbone through hydrolyzable ester bonds just like POC. POC-click elastomers showed a "first slow then fast" degradation profile and can be totally degraded within 34 weeks, which is nearly the same rate as POC and much shorter than that of CUPE cross-linked under the same condition (100 °C, 3 days) (Table 16.2). After being molded into triphasic scaffold similar to CUBPLP [112], the mechanical properties of POC-click triphasic scaffold were better than that of triphasic POC and CUPE scaffolds (data not published) for vascular applications. In addition to the superior mechanical properties of the POC-click elastomers, the residual azide groups on the surface of POC-click scaffold enabled a convenient conjugation of bioactive molecules on the polymers, which makes POC-click polymer scaffolds more amenable for biomedical applications.

16.3.1.2 Biofunctionalization of CABE Vascular Grafts

Cell-binding peptides such as cyclic RGD, RGD and their derivatives, p15, and growth factors such as REDV and VAPG have been widely used to enhance vascular graft endothelialization [59–63]. Among them, p15 is a collagen mimetic peptide that could significantly promote ECs adhesion and proliferation but are less effective for SMC adhesion and proliferation. In our recent work, p15 was conveniently conjugated onto the surface of POC-click films or scaffolds through SPAAC [70–73]—a copper-free click reaction. The p15-conjugated POC-click films showed much better HUVEC adhesion and proliferation properties compared to untreated POC-click films.

As mentioned earlier, NO plays an important role in cardiovascular application in prevention of SMC proliferation and stimulation of EC proliferation. Two different kinds of NO-releasing poly(diol citrate) elastomers, NO-releasing POC- and PDDC-DA (Table 16.1, Figure 16.8a), were developed by introducing NO-binding diol into pre-poly(diol citrates) and NO treatment after thermo-cross-linking [43–45]. Long-lasting NO-releasing poly(diol citrate) elastomers were developed by blending miscible diazeniumdiolated NO donors with photo-cross-linkable poly(diol citrates) followed by photo-cross-linking (Table 16.1 and Figure 16.8b) [41]. NO-releasing POC-DA was shown to release NO for 3 days *in vitro* and significantly reduce neointimal hyperplasia when implanted in a rat carotid artery injury model as a perivascular wrap. By using miscible diazeniumdiolated NO donor blended with rapidly photocurable methacrylated poly(diol citrates) (MA-POC or MA-PDDC), a longer duration of NO release for at least 1 week was achieved. The NO-releasing elastomers may be useful in the prevention of restenosis and thrombosis after vascular interventions such as balloon angioplasty, stent deployment, bypass grafting, and other blood-contacting surface of implant devices.

16.3.1.3 The Use of CABEs As a Vascular Graft Coating

ePTFE have been used in large-diameter (>6 mm inner diameter) blood vessel application. Their use in small-diameter blood vessels has been limited due to early graft occlusion from thrombosis. Yang *et al.* has demonstrated that the modification of ePTFE vascular grafts with POC, via a simple spin-shearing method followed by *in situ* interfacial

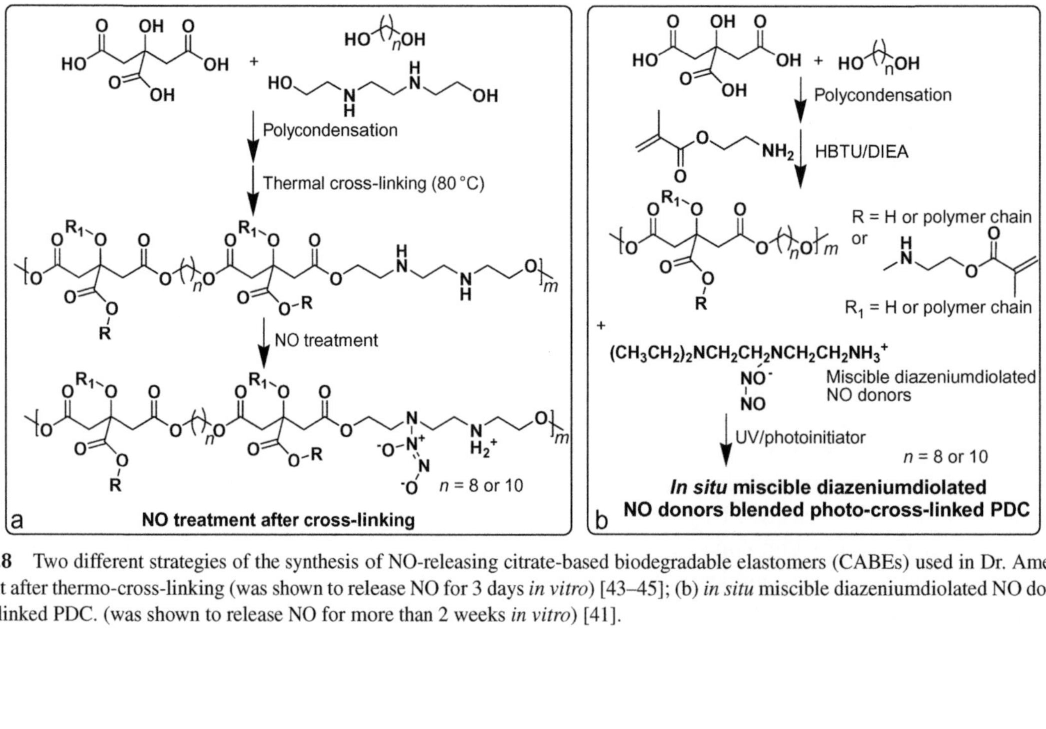

FIGURE 16.8 Two different strategies of the synthesis of NO-releasing citrate-based biodegradable elastomers (CABEs) used in Dr. Ameer's Lab. (a) NO treatment after thermo-cross-linking (was shown to release NO for 3 days *in vitro*) [43–45]; (b) *in situ* miscible diazeniumdiolated NO donors blended photo-cross-linked PDC. (was shown to release NO for more than 2 weeks *in vitro*) [41].

thermo-cross-linking, can improve the biocompatibility of ePTFE without affecting its graft compliance [17]. The POC interface conferred to the ePTFE grafts increased hydrophilicity, reduced thrombogenicity, facilitated graft endothelialization *in vitro*, and reduced macrophage infiltration *in vivo*. POC-coated ePTFE grafts were also implanted into the iliac artery in a porcine model. These grafts were found to dramatically inhibit platelet adhesion, aggregation, and activation compared to the ePTFE graft controls. This hemocompatibility may be explained by the anticoagulant and calcium-chelating properties of citrate, one of the components of POC, and the preservation of the fibril and node network of ePTFE. Moreover, POC supports the differentiation of peripheral progenitor cell into ECs, a model further characterized by Allen *et al.* [110]. Overall, POC was supported from the data to be used as a nonthrombogenic and biocompatible versatile coating, which can be widely used to improve blood-contacting devices.

16.3.2 Orthopedic Applications

16.3.2.1 Bone Tissue Engineering

Bone is a distinctive and dynamic tissue of the skeletal system that continually undergoes a coupled remodeling process defined by osteoclast bone resorption followed by new bone formation produced by osteoblasts [115]. Despite the capacity of the human skeletal system to rejuvenate itself, over 2.2 million bone-grafting procedures are performed annually worldwide to treat orthopedic pathological conditions such as fractures, tumor resections, and osteoporosis [116]. In fact, the bone has become the second most transplanted tissue in the world, and over 28 billion dollars per year was spent in total orthopedic medical costs, which are projected to significantly increase for the next decades with the demands of an aging population [117]. Autografts and allografts remain the gold standard for graft materials, but are unable to fulfill the increasing clinical demands for an effective off-the-shelf bone graft due to their limited availability, complications from donor site morbidity, and possible risk of pathogen transmission [118–120]. Synthetic counterparts are not impeded by these issues but are limited by their inability to provide mechanical compliance and mimic the native composition of bone tissue, which is composed of a 70% carbonated HA embedded in a type I collagen matrix [121]. While the incorporation of ceramic particles has been shown to improve the mechanical properties of and bone formation onto synthetic polymers, a limit in the amount of total ceramic that can be incorporated into the composite, significant inflammatory responses, long degradation times, and poor bone integration is a major roadblock for these materials [122–125]. For example, PLA-hydroxyapatite (PLLA-HA) composites can only incorporate a maximum of up to 30 wt% HA to avoid brittleness and can take up to 5 years to fully degrade, which leads to insufficient bone regeneration, significant inflammation, and poor mechanical integration. Therefore, the search for a suitable bone tissue-engineered substitute that can match the native composition of bone, provide adequate mechanical properties, minimize inflammatory responses, quickly induce bone regeneration, and fully integrate with the surrounding tissue within a year of implantation remains a significant challenge.

Citrate is traditionally known as an intermediate of the Krebs cycle (TCA cycle) for eukaryotic energy production. Previous studies have shown that a majority of the body's citrate content is located in skeletal tissues, and plays a large role in metabolism, calcium chelation, HA formation, and regulation of the thickness of bone apatite structure [30–33]. More recently, a careful solid-state nuclear magnetic resonance (NMR) study revealed that the surface of apatite nanocrystals is abundantly studded with strongly bound citrate molecules [33]. Citrate not only functions as a dissolved calcium-solubilizing agent but also is a strongly bound and integral part of the bone nanocomposite. Citrate also has a unique and innate ability to induce HA formation in SBF [34]. Surprisingly, the role of citrate is rarely mentioned in the literature related to bone development in the past 30 years including those on bone tissue engineering. The natural existence of citrate in bone and its importance in bone physiology hints that citrate should be considered in orthopedic biomaterial and scaffold design.

16.3.2.1.1 Prefabricated Implantable Bone CABE/HA Composites

In 2006, Qiu *et al.* first reported the development of a bioceramic-elastomer composite based on the citrate-based POC and HA (POC-HA) [35]. It was believed that the pendent carboxyl groups of POC could potentially aid in calcium chelation to facilitate polymer/HA interaction [34,37,126,127]. This improved calcium chelation resulting in POC-HA's ability to incorporate up to 65 wt% HA in POC-HA composites, which is not possible for traditional biodegradable lactide-based polymers (≤30 wt% HA). The enhanced amount of incorporated bioactive ceramic maximized the osteointegration of the material while maintaining suitable degradability [35]. The POC-HA composite successfully induced surface mineralization after 15 days of incubation in SBF and displayed favorable primary human osteoblast cell adhesion *in vitro*. POC-HA disks implanted into rat medial femoral condyles displayed no chronic inflammation and were well integrated with the surrounding cartilage along with mineralized chondrocytes located immediately adjacent to the implant after 6 weeks of implantation, which suggests healthy and normal bone remodeling. The composites also displayed good processability, giving them the potential to be machined and molded into bone screws for bone fixation applications [35].

Although POC-HA displayed excellent biocompatibility, osteoconductivity, and osteointegration *in vivo*, none of the investigated formulations containing other polymer/HA composites could provide sufficient mechanical strength to match that of human cortical bone [123–125,128]. To improve upon the mechanical strength of the previously reported POC-HA composites, we have recently developed a new generation of citrate-based polymer blend HA composite (CBPBHA) based on POC, CUPE [19,129], and HA. Both CUPE and POC are citrate-based cross-linkable polyester elastomers where CUPE is a urethane-doped version of POC. Although the urethane (urea)-doping chemistry sacrifices available –OH and –COOH groups in the final polymers due to the reactions between diisocyanates (like HDI) with –OH and that with –COOH, CUPE is almost 30 times stronger than POC [19,129]. We initially expected to make stronger bone composite materials using CUPE alone with HA compared to POC-HA; however, the reduced available free –COOH and –OH in CUPE may affect the polymer's ability to chelate with calcium-containing HA particles, thus influencing the material mechanical properties [37]. Therefore, our strategy for improving the mechanical strength of POC-HA was to fabricate polymer blend HA composites by blending the –COOH-rich POC with the mechanically strong CUPE to composite with HA in hopes of achieving optimal polymer/HA interactions and enhanced mechanical properties.

CBPBHA networks composed of 90% CUPE and 10% POC produced materials with a compressive strength of 116.23 ± 5.37 MPa, which falls within the range of human cortical bone (100-230 MPa) and is a significant improvement over POC-HA composites. CBPBHA promoted *in vitro* mineralization and increased C2C12 osterix (OSX) gene and alkaline phosphatase (ALP) gene expression *in vitro*. After 6-week implantation in a rabbit lateral femoral condyle defect model, CBPBHA composites elicited minimal fibrous tissue encapsulation and were well integrated with the surrounding bone tissues. The promising results in the preceding text prompted an investigation on the role of citrate supplementation in culture medium for stem cell and osteoblast culture. The results showed that citrate media supplementation *in vitro* accelerated both mesenchymal stem cell and MG-63 osteoblast phenotype progression and promoted calcified matrix formation by bone marrow stromal cells. Future studies will focus on further understanding the role of citrate in culture medium for bone stem cell differentiation and optimize the citrate contents in polymer/HA composites for orthopedic applications. CBPBHA composites represent a new generation of bone biomaterials that address the critical issues such as inflammation, osteoconductivity, and osteointegration.

16.3.2.1.2 Injectable Bone Composites

To expand the application of CABEs to bone tissue engineering, Gyawali *et al.* set out to develop an injectable, porous, and strong citric acid-based composite [38], which could be used as a delivery vehicle for cells and drugs in bone tissue engineering applications. PEGMC was combined with various percentages of HA to create PEGMC-HA composites with the help of PEG diacrylate (PEGDA) as an additional cross-linker and bicarbonates that can react with the pendent carboxylic acid groups on PEGMC to form CO_2 gas, which then forms pores to create an injectable porous bone material. The degradation profiles for PEGMC-HA networks showed increasing mass loss with lower concentrations of incorporated HA. The mechanical compressive tests showed that the PEGMC-HA networks were elastic and achieved complete recovery without any permanent deformation under cyclic load in both hydrated and dry conditions. Human fetal osteoblasts (hFOB 1.19) encapsulated in PEGMC-HA hydrogel composites were viable and functional over a 21-day culture. ECM production was measured by the total ALP and calcium content. The results show that both increased after 3 weeks of culture. Scanning electron microscopy (SEM) and energy dispersive X-ray analysis of the constructs showed that the PEGMC-HA films were covered with small cauliflower-shaped mineralized structures after 7 days of incubation in SBF. The presence of pendent groups in the PEGMC polymer allows for easy modification through the bioconjugation of biological molecules such as type I bovine collagen and resulted in enhanced cellular attachment and proliferation at the end of day 7 of subculture. An *ex vivo* study on a porcine femoral head demonstrated that PEGMC-HA is a potentially promising injectable biodegradable bone material for the treatment of osteonecrosis of the femoral head. Unlike many injectable systems, PEGMC-HA composites could also be fabricated into highly porous architectures from gas-foaming techniques *in situ* after injection into the implant site and can also be used to deliver therapeutics, as shown in the controlled release profiles using bovine serum albumin (BSA) as a model drug. Thus, unlike previous injectable materials, PEGMC-HA composites show great potential as an injectable, porous, and strong cell/drug delivery system for orthopedic applications.

16.3.2.2 Cartilage Tissue Engineering

Osteoarthritis is a joint disease that affects more than 20 million people in the United States and is characterized by articular cartilage degeneration, which ultimately leads to complete loss of cartilage tissue at the joint surface. For professional athletes and those over the age of 65, osteoarthritis is one of the most frequent causes of physical disability [130]. Due to the very limited capacity for cartilage regeneration and the varying success of techniques to repair damaged cartilage such as mosaicplasty and autologous chondrocyte implantation, the treatment of osteoarthritis and cartilage injuries has been a major challenge to orthopedic research [131–133]. Recent efforts in tissue engineering have focused on creating cartilaginous tissues *in vitro* for subsequent transplantation. However, cells grown *in vitro* under static conditions may not display the normal physiological function as cells located in the dynamic *in vivo* environment. In attempts to increase the quality of engineered constructs, cell-seeded scaffolds are often subjected to external mechanical stimuli in the form of cyclic compression and shear forces to mimic natural joint movement, which have been shown to be important in maintaining the homeostasis of cartilage glycosaminoglycan (GAG) and collagen formation [134,135]. Unfortunately, currently used materials for scaffolds in chondrocyte mechanical regimens have limited strength and elasticity and are prone to elastic deformation after cyclic compressive strains [136–138].

In 2006, Kang *et al.* published a study to address the limitations of the previous materials and assessed whether POC would be a suitable material to engineer elastomeric scaffolds for cartilage tissue engineering [139]. In this study, the authors fabricated porous POC scaffolds via a salt-leaching technique and characterized the POC scaffolds' ability to support chondrocyte attachment, proliferation, matrix synthesis, and cell differentiation. POC scaffolds were compared to 2% agarose, 4% alginate, nonwoven PGA, and nonwoven PLGA meshes in terms of recovery ratio. Of all the materials tested, only POC was able to display a 100% recovery ratio. The GAG and collagen content of chondrocytes cultured after 28 days on POC scaffolds was 36% and 26%, respectively, of those values found in bovine knee cartilage explants. Histology and immunohistochemistry evaluations confirmed that chondrocytes were able to attach to the pore walls within the scaffold, maintain their cell phenotype, and form a cartilaginous tissue during the 28 days of culture [139]. In summary, POC's elastomeric qualities showed potential as a biodegradable scaffold, which long-term cyclic and shear strains can be applied to *in vitro* to increase the GAG and collagen content of tissue-engineered cartilage. In addition, the elastomeric properties of POC constructs may be better suited to appropriately transfer local compression and shearing forces produced by joint mobilization to enhance *in vivo* cartilage regeneration.

In 2010, Jeong *et al.* published a study to compare the performance of chondrocyte-seeded scaffolds made of similar architectures to determine the influence of material on *in vitro* cartilage regeneration [140]. Three-dimensional scaffolds of the same design were fabricated using PCL, poly(glycerol sebacate) (PGS), or POC, and tissue regeneration was characterized by cell phenotype, cellular proliferation and differentiation, and matrix production. The studies showed that PGS was the least favorable material for cartilage regeneration as determined by the high dedifferentiation (Col1), hypertrophic mRNA expression (Col10), and high matrix degradation (MMP13 and MMP3) results. Although a majority of the cells seeded on PCL remained on the scaffold periphery, the PCL scaffolds showed moderate cellular activity but still caused dedifferentiation (Col1) of chondrocytes within the scaffold. Interestingly, POC provided the best support for cartilage regeneration with the highest tissue ingrowth (cell penetration), matrix production, relative mRNA expressions for chondrocyte differentiation (Col2/Col1), and DNA and sGAG content after 4 weeks of culture. This study demonstrates that POC can outperform other biodegradable elastomers for cartilage tissue engineering and warrants further *in vivo* studies.

16.3.3 Bioimaging and Drug Delivery

BPLPs are the first aliphatic BPLPs reported, which was developed in Dr. Yang's Lab based on biocompatible monomers, including citric acid and amino acids [20]. The BPLPs were synthesized simply by a polycondensation reaction of diol, citric acid, and α-amino acid [20,48]. The synthesis route is very versatile, as both organic solvent-soluble and water-soluble BPLPs (WBPLP) could be prepared by using OD and PEG, respectively. All the 20 natural amino acids and some unnatural amino acids have been used to create completely degradable polymers with intrinsic and tunable fluorescence. The backbone of BPLPs consists of ester bonds, which can be hydrolyzed in physiological conditions, and both BPLP and its degradation products are biocompatible since all the three monomers are natural or FDA-approved. The unique characteristics of BPLPs eliminated the long-term concern of fluorescent dyes and inorganic QDs, as well as their conjugation difficulties [21].

Most notably, BPLPs exhibited extraordinary fluorescence properties [20,141]. First, using different amino acids, BPLPs can emit fluorescent light from blue to near-infrared range (up to 825 nm for BPLP with α-methyl serine). Thus, BPLPs are available for both *in vitro* and *in vivo* imaging applications. Second, unlike traditional organic dyes, such as rhodamine B, BPLPs displayed less photobleaching behavior. After 3 h continuous UV excitation, BPLP-Cys only lost less than 2% of fluorescence intensity, which is significantly less than that of rhodamine B (10% loss). Third, the quantum yield of BPLPs is exceptional high. For instance, BPLP-Cys has a quantum yield of 62.3%, which is much higher than that of CdTe/ZnS QDs (20%) [142] and even higher than that of most organic dyes [143] and green fluorescent protein (GFP) [144]. These advantages of BPLPs probably result from a unique fluorescence mechanism, although it has yet to be fully understood. As Zhang *et al.* suggested [21], a six-member ring on the side of citrate backbone is the possible fluorophore of BPLPs. The long polymer chain makes the six-member ring into a planar conformation, resulting in photoluminescence without a large conjugated structure.

With the versatility of their molecular design, BPLPs have been fabricated into films, porous scaffolds, nanoparticles, micelles, and nanogels [20,21,141]. After thermo-cross-linking, BPLPs can be cross-linked into elastic films, which can be applied as medical implants. Porous tissue engineering scaffolds have also been prepared simply by salt-leaching after thermo-cross-linking. Subcutaneously implanted BPLP films and scaffolds can be observed by *in vivo* fluorescent imaging system (Figure 16.9). Recently, the fluorescence emission is a new candidate to monitor the scaffolds *in vivo* noninvasively [145]. BPLPs could help patients and doctors to locate the material better and later envision the erosion and performance simply by fluorescent imaging. Moreover, BPLP nanoparticles have been easily fabricated by nanoprecipitation or single/double emulsion approaches. Intratumor and intravenously injected BPLP nanoparticles were captured by *in vivo* fluorescent imaging system (Figure 16.9), indicating that these nanoparticles are potentially available for cancer diagnostics [141]. By adding maleic acid into the chain of WBPLP, photo-cross-linked WBPLP nanogels were also created [146]. BPLP nanoparticles and WBPLP nanogels do not need further fluorescent dye coating or conjugating to perform theranostic nanomedicine since they can emit intrinsic fluorescence and encapsulate drugs. In addition, multimodal imaging is an emerging area, since it could lead to a faster, higher-resolution, and deeper visualization by using several imaging approaches at the same time [147]. Wadajkar *et al.* utilized both oil-soluble BPLPs and WBPLPs to coat magnetic nanoparticles via double emulsion methods [148], the resulting core-shell nanoparticles showed dual-imaging (fluorescent imaging and MRI) capabilities after uptake by prostate cancer cells. It is also interesting that different cancer cells (PC3 and LNCaP cells) exhibited selective uptake behavior based on hydrophilicity/hydrophobicity of the particle surface.

Recently, our lab extended the BPLP class by doping the polymer with urethane segments [48]. The newly developed urethane-doped biodegradable photoluminescent polymers (UBPLPs) possess soft but strong mechanical properties that are suitable for off-the-shelf vascular grafts. Inherited from BPLPs, UBPLPs also showed strong manageable fluorescence. Unlike BPLPs, the chain-extended UBPLPs had controlled drug release profiles and better mechanical performance due to the modified polymer structure.

Overall, citrate-based photoluminescent biodegradable polymers are a novel class of polymers that are biocompatible, biodegradable, and promising for *in vivo* fluorescent imaging applications. BPLPs have the potential to conquer current challenges of biomedical materials in tissue engineering, drug delivery, and molecular imaging areas. Current and future research will be focused on further exploring the fluorescent mechanism and developing new polymers with better fluorescent and biological properties to meet clinical requirements. In addition, expanding the family with copolymerization, for instance, with polylactide, and creating polymer/inorganic hybrid materials for more medical applications will be another direction for the promising future of BPLPs.

16.3.4 Tissue Bioadhesive

Employing catechol chemistry, a novel family of biodegradable and strong wet-tissue adhesives, iCMBAs, was developed in our lab by introducing dopamine into the pendent groups of PEGC [22,23]. iCMBAs are superior to

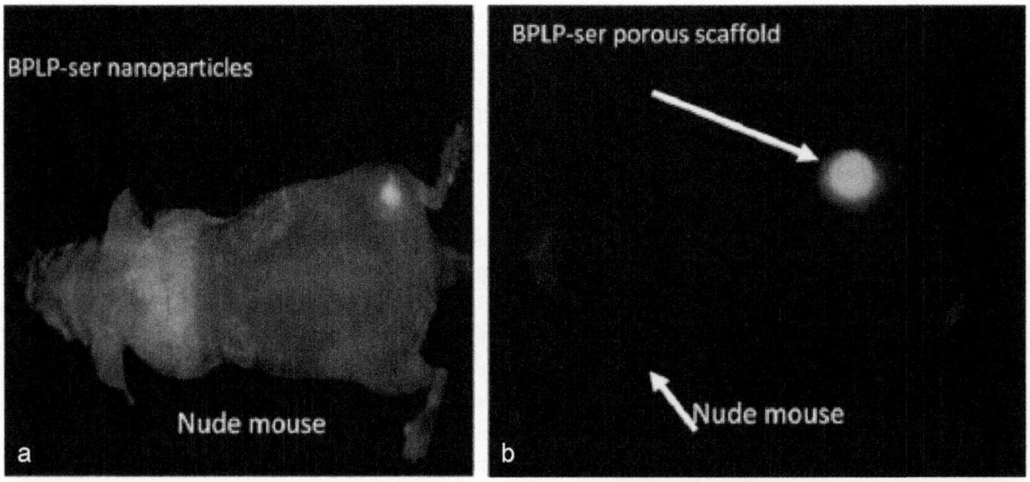

FIGURE 16.9 (a) Fluorescent image of BPLP-Ser nanoparticles injected subcutaneously in a nude mouse; (b) fluorescent image of BPLP-Ser porous scaffold implanted subcutaneously in a nude mouse [21]. (For color version of this figure, the reader is referred to the online version of this chapter.)

other PEG-based mussel-inspired bioadhesives in terms of cost-effective synthesis process and readily biodegradable properties [22]. Exhibiting excellent *in vitro* and *in vivo* cytocompatibility, iCMBAs also showed 2.5-8.0-fold stronger wet-tissue adhesion strength over clinically used fibrin glue, tunable degradability, and tissue-like elastomeric mechanical properties. Otherwise, iCMBAs were able to stop bleeding instantly and suturelessly and close the wound models (2 cm long and 0.5 cm deep) created on the backs of SD rats, which are impossible using the existing gold standard, fibrin glue, due to its weak tissue adhesion strength. Equally important, iCMBA bioadhesives facilitate wound healing and are totally biodegradable and absorbable without eliciting significant inflammatory response. All the results support that iCMBA is highly translational and could have broad impact on surgeries where surgical tissue adhesives, sealants, and hemostasis are needed.

16.3.5 Other Applications

16.3.5.1 Cell and Drug Delivery

Many tissue engineering designs using injectable, *in situ* forming systems have been reported with many advantages over previous methods. Unlike tissue engineering approaches that utilize prefabricated scaffolds, injectable systems have gained increasing interest as a unique method for delivering biomaterials into difficult to reach areas of the body using minimally invasive procedures. They also show the ability to fill and conform to any shape irrespective of the defect geometry. Furthermore, injectable systems can be used as fillers to reinforce the mechanical properties of diseased/injured tissue and as a competent carrier of cells and therapeutic agents such as drugs and growth factors [149–151]. Unfortunately, previous citrate-based composite designs required the use of organic solvents for material processing and harsh processing temperatures (>80 °C) for network formation, rendering them unable to be used in injectable strategies.

To overcome this limitation, Gyawali *et al.* recently developed a new family of *in situ* cross-linkable citrate-based polymers that can be dissolved in water and cross-linked through free radical polymerization methods based on double bond-containing PEGMC to avoid the use of harsh processing conditions required by the previous designs. Free radical cross-linking allowed for the preservation of valuable carboxyl and hydroxyl groups derived from citric acid, which could later be used to conjugate bioactive molecules into the bulk material to control cell behavior [20]. To ensure that cells and sensitive drugs/factors could be incorporated and delivered to the injury site, PEGDA and/or acrylic acid monomer was introduced into the system to create a cross-linked network with PEGMC [27]. These new citrate-based polymers could easily be injected and cross-linked through free radical polymerization. Cyclic conditioning tests showed that PEGMC networks possess a maximum tensile strength of 638 kPa with a corresponding elongation at break of 723%. In addition, PEGMC hydrogels could be compressed up to 75% strain without permanent deformation and with negligible hysteresis. PEGMC hydrogels also supported the encapsulation and proliferation of NIH 3T3 and human dermal fibroblasts. The cytotoxicity of the degradation products was comparable to the PEGDA [27]. The pH-sensitive and controlled drug release using BSA as a drug model demonstrated that PEGMC hydrogel has the potential to be used as a suitable drug delivery vehicle. In addition, PEGMC hydrogels caused minimal inflammation and were fully degraded without chronic inflammation or changes in histology within 30 days in a rodent subcutaneous model. In conclusion, PEGMC materials were synthesized in a convenient, one-pot reaction and demonstrated excellent injectability, *in situ* cross-linking, adequate functionalities, elastic mechanical properties, and controlled degradability. Collectively, the development of these platform biomaterials for injectable tissue engineering adds new members of CABE family and presents unique opportunities for many biomedical applications such as drug delivery and orthopedic tissue engineering.

16.3.5.2 Endoscopic Mucosal Resection

Gastrointestinal (GI) cancers frequently occur in industrialized countries with new cases of esophageal, gastric, and colorectal cancers affecting 3.60%, 11.4%, and 30.1%, respectively, of the developed world's population in 2008. The early stages of GI cancers (mucosal dysplasias or cancers) exhibit nonspecific symptoms, making them difficult to diagnose. This usually results in the majority of cases being diagnosed at advanced stages when bleeding, pain, or obstructions have already occurred resulting in 5-year survival rates below 30% [152,153]. Endoscopic mucosal resection (EMR) is a minimally invasive endoscopic procedure developed to remove dysplastic and malignant lesions limited to the mucosa and top part of the submucosal layer of the GI tract. Originally, EMR was accomplished by mechanical separation of the mucosal layer from the underlying muscle. However, perforation, bleeding, and damage to the muscle layer remained common occurrences. The current clinical approach to minimize EMR-associated complications is to inject a solution within the submucosal layer, which physically separates the diseased mucosal strata and provides a "safety cushion" for the subsequent underlying muscle layers [154].

Despite the recent advancements in the field, EMR has been historically limited by the available injection solution materials, which have been constrained by two design avenues: the osmolarity and viscosity of a solution are responsible for the lifting properties of EMR materials [155]. Saline is the most commonly used in clinic and is considered the "gold standard" due to its biocompatibility, low cost, and ease of use. However, it suffers from quick dissipation and requires repeated injections resulting in surgical difficulties. In order to achieve greater lift heights with longer lift durations, higher viscosity compounds such as sodium hyaluronate and fibrinogen have been employed. Sodium hyaluronate is currently being studied as a standard for comparing new compounds owing to its relative ease of injection and high viscosity, which results in better submucosal lift. However, the high cost of naturally derived solutions such as sodium hyaluronate and fibrinogen has prevented their large-scale use [155].

Albeit significant inroads have been made in viscous EMR solutions, a paradigm shift has been made toward the development of solutions, which rely on gel formation to create improved tissue elevation heights with extended lift durations. For example, photo-cross-linkable chitosan and thermoresponsive polymers have been recently reported for EMR with great enthusiasm [156,157]. Although these materials are widely available and have shown promise in creating adequate submucosal elevation heights and prolonged lift durations, there are inherent limitations to these approaches such as complex preparatory requirements and administration difficulties. Material transformation from a liquid to a gel state using photoinitiated free radical polymerization methods requires the use of an ultraviolet light, which may be difficult in hard-to-reach areas and is not widely applicable in conjunction with the current clinically used endoscopic tools. Thermoresponsive polymers utilize a liquid to gel transition when the temperature of the system is raised toward body temperature. Although this gel formation does not require the use of an ultraviolet light, the potential for clogging inside long delivery tools is common.

To address all the concerns of the current EMR solutions, Tran and Yang reported the use of a citrate-based injectable material to aid in EMR procedures and deliver therapeutics to the resected tissue [158]. This was the first injectable drug-eluting elastomeric polymer (iDEEP) system based on PEGMC and therapeutic rebamipide, which is a mucosal protective and ulcer-healing drug that stimulates prostaglandin generation and improves the speed of ulcer healing to aid in the management of EMR-induced damages [159]. PEGMC formulations used in this study demonstrated a tunable transition from a viscous flowable liquid into a cross-linked soft biodegradable hydrogel within 5 min. iDEEP-A component (composed by dissolving PEGMC, PEGDA, and tetramethylethylenediamine [TEMED, catalyst] in water), which is more viscous than saline, remained a viscous liquid until combined with the water-soluble iDEEP-B component (ammonium persulfate [redox initiator] in water) to produce a soft biodegradable hydrogel. Dividing the system into two separate components offers a huge advantage over previous designs in that the surgeon can precisely control the gel setting location and time, which avoids premature gelling inside the delivery tools. In addition, the utilization of a redox-initiated cross-linking mechanism does not require the use of additional equipment such as a UV light for the gel formation to occur.

The *in vitro* drug release profile studies of iDEEP hydrogels using rebamipide displayed an initial burst release followed by a sustained release for up to 2 weeks and could be controlled through polymer/monomer ratios. To characterize *ex vivo* submucosal lift, the upper third portions of porcine stomachs were injected with saline, sodium hyaluronate, and iDEEP; the results showed that all submucosal cushions created with iDEEP were more durable than those performed with saline and sodium hyaluronate at all time points. No significant changes in iDEEP cushion height were observed after 5 min due to gel formation. To evaluate the efficacy of iDEEP, standard EMR procedures were performed *in vivo* using a live porcine stomach model. iDEEP-A was easily injected using standard delivery tools and was able to create an adequate submucosal cushion. Using the same injection needle, an iDEEP-B solution was then injected into the same location without any clogging inside the delivery tool. After 5 min of iDEEP-B injection, the *en bloc* resection of the elevated mucosa revealed a soft biodegradable gel underneath the mucosa to provide protection for the underlying muscle layer from electrocautery damage. Although the iDEEP gel cannot be removed entirely following the EMR procedure, the remaining material can be used to deliver therapeutics. In addition, previous small animal studies have shown complete biodegradation of the hydrogel, excellent tissue compatibility, and minimal inflammation *in vivo* [27,158].

In conclusion, iDEEP is a cost-effective, readily available, and easily injectable two-component solution, which allows for biodegradable gel formation under the submucosal space without complex administration difficulties and can potentially aid in mucosal regeneration through controlled therapeutic delivery. These results suggest that iDEEP may provide a significant step toward the realization of an ideal injection material for EMR. Future studies will be dedicated to comparative long-term evaluations in living animals with pathological review to confirm the efficacy, depth of resection ability, and submucosal regeneration of the iDEEP system.

16.3.5.3 Nerve Tissue Engineering

Peripheral nerve injury remains a difficult and challenging problem in reconstructive surgery [160]. When the nerve defect or "gap" size is smaller than a few millimeters, the damaged proximal axonal stump is able to regenerate axonal sprouts toward the distal segment to reestablish motor and sensory function [161]. However, this form of neural regeneration does not always result in full functional recovery due to misdirection of the regenerating axons or inappropriate target reinnervation [162]. To complicate matters, without the presence of specific guidance, nerve ends separated by a gap size greater than 1 cm in length generally result in the backward growth of axons into the proximal nerve stump forming neuromas [163]. To increase the prospects of axonal regeneration and functional recovery, the current clinical "gold standard" for large gap repair involves the use of nerve autografts, which rely on the premise that viable Schwann cells (SC) located in the basal lamina tubes release a synergistic combination of growth factors and cell adhesion molecules to support and direct oriented axonal regeneration [164,165]. Unfortunately, autologous grafting is frequently associated with limitations including the need for multiple surgeries, donor site morbidity, distal donor site denervation, neuroma formation, and the limited availability of suitable grafts for harvesting [166].

To address the aforementioned limitations, several laboratories are actively pursuing the development of synthetic alternatives to replace nerve autografts in bridging the gap between transected nerve ends. Tissue-engineered nerve guides (TENGs) are a promising option for large gap nerve repair in that they can provide a biodegradable conduit for the delivery of therapeutic cell types, mechanical support, and chemical stimulation for axonal growth and nerve regeneration [167,168]. A variety of materials such as collagen [169], PLA [170], polyamide [171], poly(phosphoesters) [172], and poly(ethylene) [173] have been used with numerous processing strategies including TIPS and injection molding [174] techniques to fabricate synthetic alternatives to bridge neural defects. Recent years have witnessed the development of TENG with increasingly sophisticated and intricate internal structures based upon mechanisms of contact guidance and basement membrane microtube theory for nerve regeneration, which hypothesize that axon elongation requires guidance by contact with the appropriate substrate through topographical control [175,176]. By creating longitudinally oriented channels to fill the interior of the conduit, novel TENG has been produced to support the systems' natural pattern of growth [176]. The multichanneled designs are advantageous in that they provide better nerve target reinnervation, a greater surface area for cell growth, the topography necessary to direct the growth of regenerating nerve fibers (bands of Büngner), and internal support to prevent conduit collapse [162,166–168,177,178].

Our lab has recently developed a novel CUPE-multichanneled TENG. CUPE is a highly strong, soft, and biodegradable elastomer developed in our lab, which has shown excellent biocompatibility and hemocompatibility [19]. It is expected that a TENG fabricated using CUPE will have adequate strength and elasticity to withstand tension and retain sutures and will be suited for immediate implantation. In order to better recreate the native parallel channels of nerve basal lamina tubes, we have proposed the fabrication of novel porous multichanneled nerve guides with the following rationale: (1) a parallel multichannel design can better mimic the native architecture of nerve basal lamina tubes and promote nerve cell alignment through contact guidance; (2) the introduction of microporosity (<10 μm) between the channels can minimize fibrous tissue infiltration, increase permeability for cell-to-cell communication, and limit cell dispersion to enhance nerve target reinnervation; and (3) an outer sheath of the nerve guide conduit can provide the necessary mechanical strength for surgical implantation.

Porous and elastic CUPE scaffolds were developed for peripheral nerve regeneration based on the basement membrane microtube theory and designed with multiple internal longitudinally oriented channels as well as an external nonporous sheath to mimic the native endoneurial microtubular structure and epineurium, respectively. This fabrication technique allows for great flexibility in the scaffold channel geometry, porosity, and mechanical properties. CUPE-multichanneled scaffolds displayed an ultimate peak stress of 2.83 ± 0.24 MPa with corresponding elongations at break of 259.60 ± 21.49 %, which are in the range of native nerve mechanical properties. CUPE-multichanneled scaffolds were also evaluated *in vivo* for the repair of 1 cm rat sciatic nerve defects. After 8 weeks of implantation, CUPE-multichanneled scaffolds compared favorably with nerve autograft in terms of fiber density and population.

In conclusion, a novel CUPE TENG consisting of longitudinally oriented parallel microchannels was fabricated using particulate-leaching techniques and evaluated mechanically and *in vivo* for potential use in peripheral nerve tissue engineering. The scaffolds were made from CUPE, a new type of strong, soft, and hemocompatible biodegradable polyester elastomer. These studies represent the first step toward the investigation of the role of scaffold architecture on the resulting tensile, suture retention, and *in vivo* performance. Using this design, TENG can be produced with tunable strength and architecture to fit the needs of a particular application. CUPE TENG performed as well as nerve autografts in the *in vivo* evaluation studies.

16.3.5.4 Gene Delivery

In addition to providing a physical substrate for cellular growth, tissue-engineered scaffolds should also facilitate the delivery of cell-signaling factors in order to repair and/or integrate with the diseased tissues in the body [179]. The delivery of these factors should allow for both short- and long-term delivery while allowing control over dosing without compromising the biological activity of the factor. Traditionally, the physical adsorption of a protein onto a scaffold followed by the protein release during scaffold degradation has been the primary means to deliver proteins to the surrounding scaffold environment [180,181]. Unfortunately, this method does not transfer well to an *in vivo* setting due to the protein's short half-life, limited loading capacity, and significant loss of activity during entrapment. All of these are major limitations in maintaining the necessary prolonged therapeutic levels.

An alternative to direct protein delivery is to deliver genes encoding for the protein of interest, which will be internalized by the seeded cells and cause the subsequent production of the desired protein for extended periods [182,183]. In a recent report by Zhang *et al.*, POC was evaluated as a scaffold material to facilitate substrate-mediated gene delivery [179]. The rationale behind this approach was that plasmid DNA (pDNA) in its native form or in complex with a cationic polymer such as polyethylenimine (PEI) could be physically adsorbed onto the scaffold. PEI has shown the ability to disrupt the endosome through the proton sponge effect and enhance the efficiency of DNA delivery. The results show that polyplex-containing scaffolds showed a higher loading and slower initial release rates when compared to naked pDNA-containing scaffolds. HEK293 cells and porcine aortic smooth muscle cells (PASMC) seeded onto polyplex-loaded POC scaffolds demonstrated cell proliferation and transfection for up to 12 days *in vitro*. However, *in vivo* studies using a mouse intraperitoneal (IP) fat model showed that successful long-term transgene delivery was only achieved using naked pDNA-containing scaffolds, which was determined by higher levels of both luciferase and GFP expression.

Although the *in vivo* results did not match those of the *in vitro* results, these contradictions were also reported by many other groups and may be due to the interactions between carriers, host tissue, and immunity that cannot be sufficiently mirrored in an *in vitro* setting [184]. The results demonstrate that POC scaffolds can be a suitable scaffold biomaterial for substrate-mediated gene delivery and can potentially support the long-term biological cues needed to mediate tissue regeneration through nonviral gene delivery.

16.4 CONCLUSIONS

Since the creation of the first and initial citrate-based biodegradable elastomers (CABEs), poly(diol citrate) by Yang *et al.* in 2004, benefited from the facile and cost-effective synthesis process, available choices of diol comonomers (from small diol molecules to macrodiols, from hydrophobic diols to hydrophilic ones, and from saturated diols to unsaturated diols), and the multifunctionality of citric acid, CABEs have stood out and become an intensively studied and used biomaterial among the family of biodegradable polymers.

Citric acid is the key monomer for the design of CABEs and their applications. Being a natural metabolite in the body, this multifunctional monomer provides rich chemistries for the design functional biodegradable elastomers. Citric acid not only participates in degradable ester bond formation and provides pendent chemistry for biofunctionalization (BPLP and iCMBA), being inherent hemocompatible and chelating calcium ions to improve polymer/mineral interactions (CABE/HA), but also presents biochemical signals and niches to mediate tissue development (citrates in bone). CABEs have become an important branch of biodegradable polymers.

To meet the diversified needs in biological and biomedical applications, the innovation of CABEs will continue with focus on addressing some limitations on mechanical properties, degradation, and biofunctionalities required by specific applications. For example, wet mechanical strengths including adhesion strength (specifically for iCMBA) of CABEs should be improved for tissue engineering applications or orthopedic fixation device applications. Further understanding is needed on the mechanism of citrate signals for tissue development, which should be instrumental for the design of CABEs that may present dynamic citrate signals in response to tissue development such as bone regeneration. There is still much to do in the design of biodegradable polylactone biomaterials, which should constitute the next wave of innovations for CABE polymer. Conferring intriguing fluorescent properties to polylactones can give a new life to the mature polylactone syntheses/technology and will generate significant impacts on the fields that have largely benefited from the use of biodegradable polylactone materials such as drug delivery, biosensoring, bioimaging, cancer nanotechnology, and tissue engineering.

ACKNOWLEDGMENTS

This work was supported in part by NSF awards (DMR 1313553 and CMMI 1266116), NIH award (EB012575 and CA182670), and a research award from National Natural Sciences Foundation of China (31228007).

REFERENCES

[1] Q.Y. Liu, L. Jiang, R. Shi, L.Q. Zhang, Synthesis, preparation, in vitro degradation, and application of novel degradable bioelastomers—a review, Prog. Polym. Sci. 37 (2012) 715–765.
[2] R.T. Tran, Y. Zhang, D. Gyawali, J. Yang, Recent development on citric acid derived biodegradable elastomers, Recent Pat. Biomed. E 2 (2009) 216–227.
[3] H.Y. Tian, Z.H. Tang, X.L. Zhuang, X.S. Chen, X.B. Jing, Biodegradable synthetic polymers: preparation, functionalization and biomedical application, Prog. Polym. Sci. 37 (2012) 237–280.
[4] C.L. He, X.L. Zhuang, Z.H. Tang, H.Y. Tian, X.S. Chen, Stimuli-sensitive synthetic polypeptide-based materials for drug and gene delivery, Adv. Healthc. Mater. 1 (2012) 48–78.
[5] Y.D. Wang, G.A. Ameer, B.J. Sheppard, R. Langer, A tough biodegradable elastomer, Nat. Biotechnol. 20 (2002) 602–606.

[6] H. Cheng, P.S. Hill, D.J. Siegwart, N. Vacanti, A.K.R. Lytton-Jean, S.-W. Cho, A. Ye, R. Langer, D.G. Anderson, A novel family of biodegradable poly(ester amide) elastomers, Adv. Mater. 23 (2011) H95–H100.

[7] M.C. Serrano, E.J. Chung, G.A. Ameer, Advances and applications of biodegradable elastomers in regenerative medicine, Adv. Funct. Mater. 20 (2010) 192–208.

[8] J. Feng, R.-X. Zhuo, X.-Z. Zhang, Construction of functional aliphatic polycarbonate for biomedical applications, Prog. Polym. Sci. 37 (2012) 211–236.

[9] B. Amsden, Curable, biodegradable elastomers: emerging biomaterials for drug delivery and tissue engineering, Soft Matter 3 (2007) 1335–1348.

[10] Y. Li, G.A. Thouas, Q.-Z. Chen, Biodegradable soft elastomers: synthesis/properties of materials and fabrication of scaffolds, RSC Adv. 2 (2012) 8229–8242.

[11] Q.-Z. Chen, S.L. Liang, G.A. Thouas, Elastomeric biomaterials for tissue engineering, Prog. Polym. Sci. 38 (2013) 584–671.

[12] R. Rai, M. Tallawi, A. Grigore, A.R. Boccaccini, Synthesis, properties and biomedical applications of poly(glycerol sebacate) (PGS): a review, Prog. Polym. Sci. 37 (2012) 1051–1078.

[13] C.L.E. Nijst, J.P. Bruggeman, J.M. Karp, L. Ferreira, A. Zumbuehl, C.J. Bettinger, R. Langer, Synthesis and characterization of photocurable elastomers from poly(glycerol-co-sebacate), Biomacromolecules 8 (2007) 3067–3073.

[14] Q.Z. Chen, A. Bismarck, U. Hansen, S. Junaid, M.Q. Tran, S.E. Harding, N.N. Ali, A.R. Boccaccini, Characterization of a soft elastomer poly(glycerol sebacate) designed to match the mechanical properties of myocardial tissue, Biomaterials 29 (2008) 47–57.

[15] W. Wu, R.A. Allen, Y.D. Wang, Fast-degrading elastomer enables rapid remodeling of cell-free synthetic graft into a neoartery, Nat. Med. 18 (2012) 1148–1154.

[16] J. Yang, A.R. Webb, G.A. Ameer, Novel citric acid-based biodegradable elastomers for tissue engineering, Adv. Mater. 16 (2004) 511–516.

[17] J. Yang, D. Motlagh, J.B. Allen, A.R. Webb, M.R. Kibbe, O. Aalami, M. Kapadia, T.J. Carroll, G.A. Ameer, Modulating expanded polytetrafluoroethylene vascular graft host response via citric acid-based biodegradable elastomers, Adv. Mater. 18 (2006) 1493–1498.

[18] J. Yang, A.R. Webb, S.J. Pickerill, G. Hageman, G.A. Ameer, Synthesis and evaluation of poly(diolcitrate) biodegradable elastomers, Biomaterials 27 (2006) 1889–1898.

[19] J. Dey, H. Xu, J.H. Shen, P. Thevenot, S.R. Gondi, K.T. Nguyen, B.S. Sumerlin, L.P. Tang, J. Yang, Development of biodegradable crosslinked urethane-doped polyester elastomers, Biomaterials 29 (2008) 4637–4669.

[20] J. Yang, Y. Zhang, S. Gautam, L. Liu, J. Dey, W. Chen, R.P. Mason, C.A. Serrano, K.A. Schug, L.P. Tang, Development of aliphatic biodegradable photoluminescent polymers, Proc. Natl. Acad. Sci. U.S.A. 106 (2009) 10086–10091.

[21] Y. Zhang, J. Yang, Design strategies for fluorescent biodegradable polymeric biomaterials, J. Mater. Chem. B 1 (2013) 132–148.

[22] M. Mehdizadeh, H. Weng, D. Gyawali, L.P. Tang, J. Yang, Injectable citrate-based mussel-inspired tissue bioadhesives with high wet strength for sutureless wound closure, Biomaterials 33 (2012) 7972–7983.

[23] M. Mehdizadeh, J. Yang, Design strategies and applications of tissue bioadhesives, Macromol. Biosci. 13 (2013) 271–288.

[24] R.T. Tran, P. Thevenot, D. Gyawali, J.-C. Chiao, L.P. Tang, J. Yang, Synthesis and characterization of a biodegradable elastomer featuring a dual crosslinking mechanism, Soft Matter 6 (2010) 2449–2461.

[25] T. Ding, Y.Q. Xu, H. Gu, Y.R. Liang, X.M. Fang, L.Q. Zhang, Properties of poly(ethylene glycol)-based bioelastomer, J. Appl. Polym. Sci. 118 (2010) 2442–2447.

[26] T. Ding, Q.Y. Liu, R. Shi, M. Tian, J. Yang, L.Q. Zhang, Synthesis, characterization and in vitro degradation study of a novel and rapidly degradable elastomer, Polym. Degrad. Stab. 91 (2006) 733–739.

[27] D. Gyawali, P. Nair, Y. Zhang, R.T. Tran, C. Zhang, M. Samchukov, M. Makaroy, H.K.W. Kim, J. Yang, Citric acid-based in situ crosslinkable polymers for cell delivery, Biomaterials 31 (2010) 9092–9105.

[28] A.J. Francis, C.J. Dodge, J.B. Gillow, Biodegradation of metal citrate complexes and implications for toxic metal mobility, Nature 356 (1992) 140–142.

[29] L.M. Bishop, J.C. Yeager, X. Chen, J.N. Wheeler, M.D. Torelli, M.C. Benson, S.D. Burke, J.A. Pedersen, R.J. Hamers, A citric acid-derived ligand for modular functionalization of metal oxide surfaces via "click" chemistry, Langmuir 28 (2012) 1322–1329.

[30] M. Saito, A. Maruoka, T. Mori, N. Sugano, K. Hino, Experimental studies on a new bioactive bone cement: hydroxyapatite composite resin, Biomaterials 15 (1994) 156–160.

[31] H.A. Krebs, W.A. Johnson, The role of citric acid in intermediate metabolism in animal tissues, FEBS Lett. 117 (1980) K1–K10.

[32] F. Dickens, The citric acid content of animal tissues, with reference to its occurrence in bone and tumour, Biochem. J. 35 (1941) 1011–1023.

[33] Y.Y. Hu, A. Rawal, K. Schmidt-Rohr, Strongly bound citrate stabilizes the apatite nanocrystals in bone, Proc. Natl. Acad. Sci. U.S.A. 107 (2010) 22425–22429.

[34] S.H. Rhee, J. Tanaka, Effect of citric acid on the nucleation of hydroxyapatite in a simulated body fluid, Biomaterials 20 (1999) 2155–2160.

[35] H.J. Qiu, J. Yang, P. Kodali, J. Koh, G.A. Ameer, A citric acid-based hydroxyapatite composite for orthopedic implants, Biomaterials 27 (2006) 5845–5854.

[36] I. Djordjevic, E.J. Szili, N.R. Choudhury, N. Dutta, D.A. Steele, S. Kumar, Osteoblast biocompatibility on poly(octanediol citrate)/sebacate elastomers with controlled wettability, J. Biomater. Sci. Polym. Ed. 21 (2010) 1039–1050.

[37] Y. Jiao, G. Gyawali, J.M. Stark, P. Akcora, P. Nair, R.T. Tran, J. Yang, A rheological study of biodegradable injectable PEGMC/HA composite scaffolds, Soft Matter 8 (2012) 1499–1507.

[38] D. Gyawali, P. Nair, H.K.W. Kim, J. Yang, Citrate-based biodegradable injectable hydrogel composites for orthopedic applications, Biomater. Sci. 1 (2013) 52–64.

[39] D. Gyawali, R.T. Tran, K.J. Guleserian, L.P. Tang, J. Yang, Citric-acid-derived photo-cross-linked biodegradable elastomers, J. Biomater. Sci. Polym. Ed. 21 (2010) 1761–1782.

[40] D. Porter, F. Vollrath, Silk as a biomimetic ideal for structural polymers, Adv. Mater. 21 (2009) 487–492.

[41] Y. Wang, M.R. Kibbe, G.A. Ameer, Photo-crosslinked biodegradable elastomers for controlled nitric oxide delivery, Biomater. Sci. 1 (2013) 625–632.

[42] P.N. Coneski, P.A. Fulmer, J.H. Wynne, Thermal polycondensation of poly(diol citrate)s with tethered quaternary ammonium biocides, RSC Adv. 2 (2012) 12824–12834.

[43] H.C. Zhao, M.C. Serrano, D.A. Popowich, M.R. Kibbe, G.A. Ameer, Biodegradable nitric oxide-releasing poly(diol citrate) elastomer, J. Biomed. Mater. Res. A 93 (2010) 356–363.
[44] M.C. Serrano, A.K. Vavra, M. Jen, M.E. Hogg, J. Murar, J. Martinez, L.K. Keefer, G.A. Ameer, M.R. Kibbe, Poly(diol-co-citrate)s as novel elastomeric perivascular wraps for the reduction of neointimal hyperplasia, Macromol. Biosci. 11 (2011) 700–709.
[45] M.C. Jen, M.C. Serrano, R. van Lith, G.A. Ameer, Polymer based nitric oxide therapies: recent insights for biomedical applications, Adv. Funct. Mater. 22 (2012) 239–260.
[46] H.C. Zhao, G.A. Ameer, Modulating the mechanical properties of poly(diol citrates) via the incorporation of a second type of crosslink network, J. Appl. Polym. Sci. 114 (2009) 1454–1470.
[47] Y. Zhang, R.T. Tran, D. Gyawali, J. Yang, Development of photocrosslinkable urethane-doped polyester elastomers for tissue engineering, Int. J. Biomater. Res. Eng. 1 (2011) 18–31.
[48] Y. Zhang, R.T. Tran, I.S. Qattan, Y.-T. Tsai, L.P. Tang, C. Liu, J. Yang, Fluorescent imaging enabled urethane-doped citrate-based biodegradable elastomers, Biomaterials 34 (2013) 4048–4056.
[49] C. Serrano, L. Carbajal, G.A. Ameer, Novel biodegradable shape-memory elastomers with drug-releasing capabilities, Adv. Mater. 23 (2011) 2211–2215.
[50] H.-W. Sung, Y. Chang, C.-T. Chiu, C.-N. Chen, H.-C. Liang, Mechanical properties of a porcine aortic valve fixed with a naturally occurring crosslinking agent, Biomaterials 20 (1999) 1759–1772.
[51] H. Millesi, G. Zoch, R. Reihsner, Mechanical properties of peripheral nerves, Clin. Orthop. Relat. Res. 314 (1995) 76–83.
[52] G.A. Ameer, A.R. Webb, US20070071790, 2007.
[53] J. Gosline, M. Lillie, E. Carrington, P. Guerette, C. Ortlepp, K. Savage, Elastic proteins: biological roles and mechanical properties, Philos. Trans. R. Soc. Lond. B Biol. Sci. 357 (2002) 12–32.
[54] F.R. Noyes, E.S. Grood, The strength of the anterior cruciate ligament in humans and Rhesus monkeys, J. Bone Joint Surg. Am. 58 (1976) 1074–1082.
[55] K.B. Chandran, Cardiovascular Biomechanics, New York University Press, New York, USA, 1992.
[56] H. Yuan, S. Kononov, F.S. Cavalcaante, K.R. Lutchen, E.P. Lngenito, B. Suki, Effect of collagenase and elastase on the mechanical properties of lung tissue strips, J. Appl. Physiol. 89 (2000) 3–14.
[57] Q. Li, J. Wang, S. Shahani, D.D.N. Sun, B. Sharma, J.H. Elisseeff, K.W. Leong, Biodegradable and photocrosslinkable polyphosphoester hydrogel, Biomaterials 27 (2006) 1027–1034.
[58] Y. Li, C. Yang, M. Khan, S.Q. Liu, J.L. Hedrick, Y.-Y. Yang, P.-L.R. Ee, Nanostructured PEG-based hydrogels with tunable physical properties for gene delivery to human mesenchymal stem cells, Biomaterials 33 (2012) 6533–6541.
[59] A. de Mel, G. Jell, M.M. Stevens, A.M. Seifalian, Biofunctionalization of biomaterials for accelerated in situ endothelialization: a review, Biomacromolecules 9 (2008) 2969–2979.
[60] X.F. Hu, K.G. Neoh, J.Y. Zhang, E.-T. Kang, W. Wang, Immobilization strategy for optimizing VEGF's concurrent bioactivity towards endothelial cells and osteoblasts on implant surfaces, Biomaterials 33 (2012) 8082–8093.
[61] W.T. Zheng, Z.H. Wang, L.J. Song, Q. Zhao, J. Zhang, D. Li, S.F. Wang, J.H. Han, X.-L. Zheng, Z.M. Yang, D.L. Kong, Endothelialization and patency of RGD-functionalized vascular grafts in a rabbit, Biomaterials 33 (2012) 2880–2891.
[62] C. Li, A. Hill, M. Imran, In vitro and in vivo studies of ePTFE vascular grafts treated with p15 peptide, J. Biomater. Sci. Polym. Ed. 16 (2005) 875–891.
[63] C. Li, A. Hill, M. Imran, F. Tio, Peptide-coated vascular grafts: an in vivo study in sheep, Hemodial. Int. 8 (2004) 360–367.
[64] S.T. Laughlin, J.M. Baskin, S.L. Amacher, C.R. Bertozzi, In vivo imaging of membrane-associated glycans in developing Zebrafish, Science 320 (2008) 664–667.
[65] J.S. Guo, Y.B. Huang, X.B. Jing, X.S. Chen, Synthesis and characterization of functional poly(γ-benzyl-L-glutamate) (PBLG) as a hydrophobic precursor, Polymer 50 (2009) 2847–2855.
[66] C.F. Wu, Y.H. Jin, T. Schneider, D.R. Burnham, P.B. Smith, D.T. Chiu, Ultrabright and bioorthogonal labeling of cellular targets using semiconducting polymer dots and click chemistry, Angew. Chem. Int. Ed. 49 (2010) 9436–9440.
[67] J.S. Guo, Y. Wei, D.F. Zhou, P.Q. Cai, X.B. Jing, X.-S. Chen, Y.B. Huang, Chemosynthesis of poly(ε-lysine)-analogous polymers by microwave-assisted click polymerization, Biomacromolecules 12 (2011) 737–746.
[68] D. Zhao, S.W. Tan, D.Q. Yuan, W.G. Lu, Y.H. Rezenom, H.L. Jiang, L.-Q. Wang, Surface functionalization of porous coordination nanocages via click chemistry and their application, Adv. Mater. 23 (2011) 90–93.
[69] J.S. Guo, F.B. Meng, X.B. Jing, Y.B. Huang, Combination of antibiofouling and ion-interaction by click chemistry for endotoxin selective removal from protein solution. Adv. Healthc. Mater. 2 (2013) 784–789.
[70] C.A. DeForest, B.D. Polizzotti, K.S. Anseth, Sequential click reactions for synthesizing and patterning three-dimensional cell microenvironments, Nat. Mater. 8 (2009) 659–665.
[71] R. Manova, T.A. van Beek, H. Zuilhof, Surface functionalization by strain- promoted alkyne-azide click reactions, Angew. Chem. Int. Ed. 50 (2011) 5428–5430.
[72] R. Becer, R. Hoogenboom, U.S. Schubert, Click chemistry beyond metal-catalyzed cycloaddition, Angew. Chem. Int. Ed. 48 (2009) 4900–4908.
[73] S.F.M. van Dongen, P. Maiuri, E. Marie, C. Tribet, M. Piel, Triggering cell adhesion, migration or shape change with a dynamic surface coating, Adv. Mater. 25 (2013) 1687–1691.
[74] J. Hong, Q. Luo, X.M. Wan, Z.S. Petrovic, B.K. Shah, Biopolymers from vegetable oils via catalyst- and solvent-free "click" chemistry: effects of cross-linking density, Biomacromolecules 13 (2012) 261–266.
[75] J.S. Guo, F.B. Meng, X.Y. Li, M.Z. Wang, Y.J. Wu, X.B. Jing, Y.B. Huang, PEGylated click polypeptides synthesized by copper-free microwave-assisted thermal click polymerization for selective endotoxin removal from protein solutions, Macromol. Biosci. 12 (2012) 533–546.
[76] M. Gaumet, R. Gurny, F. Delie, Fluorescent biodegradable PLGA particles with narrow size distributions: preparation by means of selective centrifugation, Int. J. Pharm. 342 (2007) 222–230.
[77] P.A. Ma, S. Liu, Y.B. Huang, X.S. Chen, L.P. Zhang, X.B. Jing, Lactose mediated liver-targeting effect observed by ex vivo imaging technology, Biomaterials 31 (2010) 2646–2654.
[78] J.-H. Kim, K. Park, H.Y. Nam, S. Lee, K. Kim, I.C. Kwon, Polymers for bioimaging, Prog. Polym. Sci. 32 (2007) 1031–1053.
[79] X.H. Gao, L. Yang, J.A. Petros, F.F. Marshall, J.W. Simons, S.M. Nie, In vivo molecular and cellular imaging with quantum dots, Curr. Opin. Biotechnol. 16 (2005) 63–72.

[80] M.N. Rhyner, A.M. Smith, X.H. Gao, H. Mao, L.L. Yang, S.M. Nie, Quantum dots and multifunctional nanoparticles: new contrast agents for tumor imaging, Nanomedicine 1 (2006) 209–217.

[81] W.D. Spotnitz, S. Burks, Hemostats, sealants, and adhesives: components of the surgical toolbox, Transfusion 48 (2008) 1502–1516.

[82] W.D. Spotnitz, Fibrin sealant in the United States: clinical use at University of Virginia, Thromb. Haemost. 74 (1995) 482–485.

[83] D.H. Sierra, Fibrin sealant adhesive systems: a review of their chemistry, material properties and clinical applications, J. Biomater. Appl. 7 (1993) 309–352.

[84] N. Matsuda, N. Nakajima, T. Ttoh, T. Takakura, Development of a compliant surgical adhesive derived from novel fluorinated hexamethylene diisocyanate, ASAIO Trans. 35 (1989) 381–383.

[85] T. Matsuda, T. Takakuka, T. Itoh, Surgical adhesive, US4994542, 1991.

[86] E.J. Beckman, M. Buckley, S. Anarwal, J. Zhang, Medical adhesive and methods of tissue adhesion, US7264823, 2007.

[87] K.-D. Park, Y.-K. Joung, K.-M. Park, E.-G. Lih, In situ-forming hydrogel for tissue adhesive and biomedical use thereof, US0156164 A1, 2012.

[88] J.H. Waite, Nature's underwater adhesive specialist, Int. J. Adhes. Adhes. 7 (1987) 9–14.

[89] J.H. Waite, M.L. Tanzer, The bioadhesive of *Mytilus byssus*: a protein containing L-dopa, Biochem. Biophys. Res. Commun. 96 (1980) 1554–1561.

[90] R.L. Strausberg, R.P. Link, Protein-based medical adhesives, Trends Biotechnol. 8 (1990) 53–57.

[91] J.H. Waite, X. Qin, Polyphosphoprotein from the adhesive pads of *Mytilus edulis*, Biochemistry 40 (2001) 2887–2893.

[92] T.J. Deming, Mussel byssus and biomolecular materials, Curr. Opin. Chem. Biol. 3 (1999) 100–105.

[93] Q. Lin, D. Gourdon, C.J. Sun, N. Holten-Andersen, T.H. Anderson, J.H. Waite, J.N. Israelachvili, Adhesion mechanisms of the mussel foot proteins mfp-1 and mfp-3, Proc. Natl. Acad. Sci. U.S.A. 104 (2007) 3782–3786.

[94] H. Lee, N.F. Scherer, P.B. Messersmith, Single-molecule mechanics of mussel adhesion, Proc. Natl. Acad. Sci. U.S.A. 103 (2006) 12999–30003.

[95] C.E. Brubaker, P.B. Messersmith, Enzymatically degradable mussel-inspired adhesive hydrogel, Biomacromolecules 12 (2011) 4326–4334.

[96] J.P. Bruggeman, C.J. Bettinger, C.L.E. Nijst, D.S. Kohane, R. Langer, Biodegradable xylitol-based polymers, Adv. Mater. 20 (2008) 1922–1927.

[97] L.X. Yang, J.Z. Wei, L.S. Yan, Y.B. Huang, X.B. Jing, Synthesis of OH-group-containing, biodegradable polyurethane and protein fixation on its surface, Biomacromolecules 12 (2011) 2032–2038.

[98] L.Y. Wang, X.B. Jing, H.B. Cheng, X.L. Hu, L.X. Yang, Y.B. Huang, Rheology and crystallization of long-chain branched poly(L-lactide)s with controlled branch length, Ind. Eng. Chem. Res. 51 (2012) 10731–10741.

[99] W.S. Wang, P. Ping, H.J. Yu, X.S. Chen, X.B. Jing, Synthesis and characterization of a novel biodegradable, thermoplastic polyurethane elastomer, J. Polym. Sci. A Polym. Chem. 44 (2006) 5505–5512.

[100] P. Ping, W.S. Wang, X.S. Chen, X.B. Jing, Poly(ε-caprolactone) polyurethane and its shape-memory property, Biomacromolecules 6 (2005) 587–592.

[101] W.S. Wang, P. Ping, X.S. Chen, X.B. Jing, Biodegradable polyurethane based on random copolymer of L-lactide and ε-caprolactone and its shape-memory property, J. Appl. Polym. Sci. 104 (2007) 4182–4187.

[102] M.J.N. Pereira, B. Ouyang, C.A. Sundback, N. Lang, I. Friehs, S. Mureli, I. Pomerantseva, J. McFadden, M.C. Mochel, O. Mwizerwa, P. del Nido, D. Sarkar, P.T. Masiakos, R. Langer, L.S. Ferreira, J.M. Karp, A highly tunable biocompatible and multifunctional biodegradable elastomer, Adv. Mater. 25 (2013) 1209–1215.

[103] J. Yang, D. Motlash, A.R. Webb, G.A. Ameer, Novel biphasic elastomer scaffold for small-diameter blood vessel tissue engineering, Tissue Eng. 11 (2005) 1876–1886.

[104] J. Yang, J. Dey, Bio-polymer and scaffold sheet method for, tissue engineering, US20090093565, 2009.

[105] M.H. Davidson, Overview of prevention and treatment of atherosclerosis with lipid-altering therapy for pharmacy directors, Am. J. Manag. Care 13 (2007) S260–S269.

[106] A. Tiwari, H. Salacinski, A.M. Seifalian, G. Hamilton, New prostheses for use in bypass grafts with special emphasis on polyurethanes, Cardiovasc. Surg. 10 (2002) 191–197.

[107] L. Xue, H.P. Greisler, Biomaterials in the development and future of vascular grafts, J. Vasc. Surg. 37 (2003) 472–480.

[108] R.D. Sayers, S. Raptis, M. Berce, J.H. Miller, Long-term results of femorotibial bypass with vein or polytetrafluoroethylene, Br. J. Surg. 85 (1998) 934–938.

[109] J.K. Kirklin, D. Smith, W. Novick, D.C. Naftel, J.W. Kirklin, A.D. Pacifico, N.C. Nanda, F.R. Helmcke, R.C. Bourge, Long-term function of cryopreserved aortic homografts. A ten-year study, J. Thorac. Cardiovasc. Surg. 106 (1993) 154–165.

[110] D. Motlagh, J. Allen, R. Hoshi, J. Yang, K. Lui, G. Ameer, Hemocompatibility evaluation of poly(diol citrate) in vitro for vascular tissue engineering, J. Biomed. Mater. Res. A 101A (2007) 907–916.

[111] J. Dey, H. Xu, K.T. Nguyen, J. Yang, Crosslinked urethane doped polyester biphasic scaffolds: potential for in vivo vascular tissue engineering, J. Biomed. Mater. Res. A 95A (2010) 361–370.

[112] R.T. Tran, P. Thevenot, Y. Zhang, D. Gyawali, L.P. Tang, J. Yang, Scaffold sheet design strategy for soft tissue engineering, Materials 3 (2010) 1375–1389.

[113] T.W. Chung, D.Z. Liu, S.Y. Wang, S.S. Wang, Enhancement of the growth of human endothelial cells by surface roughness at nanometer scale, Biomaterials 24 (2003) 4655–4661.

[114] W.J. Li, C.T. Laurencin, E.J. Caterson, R.S. Tuan, F.K. Ko, Electrospun nanofibrous structure: a novel scaffold for tissue engineering, J. Biomed. Mater. Res. 60 (2002) 613–621.

[115] G.D. Roodman, Advances in bone biology: the osteoclast, Endocr. Rev. 17 (1996) 308–332.

[116] S. Nardecchia, M.C. Serrano, M.C. Gutierrez, M.T. Portoles, M.L. Ferrer, F.d. Monte, Osteoconductive performance of carbon nanotube scaffolds homogeneously mineralized by flow-through electrodeposition, Adv. Funct. Mater. 22 (2012) 4411–4420.

[117] D. Marsh, Concepts of fracture union, delayed union, and nonunion, Clin. Orthop. Relat. Res. 355 (1998) S22–S30.

[118] A.J. Salgado, O.P. Coutinho, R.L. Reis, Bone tissue engineering: state of the art and future trends, Macromol. Biosci. 4 (2004) 743–765.

[119] J.A. Goulet, L.E. Senunas, G.L. DeSilva, M.L. Greenfield, Autogenous iliac crest bone graft: complications and functional assessment, Clin. Orthop. Relat. Res. 339 (1997) 76–81.

[120] S.N. Parikh, Bone graft substitutes: past, present, future, J. Postgrad. Med. 48 (2002) 142–148.

[121] J.E. Eastoe, B. Eastoe, The organic constituents of mammalian compact bone, Biochem. J. 57 (1954) 453–459.

[122] J.A. Hunt, J.T. Callaghan, Polymer-hydroxyapatite composite versus polymer interference screws in anterior cruciate ligament reconstruction in a large animal model, Knee Surg. Sports Traumatol. Arthrosc. 16 (2008) 655–660.

[123] R.J. Kane, G.L. Converse, R.K. Roeder, Effects of the reinforcement morphology on the fatigue properties of hydroxyapatite reinforced polymers, J. Mech. Behav. Biomed. Mater. 1 (2008) 261–268.

[124] H.N. Liu, T.J. Webster, Mechanical properties of dispersed ceramic nanoparticles in polymer composites for orthopedic applications, Int. J. Nanomedicine 5 (2010) 299–313.

[125] A.J. Wagoner Johnson, B.A. Herschler, A review of the mechanical behavior of CaP and CaP/polymer composites for applications in bone replacement and repair, Acta Biomater. 7 (2011) 16–30.

[126] S.H. Rhee, J. Tanaka, Hydroxyapatite formation on cellulose cloth induced by citric acid, J. Mater. Sci. Mater. Med. 11 (2000) 449–452.

[127] A. Yokoyama, S. Yamamoto, T. Kawasaki, T. Kohgo, M. Nakasu, Development of calcium phosphate cement using chitosan and citric acid for bone substitute materials, Biomaterials 23 (2002) 1091–1101.

[128] P. Zioupos, M. Gresle, K. Winwood, Fatigue strength of human cortical bone: age, physical, and material heterogeneity effects, J. Biomed. Mater. Res. A 86 (2008) 627–636.

[129] J. Dey, R.T. Tran, J.H. Shen, L.P. Tang, J. Yang, Development and long-term in vivo evaluation of a biodegradable urethane-doped polyester elastomer, Macromol. Mater. Eng. 296 (2011) 1149–1157.

[130] M.C. Corti, C. Rigon, Epidemiology of osteoarthritis: prevalence, risk factors and functional impact, Aging Clin. Exp. Res. 15 (2003) 359–363.

[131] E.B. Hunziker, Articular cartilage repair: are the intrinsic biological constraints undermining this process insuperable? Osteoarthritis Cartilage 7 (1999) 15–28.

[132] R.F. LaPrade, M.F. Swiontkowski, New horizons in the treatment of osteoarthritis of the knee, J. Am. Med. Assoc. 281 (1999) 876–878.

[133] A.P. Newman, Articular cartilage repair, Am. J. Sports Med. 26 (1998) 309–324.

[134] M.D. Buschmann, Y.A. Gluzband, A.J. Grodzinsky, E.B. Hunziker, Mechanical compression modulates matrix biosynthesis in chondrocyte/agarose culture, J. Cell Sci. 108 (1995) 1497–1508.

[135] T.A. Kelly, C.C. Wang, R.L. Mauck, G.A. Ateshian, C.T. Hung, Role of cell-associated matrix in the development of free-swelling and dynamically loaded chondrocyte-seeded agarose gels, Biorheology 41 (2004) 223–237.

[136] C.T. Hung, R.L. Mauck, C.C. Wang, E.G. Lima, G.A. Ateshian, A paradigm for functional tissue engineering of articular cartilage via applied physiologic deformational loading, Ann. Biomed. Eng. 32 (2004) 35–49.

[137] R.L. Mauck, S.B. Nicoll, S.L. Seyhan, G.A. Ateshian, C.T. Hung, Synergistic action of growth factors and dynamic loading for articular cartilage tissue engineering, Tissue Eng. 9 (2003) 597–611.

[138] M.S. Rahman, T. Tsuchiya, Enhancement of chondrogenic differentiation of human articular chondrocytes by biodegradable polymers, Tissue Eng. 7 (2001) 781–790.

[139] Y. Kang, J. Yang, S. Khan, L. Anissian, G.A. Ameer, A new biodegradable polyester elastomer for cartilage tissue engineering, J. Biomed. Mater. Res. A 77 (2006) 331–339.

[140] C.G. Jeong, S.J. Hollister, A comparison of the influence of material on in vitro cartilage tissue engineering with PCL, PGS, and POC 3D scaffold architecture seeded with chondrocytes, Biomaterials 31 (2010) 4304–4312.

[141] J. Yang, S. Gautam, Methods and compositions of biodegradable photoluminescent polymers, US 200861074503, 2008.

[142] W.-C. Law, K.-T. Yong, I. Roy, H. Ding, R. Hu, W.W. Zhao, P.N. Paras, Aqueous-phase synthesis of highly luminescent CdTe/ZnTe core/shell quantum dots optimized for targeted bioimaging, Small 5 (2009) 1302–1310.

[143] U. Resch-Genger, M. Grabolle, S. Cavaliere-Jaricot, R. Nitschke, T. Nann, Quantum dots versus organic dyes as fluorescent labels, Nat. Methods 5 (2008) 763–775.

[144] R.Y. Tsien, The green fluorescent protein, Annu. Rev. Biochem. 67 (1998) 509–544.

[145] N. Artzi, N. Oliva, C. Puron, S. Shitreet, S. Artzi, A. bon Ramos, A. Groothuis, G. Sahagian, E.R. Edelman, In vivo and in vitro tracking of erosion in biodegradable materials using non-invasive fluorescence imaging, Nat. Mater. 10 (2011) 704–709.

[146] D. Manry, D. Gyawali, J. Yang, Size optimization of biodegradable fluorescent nanogels for cell imaging, J. High Sch. Res. (2011), In press.

[147] D.-E. Lee, H. Koo, I.-C. Sun, J.H. Ryu, K. Kim, I.C. Kwon, Multifunctional nanoparticles for multimodal imaging and theragnosis, Chem. Soc. Rev. 41 (2012) 2656–2672.

[148] A.S. Wadajkar, T. Kadapure, Y. Zhang, W.N. Cui, K.T. Nguyen, J. Yang, Dual-imaging enabled cancer-targeting nanoparticles, Adv. Healthc. Mater. 1 (2012) 450–456.

[149] J. Temenoff, A. Mikos, Injectable biodegradable materials for orthopedic tissue engineering, Biomaterials 21 (2000) 2405–2412.

[150] J. Kretlow, L. Klouda, A. Mikos, Injectable matrices and scaffolds for drug delivery in tissue engineering, Adv. Drug Deliver. Rev. 59 (2007) 263–273.

[151] J. Ifkovits, J. Burdick, Review: photopolymerizable and degradable biomaterials for tissue engineering applications, Tissue Eng. 13 (2007) 2369–2385.

[152] H. Inoue, N. Fukami, T. Yoshida, S.E. Kudo, Endoscopic mucosal resection for esophageal and gastric cancers, J. Gastroenterol. Hepatol. 17 (2002) 382–388.

[153] V. Catalano, R. Labianca, G.D. Beretta, G. Gatta, F. de Braud, E. Van Cutsem, Gastric cancer, Crit. Rev. Oncol. Hematol. 71 (2009) 127–164.

[154] M. Tada, A. Murakami, M. Karita, H. Yanai, K. Okita, Endoscopic resection of early gastric cancer, Endoscopy 25 (1993) 445–450.

[155] D. Polymeros, G. Kotsalidis, K. Triantafyllou, G. Karamanolis, J.G. Panagiotides, S.D. Ladas, Comparative performance of novel solutions for submucosal injection in porcine stomachs: an ex vivo study, Dig. Liver Dis. 42 (2010) 226–229.

[156] T. Hayashi, T. Matsuyama, K. Hanada, K. Nakanishi, M. Uenoyama, M. Fujita, M. Ishihara, M. Kikuchi, T. Ikeda, H. Tajiri, Usefulness of photocrosslinkable chitosan for endoscopic cancer treatment in alimentary tract, J. Biomed. Mater. Res. B Appl. Biomater. 71 (2004) 367–372.

[157] G. Fernandez-Esparrach, S.N. Shaikh, A. Cohen, M.B. Ryan, C.C. Thompson, Efficacy of a reverse-phase polymer as a submucosal injection solution for EMR: a comparative study (with video), Gastrointest. Endosc. 69 (2009) 1135–1139.

[158] R.T. Tran, M. Palmer, S.-J. Tang, T.L. Abell, J. Yang, Injectable drug-eluting elastomeric polymers: a novel submucosal injectable material, Gastrointest. Endosc. 75 (2012) 1092–1097.

[159] Y.J. Kim, J.H. Cheon, S.K. Lee, J.H. Kim, Y.C. Lee, Rebamipide may be comparable to H_2 receptor antagonist in healing iatrogenic gastric ulcers created by endoscopic mucosal resection: a prospective randomized pilot study, J. Korean Med. Sci. 25 (2010) 583–588.

[160] D.J. Bryan, J.B. Tang, S.A. Doherty, D.D. Hile, D.J. Trantolo, D.L. Wise, I.C. Summerhayes, Enhanced peripheral nerve regeneration through a poled bioresorbable poly(lactic-co-glycolic acid) guidance channel, J. Neural Eng. 1 (2004) 91–98.
[161] A. Bozkurt, G.A. Brook, S. Moellers, F. Lassner, B. Sellhaus, J. Weis, M. Woeltje, J. Tank, C. Beckmann, P. Fuchs, L.O. Damink, F. Schügner, I. Heschel, N. Pallua, In vitro assessment of axonal growth using dorsal root ganglia explants in a novel three-dimensional collagen matrix, Tissue Eng. 13 (2007) 2971–2979.
[162] Y.C. Huang, Y.Y. Huang, C.C. Huang, H.C. Liu, Manufacture of porous polymer nerve conduits through a lyophilizing and wire-heating process, J. Biomed. Mater. Res. B Appl. Biomater. 74 (2005) 659–664.
[163] P. Thomas, Local nerve injury: guidance factors during axonal regeneration, Muscle Nerve 12 (1989) 796–800.
[164] V. Guenard, N. Kleitman, T.K. Morrissey, R.P. Bunge, P. Aebischer, Syngeneic Schwann cells derived from adult nerves seeded in semipermeable guidance channels enhance peripheral nerve regeneration, J. Neurosci. 12 (1992) 3310–3320.
[165] W.B. Jacobs, M.G. Fehlings, The molecular basis of neural regeneration, Neurosurgery 53 (2003) 943–948, discussion 8–50.
[166] L. Flynn, P.D. Dalton, M.S. Shoichet, Fiber templating of poly (2-hydroxyethyl methacrylate) for neural tissue engineering, Biomaterials 24 (2003) 4265–4272.
[167] A.J. Krych, G.E. Rooney, B. Chen, T.C. Schermerhorn, S. Ameenuddin, L. Gross, M.J. Moore, B.L. Currier, R.J. Spinner, J.A. Friedman, M.J. Yaszemski, A.J. Windebank, Relationship between scaffold channel diameter and number of regenerating axons in the transected rat spinal cord, Acta Biomater. 5 (2009) 2551–2559.
[168] M.D. Bender, J.M. Bennett, R.L. Waddell, J.S. Doctor, K.G. Marra, Multi-channeled biodegradable polymer/CultiSpher composite nerve guides, Biomaterials 25 (2004) 1269–1278.
[169] N. Dubey, P.C. Letourneau, R.T. Tranquillo, Guided neurite elongation and Schwann cell invasion into magnetically aligned collagen in simulated peripheral nerve regeneration, Exp. Neurol. 158 (1999) 338–350.
[170] N. Rangappa, A. Romero, K.D. Nelson, R.C. Eberhart, G.M. Smith, Laminin-coated poly(L-lactide) filaments induce robust neurite growth while providing directional orientation, J. Biomed. Mater. Res. 51 (2000) 625–634.
[171] N. Terada, L.M. Bjursten, M. Papaloizos, G. Lundborg, Resorbable filament structures as a scaffold for matrix formation and axonal growth in bioartificial nerve grafts: long term observations, Restor. Neurol. Neurosci. 11 (1997) 65–69.
[172] S. Wang, A.C. Wan, X.Y. Xu, S.J. Gao, H.-Q. Mao, K.W. Leong, H. Yu, A new nerve guide conduit material composed of a biodegradable poly(phosphoester), Biomaterials 22 (2001) 1157–1169.
[173] T.R. Stevenson, V.A. Kadhiresan, J.A. Faulkner, Tubular nerve guide and epineurial repair: comparison of techniques for neurorrhaphy, J. Reconstr. Microsurg. 10 (1994) 171–174.
[174] C. Sundback, T. Hadlock, M. Cheney, J. Vacanti, Manufacture of porous polymer nerve conduits by a novel low-pressure injection molding process, Biomaterials 24 (2003) 819–830.
[175] C.A. Brayfield, K.G. Marra, J.P. Leonard, X. Tracy Cui, J.C. Gerlach, Excimer laser channel creation in polyethersulfone hollow fibers for compartmentalized in vitro neuronal cell culture scaffolds, Acta Biomater. 4 (2008) 244–255.
[176] M. Zhang, I.V. Yannas, Peripheral nerve regeneration, Adv. Biochem. Eng. Biotechnol. 94 (2005) 67–89.
[177] X.Y. Hu, J.H. Huang, Z.X. Ye, L. Xia, M. Li, B.C. Lv, X.F. Shen, Z.J. Luo, A novel scaffold with longitudinally oriented microchannels promotes peripheral nerve regeneration, Tissue Eng. Part A 15 (2009) 3297–3308.
[178] J.M. Li, T.A. Rickett, R.Y. Shi, Biomimetic nerve scaffolds with aligned intraluminal microchannels: a "sweet" approach to tissue engineering, Langmuir 25 (2009) 1813–1817.
[179] X.-Q. Zhang, H.H. Tang, R. Hoshi, L. De Laporte, H.J. Qiu, X.Y. Xu, L.D. Shea, G.A. Ameer, Sustained transgene expression via citric acid-based polyester elastomers, Biomaterials 30 (2009) 2632–2641.
[180] A.J. DeFail, C.R. Chu, N. Izzo, K.G. Marra, Controlled release of bioactive TGF-beta 1 from microspheres embedded within biodegradable hydrogels, Biomaterials 27 (2006) 1579–1585.
[181] T.P. Richardson, M.C. Peters, A.B. Ennett, D.J. Mooney, Polymeric system for dual growth factor delivery, Nat. Biotechnol. 19 (2001) 1029–1034.
[182] L. De Laporte, L.D. Shea, Matrices and scaffolds for DNA delivery in tissue engineering, Adv. Drug Deliver. Rev. 59 (2007) 292–307.
[183] J.H. Jang, T.L. Houchin, L.D. Shea, Gene delivery from polymer scaffolds for tissue engineering, Expert Rev. Med. Devices 1 (2004) 127–138.
[184] A.P. Rolland, From genes to gene medicines: recent advances in nonviral gene delivery, Crit. Rev. Ther. Drug Carrier Syst. 15 (1998) 143–198.

Chapter 17

Nucleic Acid Aptamers for Biomaterials Development

Mark R. Battig, Yong Wang

Department of Bioengineering, College of Engineering, The Pennsylvania State University, University Park, Pennsylvania, USA

Chapter Outline

17.1 INTRODUCTION

A biomaterial can be broadly defined as any material or construct that interacts with or replaces a biological system or function. For example, metals have been developed as biomaterials for use in artificial hips and long bones to maintain host mobility [1]. In addition to applications as implants, biomaterials aid the understanding of the physiology of healthy and diseased tissues. Despite their wide applications, many biomaterials have limited uses. Many implanted biomaterials (e.g., artificial organs) faced host rejection with severe tissue responses [2]. The host response to the implantation of a biomaterial is marked by adsorption of plasma proteins, acute and chronic inflammations, foreign body reactions, granulation tissue formation, and fibrous encapsulation of the implant [3,4]. The biomaterial may be incompatible with the host due to many factors such as contaminants and degradation products [3]. Thus, the development of new biomaterials aims to increase its biocompatibility and utility by accounting for various factors. However, even after identifying suitable construction reagents, biomaterials often suffer from low performance due to the inability to communicate with native tissues. In order to increase the efficacy of using biomaterials and their utility for their many applications, biomaterials have been functionalized with various ligands to enhance desired material-target interactions. Functionalization with ligands is a promising strategy to impart biomaterials with advanced properties, such as specific cell attachment or control over the release of growth factors. This chapter will discuss the development of functionalized biomaterials and their applications, with an emphasis on using nucleic acid aptamers to achieve novel functions.

17.2 BIOMATERIALS

The development of new biomaterials can be categorized into two focuses: bulk materials and nanomaterials. Both of them can be made from synthetic or biologically derived polymers. Bulk materials have been developed as coatings, cell scaffolds, and drug delivery systems. Similarly, nanomaterials have been developed as drug delivery systems, as well as imaging reagents. For example, synthetic hydrophilic polymers have been investigated as antifouling coatings to prevent plasma protein adsorption and subsequent bacteria and/or cell attachment since their properties are easily modified and easily applied to surfaces [5–7]. Antifouling

Natural and Synthetic Biomedical Polymers

properties are important considerations because the nonspecific adsorption of plasma proteins and other biomolecules (i.e., biofouling) can deteriorate biomaterial performance and lead to the failure of medical devices [8].

Despite the vast utility of biomaterials, the effectiveness of using them in their various applications is low. For example, biologically derived materials such as collagen and decellularized bladder submucosa and small intestinal submucosa have been investigated as cell scaffolds for tissue regeneration applications [9–13]. Natural materials often have inherent biological recognition properties and little to no immunogenicity [14]. Collagen and submucosa (which consists of collagen as well as other extracellular matrix proteins) contain the tripeptide cell adhesion domain sequence arginine-glycine-aspartic acid that allows for the attachment of cells to the substrate and the cultivation of new tissue, making them attractive biomaterials. However, biologically derived materials are limited in their use due to difficulties in isolating, treating, and purifying large quantities of organic material. Additionally, these cell scaffold materials are mechanically weak and are easily degraded. For tissue regeneration applications, the rate of tissue formation should equate the rate of biomaterial degradation in an ideal situation. To increase the longevity of these biomaterials, postprocessing methods such as cross-linking are needed to increase their mechanical strength and control the rate of enzymatic degradation [15–17]. To avoid many of the disadvantages found in natural materials, synthetic polymers have been used to create cell scaffolds. Unlike biologically derived biomaterials, synthetic polymers for biomaterials are easily produced on a large scale. Additionally, their mechanical strength and microstructures can be controlled with greater precision [14]. However, synthetic polymers lack biological recognition for cell attachment [14]. Therefore, synthetic biomaterials for cell scaffolds need cell adhesion ligands incorporated into them. Ligands such as peptides containing the arginine-glycine-aspartic acid sequence have been successfully incorporated into biomaterials to develop functional scaffolds [18–21]; however, these ligands adhere to cells nonspecifically and do not discriminate between different cell types. This could be an important design criterion where it is desirable to create a cell scaffold that could recruit tissue-forming cells to promote tissue growth while rejecting the attachment of bacteria to limit infections.

Delivery systems also face low efficacy. Wide arrays of synthetic and natural polymers have been used to develop macro- and nanoscale delivery systems for the extended, local delivery of growth factors to stimulate cellular activity. For example, synthetic hydrogels such as poly(ethylene glycol) (PEG) hydrogels have been used in numerous studies to deliver many kinds of growth factors [22–27] and biologically derived hydrogels such as hyaluronic acid hydrogels were used to deliver bone morphogenetic protein 2 and vascular endothelial growth factor (VEGF) to stimulate *in vivo* bone regeneration [28]. Hyaluronic acid is an anionic, nonsulfated glycosaminoglycan. It can be cross-linked to form an insoluble hydrogel through a wide variety of cross-linkers. Nanoparticles such as poly(lactide-co-glycolide) (PLGA) nanoparticles have also been used to encapsulate and deliver many kinds of therapeutics [29–38]. The submicron size of these particles allows the delivery systems to penetrate tissue to great depths and be internalized by cells [39]. These simple growth factor delivery systems have been shown to deliver a single growth factor locally, thereby raising the efficacy in treating damaged tissue in comparison to bolus injections; however, the release of therapeutics is still too rapid, resulting in the waste of expensive therapeutics and in severe side effects. Functionalization of drug delivery systems with ligands such as heparin that can bind growth factors has been explored to decrease the release rate of growth factors from biomaterials [40–48]. Heparin is a highly sulfated glycosaminoglycan bearing a strong electronegative charge density. Therefore, delivery systems functionalized with heparin can only slow the release rate of growth factors with large isoelectric points and cannot control the delivery of multiple growth factors with individual release rates. For maximum effectiveness in tissue regeneration applications, multiple growth factors should be delivered in a sequence-specific manner that mimics natural events [49,50]. Methods such as encapsulation of one species in nano- or microparticles while the second remains outside the particles but still in the bulk system and core-shell particles have been explored as dual-delivery systems [50–52]. While this approach can be applied to any growth factors, it is still difficult to control the delivery of several growth factors. Moreover, these systems do not have the capability of on-demand release of growth factors, where the release rate of growth factors can be stimulated when desired.

Metal nanoparticles and nanoparticles with fluorescence emission have been used as imaging reagents with enhanced resolution over current imaging molecules [53–58]. In addition, nanoparticles used as imaging reagents have longer residence times *in vivo* in comparison to small molecules [59]; however, they are prone to labeling tissues nonspecifically. Increases in signal-to-noise ratios could be gained if the nanoparticles could label a target specifically, where a greater signal would be attained by a greater accumulation of particles in the target while there is decreased noise by a lesser accumulation in the periphery. Additionally, the residence time can be further increased if the nanoparticle binds to the target strongly. To impart specificity to the nanoparticles, targeting ligands such as antibodies have been attached to the surface of the nanoparticles [60–62]. Nanoparticles have large surface area-to-volume ratios that allow a significant number of targeting ligands to be immobilized to increase specificity and affinity for a particular target [63]. Antibodies, however, are fragile biomolecules and make the development and use of targeting nanoparticles difficult.

Functionalization can enhance the efficacy of biomaterials in their various applications. However, current functionalization strategies can be limited. Ligands with greater specificity for a target with modular affinities are robust and stable molecules would greatly advance the development of functionalized biomaterials. Nucleic acid aptamers are promising molecules that can be used as ligand to develop biomaterials of greater efficacy for *in vivo* application because they do not face the limitations of other ligands.

17.3 NUCLEIC ACID APTAMERS

Nucleic acid aptamers are a unique class of oligonucleotides. They are either synthetic RNA or DNA that fold into well-defined nanostructures. They are unlike other oligonucleotides because these nanostructures bind a cognate partner with a high affinity and specificity. The high binding affinity and specificity of nucleic acid aptamers for their targets rivals that of antibodies for their antigens, and thus, they have been regarded as synthetic antibodies [64]. However, unlike antibodies, aptamers are robust biomolecules. Aptamers can withstand harsh chemical conditions and are therefore amenable to a wide array of chemical reactions and environments. This allows aptamers to be used in functionalizing numerous biomaterials to achieve unique goals. There are two stages to identifying an aptamer for a target. The first is upstream selection to identify binding aptamers and the second is downstream truncation to optimize the binding affinity of the aptamer to its target.

17.3.1 Upstream Selection

Nucleic acid aptamers are screened against a target through an iterative process called the systematic evolution of ligands through exponential enrichment (SELEX) [65,66]. Unlike identifying antibodies for an antigen, aptamer screening occurs *in vitro*. The advantage of an *in vitro* selection process over an *in vivo* method is that a wider variety of targets can be screened against, e.g., toxins that cannot be tolerated by the host during *in vivo* selection. The SELEX process begins with incubating a library of randomized RNA or DNA sequences with the target in controlled binding conditions (e.g., physiological conditions). A typical library consists of 10^{14}-10^{15} oligonucleotides with each oligonucleotide consisting of 80-100 bases. Molecular recognition through hydrophobic interactions, hydrogen bonds, electrostatic interactions, and steric hindrances governs the association of the target with the oligonucleotides of the library [67]. Oligonucleotides with a high affinity will associate with the target, whereas oligonucleotides with a low binding affinity will remain unbound and are removed from the pool. After partitioning the unbound oligonucleotides, the bound oligonucleotides are dissociated from the target and are amplified to create a second library pool. This process is continued by incubating the target with the new library, removing the weakly binding oligonucleotides, and amplifying the remaining oligonucleotides to create a further enriched library. After several iterations, oligonucleotides of high affinity for the target have been isolated and are called aptamers. Through SELEX, aptamers have been identified for a wide range of targets, including small molecules (e.g., metal ions [68–71], dyes [65,72], biochemicals [73–76], and molecular drugs [77–83]), amino acids [84–87] and proteins (e.g., growth factors [88–91], cell receptors [92–95], enzymes [66,96], and other proteins [97,98]), and other targets (e.g., *Escherichia coli* [99] and viruses [100–102]).

Since the introduction of the SELEX process, many modifications and improvements were implemented to identify aptamers of greater stability and greater specificity or to increase the ease of isolating an aptamer. Unmodified aptamers have short *in vivo* half-lives, on the order of a few minutes, due to enzymatic degradation and renal filtration [103]. To increase the therapeutic potential of aptamers, libraries of modified oligonucleotides with increased endonuclease resistance have been used in SELEX to yield aptamers that are more durable. Any modification of oligonucleotides must be chosen with care such that they are still recognized by the polymerases during the amplification step of SELEX. Substitutions in RNA libraries of 2′-hydroxypyrimidines with 2′-aminopyrimidine [88,104–107] or 2′-fluoropyrimidine [106,108] have been reported to be compatible with polymerases and the resulting aptamers have increased endonuclease resistance. Additionally, substitutions of the phosphodiester backbone with phosphorothioate prior to SELEX can yield aptamers with increased endonuclease resistance [109]. It is possible to screen aptamers from a pool of unmodified oligonucleotides and introduce chemical groups that confer endonuclease resistance post-SELEX (e.g., 2′-methoxypurines in RNA aptamers), but such will often decrease the binding affinity of the aptamer for its target [108]. Capping strategies (modifications of the 3′ or 5′ positions after identifying an aptamer through SELEX) have also been investigated to increase *in vivo* half-lives by increasing exonuclease resistance and/or decreasing renal clearance [110–112].

In addition to improving the oligonucleotide library for SELEX, improvements to the SELEX process have also been described. Such improvements aim to increase the selectivity of aptamers or the ease and speed of screening aptamers. For example, negative SELEX [65] and counter SELEX [77] were implemented to eliminate oligonucleotides that bound nonspecifically and oligonucleotides that bound to similar targets, respectively; capillary electrophoresis SELEX [113], FluMag-SELEX [114], non-SELEX [115], and gel electrophoresis [116,117] were developed or used to better separate aptamer-target complexes; automated SELEX [118], MonoLEX [119], and other methods [120] have been developed to ease the laborious process or shorten the time required to screen for aptamers; and microfluidics have been incorporated into SELEX to decrease sample volumes and permit high-throughput screening of aptamers [121–123].

17.3.2 Downstream Truncation

Aptamers screened from SELEX typically consist of 80-100 nucleotides. However, not all the nucleotides contribute to the high binding affinity for the target. An aptamer can be divided into three regions: a region of essential nucleotides, supporting nucleotides, and nonessential nucleotides. The essential region consists of nucleotides that are directly involved in the interaction of the aptamer with the target. Any base substitutions or removal of a base in this region will result in a significant loss of binding affinity. The second region consists of supporting nucleotides. Supporting nucleotides indirectly contribute to the binding affinity by stabilizing the secondary structure of the aptamer. Stabilization occurs through the formation of stems by intramolecular base pairing of complementary nucleotides. Base substitutions or decreases in the stem length result in modest decreases of the binding affinity. The third region is the nonessential region and consists of nucleotides that can be substituted or removed without a loss in the binding affinity for the target. It typically consists of nucleotides that serve as the primers in PCR for aptamer amplification during SELEX and nucleotides that do not participate in intra- or intermolecular binding. Because the nonessential nucleotides are irrelevant in aptamer binding, they are often removed from the aptamer. Truncation of an aptamer is desirable for many reasons. Shortened aptamers are easier and less expensive to synthesize. While the chemical synthesis process is well established, sequences of greater than 60 nucleotides or with complex secondary structures (e.g., G-quadruplexes) are difficult to synthesize. Additionally, shortened aptamers often have increased binding affinities in comparison to the original full-length oligonucleotide since the nonessential region can destabilize the aptamer structure or interfere with aptamer-target association. Methods to determine the essential region of an aptamer to guide aptamer truncation are described later.

17.3.2.1 Sequence Alignments

Concluding SELEX, a variety of aptamers are isolated that bind to the same target with varying affinities. One can identify the essential region by comparing sequence homology and analyzing common sequences though manual alignment. Reoccurring sequences often denote the nucleotides that are required for target binding. By generalizing the reoccurring sequences, a consensus sequence is revealed. One can generally conclude that the consensus sequence represents the essential region necessary for strong aptamer-target binding. Nucleotides not within the consensus sequence are likely extraneous and may be truncated from the aptamer.

17.3.2.2 In Silico Comparisons

In addition to sequence similarities, consensus motifs based on common secondary structures can be analyzed through modeling software. The secondary structure of the aptamer is important in binding the target because various folds allow the aptamer to exploit various binding mechanisms such as hydrophobicity, molecular shape complementarity, or intercalation in the case of small molecules binding double-stranded regions of the aptamer. To identify common folds or secondary structures, computer programs such as UNAFold and RNAstructure have been developed to predict the secondary structures based on minimizing free energy using thermodynamic data, which can then be used to compare aptamer structures and guide truncation [124,125]. However, many programs do not predict folds such as G-quartet structures or pseudoknots. Recently, ValFold has been developed to predict truncated aptamer sequences based not only on canonical Watson-Crick base pairings but on guanine-guanine pairings that form G-quartet motifs as well [125–128]. G-quartet motifs have been found in many aptamers and are essential for the high binding affinity of the aptamer for their respective target [96,129,130]. However, programs like ValFold are still in their infancy and optimization of their algorithm is needed to reduce computational requirements and to predict difficult folds like pseudoknots. Additionally, most software do not account for the presence of the target molecule. The target molecule may be a very important factor since aptamers often undergo conformational changes upon binding its target [131–135]. Thus, it is difficult to identify nucleotides outside of the consensus region that may bind the target when the target is introduced or that may stabilize a new secondary structure, though there has been progress in developing a process that combines aptamer folding with protein docking to account for these structural changes [136].

17.3.2.3 Experimental Methods

The advantage of using sequence comparisons and *in silico* methods for guiding aptamer truncation is their ease of use and inexpensive operating cost. The disadvantage of sequence comparisons is that there may be no consensus sequence and an aptamer of a unique sequence but high affinity could be excluded from analysis [94]. Additionally, most of these methods compare sequences and structures in a virtual environment with the target unaccounted. Thus, rigorous algorithms that can model aptamer-target interactions and aptamer refolding are needed to predict secondary structures accurately. Experimental techniques remain the most accurate method to identify the essential region of aptamers and guide aptamer truncation. Processes such as the chemical synthesis of shortened aptamers sequences, enzymatic footprinting, or partial hydrolysis or degradation of the full-length aptamer and selecting high-affinity fragments have been used to find short aptamer derivatives [90,94,137–139]. Finding optimized aptamers through these iterative processes often require months and/or are costly to accomplish. Thus, new methods that would shorten the time required to screen aptamers sequences or guide aptamer truncation would be beneficial.

A relatively new method was described that utilizes nucleic acid aptamer hybridization with complementary sequences to guide aptamer truncation [140]. In this method, complementary sequences that hybridize with either the 3′ or the 5′ end of the aptamer were analyzed for their degree of interference in binding the target. Complementary sequences that were able to hybridize with the aptamer but did not lessen the binding affinity of the aptamer for the target were presumed to have bound to the nonessential region of the aptamer. Therefore, the nucleotides from the aptamer hybridizing with the complementary sequence can be truncated. The same method can be used to probe the supportive and the essential nucleotides. Supportive nucleotides are involved in stabilizing the secondary structure of an aptamer. Therefore, inhibiting the function of these nucleotides will destabilize the aptamer structure and lessen its binding affinity. Complementary sequences that hybridize with an increasing number of supportive nucleotides will result in proportionally decreasing equilibrium dissociation constants. There will be a significant loss of affinity for the target if the complementary sequence blocks the association of an essential nucleotide with the target. This method is unlike previous methods for determining or optimizing the aptamer structure because this method provides greater information by mapping the functional regions of the aptamers and can be completed within a week.

By identifying sequences of high specificity through upstream selection and optimizing binding affinity through downstream truncation, aptamers are obtained that discriminate based on subtle molecular differences and bind their target very strongly. Methods or tools such as surface plasmon resonance, electrophoretic mobility shift assay, atomic force microscopy, and quartz crystal microbalance can be used to analyze the binding affinity of aptamers by calculating the equilibrium dissociation constant (K_D). A K_D value on the order of several nanomolar or less denotes an aptamer that binds it target very strongly, whereas a value of several micromolar or larger may denote weak or nonspecific binding.

17.4 DEVELOPMENT OF APTAMER-FUNCTIONALIZED BIOMATERIALS

Nucleic acid aptamers bind their target with high affinity. The strong association between the aptamer and the target can prevent the target from interacting with other species. For this reason, aptamers have been used as therapeutics to inhibit enzyme or protein functions [141]. Additionally, aptamer can be used as ligands to functionalize biomaterials and bestow strong molecular recognition properties. Many of these aptamer-functionalized biomaterials and their applications are discussed later.

17.4.1 Aptamer-Functionalized Hydrogels

17.4.1.1 Hydrogels for Controlled Protein Release

Hydrogels are an ideal biomaterial for many applications because they are easily synthesized, have tunable chemical and physical properties, are hydrophilic and can absorb many times their weight in water, and are stable over time. They can be synthesized from natural materials, synthetic polymers, or a combination thereof. Hydrogels are typically porous biomaterials. Therefore, they have been greatly explored for many controlled release applications. These biomaterials provide a defined three-dimensional environment for the localized delivery of specific signals (e.g., growth factors) to cells in targeted sites of the body to guide the gene expression and development of new tissues [142]. The advantage of using a delivery system rather than a bolus injection is that the therapeutic efficacy is greatly increased. However, most hydrogels have no inherent ability to strongly control the delivery of therapeutics and the release of therapeutics is still too quick to be usable in many applications. To further slow the release of therapeutics, such as growth factors, hydrogels functionalized with aptamers selected for the growth factors can be used. An aptamer selected for platelet-derived growth factor BB (PDGF-BB) was incorporated into a polyacrylamide hydrogel through free-radical polymerization to slow the release of PDGF-BB [143]. The high binding affinity of the aptamer for the target was able to greatly reduce the release rate of PDGF-BB from the hydrogel. Moreover, aptamers of lesser affinity generated through base substitutions resulted in faster release kinetics than the highest affinity aptamer when incorporated into the hydrogel. Thus, aptamer-functionalized hydrogels are able to greatly reduce the release rate of growth factors from hydrogels and the release rate can be modulated. To further show that aptamer-functionalized hydrogels are promising biomaterials for use in growth factor delivery, complementary sequences were used as molecular triggers to stimulate the release of growth factors [144]. The hydrogels were able to release additional proteins when triggered by the complementary sequences and the hydrogels were able to be triggered multiple times. The ability of the hydrogel to release the growth factors on demand is an important property for theranostics. As aptamers are highly specific, multiple aptamers can be incorporated into a hydrogel to control the delivery of multiple growth factors. Aptamers selected for VEGF and PDGF-BB were incorporated into a hydrogel and the ability to trigger the individual release of a growth factor was explored [145]. The length of the aptamers and their complementary sequences were varied to see its effect on growth factor release in one-protein systems. After optimizing the triggered release for each protein, a two-protein system was developed and subjected to treatment of the complementary sequences. It was demonstrated that one complementary sequence would only inactivate one aptamer to release one growth factor [145]. Because aptamers are specific to their target and each aptamer has a unique complementary sequence, it is unsurprising that a complementary sequence triggered the release of only one growth factor and not both. Thus, rationally designed aptamer and complementary sequences can be applied to control and trigger the delivery of multiple growth factor from hydrogels in a highly regulated manner.

To further investigate the feasibility of using aptamer-functionalized hydrogels for controlled growth factor delivery, PEGylated complementary sequences were tested for their ability to trigger protein release [146]. PEGylated complementary sequences (complementary sequences conjugated to PEG) were investigated since conjugated oligonucleotides are more resistant to nuclease degradation, whereas unmodified oligonucleotides have a half-life of several minutes *in vivo*. Therefore, PEGylated complementary sequences would be more practical than unmodified complementary sequences in triggering growth factor release *in vivo*. The PEGylated complementary sequences were able to trigger PDGF-BB release and the amount of PDGF-BB released was dependent on the length of PEG attached to the complementary sequence [146]. Thus, PEGylation is a method to protect exogenous oligonucleotides *in vivo* and the length of PEG can be tailored to achieve desired on-demand release profiles.

17.4.2 Aptamer-Functionalized Coatings

A major obstacle facing medical devices postimplantation is rejection by the host. One method to encourage implant integration with the host is to promote the attachment of specific cells onto the implant. Aptamers screened for osteoblasts have been attached to tissue culture plates via biotin-streptavidin association [147]. Aptamer-coated plates were able to quickly bind osteoblasts from solution. However, the system relied on precoating the plates with streptavidin. A simplified method would be to directly incorporate aptamers into the coating during material synthesis. Aptamers bearing free carbon-carbon double bonds have been incorporated into polyacrylamide and PEG hydrogels via free-radical polymerization [5,6]. Thin films of polyacrylamide or PEG can be applied to substrates and prevent cell attachment that would otherwise attach to the base substrate. When an aptamer specific to a cell line is incorporated into the hydrogel layer, the hydrogel will allow the target cell line to adhere to the hydrogel while rejecting the attachment of undesired cells. Furthermore, the attachment of cells to the aptamers can be reversed using complementary sequences to release the cells in a nondestructive manner [148].

These studies utilized unmodified aptamers and complementary sequences to capture and release cells. However, for real applications, modified aptamers and complementary sequences that resist degradation by nucleases should be used to maintain the functionality of the biomaterial.

17.4.3 Aptamer-Functionalized Nanomaterials

17.4.3.1 Polymer Nanoparticles

Polymer nanoparticles such as PLA have been functionalized with aptamers to impart targeting capabilities to develop therapeutic delivery systems for the treatment of cancer systemically. For example, nanoparticles of PEGylated PLA with a terminal carboxylic acid functional group were used to encapsulate rhodamine-labeled dextran [149]. The carboxylic acid group was used to conjugate an RNA aptamer that binds to the prostate-specific membrane antigen (PSMA), a well-known transmembrane protein that is overexpressed on prostate cancer epithelial cells. Prostate LNCaP epithelial cells, which express PSMA, showed a preferred uptake of over 75-fold in comparison to the control cell line *in vitro* [149]. When tested *in vivo*, anti-PSMA aptamer-functionalized PLGA nanoparticles showed a near-fourfold increase in tumor accumulation in comparison to nonfunctionalized nanoparticles at 24 h when injected systemically [150]. The strategy provided for creating targeting drug delivery nanoparticles is amenable to encapsulating different drug molecules (e.g., paclitaxel, cisplatin, docetaxel, or small interfering RNA) [151–154] or attaching different aptamers. An antinucleolin aptamer and an aptamer selected against mucin 1 were conjugated to PLGA to target nucleolin receptors and mucin 1 proteins on tumor cells, respectively [154–156].

In addition to using aptamer-functionalized nanomaterials as targeting drug delivery systems, these biomaterials can be utilized to diminish the hazards of toxins. For example, PLGA nanoparticles with selenomethionine encapsulated within the particles and an aptamer with a strong affinity for mercury covering the particle were developed to reduce mercury toxicity *in vivo* [157]. Administration of these functionalized nanoparticles reduced accumulation of mercury in the kidneys and the brain approximately 60% and enhanced urinary excretion of mercury while restoring lost locomotive and cognitive activity in rats [157]. Thus, aptamer-functionalized polymer nanoparticles are promising tools for anticancer therapy and as antidotes to neutralize systemic toxicity of hazardous toxins for human health applications.

17.4.3.2 Quantum Dots

To increase the resolution and residence time of imaging reagents, particles such as quantum dots have been conjugated with aptamers to impart specificity and prolong the attachment of quantum dots to the target [158–163]. The large surface area-to-volume ratio found in quantum dots and other nanoparticles permits the attachment of a large quantity of aptamers to the surface. The large concentration of aptamers on the surface increases the binding affinity of the quantum dots for the target manyfold [164]. Moreover, the conjugation of aptamers to quantum dots protects the aptamers from enzymatic degradation, making aptamers suitable ligands for *in vivo* imaging [165].

Aptamers selected against hepatocellular carcinoma using subtractive SELEX between cells of high metastatic potential and low metastatic potential have been conjugated to quantum dots that can discriminate between the two cell lines with high specificity [161]. Discriminating between high and low metastasis potentials can reveal cell-specific molecules for the prediction of cancer metastasis and treatment of recurrences. Labeling of human hepatocellular carcinoma tissue with the aptamer and quantum dots revealed selective binding, whereas the control aptamer did not. Moreover, the aptamer-quantum dot conjugate was able to selectively label hepatocellular carcinoma cells from whole blood samples. Thus, aptamers conjugated to quantum dots are promising probes to analyze metastatic cells as well as tumor tissues.

A multifunctional nanoparticle system was developed utilizing quantum dots as a carrier and imaging reagent; doxorubicin conjugated to the quantum dot through a cleavable, pH-sensitive bond as an anticancer drug; and an aptamer targeting mucin 1, for the enhanced targeting and treatment of drug-resistant cancers [163]. Intravenous administration of quantum dots in nude mice showed preferred uptake by the tumor of aptamer-conjugated quantum dots over the bare quantum dots. Both cases showed accumulation of quantum dots in the liver and the kidneys due to the enhanced permeation and retention effect, suggesting that the toxicity of quantum dots needs to be considered for *in vivo* imaging applications. By using nontoxic imaging reagents in conjunction with aptamers for targeted delivery, highly specific systems with simultaneous imaging and therapeutic properties can be developed for advanced monitoring and treatment of cancers.

17.4.3.3 Dendrimers

Similar to the nanoparticles mentioned earlier, dendrimers have been developed as new biomaterials with unique properties. Dendrimers are nanometer-scale macromolecules with a radial series of branches extending from an inner core. These hyperbranched macromolecules are synthesized in a stepwise fashion using either a convergent or a divergent method. The high level of control of the dendrimer architecture (monodispersity in size and weight), high ligand density, multivalency and surface functionality, lack of immunogenicity, good biocompatibility, ability to cross cell membranes, and high drug-loading capacity make dendrimers favorable nanoscaffolds for various biomedical applications. To utilize dendrimers for drug delivery applications, the dendrimers must be quickly delivered to and retained by the target site to avoid the loss of drug molecules. To direct dendrimers to a target site, dendrimers functionalized with aptamers were developed [166–169]. The first aptamer-dendrimer biomaterial was synthesized using an aptamer selected for CCRF-CEM, a human T lymphocytic leukemia cell line, and generation-five poly(amidoamine) (PAMAM) dendrimer [166]. The aptamer was successfully conjugated to the dendrimer without loss of specificity or affinity. In addition, the size of the complex was approximately 8 nm, making it feasible for use in *in vivo* application because its size avoids renal clearance for long systemic circulation [166]. This concept was further extended to develop a dendrimer that was functionalized with bivalent aptamers, resulting in bivalent aptamer-dendrimer complexes that were able to bind to their target with greater affinity than their monovalent counterpart [167].

To create a cancer-targeting complex, an aptamer that recognizes the PSMA was conjugated to generation-four PAMAM dendrimers, which was then loaded with doxorubicin [168]. *In vivo* studies in prostate tumor models with the aptamer-dendrimer bioconjugate revealed enhanced aptamer stability, targeting specificity, and excellent antitumor efficacy [168].

17.4.3.4 Micelles and Liposomes

Micelles and liposomes have also been investigated for drug delivery applications for their ability to store and protect drug molecules within itself. They are spontaneously formed from the self-assembly of long organic or polymeric amphiphilic molecules through the hydrophobic effect. The entropic penalty of "burying" the hydrophobic tail into an oil-like core is less than the entropic penalty of surrounding the amphiphile with water molecules and drives the formation of the micelle. In aqueous solutions, the hydrophilic heads form a shell contacting the solvent and the hydrophobic tails form the center of the micelle. Micelles often adopt a spherical structure since the least amount of energy is needed to maintain this shape. Liposomes are similar to micelles, but liposomes have a bilayer structure, where aqueous solution is encapsulated by the hydrophobic membrane. Small drug molecules can be encapsulated within micelles or liposomes during their formation, and the drugs can be retained for long periods since these molecules cannot easily pass by the amphiphilic molecules. In order for micelles or liposomes to be successful in drug delivery applications, the carriers need to be easily and uniformly synthesized, need a high retention of the drugs until reaching the target, and need preferential and sufficient uptake by the target [170]. The development of micelle and liposome assembly techniques has resulted in the formations of vesicles that are well defined with desired sizes and shapes [171–175]. However, the ability to deliver the drug-laden micelle or liposome specifically to the target site was limited. In order to expand their use from drug containers to targeting drug delivery vehicles, new micelles and liposomes were developed with aptamers decorating the surfaces [176–178]. Micelles or liposomes displaying aptamers are created using amphiphiles with nucleic acid aptamers covalently bonded to the hydrophilic head group. As nucleic acid aptamers are hydrophilic, they will extend into the aqueous environment from the surface of the micelle or liposome. Aptamers conjugated to micelles or liposome often retain their high binding functionality. Interestingly, when an aptamer specific for Ramos cells (a B-cell lymphoma cell line) was incorporated into a micelle, it was found that the binding functionality of the aptamer increased at physiological conditions, perhaps due to a combination of multivalency from multiple aptamers binding the target to the micelle and the fluidic nature of the nanomaterial [179]. Other merits of the aptamer-functionalized micelle include a rapid targeting ability, high sensitivity, high specificity, and the creation of a dual drug delivery pathway [179]. Micelles and liposomes can fuse to cells and introduce therapeutics that could not penetrate the cell membrane, directly into the cell. Additionally, aptamer-functionalized micelles can be destroyed using ultrasound for triggered drug delivery after the micelle has reached its intended target [180]. The plasma residence time of aptamers also increases when the aptamer is conjugated to liposomes [181]. The efficacy of drug delivery from aptamer-functionalized liposomes can also be decreased using complementary sequences, an ability not inherent in bare liposomes [182]. After the drug-delivering liposome binds its target via the aptamer, sequences complementary to the aptamer can be introduced into a system to inactivate the aptamer and free the liposome from the target, thereby removing the drug delivery system from the target site.

17.5 CONCLUSION

Nucleic acid aptamers are a unique class of oligonucleotides that bind to target molecules with high affinity and specificity. Their strong interactions with their target make aptamers exceptional ligands to functionalize biomaterials. Aptamer-functionalized hydrogels and nanoparticles have been developed to overcome many limitations to create biomaterials for regulated growth factor delivery, cell capture and release, and targeting imaging reagents. These biomaterials with innovative functions are promising materials for the advancement of biomedical applications.

ACKNOWLEDGEMENT

We greatly acknowledge financial support from the NSF (DMR 1332351) and the Penn State Start-up Fund.

REFERENCES

[1] M.K. Sewell-Loftin, Y. Chun, A. Khademhosseini, W.D. Merryman, EMT-inducing biomaterials for heart valve engineering: taking cues from developmental biology, J. Cardiovasc. Transl. Res. 4 (5) (2011) 658–671.
[2] M.L. Oprea, H. Schnöring, J.S. Sachweh, H. Ott, J. Biertz, J.F. Vazquez-Jimenez, Allergy to pacemaker silicone compounds: recognition and surgical management, Ann. Thorac. Surg. 87 (4) (2009) 1275–1277.
[3] J.M. Anderson, Biological responses to materials, Annu. Rev. Mater. Res. 31 (1) (2001) 81–110.
[4] J.M. Anderson, A. Rodriguez, D.T. Chang, Foreign body reaction to biomaterials, Semin. Immunol. 20 (2) (2008) 86–100.
[5] S. Li, N. Chen, Z. Zhang, Y. Wang, Endonuclease-responsive aptamer-functionalized hydrogel coating for sequential catch and release of cancer cells, Biomaterials 34 (2) (2013) 460–469.
[6] N. Chen, Z. Zhang, B. Soontornworajit, J. Zhou, Y. Wang, Cell adhesion on an artificial extracellular matrix using aptamer-functionalized PEG hydrogels, Biomaterials 33 (5) (2012) 1353–1362.
[7] A.M. Telford, M. James, L. Meagher, C. Neto, Thermally cross-linked PNVP films as antifouling coatings for biomedical applications, ACS Appl. Mater. Interfaces 2 (8) (2010) 2399–2408.
[8] H. Chen, L. Yuan, W. Song, Z. Wu, D. Li, Biocompatible polymer materials: role of protein–surface interactions, Prog. Polym. Sci. 33 (11) (2008) 1059–1087.
[9] K.M. Brouwer, W.F. Daamen, N. van Lochem, D. Reijnen, R.M.H. Wijnen, T.H. van Kuppevelt, Construction and in vivo evaluation of a dual layered collagenous scaffold with a radial pore structure for repair of the diaphragm, Acta Biomater. 9 (6) (2013) 6844–6851.
[10] C.B. Kwok, F.C. Ho, C.W. Li, A.H.W. Ngan, D. Chan, B.P. Chan, Compression-induced alignment and elongation of human mesenchymal stem cell (hMSC) in 3D collagen constructs is collagen concentration dependent, J. Biomed. Mater. Res. 101A (6) (2013) 1716–1725.
[11] S. Wu, Y. Liu, S. Bharadwaj, A. Atala, Y. Zhang, Human urine-derived stem cells seeded in a modified 3D porous small intestinal submucosa scaffold for urethral tissue engineering, Biomaterials 32 (5) (2011) 1317–1326.
[12] B. Kim, J. Choi, J. Kim, Y. Choi, Y. Cho, Recellularization of decellularized human adipose-tissue-derived extracellular matrix sheets with other human cell types, Cell Tissue Res. 348 (3) (2012) 559–567.
[13] M.T. Wolf, K.A. Daly, E.P. Brennan-Pierce, S.A. Johnson, C.A. Carruthers, A. D'Amore, S.P. Nagarkar, S.S. Velankar, S.F. Badylak, A hydrogel derived from decellularized dermal extracellular matrix, Biomaterials 33 (29) (2012) 7028–7038.
[14] B.S. Kim, C.E. Baez, A. Atala, Biomaterials for tissue engineering, World J. Urol. 18 (1) (2000) 2–9.
[15] C.E. Schmidt, J.M. Baier, Acellular vascular tissues: natural biomaterials for tissue repair and tissue engineering, Biomaterials 21 (22) (2000) 2215–2231.

[16] E. Song, S. Yeon Kim, T. Chun, H.J. Byun, Y.M. Lee, Collagen scaffolds derived from a marine source and their biocompatibility, Biomaterials 27 (15) (2006) 2951–2961.
[17] J. Haag, S. Baiguera, P. Jungebluth, D. Barale, C. Del Gaudio, F. Castiglione, A. Bianco, C.E. Comin, D. Ribatti, P. Macchiarini, Biomechanical and angiogenic properties of tissue-engineered rat trachea using genipin cross-linked decellularized tissue, Biomaterials 33 (3) (2012) 780–789.
[18] L. Schukur, P. Zorlutuna, J.M. Cha, H. Bae, A. Khademhosseini, Directed differentiation of size-controlled embryoid bodies towards endothelial and cardiac lineages in RGD-modified poly(Ethylene Glycol) hydrogels, Adv. Healthc. Mater. 2 (1) (2013) 195–205.
[19] F. Yang, C.G. Williams, D.A. Wang, H. Lee, P.N. Manson, J. Elisseeff, The effect of incorporating RGD adhesive peptide in polyethylene glycol diacrylate hydrogel on osteogenesis of bone marrow stromal cells, Biomaterials 26 (30) (2005) 5991–5998.
[20] I.L. Kim, S. Khetan, B.M. Baker, C.S. Chen, J.A. Burdick, Fibrous hyaluronic acid hydrogels that direct MSC chondrogenesis through mechanical and adhesive cues, Biomaterials 34 (22) (2013) 5571–5580.
[21] L.A. Smith Callahan, A.M. Ganios, E.P. Childers, S.D. Weiner, M.L. Becker, Primary human chondrocyte extracellular matrix formation and phenotype maintenance using RGD-derivatized PEGDM hydrogels possessing a continuous Young's modulus gradient, Acta Biomater. 9 (4) (2013) 6095–6104.
[22] F. Yang, J. Wang, J. Hou, H. Guo, C. Liu, Bone regeneration using cell-mediated responsive degradable PEG-based scaffolds incorporating with rhBMP-2, Biomaterials 34 (5) (2013) 1514–1528.
[23] A.S. Salimath, E.A. Phelps, A.V. Boopathy, P.L. Che, M. Brown, A.J. García, M.E. Davis, Dual delivery of hepatocyte and vascular endothelial growth factors via a protease-degradable hydrogel improves cardiac function in rats, PLoS One 7 (11) (2012) e50980.
[24] O. Oliviero, M. Ventre, P.A. Netti, Functional porous hydrogels to study angiogenesis under the effect of controlled release of vascular endothelial growth factor, Acta Biomater. 8 (9) (2012) 3294–3301.
[25] S. Sokic, G. Papavasiliou, FGF-1 and proteolytically mediated cleavage site presentation influence three-dimensional fibroblast invasion in biomimetic PEGDA hydrogels, Acta Biomater. 8 (6) (2012) 2213–2222.
[26] C.C. Lin, K.S. Anseth, PEG hydrogels for the controlled release of biomolecules in regenerative medicine, Pharm. Res. 26 (3) (2009) 631–643.
[27] J.A. Burdick, M.N. Mason, A.D. Hinman, K. Thorne, K.S. Anseth, Delivery of osteoinductive growth factors from degradable PEG hydrogels influences osteoblast differentiation and mineralization, J. Control. Release. 83 (1) (2002) 53–63.
[28] J. Patterson, R. Siew, S.W. Herring, A.S.P. Lin, R. Guldberg, P.S. Stayton, Hyaluronic acid hydrogels with controlled degradation properties for oriented bone regeneration, Biomaterials 31 (26) (2010) 6772–6781.
[29] H. Wang, Y. Jia, W. Hu, H. Jiang, J. Zhang, L. Zhang, Effect of preparation conditions on the size and encapsulation properties of mPEG-PLGA nanoparticles simultaneously loaded with vincristine sulfate and curcumin, Pharm. Dev. Technol. 18 (3) (2012) 694–700.
[30] L. Zou, X. Song, T. Yi, S. Li, H. Deng, X. Chen, Z. Li, Y. Bai, Q. Zhong, Y. Wei, Administration of PLGA nanoparticles carrying shRNA against focal adhesion kinase and CD44 results in enhanced antitumor effects against ovarian cancer, Cancer Gene Ther. 20 (2013) 242–250.
[31] M. Cetin, A. Atila, S. Sahin, I. Vural, Preparation and characterization of metformin hydrochloride loaded-Eudragit (R) RSPO and Eudragit (R) RSPO/PLGA nanoparticles, Pharm. Dev. Technol. 18 (2013) 570–576.
[32] M. Callewaert, S. Dukic, L. Van Gulick, M. Vittier, V. Gafa, M.C. Andry, M. Molinari, V.G. Roullin, Etoposide encapsulation in surface-modified poly(lactide-co-glycolide) nanoparticles strongly enhances glioma antitumor efficiency, J. Biomed. Mater. Res. 101A (5) (2013) 1319–1327.
[33] X. Zhang, G. Chen, L. Wen, F. Yang, A.L. Shao, X. Li, W. Long, L. Mu, Novel multiple agents loaded PLGA nanoparticles for brain delivery via inner ear administration: in vitro and in vivo evaluation, Eur. J. Pharm. Sci. 48 (4–5) (2013) 595–603.
[34] R. Saadati, S. Dadashzadeh, Z. Abbasian, H. Soleimanjahi, Accelerated blood clearance of PEGylated PLGA nanoparticles following repeated injections: effects of polymer dose, PEG coating, and encapsulated anticancer drug, Pharm. Res. 30 (4) (2013) 985–995.
[35] L.E. Hill, T.M. Taylor, C. Gomes, Antimicrobial efficacy of poly (DL-lactide-co-glycolide) (PLGA) nanoparticles with entrapped cinnamon bark extract against *Listeria monocytogenes* and *Salmonella typhimurium*, J. Food Sci. 78 (4) (2013) N626–N632.
[36] J. Panyam, M.M. Dali, S.K. Sahoo, W. Ma, S.S. Chakravarthi, G.L. Amidon, R.J. Levy, V. Labhasetwar, Polymer degradation and in vitro release of a model protein from poly(D,L-lactide-co-glycolide) nano- and microparticles, J. Control. Release 92 (1–2) (2003) 173–187.
[37] C. Fonseca, S. Simões, R. Gaspar, Paclitaxel-loaded PLGA nanoparticles: preparation, physicochemical characterization and in vitro anti-tumoral activity, J. Control. Release 83 (2) (2002) 273–286.
[38] T. Govender, S. Stolnik, M.C. Garnett, L. Illum, S.S. Davis, PLGA nanoparticles prepared by nanoprecipitation: drug loading and release studies of a water soluble drug, J. Control. Release 57 (2) (1999) 171–185, Ref Type: Abstract.
[39] J. Panyam, V. Labhasetwar, Biodegradable nanoparticles for drug and gene delivery to cells and tissue, Adv. Drug Deliv. Rev. 64 (Suppl.) (2012) 61–71.
[40] S.E. Sakiyama-Elbert, J.A. Hubbell, Development of fibrin derivatives for controlled release of heparin-binding growth factors, J. Control. Release 65 (3) (2000) 389–402.
[41] S.E. Sakiyama-Elbert, J.A. Hubbell, Controlled release of nerve growth factor from a heparin-containing fibrin-based cell ingrowth matrix, J. Control. Release 69 (1) (2000) 149–158.
[42] H.S. Yang, W.G. La, Y.M. Cho, W. Shin, G.D. Yeo, B.S. Kim, Comparison between heparin-conjugated fibrin and collagen sponge as bone morphogenetic protein-2 carriers for bone regeneration, Exp. Mol. Med. 44 (2012) 350–355.
[43] F. Zomer Volpato, J. Almodóvar, K. Erickson, K.C. Popat, C. Migliaresi, M.J. Kipper, Preservation of FGF-2 bioactivity using heparin-based nanoparticles, and their delivery from electrospun chitosan fibers, Acta Biomater. 8 (4) (2012) 1551–1559.
[44] X. Li, J. Wang, G. Su, Z. Zhou, J. Shi, L. Liu, M. Guan, Q. Zhang, Spatiotemporal control over growth factor delivery from collagen-based membrane, J. Biomed. Mater. Res. 100A (2) (2012) 396–405.
[45] S.T.M. Nillesen, P.J. Geutjes, R. Wismans, J. Schalkwijk, W.F. Daamen, T.H. van Kuppevelt, Increased angiogenesis and blood vessel maturation in acellular collagen–heparin scaffolds containing both FGF2 and VEGF, Biomaterials 28 (6) (2007) 1123–1131.
[46] S. Cai, Y. Liu, X. Zheng Shu, G.D. Prestwich, Injectable glycosaminoglycan hydrogels for controlled release of human basic fibroblast growth factor, Biomaterials 26 (30) (2005) 6054–6067.

[47] A. Zieris, K. Chwalek, S. Prokoph, K.R. Levental, P.B. Welzel, U. Freudenberg, C. Werner, Dual independent delivery of pro-angiogenic growth factors from starPEG-heparin hydrogels, J. Control. Release 156 (1) (2011) 28–36.
[48] A.D. Baldwin, K.L. Kiick, Polysaccharide-modified synthetic polymeric biomaterials, Pept. Sci. 94 (1) (2010) 128–140.
[49] R. Dimitriou, E. Tsiridis, P.V. Giannoudis, Current concepts of molecular aspects of bone healing, Injury 36 (12) (2005) 1392–1404.
[50] T.P. Richardson, M.C. Peters, A.B. Ennett, D.J. Mooney, Polymeric system for dual growth factor delivery, Nat. Biotechnol. 19 (11) (2001) 1029–1034.
[51] D.H. Choi, C.H. Park, I.H. Kim, H.J. Chun, K. Park, D.K. Han, Fabrication of core–shell microcapsules using PLGA and alginate for dual growth factor delivery system, J. Control. Release 147 (2) (2010) 193–201.
[52] D.H.R. Kempen, L. Lu, A. Heijink, T.E. Hefferan, L.B. Creemers, A. Maran, M.J. Yaszemski, W.J.A. Dhert, Effect of local sequential VEGF and BMP-2 delivery on ectopic and orthotopic bone regeneration, Biomaterials 30 (14) (2009) 2816–2825.
[53] R. Toy, E. Hayden, A. Camann, Z. Berman, P. Vicente, E. Tran, J. Meyers, J. Pansky, P.M. Peiris, H. Wu, A. Exner, D. Wilson, K.B. Ghaghada, E. Karathanasis, Multimodal in vivo imaging exposes the voyage of nanoparticles in tumor microcirculation, ACS Nano 7 (4) (2013) 3118–3129.
[54] A. Astolfo, E. Schültke, R.H. Menk, R.D. Kirch, B.H.J. Juurlink, C. Hall, L.A. Harsan, M. Stebel, D. Barbetta, G. Tromba, F. Arfelli, In vivo visualization of gold-loaded cells in mice using x-ray computed tomography, Nanomedicine 9 (2) (2013) 284–292.
[55] D.H. Hu, Z.H. Sheng, P.F. Zhang, D.Z. Yang, S.H. Liu, P. Gong, D.Y. Gao, S.T. Fang, Y.F. Ma, L.T. Cai, Hybrid gold-gadolinium nanoclusters for tumor-targeted NIRF/CT/MRI triple-modal imaging in vivo, Nanoscale 5 (4) (2013) 1624–1628.
[56] O. Rabin, J. Manuel Perez, J. Grimm, G. Wojtkiewicz, R. Weissleder, An X-ray computed tomography imaging agent based on long-circulating bismuth sulphide nanoparticles, Nat. Mater. 5 (2) (2006) 118–122.
[57] D. Kim, S. Park, J.H. Lee, Y.Y. Jeong, S. Jon, Antibiofouling polymer-coated gold nanoparticles as a contrast agent for in vivo X-ray computed tomography imaging, J. Am. Chem. Soc. 129 (24) (2007) 7661–7665.
[58] J.F. Hainfeld, D.N. Slatkin, T.M. Focella, H.M. Smilowitz, Gold nanoparticles: a new X-ray contrast agent, Br. J. Radiol. 79 (939) (2006) 248–253.
[59] S.M. Moghimi, A.C. Hunter, J.C. Murray, Long-circulating and target-specific nanoparticles: theory to practice, Pharmacol. Rev. 53 (2) (2001) 283–318.
[60] Y.C. Kuo, H.F. Ko, Targeting delivery of saquinavir to the brain using 83-14 monoclonal antibody-grafted solid lipid nanoparticles, Biomaterials 34 (20) (2013) 4818–4830.
[61] L.C. Cheng, H.M. Chen, T.C. Lai, Y.C. Chan, R.S. Liu, Targeting polymeric fluorescent nanodiamond-gold/silver multi-functional nanoparticles as a light-transforming hyperthermia reagent for cancer cells, Nanoscale 5 (9) (2013) 3931–3940.
[62] X. Gao, Y. Cui, R.M. Levenson, L.W.K. Chung, S. Nie, In vivo cancer targeting and imaging with semiconductor quantum dots, Nat. Biotechnol. 22 (8) (2004) 969–976.
[63] J. Wang, Nanomaterial-based amplified transduction of biomolecular interactions, Small 1 (11) (2005) 1036–1043.
[64] S.D. Jayasena, Aptamers: an emerging class of molecules that rival antibodies in diagnostics, Clin. Chem. 45 (9) (1999) 1628–1650.
[65] A.D. Ellington, J.W. Szostak, In vitro selection of RNA molecules that bind specific ligands, Nature 346 (6287) (1990) 818–822.
[66] C. Tuerk, L. Gold, Systematic evolution of ligands by exponential enrichment—RNA ligands to bacteriophage-T4 DNA-polymerase, Science 249 (4968) (1990) 505–510.
[67] G. Aquino-Jarquin, J.D. Toscano-Garibay, RNA aptamer evolution: two decades of SELEction, Int. J. Mol. Sci. 12 (12) (2011) 9155–9171.
[68] Y. Miyake, H. Togashi, M. Tashiro, H. Yamaguchi, S. Oda, M. Kudo, Y. Tanaka, Y. Kondo, R. Sawa, T. Fujimoto, T. Machinami, A. Ono, MercuryII-mediated formation of thymine-HgII-thymine base pairs in DNA duplexes, J. Am. Chem. Soc. 128 (7) (2006) 2172–2173.
[69] T. Li, S. Dong, E. Wang, A lead(II)-driven DNA molecular device for turn-on fluorescence detection of lead(II) ion with high selectivity and sensitivity, J. Am. Chem. Soc. 132 (38) (2010) 13156–13157.
[70] J. Ciesiolka, J. Gorski, M. Yarus, Selection of an RNA domain that binds Zn^{2+}, RNA 1 (5) (1995) 538–550.
[71] H.P. Hofmann, S. Limmer, V. Hornung, M. Sprinzl, Ni^{2+}-binding RNA motifs with an asymmetric purine-rich internal loop and a G-A base pair, RNA 3 (11) (1997) 1289–1300.
[72] D. Grate, C. Wilson, Laser-mediated, site-specific inactivation of RNA transcripts, Proc. Natl. Acad. Sci. U.S.A. 96 (11) (1999) 6131–6136.
[73] M. Sassanfar, J.W. Szostak, An RNA motif that binds ATP, Nature 364 (6437) (1993) 550–553.
[74] J.R. Lorsch, J.W. Szostak, In vitro selection of RNA aptamers specific for cyanocobalamin, Biochemistry 33 (4) (1994) 973–982.
[75] C. Wilson, J.W. Szostak, In vitro evolution of a self-alkylating ribozyme, Nature 374 (6525) (1995) 777–782.
[76] C. Mannironi, A. Di Nardo, P. Fruscoloni, G.P. Tocchini-Valentini, In vitro selection of dopamine RNA ligands, Biochemistry 36 (32) (1997) 9726–9734.
[77] R.D. Jenison, S.C. Gill, A. Pardi, B. Polisky, High-resolution molecular discrimination by RNA, Science 263 (5152) (1994) 1425–1429.
[78] Z. Kiani, M. Shafiei, P. Rahimi-Moghaddam, A.A. Karkhane, S.A. Ebrahimi, In vitro selection and characterization of deoxyribonucleic acid aptamers for digoxin, Anal. Chim. Acta 748 (2012) 67–72.
[79] Y. Wang, R.R. Rando, Specific binding of aminoglycoside antibiotics to RNA, Chem. Biol. 2 (5) (1995) 281–290.
[80] R. Stoltenburg, N. Nikolaus, B. Strehlitz, Capture-SELEX: selection of DNA aptamers for aminoglycoside antibiotics, J. Anal. Methods Chem. 22 (22) (2012) 1–14.
[81] M.N. Win, J.S. Klein, C.D. Smolke, Codeine-binding RNA aptamers and rapid determination of their binding constants using a direct coupling surface plasmon resonance assay, Nucleic Acids Res. 34 (19) (2006) 5670–5682.
[82] M.N. Stojanovic, P. de Prada, D.W. Landry, Fluorescent sensors based on aptamer self-assembly, J. Am. Chem. Soc. 122 (46) (2000) 11547–11548.
[83] A. Okazawa, H. Maeda, E. Fukusaki, Y. Katakura, A. Kobayashi, In vitro selection of hematoporphyrin binding DNA aptamers, Bioorg. Med. Chem. Lett. 10 (23) (2000) 2653–2656.
[84] A. Geiger, P. Burgstaller, H. vonderEltz, A. Roeder, M. Famulok, RNA aptamers that bind L-arginine with sub-micromolar dissociation constants and high enantioselectivity, Nucleic Acids Res. 24 (6) (1996) 1029–1036.

[85] C. Lozupone, S. Changayil, I. Majerfeld, M. Yarus, Selection of the simplest RNA that binds isoleucine, RNA (Cambridge) 9 (11) (2003) 1315–1322.

[86] I. Majerfeld, D. Puthenvedu, M. Yarus, RNA affinity for molecular L-histidine; genetic code origins, J. Mol. Evol. 61 (2) (2005) 226–235.

[87] X. Yang, T. Bing, H. Mei, C. Fang, Z. Cao, D. Shangguan, Characterization and application of a DNA aptamer binding to L-tryptophan, Analyst 136 (3) (2011) 577–585.

[88] L.S. Green, D. Jellinek, C. Bell, L.A. Beebe, B.D. Feistner, S.C. Gill, F.M. Jucker, N. Janjic, Nuclease-resistant nucleic acid ligands to vascular permeability factor/vascular endothelial growth factor, Chem. Biol. 2 (10) (1995) 683–695.

[89] H. Hasegawa, K. Sode, K. Ikebukuro, Selection of DNA aptamers against VEGF165 using a protein competitor and the aptamer blotting method, Biotechnol. Lett. 30 (5) (2008) 829–834.

[90] L.S. Green, D. Jellinek, R. Jenison, A. Ostman, C.H. Heldin, N. Janjic, Inhibitory DNA ligands to platelet-derived growth factor B-chain, Biochemistry 35 (45) (1996) 14413–14424.

[91] D. Jellinek, C.K. Lynott, D.B. Rifkin, N. Janjic, High-affinity RNA ligands to basic fibroblast growth factor inhibit receptor binding, Proc. Natl. Acad. Sci. U.S.A. 90 (23) (1993) 11227–11231.

[92] D. Shangguan, Z.C. Cao, Y. Li, W. Tan, Aptamers evolved from cultured cancer cells reveal molecular differences of cancer cells in patient samples, Clin. Chem. 53 (6) (2007) 1153–1155.

[93] P. Mallikaratchy, Z. Tang, S. Kwame, L. Meng, D. Shangguan, W. Tan, Aptamer directly evolved from live cells recognizes membrane bound immunoglobin heavy mu chain in Burkitt's lymphoma cells, Mol. Cell. Proteomics 6 (12) (2007) 2230–2238.

[94] S.E. Lupold, B.J. Hicke, Y. Lin, D.S. Coffey, Identification and characterization of nuclease-stabilized RNA molecules that bind human prostate cancer cells via the prostate-specific membrane antigen, Cancer Res. 62 (14) (2002) 4029–4033.

[95] S. Shigdar, L. Qiao, S.F. Zhou, D. Xiang, T. Wang, Y. Li, L.Y. Lim, L. Kong, L. Li, W. Duan, RNA aptamers targeting cancer stem cell marker CD133, Cancer Lett. 330 (1) (2013) 84–95.

[96] L.C. Bock, L.C. Griffin, J.A. Latham, E.H. Vermaas, J.J. Toole, Selection of single-stranded DNA molecules that bind and inhibit human thrombin, Nature 355 (6360) (1992) 564–566.

[97] V.J.B. Ruigrok, E. van Duijn, A. Barendregt, K. Dyer, J.A. Tainer, R. Stoltenburg, B. Strehlitz, M. Levisson, H. Smidt, J. van der Oost, Kinetic and stoichiometric characterisation of streptavidin-binding aptamers, Chembiochem 13 (6) (2012) 829–836.

[98] B. Shui, A. Ozer, W. Zipfel, N. Sahu, A. Singh, J.T. Lis, H. Shi, M.I. Kotlikoff, RNA aptamers that functionally interact with green fluorescent protein and its derivatives, Nucleic Acids Res. 40 (5) (2012) e39.

[99] J.G. Bruno, M.P. Carrillo, T. Phillips, In Vitro antibacterial effects of antilipopolysaccharide DNA aptamer-C1qrs complexes, Folia Microbiol. 53 (4) (2008) 295–302.

[100] H.R. Liang, G.Q. Hu, T. Zhang, Y.J. Yang, L.L. Zhao, Y.L. Qi, H.L. Wang, Y.W. Gao, S.T. Yang, X.Z. Xia, Isolation of ssDNA aptamers that inhibit rabies virus, Int. Immunopharmacol. 14 (3) (2012) 341–347.

[101] P. Punnarak, M. Santos, S. Hwang, H. Kondo, I. Hirono, Y. Kikuchi, T. Aoki, RNA aptamers inhibit the growth of the fish pathogen viral hemorrhagic septicemia virus (VHSV), Mar. Biotechnol. 14 (6) (2012) 752–761.

[102] M. Wongphatcharachai, P. Wang, S. Enomoto, R.J. Webby, M.R. Gramer, A. Amonsin, S. Sreevatsan, Neutralizing DNA aptamers against swine influenza H3N2 viruses, J. Clin. Microbiol. 51 (1) (2013) 46–54.

[103] F. Tolle, G. Mayer, Dressed for success—applying chemistry to modulate aptamer functionality, Chem. Sci. (Cambridge) 4 (1) (2013) 60–67.

[104] D. Jellinek, L.S. Green, C. Bell, C.K. Lynott, N. Gill, C. Vargeese, G. Kirschenheuter, D.P.C. McGee, P. Abesinghe, Potent 2'-amino-2'-deoxypyrimidine RNA inhibitors of basic fibroblast growth factor, Biochemistry 34 (36) (1995) 11363–11372.

[105] D. O'Connell, A. Koenig, S. Jennings, B. Hicke, H.L. Han, T. Fitzwater, Y.F. Chang, N. Varki, D. Parma, A. Varki, Calcium-dependent oligonucleotide antagonists specific for L-selectin, Proc. Natl. Acad. Sci. U.S.A. 93 (12) (1996) 5883–5887.

[106] N.C. Pagratis, C. Bell, Y.F. Chang, S. Jennings, T. Fitzwater, D. Jellinek, C. Dang, Potent 2'-amino-, and 2'-fluoro-2'-deoxyribonucleotide RNA inhibitors of keratinocyte growth factor, Nat. Biotechnol. 15 (1) (1997) 68–73.

[107] Y. Lin, D. Nieuwlandt, A. Magallanez, B. Feistner, S.D. Jayasena, High-affinity and specific recognition of human thyroid stimulating hormone (hTSH) by in vitro-selected 2'-amino-modified RNA, Nucleic Acids Res. 24 (17) (1996) 3407–3414.

[108] J. Ruckman, L.S. Green, J. Beeson, S. Waugh, W.L. Gillette, D.D. Henninger, L. Claesson-Welsh, N. Janjic, 2'-Fluoropyrimidine RNA-based aptamers to the 165-amino acid form of vascular endothelial growth factor (VEGF165): inhibition of receptor binding and VEGF-induced vascular permeability through interactions requiring the exon 7-encoded domain, J. Biol. Chem. 273 (32) (1998) 20556–20567.

[109] D.J. King, D.A. Ventura, A.R. Brasier, D.G. Gorenstein, Novel combinatorial selection of phosphorothioate oligonucleotide aptamers, Biochemistry 37 (47) (1998) 16489–16493.

[110] H. Dougan, D.M. Lyster, C.V. Vo, A. Stafford, J.I. Weitz, J.B. Hobbs, Extending the lifetime of anticoagulant oligodeoxynucleotide aptamers in blood, Nucl. Med. Biol. 27 (3) (2000) 289–297.

[111] L. Beigelman, J. Matulic-Adamic, P. Haeberli, N. Usman, B. Dong, R.H. Silverman, S. Khamnei, P.F. Torrence, Synthesis and biological activities of a phosphorodithioate analog of 2', 5'-oligoadenylate, Nucleic Acids Res. 23 (19) (1995) 3989–3994.

[112] J.M. Healy, S.D. Lewis, M. Kurz, R.M. Boomer, K.M. Thompson, C. Wilson, T.G. McCauley, Pharmacokinetics and biodistribution of novel aptamer compositions, Pharm. Res. 21 (12) (2004) 2234–2246.

[113] S.D. Mendonsa, M.T. Bowser, In vitro selection of high-affinity DNA ligands for human IgE using capillary electrophoresis, Anal. Chem. 76 (18) (2004) 5387–5392.

[114] R. Stoltenburg, C. Reinemann, B. Strehlitz, FluMag-SELEX as an advantageous method for DNA aptamer selection, Anal. Bioanal. Chem. 383 (1) (2005) 83–91.

[115] M. Berezovski, M. Musheev, A. Drabovich, S.N. Krylov, Non-SELEX selection of aptamers, J. Am. Chem. Soc. 128 (5) (2006) 1410–1411.

[116] D. Smith, G.P. Kirschenheuter, J. Charlton, D.M. Guidot, J.E. Repine, In vitro selection of RNA-based irreversible inhibitors of human neutrophil elastase, Chem. Biol. 2 (11) (1995) 741–750.

[117] R.Y.L. Tsai, R.R. Reed, Identification of DNA recognition sequences and protein interaction domains of the multiple-Zn-finger protein Roaz, Mol. Cell. Biol. 18 (11) (1998) 6447–6456.

[118] J.C. Cox, P. Rudolph, A.D. Ellington, Automated RNA selection, Biotechnol. Prog. 14 (6) (1998) 845–850.

[119] A. Nitsche, A. Kurth, A. Dunkhorst, O. Panke, H. Sielaff, W. Junge, D. Muth, F. Scheller, W. Stocklein, C. Dahmen, G. Pauli, A. Kage, One-step selection of Vaccinia virus-binding DNA aptamers by MonoLEX, BMC Biotechnol. 7 (1) (2007) 48.
[120] L. Peng, B.J. Stephens, K. Bonin, R. Cubicciotti, M. Guthold, A combined atomic force/fluorescence microscopy technique to select aptamers in a single cycle from a small pool of random oligonucleotides, Microsc. Res. Tech. 70 (4) (2007) 372–381.
[121] C. Minseon, Y. Xiao, J. Nie, S. Ron, A.T. Csordas, S.S. Oh, J.A. Thomson, H.T. Soh, Quantitative selection of DNA aptamers through microfluidic selection and high-throughput sequencing, Proc. Natl. Acad. Sci. U.S.A. 107 (35) (2010) 15373–15378.
[122] C.J. Huang, H.I. Lin, S.C. Shiesh, G.B. Lee, Integrated microfluidic system for rapid screening of CRP aptamers utilizing systematic evolution of ligands by exponential enrichment (SELEX), Biosens. Bioelectron. 25 (7) (2010) 1761–1766.
[123] A. Jolma, T. Kivioja, J. Toivonen, L. Cheng, G. Wei, M. Enge, M. Taipale, J.M. Vaquerizas, J. Yan, M.J. Sillanpää, M. Bonke, K. Palin, S. Talukder, T.R. Hughes, N.M. Luscombe, E. Ukkonen, J. Taipale, Multiplexed massively parallel SELEX for characterization of human transcription factor binding specificities, Genome Res. 20 (6) (2010) 861–873.
[124] N.R. Markham, M. Zuker, UNAFold: software for nucleic acid folding and hybridization, Methods Mol. Biol. (Clifton, NJ) 453 (2008) 3–31.
[125] J.S. Reuter, D.H. Mathews, RNAstructure: software for RNA secondary structure prediction and analysis, BMC Bioinformatics 11 (2010) 129.
[126] J. Akitomi, S. Kato, Y. Yoshida, K. Horii, M. Furuichi, ValFold: program for the aptamer truncation process, Bioinformation 7 (1) (2011) 38–40.
[127] D.H. Mathews, J. Sabina, M. Zuker, D.H. Turner, Expanded sequence dependence of thermodynamic parameters improves prediction of RNA secondary structure, J. Mol. Biol. 288 (5) (1999) 911–940.
[128] J. SantaLucia, A unified view of polymer, dumbbell, and oligonucleotide DNA nearest-neighbor thermodynamics, Proc. Natl. Acad. Sci. U.S.A. 95 (4) (1998) 1460–1465.
[129] J. Wu, C. Wang, X. Li, Y. Song, W. Wang, C. Li, J. Hu, Z. Zhu, J. Li, W. Zhang, Z. Lu, C.J. Yang, Identification, characterization and application of a G-quadruplex structured DNA aptamer against cancer biomarker protein anterior gradient homolog 2, PLoS One 7 (9) (2012) e46393.
[130] Y. Nonaka, K. Sode, K. Ikebukuro, Screening and improvement of an anti-VEGF DNA aptamer, Molecules (Basel, Switzerland) 15 (1) (2010) 215–225.
[131] A. Haller, R.B. Altman, M.F. Soulière, S.C. Blanchard, R. Micura, Folding and ligand recognition of the TPP riboswitch aptamer at single-molecule resolution, Proc. Natl. Acad. Sci. U.S.A. 110 (11) (2013) 4188–4193.
[132] U. Forster, J.E. Weigand, P. Trojanowski, B. Suess, J. Wachtveitl, Conformational dynamics of the tetracycline-binding aptamer, Nucleic Acids Res. 40 (4) (2012) 1807–1817.
[133] J. Zhou, A.V. Ellis, H. Kobus, N.H. Voelcker, Aptamer sensor for cocaine using minor groove binder based energy transfer, Anal. Chim. Acta 719 (2012) 76–81.
[134] J.R. Williamson, Induced fit in RNA-protein recognition, Nat. Struct. Biol. 7 (10) (2000) 834.
[135] J.C. Manimala, S.L. Wiskur, A.D. Ellington, E.V. Anslyn, Tuning the specificity of a synthetic receptor using a selected nucleic acid receptor, J. Am. Chem. Soc. 126 (50) (2004) 16515–16519.
[136] W.M. Rockey, F.J. Hernandez, S.Y. Huang, S. Cao, C.A. Howell, Rational truncation of an RNA aptamer to prostate-specific membrane antigen using computational structural modeling, Nucleic Acid Ther. 21 (5) (2011) 299–314.
[137] N.O. Fischer, J.B.H. Tok, T.M. Tarasow, Massively parallel interrogation of aptamer sequence, structure and function, PLoS One 3 (7) (2008) e2720.
[138] N.M. Sayer, M. Cubin, A. Rhie, M. Bullock, A. Tahiri-Alaoui, W. James, Structural determinants of conformationally selective, prion-binding aptamers, J. Biol. Chem. 279 (13) (2004) 13102–13109.
[139] M. Legiewicz, M. Yarus, A more complex isoleucine aptamer with a cognate triplet, J. Biol. Chem. 280 (20) (2005) 19815–19822.
[140] J. Zhou, B. Soontornworajit, M.P. Snipes, Y. Wang, Structural prediction and binding analysis of hybridized aptamers, J. Mol. Recognit. 24 (1) (2011) 119–126.
[141] S.M. Nimjee, C.P. Rusconi, B.A. Sullenger, Aptamers: an emerging class of therapeutics, Annu. Rev. Med. 56 (2005) 555–583.
[142] A.J. Putnam, D.J. Mooney, Tissue engineering using synthetic extracellular matrices, Nat. Med. 2 (7) (1996) 824–826.
[143] B. Soontornworajit, J. Zhou, M.T. Shaw, T.H. Fan, Y. Wang, Hydrogel functionalization with DNA aptamers for sustained PDGF-BB release, Chem. Commun. 46 (11) (2010) 1857–1859.
[144] B. Soontornworajit, J. Zhou, Y. Wang, A hybrid particle-hydrogel composite for oligonucleotide-mediated pulsatile protein release, Soft Matter 6 (17) (2010) 4255–4261.
[145] M.R. Battig, B. Soontornworajit, Y. Wang, Programmable release of multiple protein drugs from aptamer-functionalized hydrogels via nucleic acid hybridization, J. Am. Chem. Soc. 134 (30) (2012) 12410–12413.
[146] B. Soontornworajit, J. Zhou, M.P. Snipes, M.R. Battig, Y. Wang, Affinity hydrogels for controlled protein release using nucleic acid aptamers and complementary oligonucleotides, Biomaterials 32 (28) (2011) 6839–6849.
[147] K. Guo, H.P. Wendel, L. Scheideler, G. Ziemer, A.M. Scheule, Aptamer-based capture molecules as a novel coating strategy to promote cell adhesion, J. Cell. Mol. Med. 9 (3) (2005) 731–736.
[148] Z. Zhang, N. Chen, S. Li, M.R. Battig, Y. Wang, Programmable hydrogels for controlled cell catch and release using hybridized aptamers and complementary sequences, J. Am. Chem. Soc. 134 (38) (2012) 15716–15719.
[149] O.C. Farokhzad, S. Jon, A. Khademhosseini, T.N. Tran, D.A. LaVan, R. Langer, Nanoparticle-aptamer bioconjugates: a new approach for targeting prostate cancer cells, Cancer Res. 64 (21) (2004) 7668–7672.
[150] J. Cheng, B.A. Teply, I. Sherifi, J. Sung, G. Luther, F.X. Gu, E. Levy-Nissenbaum, A.F. Radovic-Moreno, R. Langer, O.C. Farokhzad, Formulation of functionalized PLGA–PEG nanoparticles for in vivo targeted drug delivery, Biomaterials 28 (5) (2007) 869–876.
[151] S. Dhar, F.X. Gu, R. Langer, O.C. Farokhzad, S.J. Lippard, Targeted delivery of cisplatin to prostate cancer cells by aptamer functionalized Pt(IV) prodrug-PLGA-PEG nanoparticles, Proc. Natl. Acad. Sci. U.S.A. 105 (45) (2008) 17356–17361.
[152] N. Kolishetti, S. Dhar, P.M. Valencia, L.Q. Lin, R. Karnik, Engineering of self-assembled nanoparticle platform for precisely controlled combination drug therapy, Proc. Natl. Acad. Sci. U.S.A. 107 (42) (2010) 17939–17944.
[153] S. Dhar, N. Kolishetti, S.J. Lippard, O.C. Farokhzad, Targeted delivery of a cisplatin prodrug for safer and more effective prostate

cancer therapy in vivo, Proc. Natl. Acad. Sci. U.S.A. 108 (5) (2011) 1850–1855.

[154] J. Yang, S.X. Xie, Y. Huang, M. Ling, J. Liu, Prostate-targeted biodegradable nanoparticles loaded with androgen receptor silencing constructs eradicate xenograft tumors in mice, Nanomedicine (London, England) 7 (9) (2012) 1297–1309.

[155] J. Guo, X. Gao, L. Su, H. Xia, G. Gu, Z. Pang, X. Jiang, L. Yao, J. Chen, H. Chen, Aptamer-functionalized PEG–PLGA nanoparticles for enhanced anti-glioma drug delivery, Biomaterials 32 (31) (2011) 8010–8020.

[156] C. Yu, Y. Hu, J. Duan, W. Yuan, C. Wang, H. Xu, X.D. Yang, Novel aptamer-nanoparticle bioconjugates enhances delivery of anticancer drug to MUC1-positive cancer cells in vitro, PLoS One 6 (9) (2011) e24077.

[157] X. Hu, K.L. Tulsieram, Q. Zhou, L. Mu, J. Wen, Polymeric nanoparticle–aptamer bioconjugates can diminish the toxicity of mercury in vivo, Toxicol. Lett. 208 (1) (2012) 69–74.

[158] R.M. Kong, X.B. Zhang, Z. Chen, W. Tan, Aptamer-assembled nanomaterials for biosensing and biomedical applications, Small (Weinheim an der Bergstrasse, Germany) 7 (17) (2011).

[159] V. Bagalkot, L. Zhang, E. Levy-Nissenbaum, S. Jon, P.W. Kantoff, R. Langer, O.C. Farokhzad, Quantum dot–aptamer conjugates for synchronous cancer imaging, therapy, and sensing of drug delivery based on bi-fluorescence resonance energy transfer, Nano Lett. 7 (10) (2007) 3065–3070.

[160] S. Lian, P. Zhang, P. Gong, D. Hu, B. Shi, C. Zeng, L. Cai, A universal quantum dots-aptamer probe for efficient cancer detection and targeted imaging, J. Nanosci. Nanotechnol. 12 (10) (2012) 7703–7708, Ref Type: Abstract.

[161] F.B. Wang, Y. Rong, M. Fang, J.P. Yuan, C.W. Peng, S.P. Liu, Y. Li, Recognition and capture of metastatic hepatocellular carcinoma cells using aptamer-conjugated quantum dots and magnetic particles, Biomaterials 34 (15) (2013) 3816–3827.

[162] D. Hu, P. Zhang, P. Gong, S. Lian, Y. Lu, A fast synthesis of near-infrared emitting CdTe/CdSe quantum dots with small hydrodynamic diameter for in vivo imaging probes, Nanoscale 3 (11) (2011) 4724–4732.

[163] R. Savla, O. Taratula, O. Garbuzenko, T. Minko, Tumor targeted quantum dot-mucin 1 aptamer-doxorubicin conjugate for imaging and treatment of cancer, J. Control. Release 153 (1) (2011) 16–22.

[164] Y. Kim, Z. Cao, W. Tan, Molecular assembly for high-performance bivalent nucleic acid inhibitor, Proc. Natl. Acad. Sci. U.S.A. 105 (15) (2008) 5664–5669.

[165] Y. Wu, J.A. Phillips, H. Liu, R. Yang, W. Tan, Carbon nanotubes protect DNA strands during cellular delivery, ACS Nano 2 (10) (2008) 2023–2028.

[166] J. Zhou, B. Soontornworajit, J. Martin, B.A. Sullenger, E. Gilboa, Y. Wang, A hybrid DNA aptamer–dendrimer nanomaterial for targeted cell labeling, Macromol. Biosci. 9 (9) (2009) 831–835.

[167] J. Zhou, B. Soontornworajit, Y. Wang, A temperature-responsive antibody-like nanostructure, Biomacromolecules 11 (8) (2010) 2087–2093.

[168] I.H. Lee, S. An, M.K. Yu, H.K. Kwon, S.H. Im, S. Jon, Targeted chemoimmunotherapy using drug-loaded aptamer–dendrimer bioconjugates, J. Control. Release 155 (3) (2011) 435–441.

[169] X. Wu, B. Ding, J. Gao, H. Wang, W. Fan, Second-generation aptamer-conjugated PSMA-targeted delivery system for prostate cancer therapy, Int. J. Nanomed. 6 (2011) 1747–1756.

[170] G. Gregoriadis, Engineering liposomes for drug delivery: progress and problems, Trends Biotechnol. 13 (12) (1995) 527–537.

[171] L. Lesoin, O. Boutin, C. Crampon, E. Badens, CO_2/water/surfactant ternary systems and liposome formation using supercritical CO_2: a review, Colloids Surf. A 377 (1–3) (2011) 1–14.

[172] K. Morigaki, P. Walde, Fatty acid vesicles, Curr. Opin. Colloid Interface Sci. 12 (2) (2007) 75–80.

[173] C. Wang, Z. Wang, X. Zhang, Amphiphilic building blocks for self-assembly: from amphiphiles to supra-amphiphiles, Acc. Chem. Res. 45 (4) (2012) 608–618.

[174] M.C. Jones, J.C. Leroux, Polymeric micelles—a new generation of colloidal drug carriers, Eur. J. Pharm. Biopharm. 48 (2) (1999) 101–111.

[175] G. Riess, Micellization of block copolymers, Prog. Polym. Sci. 28 (7) (2003) 1107–1170.

[176] H. Kang, M.B. O'Donoghue, H. Liu, W. Tan, A liposome-based nanostructure for aptamer directed delivery, Chem. Commun. 46 (2) (2010) 249–251.

[177] Y.K. Jung, T.W. Kim, H.G. Park, H.T. Soh, Specific colorimetric detection of proteins using bidentate aptamer-conjugated polydiacetylene (PDA) liposomes, Adv. Funct. Mater. 20 (18) (2010) 3092–3097.

[178] A.P. Mann, R.C. Bhavane, A. Somasunderam, B.L. Montalvo-Ortiz, K.B. Ghaghada, D. Volk, R. Nieves-Alicea, K.S. Suh, M. Ferrari, A. Annapragada, D.G. Gorenstein, T. Tanaka, Thioaptamer conjugated liposomes for tumor vasculature targeting, Oncotarget 2 (4) (2011) 298–304.

[179] Y. Wu, K. Sefah, H. Liu, R. Wang, W. Tan, DNA aptamer–micelle as an efficient detection/delivery vehicle toward cancer cells, Proc. Natl. Acad. Sci. U.S.A. 107 (1) (2010) 5–10.

[180] C.H. Wang, S.T. Kang, Y.H. Lee, Y.L. Luo, Y.F. Huang, C.K. Yeh, Aptamer-conjugated and drug-loaded acoustic droplets for ultrasound theranosis, Biomaterials 33 (6) (2012) 1939–1947.

[181] M.C. Willis, B. Collins, T. Zhang, L.S. Green, D.P. Sebesta, C. Bell, E. Kellogg, S.C. Gill, A. Magallanez, S. Knauer, R.A. Bendele, P.S. Gill, N. Janjic, Liposome-anchored vascular endothelial growth factor aptamers, Bioconjug. Chem. 9 (5) (1998) 573–582.

[182] Z. Cao, R. Tong, A. Mishra, W. Xu, G. Wong, J. Cheng, Y. Lu, Reversible cell-specific drug delivery with aptamer-functionalized liposomes, Angew. Chem. Int. Ed. Engl. 48 (35) (2009) 6494–6498.

Chapter 18

Biomedical Applications of Nondegradable Polymers

Anuradha Subramaniam, Swaminathan Sethuraman

School of Chemical & Biotechnology, Centre for Nanotechnology & Advanced Biomaterials (CeNTAB), SASTRA University, Thanjavur, India

Chapter Outline

18.1 INTRODUCTION

Synthetic polymers have been used extensively to restore the function of organ and found to have a history in medicine. Synthetic polymers are broadly classified into degradable and nondegradable based on the chemical reactivity [1]. Although the biodegradable polymers have gained much attention in today's medicine, exploration of biomaterials in treating various health problems originated with the use of nondegradable polymers. This extended association with the biomedical field progresses the performance of the nondegradable polymers, thereby improving the quality of patient's lives through dental fillings, contact lens, vascular prosthesis, orthopedic implants, and artificial hips [2]. Nondegradable polymers can also be called as biostable or biointegrable polymers as these offer enduring support over time along with the ultimate performance during the patient's lifetime [3]. This can also surmount the problems of asynchronous degradation with new tissue regeneration and harmful end products of degraded polymers [4].

Among the nonresorbable materials, polymers have received much attention as compared to metals or ceramics due to the ease of processing into various shapes, with the existence of materials with broad range of physical, mechanical, chemical, and thermal properties [5]. Further, all these material properties can be tailored by changing the structure, crystallinity, molecular architecture, and additives. Poly(methyl methacrylate) (PMMA) was first used as denture base material in dentistry in 1937 [6,7]. PMMA has been used to repair skull defects since 1940s, and the foremost successful intraocular lens was manufactured in 1949 [8–10]. In the late 1950s, PMMA has been identified as bone cement in fixation of femur and acetabulum implant, and in 1958, poly(tetrafluoroethylene) (PTFE)-based bearing materials were developed. PTFE was replaced by ultrahigh-molecular-weight polyethylene (UHMWPE) due to its limitations [11]. In the 1940s silicone elastomers has been used to repair bile duct and in the 1950s, Dacron sutures were used in surgery where high strength and long-term performance were required [12]. In the 1960s silicone gel was developed for breast implant, and woven and knitted PET was clinically identified as synthetic vascular prosthesis [13,14]. PET has been used as a sewing cuff around the heart valve to encourage tissue ingrowth, and polyurethane was also used for various biomedical applications [15]. Thus, nondegradable polymers have been clinically used for a long time, and this chapter portrays the history of nondegradable polymers in medicine with detailed performance of a wide range of polymers as biomaterials based on the properties. Moreover, detailed knowledge of the properties of different biostable polymers, their response *in vivo*, and the future trends of nondegradable polymers to address current problems for the development of next-generation biomaterials are also discussed.

18.2 NONDEGRADABLE POLYMERS AS BIOMATERIALS

Nondegradable polymers have been extensively used in medicine as drug delivery systems, fillings, orthopedic implants, ocular lens, heart valves, bone cements, vascular grafts, and tissue engineering scaffolds. Drug release from nondegradable carriers has been controlled through diffusion alone. Advantages such as ease of processing, tissue/blood compatibility, biological stability, robustness, and excellent mechanical properties during *in vivo* application can promote its potential use as biomaterials over metals and ceramics. Table 18.1 shows the list of several commercialized nondegradable products with its applications. Figure 18.1 shows the chemical structure of the nondegradable polymers that are discussed in the subsequent sections of this chapter.

18.2.1 Poly(Ethylene)

Based on the molecular weight, versatile biomedical polymer, polyethylene, is classified as low-density polyethylene, high-density polyethylene (HDPE), and UHMWPE (Figure 18.1a). Among these, HDPE and UHMWPE have gained much attention for the use of biomaterials due to chemical inertness, mechanical strength, limited tissue reaction, and biostability [16]. HDPE has a long history as bone and cartilage substitutes and has been used clinically since its mechanical properties mimic the natural bones [17]. MedPor® is a sintered porous scaffold developed using HDPE microbeads, which tend to allow tissue ingrowth, thereby treating craniofacial defects, external ear reconstruction, and chin, malar, and nasal augmentations as implants [18]. HAPEX™ (hydroxyapatite/HDPE composite), developed by Bonfield *et al.* in the early 1980s, mimics the bone composition as inorganic fillers with organic matrix [19]. He also had an attempt to improve the mechanical stiffness by chemical coupling like silane and acrylic acid grafting, named as HAPEX™, and found to replace

TABLE 18.1 Nondegradable Polymers That Have Been Used for Various Clinical Applications

Polymer	Commercial Name	Company	Applications
Expanded poly(tetrafluoro ethylene) (ePTFE)	GORE-TEX®	W.L. Gore & Associates, Flagstaff, USA	Regenerative membrane, osteoconductive membrane, large diameter aortic, and carotid vascular grafts, tension-free repair of ventral incisional hernia; orbital reconstruction, facial reconstruction, rhinoplasty
Poly(ethylene terephthalate) (PET)	DACRON®	Dupont	Surgical suture membrane, vascular grafts, anterior cruciate ligament prosthesis
Poly(tetrafluoro ethylene) (PTFE)	TEFLON®	Dupont	Replace urinary tissues, vascular graft, guided tissue regenerations, orbital reconstruction, facial reconstruction, rhinoplasty
Poly(methyl methacrylate) (PMMA)	PALACOS®, OSTEOPAL®	Merck	Intraocular lens, bone cements
PDMS	SLILASTIC®	Dow chemicals	Drug delivery implants, gas exchange membranes, intraocular lens
Ethylene-*co*-inylacetate (EVA)	ELVAX®	DuPont	Drug delivery implants
Poly(ether-urethanes) (PU)	BIOSPAN®	Polymer Technology Group, Inc.	Heart valves, vascular grafts, and other blood-contacting devices
Silastic® membrane for delivery of Levonorestrel	NORPLANT®	Wyeth-Ayerst	Contraceptive agent
Poly(propylene)	ACTISITE®	Procter & Gamble	Periodontal drug delivery system
Ethylene vinyl acetate	OCUSERT®	Alza	Treat glaucoma

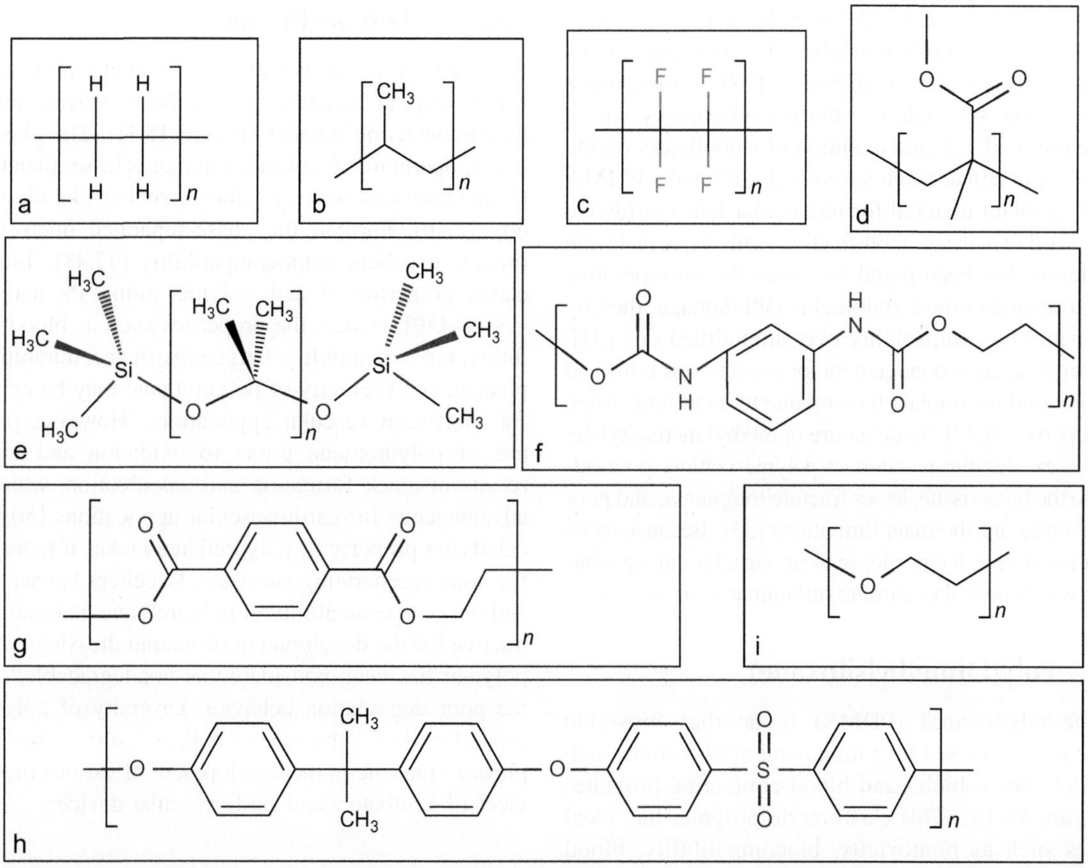

FIGURE 18.1 Chemical structures of (a) poly(ethylene); (b) poly(propylene); (c) poly(tetrafluoroethylene); (d) poly(methyl methacrylate); (e) poly(dimethylsiloxane); (f) polyurethanes; (g) poly(ethylene terphthlate); (h) poly(sulfone); (i) poly(ethyleneoxide). (For color version of this figure, the reader is referred to the online version of this chapter.)

minor load-bearing bones like the cheek bone and the middle eardrum [20]. Nath *et al.* have improved the mechanical properties like stiffness and hardness by the combination of alumina particles with hydroxylapatite in HDPE matrix for load-bearing hard tissue applications such as the bone, teeth, and joints [20]. Similarly cross-linked UHMWPE or incorporation of additives like vitamin E to UHMWPE matrix has been used for total hip and knee replacements due to its improved wear resistance, thereby protecting the implant from wear-mediated osteolysis over conventional UHMWPE [11,21].

18.2.2 Poly(Propylene)

Poly(propylene) is one of the biocompatible, biostable polymer used widely in clinical applications ranging from sutures to load-bearing implants (Figure 18.1b). Poly(propylene) has fiber-forming characteristics and has been used in the treatment of ventral incisional hernia [22,23]. Poly(propylene) has excellent stiffness and strength when compared to polyethylene [24]. Superior mechanical performance in fatigue and temperature resistance offers sufficient mechanical property even at body temperature and able to bear millions of loading-unloading cycle in orthopeadic implants [24]. This can also be used in the root canal treatment by delivering the antimicrobial agent, tetracycline [25].

18.2.3 Poly(Tetrafluoroethylene)

PTFE is a fluorocarbon-based polymer also called as Teflon (Figure 18.1c). Chemical and biological inertness, nondegradability, hydrophobicity, high crystallinity, hemocompatibility, biocompatibility, low elastic modulus, and tensile strength are the ideal properties that can make the material to more suitable for artificial vascular graft, catheter, and sutures [23]. Expanded PTFE (ePTFE) allows tissue to grow inside, and the surface of ePTFE grafts has been modified to enhance the endothelial adhesion, thereby achieving strong thromboresistance property to protect from occlusion in smaller-diameter arteries [26].

18.2.4 Poly(Methyl Methacrylate)

Though it is a hard brittle material, PMMA have an extensive record in medical applications as dental fillings, intraocular lens, and bone cements since the 1940s (Figure 18.1c). This also has an extended history in dentistry due to its ease of handling, processing, and polishing characteristics, and known biocompatibility in oral environment [27]. Bioinertness, nondegradability, UV light resistance, transparency with a refractive index of 1.5, and retention of smoothness on the surface when in contact with vascular tissues make PMMA a standard implant material for intraocular lens worldwide [28–30]. Further surface modification with heparin in the PMMA lenses has been found to reduce the postoperative inflammation and corneal endothelial cell damage, thereby offering better biocompatibility over unmodified one [31]. PMMA can be used as bone cement for osteoporotic bone and also can be used for implant fixation due to its instant strong fixing property [28,32]. Toxic nature of methyl methacrylate, brittleness, exothermic reaction of polymerization, poor adherence to the bone tissue, lesser fracture toughness, and poor wear resistance are the main limitations [33]. Because of its low wear resistance, it can release wear particles during compression, which provokes chronic inflammatory response.

18.2.5 Poly(Dimethylsiloxane)

Poly(dimethylsiloxane) (PDMS) is another biostable synthetic polymer used for biomedical applications such as drug delivery vehicles and blood-contacting biomaterials (Figure 18.1e). This elastomeric polymer has ideal properties such as nontoxicity, biocompatibility, blood compatibility, elasticity, transparency, and durability [34–36]. Additionally, flexibility of polymer backbone in PDMS exposes its methyl group at numerous interfaces, which are low interacting substitute, thereby creating minimum level of interaction at their surface [37]. This bioinertness inhibits the microbial growth, thereby making it more attractive for biomedical applications [38]. High gas permeability of PDMS allows the diffusion of gases like oxygen and carbon dioxide and makes it more appropriate in medical applications like wound dressing and contact lenses [39]. However, the hydrophobic nature of PDMS does not encourage cell adhesion, which is a very critical requirement for wound healing process and angiogenesis [40–42]. Different approaches such as plasma treatment and laser treatment have been used to modify the silicone surface to improve its wettability and cell-adhesive property. However, these techniques do not form a stable cell-adhesive property due to surface reorganization [43]. Hence, cell-adhesive proteins like fibronectin, laminin, and collagen have been coated onto the PDMS surface for long-lived cell-adhesive property [44]. Applying such proteins provoke immune responses, disease transmission, and poor stability due to proteolytic action, which restricts its applications in medical applications. Modified PDMS surface has been developed and used as stable cell-adhesive surface using peptide conjugation [45].

18.2.6 Polyurethanes

Polyurethane is another class of synthetic polymers, used extensively in vascular grafts, heart valves, and other blood-contacting devices (Figure 18.1f). The glass transition temperature (T_g) of this polymer is lesser than the room temperature and hence exhibits a rubber-like elastic property [2,46]. Further, the phase-separated microstructures provide excellent hemocompatibility [47,48]. Tensile and elastic properties of such polymer mimic the native blood vessels [49]. Hence, the properties such as blood compatibility, biocompatibility, fatigue resistance, durability, compliance, and elasticity of polyurethane may be appropriate for short-term vascular applications. However, prolonged use of polyurethane prone to oxidation and hydrolysis results in crack formation and calcification, which is not advantageous for cardiovascular applications [50,51]. The calcifying property of polyurethane makes it more suitable for bone regenerating substrate. Excellent barrier property and oxygen permeability of polyurethane makes it more attractive for the development of wound dressings [52]. This polymer has been considered as nondegradable, owing to the poor degradation behavior. Diversity of polyurethane properties due to the ease of bulk and surface modification plays a vital role in the development of various medical devices like catheters and cardiovascular devices.

18.2.7 Poly(Ethylene Terphthlate)

Poly(ethylene terephthalate) is thermoplastic polyester, commercialized as Dacron®. Dacron has many medical applications like sutures and vascular grafts (Figure 18.1g). The properties such as hardness; stiffness; bio-, chemical, and dimensional stability; and biocompatibility make the material more promising for biomedical applications. Biostability of Dacron is mainly due to the presence of hydrophobic aromatic groups with high crystallinity which restricts hydrolytic breakdown [53]. However, absence of bioactiveness limits its potential use in tissue engineering [54]. Surface functionalization has been used to improve polymer-cell interaction. Another problem in the use of knitted Dacron is that blood has to be clotted before the implantation of knitted Dacron and this has been overcome by the impregnating albumin, collagen and gelatin to make it more impermeable [55].

18.2.8 Poly(Sulfone)

Poly(sulfone) has the characteristic properties such as chemical inertness, excellent strength, stiffness, high resistance to radiations and temperature, dimensional stability, and biocompatibility [2,16,56]. These properties resemble the properties of light metals [57]. Further, the properties of poly(sulfone) do not change during sterilization (Figure 18.1h). However, hydrophobic nature of poly(sulfone) limits its potential use in tissue engineering applications and blood-contacting applications [58]. Membrane made of poly(sulfone) has been used widely in hemodialysis applications by allowing the diffusion of toxins from blood without compromising the blood proteins [59,60]. It has also been used in the design of bioartificial liver (BAL) [61]. Poly(sulfone) capillary fiber has been proved to be a promising device for the extended drug release in intraocular applications [62].

18.2.9 Poly(Ethyleneoxide)

Another synthetic, water soluble polymer, poly(ethyleneoxide) (PEO), is approved by FDA for various medical applications (Figure 18.1i). PEO has hydrogel-forming characteristics, which can be suitable for developing drug delivery vehicles and regenerating scaffolds [63]. This polymer also restricts the protein adhesion, thereby reducing the complement activation system, immunogenicity, and antigenicity in the body [64,65]. Hence, this can be an excellent substrate for cell-based tissue engineering applications. Nontoxicity and solubility in both water and organic solvents make this polymer more attractive for biomedical applications. PEO has also been used in the development of amphiphilic copolymers and plays a vital role in micellization process (Pluronics, PEO-PPO-PEO). Further addition of PEO on to the biomaterial surface has received more attention to provide surface compatibility by avoiding adhesion of plasma proteins, bacteria, and also platelets through steric repulsion.

18.3 CHARACTERIZATION

18.3.1 Surface Characterization

Biomaterial surface properties such as morphology, roughness, charge density, chemistry, and wettability play a vital role on the success of biomaterial devices or implants [66–68]. For example, nondegradable polymers are mostly hydrophobic, which has been extensively used in cardiovascular applications. Though the materials were found to be biocompatible, and mechanically stable, blood compatibility remains a great challenge in the development of cardiovascular device because hydrophobic surface enhances the fibrinogen (rod-shaped blood plasma protein) adhesion, thereby stimulating the platelet aggregation and thrombosis [69]. Further, this hydrophobic surface also limits the endothelial adhesion, which is a chief factor to eliminate platelet aggregation. Hence, most of these synthetic biomaterials have been subjected to surface modification in order to stimulate biological performance (endothelialization). Similarly Bison *et al.* grafted acrylic acid onto the PET film in order to immobilize types I and III collagen, thereby favoring the long-term cultivation of human bladder smooth muscle cell [70]. The polyacrylic acid-grafted PET surface immobilized with galactose supports the hepatocyte interaction and retains its phenotype [71]. Various characterization techniques are used to study biomaterials in order to tailor the surface for better biological responses. Surface characterization is mainly based on three major classifications: microscopic methods, spectroscopic methods, and thermodynamic methods. Microscopic methods have been mainly used to characterize the morphology, cell-scaffold interaction, topography like roughness, various features (ridges, grooves, and pores), mapping of surface composition, and biomolecules (protein, lipid) interaction with the surface using scanning electron microscope, scanning tunneling microscope, atomic force microscope, transmission electron microscope, and confocal microscope [68].

Spectroscopic methods are mainly useful to identify the elements, functional groups, and chemical structures close to the biomaterial surface. Techniques like X-ray photoelectron spectroscopy, attenuated total reflection fourier transform infrared spectroscopy, and secondary ion mass spectrometry have been used widely in determining the elemental composition and charged species at biomaterial surface. Other important characteristics obtained from the biomaterial surface are surface energy and relative wettability using thermodynamic method such as contact angle experiments.

18.3.2 Biostability

Biostability of implants is greatly affected by adsorption and absorption of active compounds in physiological solutions [72]. Hence, characterization of *in vivo* biostability of polymeric biomaterials was considered an important parameter for long-term implantable medical devices. Choice of the biostable material is mainly dependent on the function of implant such that materials for artificial joints should withstand abrasive wear. Excellent elastic polymer polyurethane has been widely used for the development of vascular grafts as compared to ePTFE and Dacron [73]. However, poor biostability as well as compliance mismatch limits its clinical applications [74]. Tissue response against foreign body-like polyurethane facilitated the autoxidation through the release of hydrogen peroxide, thereby reducing the biostability [75]. Polycarbonate and polyurethane have been identified to possess excellent biostability against hydrolytic degradation, environmental stress cracking, and metal ion oxidation than polyether polyurethane [76]. Ward *et al.* had improved the biostasbility by replacing polyether with oxidation-resistant polysiloxane [75].

18.3.2.1 In Vivo *Tissue Response*

Tissue compatibility of implants is used to evaluate whether the medical device can provoke any significant harmful reactions to the patient. This test has been used to determine toxicities like acute, subacute, and chronic and genotoxicity, carcinogenicity, and also immune response. Foreign body tissue reaction for silicone-modified polyurethanes has validated that the silicone modification provokes apoptosis of adherent macrophages *in vivo* even at 4 days using subcutaneous cage implant model [77]. Similarly, PDMS endcapped polyether, polyurethane, was found to have lower foreign body giant cell density as compared to unmodified due to the presence of hydrophobic surface [78]. Sherman *et al.* have found that the polyisobutylene/PDMS networks elicited least tissue and cellular response as compared to standard polyethylene through intraperitoneal implantation [79]. *In vivo* response to polymeric systems such as poly(phosphazenes) and a blend of poly(lactide-*co*-glycolide)-poly(thiophene) have been evaluated using a subcutaneous rat model [80,81]. Implantation of bone cement based on modified PMMA with γ-methacryloxyp ropyltrimethoxysilane and calcium acetate in rabbit model has shown acute inflammation after a week, in turn promoting osteoconduction after 4 weeks through the formation of hydroxyapatite [82]. PET vascular grafts were tested for carcinogenicity through long-term subcutaneous implantation and proved to be noncarcinogenic for BDF1 mice and Syrian golden hamsters [79].

18.4 FUTURE PROSPECTS

Investigation of new synthetic biomaterials is still in the process due to various factors such as need for better *in vivo* efficacy, superior mechanical and biological properties, nontoxic material, requirement of cost-effective material, and threat of health hazards of existing biomaterials. Engineering new surfaces on the existing biomaterials will also receive much attention to establish excellent biological response both *in vitro* and *in vivo*. Research development in biomaterials will be expected to evaluate the long-term complications of biostable polymer and to identify a material with improved performance *in vivo*. Though materials like PET, PTFE, and PU have been used for the development of vascular graft, developing small-diameter vascular grafts still remains a challenge due to the absence of antithrombogenic property. Hence, development of synthetic vascular graft with antithrombotic property, and mechanical property similar to native blood vessels has to be addressed in future. Incorporation of biomolecular cues into the biointegrable polymers could dictate the cell fate and facilitate the evolution of next-generation synthetic polymers for biomedical applications.

REFERENCES

[1] V.P. Shastri, Non-degradable biocompatible polymers in medicine: past, present, and future, *Curr. Pharm. Biotechnol.* 4 (5) (2003) 331–337.

[2] S.H. Teoh, Z.G. Tang, G.W. Hastings, Thermoplastic polymers in biomedical applications: structures, properties and processing, in: J. Black, G. Hastings (Eds.), Handbook of Biomaterial Properties, Chapman & Hall, London, 1998, pp. 270.

[3] L. Fassina, L. Visai, L. Asti, F. Benazzo, P. Speziale, M.C. Tanzi, G. Magenes, Calcified matrix production by SAOS-2 cells inside a polyurethane porous scaffold, using a perfusion bioreactor, *Tissue Eng.* 11 (5–6) (2005) 685–700.

[4] S. Fare, P. Petrini, M.C. Tanzi, A. Bigi, N. Roveri, Biointegrable 3D polyurethane/α-TCP composites for bone reconstruction, in: D. Mantovani (Ed.), Advanced Materials for Biomedical Application, Canadian Institute of Mining, Metallurgy and Petroleum, Montreal, Cananda, 2002, pp. 17–26.

[5] L. Eschbach, Nonresorbable polymers in bone surgery, *Injury* 31 (4) (2000) S-D22–S-D27.

[6] F.A. Peyton, History of resins in dentistry, *Dent. Clin. North Am.* 19 (2) (1975) 211–222.

[7] X. Zhang, X. Zhang, B. Zhu, K. Lin, J. Chang, Mechanical and thermal properties of denture reinforced with silanized aluminum borate whiskers, *Dent. Mater. J.* 31 (6) (2012) 903–908.

[8] A. Sharma, Neuroprosthetic rehabilitation of acquired skull defects, *Int. J. Periodontics Restorative Dent.* 1 (1) (2011) 65–70.

[9] S.M. Kurtz, M.L. Villarraga, K. Zhao, A.A. Edidin, Static and fatigue mechanical behavior of bone cement with elevated barium sulfate content for treatment of vertebral compression fractures, *Biomaterials* 26 (2005) 3699–3712.

[10] D.J. Apple, Sir Harold Ridley and His Fight for Sight: He Changed the World So That We May Better See It, Slack Incorp., USA, 2006.

[11] M.K. Musib, A review of the history and role of UHMWPE as a component in total joint replacements, *Int. J. Biol. Eng.* 1 (1) (2011) 6–10.

[12] F.H. Lahey, Comments made following the speech "Results from using Vitallium tubes in biliary surgery", read by Pearse, H.E. before the American Surgical Association, Hot Springs, VA, *Ann. Surg.* 124 (1946) 1027.

[13] A. Colas, J. Curtis, Silicone biomaterials: history and chemistry & medical applications of silicones, in: B.D. Ratner, A.S. Hoffman, F.J. Schoen, J.E. Lemons (Eds.), Biomaterial Science: An Introduction to Materials in Medicine, second ed., Elsevier Academic Press, San Diego, CA, 2004.

[14] A. Metzger, Polyethylene terephthalate and the pillar palatal implant: its historical usage and durability in medical applications, *J. Am. Podiatry Assoc.* 65 (1975) 1–12.

[15] J.F. Mano, R.A. Sousa, L.F. Boesel, N.M. Neves, R.L. Reis, Bioinert, biodegradable and injectable polymeric matrix composites for hard tissue replacement: State of the art and recent developments, *Compos. Sci. Technol.* 64 (6) (2004) 789–817.

[16] L.R. Rubin, Polyethylene as a bone and cartilage substitute: a 32 year retrospective, in: L.R. Rubin (Ed.), Biomaterials in Reconstructive Surgery, C V Mosby, St. Louis, 1983, pp. 474–493.

[17] K. Sevin, I. Askar, A. Saray, E. Yormuk, Exposure of high density porous polyethylene (Medpor) used for contour restoration and treatment, *Br. J. Oral Maxillofac. Surg.* 38 (1) (2000) 44–49.

[18] W. Bonfield, From concept to patient-engineering solutions to medical problems, in: Engineers and Society: The 1997 CSE International Lecture, The Royal Academy of Engineering, London, 1997, p. 5.

[19] C.X. Dong, S.J. Zhu, M. Mizuno, M. Hashimoto, Fatigue behavior of HDPE composite reinforced with silane modified TiO2, *J. Mater. Sci. Technol.* 27 (1) (2011) 659–677.

[20] S. Nath, S. Bodhak, B. Basu, HDPE-Al2O3-Hap composites for biomedical applications: processing and characterizations, *J. Biomed. Mater. Res. B Appl. Biomater.* 88B (2009) 1–11.
[21] E.M.B. Prever, A. Bistolfi, P. Bracco, L. Costa, UHMWPE for arthoplasty: past or future? *J. Orthopaedic Traumatol.* 10 (1) (2009) 1–8.
[22] J.R. San Pio, T.E. Damsgaard, O. Momsen, I. Villadsen, J. Larsen, Repair of giant incisional hernias with polypropylene mesh: a retrospective study, *Scand. J. Plast. Reconstr. Surg. Hand Surg.* 37 (2) (2003) 102–106.
[23] S.R. Steele, P. Lee, M.J. Martin, P.S. Mullenix, E.S. Sullivan, Is parastomal hernia repair with polypropylene mesh safe? *Am. J. Surg.* 185 (5) (2003) 436–440.
[24] Y. Liu, M. Wang, Fabrication and characteristics of hydroxyapatite reinforced polypropylene as a bone analogue biomaterial, *J. Appl. Polym. Sci.* 106 (4) (2007) 2780–2790.
[25] J.M. Goodson, Life on a string: development of the tetracycline fiber delivery system, *Technol. Health Care* 4 (3) (1996) 269–282.
[26] A. Mel, B.G. Cousins, A.M. Seifalian, Surface modification of biomaterials: a quest for blood compatibility, *Int. J. Biomater.* 2012 (2012) 707863.
[27] T.G. Tihan, M.D. Ionita, R.G. Popescu, D. Iordachescu, Effect of hydrophilic-hydrophobic balance on biocompatibility of poly(methyl methacrylate) (PMMA)-hydroxyapatite (HA) composites, *Mater. Chem. Phys.* 118 (2–3) (2009) 265–269.
[28] I.M. Carvalho Costa, C.P. Salaro, M.C. Costa, Polymethylmethacrylate facial implant: a successful personal experience in Brazil for more than 9 years, *Dermatol. Surg.* 35 (8) (2009) 1221–1227.
[29] M. Stickler, T. Rhein, Polymethacrylates, in: B. Elvers, S. Hawkins, G. Schultz (Eds.), Ullmann's Encyclopedia of Industrial Chemistry, fifth ed., VHS, New York, 1992, p. 473, A21.
[30] C. Batich, P. Leamy, Biopolymers, in: M. Kutz (Ed.), Standard Handbook of Biomedical Engineering and Design, McGraw-Hill Professional, New York, 2003, p. 11.3.
[31] B. Dick, K. Greiner, G. Magdowski, N. Pfeiffer, Long term stability of heparin coated PMMA intraocular lenses. Results of an in vivo study, *Ophthalmologe* 94 (12) (1997) 920–924.
[32] V.L. Schade, T.S. Roukis, Role of polymethylmethacrylate antibiotic-loaded cement in addition to debridement for the treatment of soft tissue and osseous infections of the foot and ankle, *J. Foot Ankle Surg.* 49 (1) (2010) 55–62.
[33] Y.Y. Cheng, W.L. Cheung, T.W. Chow, Strain analysis of maxillary complete denture with three-dimensional finite element method, *J. Prosthet. Dent.* 103 (5) (2010) 309–318.
[34] L. Wong, C. Ho, Surface molecular property modifications for poly(dimethylsiloxane) (PDMS) based microfluidic devices, *Microfluid. Nanofluidics* 7 (3) (2009) 291–306.
[35] M.H. Wu, Simple poly(dimethylsiloxane) surface modification to control cell adhesion, *Surf. Interface Anal.* 41 (2009) 11–16.
[36] S. Pinto, P. Alves, C.M. Matos, A.C. Santos, L.R. Rodrigues, J.A. Teixeira, M.H. Gil, Poly(dimethyl siloxane) surface modification by low pressure plasma to improve its characteristics towards biomedical applications, *Colloids Surf. B Biointerfaces* 81 (1) (2010) 20–26.
[37] M.J. Owen, Why Silicones behave funny, *Chemtech* 11 (1981) 288.
[38] H. Chen, M.A. Brook, H. Sheardown, Silicon elastomers for reduced protein adsorption, *Biomaterials* 25 (12) (2004) 2273–2282.
[39] R.R. Bott, M.S. Gebert, P.C. Klykken, I. Mazeaud, X. Thomas, Preparation for topical use and treatment, Patent US20030180281, Sep 25, 2003.
[40] J.Y. Park, D. Ahn, Y.Y. Choi, C.M. Hwang, S. Takayama, S.H. Lee, S. Lee, Surface chemistry modifications of PDMS elastomers with boiling water improves cellular adhesion, *Sens. Actuators B* 173 (2012) 765–771.
[41] N.Q. Balaban, U.S. Schwarz, D. Riveline, P. Goichberg, G. Tzur, I. Sabanay, D. Mahalu, S. Safran, A. Bershadsky, L. Addadi, B. Geiger, Force and focal adhesion assembly: a close relationship studied using elastic micropatterned substrates, *Nat. Cell Biol.* 3 (5) (2001) 466–472.
[42] E. Zamir, M. Katz, Y. Posen, N. Erez, K.M. Yamada, B.Z. Katz, S. Lin, D.C. Lin, A. Bershadsky, Z. Kam, B. Geiger, Dynamics and segregation of cell-matrix adhesions in cultured fibroblasts, *Nat. Cell Biol.* 2 (4) (2000) 191–196.
[43] X.Q. Brown, K. Ookawa, J.Y. Wong, Evaluation of polydimethylsiloxane scaffolds with physiologicallyrelevant elastic moduli: interplay of substrate mechanics and surface chemistry effects on vascular smooth muscle cell response, *Biomaterials* 26 (16) (2005) 3123–3129.
[44] L. Wang, B. Sun, K.S. Ziemer, G.A. Barabino, R.L. Carrier, Chemical and physical modifications to poly(dimethylsiloxane) surfaces affect adhesion of Caco-2 cells, *J. Biomed. Mater. Res. A* 93 (4) (2010) 1260–1271.
[45] B. Li, J. Chen, H. Wang, RGD peptide-congugated poly(dimethylsiloxane) promotes adhesion, proliferation, and collagen secretion of human fibroblasts, *J. Biomed. Mater. Res.* 79A (2006) 989–998.
[46] R.J. Zdrahala, I.J. Zdrahala, Biomedical applications of polyurethanes: a review of pase promises, present realities and vibrant future, *J. Biomater. Appl.* 14 (1) (1999) 67–90.
[47] T.G. Grasel, S.L. Cooper, Surface properties and blood compatibility of polyurethaneureas, *Biomaterials* 7 (5) (1986) 315–328.
[48] S. Hsu, C. Chen, C. Chao, C. Chen, J. Chiu, Cell migration on nanophase-separated poly(carbonate urethane)s, *J. Med. Biol. Eng.* 27 (1) (2007) 15–21.
[49] A. Ratcliffe, Tissue engineering of vascular grafts, *Matrix Biol.* 19 (4) (2000) 353–357.
[50] I. Alferiev, S.J. Stachelek, Z. Lu, A.L. Fu, T.L. Sellaro, J.M. Connolly, R.W. Bianco, M.S. Sacks, R.J. Levy, Prevention of polyurethane valve cusp calcification with covalently attached bisphosphonate diethylamino moieties, *J. Biomed. Mater. Res. A* 66 (2) (2003) 385–395.
[51] I. Alferiev, N. Vyavahare, C. Song, J. Connolly, J.T. Hinson, Z. Lu, S. Tallapragada, R. Bianco, R. Levy, Bisphosphonate derivatized polyurethanes resist calcification, *Biomaterials* 22 (19) (2001) 2683–2693.
[52] M.S. Khil, D.I. Cha, H.Y. Kim, I.S. Kim, N. Bhattarai, Electrospun nanofibrous polyurethane membrane as wound dressing, *J. Biomed. Mater. Res. B Appl. Biomater.* 67 (2) (2003) 675–679.
[53] R.S. Greco (Ed.), Implantation Biology: The host response and Biomedical Devices, CRC press, Boca Raton, FL, 1994.
[54] Y. Liu, T. He, C. Gao, Surface modification of poly(ethylene terephthalate) via hydrolysis and layer-by-layer assembly of chitosan and chondroitin sulfate to construct cytocompatible layer for human endothelial cells, *Colloids Surf. B Biointerfaces* 46 (2) (2005) 117–126.
[55] J. Chlupac, E. Filova, L. Bacakova, Blood vessel replacement: 50 years of development and tissue engineering paradigms in vascular surgery, *Physiol. Res.* 58 (2) (2009) S119–S139.
[56] H. Toiserkani, G. Yilmaz, Y. Yagci, L. Torun, Functionalization of polysulfones by click chemistry, *Macromol. Chem. Phys.* 211 (22) (2010) 2389–2395.

[57] M. Wang, C.Y. Yue, B. Chua, Production and evaluation of hydroxyapatite reinforced polysulfone for tissue replacement, *J. Mater. Sci. Mater. Med.* 12 (9) (2001) 821–826.

[58] J.Y. Park, M.H. Acar, A. Akthakul, W. Kuhlman, A.M. Mayes, Polysulfone-graft-poly(ethylene glycol) graft copolymers for surface modification of polysulfone membranes, *Biomaterials* 27 (6) (2006) 856–865.

[59] L. De Bartolo, S. Salerno, E. Curcio, A. Piscioneri, M. Rende, S. Morelli, F. Tasselli, A. Bader, E. Drioli, Human hepatocyte functions in a crossed hollow fiber membrane bioreactor, *Biomaterials* 30 (13) (2009) 2531–2543.

[60] K. Heilmann, T. Keller, Polysulfone: the development of a membrane for convective therapies, *Contrib. Nephrol.* 175 (2011) 15–26.

[61] B. Carpentier, A. Gautier, C. Legallais, Artificial and bioartificial liver devices: present and future, *Gut* 58 (12) (2009) 1690–1702.

[62] M.H. Rahimy, G.A. Peyman, S.Y. Chin, R. Golshani, C. Aras, H. Borhani, H. Thompson, Polysulfone capillary fiber for intraocular drug delivery: in vitro and in vivo evaluations, *J. Drug Target.* 2 (4) (1994) 2455–2480.

[63] J. Zhu, R.E. Marchant, Design properties of hydrogel tissue-engineering scaffolds, *Expert Rev. Med. Devices* 8 (5) (2011) 607–626.

[64] J.M. Harris, Poly(ethylene glycol) chemistry: biotechnical and biomedical applications, Plenum Press, New York, 1992.

[65] J.M. Harris, S. Zalipsky, Poly(Ethylene Glycol): Chemistry and Biological Application, American Chemical Society, Washington, DC, 1997.

[66] Surface and interfacial aspects of biomedical polymers, in: J.D. Andrade (Ed.), Surface Chemistry and Physics, Plenum Press, New York, 1985.

[67] T. Cunningham, F.M. Serry, L.M. Ge, D. Gotthard, D.J. Dawson, Atomic force profilometry and long scan atomic force microscopy: new techniques for characterization of surfaces, *Surf. Eng.* 16 (4) (2000) 295–298.

[68] M. Merrett, R.M. Cornelius, W.G. Mccluung, L.D. Unsworth, H. Sheardown, Surface analysis methods for characterizing polymeric biomaterials, *J. Biomater. Sci. Polym. Ed.* 13 (6) (2002) 593–621.

[69] S. Sarkar, K.M. Sales, G. Hamilton, A.M. Seifalian, Addressing thrombogenicity in vascular graft construction, *J. Biomed. Mater. Res. B Appl. Biomater.* 82 (1) (2007) 100–108.

[70] I. Bison, M. Kosinski, S. Ruault, B. Gupta, J. Hilborn, F. Wurm, P. Frey, Acrylic acid grafting and collagen immobilization on poly(ethylene terephthalate) surfaces for adherence and growth of human bladder smooth muscle cells, *Biomaterials* 23 (15) (2002) 3149–3158.

[71] L. Ying, C. Yin, R.X. Zhuo, K.W. Leong, H.Q. Mao, E.T. Kang, K.G. Neoh, Immobilization of galactose ligands on acrylic acid graft-copolymerized poly(ethylene terephthalate) film and its application to hepatocyte culture, *Biomacromolecules* 4 (1) (2003) 157–165.

[72] S.D. Bruck, Biostability of materials and implants, *J. Long Term Eff. Med. Implants* 1 (1) (1991) 89–106.

[73] G.J. Wilson, D.C. MacGregor, P. Klement, J.P. Dereume, B.A. Weber, A.G. Binnington, L. Pinchuk, The composite Corethane/Dacron vascular prosthesis. Canine in vivo evaluation of 4 mm diameter grafts with 1 year follow-up, *ASAIO Trans.* 37 (3) (1991) M475–M476.

[74] M.R. Kapadia, D.A. Popowich, M.R. Kibbe, Modified prosthetic vascular conduits, *Circulation* 117 (2008) 1873–1882.

[75] B. Ward, J. Anderson, M. Ebert, R. McVenes, K. Stokes, In vivo biostability of polysiloxane polyether polyurethanes: resistance to metal ion oxidation, *J. Biomed. Mater. Res. A* 77 (2006) 380–389.

[76] K. Stokes, R. McVenes, J.M. Anderson, Polyurethane elastomer biostability, *J. Biomater. Appl.* 9 (4) (1995) 321–354.

[77] E.M. Christenson, M. Dadsetan, A. Hiltner, Biostability and macrophage-mediated foreign body reaction of silicone-modified polyurethanes, *J. Biomed. Mater. Res. A* 74 (2) (2005) 141–155.

[78] A.B. Mathur, T.O. Collier, W.J. Kao, M. Wiggins, M.A. Schubert, A. Hiltner, J.M. Anderson, In vivo biocompatibility and biostability of modified polyurethanes, *J. Biomed. Mater. Res.* 36 (2) (1997) 246–257.

[79] P. Blagoeva, I. Stoichev, R. Balanski, L. Purvanova, T.S. Mircheva, A. Smilov, The testing for carcinogenicity of a polyethylene terephthalate vascular prosthesis, *Khirurgila (Soflia)* 43 (6) (1990) 98–105.

[80] S. Sethuraman, L.S. Nair, S. El-Amin, R. Farrar, M.T. Nguyen, A. Singh, H.R. Allocock, Y.E. Greish, P.W. Brown, C.T. Laurencin, In vivo biodegradability and biocompatibility evaluation of novel alanine ester based polyphosphazenes in a rat model, *J. Biomed. Mater. Res. A* 77 (4) (2006) 679–687.

[81] A. Subramanian, U.M. Krishnan, S. Sethuraman, In vivo biocompatibility of PLGA-polyhexylthiophene nanofiber scaffolds in a rat model, *Biomed Res. Int.* 2013 (2013) 390518.

[82] A. Sugino, C. Ohtsuki, T. Miyazaki, In vivo response of bioactive PMMA-based bone cement modified with alkoxysilane and calcium acetate, *J. Biomater. Appl.* 23 (3) (2008) 213–228.

Chapter 19

Polymeric Biomaterials for Implantable Prostheses

Tram T. Dang*,†,1, Mehdi Nikhkhah*,‡,1, Adnan Memic*,‡,§, Ali Khademhosseini*,‡,¶

*Center for Biomedical Engineering, Department of Medicine, Brigham and Women's Hospital, Harvard Medical School, Boston, Massachusetts, USA
†David H. Koch Institute for Integrative Cancer Research, Massachusetts Institute of Technology, Cambridge, Massachusetts, USA
‡Harvard-MIT Division of Health Sciences and Technology, Massachusetts Institute of Technology, Cambridge, Massachusetts, USA
§Center of Nanotechnology, King Abdulaziz University, Jeddah, Saudi Arabia
¶Wyss Institute for Biologically Inspired Engineering, Harvard University, Boston, Massachusetts, USA

Chapter Outline

ABBREVIATIONS

CF/PEEK carbon fiber-reinforced polyetheretherketone
DES drug-eluting stent
ECM extracellular matrix
ePTFE expanded PTFE
FDA Food and Drug Administration
OOKP osteo-odonto-keratoprosthesis
PBMA polybutyl methacrylate
PCL polycaprolactone
PCU polycarbonate urethane
PDLLA poly-DL-lactide acid
PDMS Polydimethylsiloxane
PEEK polyetheretherketone
PEG polyethylene glycol
PET poly(ethylene terephthalate)
PEU polyether urethane
PEVA polyethylene-*co*-vinyl acetate
pHEMA poly(2-hydroxyethyl methacrylate)
PLLA poly-L-lactide acid
PMMA polymethyl methacrylate
PNIPAAm poly(*N*-isopropylacrylamide)
POSS polyhedral oligomeric silsesquioxane
PTFE polytetrafluoroethylene
PU polyurethane
PVA polyvinyl alcohol
QAS quaternary ammonium salts
SIBS poly(styrene-*b*-isobutylene-*b*-styrene)
SIPN semi-interpenetrating polymer network
SMP shape-memory polymer
SR silicone rubber
UHMWPE ultra-high-molecular-weight polyethylene

[1]These authors contributed equally to this work.

19.1 INTRODUCTION

Implantable prostheses have a great potential to enhance human health by improving the quality of life and extending longevity. As reported in 2000, the worldwide market for medical devices including implantable prostheses was over $300 billion, comprising approximately 8% of the worldwide healthcare expenditure and catering to over 20 million patients [1]. "Prosthesis," derived from the Greek words "pro-" for "instead of" and "thesis" for "placing," is defined as the replacement of all or parts of a damaged or diseased anatomical organ by an artificial device [2]. Implantable prostheses are devices that require surgical procedures for their placement inside the human body when they are deemed medically more beneficial than repairing the ailing native tissues [2]. For example, when patients, especially the elderly population, suffer from degenerative osteoarthritis, total hip replacement can help to alleviate pain and improve mobility more effectively than joint repairing or excision [3]. Consequently, artificial hip joint is one of the earliest prostheses to be clinically accepted in modern medicine [4] and has become one of the most implanted medical devices in the United States [5].

Fabrication of implantable prostheses involves the use of several major classes of materials such as ceramics, metals, and polymers. Polymeric biomaterials offer the advantage that they can be synthesized or processed to have different compositions and structures, resulting in controllable chemical, physical, and biological characteristics to meet specific prosthesis requirements [6]. Synthetic polymers, rather than naturally derived materials, have historically been used for the fabrication of medical devices. Post-World War II, pioneering surgeons utilized commercially available polymers including nylon, polyesters, polytetrafluoroethylene (PTFE), polymethyl methacrylate (PMMA), polyethylene, and silicones to heuristically fabricate device components for clinical applications [7,8]. For instance, these materials can be employed as the subcomponent of a device such as the occluder of the titled disk heart valve or as the bulk of the entire device as in the case of polymeric vascular grafts. Traditionally, polymeric biomaterials were selected for device fabrication based on their availability, biological inertness, and mechanical and structural compatibility. Recently, the field of polymeric biomaterials has evolved toward the design and development of new materials with enhanced biological or mechanical functionality for specific purposes.

This chapter addresses the application of polymeric biomaterials in the context of implantable devices intended for long-term functionality and permanent existence in the recipients. Basic concepts of biocompatibility as well as mechanical and structural compatibility are discussed to provide appropriate background for the understanding of polymer usage in cardiovascular, orthopedic, ophthalmologic, and dental prostheses. Furthermore, emerging classes of rationally designed polymers with potential applications in implantable devices are also highlighted.

19.2 BIOCOMPATIBILITY OF POLYMERIC PROSTHESES

When prostheses are implanted in recipients, in some cases, the devices last for a long time while others encounter complications leading to their rapid failure. Complications often occur due to the adverse biomaterial-host interaction between the implants and the surrounding biological environment of the host tissues [9]. The mutual effects of the implant on the host tissues and of the host on the implant are critical in determining whether subsequent device failure will occur [9]. The relationship between an implanted material and its host tissues constitutes the concept of biocompatibility, which was historically defined by David F. William as the ability of a material to perform with an appropriate host response in a specific application [10]. Though this definition is a general concept, it underlines three important aspects of a biomaterial, specifically the following: (a) the material should exhibit functional performance and not merely exist in the host tissue, (b) the response has to be appropriate for the desired application, and (c) the nature of the response may vary depending on the specific application [11,12]. A lack of biocompatibility can result in complications of implanted prostheses such as loosening of orthopedic joint prostheses [13], fibrotic overgrowth of mammary prostheses [14], and thrombogenic reactions on vascular grafts [15,16].

19.2.1 Effects of Implantable Prostheses on Host Tissues and Immune System

Implantation of a prosthesis can activate the host's blood coagulation mechanism, trigger immunologic reactions, or introduce foreign organisms resulting in common complications such as thrombosis, formation of fibrotic hyperplasia, and bacterial infection [9,17]. Figure 19.1a illustrates the typical sequence of biological responses following surgical implantation of a device. Immediately after tissue injury, plasma proteins adsorb onto the device surface via electrostatic bonds and noncovalent, nonspecific van der Waals forces [17]. Formation of such protein films directs subsequent events such as complement activation, fibrin formation, coagulation cascade, and the recruitment of inflammatory cells [17,20–23].

Bleeding from injured blood vessels can be arrested by the activation of the hemostatic mechanism [24]. However, an artificial device surface placed in contact with blood can disturb the balance of this mechanism [24]. Excessive blood clotting with increased risk of embolization can occur due to the adhesion of platelets and leukocytes to device surface and the formation of fibrin network following the

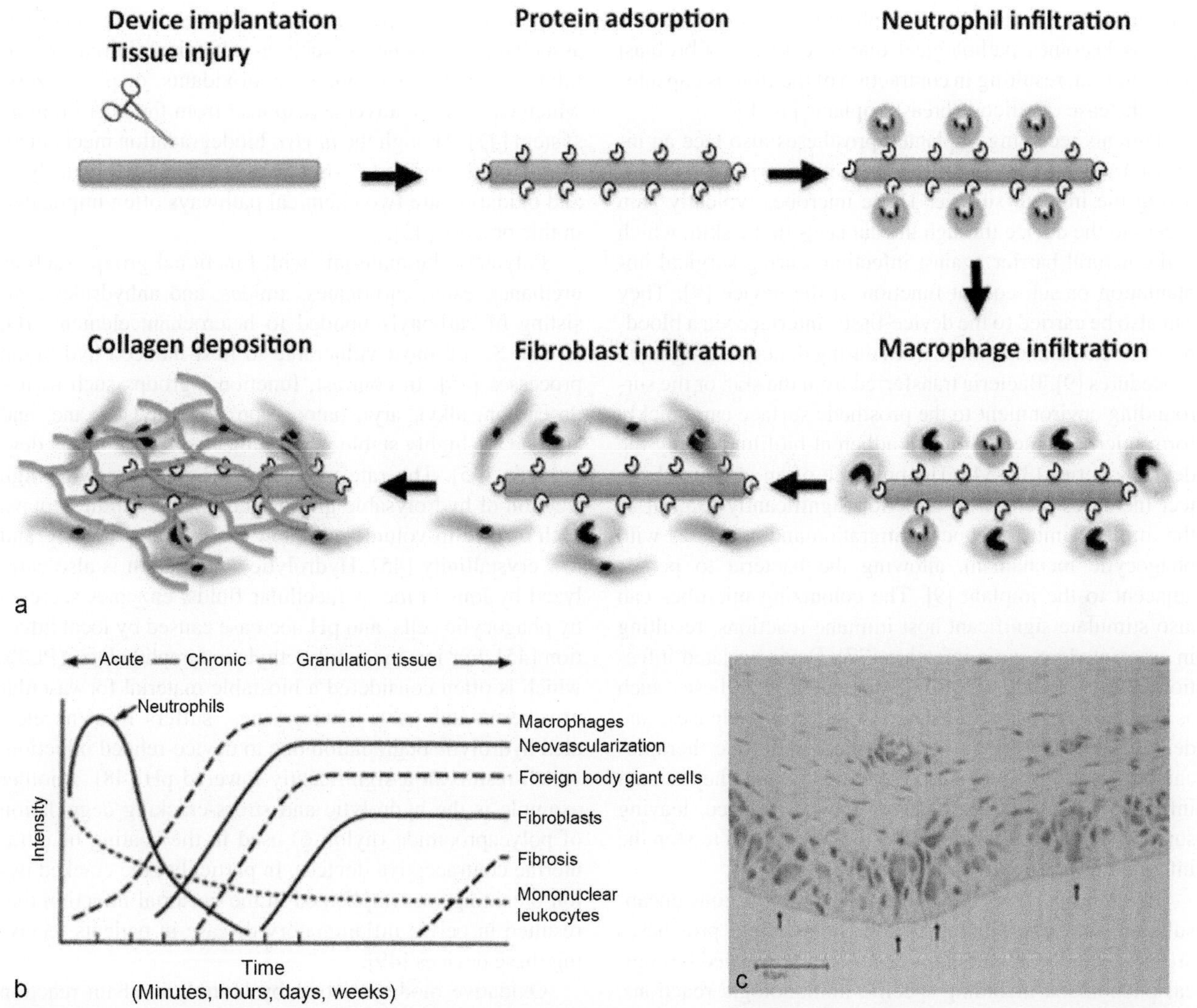

FIGURE 19.1 Host response to an implanted prosthesis. (a) Schematic illustration of the biological events occurring after the surgical implantation of a device. (b) The approximate kinetics of cellular infiltration in the host response. (c) Hematoxylin and eosin stain of the fibrous capsules surrounding silicone breast implants with arrows pointed at the multinucleated foreign body giant cells. (For color version of this figure, the reader is referred to the online version of this chapter.) Panel (a): Adapted from Anderson and Shive [18] with permission from Elsevier, copyright (1997). Panel (b): Adapted from Siggelkow *et al.* [19] with permission from Elsevier, copyright (2003).

adsorption of plasma proteins [24]. This complication is largely associated with cardiovascular devices such as intravascular stents and heart valves, which are in direct contact with blood [25].

In addition to the blood-material interactions, implantation of a prosthesis *in vivo* triggers a temporal recruitment of inflammatory cells as shown in Figure 19.1b [18,26,27]. Typically, during the first few hours or days after implantation, neutrophils, a type of short-lived white blood cells, migrate through the leaky vasculature to the injured tissues, where they engulf and degrade any foreign materials such as cellular or polymeric debris in a process called phagocytosis [26,27]. Next, macrophages replace neutrophils as the dominant phagocytic cell type, which can persist for weeks or months [26,27]. Macrophages consist of different subphenotypes that can play either proinflammatory or regenerative roles in the wound healing response associated with implanted materials [28,29]. When macrophages fail to engulf the implanted materials, they can fuse together at the device-implant interface forming multinucleated foreign body giant cells (Figure 19.1c) [19,21,23,30]. Macrophages also release growth factors such as fibroblast growth factor and transforming growth factor-beta that attract fibroblasts [31]. Subsequently, fibroblasts synthesize collagen and other matrix molecules, forming multiple fibrotic cell layers surrounding the implanted objects [26,27]. Stabilized fibrous encapsulation is a normal end-stage healing response to an implanted prosthesis intended for long-term functionality. Complications only occur if this process becomes pathological due to extensive fibroblast proliferation, resulting in contraction of the fibrous capsules as in the case of silicone breast implants [14,17].

Patients receiving implanted prostheses also face an increased risk of infection caused by microorganisms colonizing the implant surface. These microbes typically gain access to the device through the damages in the skin, which is the natural barrier against infection, during surgical implantation or subsequent function of the device [9]. They can also be carried to the device-tissue interface via a blood-borne route from distant sources during dental or urogenital procedures [9]. Bacteria transferred from the skin or the surrounding environment to the prosthetic surface can quickly form microcolonies within an adherent biofilm coating the device exterior [32–34]. The presence of an implanted object increases the risk of infection significantly [35,36] as the implant limits phagocyte migration and interferes with phagocytic mechanism, allowing the bacteria to persist adjacent to the implant [9]. The colonizing microbes can also stimulate significant host immune reactions, resulting in associated systemic infection [37]. Device-related infection affects a wide range of implantable prostheses such as vascular grafts, artificial hip joints, contact lenses, and dental devices. Although conventional antibiotic therapies can relieve the symptoms of systemic infection, they remain infective against biofilms on the implant surface, leaving surgical removal of the device as the only option to stop the infection [38,39].

In addition to blood-material interaction, fibrous encapsulation, and microbial infection, implantable prostheses are associated with a multitude of other interrelated biological phenomena including specific immunologic reactions, complement activation, systemic toxicity, hypersensitivity, and implant-associated tumor formation [9]. For more detailed discussion on these biological processes, the readers are referred to specialized textbooks [9,17] and reviews [12,30,40].

19.2.2 Biochemical Effects of Host Environment on Polymeric Implants

Implanted biomaterials or medical devices are subjected to the surrounding host environment, which contains biochemical molecules such as enzymes [41,42], free radicals, peroxides [43], and hydrogen ions secreted by inflammatory cells and infecting microbes [44–46]. The phagocytic mechanism of inflammatory cells such as neutrophils and macrophages has naturally evolved as a defense strategy for the body to ensure the removal of undesired foreign objects. Therefore, the potent biochemical actions of the secreted species can result in the unintended breakdown of solid-phase polymeric components of implanted devices over an extended period of time (months or years) [45]. This degradation process can also release biomaterial-associated components such as chemical initiators, inhibitors, residual monomers, antioxidants, or plasticizers, which can invoke adverse response from the host immune system [47]. Though the *in vivo* biodegradation mechanism of polymeric implants is not firmly understood, hydrolysis and oxidation are two chemical pathways often implicated in this process [45].

Polymeric biomaterials with functional groups such as urethanes, ester, carbonates, amides, and anhydrides consisting of carbonyls bonded to heterochain elements (O, N, and S) are most vulnerable to host-induced hydrolytic processes [45]. In contrast, functional groups such as hydrocarbon, alkyl, aryl, halocarbon, dimethylsiloxane, and sulfone are highly stable when subjected to hydrolytic degradation [45]. The rate of hydrolysis increases with high fraction of hydrolysable groups and/or hydrophilic groups, high surface-to-volume ratio, low cross-linking density, and low crystallinity [45]. Hydrolytic degradation is also catalyzed by ions in the extracellular fluids, enzymes secreted by phagocytic cells, and pH decrease caused by local infection [45]. For instance, poly(ethylene terephthalate) (PET), which is often considered a biostable material for vascular prostheses and valve sewing rings, suffers from accelerated hydrolytic degradation due to device-related infection, which results in a significantly lowered pH [48]. Another example is the hydrolytic and stress-cracking degradation of polycaproamide (nylon 6) used in the coating of intrauterine contraceptive devices. In particular, the cracked nylon 6 coating was implicated in the bacterial infection that resulted in pelvic inflammatory disease in patients receiving these devices [49].

Oxidative biodegradation by homolytic chain reaction or heterolytic mechanism is another biochemical pathway for the unintended *in vivo* breakdown of polymeric implants. These reactions are initiated and propagated by oxidizing molecular species such as superoxide anion radicals and their derivatives (hydrogel peroxide and hypochlorite) that are generated from activated phagocytic cells in high concentrations at the device-tissue interface [43,45]. In these processes, polymeric sites that facilitate the removal of an ion or atom and provide resonance stabilization for the resultant ion or radical are most vulnerable to host-induced oxidative attack [45]. Functional groups susceptible to oxidative degradation include branched aliphatic hydrocarbon, allylic hydrocarbon, ether, and aromatic ring-containing polymers [45]. Furthermore, oxidative attack on the polymeric surface might also induce shallow brittle microcracks, which is subsequently propagated to deeper and wider cracks [45]. For example, polyether urethane (PEU) have been reliably used as connectors and insulators for cardiac pacemakers. However, in some cases, pacing leads displayed surface cracks, which were linked to oxidative loss of ether functional groups [50–52].

Given the complex biological and chemical interaction between the polymeric implants and the *in vivo* environment of the host, the physiochemical properties of the polymers must be carefully considered in the selection of polymeric biomaterials that are biochemically appropriate for specific applications in implantable prostheses. Overall, the ideal material should minimize its adverse effects on the host tissues and immune system while resisting hydrolytic and oxidative degradation to achieve successful device integration and the intended long-term functionality. In addition, the choice of materials depends not only on the functions of the prostheses but also on other factors such as the site of implantation, the age group of recipients, and the intended period of use.

19.3 STRUCTURAL COMPATIBILITY AND MECHANICAL DURABILITY OF POLYMERIC PROSTHESES

The tissues in the body can be classified into hard and soft types. Soft tissues exhibit significantly lower Young's modulus (stiffness) and are generally weaker (with lower tensile strength) compared to hard tissues. The skin, tendon, and arterial tissues are considered soft tissues, while the hard tissues include the bone, tooth, and enamel [53]. Successful development and optimal performance of an implantable prosthesis heavily rely on surface and structural compatibility of the implant in both hard and soft tissues. Surface biocompatibility usually refers to the tissue-material interaction as discussed in Section 19.2 [9,12]. Alternatively, structural compatibility refers to optimal matching of the implant to the host tissue in terms of mechanical and structural properties [54]. In particular, structural compatibility refers to a set of bulk properties of the implant material including stiffness, strength to high mechanical loading condition, load transition capacity within the interface of the native tissue, fatigue strength (for approximately 10^8 cycles of loading in a lifetime), and creep and wear resistance within the temperature and corrosive environment of the human body [54,55]. Any mismatch between the mechanical and structural properties of the implant and the surrounding tissues results in poor functionality of the implant and further complications in the host tissue. For instance, in the case of metallic or ceramic implants for bone replacement, high modulus of the implant often leads to stress shielding, in which the bone is insufficiently loaded compared to the surrounding tissue. Such phenomenon ultimately leads to increased bone remodeling and porosity [55–59]. In the case of cardiovascular applications, the mismatch between the compliance of the prosthetic graft and surrounding native blood vessels ultimately results in the failure of the graft [60–63].

One of the major advantages of polymeric materials compared to ceramics or metals for the development of implantable prostheses is that polymers can be made with different compositions and architectures (i.e., linear and star-branched) to match the desired mechanical properties of the native soft and hard tissues [6]. For example, in the case of soft tissues such as blood vessels, polymeric biomaterials are attractive material candidates since the radial compliance of these materials can be tailored to match the mechanical properties of the native structures [60]. Alternatively, for hard tissue applications, mechanical properties of polymeric materials can be reinforced through blending of fibers with variable geometric features and organization to obtain desired mechanical properties [54].

Polymeric prostheses should exhibit suitable mechanical properties including compatible modulus and tensile and fatigue strength in response to cyclic loading and repetitive stress since the majority of implants are subjected to these types of stress [54,55,64]. For instance, cardiovascular prostheses are subjected to pulsatile pressure and orthopedic implants are subjected to static and dynamic loading depending on the postural position and activity (i.e., running) of the implant recipients. An ideal polymeric prosthesis should also exhibit high durability to degradation along with excellent wear resistance to minimize the debris generated through the interaction of the implant with neighboring tissue and biological fluids (i.e., blood). Creep resistance of polymeric biomaterials is another crucial factor, which needs to be carefully considered to achieve the optimal functionality of the implant. This is particularly important since the normal body temperature is a significant fraction of the melting point of the majority of polymeric biomaterials [6,55]. In addition to bulk properties of the polymer biomaterials, surface properties including topography and roughness of the implant surface also play a major role in optimal performance of the implant [6,65,66].

In general, selection of a mechanically appropriate polymeric biomaterial for the development of prosthetic implants is highly dependent on the endpoint application of the implant [6,64]. For instance, in the case of orthopedic and dental prostheses, polymeric biomaterials should exhibit high resistance to fatigue and corrosion, respectively. Alternatively, for cardiovascular devices, wear and tear resistance, fatigue resistance, and the porosity of the polymeric biomaterials are among the major design factors dictating the choice of materials. Therefore, it is crucial to consider detailed structural and mechanical factors in order to design application-specific implants with optimal and long-term performance.

19.4 APPLICATIONS OF POLYMERIC BIOMATERIALS IN IMPLANTABLE PROSTHESES

Table 19.1 summarizes some of the existing usage of polymeric biomaterials in a variety of implantable prostheses for cardiovascular, orthopedic, ophthalmologic, and dental applications.

TABLE 19.1 Existing Usage of Polymeric Biomaterials in Implantable Prostheses

Device/Prosthesis	Material	Company/Manufacturer	Reference
Heart valve	POSS-PCU	–	[67]
	PU, PCU, and PEU		[68–74]
	Hexsyn rubber	–	[75]
	PTFE	–	[76]
	PVA	–	[77]
Vascular graft	PTFE/ePTFE	–	[78–87]
	Dacron®	–	[88]
	PU/PCU, poly(carbonate-urea) urethane, and poly(ether-urethane-urea)	–	[89–94]
Stent	PEVA-coated stainless steel (Cypher®)	Johnson & Johnson, USA	[95]
	SIBS-coated stainless steel (Taxus®)	Boston Scientific, USA	[96]
	PLLA (Igaki-Tamai)		[97]
	PLLA and PDLLA	Abbott, USA	[98]
	Poly(tyrosine carbonate)	REVA Medical	[95,99]
Hip/knee joints	Several versions of UHMWPE	Zimmer, DePuy, and others	[54,100–102]
Vertebral disks	UHMWPE	DePuy, NuVasive, and others	[103]
	PU	Medtronic	[104]
Menisci	Collagen	Menaflex® ReGen Biologics (FDA approval reversed)	[105–107]
	PU	Actifit® by Orteq	[105–107]
	PVA	–	[108]
Tendons/ligaments	PET fibers	OrthoCoupler™ by Surgical Energetics	[109]
Intraocular lenses	PMMA and other acrylic polymers	Numerous manufacturers	[110]
	Collagen	Numerous manufacturers	[111,112]
	Poloxamer	–	[113]
Artificial cornea	PMMA	Numerous manufacturers	[114]
	pHEMA	AlphaCor™	[115,116]
	PTFE	–	[117]
	PVA	–	[118]
Dental implants	PMMA	–	[54,119]
	PTFE	–	[54]

POSS, polyhedral oligomeric silsesquioxane; PCU, polycarbonate urethane; PU, polyurethane; PEU, poly(ether urethanes); PTFE, polytetrafluoroethylene; PVA, polyvinyl alcohol; PEVA, polyethylene-*co*-vinyl acetate; SIBS, poly(styrene-*b*-isobutylene-*b*-styrene); PLLA, poly-L-lactide acid; PDLLA, poly-DL-lactide acid; UHMWPE, ultra-high-molecular-weight polyethylene; PET, poly(ethylene terephthalate); PMMA, polymethyl methacrylate; pHEMA, poly(2-hydroxyethyl methacrylate).

19.4.1 Cardiovascular Applications

19.4.1.1 Polymeric Heart Valves

Replacement of heart valves was introduced into clinical practice around 1960. In the past few years, heart valves have been among the most applied cardiovascular prostheses, which have had a dramatic effect on patients suffering from valvular diseases [120,121]. Currently, clinically accepted heart valves include mechanical valves and bioprosthetic valves of porcine or bovine origins. Mechanical valves evolved from cage ball models to a monoleaflet configuration (tilting disk) to a more advanced bileaflet or trileaflet design (Figure 19.2a) [123,124]. On the other hand, bioprosthetic valves can be categorized into stented, stentless, or percutaneous valves [123,125]. Mechanical valves exhibit long-term durability and functionality but they are at high risk for thrombogenicity. Alternatively, prosthetic valves maintain enhanced hemodynamic function; however, since they are subjected to high stress, they are at increased risk for early failure [123].

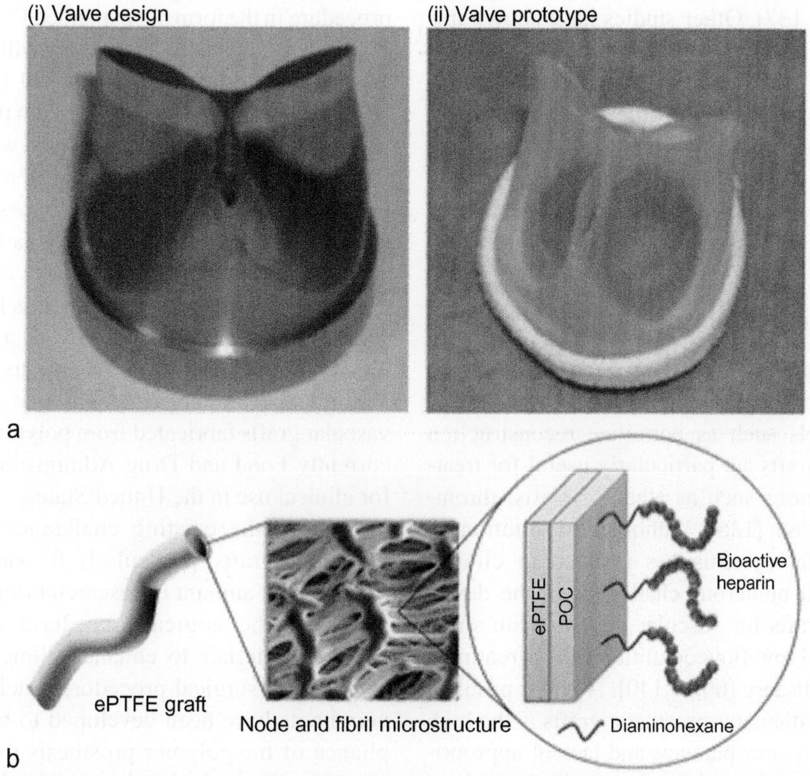

FIGURE 19.2 Polymeric prostheses for cardiovascular applications. (a) Design (i) and prototype (ii) of a synthetic heart valve fabricated with POSS-PCU nanocomposite polymer. (b) Schematics of an ePTFE vascular graft with heparin immobilized on the surface for anticoagulant effects. (For color version of this figure, the reader is referred to the online version of this chapter.) Panel (a): Adapted from Kidane *et al.* [67] with permission from Elsevier, copyright (2009). Panel (b): Adapted from Hoshi *et al.* [122] with permission from Elsevier, copyright (2013).

The use of polymeric biomaterials has been widely explored in the modification of bioprosthetic valves and in the construction of mechanical valves. In bioprosthetic valves, a polymeric coating is often used to cover the base ring of the valve to successfully suture the valve to the surrounding host tissues [126]. Examples of these polymers include PET, PTFE (Teflon®), and polyester [126]. In addition, different parts of the mechanical valve can be fabricated using polymeric materials along with the metallic support ring. In cage ball valves, rigid silicone rubber (SR) has been widely used to make the moving ball [124]. Furthermore, the occluder in tilted disk monoleaflet valves can also be made from polymeric biomaterials such as ultra-high-molecular-weight polyethylene (UHMWPE) [127].

With the recent advances in the development of novel polymeric biomaterials along with surface modification techniques, significant research effort has been devoted toward fabrication of fully polymeric valves as an attractive alternative to mechanical and bioprosthetic valves [128]. Development of these types of valves is specifically appealing, since polymeric materials offer significant flexibility in terms of material properties and manufacturing process to achieve reduced thrombogenicity and improved durability and biocompatibility [127,128]. However, fully polymeric heart valves are still at the preclinical research and assessment stage. The major challenges for the application of these valves were pathological calcification, poor mechanical stability and durability, and poor flow dynamics [76,128,129]. To date, numerous polymeric biomaterials including Hexsyn rubber [75], PTFE [76], polyvinyl alcohol (PVA) [77], and polyurethane (PU) and its modified forms including polyester urethane, polycarbonate urethane (PCU), poly(ether urethanes) [68–74,130–132], and polyhedral oligomeric silsesquioxane-PCU (POSS-PCU) [67] (Table 19.1) have been investigated for the development of heart valves. However, among these polymeric biomaterials, PU has been most widely accepted as a suitable biomaterial for heart valve applications, mainly because of its unique characteristics including biphasic hard crystalline and soft elastomeric properties [128]. PU offers several advantages including flexibility, hemocompatibility, mechanical robustness, and low risk of thrombogenicity [71,128,133,134]. The major drawbacks of PU are the potential risk of calcification [68] and high degradation rate [135]. So far, numerous efforts have been focused on improving the material properties of PU by enhancing the mechanical properties of the soft segment. For instance, PCU, an alternative form of PU, exhibits enhanced resistance to oxidative stability and lower biodegradation rate compared to PU [136,137]. Other studies have shown improved structural and material modification of PU through the incorporation of other polymeric materials such as polydimethylsiloxane (PDMS) into the backbone of PU to enhance its mechanical properties [138,139]. For a more comprehensive review on the development of fully polymeric heart valves, readers are referred to specialized literature [128].

19.4.1.2 Polymeric Vascular Grafts

Polymeric vascular grafts have long been accepted as suitable candidates for bypass surgery of large-diameter (>6 mm) blood vessels such as aortoiliac reconstruction [61]. These types of grafts are particularly useful for treatment of vascular diseases such as atherosclerosis, thrombosis, and inflammation [126]. Although substitution of large-diameter vascular systems has resulted in clinical success, there are still numerous challenges in the development of suitable grafts for vascular systems with small diameter (<6 mm) and low flow condition in the treatment of peripheral arterial disease [60,61,140]. A major problem associated with small-diameter prosthetic grafts is the high risk of thrombogenicity, low patency, and lack of appropriate hemodynamic property, mainly due to the low blood flow rate and compliance mismatch between the graft and the host vasculature [60,62,63,141]. Therefore, autologous veins still remain the best candidate for bypass grafting of small-diameter blood vessels [60,63].

As discussed earlier in this chapter, polymeric biomaterials should exhibit excellent surface and structural compatibility depending on the endpoint application of prostheses. In particular, vascular grafts should exhibit long-term biostability, high endurance, and excellent resistance to cyclic loading condition, nonthrombogenicity, and excellent patency over a long period of lifetime [6,142]. In addition, the mechanical properties of the vascular such as elasticity should be comparable to those of the *in vivo* host vasculature. Particularly, any mismatch between the radial stiffness of the graft and the native vasculature can result in alteration of the flow pattern (i.e., turbulent vs. laminar) and shear stress, affecting graft patency and performance, and may cause hyperplasia at the anastomosis locations [62,141]. Usually in the case of large and high-pressure blood vessels, the stiffness of the graft plays a key role in the selection of the biomaterial candidate, while in the case of small blood vessels, radial compliance of the vascular graft is the most critical factor [60,141].

To date, PTFE and PET (Dacron®) have been the two most commonly used polymeric biomaterials for large blood vessel substitution [61,143]. PTFE has a crystalline architecture with tensile strength of 14 MPa and Young's modulus of around 0.5 GPa [60]. PTFE was first introduced into clinical practice in 1970s for lower extremity bypass procedure in the form of "expanded PTFE" (ePTFE) [63,78]. Dacron® is a crystalline polyester with a tensile modulus of 14 GPa and tensile strength of ~170-180 MPa [60]. Though it was historically developed for textile industry, Dacron®, in the form of knitted architecture with high strength and flexibility, is currently used extensively for prosthetic vascular grafts [126,143]. Other types of polymeric biomaterials such as PUs, PCU, poly(carbonate-urea)urethane, and poly(ether-urethane-urea) have also been explored for application in vascular bypass grafts [89–93]. For instance, some studies have suggested PU as a biomaterial for small blood vessel substitution due to its high elasticity [61]. Vectra® (BARD Peripheral Vascular), one of the synthetic vascular grafts fabricated from poly(ether-urethane-urea), is currently Food and Drug Administration (FDA) approved for clinical use in the United States.

Due to the existing challenges in the development polymeric grafts, particularly for small-diameter vessels, a significant amount of research efforts have been devoted to modify the surgical procedures and the properties of existing materials to enhance clinical outcomes. For instance, new surgical procedures such as vein patches and vein boots have been developed to better match the compliance of the polymer prosthesis to the host vasculature [79,80]. Alternatively, chemical and biological modification approaches have also been extensively used to enhance the properties of the developed grafts. Common strategies include immobilization of growth factors [81,82], heparin coating (Figure 19.2b) [83,94,122], silicone-based surface modification (Aria® graft, Thoratec) [144], carbon-impregnated ePTFE grafts [84] (IMPRA® Carboflo® vascular grafts, BARD Peripheral Vascular) [85], anticoagulant protein and antagonist coating (i.e., hirudin) [86,88], and sirolimus and paclitaxel coating [87,94]. For a more comprehensive review on polymeric vascular grafts, readers are referred to specialized literature [60,61,63].

19.4.1.3 Polymeric Stents

Stents are tubular-shaped implants that are placed inside major blood vessels to mechanically support the stenotic vessels or obstructed conduits [145,146]. An ideal stent should be trackable, thromboresistant, reliably expandable, and flexible [147]. Although bare-metal stents were used for thousands of surgeries since 1990 [148], the long-term clinical outcome of these types of stents was significantly affected by the risk of thrombosis and restenosis [95]. To address these limitations, polymeric biomaterials have been extensively explored as an alternative for metals in stent design. Polymers can constitute either the components of primarily metallic stents or the entire stent scaffolds. For instance, in metallic drug-eluting stents (DESs), a permanent or degradable polymer containing an active pharmaceutical agent is coated on the metallic scaffolds [95,149]. The polymeric coating allows the gradual release of the antifibrotic drug [95,150–153]. Metallic DESs have shown a great promise in decreasing restenosis and reducing early occurrence of thrombosis [154,155]. Cypher® (Johnson & Johnson, USA) and Taxus® (Boston Scientific, USA) stents were among the first generation of DESs, approved by FDA, which were coated with permanent polymers such as a blend of polyethylene-*co*-vinyl acetate (PEVA) and polybutyl methacrylate (PBMA) [95] and poly(styrene-*b*-isobutylene-*b*-styrene) [96], respectively. Furthermore, Endeavor® (Medtronic, USA) [156] and XIENCE V® stents (Abbott, USA) [157] were among the next generation of DESs, which carry phosphorylcholine and poly(vinylidene fluoride-*co*-hexafluoropropylene) polymeric coatings, respectively.

In the recent years, there has been significant effort toward the development of fully polymeric stents in which polymeric biomaterials are used as the main scaffold and the carrier matrix for drug delivery. These types of stents usually allow for proper vascular remodeling [95]; however, their poor mechanical properties have limited their application in clinical trials [95]. For instance, the Igaki-Tamai stent is a self-expanding stent fabricated from poly-L-lactide acid (PLLA) with a zigzag-helical scaffold structure. A clinical trial of this stent on 15 patients in 2000 revealed a low percentage of restenosis in a 6-month follow-up and full absorption of the stent in a 4-year follow-up [97,158]. The BVS stent (Abbott, USA) is another example of fully polymeric DESs in clinical trials. The body of this stent is fabricated from PLLA coated with a layer of poly-DL-lactide acid, which encapsulates the antiproliferative drug everolimus [98]. The clinical trial of this stent on 30 patients in a 2-year follow-up showed complete degradation of the stent within 18 months of placement with no cardiac complications or thrombosis [159]. In 2007, REVA Medical launched another type of polymeric DESs, which are balloon-expandable stents made with trypsin-derived polycarbonate polymer. REVA's ReZolve® stent, which are under clinical evaluation, comprises a locking mechanism element, which prevents deformation and recoiling upon expansion and implantation [95,99]. Fully polymeric stents are envisioned to substitute the permanent metallic stents; however, the major challenges such as appropriate biodegradation and mechanical robustness need to be addressed to advance absorbable stents further in clinical applications. For a more comprehensive review on DESs and polymeric stents, readers are referred to specialized literature [95,160].

19.4.2 Orthopedic Applications

In 2008, the orthopedic market was estimated to be over $14 billion in the United States alone, and it has been projected to double in less than 10 years [161]. The shift toward hard-on-soft materials in the engineering of orthopedic implants started with the pioneering work of Sir John Charnley in the 1960s and 1970s [4]. Since then, polymers and polymer composites have gained increasing prominence in the development of orthopedic prosthetic devices such as reconstructive joint replacements, spinal implants, and orthobiologics [100].

19.4.2.1 Artificial Joint Replacement

Two of the most common joint replacements are that of the knee and the hip. Total hip replacements alone are performed over 230,000 times annually in the United States [5]. Currently, clinical success rates for total hip replacements stand at 93% for 10 years, 85% for 15 years, and roughly 77% for 25 years postsurgery [162,163]. The success rate for polyethylene knee replacement has been slightly lower, especially in young adults [54,164]. Favorable clinical outcome of an artificial joint replacement depends on mechanical properties such as high strength, elastic modulus, fracture toughness, and resistance to wear, fatigue, and deformation [165,166]. Additionally, biocompatibility and wetting for improved body-implant lubrication are also critical factors [166,167].

To date, polymeric biomaterials have been utilized to fabricate components of hip and knee replacement with varying extent of clinical success. For instance, UHMWPE has been utilized in numerous attempts to improve wear resistance in knee replacement. However, even UHMWPE composites with improved stiffness and strength have not been as successful in knee replacement as that of the hip, except for the case of carbon fiber-reinforced UHMWPE implants [167]. For hip replacement, PTFE was initially used but soon discontinued due to the unacceptably high wear and distortion resulting in granuloma [54]. Furthermore, polymeric materials such as carbon fiber-reinforced polyetheretherketone (CF/PEEK) [168] have been used in the fabrication of acetabular components of hip joint cups (Figure 19.3a). Short-term function of UHMWPE acetabular cups has proven to be satisfactory; however, achieving long-term functionality still remains challenging [101]. Particularly, the formation of polyethylene wear particles and material degradation caused by oxidative stress have been tackled by improved fabrication and sterilization methods [169]. More recently, vitamin E-impregnated UHMWPE has been used to decrease oxidative damage, improve biocompatibility, and lower bacterial adhesiveness [102,170]. Additionally, CF/PEEK has been proposed for coatings in artificial hips with bioactive surface designs that provided minimal stress shielding [171].

19.4.2.2 Vertebral Applications

The application of polymers in treating vertebral injury dates back to the work performed by Galibert *et al.* in 1984

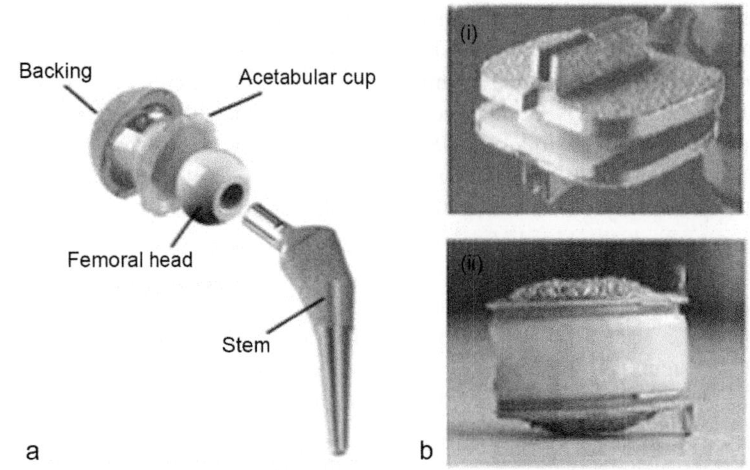

FIGURE 19.3 Polymeric prostheses for orthopedic applications. (a) Major components of an artificial hip joint. (b) Two examples of artificial cervical disks with polymeric components. (i) ProDisc-C disk with an insert fabricated from UHMWPE. (ii) Bryan disk with a nucleus fabricated from polyurethane. (For color version of this figure, the reader is referred to the online version of this chapter.) Panel (a): Adapted from Mattei *et al.* [166] with permission from Elsevier, copyright (2011). Panel (b): Adapted from Link *et al.* [104] with permission from Elsevier, copyright (2004).

in vertebroplasty that involved procedures for stabilizing compression fractures of the vertebrae with bone cement injections [172]. However, for more serious injuries, artificial polymeric implants are often required to replace degenerative intervertebral disks. These prostheses are designed to allow for spinal mobility with structural integrity while assuring protection of the spinal cord [173,174]. In severe cases of vertebral injury, spinal fusion and disk replacement are the two most common treatment approaches currently used [173,174]. Spinal fusions involve removing the vertebrae and restoring the site with synthetic bone grafts to immobilize the joints between two vertebras. In this regard, several groups have developed bioglass/PU composites while others have utilized bioglass/polystyrene materials for vertebral body replacement and bone grafting purposes [175,176]. Alternatively, vertebral prosthesis such as brackets and cages can be used to promote bone growth and subsequent osseointegration with the prosthesis (Figure 19.3b). In a few cases, polymeric composites such as CF/PEEK and carbon fiber-reinforced polystyrene have also been proposed for vertebral prostheses [177].

Several implants are commercially available for total or partial disk replacements. For instance, currently, two polymer-based cervical and two lumbar disk prostheses approved by the FDA are being widely used for disk replacement applications [103]. The first artificial disk (DePuy Inc.), approved by the FDA in 2004, was based on a hard-on-soft technology, which employed a CoCrMo alloy in conjunction with UHMWPE. Alternatively, ProDisc-C™ (Figure 19.3b,i), approved in 2006 and 2007 for both lumbar and cervical replacements, respectively, was based on similar types of polymer composites. More recently, Medtronic developed Bryan™ prosthesis using titanium alloys and PU polymer (Figure 19.3b,ii) [104]. Other types of polymeric biomaterials including poly(2-hydroxyethyl methacrylate) (pHEMA)/PMMA, semi-interpenetrating polymer network (SIPN) hydrogels reinforced with PET fibers, and polyethylene composite reinforced with hydroxyapatite (HAPEX™) have also been proposed for annulus substitution and anchoring endplates [178]. Similarly, pHEMA/polycaprolactone (PCL) SIPN hydrogels reinforced with PET fibers have shown promises for intervertebral disk prosthesis [179]. In addition, a variety of polymeric biomaterials such as polyethylene, PU, SR, and PET/SR have been proposed for fabrication of artificial disks [54]. To date, long-term clinical application of the artificial disks remains challenging due to the difficulty of replicating the properties of the native vertebral disk to function in a complex biomechanical environment [54,174].

19.4.2.3 Orthobiologics

There are several polymeric implants and prosthesis, which are being used for orthobiologic applications in clinical practice. One of the most common applications is synthetic menisci [105,106]. Menisci are cartilage tissues, which serve to disperse friction in the knee joint. Traumatic injuries and degenerative tears, perhaps due to tissue deterioration with age, usually necessitate meniscal replacement. Currently, the two partial meniscal substitutes approved by the FDA are collagen meniscus implants (CMI®) and polymeric PU implants (Actifit®) [105,107]. Recent studies have demonstrated that nonporous PVA hydrogel-based implants are promising for permanent meniscal replacements [108], while hyaluronic acid/PCL composite matrix reinforced with polylactic acid fibers is porous alternatives for total or partial artificial menisci replacement [180]. Another example of orthobiologic applications is artificial tendons. Though currently there are no permanent and clinically reliable artificial tendons, significant amount of research has been devoted toward this direction [181]. For instance, OrthoCoupler™ devices have been designed from bundles of fine polyester (PET) fibers to imitate tendon connections [109]. Alternatively, others have used self-assembled layer-by-layer of chitosan and hyaluronic acid on PET artificial ligaments to improve biocompatibility and osteoblast proliferation [182]. Finally, orthobiologics also include bone fracture repair by either bioinert internal fixtures or biodegradable plates and pins [105]. Several polymeric composites that are thermoset and thermoplastic have been used for this application, including CF/PEEK composites due to their biological inertness, high mechanical strength, and good fatigue resistance [183].

19.4.3 Ophthalmologic Applications

Polymeric biomaterials have also been widely used in ophthalmologic application. Cataracts are one of the leading causes of blindness in the aging population [184], and cataract surgery is one of the most common ophthalmologic procedures [185]. Over the last few decades, cataract surgery techniques have changed enormously. One of the major contributors to this progress is the advent of advanced materials in the design of intraocular lenses (IOLs) [186]. Material selection for such lenses depends on optical calculation performed prior to implantation so that the chosen materials have optimal properties such as refractive index [187]. Close attention should be given to refractive targeting, biometry, and IOL power calculation formulae to achieve the best postoperative results [185]. Currently, several manufacturers offer hundreds of different types of IOLs fabricated from polymeric biomaterials such as plastics, acrylics, and silicone [110]. Initially, in most commercially available IOLs, the optical components were made of PMMA or polypropylene; however, more recently, some manufacturers have introduced collagen copolymers [111,112,187]. In general, cataract surgery is very safe but IOL complications still exist, most commonly due to the same oxidative stress that cause cataracts or due to incomplete IOL integration, protein adsorption, infection, and inflammation [188]. More recently, Kwon *et al.* have suggested injectable IOLs based on poloxamer hydrogels, which exhibited no inflammatory or toxic effect in a rabbit animal model [113].

Another ophthalmologic application of polymeric prostheses is the artificial cornea or keratoprosthesis. This device has been developed for more than a century, starting with the work of Heusser who was the first to implant artificial corneas in patients in 1859 [189]. However, it was not until after World War II that researchers noticed that PMMA particles were well tolerated in patient eyes [190]. Ever since, other polymeric materials such as nylon, PU, and PTFE [117,189] have also been explored for keratoprosthetic application. Today, artificial corneas are the last resort for patients after multiple corneal transplants have or are expected to fail [191]. Existing commercially available keratoprosthetic devices include the osteo-odonto-keratoprosthesis and Boston keratoprosthesis [192]. Though PMMA remains the most common polymer used in these existing keratoprostheses [114,189], more recently, hydrogel-based devices such as the AlphaCor made from pHEMA have also been FDA approved [115]. These devices are fabricated from softer biopolymeric materials with improved biocompatibility thus allowing for better host tissue regeneration [193].

For successful integration of the prosthesis, i.e., minimizing infection and other postoperative complications, it is important that the keratoprosthesis allows for formation of epithelial cell layers that act as a barrier to external contamination while allowing for nutrition diffusion to retinal cells [194]. To address this challenge, several new approaches have been developed ranging from novel hydrogels and synthetic polymers to collagen-based bio-copolymers and IPNs [116]. For instance, collagen-based networks cross-linked with dendrimers and PMDS in combination with poly(*N*-isopropylacrylamide) (PNIPAAm) have been developed for keratoprosthetic application [195]. These PDMS/PNIPAAm IPNs allow for blending and complementarity of features including improved biocompatibility and glucose/nutrient permeability while preserving mechanical strength and transparency. Additionally, collagen-based hydrogels composed of collagen, chitosan, and glycosaminoglycans have also exhibited great potential in animal models [196].

Another ophthalmologic application of polymeric biomaterials is the development of ocular prosthesis and biologically inspired compound eyes [197,198]. Such prostheses, commonly fabricated from porous polyethylene, are designed to serve as nonfunctional artificial substitutes for enucleated eyeballs [199].

19.4.4 Dental Applications

Dental treatment is one of the most common medical procedures in humans [54]. It can range from simple procedures such as cavity filling to replacing tooth segments or complete teeth. A vast majority of materials used in dental applications have polymeric components, especially those used for restorative and prosthetic purposes [200]. However, the field of polymeric biomaterials in dentistry still has a large void, allowing for substantial advancement in this area [201].

The materials selected for dental prostheses must resemble the physical, mechanical, and aesthetic characteristics of natural teeth [54]. PMMA polymers constitute the most frequently used group of materials in dentistry. These materials are easily adapted to specific prosthetic purposes while exhibiting excellent biocompatibility. In addition to their good mechanical properties and wear resistance, the advantage of PMMA polymers in dentistry is that they are tasteless, odorless, nontoxic, nonirritating, and resistant to numerous microbial agents [202].

In recent years, dental research has been focused on dental implants and artificial teeth rooted in a patient's jaw allowing for a permanent denture, as alternatives to bridges or false teeth. A wide array of materials including polymers such as UHMWPE, PTFE, and PET have been used in many types of existing dental implants [54,119]. Porous polymeric surfaces are now designed to facilitate bone integration [54]. Other dental applications of polymeric biomaterials have been for the development of a dental bridge, meant as a partial denture or false teeth. In extreme cases, removable dentures fabricated from PMMA are used to overcome the loss of all teeth [203].

Some obstacles still remain in the design of optimal orthodontic prostheses due to their poor thermal conductivity and effect on taste. However, recent publications indicate that in the case of bridges, such issues could be resolved with microwires incorporated into PMMA polymers thereby increasing their diffusivity and resulting in better performance [204].

19.5 EMERGING CLASSES OF POLYMERIC BIOMATERIALS FOR IMPLANTABLE PROSTHESES

19.5.1 Antifouling Polymeric Coatings

After implantation, the surface of the prosthesis is the first component to come into contact with the surrounding biological milieu. Therefore, surface characteristics play an important role in controlling the course of subsequent biological reactions. Antifouling materials are materials that can resist protein adsorption or microbial adhesion [205,206]. Hence, they have potential applications as surface coatings on implantable devices such as heart valves and hip joint prostheses to minimize biofilm formation and subsequent device-associated infections.

19.5.1.1 Passive Antifouling Materials

Polyethylene glycols (PEGs) are the earliest polymeric biomaterials documented in the literature for its ability to impart protein resistance to a surface [207–209]. PEGs have traditionally been immobilized on metal, glass, or polymer surfaces via block or graft copolymerization, direct covalent attachment, chemical adsorption, and physical adsorption [206,210]. Steric repulsion forces between hydrophilic flexible PEG chains facilitate retention of interfacial water molecules, rendering PEG coatings highly resistant to protein adsorption [211–213]. However, PEG-based coatings have the tendency to auto-oxidize in the presence of oxygen [214], thus losing their protein resistance property and limiting their utility as nonfouling surfaces for long-term implant applications.

Recently, new classes of synthetic and natural polymers such as zwitterionic polymers [215–221], peptoids [222], carbohydrate and glycerol derivatives [223–225], and poly-L-lysine-graft-dextran [226] have emerged as candidate materials for the development of non-PEG protein-resistant surfaces. For example, Jiang and coworkers demonstrated that zwitterionic materials such as poly(sulfobetaine methacrylate) were not only highly resistant to nonspecific protein adsorption [227–230] but also able to significantly decrease bacterial adhesion and biofilm formation [231].

19.5.1.2 Active Antifouling Materials

Progress has also been made in the past decade in the development of antifouling polymeric materials with active self-cleansing and self-defending properties [206]. Stimuli-responsive polymeric materials have been explored to design self-cleansing coatings that can release adherent microbes upon exposure to temperature or pH changes [232–235]. For instance, thermoresponsive PNIPAAm [232,233] and its derivatives [236,237] facilitate bacteria detachment from their surfaces when their temperatures decrease below their lower critical solution temperatures. This process causes the materials to switch from a water-insoluble, collapsed hydrophobic structure to a soluble, extended hydrophilic configuration [206,235,238].

Another approach to prevent biofouling involves the gradual release of active antimicrobial molecules from polymeric coatings in a self-defending strategy to inhibit bacterial colonization and biofilm formation. Silver nanoparticles [239–241] and small-molecule antibiotics [242] have been incorporated in multilayer coatings to achieve controlled release kinetics. Nonetheless, the major shortcoming of this release-based approach is the rapid exhaustion of the active species, which results in diminished antimicrobial activity over time [206]. To overcome this disadvantage, polymeric materials with covalently attached antimicrobial moieties have been developed. PEG-acrylate derivatives of the antibiotic vancomycin can be polymerized onto a device surface to allow a loading dose several thousand times larger than that achieved with monolayer vancomycin coupling approaches [243,244]. Alternatively, polymers with pendant quaternary polycationic groups [245,246] or covalently attached antimicrobial peptides [247,248] can retain their antimicrobial characteristics even after repeated usage.

Lastly, a few studies have demonstrated that combining two or more of the aforementioned techniques can result in polymeric coatings with dual self-cleansing and self-defending functionalities or more lasting antimicrobial effect. Cheng *et al.* reported a pCBMA-based polycationic coating that can kill bacteria and then switch to a zwitterionic surface upon hydrolysis to release dead bacteria [249]. Polymeric multilayers containing both immobilized quaternary ammonium salts (QAS) and silver ions/nanoparticles have also been fabricated to achieve an extended microbicidal effect due to the presence of the QAS even after the exhaustion of the silver species [250].

19.5.2 Polymeric Surfaces to Direct Biological Responses

While antifouling materials are designed to achieve biological inertness by preventing adhesion of undesirable proteins or microbes, another new class of biomaterials has been rationally engineered to actively control or direct cellular responses. Biomaterial scientists have recently explored biochemical signals and surface topological cues with the long-term goal of achieving improved biofunctionality of the prostheses or expedite device-tissue integration.

19.5.2.1 Surface with Biochemical Cues

Bioactive molecules can be immobilized on polymeric surfaces by covalent chemical attachment or physical adsorption [251]. Solid polymer surfaces with reactive functional groups (OH, NH_2, COOH, SH, and $CHCH_2$) present an opportunity for covalent conjugation with biomolecules either directly or via a spacer linker [251]. Relatively inert polymers can also be treated with surface modification techniques to introduce reactive functional groups and allow chemical functionalization with desired biochemical moieties [251]. Alternatively, biomolecules can also be physically immobilized on polymeric surfaces via van der Waal forces, affinity binding, or electrostatic interaction [251]. Notably, these biomolecules must be presented on the material surfaces with stabilized conformations and correct orientations to preserve their bioactivity [252–255].

Polymeric biorecognition surfaces have been constructed by immobilizing components of the extracellular matrix (ECM) that are critical for specific cell functions [256]. For example, oligopeptide motifs such as the Arg-Gly-Asp (RGD) sequence identified in a number of ECM proteins and the integrin $\alpha_2\beta_1$ binding domain found in collagen type I have been used to engineer surfaces with increased cell adhesion properties [252,257,258]. A graft copolymer of poly-L-lysine backbone with pendant PEG side chains, which were partially functionalized with RGD sequences, was able to reduce serum proteins adsorption to treated polystyrene surface and facilitate the binding of human dermal fibroblasts [259]. Alternatively, ECM proteins such as fibronectin or osteopontin have also been immobilized on material surface to promote multiple responses mediated by transmembrane receptors. Osteopontin has been identified as an important mediator in the recruitment of immune cells during wound healing and an inhibitor of mineral deposition during pathological calcification of blood vessels and heart valves [260]. This protein has been immobilized on the surface of pHEMA to promote endothelial cell adhesion and spreading [254]. Such osteopontin-based materials might have potential therapeutic value in inhibiting pathological calcification and promoting endothelialization on prosthetic heart valves and blood vessels. In addition, polymeric coatings with immobilized fibronectin and laminin or PVA hydrogels conjugated with amniotic membrane are potentially useful for keratoprosthetic devices due to their ability to enhance corneal epithelialization [118,261,262].

19.5.2.2 Surface with Topographical Cues

Recent advances in micro- and nanofabrication technologies have provided increasingly sophisticated tools to design material surfaces with controlled topography at different length scales. Polymeric surfaces with topographical features such as grooves, ridges, fiber, pillar, islands, and pits have been fabricated by various techniques such as soft lithography [263], electrospinning [264,265], and nonlithographic templating [266,267] for micrometer-length scale or e-beam lithography [268], polymer demixing [269,270], and block copolymer phase separation [271–273] for nanometer dimensions.

Microscale features have been reported to affect cellular morphology, adhesion, migration, and differentiation [263]. For example, contact guidance is a well-established phenomenon in which cells become elongated, align, and migrate in the direction of microgrooves with comparable dimension to that of individual cells [274,275]. On the other hand, the effects of nanoscale features on protein adsorption and cellular response are currently less well understood [276–278]. Nonetheless, nanoscale features are of relevant length scales to the components of the ECM such as the basement membrane [278,279] and the interconnecting nanopores [277,280,281] and can potentially influence the conformation and orientation of adsorbed proteins [276,282,283]. In addition, the effects of surface micro- and nanotopography on cellular behavior can vary widely depending on the cell types [276,277,284]. For example, fibroblasts demonstrate full alignment in microgrooves while only partial alignment is observed for macrophages and no ordered orientation observed for keratinocytes or neutrophils [285].

Existing studies on polymeric surfaces with biochemical and/or topographical cues have only focused on gaining initial understanding of the cell-material interaction on simple two-dimensional flat substrate fabricated from a limited range of polymers. Future opportunities exist for transferring these knowledge and material design strategies to more device-specific materials or addressing complex three-dimensional geometry and curvature of clinically

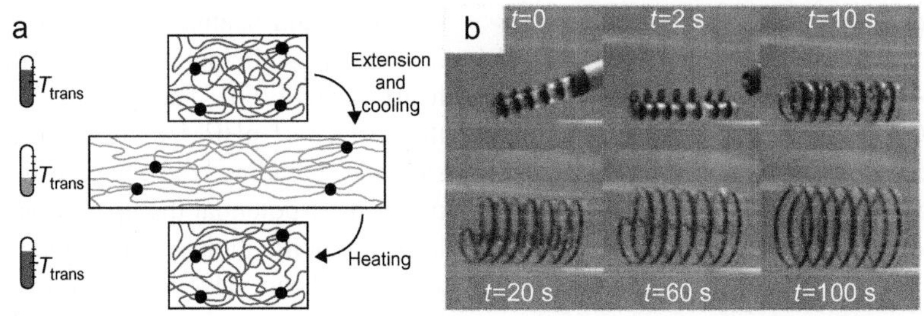

FIGURE 19.4 Molecular mechanism and macroscopic effect of a shape-memory polymer. (a) Schematic representation of the thermally induced shape-memory effect of a polymer network with $T_{trans}=T_g$. (b) Shape recovery of a stent with $T_g=52°$ in water at 37 °C. The stent gradually changed from its temporary compacted shape at $t=0$ s to its permanent expanded shape at $t=100$ s. (For color version of this figure, the reader is referred to the online version of this chapter.) Panel (a): Adapted from Lendlein and Kelch [290] with permission from Elsevier, copyright (2002). Panel (b): Adapted from Yakacki *et al.* [291] with permission from Elsevier, copyright (2007).

relevant prostheses [256]. For example, Berry and coworkers have recently proposed a simple polymer demixing approach to fabricate three-dimensional nylon tubing with internal nanotopography to influence fibroblast morphology, size, and adhesion [286,287]. This approach is potentially useful in enhancing endothelialization of prostheses with polymeric tubular structures such as vascular grafts or stents.

19.5.3 Shape-Memory Polymers

Shape-memory polymers (SMPs) are polymeric materials that have the ability to switch from a temporary to a permanent shape in response to external stimuli such as heat, infrared radiation, electrical signals, magnetic field, or immersion in water [288,289]. For example, Figure 19.4 illustrates the molecular mechanism and macroscopic effect of a covalently cross-linked polymer network with temperature-dependent shape-memory effect [288,290]. During the fabrication or "programming" of a SMP (Figure 19.4a), the polymer network is first formed into the initial, permanent shape when the material is maintained above the transition temperature (T_{trans}) of its polymeric chain segments, allowing them to remain flexible. The material is then deformed into the temporary shape and this configuration can be fixed by decreasing its temperature below T_{trans}, causing the chain segments to be rigid. Upon application of the appropriate external stimulus, i.e, heating above T_{trans}, the material recovers its original shape [290,292]. Compared to traditional shape-memory alloys, SMPs offer the advantages of lower density, lower cost of raw materials and fabrication, and the ease of processing and tailoring [289]. However, one important design consideration for SMP-based medical devices is to ensure that the shape-memory stimulus meets safety requirements to avoid undesired trigger and subsequent damages to surrounding native healthy tissues [289]. Nonetheless, the versatility of SMPs renders them attractive for potential applications in a new generation of implantable prostheses such as ocular implants [293], neural electrode implants [294], cochlear implants [295], and removable stents [296–298]. For example, future SMP-based stents (Figure 19.4b) have the potential advantage of anchoring themselves during intravascular intervention by gradual expansion when exposed to body temperature [289,291]. They might also be replaced if inadequately positioned, simply by softening and decreasing their diameters [289]. In general, further material development and performance characterization are needed to achieve mechanical properties matching vasculature requirements and long-term material stability before SMP-based stents can reach their clinical success [289].

19.6 CONCLUSION AND PERSPECTIVES

Polymeric biomaterials have been widely used in many implantable prostheses and play a critical role in improving human health. The first generations of polymers selected for implantable device have been intended to achieve biological inertness and mechanical and structural compatibility. Recently, the paradigm shift toward bioactive polymers might lead to the development of new material classes that not only are biocompatible but also can respond to molecular, cellular, and mechanical cues. Most of the studies conducted for the emerging materials so far have been limited to early synthetic approaches and *in vitro* evaluation. However, new polymeric biomaterials intended for long-term prostheses must be subjected to predictive testing such as evaluation in animal models, accelerated aging, and statistical projections before reaching its clinical application. The selection of appropriate polymeric biomaterials can only be iteratively improved with information from postimplant monitoring of devices in existing clinical use.

ACKNOWLEDGMENTS

The authors acknowledge funding from the Presidential Early Career Award for Scientists and Engineers (PECASE), the Office of Naval Research Young National Investigator Award, the National Science Foundation

CAREER Award (DMR 0847287), the National Science Foundation (1240443) and the National Institutes of Health (HL092836, AR057837, DE021468, DE019024, EB012597, HL099073, EB008392), and MIT Portugal Program (MPP-09Call-Langer-47). T. T. D. was supported by the Sung Wan Kim Postdoctoral Fellowship from the Controlled Release Society Foundation. A. M. would like to thank the Strategic Technologies Program of King Abdulaziz City for Science and Technology (KACST), 10-NAN1081-3 for their support.

REFERENCES

[1] M.J. Lysaght, J.A. O'Loughlin, Demographic scope and economic magnitude of contemporary organ replacement therapies, ASAIO J. 46 (2000) 515–521.

[2] D. Williams, Implantable prostheses, Phys. Med. Biol. 25 (1980) 611–636.

[3] P.F. Gomez, J.A. Morcuende, Early attempts at hip arthroplasty: 1700s to 1950s, Iowa Orthop. J. 25 (2005) 25–29.

[4] W. Waugh, John Charnley: The Man and the Hip, Springer-Verlag, London, New York, 1990.

[5] D.A. McIntyre, The eleven most implanted medical devices in America, 24/7 Wall Street (2011, April 9, 2013). http://247wallst.com/healthcare-economy/2011/07/18/the-eleven-most-implanted-medical-devices-in-america/.

[6] N. Angelova, D. Hunkeler, Rationalizing the design of polymeric biomaterials, Trends Biotechnol. 17 (1999) 409–421.

[7] B.D. Ratner, The history of biomaterials, in: B.D. Ratner, A.S. Hoffman, F.J. Schoen, J.E. Lemons (Eds.), Biomaterials Science: An Introduction to Materials in Medicine, third ed., Academic Press, Elsevier, Waltham, MA, USA, 2013.

[8] B.D. Ratner, A.S. Hoffman, F.J. Schoen, J.E. Lemons, Biomaterials Science: An Introduction to Materials in Medicine, third ed., Academic Press, Elsevier, Waltham, MA, USA, 2013.

[9] F.J. Schoen, Introduction: biological responses to biomaterials, in: B.D. Ratner, A.S. Hoffman, F.J. Schoen, J.E. Lemons (Eds.), Biomaterials Science: An Introduction to Materials in Medicine, third ed., Academic Press, Elsevier, Waltham, MA, USA, 2013.

[10] D.F. Williams, Definitions in Biomaterials, Elsevier, Amsterdam, 1987.

[11] D.F. Williams, The Williams Dictionary of Biomaterials, Liverpool University Press, Liverpool, 1999.

[12] D.F. Williams, On the mechanisms of biocompatibility, Biomaterials 29 (2008) 2941–2953.

[13] Y. Abu-Amer, I. Darwech, J.C. Clohisy, Aseptic loosening of total joint replacements: mechanisms underlying osteolysis and potential therapies, Arthritis Res. Ther. 9 (2007) S6.

[14] J. Destouet, B. Monsees, R. Oser, J. Nemecek, V. Young, T. Pilgram, Screening mammography in 350 women with breast implants: prevalence and findings of implant complications, Am. J. Roentgenol. 159 (1992) 973–978.

[15] A.W. Bridges, A.J. Garcia, Anti-inflammatory polymeric coatings for implantable biomaterials and devices, J. Diabetes Sci. Technol. 2 (2008) 984–994.

[16] M.B. Gorbet, M.V. Sefton, Biomaterial-associated thrombosis: roles of coagulation factors, complement, platelets and leukocytes, Biomaterials 25 (2004) 5681–5703.

[17] D. Chauvel-Lebret, P. Auroy, M. Bonnaure-Mallet, Biocompatibility of elastomers, in: S. Dumitriu (Ed.), Polymeric Biomaterials, second ed., Marcel Dekker, Inc., New York, 2002.

[18] J.M. Anderson, M.S. Shive, Biodegradation and biocompatibility of PLA and PLGA microspheres, Adv. Drug Deliv. Rev. 28 (1997) 5–24.

[19] W. Siggelkow, A. Faridi, K. Spiritus, U. Klinge, W. Rath, B. Klosterhalfen, Histological analysis of silicone breast implant capsules and correlation with capsular contracture, Biomaterials 24 (2003) 1101–1109.

[20] E. Piskin, Biodegradable polymers as biomaterials, J. Biomater. Sci. Polym. Ed. 6 (1995) 775–795.

[21] M. Shen, I. Garcia, R.V. Maier, T.A. Horbett, Effects of adsorbed proteins and surface chemistry on foreign body giant cell formation, tumor necrosis factor alpha release and procoagulant activity of monocytes, J. Biomed. Mater. Res. A 70 (2004) 533–541.

[22] C.J. Wilson, R.E. Clegg, D.I. Leavesley, M.J. Pearcy, Mediation of biomaterial-cell interactions by adsorbed proteins: a review, Tissue Eng. 11 (2005) 1–18.

[23] T. Collier, J. Anderson, Protein and surface effects on monocyte and macrophage adhesion, maturation, and survival, J. Biomed. Mater. Res. 60 (2002) 487–496.

[24] S.R. Hanson, E.I. Tucker, Blood coagulation and blood-materials interactions, in: B.D. Ratner, A.S. Hoffman, F.J. Schoen, J.E. Lemons (Eds.), Biomaterial Science: An Introduction to Materials in Medicine, third ed., Academic Press, Elsevier, Waltham, MA, USA, 2013.

[25] S.G. Wise, A. Waterhouse, P. Michael, M.K. Ng, Extracellular matrix molecules facilitating vascular biointegration, J. Funct. Biomater. 3 (2012) 569–587.

[26] J.M. Anderson, Inflammatory response to implants, ASAIO J. 34 (1988) 101–107.

[27] J.M. Anderson, Inflammation, wound healing, and the foreign body response, in: B.D. Ratner, A.S. Hoffman, F.J. Schoen, J.E. Lemons (Eds.), Biomaterial Science: An Introduction to Materials in Medicine, third ed., Academic Press, Elsevier, Waltham, MA, USA, 2013.

[28] D.M. Mosser, J.P. Edwards, Exploring the full spectrum of macrophage activation, Nat. Rev. Immunol. 8 (2008) 958–969.

[29] S. Franz, S. Rammelt, D. Scharnweber, J.C. Simon, Immune responses to implants—a review of the implications for the design of immunomodulatory biomaterials, Biomaterials 32 (2011) 6692–6709.

[30] J.M. Anderson, A. Rodriguez, D.T. Chang, Foreign body reaction to biomaterials, Semin. Immunol. 20 (2008) 86–100.

[31] S. Barrientos, O. Stojadinovic, M.S. Golinko, H. Brem, M. Tomic-Canic, Growth factors and cytokines in wound healing, Wound Repair Regen. 16 (2008) 585–601.

[32] T.J. Marrie, J. Nelligan, J. Costerton, A scanning and transmission electron microscopic study of an infected endocardial pacemaker lead, Circulation 66 (1982) 1339–1341.

[33] J. Costerton, L. Montanaro, C. Arciola, Biofilm in implant infections: its production and regulation, Int. J. Artif. Organs 28 (2005) 1062–1068.

[34] R.M. Donlan, R. Donlan, Biofilms: microbial life on surfaces, Emerg. Infect. Dis. 8 (2002) 881–890.

[35] S.D. Elek, P. Conen, The virulence of Staphylococcus pyogenes for man. A study of the problems of wound infection, Br. J. Exp. Pathol. 38 (1957) 573–586.

[36] E.T.J. Rochford, D.J. Jaekel, N.J. Hickock, R.G. Richards, T.F. Moriarty, A.H.C. Poulsson, Bacterial interactions with polyaryletheretherketone, in: S.M. Kurtz (Ed.), PEEK Biomaterials Handbook, Elsevier, USA, 2012, pp. 93–119.

[37] P. Stoodley, L. Hall-Stoodley, B. Costerton, P. DeMeo, M. Shirtliff, E. Gawalt, S. Kathju, Biofilms, biomaterials, and device-related infections, in: B.D. Ratner, A.S. Hoffman, F.J. Schoen, J.E. Lemons (Eds.), Biomaterial Science: Introduction to Materials in Medicine, third ed., Academic Press, Elsevier, Waltham, MA, USA, 2013.

[38] A. Trampuz, W. Zimmerli, Diagnosis and treatment of implant-associated septic arthritis and osteomyelitis, Curr. Infect. Dis. Rep. 10 (2008) 394–403.

[39] J. Del Pozo, R. Patel, The challenge of treating biofilm-associated bacterial infections, Clin. Pharmacol. Ther. 82 (2007) 204–209.

[40] D. Williams, Tissue-biomaterial interactions, J. Mater. Sci. 22 (1987) 3421–3445.

[41] T.T. Dang, K.M. Bratlie, S.R. Bogatyrev, X.Y. Chen, R. Langer, D.G. Anderson, Spatiotemporal effects of a controlled-release anti-inflammatory drug on the cellular dynamics of host response, Biomaterials 32 (2011) 4464–4470.

[42] K.M. Bratlie, T.T. Dang, S. Lyle, M. Nahrendorf, R. Weissleder, R. Langer, D.G. Anderson, Rapid biocompatibility analysis of materials via in vivo fluorescence imaging of mouse models, PLoS One 5 (2010) e10032.

[43] W.F. Liu, M. Ma, K.M. Bratlie, T.T. Dang, R. Langer, D.G. Anderson, Real-time in vivo detection of biomaterial-induced reactive oxygen species, Biomaterials 32 (2011) 1796–1801.

[44] G. Zaikov, Quantitative aspects of polymer degradation in the living body, J. Macromol. Sci. Rev. Macromol. Chem. Phys. 25 (1985) 551–597.

[45] A.J. Coury, Chemical and biochemical degradation of polymers intended to be biostable, in: B.D. Ratner, A.S. Hoffman, F.J. Schoen, J.E. Lemons (Eds.), Biomaterials Science: An Introduction to Materials in Medicine, third ed., Academic Press, Elsevier, Waltham, MA, USA, 2013.

[46] B.D. Ratner, Introduction: the body fights back—degradation of materials in the biological environment, in: B.D. Ratner, A.S. Hoffman, F.J. Schoen, J.E. Lemons (Eds.), Biomaterials Science: An Introduction to Materials in Medicine, third ed., Academic Press, Elsevier, Waltham, MA, USA, 2013.

[47] A. Hensten, N. Jacobsen, Systemic toxicity and hypersensitivity, Biomaterials Science: An Introduction to Materials in Medicine, third ed., Academic Press, Elsevier, Waltham, MA, USA, 2013.

[48] K. Gumargalieva, Y.V. Moiseev, T. Daurova, O. Voronkova, Effect of infections on the degradation of polyethylene terephthalate implants, Biomaterials 3 (1982) 177–180.

[49] J.A. Hudson, A. Crugnola, The in vivo biodegradation of nylon 6 utilized in a particular IUD, J. Biomater. Appl. 1 (1986) 487–501.

[50] A. Coury, K. Stokes, P. Cahalan, P. Slaikeu, Biostability considerations for implantable polyurethanes, Life Support Syst. 5 (1987) 25–39.

[51] D.J. Martin, L.A. Poole Warren, P.A. Gunatillake, S.J. McCarthy, G.F. Meijs, K. Schindhelm, New methods for the assessment of in vitro and in vivo stress cracking in biomedical polyurethanes, Biomaterials 22 (2001) 973–978.

[52] K. Stokes, A. Coury, P. Urbanski, Autooxidative degradation of implanted polyether polyurethane devices, J. Biomater. Appl. 1 (1986) 411–448.

[53] J. Black, G. Hastings, Handbook of Biomaterial Properties, Chapman and Hall, London, UK, 1998.

[54] S. Ramakrishna, J. Mayer, E. Wintermantel, K.W. Leong, Biomedical applications of polymer-composite materials: a review, Compos. Sci. Technol. 61 (2001) 1189–1224.

[55] J.R. Davis, Handbook of Materials for Medical Devices, ASM International, Ohio, 2003.

[56] R. Huiskes, Some fundamental aspects of human joint replacement. Analyses of stresses and heat conduction in bone-prosthesis structures, Acta Orthop. Scand. Suppl. 185 (1980) 1–208.

[57] E. Schneider, C. Kinast, J. Eulenberger, D. Wyder, G. Eskilsson, S.M. Perren, A comparative study of the initial stability of cementless hip prostheses, Clin. Orthop. Relat. Res. 248 (1989) 200–209.

[58] D.R. Sumner, J.O. Galante, Determinants of stress shielding—design versus materials versus interface, Clin. Orthop. Relat. Res. 274 (1992) 202–212.

[59] M. Long, H.J. Rack, Titanium alloys in total joint replacement—a materials science perspective, Biomaterials 19 (1998) 1621–1639.

[60] R.Y. Kannan, H.J. Salacinski, P.E. Butler, G. Hamilton, A.M. Seifalian, Current status of prosthetic bypass grafts: a review, J. Biomed. Mater. Res. B. 74B (2005) 570–581.

[61] M.R. Kapadia, D.A. Popowich, M.R. Kibbe, Modified prosthetic vascular conduits, Circulation 117 (2008) 1873–1882.

[62] W.M. Abbott, J. Megerman, J.E. Hasson, G. Litalien, D.F. Warnock, Effect of compliance mismatch on vascular graft patency, J. Vasc. Surg. 5 (1987) 376–382.

[63] S. Venkatraman, F. Boey, L.L. Lao, Implanted cardiovascular polymers: natural, synthetic and bio-inspired, Prog. Polym. Sci. 33 (2008) 853–874.

[64] V.P. Shastri, Non-degradable biocompatible polymers in medicine: past, present and future, Curr. Pharm. Biotechnol. 4 (2003) 331–337.

[65] J.S. Hanker, B.L. Giammara, Biomaterials and biomedical devices, Science 242 (1988) 885–892.

[66] A. Kurella, N.B. Dahotre, Review paper: surface modification for bioimplants: the role of laser surface engineering, J. Biomater. Appl. 20 (2005) 5–50.

[67] A.G. Kidane, G. Burriesci, M. Edirisinghe, H. Ghanbari, P. Bonhoeffer, A.M. Seifalian, A novel nanocomposite polymer for development of synthetic heart valve leaflets, Acta Biomater. 5 (2009) 2409–2417.

[68] G.M. Bernacca, T.G. Mackay, R. Wilkinson, D.J. Wheatley, Calcification and fatigue failure in a polyurethane heart-valve, Biomaterials 16 (1995) 279–285.

[69] G.M. Bernacca, T.G. Mackay, R. Wilkinson, D.J. Wheatley, Polyurethane heart valves: fatigue failure, calcification, and polyurethane structure, J. Biomed. Mater. Res. 34 (1997) 371–379.

[70] S.H. Daebritz, B. Fausten, B. Hermanns, A. Franke, J. Schroeder, J. Groetzner, R. Autschbach, B.J. Messmer, J.S. Sachweh, New flexible polymeric heart valve prostheses for the mitral and aortic positions, Heart Surg. Forum 7 (2004) E525–E532.

[71] M. Kuetting, J. Roggenkamp, U. Urban, T. Schmitz-Rode, U. Steinseifer, Polyurethane heart valves: past, present and future, Expert Rev. Med. Devices 8 (2011) 227–233.

[72] M. Butterfield, D.J. Wheatley, D.F. Williams, J. Fisher, A new design for polyurethane heart valves, J. Heart Valve Dis. 10 (2001) 105–110.

[73] G.M. Bernacca, I. Straub, D.J. Wheatley, Mechanical and morphological study of biostable polyurethane heart valve leaflets explanted from sheep, J. Biomed. Mater. Res. 61 (2002) 138–145.

[74] D.J. Wheatley, L. Raco, G.M. Bernacca, I. Sim, P.R. Belcher, J.S. Boyd, Polyurethane: material for the next generation of heart valve prostheses? Eur. J. Cardiothorac. Surg. 17 (2000) 440–447.

[75] R. Kiraly, R. Yozu, D. Hillegass, H. Harasaki, S. Murabayashi, J. Snow, Y. Nose, Hexsyn trileaflet valve—application to temporary blood pumps, Artif. Organs 6 (1982) 190–197.

[76] F. Nistal, V. Garciamartinez, E. Arbe, D. Fernandez, E. Artinano, F. Mazorra, I. Gallo, In vivo experimental assessment of polytetrafluoroethylene trileaflet heart-valve prosthesis, J. Thorac. Cardiovasc. Surg. 99 (1990) 1074–1081.

[77] H.J. Jiang, G. Campbell, D. Boughner, W.K. Wan, M. Quantz, Design and manufacture of a polyvinyl alcohol (PVA) cryogel tri-leaflet heart valve prosthesis, Med. Eng. Phys. 26 (2004) 269–277.

[78] C.D. Campbell, D.H. Brooks, M.W. Webster, H.T. Bahnson, Use of expanded microporous polytetrafluoroethylene for limb salvage: a preliminary report, Surgery 79 (1976) 485–491.

[79] P.A. Stonebridge, R.J. Prescott, C.V. Ruckley, Randomized trial comparing infrainguinal polytetrafluoroethylene bypass grafting with and without vein interposition cuff at the distal anastomosis, J. Vasc. Surg. 26 (1997) 543–550.

[80] G.D. Griffiths, J. Nagy, D. Black, P.A. Stonebridge, G. Joint Vasc Res, Randomized clinical trial of distal anastomotic interposition vein cuff in infrainguinal polytetrafluoroethylene bypass grafting, Br. J. Surg. 91 (2004) 560–562.

[81] H.P. Greisler, D.J. Cziperle, D.U. Kim, J.D. Garfield, D. Petsikas, P.M. Murchan, E.O. Applegren, W. Drohan, W.H. Burgess, Enhanced endothelialization of expanded polytetrafluoroethylene grafts by fibroblast growth factor type-1 pretreatment, Surgery 112 (1992) 244–255.

[82] B.H. Walpoth, P. Zammaretti, M. Cikirikcioglu, E. Khabiri, M.K. Djebaili, J.C. Pache, J.C. Tille, Y. Aggoun, D. Morel, A. Kalangos, J.A. Hubbell, A.H. Zisch, Enhanced intimal thickening of expanded polytetrafluoroethylene grafts coated with fibrin or fibrin-releasing vascular endothelial growth factor in the pig carotid artery interposition model, J. Thorac. Cardiovasc. Surg. 133 (2007) 1163–1170.

[83] M. Bosiers, K. Deloose, J. Verbist, H. Schroe, G. Lauwers, W. Lansink, P. Peeters, Heparin-bonded expanded polytetrafluoroethylene vascular graft for femoropopliteal and femorocrural bypass grafting: 1-year results, J. Vasc. Surg. 43 (2006) 313–318.

[84] D.L. Akers, Y.H. Du, R.F. Kempczinski, The effect of carbon coating and porosity on early patency of expanded polytetrafluoroethylene grafts—an experimental study, J. Vasc. Surg. 18 (1993) 10–15.

[85] X. Kapfer, W. Meichelboeck, F.M. Groegler, Comparison of carbon-impregnated and standard ePTFE prostheses in extra-anatomical anterior tibial artery bypass: a prospective randomized multicenter study, Eur. J. Vasc. Endovasc. Surg. 32 (2006) 155–168.

[86] M. Heise, G. Schmidmaier, I. Husmann, C. Heidenhain, J. Schmidt, P. Neuhaus, U. Settmacher, PEG-hirudin/iloprost coating of small diameter ePTFE grafts effectively prevents pseudointima and intimal hyperplasia development, Eur. J. Vasc. Endovasc. Surg. 32 (2006) 418–424.

[87] I. Baek, C.Z. Bai, J. Hwang, J. Park, J.S. Park, D.J. Kim, Suppression of neointimal hyperplasia by sirolimus-eluting expanded polytetrafluoroethylene (ePTFE) haemodialysis grafts in comparison with paclitaxel-coated grafts, Nephrol. Dial. Transplant. 27 (2012) 1997–2004.

[88] M.C. Wyers, M.D. Phaneuf, E.M. Rzucidlo, M.A. Contreras, F.W. LoGerfo, W.C. Quist, In vivo assessment of a novel Dacron surface with covalently bound recombinant hirudin, Cardiovasc. Pathol. 8 (1999) 153–159.

[89] T. Okoshi, G. Soldani, M. Goddard, P.M. Galletti, Very small diameter polyurethane vascular prosthesis with rapid endothelialization for coronary-artery bypass grafting, J. Thorac. Cardiovasc. Surg. 105 (1993) 791–795.

[90] R.D.M. Allen, E. Yuill, B.J. Nankivell, D.M.A. Francis, Australian multicentre evaluation of a new polyurethane vascular access graft, Aust. N.Z. J. Surg. 66 (1996) 738–742.

[91] A.M. Seifalian, H.J. Salacinski, A. Tiwari, A. Edwards, S. Bowald, G. Hamilton, In vivo biostability of a poly(carbonate-urea)urethane graft, Biomaterials 24 (2003) 2549–2557.

[92] H. Martz, R. Paynter, S. Benslimane, G. Beaudoin, R. Guidoin, J. Borzone, H. Bensimhon, R. Satin, N. Sheiner, Hydrophilic microporous polyurethane versus expanded PTFE grafts as substitutes in the carotid arteries of dogs—a limited study, J. Biomed. Mater. Res. 22 (1988) 63–69.

[93] S.K. Kakkos, R. Haddad, G.K. Haddad, D.J. Reddy, T.J. Nypaver, J.C. Lin, A.D. Shepard, Results of aggressive graft surveillance and endovascular treatment on secondary patency rates of Vectra Vascular Access Grafts, J. Vasc. Surg. 45 (2007) 974–980.

[94] Y. Ishii, S.I. Sakamoto, R.T. Kronengold, R. Virmani, E.A. Rivera, S.M. Goldman, E.J. Prechtel, J.G. Hill, R.J. Damiano, A novel bioengineered small-caliber vascular graft incorporating heparin and sirolimus: excellent 6-month patency, J. Thorac. Cardiovasc. Surg. 135 (2008) 1237–1246.

[95] N. Grabow, D.P. Martin, K.-P. Schmitz, K. Sternberg, Absorbable polymer stent technologies for vascular regeneration, J. Chem. Technol. Biotechnol. 85 (2010) 744–751.

[96] S.V. Ranade, K.M. Miller, R.E. Richard, A.K. Chan, M.J. Allen, M.N. Helmus, Physical characterization of controlled release of paclitaxel from the TAXUS(TM) Express(2TM) drug-eluting stent, J. Biomed. Mater. Res. A 71A (2004) 625–634.

[97] H. Tamai, K. Igaki, E. Kyo, K. Kosuga, A. Kawashima, S. Matsui, H. Komori, T. Tsuji, S. Motohara, H. Uehata, Initial and 6-month results of biodegradable poly-l-lactic acid coronary stents in humans, Circulation 102 (2000) 399–404.

[98] J.A. Ormiston, M.W.I. Webster, G. Armstrong, First-in-human implantation of a fully bioabsorbable drug-eluting stent: the BVS poly-L-lactic acid everolimus-eluting coronary stent, Catheter. Cardiovasc. Interv. 69 (2007) 128–131.

[99] J. Kohn, J. Zeltinger, Degradable, drug-eluting stents: a new frontier for the treatment of coronary artery disease, Expert Rev. Med. Devices 2 (2005) 667–671.

[100] W. Wang, Y. Ouyang, C.K. Poh, Orthopaedic implant technology: biomaterials from past to future, Ann. Acad. Med. Singapore 40 (2011) 237–244.

[101] L. Costa, M.P. Luda, L. Trossarelli, E.M. Brach del Prever, M. Crova, P. Gallinaro, In vivo UHMWPE biodegradation of retrieved prosthesis, Biomaterials 19 (1998) 1371–1385.

[102] G. Banche, P. Bracco, A. Bistolfi, V. Allizond, M. Boffano, L. Costa, A. Cimino, A.M. Cuffini, E.M. Del Prever, Vitamin E blended UHMWPE may have the potential to reduce bacterial adhesive ability, J. Orthop. Res. 29 (2011) 1662–1667.

[103] S.R. Golish, P.A. Anderson, Bearing surfaces for total disc arthroplasty: metal-on-metal versus metal-on-polyethylene and other biomaterials, Spine J. 12 (2012) 693–701.

[104] H.D. Link, P.C. McAfee, L. Pimenta, Choosing a cervical disc replacement, Spine J. 4 (2004) S294–S302.

[105] A.C. Vrancken, P. Buma, T.G. van Tienen, Synthetic meniscus replacement: a review, Int. Orthop. 37 (2013) 291–299.

[106] T.G. van Tienen, G. Hannink, P. Buma, Meniscus replacement using synthetic materials, Clin. Sports Med. 28 (2009) 143–156.

[107] S. Spencer, A. Saithna, M. Carmont, M. Dhillon, P. Thompson, T. Spalding, Meniscal scaffolds: early experience and review of the literature, Knee 19 (2012) 760–765.

[108] M. Kobayashi, Y.S. Chang, M. Oka, A two year in vivo study of polyvinyl alcohol-hydrogel (PVA-H) artificial meniscus, Biomaterials 26 (2005) 3243–3248.

[109] A. Melvin, A. Litsky, J. Mayerson, K. Stringer, D. Melvin, N. Juncosa-Melvin, An artificial tendon to connect the quadriceps muscle to the tibia, J. Orthop. Res. 29 (2011) 1775–1782.

[110] C.K. Wei, S.M. Wang, J.C. Lin, A study of patient satisfaction after cataract surgery with implantation of different types of intraocular lenses, BMC Res. Notes 5 (2012) 592.

[111] D.P. DeVore, Long-term compatibility of intraocular lens implant materials, J. Long Term Eff. Med. Implants 1 (1991) 205–216.

[112] G.D. Kymionis, M.A. Grentzelos, A.E. Karavitaki, P. Zotta, S.H. Yoo, I.G. Pallikaris, Combined corneal collagen cross-linking and posterior chamber toric implantable collamer lens implantation for keratoconus, Ophthalmic Surg. Lasers Imaging 42 (2011) e22–e25.

[113] J.W. Kwon, Y.K. Han, W.J. Lee, C.S. Cho, S.J. Paik, D.I. Cho, J.H. Lee, W.R. Wee, Biocompatibility of poloxamer hydrogel as an injectable intraocular lens: a pilot study, J. Cataract Refract. Surg. 31 (2005) 607–613.

[114] M.G. Doane, C.H. Dohlman, G. Bearse, Fabrication of a keratoprosthesis, Cornea 15 (1996) 179–184.

[115] T.V. Chirila, C.R. Hicks, P.D. Dalton, S. Vijayasekaran, X. Lou, Y. Hong, A.B. Clayton, B.W. Ziegelaar, J.H. Fitton, S. Platten, G.J. Crawford, I.J. Constable, Artificial cornea, Prog. Polym. Sci. 23 (1998) 447–473.

[116] T.V. Chirila, S. Vijayasekaran, R. Horne, Y.C. Chen, P.D. Dalton, I.J. Constable, G.J. Crawford, Interpenetrating polymer network (IPN) as a permanent joint between the elements of a new type of artificial cornea, J. Biomed. Mater. Res. 28 (1994) 745–753.

[117] J.C. Barber, Keratoprosthesis: past and present, Int. Ophthalmol. Clin. 28 (1988) 103–109.

[118] Y. Uchino, S. Shimmura, H. Miyashita, T. Taguchi, H. Kobayashi, J. Shimazaki, J. Tanaka, K. Tsubota, Amniotic membrane immobilized poly(vinyl alcohol) hybrid polymer as an artificial cornea scaffold that supports a stratified and differentiated corneal epithelium, J. Biomed. Mater. Res. B 81 (2007) 201–206.

[119] N. Bjork, K. Ekstrand, I.E. Ruyter, Implant-fixed, dental bridges from carbon/graphite fibre reinforced poly(methyl methacrylate), Biomaterials 7 (1986) 73–75.

[120] V.E. Friedewald, R.O. Bonow, J.S. Borer, B.A. Carabello, P.P. Kleine, C.W. Akins, W.C. Roberts, The editor's roundtable: cardiac valve surgery, Am. J. Cardiol. 99 (2007) 1269–1278.

[121] P. Zilla, J. Brink, P. Human, D. Bezuidenhout, Prosthetic heart valves: catering for the few, Biomaterials 29 (2008) 385–406.

[122] R.A. Hoshi, R. Van Lith, M.C. Jen, J.B. Allen, K.A. Lapidos, G. Ameer, The blood and vascular cell compatibility of heparin-modified ePTFE vascular grafts, Biomaterials 34 (2013) 30–41.

[123] P. Pibarot, J.G. Dumesnil, Prosthetic heart valves selection of the optimal prosthesis and long-term management, Circulation 119 (2009) 1034–1048.

[124] V.L. Gott, D.E. Alejo, D.E. Cameron, Mechanical heart valves: 50 years of evolution, Ann. Thorac. Surg. 76 (2003) S2230–S2239.

[125] P. Bloomfield, Choice of heart valve prosthesis, Heart 87 (2002) 583–589.

[126] M.I. Shtilman, Polymeric Biomaterials. Part I. Polymer Implants, Ridderprint, Ridderkerk, Netherlands, 2003.

[127] G.S. Bhuvaneshwar, A.V. Ramani, K.V. Chandran, Polymeric occluders in tilting disc heart valve prostheses, in: S. Dumitriu (Ed.), Polymeric Biomaterials, second ed., Marcel Dekker Inc., New York, 2002.

[128] H. Ghanbari, H. Viatge, A.G. Kidane, G. Burriesci, M. Tavakoli, A.M. Seifalian, Polymeric heart valves: new materials, emerging hopes, Trends Biotechnol. 27 (2009) 359–367.

[129] S.L. Hilbert, V.J. Ferrans, Y. Tomita, E.E. Eidbo, M. Jones, Evaluation of explanted polyurethane trileaflet cardiac valve prosthesis, J. Thorac. Cardiovasc. Surg. 94 (1987) 419–429.

[130] S.H. Daebritz, J.S. Sachweh, B. Hermanns, B. Fausten, A. Franke, J. Groetzner, B. Klosterhalfen, B.J. Messmer, Introduction of a flexible polymeric heart valve prosthesis with special design for mitral position, Circulation 108 (2003) 134–139.

[131] T.G. Mackay, D.J. Wheatley, G.M. Bernacca, A.C. Fisher, C.S. Hindle, New polyurethane heart valve prosthesis: design, manufacture and evaluation, Biomaterials 17 (1996) 1857–1863.

[132] V.G. Sister, V.N. Iurechko, Experimental study of the hydrodynamics of polyurethane tricuspid heart valves, Med. Tekh. 6 (2006) 8–14.

[133] J.W. Boretos, W.S. Pierce, Segmented polyurethane—a new elastomer for biomedical applications, Science 158 (1967) 1481–1482.

[134] D. Cosgrove, R. Frater Wheatley, A. Ritchie, Polyurethane: material for the next generation of heart valve prostheses? Eur. J. Cardiothorac. Surg. 17 (2000) 440–448.

[135] M. Szycher, Biostability of polyurethane elastomers: a critical review, J. Biomater. Appl. 3 (1988) 297–402.

[136] Y.W. Tang, R.S. Labow, J.P. Santerre, Enzyme-induced biodegradation of polycarbonate-polyurethanes: dependence on hard-segment chemistry, J. Biomed. Mater. Res. 57 (2001) 597–611.

[137] H.J. Salacinski, M. Odlyha, G. Hamilton, A.M. Seifalian, Thermo-mechanical analysis of a compliant poly(carbonate-urea)urethane after exposure to hydrolytic, oxidative, peroxidative and biological solutions, Biomaterials 23 (2002) 2231–2240.

[138] M. Dabagh, M.J. Abdekhodaie, M.T. Khorasani, Effects of polydimethylsiloxane grafting on the calcification, physical properties, and biocompatibility of polyurethane in a heart valve, J. Appl. Polym. Sci. 98 (2005) 758–766.

[139] A. Simmons, J. Hyvarinen, R.A. Odell, D.J. Martin, P.A. Gunatillake, K.R. Noble, L.A. Poole-Warren, Long-term in vivo biostability of poly(dimethylsiloxane)/poly(hexamethylene oxide) mixed macrodiol-based polyurethane elastomers, Biomaterials 25 (2004) 4887–4900.

[140] F.J. Veith, S.K. Gupta, E. Ascer, S. Whiteflores, R.H. Samson, L.A. Scher, J.B. Towne, V.M. Bernhard, P. Bonier, W.R. Flinn, P. Astelford, J.S.T. Yao, J.J. Bergan, 6-Year prospective multicenter randomized comparison of autologous saphenous vein and expanded polytetrafluoroethylene grafts in infrainguinal arterial reconstructions, J. Vasc. Surg. 3 (1986) 104–114.

[141] L.J. Brossollet, Mechanical issues in vascular grafting—a review, Int. J. Artif. Organs 15 (1992) 579–584.

[142] S.G. Freidman, A History of Vascular Surgery, Blackwell Publishing Inc., Malden, MA, 2005.

[143] N.D. Ku, R.C. Allen, Vascular grafts, in: J. Bronzino (Ed.), The Biomedical Engineering Handbook, vol. 2, CRC Press, Boca Raton, FL, 2000.

[144] D.J. Farrar, Development of a prosthetic coronary artery bypass graft, Heart Surg. Forum 3 (2000) 36–40.

[145] U. Sigwart, J. Puel, V. Mirkovitch, F. Joffre, L. Kappenberger, Intravascular stents to prevent occlusion and restenosis after transluminal angioplasty, N. Engl. J. Med. 316 (1987) 701–706.

[146] J.C. Palmaz, D.T. Kopp, H. Hayashi, R.A. Schatz, G. Hunter, F.O. Tio, O. Garcia, R. Alvarado, C. Rees, S.C. Thomas, Normal and stenotic renal-arteries—experimental balloon-expandable intraluminal stenting, Radiology 164 (1987) 705–708.

[147] E. Mariano, G.M. Sangiorgi, Coronary stents, in: P. Pavone, M. Fioranelli, D.A. Dowe (Eds.), CT Evaluation of Coronary Artery Disease, first ed., Springer-Verlag, Milan, Italy, 2009, pp. 113–128.

[148] D.L. Fischman, M.B. Leon, D.S. Baim, R.A. Schatz, M.P. Savage, I. Penn, K. Detre, L. Veltri, D. Ricci, M. Nobuyoshi, M. Cleman, R. Heuser, D. Almond, P.S. Teirstein, R.D. Fish, A. Colombo, J. Brinker, J. Moses, A. Shaknovich, J. Hirshfeld, S. Bailey, S. Ellis, R. Rake, S. Goldberg, A randomized comparison of coronary-stent placement and balloon angioplasty in the treatment of coronary artery disease, N. Engl. J. Med. 331 (1994) 496–501.

[149] T.C. Woods, A.R. Marks, Drug-eluting stents, Annu. Rev. Med. 55 (2004) 169–178.

[150] A. Kraitzer, Y. Kloog, M. Zilberman, Approaches for prevention of restenosis, J. Biomed. Mater. Res. B. 85B (2008) 583–603.

[151] M.N. Babapulle, M.J. Eisenberg, Coated stents for the prevention of restenosis: part I, Circulation 106 (2002) 2734–2740.

[152] M.N. Babapulle, M.J. Eisenberg, Coated stents for the prevention of restenosis: part II, Circulation 106 (2002) 2859–2866.

[153] E. Regar, G. Sianos, P.W. Serruys, Stent development and local drug delivery, Br. Med. Bull. 59 (2001) 227–248.

[154] S. Venkatraman, F. Boey, Release profiles in drug-eluting stents: issues and uncertainties, J. Control. Release 120 (2007) 149–160.

[155] I. Iakovou, T. Schmidt, E. Bonizzoni, L. Ge, G.M. Sangiorgi, G. Stankovic, F. Airoldi, A. Chieffo, M. Montorfano, M. Carlino, I. Michev, N. Corvaja, C. Briguori, U. Gerckens, E. Grube, A. Colombo, Incidence, predictors, and outcome of thrombosis after successful implantation of drug-eluting stents, JAMA 293 (2005) 2126–2130.

[156] D.E. Kandzari, M.B. Leon, J.J. Popma, P.J. Fitzgerald, C. O'Shaughnessy, M.W. Ball, M. Turco, R.J. Applegate, P.A. Gurbel, M.G. Midei, S.S. Badre, L. Mauri, K.P. Thompson, L.A. LeNarz, R.E. Kuntz, E.I. Investigators, Comparison of zotarolimus-eluting and sirolimus-eluting stents in patients with native coronary artery disease—a randomized controlled trial, J. Am. Coll. Cardiol. 48 (2006) 2440–2447.

[157] M.A.M. Beijk, J.J. Piek, XIENCE V everolimus-eluting coronary stent system: a novel second generation drug-eluting stent, Expert Rev. Med. Devices 4 (2007) 11–21.

[158] T. Tsuji, The influence of biodegradation of the biodegradable stent on vessel in long-term period: serial intravascular ultrasound analysis of the Igaki-Tamai biodegradable stent, American heart association abstracts from scientific sessions, Circulation 106 (2002) 356.

[159] P.W. Serruys, J.A. Ormiston, Y. Onuma, E. Regar, N. Gonzalo, H.M. Garcia-Garcia, K. Nieman, N. Bruining, C. Dorange, K. Miquel-Hebert, S. Veldhof, M. Webster, L. Thuesen, D. Dudek, A bioabsorbable everolimus-eluting coronary stent system (ABSORB): 2-year outcomes and results from multiple imaging methods, Lancet 373 (2009) 897–910.

[160] J.A. Ormiston, P.W. Serruys, Bioabsorbable coronary stents, Circ. Cardiovasc. Interv. 2 (2009) 255–260.

[161] P.H. Long, Medical devices in orthopedic applications, Toxicol. Pathol. 36 (2008) 85–91.

[162] H. Bougherara, R. Zdero, A. Dubov, S. Shah, S. Khurshid, E.H. Schemitsch, A preliminary biomechanical study of a novel carbon-fibre hip implant versus standard metallic hip implants, Med. Eng. Phys. 33 (2011) 121–128.

[163] J.J. Callaghan, J.C. Albright, D.D. Goetz, J.P. Olejniczak, R.C. Johnston, Charnley total hip arthroplasty with cement. Minimum twenty-five-year follow-up, J. Bone Joint Surg. Am. 82 (2000) 487–497.

[164] A.J. Carr, O. Robertsson, S. Graves, A.J. Price, N.K. Arden, A. Judge, D.J. Beard, Knee replacement, Lancet 379 (2012) 1331–1340.

[165] M.N. Rahaman, A. Yao, B.S. Bal, J.P. Garino, M.D. Ries, Ceramics for prosthetic hip and knee joint replacement, J. Am. Ceram. Soc. 90 (2007) 1965–1988.

[166] L. Mattei, F. Di Puccio, B. Piccigallo, E. Ciulli, Lubrication and wear modelling of artificial hip joints: a review, Tribol. Int. 44 (2011) 532–549.

[167] M.S. Scholz, J.P. Blanchfield, L.D. Bloom, B.H. Coburn, M. Elkington, J.D. Fuller, M.E. Gilbert, S.A. Muflahi, M.F. Pernice, S.I. Rae, J.A. Trevarthen, S.C. White, P.M. Weaver, I.P. Bond, The use of composite materials in modern orthopaedic medicine and prosthetic devices: a review, Compos. Sci. Technol. 71 (2011) 1791–1803.

[168] R.E. Field, K. Rajakulendran, V.K. Eswaramoorthy, N. Rushton, Three-year prospective clinical and radiological results of a new flexible horseshoe acetabular cup, Hip Int. 22 (2011) 598–606.

[169] F.J. Medel, P. Pena, J. Cegonino, E. Gomez-Barrena, J.A. Puertolas, Comparative fatigue behavior and toughness of remelted and annealed highly crosslinked polyethylenes, J. Biomed. Mater. Res. B 83 (2007) 380–390.

[170] P. Bracco, E. Oral, Vitamin E-stabilized UHMWPE for total joint implants: a review, Clin. Orthop. Relat. Res. 469 (2011) 2286–2293.

[171] I. Nakahara, M. Takao, S. Bandoh, N. Bertollo, W.R. Walsh, N. Sugano, Novel surface modifications of carbon fiber-reinforced polyetheretherketone hip stem in an ovine model, Artif. Organs 36 (2012) 62–70.

[172] H. Deramond, C. Depriester, P. Galibert, D. Le Gars, Percutaneous vertebroplasty with polymethylmethacrylate. Technique, indications, and results, Radiol. Clin. North Am. 36 (1998) 533–546.

[173] J.r. Krämer, Intervertebral Disk Diseases: Causes, Diagnosis, Treatment, and Prophylaxis, third ed., Thieme, Stuttgart, NY, 2009.

[174] Q.B. Bao, G.M. McCullen, P.A. Higham, J.H. Dumbleton, H.A. Yuan, The artificial disc: theory, design and materials, Biomaterials 17 (1996) 1157–1167.

[175] M. Marcolongo, P. Ducheyne, J. Garino, E. Schepers, Bioactive glass fiber/polymeric composites bond to bone tissue, J. Biomed. Mater. Res. 39 (1998) 161–170.

[176] A. Ignatius, K. Unterricker, K. Wenger, M. Richter, L. Claes, P. Lohse, H. Hirst, A new composite made of polyurethane and glass ceramic in a loaded implant model: a biomechanical and histological analysis, J. Mater. Sci. Mater. Med. 8 (1997) 753–756.

[177] P. Ciappetta, S. Boriani, G.P. Fava, A carbon fiber reinforced polymer cage for vertebral body replacement: technical note, Neurosurgery 41 (1997) 1203–1206.

[178] A. Gloria, R. De Santis, L. Ambrosio, F. Causa, K.E. Tanner, A multi-component fiber-reinforced PHEMA-based hydrogel/HAPEX device for customized intervertebral disc prosthesis, J. Biomater. Appl. 25 (2011) 795–810.

[179] L. Ambrosio, R. De Santis, L. Nicolais, Composite hydrogels for implants, Proc. Inst. Mech. Eng. H 212 (1998) 93–99.

[180] E. Kon, G. Filardo, M. Tschon, M. Fini, G. Giavaresi, L. Marchesini Reggiani, C. Chiari, S. Nehrer, I. Martin, D.M. Salter, L. Ambrosio, M. Marcacci, Tissue engineering for total meniscal substitution: animal study in sheep model—results at 12 months, Tissue Eng. A 18 (2012) 1573–1582.

[181] A. Melvin, A. Litsky, J. Mayerson, D. Witte, D. Melvin, N. Juncosa-Melvin, An artificial tendon with durable muscle interface, J. Orthop. Res. 28 (2010) 218–224.
[182] H. Li, Y. Ge, P. Zhang, L. Wu, S. Chen, The effect of layer-by-layer chitosan-hyaluronic acid coating on graft-to-bone healing of a poly(ethylene terephthalate) artificial ligament, J. Biomater. Sci. Polym. Ed. 23 (2012) 425–438.
[183] S.M. Kurtz, J.N. Devine, PEEK biomaterials in trauma, orthopedic, and spinal implants, Biomaterials 28 (2007) 4845–4869.
[184] C.R. Dawson, I.R. Schwab, Epidemiology of cataract—a major cause of preventable blindness, Bull. World Health Organ. 59 (1981) 493–501.
[185] P.T. Ashwin, S. Shah, J.S. Wolffsohn, Advances in cataract surgery, Clin. Exp. Optom. 92 (2009) 333–342.
[186] D.J. Apple, J. Sims, Harold Ridley and the invention of the intraocular lens, Surv. Ophthalmol. 40 (1996) 279–292.
[187] J.K. Wang, C.Y. Hu, S.W. Chang, Intraocular lens power calculation using the IOLMaster and various formulas in eyes with long axial length, J. Cataract Refract. Surg. 34 (2008) 262–267.
[188] J.T. Banta, P.J. Rosenfeld, Cataract surgery and intraocular lens selection in patients with age-related macular degeneration: pearls for success, Int. Ophthalmol. Clin. 52 (2012) 73–80.
[189] D. Myung, P.E. Duhamel, J.R. Cochran, J. Noolandi, C.N. Ta, C.W. Frank, Development of hydrogel-based keratoprostheses: a materials perspective, Biotechnol. Prog. 24 (2008) 735–741.
[190] W. Stone Jr., E. Herbert, Experimental study of plastic material as replacement for the cornea: a preliminary report, Am. J. Ophthalmol. 36 (1953) 168–173.
[191] D.T.H. Tan, J.K.G. Dart, E.J. Holland, S. Kinoshita, Corneal transplantation, Lancet 379 (2012) 1749–1761.
[192] H.F. Chew, B.D. Ayres, K.M. Hammersmith, C.J. Rapuano, P.R. Laibson, J.S. Myers, Y.P. Jin, E.J. Cohen, Boston keratoprosthesis outcomes and complications, Cornea 28 (2009) 989–996.
[193] D.J. Carlsson, F. Li, S. Shimmura, M. Griffith, Bioengineered corneas: how close are we? Curr. Opin. Ophthalmol. 14 (2003) 192–197.
[194] C. Liu, K. Hille, D. Tan, C. Hicks, J. Herold, Keratoprosthesis surgery, Dev. Ophthalmol. 41 (2008) 171–186.
[195] M.A. Princz, H. Sheardown, Heparin-modified dendrimer cross-linked collagen matrices for the delivery of basic fibroblast growth factor (FGF-2), J. Biomater. Sci. Polym. Ed. 19 (2008) 1201–1218.
[196] Y.X. Huang, Q.H. Li, An active artificial cornea with the function of inducing new corneal tissue generation in vivo—a new approach to corneal tissue engineering, Biomed. Mater. 2 (2007) S121–S125.
[197] K.H. Jeong, J. Kim, L.P. Lee, Biologically inspired artificial compound eyes, Science 312 (2006) 557–561.
[198] P.L. Custer, R.H. Kennedy, J.J. Woog, S.A. Kaltreider, D.R. Meyer, Orbital implants in enucleation surgery: a report by the American Academy of Ophthalmology, Ophthalmology 110 (2003) 2054–2061.
[199] S.B. Patil, R. Meshramkar, B.H. Naveen, N.P. Patil, Ocular prosthesis: a brief review and fabrication of an ocular prosthesis for a geriatric patient, Gerodontology 25 (2008) 57–62.
[200] D.G. Purton, J.A. Payne, Comparison of carbon fiber and stainless steel root canal posts, Quintessence Int. 27 (1996) 93–97.
[201] N.B. Cramer, J.W. Stansbury, C.N. Bowman, Recent advances and developments in composite dental restorative materials, J. Dent. Res. 90 (2011) 402–416.
[202] R. Gautam, R.D. Singh, V.P. Sharma, R. Siddhartha, P. Chand, R. Kumar, Biocompatibility of polymethylmethacrylate resins used in dentistry, J. Biomed. Mater. Res. B 100 (2012) 1444–1450.
[203] J.P. Louis, M. Dabadie, Fibrous carbon implants for the maintenance of bone volume after tooth avulsion: first clinical results, Biomaterials 11 (1990) 525–528.
[204] P.B. Messersmith, A. Obrez, S. Lindberg, New acrylic resin composite with improved thermal diffusivity, J. Prosthet. Dent. 79 (1998) 278–284.
[205] A. Hucknall, S. Rangarajan, A. Chilkoti, In pursuit of zero: polymer brushes that resist the adsorption of proteins, Adv. Mater. 21 (2009) 2441–2446.
[206] I. Banerjee, R.C. Pangule, R.S. Kane, Antifouling coatings: recent developments in the design of surfaces that prevent fouling by proteins, bacteria, and marine organisms, Adv. Mater. 23 (2010) 690–718.
[207] B.D. Ratner, A.S. Hoffman, Nonfouling surfaces, in: B.D. Ratner, A.S. Hoffman, F.J. Schoen, J.E. Lemons (Eds.), Biomaterials Science: An Introduction to Materials in Medicine, third ed., Academic Press, Elsevier, Waltham, MA, USA, 2013.
[208] K.L. Prime, G.M. Whitesides, Self-assembled organic monolayers: model systems for studying adsorption of proteins at surfaces, Science 252 (1991) 1164–1167.
[209] W.R. Gombotz, W. Guanghui, T.A. Horbett, A.S. Hoffman, Protein adsorption to poly (ethylene oxide) surfaces, J. Biomed. Mater. Res. 25 (1991) 1547–1562.
[210] G.R. Llanos, M.V. Sefton, Does polyethylene oxide possess a low thrombogenicity? J. Biomater. Sci. Polym. Ed. 4 (1993) 381–400.
[211] S. Chen, L. Li, C. Zhao, J. Zheng, Surface hydration: principles and applications toward low-fouling/nonfouling biomaterials, Polymer 51 (2010) 5283–5293.
[212] S. Jeon, J. Andrade, Protein—surface interactions in the presence of polyethylene oxide: II. Effect of protein size, J. Colloid Interface Sci. 142 (1991) 159–166.
[213] S. Jeon, J. Lee, J. Andrade, P. De Gennes, Protein–surface interactions in the presence of polyethylene oxide: I. Simplified theory, J. Colloid Interface Sci. 142 (1991) 149–158.
[214] D.A. Herold, K. Keil, D.E. Bruns, Oxidation of polyethylene glycols by alcohol dehydrogenase, Biochem. Pharmacol. 38 (1989) 73–76.
[215] S. Chen, J. Zheng, L. Li, S. Jiang, Strong resistance of phosphorylcholine self-assembled monolayers to protein adsorption: insights into nonfouling properties of zwitterionic materials, J. Am. Chem. Soc. 127 (2005) 14473–14478.
[216] S. Chen, L. Liu, S. Jiang, Strong resistance of oligo (phosphorylcholine) self-assembled monolayers to protein adsorption, Langmuir 22 (2006) 2418–2421.
[217] V.A. Tegoulia, W. Rao, A.T. Kalambur, J.F. Rabolt, S.L. Cooper, Surface properties, fibrinogen adsorption, and cellular interactions of a novel phosphorylcholine-containing self-assembled monolayer on gold, Langmuir 17 (2001) 4396–4404.
[218] M. Tanaka, T. Sawaguchi, Y. Sato, K. Yoshioka, O. Niwa, Synthesis of phosphorylcholine–oligoethylene glycol–alkane thiols and their suppressive effect on non-specific adsorption of proteins, Tetrahedron Lett. 50 (2009) 4092–4095.
[219] W. Feng, J.L. Brash, S. Zhu, Non-biofouling materials prepared by atom transfer radical polymerization grafting of 2-methacryloloxyethyl phosphorylcholine: separate effects of graft density and chain length on protein repulsion, Biomaterials 27 (2006) 847–855.
[220] Y. Xu, M. Takai, K. Ishihara, Protein adsorption and cell adhesion on cationic, neutral, and anionic 2-methacryloyloxyethyl phosphorylcholine copolymer surfaces, Biomaterials 30 (2009) 4930–4938.

[221] A. Yamasaki, Y. Imamura, K. Kurita, Y. Iwasaki, N. Nakabayashi, K. Ishihara, Surface mobility of polymers having phosphorylcholine groups connected with various bridging units and their protein adsorption-resistance properties, Colloids Surf. B 28 (2003) 53–62.
[222] A.R. Statz, R.J. Meagher, A.E. Barron, P.B. Messersmith, New peptidomimetic polymers for antifouling surfaces, J. Am. Chem. Soc. 127 (2005) 7972–7973.
[223] C. Siegers, M. Biesalski, R. Haag, Self-assembled monolayers of dendritic polyglycerol derivatives on gold that resist the adsorption of proteins, Chemistry 10 (2004) 2831–2838.
[224] M. Wyszogrodzka, R. Haag, Synthesis and characterization of glycerol dendrons, self-assembled monolayers on gold: a detailed study of their protein resistance, Biomacromolecules 10 (2009) 1043–1054.
[225] M. Metzke, J.Z. Bai, Z. Guan, A novel carbohydrate-derived side-chain polyether with excellent protein resistance, J. Am. Chem. Soc. 125 (2003) 7760–7761.
[226] C. Perrino, S. Lee, S.W. Choi, A. Maruyama, N.D. Spencer, A biomimetic alternative to poly (ethylene glycol) as an antifouling coating: resistance to nonspecific protein adsorption of poly (L-lysine)-graft-dextran, Langmuir 24 (2008) 8850–8856.
[227] Z. Zhang, S. Chen, Y. Chang, S. Jiang, Surface grafted sulfobetaine polymers via atom transfer radical polymerization as superlow fouling coatings, J. Phys. Chem. B 110 (2006) 10799–10804.
[228] Y. Chang, S. Chen, Z. Zhang, S. Jiang, Highly protein-resistant coatings from well-defined diblock copolymers containing sulfobetaines, Langmuir 22 (2006) 2222–2226.
[229] Y. Chang, S. Chen, Q. Yu, Z. Zhang, M. Bernards, S. Jiang, Development of biocompatible interpenetrating polymer networks containing a sulfobetaine-based polymer and a segmented polyurethane for protein resistance, Biomacromolecules 8 (2007) 122–127.
[230] S. Chen, S. Jiang, An new avenue to nonfouling materials, Adv. Mater. 20 (2008) 335–338.
[231] G. Cheng, Z. Zhang, S. Chen, J.D. Bryers, S. Jiang, Inhibition of bacterial adhesion and biofilm formation on zwitterionic surfaces, Biomaterials 28 (2007) 4192–4199.
[232] L.K. Ista, V.H. Pérez-Luna, G.P. López, Surface-grafted, environmentally sensitive polymers for biofilm release, Appl. Environ. Microbiol. 65 (1999) 1603–1609.
[233] L. Ista, G. Lopez, Lower critical solubility temperature materials as biofouling release agents, J. Ind. Microbiol. Biotechnol. 20 (1998) 121–125.
[234] L.K. Ista, S. Mendez, G.P. Lopez, Attachment and detachment of bacteria on surfaces with tunable and switchable wettability, Biofouling 26 (2010) 111–118.
[235] L.K. Ista, S. Mendez, V.H. Pérez-Luna, G.P. López, Synthesis of poly (N-isopropylacrylamide) on initiator-modified self-assembled monolayers, Langmuir 17 (2001) 2552–2555.
[236] D. Cunliffe, C. de las Heras Alarcón, V. Peters, J.R. Smith, C. Alexander, Thermoresponsive surface-grafted poly (N-isopropylacrylamide) copolymers: effect of phase transitions on protein and bacterial attachment, Langmuir 19 (2003) 2888–2899.
[237] C. de Las Heras Alarcón, B. Twaites, D. Cunliffe, J.R. Smith, C. Alexander, Grafted thermo-and pH responsive co-polymers: surface-properties and bacterial adsorption, Int. J. Pharm. 295 (2005) 77–91.
[238] H. Fu, X. Hong, A. Wan, J.D. Batteas, D.E. Bergbreiter, Parallel effects of cations on PNIPAM graft wettability and PNIPAM solubility, ACS Appl. Mater. Interfaces 2 (2010) 452–458.
[239] S. Joly, R. Kane, L. Radzilowski, T. Wang, A. Wu, R. Cohen, E. Thomas, M. Rubner, Multilayer nanoreactors for metallic and semiconducting particles, Langmuir 16 (2000) 1354–1359.
[240] D. Lee, R.E. Cohen, M.F. Rubner, Antibacterial properties of Ag nanoparticle loaded multilayers and formation of magnetically directed antibacterial microparticles, Langmuir 21 (2005) 9651–9659.
[241] T.C. Wang, M.F. Rubner, R.E. Cohen, Polyelectrolyte multilayer nanoreactors for preparing silver nanoparticle composites: controlling metal concentration and nanoparticle size, Langmuir 18 (2002) 3370–3375.
[242] H.F. Chuang, R.C. Smith, P.T. Hammond, Polyelectrolyte multilayers for tunable release of antibiotics, Biomacromolecules 9 (2008) 1660–1668.
[243] M.C. Lawson, R. Shoemaker, K.B. Hoth, C.N. Bowman, K.S. Anseth, Polymerizable vancomycin derivatives for bactericidal biomaterial surface modification: structure–function evaluation, Biomacromolecules 10 (2009) 2221–2234.
[244] M.C. Lawson, C.N. Bowman, K.S. Anseth, Vancomycin derivative photopolymerized to titanium kills S. epidermidis, Clin. Orthop. Relat. Res. 461 (2007) 96–105.
[245] C.J. Waschinski, J. Zimmermann, U. Salz, R. Hutzler, G. Sadowski, J.C. Tiller, Design of contact-active antimicrobial acrylate-based materials using biocidal macromers, Adv. Mater. 20 (2008) 104–108.
[246] J.C. Tiller, C.-J. Liao, K. Lewis, A.M. Klibanov, Designing surfaces that kill bacteria on contact, Proc. Natl. Acad. Sci. U.S.A. 98 (2001) 5981–5985.
[247] K. Glinel, A.M. Jonas, T. Jouenne, J. Leprince, L. Galas, W.T. Huck, Antibacterial and antifouling polymer brushes incorporating antimicrobial peptide, Bioconjug. Chem. 20 (2008) 71–77.
[248] M. Bagheri, M. Beyermann, M. Dathe, Immobilization reduces the activity of surface-bound cationic antimicrobial peptides with no influence upon the activity spectrum, Antimicrob. Agents Chemother. 53 (2009) 1132–1141.
[249] G. Cheng, H. Xue, Z. Zhang, S. Chen, S. Jiang, A switchable biocompatible polymer surface with self-sterilizing and nonfouling capabilities, Angew. Chem. Int. Ed. Engl. 47 (2008) 8831–8834.
[250] Z. Li, D. Lee, X. Sheng, R.E. Cohen, M.F. Rubner, Two-level antibacterial coating with both release-killing and contact-killing capabilities, Langmuir 22 (2006) 9820–9823.
[251] A.S. Hoffman, J.A. Hubbell, Surface-immobilized biomolecules, in: B.D. Ratner, A.S. Hoffman, F.J. Schoen, J.E. Lemons (Eds.), Biomaterial Science: An Introduction to Materials in Medicine, third ed., Academic Press, Elsevier, Waltham, MA, USA, 2013.
[252] B.D. Ratner, S.J. Bryant, Biomaterials: where we have been and where we are going, Annu. Rev. Biomed. Eng. 6 (2004) 41–75.
[253] B.D. Ratner, A paradigm shift: biomaterials that heal, Polym. Int. 56 (2007) 1183–1185.
[254] S.M. Martin, R. Ganapathy, T.K. Kim, D. Leach-Scampavia, C.M. Giachelli, B.D. Ratner, Characterization and analysis of osteopontin-immobilized poly (2-hydroxyethyl methacrylate) surfaces, J. Biomed. Mater. Res. A 67 (2003) 334–343.
[255] L. Liu, S. Chen, C.M. Giachelli, B.D. Ratner, S. Jiang, Controlling osteopontin orientation on surfaces to modulate endothelial cell adhesion, J. Biomed. Mater. Res. A 74A (2005) 23–31.
[256] N.M. Alves, I. Pashkuleva, R.L. Reis, J.F. Mano, Controlling cell behavior through the design of polymer surfaces, Small 6 (2010) 2208–2220.
[257] R.G. LeBaron, K.A. Athanasiou, Extracellular matrix cell adhesion peptides: functional applications in orthopedic materials, Tissue Eng. 6 (2000) 85–103.

[258] J.P. Ranieri, R. Bellamkonda, E.J. Bekos, T.G. Vargo, J.A. Gardella Jr., P. Aebischer, Neuronal cell attachment to fluorinated ethylene propylene films with covalently immobilized laminin oligopeptides YIGSR and IKVAV. II, J. Biomed. Mater. Res. 29 (1995) 779–785.
[259] S. VandeVondele, J. Vörös, J.A. Hubbell, RGD-grafted poly-l-lysine-graft-(polyethylene glycol) copolymers block non-specific protein adsorption while promoting cell adhesion, Biotechnol. Bioeng. 82 (2003) 784–790.
[260] C.M. Giachelli, S. Steitz, Osteopontin: a versatile regulator of inflammation and biomineralization, Matrix Biol. 19 (2000) 615–622.
[261] M. Ohji, L. Mandarino, N. SundarRaj, R.A. Thoft, Corneal epithelial cell attachment with endogenous laminin and fibronectin, Invest. Ophthalmol. Vis. Sci. 34 (1993) 2487–2492.
[262] X. Duan, C. McLaughlin, M. Griffith, H. Sheardown, Biofunctionalization of collagen for improved biological response: scaffolds for corneal tissue engineering, Biomaterials 28 (2007) 78–88.
[263] M. Nikkhah, F. Edalat, S. Manoucheri, A. Khademhosseini, Engineering microscale topographies to control the cell-substrate interface, Biomaterials 33 (2012) 5230–5246.
[264] S. Sant, C.M. Hwang, S.H. Lee, A. Khademhosseini, Hybrid PGS–PCL microfibrous scaffolds with improved mechanical and biological properties, J. Tissue Eng. Regen. Med. 5 (2011) 283–291.
[265] F. Tian, H. Hosseinkhani, M. Hosseinkhani, A. Khademhosseini, Y. Yokoyama, G.G. Estrada, H. Kobayashi, Quantitative analysis of cell adhesion on aligned micro- and nanofibers, J. Biomed. Mater. Res. A 84 (2008) 291–299.
[266] M.T. Eliason, E.O. Sunden, A.H. Cannon, S. Graham, A.J. García, W.P. King, Polymer cell culture substrates with micropatterned carbon nanotubes, J. Biomed. Mater. Res. A 86 (2008) 996–1001.
[267] T.T. Dang, Q. Xu, K.M. Bratlie, E.S. O'Sullivan, X.Y. Chen, R. Langer, D.G. Anderson, Microfabrication of homogenous, asymmetric cell-laden hydrogel capsules, Biomaterials 30 (2009) 6896–6902.
[268] M.J. Dalby, N. Gadegaard, R. Tare, A. Andar, M.O. Riehle, P. Herzyk, C.D. Wilkinson, R.O. Oreffo, The control of human mesenchymal cell differentiation using nanoscale symmetry and disorder, Nat. Mat. 6 (2007) 997–1003.
[269] M. Dalby, S. Childs, M. Riehle, H. Johnstone, S. Affrossman, A. Curtis, Fibroblast reaction to island topography: changes in cytoskeleton and morphology with time, Biomaterials 24 (2003) 927–935.
[270] M.J. Dalby, S.J. Yarwood, M.O. Riehle, H.J. Johnstone, S. Affrossman, A.S. Curtis, Increasing fibroblast response to materials using nanotopography: morphological and genetic measurements of cell response to 13-nm-high polymer demixed islands, Exp. Cell Res. 276 (2002) 1–9.
[271] K.A. Lau, J. Bang, C.J. Hawker, D.H. Kim, W. Knoll, Modulation of protein–surface interactions on nanopatterned polymer films, Biomacromolecules 10 (2009) 1061–1066.
[272] S.-M. Park, O.-H. Park, J.Y. Cheng, C.T. Rettner, H.-C. Kim, Patterning sub-10 nm line patterns from a block copolymer hybrid, Nanotechnology 19 (2008) 455304.
[273] I. Hamley, Nanostructure fabrication using block copolymers, Nanotechnology 14 (2003) R39.
[274] C. Wilkinson, M. Riehle, M. Wood, J. Gallagher, A. Curtis, The use of materials patterned on a nano- and micro-metric scale in cellular engineering, Mater. Sci. Eng. C 19 (2002) 263–269.
[275] A. Curtis, C. Wilkinson, Topographical control of cells, Biomaterials 18 (1997) 1573–1583.
[276] M.S. Lord, M. Foss, F. Besenbacher, Influence of nanoscale surface topography on protein adsorption and cellular response, Nano Today 5 (2010) 66–78.
[277] E.K.F. Yim, K.W. Leong, Significance of synthetic nanostructures in dictating cellular response, Nanomedicine 1 (2005) 10–21.
[278] M.M. Stevens, J.H. George, Exploring and engineering the cell surface interface, Science 310 (2005) 1135–1138.
[279] I. Wheeldon, A. Farhadi, A.G. Bick, E. Jabbari, A. Khademhosseini, Nanoscale tissue engineering: spatial control over cell-materials interactions, Nanotechnology 22 (2011) 212001.
[280] G. Abrams, S. Goodman, P. Nealey, M. Franco, C. Murphy, Nanoscale topography of the basement membrane underlying the corneal epithelium of the rhesus macaque, Cell Tissue Res. 299 (2000) 39–46.
[281] H. Gong, J. Ruberti, D. Overby, M. Johnson, T.F. Freddo, A new view of the human trabecular meshwork using quick-freeze, deep-etch electron microscopy, Exp. Eye Res. 75 (2002) 347–358.
[282] M. Lord, B. Cousins, P. Doherty, J. Whitelock, A. Simmons, R. Williams, B. Milthorpe, The effect of silica nanoparticulate coatings on serum protein adsorption and cellular response, Biomaterials 27 (2006) 4856–4862.
[283] D.S. Sutherland, M. Broberg, H. Nygren, B. Kasemo, Influence of nanoscale surface topography and chemistry on the functional behaviour of an adsorbed model macromolecule, Macromol. Biosci. 1 (2001) 270–273.
[284] R. Flemming, C. Murphy, G. Abrams, S. Goodman, P. Nealey, Effects of synthetic micro- and nano-structured surfaces on cell behavior, Biomaterials 20 (1999) 573–588.
[285] J. Meyle, K. Gültig, W. Nisch, Variation in contact guidance by human cells on a microstructured surface, J. Biomed. Mater. Res. 29 (1995) 81–88.
[286] C.C. Berry, M.J. Dalby, D. McCloy, S. Affrossman, The fibroblast response to tubes exhibiting internal nanotopography, Biomaterials 26 (2005) 4985–4992.
[287] C. Berry, M. Dalby, R. Oreffo, D. McCloy, S. Affrossman, The interaction of human bone marrow cells with nanotopographical features in three dimensional constructs, J. Biomed. Mater. Res. A 79 (2006) 431–439.
[288] A. Lendlein, M. Behl, B. Hiebl, C. Wischke, Shape-memory polymers as a technology platform for biomedical applications, Expert Rev. Med. Devices 7 (2010) 357–379.
[289] M.C. Serrano, G.A. Ameer, Recent insights into the biomedical applications of shape-memory polymers, Macromol. Biosci. 12 (2012) 1156–1171.
[290] A. Lendlein, S. Kelch, Shape-memory polymers, Angew. Chem. Int. Ed. Engl. 41 (2002) 2034–2057.
[291] C.M. Yakacki, R. Shandas, C. Lanning, B. Rech, A. Eckstein, K. Gall, Unconstrained recovery characterization of shape-memory polymer networks for cardiovascular applications, Biomaterials 28 (2007) 2255.
[292] M. Behl, A. Lendlein, Shape-memory polymers, Mater. Today 10 (2007) 20–28.
[293] J.H. Shadduck, Implants for treating ocular hypertension, methods of use and methods of fabrication, US Patent Application 10/770, 436 (September 30, 2004).
[294] A.A. Sharp, H.V. Panchawagh, A. Ortega, R. Artale, S. Richardson-Burns, D.S. Finch, K. Gall, R.L. Mahajan, D. Restrepo, Toward a self-deploying shape memory polymer neuronal electrode, J. Neural Eng. 3 (2006) L23–L30.
[295] F.A. Spelman, B.M. Clopton, A. Voie, C.N. Jolly, K. Huynh, J. Boogaard, J.W. Swanson, Cochlear implant with shape memory material and method for implanting the same, US Patent 5800500 (September 1, 1998).

[296] I. Bellin, S. Kelch, R. Langer, A. Lendlein, Polymeric triple-shape materials, Proc. Natl. Acad. Sci. U.S.A. 103 (2006) 18043–18047.
[297] L. Xue, S. Dai, Z. Li, Biodegradable shape-memory block co-polymers for fast self-expandable stents, Biomaterials 31 (2010) 8132–8140.
[298] G.M. Baer, W. Small IV, T.S. Wilson, W.J. Benett, D.L. Matthews, J. Hartman, D.J. Maitland, Fabrication and in vitro deployment of a laser-activated shape memory polymer vascular stent, Biomed. Eng. Online 6 (2007) 43.

Chapter 20

Polymeric Materials in Drug Delivery

Bing Gu, Diane J. Burgess

Department of Pharmaceutical Sciences, School of Pharmacy, University of Connecticut, Storrs, Connecticut, USA

Chapter Outline

20.1 INTRODUCTION

Both natural and synthetic polymers are widely used in drug delivery systems. Polymers have inherent flexibility in that they can be synthesized and modified to provide versatile properties to meet the desired controlled drug release profile and biocompatibility. Material performance is highly linked to strength [1], porosity [2], particle size [3], amorphousity [4], biocompatibility [5], and dissolution properties [6], etc. In order to understand, predict, and improve the performance of polymeric drug delivery systems, various characterization techniques are required, and these are detailed in the following sections of this chapter.

The most commonly used polymers in drug formulations include polyesters (such as poly(lactic-*co*-glycolic acid) (PLGA), polylactic acid, polyglycolic acid, and poly(ε-caprolactone)), polyanhydrides (such as poly(sebacic acid), poly(adipic acid), poly(terephthalic acid), and their copolymers), polyamides (such as poly(imino carbonates) and polyamino acids), polyethers (such as polyethylene glycol (PEG) and polypropylene glycol), cellulose derivatives (such as carboxymethyl cellulose, hydroxypropyl methyl cellulose (HPMC), ethyl cellulose, hydroxypropyl methyl cellulose acetate succinate (HPMC-AS), and cellulose acetate succinate (CAS)), protein-based polyesters (such as collagen, gelatin, and albumin), and polysaccharides (such as chitosan, alginate, carrageenan, cyclodextrin, hyaluronic acid, and agarose) [7]. Due to their biodegradability, polyester-based polymers are the most widely investigated polymers for drug delivery. Of the polyesters, PLGA is the most commonly used since it is biodegradable and biocompatible and has been used in commercial sutures since the 1970s [8]. Most of the marketed parenteral sustained release formulations including Lupron Depot®, Sandostatin® Depot, and Risperdal® Consta® are PLGA-based [9]. The degradation period of PLGA can last from days to months, which is modulated by adjusting the ratio of lactic and glycolic acid [10]. Fast degradation and rapid erosion are normally observed for polyanhydride-based materials. The polyanhydride (polifeprosan 20, a copolymer of aliphatic (sebacic acid) and aromatic (carboxyphenoxy) propane) is used in Gliadel® implantable disks (for treatment of malignant melanoma) [11]. The application of polyamides is limited by their antigenic potential and uncontrollable release due to enzyme-dependent degradation. PEG is used to prolong the circulation time of nanoparticles and proteins in the blood by decreasing plasma protein adsorption. For example, the FDA-approved product, Doxil®, is a pegylated liposome formulation used for cancer therapy. HPMC-AS and CAS are used as enteric coatings for tablets to achieve controlled drug release in the small intestine. These polymers do not undergo degradation in the low pH conditions experienced in the stomach, and therefore, drug is not released until the dosage form reaches the intestine. The natural polymer chitosan has good biodegradability and biocompatibility and is used for preparation of hydrogels, nanoparticles, and bioadhesive microspheres. Cyclodextrins are natural cyclic oligosaccharides composed of five or more α-D-glucopyranoside units forming a ring structure. Drug molecules can be included into the rings of cyclodextrins to form complexes with increased solubility, enhanced bioavailability, and reduced gastrointestinal irritation. There are currently more than 30 pharmaceutical products with cyclodextrins available on the market [12]. The excellent biocompatibility and biodegradability of collagen make it another widely used polymer in drug delivery systems. Formulations prepared using collagen include minipellets and tablets for protein delivery, combined hydrogel-liposome formulations for sustained drug release, and implantable matrices for bone healing [13].

Different polymer properties are required for different drug delivery applications, and these necessitate a diverse array of characterization methods. For example, hydrogels and other implants should be biocompatible and have sufficient mechanical strength to withstand forces involved during implantation as well as in the *in vivo* environment [14]. Dynamic mechanical analysis (DMA) and rotational rheometers play important roles in determining complex mechanical properties. Although consisting of the same monomeric units, polymers with different molecular weights can have considerably different physicochemical as well as mechanical properties. For example, microspheres prepared with high-molecular-weight PLGA have higher encapsulation efficiencies and longer drug release profiles compared to microspheres prepared with low-molecular-weight PLGA [15]. Accordingly, it is important to understand and measure the molecular weight and weight distribution of polymers. Gel permeation chromatography (GPC) and matrix-assisted laser desorption/ionization (MALDI) mass spectroscopy are typically used to determine polymer molecular weight. In the case of particulate polymer drug delivery systems, particle size is directly related to the *in vivo* performance of the formulation in terms of drug release rates as well as drug targeting. Accordingly, it is important to determine the particle size and size distribution of these delivery systems. Particle size is usually measured by light scattering techniques; however, other methods such as light and electron microscopy as well as light obscuration techniques may also be used depending on the size range of the formulation.

Upon introduction *in vivo*, the interface between the delivery system and the biological tissue and/or fluid is critically important to the *in vivo* performance [5]. Accordingly, surface properties including surface chemical composition and surface area must be well characterized. X-ray photoelectron spectroscopy (XPS) is a widely used technique to obtain surface elemental composition, and Brunauer, Emmett, and Teller (BET) measurement is used to provide information on surface area. Surface morphology is typically assessed via light, electron, and atomic force microscopy (AFM). The amorphous and crystalline nature of materials can be determined from X-ray diffraction (XRD) and density measurements.

In vitro and *in vivo* biocompatibility tests are required by government agencies for drug delivery system and biomedical device approval [16]. *In vitro* dissolution testing is used to assess *in vivo* performance and is important during formulation development as well as for product quality assurance. Standardized dissolution methods are under development for novel polymeric formulations such as microspheres, nanoparticles, and *in situ* forming gels [17].

20.2 MECHANICAL/THERMAL PROPERTIES OF POLYMERS

The mechanical properties of polymers can be described as "strong," "tough," "soft," "viscous," or "ductile." The mechanical properties of polymeric formulations are important as they relate to product performance. For example, subcutaneous (s.c.) implants should have similar tensile strength to the s.c. soft tissue to achieve biocompatibility yet have sufficient mechanical strength to endure the *in vivo* environment [18]. Polymer coatings on formulations such as tablets and drug-eluting stents should have good adhesive properties and appropriate mechanical strength as per performance requirements [19]. Polymer mechanical properties are usually analyzed through stress-strain or viscoelastic measurements. Simple stress-strain curves can be obtained from tensile testing (such as that shown in Figure 20.1a). DMA or oscillatory shear rheometers are usually used to measure complex viscoelastic properties of polymers. Besides

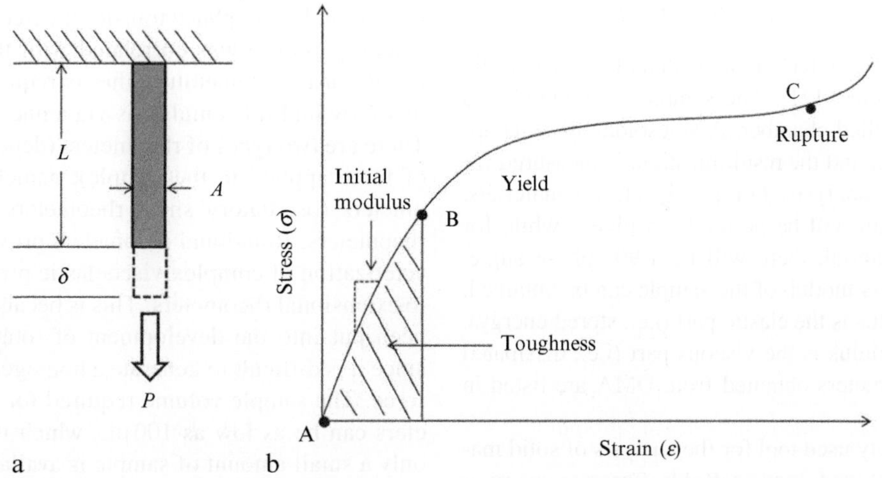

FIGURE 20.1 A typical tensile test (a) and stress-strain curve (b).

mechanical properties, it is also important to determine thermal transitions such as glass transition, crystallization, and melting temperatures. For example, both the stability and dissolution profiles of PLGA microspheres are related to the microsphere glass transition temperature [20]. Differential scanning calorimetry (DSC) can be used to determine these transition temperatures.

20.2.1 Stress-Strain

The stress-strain test (tensile test) is the most common mechanical test used for polymers. This test is conducted by fixing the polymer sample at one end to a loading frame, and a force is applied at the other end to achieve a controlled displacement δ (see Figure 20.1a). A stress-strain curve (Figure 20.1b) is obtained and analyzed to provide the mechanical properties. The stress (σ) and strain (ε) are defined using the following equations:

$$\sigma = \frac{P}{A}, \quad \varepsilon = \frac{\delta}{L}$$

Figure 20.1a is a schematic diagram representing a typical stress-strain curve obtained via tensile testing. The curve can be divided into different regions: A-B is an elastic region, in which linear stress-strain deformation can be observed; B-C is plastic region indicated by nonlinearity; point C indicates rupture of the material when the displacement increases significantly. The strain energy is the area under the stress-strain curve, which can be used to describe the toughness of a polymer. The slope of the elastic region is called the Young's modulus (E), a quantity used to assess the stiffness of the material. Yield stress is the maximum stress that can be applied to a material without causing permanent deformation. Yield stress usually marks the end of the elastic region (A-B region in Figure 20.1b). For example, PVA hydrogel coatings that have been developed for glucose biosensors have a Young's modulus between 20 and 40 MPa in the dry state. In the hydrated state, these coatings have a Young's modulus of 0.5-1 MPa that is comparable to human soft tissue [1].

20.2.2 Viscoelastic

Polymers can exhibit both viscous and elastic properties when undergoing deformation. Elasticity is stored energy, whereas viscosity is dissipated energy. In models used to describe the viscoelastic properties of polymers, a spring is usually used to represent elastic behavior, and a piston is used to describe viscous behavior. Various models with different combinations of "springs" and "pistons" are used to describe viscoelastic properties. These models include the Maxwell model where the piston and spring are in series and the Kelvin-Voigt model where the piston and spring are in parallel (as shown in Figure 20.2). The viscoelastic properties of polymers are typically complex and involve more than one Maxwell and/or Voigt model.

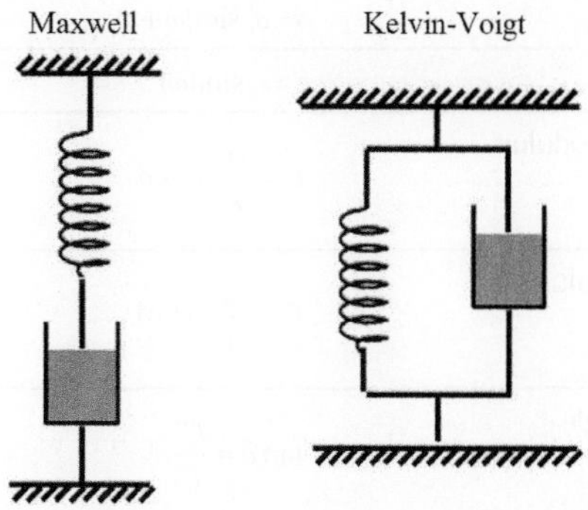

FIGURE 20.2 Maxwell and Kelvin-Voigt viscoelastic model.

20.2.2.1 Dynamic Mechanical Analysis

DMA is the most useful technique to study the viscoelastic properties of polymers [21]. The sample is mounted in a temperature-controlled chamber. A sinusoidal stress is applied to the sample, and the resulting strain is measured for complex modulus analysis. For purely elastic materials, the stress and strain will be perfectly in phase, while for purely viscous material, there will be a 90° phase angle. The storage and loss moduli of the sample can be obtained. The storage modulus is the elastic part (i.e., stored energy), while the loss modulus is the viscous part (i.e., dissipated energy). The parameters obtained from DMA are listed in Table 20.1.

DMA is a widely used tool for the analysis of solid materials such as films, rods, and scaffolds. Creep recovery or stress relaxation can be performed by holding the sample at either constant stress or strain. DMA can also be used to measure transition temperatures such as the glass transition temperature (T_g) by determining mechanical changes when conducting a temperature-sweeping test. T_g can be recognized as a significant decrease in the storage modulus and a significant increase in the loss modulus in a modulus-temperature curve. DMA is a very sensitive technique for characterization of T_g (approximately 100 times more sensitive compared to DSC) [22]. However, DMA normally requires large sample sizes.

20.2.2.2 Oscillatory Shear Rheometer

For polymeric dispersed systems such as hydrogels, suspensions, and emulsions, dynamical mechanical properties can be obtained using a rheometer. Rheological properties such as viscosity can be used to understand the system microstructure and formulation parameters. For example, the phase transition detected from the bulk viscosity profile was correlated with the ultralow interfacial tension temperature that is required for formation of w/o/w multiple emulsions via a one-step process [23]. There are two types of rheometers (depending on the type of force applied to the sample), namely, rotational rheometers (oscillatory shear rheometers) and extensional rheometers. Rotational rheometers provide a better characterization of complex viscoelastic properties compared to extensional rheometers. This is because more effort has been put into the development of rotational rheometers since it is difficult to generate a homogeneous extensional force. The sample volume required for rotational rheometers can be as low as 100 μL, which is beneficial when only a small amount of sample is available. These instruments have different geometries (concentric cylinder, cone and plate, and parallel plates) that are used for materials of varying viscosity. One part of the instrument, for example, the cone, is connected to a rotor, while the other part, for example, the plate, is stationary. The sample is placed between these two parts (as shown in Figure 20.3) and shear stress is applied. In order to distinguish system microstructural change from bulk rheological properties, the shear stress applied should be conducted in a region where the viscoelastic properties are independent of applied stress and strain. This region is called the linear viscoelastic region (LVR), which can be defined by performing a frequency-sweeping test. The length of LVR is an indication of the structural stability of the system [21]. The viscoelastic measurement is then conducted by applying a sinusoidal shear stress to the material at a frequency within the LVR and measuring the corresponding strain. Shear viscosity, viscoelastic modulus, and phase angle of the material can be analyzed from the data. Oscillatory shear rheometry is a superior technique to characterize hydrogel systems compared to DMA. For example, an oscillatory shear rheometer provided a better correlation between the storage modulus and the equilibrium-swollen volume fraction of dextran methacrylate hydrogels compared to DMA [24].

TABLE 20.1 Parameters Measured in a Dynamic Mechanical Analysis (ω Is the Frequency of Oscillatory Stress, t Is the Time, and δ Is the Phase Angle Between Stress and Strain)

Stress	$\sigma = \sigma_0 \sin(t\omega + \delta)$
Strain	$\varepsilon = \varepsilon_0 \sin(t\omega)$
Storage modulus	$E' = \frac{\sigma_0}{\varepsilon_0} \cos\delta$
Loss modulus	$E'' = \frac{\sigma_0}{\varepsilon_0} \sin\delta$
Phase angle	$\tan\delta = \frac{E''}{E'}$

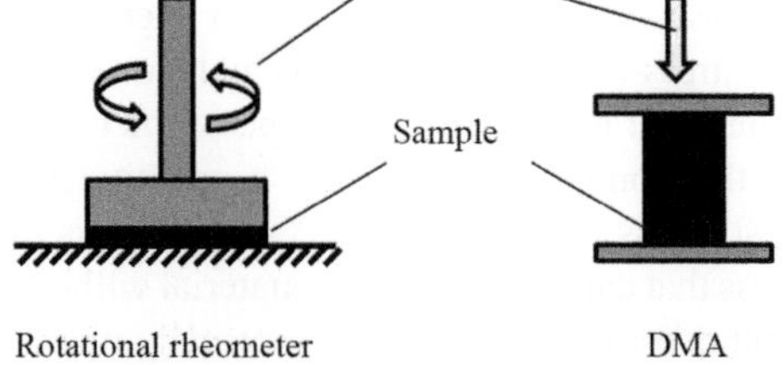

FIGURE 20.3 Illustration of how forces are applied to samples in a rotational rheometer and a dynamic mechanical analyzer.

20.2.3 Differential Scanning Calorimetry

DSC is a powerful tool for the measurement of temperature and heat flow associated with phase transition temperatures of materials. A sample and an inert reference are heated/cooled at a chosen rate in a controlled environment. In order to maintain the same temperature between the sample and reference, heat flow is applied to the sample. The heat flow is recorded to reflect the thermal events within the sample such as glass transition, melting, and crystallization. DSC measurements can be conducted at either a constant heating rate (linear temperature-time change) or a variable heating rate (modulated temperature). A linear heating rate DSC can be conducted with a temperature increase of 300 °C/min using a small sample size (as small as 0.5 mg). This type of measurement is called fast-scanning DSC (FSDSC). FSDSC can measure heat capacity, T_g, low levels of amorphousity in crystalline samples, and drug solubility in polymers [25]. DSC is very useful in understanding the state of drug as well as any drug polymer interactions within drug delivery systems [26]. Modulated DSC (MDSC) can be performed in order to improve the sensitivity and resolution of the technique. Conditions that should be carefully chosen when performing MDSC include sample size (5-20 mg for polymers), modulation period, underlying heating rate, and modulation amplitude. An advantage of MDSC is that the slow heating rate of MDSC (~2 °C/min) allows easy separation between glass transition temperatures and kinetic transition temperatures that would otherwise overlap at higher heating rates [27,28]. However, MDSC measurements usually take hours that may lead to potential sample changes. As a complimentary thermal measurement to DSC, thermal gravimetrical analysis (TGA) measures the rate of change in the mass of a sample as a function of temperature or time. A TGA instrument is usually composed of a sample pan, a precision balance, and a programmable furnace. The furnace can heat the sample at various rates for temperatures up to 2000 °C, and the mass change during the heating process can be obtained. Thus, TGA can provide information about both physical phenomena (vaporization, sublimation, absorption, and desorption) and chemical changes (desolvation, decomposition, oxidation, and combustion) of the materials [29].

20.3 SURFACE AND MORPHOLOGICAL CHARACTERIZATION OF POLYMERS

20.3.1 Morphology

Two aspects are considered when studying the morphology of polymers: (1) the macroscopic shape and structure and (2) the crystallinity and amorphousity of the material. The structure can affect the properties of the drug delivery system, e.g., increasing the porosity of PLGA microspheres can lead to an increase the initial burst release of drug [20]. The shape of nanoparticles is related to their intracellular uptake and cytotoxicity. In the case of PLGA, which is a copolymer of lactic and glycolic acid, the ratio of the two monomers dictates the crystallinity/amorphousity of the copolymer. For example, a 50/50 ratio of lactic/glycolic acid results in an amorphous polymer, whereas a copolymer composed primarily of lactic acid will have crystalline regions. The crystallinity increases with increase in the lactic acid ratio, and this affects the release properties of drug from PLGA delivery systems such as microspheres, implants, scaffolds, and coatings [30]. Microscopic methods are usually used to determine the shape and structure of polymeric drug delivery systems. X-ray diffraction is the best method to determine the crystallinity/amorphousity of polymeric drug delivery systems.

20.3.1.1 Microscopic Methods

Microscopic methods are the most direct way of visualizing the morphology of materials. Images can be generated using microscopy for samples that cannot be seen with the naked eye. Different types of microscopes include light-, electron-, and scanning probe-based systems. Optical microscopes are based on diffraction phenomena caused by interaction of matter with photons so that the resolution limit of an optical microscope is the wavelength of light. Fluorescent microscopy is a special type of optical microscopy where a fluorescent sample is illuminated with light of certain wavelength (sample-dependent), and the emitted photons are measured. In the case of confocal fluorescent microscopy, point illumination is conducted using a focused laser with a certain wavelength to construct a three-dimensional fluorescent structure. Another extensively used type optical microscopy in pharmaceutical research is polarized light microscopy (PLM), which takes advantage of polarized light by limiting light oscillation in one direction. A polarizer is usually beneath the sample to create plane polarized light with the correct vibrational direction. The analyzer is above the sample and can also limit the light oscillation in one direction. The angle between the analyzer and polarizer can be adjusted to analyze the light passing through the sample. PLM can provide optical crystallography properties of material such as refractive indices both qualitatively and quantitatively. PLM can be used to determine the effect of various polymer excipients on the crystallinity/amorphousity of drug molecules such as polymer-stabilized suspensions [31].

Electron microscopy is based on the interaction between electrons and the sample. Electron microscopes have extremely higher resolution since the wavelength of electrons is considerably smaller than that of photons. Scanning electron microscopy (SEM) allows visualization down to 1 nm and smaller. Transmission electron microscopy (TEM) allows visualization down to 0.1 nm and smaller. At present, atomic resolution (<50 pm) can be achieved using spherical aberration-corrected electron microscopy, which is able to detect individual atoms and crystalline defects [32]. TEM and SEM are used to characterize polymeric drug delivery systems. SEM obtains images by scanning the sample surface with a focused beam of electrons. The most common detection mode is analyzing the second electrons generated by ionizing the surface molecules. The number of second electrons depends on the surface and angle exposed to the electrons. Three-dimensional images of the surface can be generated. Traditional SEM can only be conducted for electronically conductive samples. Therefore, nonconductive samples need to be coated with a thin layer of conductive material to avoid scanning faults or image artifacts. Nonconductive samples can also be imaged without coating using environmental scanning electron microscopy. The images from TEM are obtained by transmitting a beam of electrons through an ultrathin sample. The image is then obtained from the electrons transmitted through the sample.

Scanning tunneling microscopy (STM) and AFM are commonly used scanning probe-based microscopy techniques that were developed in the 1980s [33]. STM usually has a probe made of tungsten (W) or praseodymium (Pr)/iridium (Ir) alloy with only one atom at the end of the probe tip. A voltage difference is applied between the tip and the surface of the sample under high vacuum. The electrons can tunnel through the vacuum between the tip and the sample resulting in a current that is measured and represented in the form of an image. By manipulating the voltage difference, STM can also be used for controlled polymerization on the surface at the nanoscale [34]. AFM is another type of probe-based microscopy that provides imaging, measuring, and manipulating matter at the nanoscale. AFM consists of a cantilever with a sharp tip at the end to scan the sample surface. Deflection of the cantilever can be measured as output of the image and surface force [35]. Deflection of the cantilever can also be used for nanoscale mechanical investigation of the material. A combination of AFM and SEM has been used for nanomechanical testing of various electrospun nanofibers. Complete mechanical behavior of PVA, nylon-6, and collagen fibrils at the nanoscale has been recorded using this technique [36]. Various types of microscopy and application examples of each are listed in Table 20.2.

20.3.1.2 Crystalline and Amorphous Characterizations

XRD cryptography is a very useful method to determine polymer crystallinity. The sample is bombarded with X-rays while rotating. Interaction of the X-rays with the material generates diffraction patterns that are used to describe sample crystallinity. XRD follows Bragg's law in that the reflected X-rays

TABLE 20.2 Comparison of Different Types of Microscopy

Mechanism	Types of Microscopy	Application Examples
Optical microscope	Fluorescent microscope	Phase separation of polymer blends [37]
	Confocal microscope	Phase separation of polymer blends [38,39]
		Protein/polymer colocalization in polymeric films [40]
		Polymer blend in microparticles [41]
	Polarized light microscope	Polymorphism of polyvinylidene fluoride thin films [42]
		Crystalline structure of microspheres and drug loaded within [43]
Electron microscope	Transmission electron microscope	Self-assembly of poly(amidoamine) nanoparticles [44]
		Optically switchable spiropyran-based polymer nanoparticles [45]
	Scanning electron microscope	Morphology of PVA hydrogel/PLGA microsphere composites [2]
		Defects in polymer coatings of stents [46]
		Morphology of poly(*N*-isopropylacrylamide)-based hydrogel at the water/oil interphase [47]
Scanning probe microscope	Scanning tunneling microscope	Controlled localization of 2D surface polymerization [48]
		Direct visualization of the formation of single-molecule conjugated copolymers [49]
	Atomic force microscope	Polymer adsorption onto the nanocrystals [50]
		In situ tensile testing of polymer fibers [36]

from different crystal layers with long range order undergo constructive interference. This causes high-intensity peaks in the spectrum. For materials without long range order such as amorphous systems, no peaks are observed. In order to obtain detailed crystallographic structure of a crystal material, circular diffraction patterns of bright dots with various intensities can also be collected and analyzed. Circular diffraction patterns can be used to evaluate amorphous systems such as collagen and keratin that usually do not exhibit characteristic X-ray patterns [51].

Besides XRD, density measurements can also be used to characterize the crystallinity of a polymer. The crystalline region is more densely packed than the amorphous region leading to higher density of crystalline polymers. The density of a polymer can be determined using a density-gradient method [52]. A density-gradient column can be prepared by adding a liquid with continuous increase of density from the meniscus to the base. The sample can then be suspended in the column at a certain height that matches the specific density. Direct density measurement is another more convenient method of measuring the density. It is performed by measuring the weight and volume of a polymer directly. The sample is weighed both in air and in a liquid of known density in order to calculate the volume. The sample volume is equal to the loss of weight in the liquid divided by the density of liquid. All these measurements are sensitive to air bubbles trapped in the polymer, the moisture content, and any inhomogeneities in the polymer.

20.3.2 Molecular Weight and Particle Size

20.3.2.1 Molecular Weight

The average molecular weight is typically used to describe the polymer size. There are different methods of calculating the average molecular weight: number average molecular weight (M_n), weight average molecular weight (M_w), viscosity average molecular weight (M_v), and Z-average molecular weight (M_z). Calculations for different average molecular weights are given in the equations in the succeeding text. M_i is the molecular weight of each polymer chain, N_i is the number of polymer chains with molecular weight of M_i, and a is the Mark-Houwink parameter that is discussed in the succeeding text:

$$M_n = \frac{\sum M_i N_i}{\sum N_i}, \quad M_w = \frac{\sum M_i^2 N_i}{\sum M_i N_i},$$

$$M_z = \frac{\sum M_i^3 N_i}{\sum M_i^2 N_i}, \quad M_v = \left[\frac{\sum M_i^{1+a} N_i}{\sum M_i N_i}\right]^{1/a}$$

Techniques based on various principles provide different molecular weights. For example, methods measuring responses depending on the number of polymer molecules regardless of shape or molecular weight can be used to determine M_n. These techniques include end group analysis, osmotic pressure measurement, and light scattering. In most cases, it is more helpful to know the molecular weight distribution of a polymer by plotting molecular weight on the x-axis and the amount of polymer on the y-axis. We will discuss the techniques commonly used to obtain the distribution of molecular weight such as size exclusion chromatography (SEC) and MALDI mass spectroscopy.

SEC is conducted by forcing the polymer solution through a column filled with a cross-linked porous matrix to separate polymers with different molecular weights. The most widely used SEC for determining the molecular weight distribution of polymers is GPC. GPC is a variant of high-performance liquid chromatography (HPLC). Organic solvents such as tetrahydrofuran or toluene are usually used as the mobile phase in GPC systems. The elution volume (or retention time) is proportional to the molecular size (namely, the hydrodynamic volume (V_η)) of the polymer. V_η is proportional to the product of intrinsic viscosity ($[\eta]$) and M_v, which can be related using the Mark-Houwink equation shown later. K and a are the Mark-Houwink parameters:

$$[\eta] = KM_v^a$$

Based on a calibration curve prepared by measuring the elution volume of a series monodispersed polymer standards with known molecular weights, the molecular weight of a polymer sample can then be determined. Based on this principle, GPC determines the distribution of M_v, while M_n, M_w, and M_z can also be calculated using other data. As GPC only measures molar masses of polymers indirectly, some researchers regard this technique as semiquantitative [53]. GPC can be coupled with multidimensional polymer HPLC techniques to obtain the quantitative molar mass of complex polymer systems [54].

In a MALDI mass spectroscopy, the polymer is dispersed in a matrix of a compound with high UV absorption such as trans-cinnamic acid or 2,5-dihydroxybenzoic acid. A UV laser with a wavelength of 330-360 nm is then applied to the matrix so that the polymers become charged ions. Following this, the charged polymers are vaporized between two electrodes, and the electrical force accelerates the charged polymer molecules causing them fly to the detector. Polymer molecules with different mass have different rates of acceleration, and therefore, they hit the detector in the order of mass. Based on this, MALDI mass spectroscopy gives absolute values of mass. In a MALDI mass spectrum, the x-axis of the diagram is molecular weight and y-axis is the number of molecules [55]. MALDI mass spectroscopy is widely used for molecular weight determination of synthetic polymers such as homopolymers, copolymers, and polymer blends [55]. It is also very useful in characterizing covalent modifications of pharmaceutical formulations using polymers [56]. Another interesting application of MALDI mass spectroscopy is to generate images of two-dimensional spatial distribution of drug molecules in thin samples such as sections of implants [57].

20.3.2.2 Particle Size

The particle size of different drug delivery systems can range from nanometers to micrometers. Particle size affects the deposition, drug release rate and mechanism, and therefore the bioavailability of drug in various delivery systems. For example, the particle size of pulmonary microspheres is critical to deposition in lung [58]. Fabricating highly insoluble drug particles in the nanosize range can significantly increase bioavailability [59]. The microscopic methods described earlier are direct ways to visualize particles. In this section, indirect particle size measurement methods based on light scattering, laser diffraction, or light obscuration are discussed.

Light scattering can be classified as static light scattering (SLS) and dynamic light scattering (DLS). SLS measures the time-averaged intensity of scattered light, while DLS measures fluctuations in intensity of scattered light with time. Very small sample sizes and concentrations are required for light scattering measurement, and the particle radius measured can range from 2 to 500 nm. The sample is illuminated with a monochromatic laser light, and the scattered light intensity is measured at the detector at various angles. Rayleigh-Debye-Zimm theory is the theoretical foundation for SLS. Parameters that can be obtained from SLS are number average molecular weight (M_n), weight average molecular weight (M_w), root-mean-square radius, and the second virial coefficient. Solvent refractive index, sample concentration, and refractive index increment are required to perform SLS data analysis.

When monochromatic light hits the particles, time-dependent fluctuation in the scattered light intensity is observed as long as the particle size is smaller than the wavelength of the incident light. This is due to the Brownian motion of the particles and can be described using the Stokes-Einstein equation shown in the succeeding text:

$$D = \frac{k_B T}{6\pi\eta r}$$

Based on this, DLS can provide the translational diffusion coefficient and hydrodynamic radius of the sample. DLS measurement requires knowledge of solvent viscosity, solvent refractive index, and sample temperature. A comprehensive review of the theory/application of SLS and DLS in polymer systems is given by Schärtl [60].

Light obscuration is used to obtain the particle size of micron-sized particles. In this technique, a dilute stream of particles in suspension is passed between a laser light source and a detector. The particles cause blockage of light at the detector. The reduction of light intensity is related to the particle size. Light obscuration methods are used for particles in the range of 0.5-500 μm.

20.3.3 Surface Characterization

20.3.3.1 Spectroscopic Methods

The most widely used spectroscopic method for surface analysis is XPS. XPS is used to determine the elemental composition of solid surfaces. Although there is no special requirement for sample preparation, surface contamination through storage/transportation must be prevented. The principle of XPS is based on electron emission from the surface material in response to irradiation with monochromatic X-rays (usually 1253.6 eV, MgKa, or 1486.6 eV, AlKa). Different photoelectrons are emitted from different elements, and a percentage of the different surface elements can be obtained. The emitted electrons can only penetrate 100 Å, and therefore, only the elements at the surface can be quantified. Two types of scans can be performed from XPS, namely, low-resolution scans (1000 eV wide) and high-resolution scans (20 eV wide). In addition to surface elemental composition, high-resolution scans can provide information on aromatic groups or unsaturated bonds. For example, high-resolution XPS has been used to determine C-O group change in the surface of polyurethane after modification with poly(ethylene oxide) [61]. Angle-resolved XPS (ARXPS) performs measurements at different emission angles, which can provide compositional variation of the sample as a function of depth. ARXPS is useful to examine the layer-by-layer structure as well as the transition from surface to bulk. XPS is a relatively nondestructive technique although some care is needed to rule out the surface chemistry change by limiting the X-ray exposure time [62,63]. A destructive method, argon etching of the material, can be used in order to determine composition levels deeper than 100 Å [64]. Recent advances include depth XPS profile sputtering using a C_{60} source, and the sputtering rate decreases with the total ion current [65]. Surface reorientation during the process must be taken into consideration while using XPS technique to characterize hydrated or freeze-dried surfaces.

Other spectroscopic techniques for surface analysis include Auger electron spectroscopy (AES), secondary ion mass spectroscopy (SIMS), and attenuated total reflection Fourier transform infrared spectroscopy (ATR-FTIR). The use of AES in drug delivery systems is very limited because the high-energy beam in AES can burn organic materials leading to surface change of these materials [62]. SIMS can provide detailed surface molecular information by bombarding the surface with a focused ion beam (approximately 5-25 keV) and collecting/analyzing the ejected secondary particles around the surface. SIMS can measure the surface composition to a depth of 1-2 nm. SIMS can be used to predict the surface properties of the material such as wettability, adhesiveness, and reactivity by measuring the covalent binding of molecules at the surface [66]. Another interesting application of SIMS is for imaging purposes as it can provide a detailed surface pattern of materials with a submicron spatial resolution.

20.3.3.2 Surface Area Measurement

Although particle size measurement methods mentioned in 3.2 earlier provide information on surface area, those methods are based on the assumption that the particles are smooth solid spheres without taking particle morphology into consideration. Unlike those methods, BET measurements are based on monolayer adsorption of gas molecules on the surface of solid materials. These methods can also be used to reflect the surface morphology of particles (e.g., porosity). The BET theory was first published in 1938 by Brunauer *et al.* [67]. The BET theory is expressed in the equation in the succeeding text:

$$\frac{1}{v\left[\left(\frac{p_0}{p}-1\right)\right]}=\frac{c-1}{v_{\mathrm{m}}c}\left(\frac{p}{p_0}\right)+\frac{1}{v_{\mathrm{m}}c}$$

where v is the volume of gas, p is the equilibrium pressure, p_0 is the saturation pressure, v_{m} is the quantity of monolayer adsorbed gas, and c is the BET constant. By defining a relative pressure, φ $(=P/P_0)$, the equation is further transformed to

$$\frac{\varphi}{v\left[(1-\varphi)\right]}=\frac{c-1}{v_{\mathrm{m}}c}\varphi+\frac{1}{v_{\mathrm{m}}c}$$

The monolayer adsorption v_{m} and BET constant c are determined from the $\frac{\varphi}{v\left[(1-\varphi)\right]}$ versus φ plot. The specific surface area is calculated using the cross-sectional area of the gas molecule, the monolayer adsorption volume, and the weight of the sample. Krypton is usually used as the gas for BET measurement as it has a lower saturation pressure (2.7 mmHg) compared to nitrogen (760 mmHg), making it a more sensitive adsorbing gas. This method was used to determine the correlation of surface area and porous structure of PLGA microspheres prepared using different polymers. Sample sizes larger than 0.3 g are recommended to obtain a consistent and reproducible result, and for surface areas less than 0.5 m^2/g, a sample size of 0.5 g or greater is needed [68].

20.4 BIOCOMPATIBILITY TESTING OF POLYMERIC MATERIALS

20.4.1 Regulatory Guidelines for Biocompatibility Testing

Although FDA does not have specific guidance for biocompatibility testing of polymers, the International Standard ISO-10993 is recommended by the FDA for biological evaluation of biomedical devices. ISO-10993 entails 20 detailed standards to evaluate the biocompatibility of a material device prior to clinical testing. These tests include both *in vitro* and *in vivo* assays as well as physicochemical characterization of the materials. The Office of Device Evaluation General Program Memorandum #G95-1 entitled "Use of International Standard ISO-10993, 'Biological Evaluation of Medical Devices Part 1: Evaluation and Testing'" dated May 1, 1995 provides the application of ISO-10993 for biocompatibility testing of biomedical devices. On April 23rd, 2013, the Food and Drug Administration (FDA) announced the availability of the draft guidance entitled "Use of International Standard ISO-10993, 'Biological Evaluation of Medical Devices Part 1: Evaluation and Testing,'" to update and clarify the previous guidance. The final version of this draft will replace #G95-1. Table 20.3 is taken from the draft and indicates the initial evaluation and consideration for biological testing of biomedical devices.

The US Pharmacopeia (USP) describes the *in vitro* and *in vivo* biological reactivity tests of elastomerics, plastics, and other polymeric materials with direct or indirect patient contact. Agar diffusion, direct contact, and elution tests are *in vitro* tests using L-929 mammalian fibroblast cells. Systemic injection, intracutaneous, and implantation tests are *in vivo* tests. A brief summary of these tests is listed in Table 20.4. For detailed information, readers are referred to USP chapters 87 and 88.

20.4.2 Biodegradable Polymers

Biodegradability is one of the most important properties of biocompatible polymers. The applications of biodegradable polymers in controlled drug delivery include implants, drug-eluting stents, nanoparticles, microspheres, and polymeric scaffolds. Some examples of natural biodegradable polymers are fibrin, collagen, chitosan, gelatin, etc., and synthetic biodegradable polymers include polyesters, polyanhydrides, and polyamides [69,70]. Biodegradation tests must be performed for biodegradable polymers. These tests focus on the tissue response to the polymer and its degradation products, the period and mechanism of degradation, and evaluation of degradation products and leachables (residues from polymerization, additives, and degraded low-molecular-weight fragments) in the tissue. Chemical or enzymatic hydrolysis is the major mechanism of polymer degradation. Many toxicity assays involve the use of extracts of materials. Cell culture medium is used to extract the leachables from the materials. The extracts are highly dependent on the morphology (especially the surface area) of the material and the process of extraction.

20.4.3 *In Vitro* Cytotoxicity Assessment

Cytotoxicity is the most important parameter to evaluate the biocompatibility of biodegradable polymers *in vitro*. Various assays are available to measure cell viability after exposure to polymer extracts, such as lactate dehydrogenase leakage assay (LDH), neutral red uptake assay (NRU), and trypan blue assay. Table 20.5 summarizes the different types

TABLE 20.3 Evaluation and Consideration for Biological Testing of Biomedical Devices

Device Categories			Biological Effect							
Nature of Body Contact		Contact Duration A-Limited (24 h) B-Prolonged (24 h to 30 days) C-Permanent (>30 days)								
Category	Contact		Cytotoxicity	Sensitization	Irritation or Intracutaneous Reactivity	Systemic Toxicity (Acute)	Subchronic Toxicity (Subacute Toxicity)	Genotoxicity	Implantation	Hemocompatibility
Surface devices	Intact skin	A	x	x	x					
		B	x	x	x					
		C	x	x	x					
	Mucosal membrane	A	x	x	x					
		B	x	x	x	o	o		o	
		C	x	x	x	o	x	x	o	
	Breached or compromised surfaces	A	x	x	x	o				
		B	x	x	x	o	o		o	
		C	x	x	x	o	x	x	o	
External communicating devices	Blood path, indirect	A	x	x	x	x				x
		B	x	x	x	x	o			x
		C	x	x	o	x	x	x	o	x
	Tissue/bone/dentin[a]	A	x	x	x	o				
		B	x	x	x	x	x	x	x	
		C	x	x	x	x	x	x	x	
	Circulating blood	A	x	x	x	x		o[b]		x
		B	x	x	x	x	x	x	x	x
		C	x	x	x	x	x	x	x	x
Implant devices	Tissue/bone	A	x	x	x	o				
		B	x	x	x	x	x	x	x	
		C	x	x	x	x	x	x	x	
	Blood	A	x	x	x	x	x		x	x
		B	x	x	x	x	x	x	x	x
		C	x	x	x	x	x	x	x	x

x, ISO evaluation tests for consideration.

o, Additional tests should be addressed in the submission, by either inclusion of the testing or a rational for the omission.

[a]*Tissue includes tissue fluids and subcutaneous spaces.*

[b]*For all devices used in extracorporeal circuits.*

TABLE 20.4 *In Vitro* and *In Vivo* Biological Reactivity Tests from USP

Test Type		Cell Line or Animal	Procedure	Interpretation of Results
In vitro	Agar diffusion test	L-929 mammalian fibroblast cells	Prepare monolayer cell culture with medium containing agar, place the flat surface of samples and controls in contact with the agar surface	Size of the zone around or under sample
	Direct contact test	L-929 mammalian fibroblast cells	Prepare monolayer cell culture and place the flat surface of samples and controls in contact with the agar surface	Size of the zone around or under sample
	Elution test	L-929 mammalian fibroblast cells	Prepare monolayer cell culture and replace the culture medium with extracts of samples and controls for incubation	Cell morphology (round or loosely attached) and cell lysis
In vivo	Systemic injection	Albino mice	Inject extracts of materials i.v. or i.p.	Abnormal behavior, weight loss, or death of the animals
	Intracutaneous test	Thin-skinned albino rabbits	Inject extracts of materials intracutaneously at one side of the animal and a blank at the other side	Skin reaction such as erythema, eschar, and edema formation
	Implantation	Adult rabbits	Implant four strips of the sample/control into the paravertebral muscle at each side of the animal	Capsule width of encapsulation

TABLE 20.5 Different Types and Mechanisms of Cytotoxicity Assays

Assay Type	Mechanism	Assay Name
Altered cell permeability assays	The loss of cell membrane integrity can result in the release of certain soluble, cytosolic enzymes	Lactate dehydrogenase release assay (tetrazolium salt-based assay)
		Glucose-6-phosphate dehydrogenase release assay (resazurin-based assay)
		Glyceraldehyde-3-phosphate dehydrogenase release assay
Cell viability assays	Stains dead cells	Trypan blue and alamar blue assay
	Total cell protein quantification	Sulforhodamine B assay
Cell survival assays	Stains survival cells	Neutral red uptake assay

and mechanisms of various cytotoxicity assays. Altered cell permeability assays (such as LDH) are more sensitive in detecting short-term effects (1 h), while cell viability assays (such as trypan blue) and cell survival assays (such as NRU) are more suitable for long-term effects (12-24 h) [71]. The most widely used assay is the MTT assay based on cell membrane permeability change. MTT is a tetrazolium dye that can react with NADH produced during lactate oxidation catalyzed by LDH. The colored product of this reaction is formazan that has UV absorption at 490-520 nm. The amount of formazan produced is proportional to the LDH concentration that is related to the amount of dead cells.

20.4.4 *In Vivo* Biocompatibility Evaluation

The biocompatibility of polymeric drug delivery systems focuses on their *in vivo* biological responses following administration of the formulation. Besides the efficacy of the drug loaded in the delivery systems, the effect of the polymeric materials should also be investigated. Depending on different formulations, biocompatibility tests include sensitization and irritation reactivity evaluation, subchronic toxicity (subacute toxicity), systemic toxicity (acute toxicity), implantation, genotoxicity, hemocompatibility, carcinogenicity, reproductive and developmental toxicity, and immune response. Sensitization tests are usually conducted using guinea pigs as a model animal to determine the skin reaction following exposure or contact with devices, materials, or extracts. Irritation is described as an inflammatory reaction to chemicals (leachables in the case of polymers). Intracutaneous reactivity tests determine the local tissue reaction to extracts of materials following intracutaneous injection. Sensitization, irritation, and intracutaneous reactions mostly focus on the response to the extracts of the materials. Therefore, the solvent selection and the extraction process must be considered for different polymers. Systemic toxicity evaluates the effects of materials on target tissues and organs, such as the liver, spleen, and kidney. Acute toxicity is considered an adverse event within 24 h after the administration of the material. Subacute toxicity is the study of any adverse events within 14-28 days following administration. The subchronic effects are usually within 90 days but not exceeding 10% of the animals' life span, while chronic effects occur at time frames over 10% of the animals' life span. When the material has the potential of genotoxicity, carcinogenicity, and reproductive/developmental toxicity, these aspects must be investigated. Hemocompatibility tests study the effect on blood/blood components by blood-contacting materials. Thrombosis, coagulation, platelets, hematology, and immunology are evaluated in these tests.

Implantation tests assess the local pathological effects induced by the biomaterials. Parameters such as number and distribution of inflammatory cells, thickness of fibrous capsule, and the presence of necrosis are determined at various times following implantation. Control of the inflammatory reaction and foreign body response is a big issue for implantable devices and drug delivery systems [5]. Upon implantation, both acute and chronic inflammations are encountered. Acute inflammation has been shown to be dependent on the extent of the initial trauma, whereas chronic inflammation has been shown to depend on the implant size [72]. Dexamethasone-loaded PLGA microsphere/PVA hydrogel composite coatings for biosensors have been developed for continuous release of dexamethasone at the implantation site to control negative tissue responses [73]. These composite coatings were able to prevent inflammation and fibrosis for periods of up to three months [74].

20.5 *IN VITRO* DISSOLUTION TESTING METHODS FOR POLYMERIC FORMULATIONS

Drug efficacy is dependent on the concentration at the site of action, and therefore, monitoring of drug release from controlled release formulations is critical. Pharmacokinetic (PK) studies are performed to determine drug plasma concentrations versus time as well as drug concentrations at the local site of action. The properties of absorption, distribution, metabolism, and elimination of the drug can be analyzed from the PK curve. Polymers used in drug delivery systems can affect the drug release as well as the PK characteristics of the formulation. For example, pegylated formulations (protein, liposome, or nanoparticles) have elongated circulating times in the blood that is beneficial for tumor targeting due to the enhanced permeation and retention effect compared to nonpegylated formulations [75]. Microspheres prepared using PLGA can maintain effective therapeutic plasma drug concentrations from days to months [9]. One of the major purposes of *in vitro* dissolution testing is to predict the *in vivo* performance of formulations from *in vitro* data. This is an important tool for both formulation development and quality control in the pharmaceutical industry. Initially dissolution methods were developed for immediate release (IR) and modified release solid oral dosage forms [76,77]. In the past several decades, more and more complex drug delivery systems (such as microspheres, hydrogels, and nanoparticles) have emerged. However, till now, there have been no standardized *in vitro* release testing methods for these complex parenteral products [78,17,79]. Therefore, there is a need for the development of dissolution testing methods for these dosage forms. The regulatory guidelines for dissolution methods mainly focus on solid oral dosage forms such as tablets. Traditional dissolution methods for novel formulations such as microspheres and liposomes include sample and separate and dialysis methods. Sample and separate methods can lead to aggregation of microparticles and incomplete exclusion of nanoparticles, while the dialysis methods have the potential of violating sink conditions [80]. An overview of the current regulatory guidance for dissolution testing is provided along with a summary of the progress that has been made in developing standardized dissolution methods for the novel polymeric formulations.

20.5.1 Regulatory Guidelines and *In Vitro-In Vivo* Correlation

Polymers are widely used in various pharmaceutical formulations, such as traditional oral formulations (tablets and capsule) and complex parenteral formulations (such as microspheres, nanoparticles, implants, and *in situ* forming gels). Dissolution testing is used in formulation development, quality control for batch release of product, and *in vitro-in vivo* correlations. FDA has a database listing dissolution methods for around 1000 formulations such as tablets, capsules, granules, and injectable suspensions. These methods provide information about the standardized USP apparatus used, testing speed, type and volume of release medium, and the recommended sampling points. For detailed information about the seven standardized dissolution apparatuses, method development, and validation of compendial methods, the reader is referred to USP Chapters 711,

724, 1087, 1090, and 1092. The FDA guidance entitled "Dissolution testing of IR solid oral dosage forms" lists methods for comparison of *in vitro* release profiles [77].

In vitro-in vivo correlation (IVIVC) is defined as a predictive mathematical model describing the relationship between an *in vitro* property (usually the rate or extend of a drug dissolution or release) and a relevant *in vivo* response (e.g., plasma drug concentration or amount of drug absorbed). Once an IVIVC is established, dissolution testing can be used to waive *in vivo* bioequivalence studies for scale up and postapproval changes as well as for generic drugs, according to FDA guidelines [76,81]. There are five types of IVIVC, level A, level B, level C, multiple level C, and level D. Out of these, only level A, a point-to-point correlation, can be used for bioequivalence decisions. To achieve a point-to-point correlation, the deconvoluted PK data are plotted against the drug release data from a dissolution study. An ideal level A correlation is obtained if the *y*-intercept is zero and the slope of the line is equal to one. Timescale and time shifting are also allowed for the development of IVIVC. The FDA requires two or more formulations with differing release rates to achieve an IVIVC [76].

20.5.2 *In Vitro* Dissolution Testing for Microspheres

Table 20.6 lists FDA-approved PLGA-based parenteral microsphere products available on the US market. These formulations are based on the degradation of PLGA to achieve sustained release *in vivo* from 1 month to 3 months. According to the FDA dissolution method database, only Trelstar® Depot has a standardized dissolution method listed using USP apparatus 2. FDA recommends using USP apparatus 2 or USP apparatus 4 to develop dissolution methods for four of the products, while for the other two products, there are no dissolution methods listed. Recent workshop reports have recommended the use of USP apparatus 4 for microsphere products [82].

Drug release from PLGA microspheres is based on polymer degradation. Two mechanisms are usually followed, bulk degradation and surface degradation. Degradation of PLGA is pH- and temperature-sensitive. The USP apparatus 4 method is a flow-through cell method composed of flow-through cells, a piston pump, and a release media reservoir with accurate temperature control. There are different types of flow-through cells available for tablets or implants. An adaptation of USP apparatus 4 for microspheres was made to the flow-through cell by dispersing the microspheres with glass beads inside the cell (as shown in Figure 20.4) [83]. The glass beads were added to prevent aggregation of the microspheres and to achieve laminar flow inside the cell [83].

Risperdal® Consta® was used as a model microsphere formulation to assess the utility of USP apparatus 4 for microsphere *in vitro* release testing [84]. Following validation for robustness and reproducibility, this method appears to be suitable for possible compendial adaptation and is useful for microsphere product development, quality assurance, and regulatory approval. The modified USP 4 method was also shown to be useful for investigation of *in vitro*

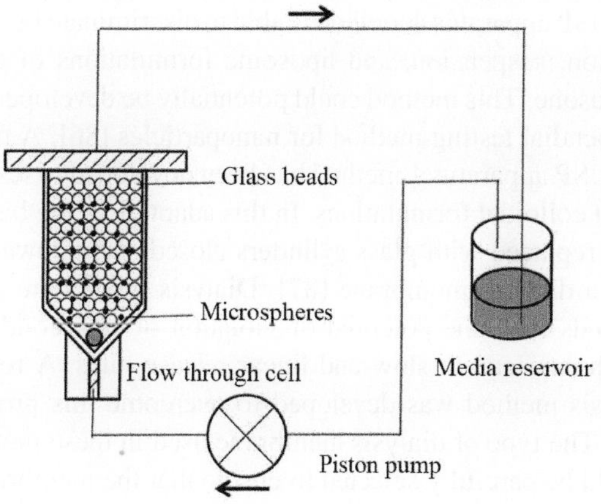

FIGURE 20.4 Illustration of USP apparatus 4 for *in vitro* release testing of microspheres.

TABLE 20.6 FDA-Approved PLGA-Based Parenteral Microspheres and Dissolution Methods

Product	API	Route of Administration	Polymer	Standardized Dissolution Method
Lupron Depot®	Leuprolide	i.m.	PLGA	Under development
Sandostatin® Depot	Octreotide	i.m.	PLGA	Under development
Trelstar® Depot	Triptorelin	i.m.	PLGA	USP II
Somatuline® Depot	Lanreotide	s.c.	PLGA	N/A
Nutropin Depot®	Somatropin	s.c.	PLGA	N/A
Risperdal® Consta®	Risperidone	i.m.	PLGA	Under development
Vivitrol®	Naltrexone	i.m.	PLGA	Under development

release of protein-loaded microspheres [85]. By inserting an *in situ* fiber-optic probe in the media reservoir of USP apparatus 4, a complete characterization of the initial burst release of drug from the microspheres was conducted and a 16% higher cumulative release was obtained compared to data obtained using a sample and separate method [83].

20.5.3 *In Vitro* Dissolution Testing for Nanoparticles

Although there are various polymeric/lipid nanoparticles under development, there is currently no standardized dissolution method for polymeric nanoparticles. The major obstacle to dissolution method development for nanoparticles is separation of the drug and the formulation. Dialysis is widely used as an *in vitro* release method for these formulations. A modified USP apparatus 4 dialysis adapter has been developed in an effort to standardize the dialysis method for *in vitro* release testing of nanoparticulate drug delivery systems. A dialysis adapter was designed as a sample holder inserted into the flow-through cell of USP apparatus 4 as shown in Figure 20.5. This method has been used for dissolution testing of suspensions and liposome formulations. The USP apparatus 4 method is able to discriminate between solution, suspension, and liposome formulations of dexamethasone. This method could potentially be developed as a compendial testing method for nanoparticles [86]. A modified USP apparatus 1 method has been developed to test different colloidal formulations. In this adaptation, the baskets were replaced with glass cylinders closed at the lower end with a dialysis membrane [87]. Dialysis membrane-based methods have the potential of violation of sink condition, which can lead to slow and linear release rates. A reverse dialysis method was developed to overcome this problem [80]. The type of dialysis membrane used in these methods should be carefully selected to ensure that the membrane is compatible with the release medium and the drug. A test to determine drug diffusion through the membrane is recommended to make sure that drug diffusion through the membrane is not the rate-limiting step during a dissolution test of a formulation.

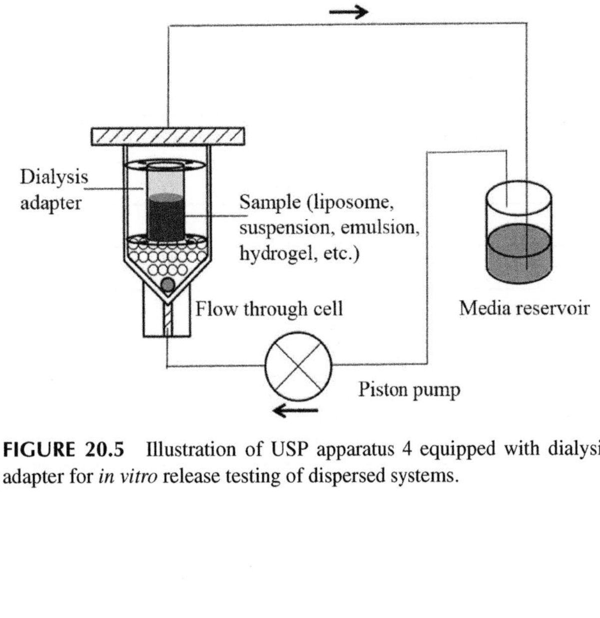

FIGURE 20.5 Illustration of USP apparatus 4 equipped with dialysis adapter for *in vitro* release testing of dispersed systems.

20.5.4 *In Vitro* Dissolution Testing for *In Situ* Gel Formulations

Similar to microsphere/nanoparticle formulations, dissolution testing of hydrogel formulations relies on separating the gel from the dissolution medium. This test can be conducted by inserting the formulation directly into the release medium. However, the shape of the formed gel is not controllable, and therefore, the reproducibility of the method is sacrificed. In some reports, preformed gels are used to investigate release from thermosensitive hydrogels. A layer of hydrogel was first formed at the bottom of a vial, and the release medium was then carefully added on top of the hydrogel [88,89]. This method ensures that the surface area of the hydrogel is in contact with the release medium and can provide reproducible results. However, this method relies on a preformed hydrogel and cannot be easily transformed to a standardized compendial method. The USP apparatus 4 equipped with dialysis adapters described in Section 20.5.3 can potentially be used for dissolution testing of this type of formulations. The dialysis membrane can act as a holder for the sample to achieve *in situ* gelling of the formulations. Other researchers have designed holders such as nylon pouches and Teflon cavities in USP apparatus 2 systems in order to control the shape and standardize the dissolution method [90].

20.6 CONCLUSIONS

Various characterization methods both *in vitro* and *in vivo* can provide information to understand, predict, and improve the performance of drug delivery systems. Selection of methods depends on the material properties and their applications. Viscoelastic properties can be measured using both DMA and oscillatory shear rheometry. DSC is a most useful method of measuring thermal transitions. Various microscopic methods are available to obtain the microstructure and shape of the materials. Amorphous and crystalline materials have different packing patterns of molecules, and these properties can be determined from XRD or density measurements. Surface properties such as surface elemental composition and material porosity can be obtained from various spectroscopic methods as well as from BET measurements. The biocompatibility of the material can be determined from both *in vitro* and *in vivo* assays. *In vitro* dissolution testing can be utilized to correlate with the *in vivo* performance of polymeric drug delivery systems. All these characterization methods can provide valuable information for drug delivery system design, optimization, manufacturing, and quality control. The regulatory guidelines provide instructions for performing some of these tests. However, more detailed development of testing procedures such as *in vitro* dissolution methods is required for novel polymeric parenteral formulations.

REFERENCES

[1] I. Galeska, et al., Controlled release of dexamethasone from PLGA microspheres embedded within polyacid-containing PVA hydrogels, AAPS J. 7 (1) (2005) E231–E240.
[2] S. Vaddiraju, et al., Microsphere erosion in outer hydrogel membranes creating macroscopic porosity to counter biofouling-induced sensor degradation, Anal. Chem. 84 (20) (2012) 8837–8845.
[3] S. Verma, R. Gokhale, D.J. Burgess, A comparative study of top-down and bottom-up approaches for the preparation of micro/nanosuspensions, Int. J. Pharm. 380 (1–2) (2009) 216–222.
[4] A. Bohr, et al., Preparation of microspheres containing low solubility drug compound by electrohydrodynamic spraying, Int. J. Pharm. 412 (1–2) (2011) 59–67.
[5] J.M. Morais, F. Papadimitrakopoulos, D.J. Burgess, Biomaterials/tissue interactions: possible solutions to overcome foreign body response, AAPS J. 12 (2) (2010) 188–196.
[6] J. Shen, D. Burgess, In vitro dissolution testing strategies for nanoparticulate drug delivery systems: recent developments and challenges, Drug Deliv. Trans. Res. 3 (5) (2013) 409–415.
[7] O. Pillai, R. Panchagnula, Polymers in drug delivery, Curr. Opin. Chem. Biol. 5 (4) (2001) 447–451.
[8] J.M. Lu, et al., Current advances in research and clinical applications of PLGA-based nanotechnology, Expert. Rev. Mol. Diagn. 9 (4) (2009) 325–341.
[9] Y. Wang, D.J. Burgess, Microsphere technology, in: J.C. Wright, D.J. Burgess (Eds.), Long Acting Injections and Implants, CRC Press, Springer, 2012, pp. 195–220.
[10] H.K. Makadia, S.J. Siegel, Poly lactic-co-glycolic acid (PLGA) as biodegradable controlled drug delivery carrier, Polymers 3 (3) (2011) 1377–1397.
[11] C. Buonerba, et al., A comprehensive outlook on intracerebral therapy of malignant gliomas, Crit. Rev. Oncol. Hematol. 80 (1) (2011) 54–68.
[12] M.E. Brewster, T. Loftsson, Cyclodextrins as pharmaceutical solubilizers, Adv. Drug Deliv. Rev. 59 (7) (2007) 645–666.
[13] W. Friess, Collagen–biomaterial for drug delivery, Eur. J. Pharm. Biopharm. 45 (2) (1998) 113–136.
[14] R.E. McMahon, et al., Development of nanomaterials for bone repair and regeneration, J. Biomed. Mater. Res. B Appl. Biomater. 101 (2) (2013) 387–397.
[15] B.S. Zolnik, D.J. Burgess, Evaluation of in vivo–in vitro release of dexamethasone from PLGA microspheres, J. Control. Release 127 (2) (2008) 137–145.
[16] FDA, Use of International Standard ISO-10993, Biological evaluation of medical devices part 1: Evaluation and testing, C.f.D.E.a.R.C. U.S. Food and Drug Administration, Editor. Washington, DC, 1995.
[17] M.N. Martinez, et al., Breakout session summary from AAPS/CRS joint workshop on critical variables in the in vitro and in vivo performance of parenteral sustained release products, J. Control. Release 142 (1) (2010) 2–7.
[18] Y. Ikada, Challenges in tissue engineering, J. R. Soc. Interface. 3 (10) (2006) 589–601.
[19] S.C. Porter, L.A. Felton, Techniques to assess film coatings and evaluate film-coated products, Drug Dev Ind Pharm 36 (2) (2010) 128–142.
[20] A. Rawat, D.J. Burgess, Effect of ethanol as a processing co-solvent on the PLGA microsphere characteristics, Int. J. Pharm. 394 (1–2) (2010) 99–105.
[21] D.S. Jones, et al., Pharmaceutical applications of dynamic mechanical thermal analysis, Adv. Drug Deliv. Rev. 64 (5) (2012) 440–448.
[22] R. Chartoff, Thermoplastic polymers, in: E.A. Turi (Ed.), second ed., Thermal Characterization of Polymeric Materials, 1, Academic Press, New York, 1997, pp. 483–743.
[23] J.M. Morais, P.A. Rocha-Filho, D.J. Burgess, Relationship between rheological properties and one-step W/O/W multiple emulsion formation, Langmuir 26 (23) (2010) 17874–17881.
[24] T.K.L. Meyvis, et al., A comparison between the use of dynamic mechanical analysis and oscillatory shear rheometry for the characterisation of hydrogels, Int. J. Pharm. 244 (1–2) (2002) 163–168.
[25] J.L. Ford, T.E. Mann, Fast-Scan DSC and its role in pharmaceutical physical form characterisation and selection, Adv. Drug Deliv. Rev. 64 (5) (2012) 422–430.
[26] M.H. Chiu, E.J. Prenner, Differential scanning calorimetry: an invaluable tool for a detailed thermodynamic characterization of macromolecules and their interactions, J Pharm Bioallied Sci 3 (1) (2011) 39–59.
[27] N.J. Coleman, D.Q.M. Craig, Modulated temperature differential scanning calorimetry: a novel approach to pharmaceutical thermal analysis, Int. J. Pharm. 135 (1–2) (1996) 13–29.
[28] B. Keymolen, et al., Investigation of the polymorphic transformations from glassy nifedipine, Thermochim. Acta 397 (1–2) (2003) 103–117.
[29] A.W. Coats, J.P. Redfern, Thermogravimetric analysis. A review, Analyst 88 (1053) (1963) 906–924.
[30] M.D. Kofron, et al., The implications of polymer selection in regenerative medicine: a comparison of amorphous and semi-crystalline polymer for tissue regeneration, Adv. Funct. Mater. 19 (9) (2009) 1351–1359.
[31] B. Van Eerdenbrugh, L.S. Taylor, An ab initio polymer selection methodology to prevent crystallization in amorphous solid dispersions by application of crystal engineering principles, CrystEngComm 13 (20) (2011) 6171–6178.
[32] H. Sawada, et al., STEM imaging of 47-pm-separated atomic columns by a spherical aberration-corrected electron microscope with a 300-kV cold field emission gun, J. Electron Microsc. 58 (2009) 357–361.
[33] D. Baird, A. Shew, Probing the history of scanning tunneling microscopy, in: D. Baird, A. Nordmann, J. Schummer (Eds.), Discovering the Nanoscale, IOS Press, Amsterdam, 2004.
[34] F.R. Zypman, Scanning tunneling microscope spectroscopy of polymers, Scanning 24 (3) (2002) 154–156.
[35] C. Steffens, et al., Atomic force microscopy as a tool applied to nano/biosensors, Sensors (Basel) 12 (6) (2012) 8278–8300.
[36] F. Hang, et al., In situ tensile testing of nanofibers by combining atomic force microscopy and scanning electron microscopy, Nanotechnology 22 (36) (2011) 365708.
[37] J. Chappell, D.G. Lidzey, Phase separation in polyfluorene-polymethylmethacrylate blends studied using UV near-field microscopy, J. Microsc. 209 (Pt. 3) (2003) 188–193.

[38] M. Doroshenko, et al., Monitoring the dynamics of phase separation in a polymer blend by confocal imaging and fluorescence correlation spectroscopy, Macromol. Rapid Commun. 33 (18) (2012) 1568–1573.
[39] T. Crisenza, et al., Direct 3D visualization of the phase-separated morphology in chlorinated polyethylene/nylon terpolyamide based thermoplastic elastomers, Macromol Rapid Commun. 33 (2) (2012) 114–119.
[40] K.L. Liu, et al., A new insight for an old system: protein-PEG colocalization in relation to protein release from PCL/PEG blends, Mol. Pharm. 8 (6) (2011) 2173–2182.
[41] M.A. Alhnan, A.W. Basit, Engineering polymer blend microparticles: an investigation into the influence of polymer blend distribution and interaction, Eur. J. Pharm. Sci. 42 (1–2) (2011) 30–36.
[42] R. Song, et al., Modification of polymorphisms in polyvinylidene fluoride thin films via water and hydrated salt, J. Colloid Interface Sci. 401 (2013) 50–57.
[43] K.J. Zhu, et al., Preparation, characterization and in vitro release properties of ibuprofen-loaded microspheres based on polylactide, poly(ε-caprolactone) and their copolymers, J. Microencapsul. 22 (1) (2005) 25–36.
[44] C.-J. Chen, et al., Self-assembly cationic nanoparticles based on cholesterol-grafted bioreducible poly(amidoamine) for siRNA delivery, Biomaterials 34 (21) (2013) 5303–5316.
[45] M.-Q. Zhu, et al., Spiropyran-based photochromic polymer nanoparticles with optically switchable luminescence, J. Am. Chem. Soc. 128 (13) (2006) 4303–4309.
[46] Y. Otsuka, et al., Scanning electron microscopic analysis of defects in polymer coatings of three commercially available stents: comparison of BiodivYsio, Taxus and Cypher stents, J. Invasive Cardiol. 19 (2) (2007) 71–76.
[47] K. Geisel, L. Isa, W. Richtering, Unraveling the 3D localization and deformation of responsive microgels at oil/water interfaces: a step forward in understanding soft emulsion stabilizers, Langmuir 28 (45) (2012) 15770–15776.
[48] S. Clair, et al., Tip- or electron beam-induced surface polymerization, Chem. Commun. (Camb.) 47 (28) (2011) 8028–8030.
[49] H. Sakaguchi, et al., Direct visualization of the formation of single-molecule conjugated copolymers, Science 310 (5750) (2005) 1002–1006.
[50] Y. Li, et al., A strategy for the improvement of the bioavailability and anti-osteoporosis activity of BCS IV flavonoid glycosides through the formulation of their lipophilic aglycone into nanocrystals, Mol. Pharm. 10 (7) (2013) 2534–2542.
[51] J. Skakle, Applications of X-ray powder diffraction in materials chemistry, Chem. Rec. 5 (5) (2005) 252–262.
[52] C.H. Lindsley, Preparation of density-gradient columns, J. Polym. Sci. 46 (148) (1960) 543–545.
[53] D. Berek, Size exclusion chromatography—a blessing and a curse of science and technology of synthetic polymers, J. Sep. Sci. 33 (3) (2010) 315–335.
[54] D. Berek, Strategies in two-dimensional liquid chromatographic separation of complex polymer systems, Macromol. Symp. 174 (1) (2001) 413–434.
[55] M.W.F. Nielen, Maldi time-of-flight mass spectrometry of synthetic polymers, Mass Spectrom. Rev. 18 (5) (1999) 309–344.
[56] A.P. Kafka, et al., The application of MALDI TOF MS in biopharmaceutical research, Int. J. Pharm. 417 (1–2) (2011) 70–82.
[57] F. Kreye, et al., MALDI-TOF MS imaging of controlled release implants, J. Control. Release 161 (1) (2012) 98–108.
[58] F. Ungaro, et al., Engineered PLGA nano- and micro-carriers for pulmonary delivery: challenges and promises, J. Pharm. Pharmacol. 64 (9) (2012) 1217–1235.
[59] R. Shegokar, R.H. Müller, Nanocrystals: industrially feasible multifunctional formulation technology for poorly soluble actives, Int. J. Pharm. 399 (1–2) (2010) 129–139.
[60] W. Schärtl, Light scattering from polymer solutions and nanoparticle dispersions, Springer Laboratory, in: I. Alig, et al. (Eds.), Springer, New York, 2007.
[61] J.G. Archambault, J.L. Brash, Protein resistant polyurethane surfaces by chemical grafting of PEO: amino-terminated PEO as grafting reagent, Colloids Surf. B: Biointerfaces 39 (1–2) (2004) 9–16.
[62] B.D. Ratner, Advances in the analysis of surfaces of biomedical interest, Surf. Interface Anal. 23 (7–8) (1995) 521–528.
[63] R.N.S. Sodhi, Application of surface analytical and modification techniques to biomaterial research, J. Electron Spectrosc. Relat. Phenom. 81 (3) (1996) 269–284.
[64] J.L. Grant, D.S. Dunn, D.J. McClure, Surface Characterization of Sputter Etched Polymer Films. MRS Online Proceedings Library, 119 (1988) 297–302.
[65] B.-Y. Yu, et al., Depth profiling of organic films with X-ray photoelectron spectroscopy using C60+ and Ar+ Co-sputtering, Anal. Chem. 80 (9) (2008) 3412–3415.
[66] D. Léonard, H.J. Mathieu, Characterisation of biomaterials using ToF-SIMS, Fresenius J. Anal. Chem. 365 (1–3) (1999) 3–11.
[67] S. Brunauer, P.H. Emmett, E. Teller, Adsorption of gases in multimolecular layers, J. Am. Chem. Soc. 60 (2) (1938) 309–319.
[68] A.G. Hausberger, P.P. DeLuca, Characterization of biodegradable poly(d, l-lactide-co-glycolide) polymers and microspheres, J. Pharm. Biomed. Anal. 13 (6) (1995) 747–760.
[69] H. Seyednejad, et al., Functional aliphatic polyesters for biomedical and pharmaceutical applications, J. Control. Release 152 (1) (2011) 168–176.
[70] M. Navarro, et al., Biomaterials in orthopaedics, J. R. Soc. Interface. 5 (27) (2008) 1137–1158.
[71] V.N. Sumantran, Cellular chemosensitivity assays: an overview, Methods Mol. Biol. 731 (2011) 219–236.
[72] Y. Wang, B. Gu, D.J. Burgess, Unpublished data.
[73] S.D. Patil, F. Papadimitrakopoulos, D.J. Burgess, Dexamethasone-loaded poly(lactic-co-glycolic) acid microspheres/poly(vinyl alcohol) hydrogel composite coatings for inflammation control, Diabetes Technol. Ther. 6 (6) (2004) 887–897.
[74] U. Bhardwaj, et al., PLGA/PVA hydrogel composites for long-term inflammation control following s.c. implantation, Int. J. Pharm. 384 (1–2) (2010) 78–86.
[75] J.M. Harris, N.E. Martin, M. Modi, Pegylation: a novel process for modifying pharmacokinetics, Clin. Pharmacokinet. 40 (7) (2001) 539–551.
[76] FDA, Extended release oral dosage forms: development, evaluation, and application of in vitro/in vivo correlations, C.f.D.E.a.R.C. U.S. Food and Drug Administration, Editor. Washington, DC, 1997.
[77] FDA, Dissolution testing of immediate release solid oral dosage forms, C.f.D.E.a.R.C. U.S. Food and Drug Administration, Editor. Washington, DC, 1997.
[78] J. Shen, D.J. Burgess, Accelerated in-vitro release testing methods for extended-release parenteral dosage forms, J. Pharm. Pharmacol. 64 (7) (2012) 986–996.

[79] M.J. Rathbone, et al., CRS/AAPS joint workshop on critical variables in the in vitro and in vivo performance of parenteral sustained-release products, Dissolution Technol. 16 (2) (2009) 2.
[80] N. Chidambaram, D.J. Burgess, A novel in vitro release method for submicron sized dispersed systems, AAPS PharmSci 1 (3) (1999) E11.
[81] FDA, SUPAC-MR: Modified release solid oral dosage forms. Scale-up and postapproval changes: Chemistry, manufacturing, and controls; In vitro dissolution testing and in vivo bioequivalence document, C.f.D.E.a.R.C. U.S. Food and Drug Administration, Editor. Washington, DC, 1997.
[82] D.J. Burgess, et al., Assuring quality and performance of sustained and controlled release parenterals: EUFEPS workshop report, AAPS J. 6 (1) (2004) 100–111.
[83] B.S. Zolnik, J.-L. Raton, D.J. Burgess, Application of USP apparatus 4 and in situ fiber optic analysis to microsphere release testing, Dissolution Technol. 12 (2) (2005) 4.
[84] A. Rawat, et al., Validation of USP apparatus 4 method for microsphere in vitro release testing using Risperdal® Consta®, Int. J. Pharm. 420 (2) (2011) 198–205.
[85] A. Rawat, D.J. Burgess, USP apparatus 4 method for in vitro release testing of protein loaded microspheres, Int. J. Pharm. 409 (1–2) (2011) 178–184.
[86] U. Bhardwaj, D.J. Burgess, A novel USP apparatus 4 based release testing method for dispersed systems, Int. J. Pharm. 388 (1–2) (2010) 287–294.
[87] M.M.A. Abdel-Mottaleb, A. Lamprecht, Standardized in vitro drug release test for colloidal drug carriers using modified USP dissolution apparatus I, Drug Dev. Indus. Pharm. 37 (2) (2011) 178–184.
[88] Y. Liu, et al., Controlled delivery of recombinant hirudin based on thermo-sensitive Pluronic® F127 hydrogel for subcutaneous administration: in vitro and in vivo characterization, J. Control. Release 117 (3) (2007) 387–395.
[89] L. Zhang, et al., Development and in-vitro evaluation of sustained release Poloxamer 407 (P407) gel formulations of ceftiofur, J. Control. Release 85 (1–3) (2002) 73–81.
[90] S. Bedi, A novel shear thinning and thixotropic PLGA microparticulate suspension system for controlled drug delivery, in: College of Pharmacy, The University of Tennessee Health Science Center, Memphis, 2011, p. 205.

Chapter 21

Polymeric Biomaterials in Tissue Engineering and Regenerative Medicine

Xiaoyan Tang[*,†,‡,§,1], Shalumon Kottappally Thankappan[*,†,‡,1], Paul Lee[¶], Sahar E. Fard[¶], Matthew D. Harmon[*,†,‡,§], Katelyn Tran[¶], Xiaojun Yu[¶,2]

*Institute for Regenerative Engineering, University of Connecticut Health Center, Farmington, Connecticut, USA

†Raymond and Beverly Sackler Center for Biological, Physical and Engineering Sciences, Farmington, Connecticut, USA

‡Department of Orthopaedic Surgery, University of Connecticut Health Center, Farmington, Connecticut, USA

§Department of Chemical, Materials and Biomolecular Engineering, University of Connecticut, Storrs, Connecticut, USA

¶Department of Chemistry, Chemical Biology, and Biomedical Engineering, Stevens Institute of Technology, Hoboken, New Jersey, USA

2Corresponding author: e-mail:xyu@stevens.edu

Chapter Outline

21.1 INTRODUCTION

According to Langer and Vaccanti, "Tissue engineering is an interdisciplinary field that applies the principles of engineering and life sciences toward the development of biological substitutes that restore, maintain, or improve tissue function or a whole organ" [1]. Later in 1999, Laurencin defined tissue engineering as "The application of biological, chemical, and engineering principals toward the repair, restoration, or regeneration of living tissue using biomaterials, cells, and factors alone or in combination" [2]. It is an interdisciplinary field combining biological, biochemical, and physicochemical factors to reproduce the biological functions. In the late 2000s, tissue engineering strategies for tissue repair were transformed by the combination of stem cells with scaffold-based therapy to form a new integrated field called "regenerative medicine" or "regenerative engineering."

[1]Both authors have equally contributed.

This is a rapidly growing multidisciplinary field seeking stem cell therapy to aid in the regeneration process of a diseased organ or injury. In 2010, Laurencin redefined tissue engineering in terms of regenerative engineering as "the integration of materials science and tissue engineering with stem and developmental cell biology and regenerative medicine toward the regeneration of complex tissues, organs, or organ systems" [3]. Significant developments have been reported in the past two decades in the area of tissue engineering and regenerative medicine for the repair of damaged tissues such as cartilage, skin, bladder, muscle, bone, and blood vessels using different biomaterials.

The primary requirement for a biomaterial is biocompatibility. During the past two decades, researchers have proposed a wide variety of biomaterials ranging from metals through ceramics to polymers. Among them, polymers possess significant importance due to their chemical tunability to allow scaffolds with appropriate physical, biological, and mechanical properties. Additional important factors in tissue engineering are the scaffold architecture, biodegradability, and the physical stability of scaffold to meet the complex functionalities possessed by each tissue type. Moreover, a scaffold should have the capability to promote greater material-cell interaction and improve cell adhesion and proliferation. Also, a scaffold should allow adequate transfer of gas, nutrients, and growth factors required for cellular development. Proper exchange of nutrients and growth factors will be possible through scaffold porosity and pore interconnectivity and hence scaffold morphology plays a key role in tissue repair and regeneration. Another important factor to be mentioned is the degradability of the scaffold. Ideally, the degradation rate of the polymeric scaffold should be in accordance with regeneration of the tissue.

A large number of natural and synthetic polymers are currently used in tissue engineering and regenerative medicine. It is considered that natural polymers were among the first used for clinical applications [4]. Natural polymers include collagen, gelatin, elastin, actin, keratin, albumin, chitosan, alginic acid, chitin, cellulose, silk, and hyaluronic acid. Many natural polymeric materials are more easily accepted by biological systems where they can be metabolically processed through established pathways. However, natural biomaterials have some disadvantages including possible immunogenicity, structural complexity, and inferior biomechanical properties.

Due to their availability and controllable degradation rate, synthetic biomaterials are also considered to be potential candidates in tissue engineering and regenerative medicine. Compared to natural polymers, synthetic polymers can easily be tailored into any form suitable for tissue engineering applications. Such materials should provide a three-dimensional (3D) structure that not only plays a supportive role for the tissue but also interacts with cells to control their function and differentiation [5]. Major synthetic polymers used for tissue engineering/regenerative medicine include poly(lactic acid) (PLA), poly(glycolic acid) (PGA), poly(lactide-*co*-glycolic acid) (PLGA), polyhydroxybutyrate (PHB), poly(hydroxyvalerate) (PHV), poly(hydroxybutyrate-valerate) (PHBV), poly(dioxanone) (PDS), polycaprolactone (PCL), polyurethanes (PUs), polyphosphazenes, polyanhydrides, polyacetals, poly(propylene fumarate) (PPF), and poly(ethylene glycol) (PEG).

This chapter reviews major natural and synthetic polymers, based on their diverse applications in tissue engineering/regenerative medicine. It also gives an outline of the advantages of biodegradable polymers over conventionally used allografts and autografts.

21.2 NATURAL POLYMERS IN TISSUE ENGINEERING AND REGENERATIVE MEDICINE

The major natural polymers currently in use are proteins and polysaccharides. Due to diverse physical and chemical properties, natural polymers play a key role in tissue engineering and regenerative medicine. In the following section, we describe some frequently used protein polymers for tissue engineering and regenerative medicine applications.

21.2.1 Proteins as Biomaterials

Proteins or amino acids are the major constituents present in natural tissues and they are well known for their controlled natural degradation ability. Such protein-based materials are especially useful in suturing applications, for scaffold materials, as drug delivery vehicles, and much more.

21.2.1.1 Collagen

Collagen is one of the body's extracellular matrix (ECM) proteins, abundantly found in musculoskeletal tissues, and has been extensively used as scaffold material for tissue engineering [6,7]. There are nearly 22 different types of collagen that exist in the human body, and, among them, types I-IV are the most common. Collagen can be extracted from mammals such as rats, bovines, and humans and it has been used successfully in many different applications ranging from tissue regeneration to cosmetic surgery. Since a major part of the ECM contains a fibrous collagenous component, scaffolds made up of collagen are considered to be ideal for tissue engineering. A scaffold's features can be varied by applying different concentrations of collagen [8]. Recent studies demonstrate that nanofibrous collagen scaffolds display physical and mechanical properties similar to normal tissue [9]. Another study suggests the possibility of applying nanostructured scaffolds in tissue engineering for better cell adhesion and growth *in vitro* [10].

Schauss *et al.* [11] reported the use of hydrolyzed type II collagen with chondroitin sulfate (CS) and hyaluronic acid (BioCell Collagen) for osteoarthritis as oral supplements. Another FDA-approved scaffold "Collagrafts," a composite form of fibrillar collagen, hydroxyapatite, and tricalcium phosphate, was found to be very effective as a biodegradable bone graft substitute [12,13]. There are numerous collagen scaffolds available on the market for tissue engineering applications. Figure 21.1 represents photographs of some commercially available collagen scaffolds in the market. A recent study revealed that a collagen-chitosan-PLA-derived scaffold could help better cell adhesion and proliferation for *blood outgrowth endothelial cells* [14]. Compared to synthetic polymers, collagen-hydroxyapatite (HA) composite fibers are found not only to be a good template for enhanced osteoblast differentiation but also as a nucleating agent for HA deposition [15]. Another advantage of collagen is its processability. Collagen scaffolds can be fashioned into desired shapes such as tubes [16,17], sponges [18,19], powders [20], and sheets [21] for tissue engineering. Due to the structural integrity and fibrous nature of collagen, it has proven its efficiency in numerous tissue engineering applications such as muscle [22], cardiovascular [23], skin [24], cartilage [25], tendon [26], and ligament [27].

21.2.1.2 Gelatin

Gelatin is one of the most well-known biomaterials used in various applications ranging from food products to tissue engineering. Worldwide production of gelatin is roughly ~600 million pounds in a year. Being one of the main components of skin, bone, cartilage, and connective tissues, it is less immunogenic compared to collagen and has potential for promoting cell attachment, differentiation, and proliferation [28,29]. Gelatin is derived by denaturization and partial hydrolysis of collagen [30]. Due to its biodegradability, biocompatibility, and relatively low cost, gelatin has widely contributed to applications in the pharmaceutical and medical fields [31]. It is fairly easily processed by dissolving in water and also able to be dissolved in various organic solvents. One of the commercially available gelatin-derived products is Gelfoam® by Pfizer. It is normally used as a bandage substitute during surgery where it is used as a hemostatic [32]. Due to its potential porosity, gelatin is able to absorb 45 times its weight in liquid and has shown to be completely absorbed into the body in 4-6 weeks [33].

Kang *et al.* [34] reported the development of a porous gelatin scaffold using water as porogens. Elimination of ice particles by freeze-drying created uniform pores in the scaffold matrix suitable for tissue engineering. In another study, a combination of gelatin with chitosan and HA resulted in a 3D porous scaffold; further culture with rat calvaria osteoblasts resulted in a promising candidate for bone tissue engineering [35]. A reliable method for altering the gelatin degradability is by combining with synthetic polymers [36]. In one study, 70% PCL was selected as a synthetic counterpart with gelatin, in which PCL\gelatin-aligned fibers showed enhanced neural stem cell (NSC) growth compared to random fibers. One of the advantages of gelatin-based scaffolds is their feasibility to use with a wide variety of natural and synthetic polymers to facilitate successful development of tissue engineering scaffolds [37–41].

21.2.1.3 Elastin

Elastin is an elastic insoluble protein in connective tissue, with cross-linked tropoelastin as a major component. It is responsible for the contraction of skin, lung, and vascular tissues in our body [42]. The insolubility and enhanced immune response [43] of native elastin limits its biomedical application and demands new materials for tissues. The ability of elastin to maintain minimum platelet interaction makes it a suitable material for making biological coatings for synthetic vascular grafts [44,45]. Currently used synthetic vascular biomaterials like poly(ethylene terephthalate) (Dacron) and poly(tetrafluoroethylene) show

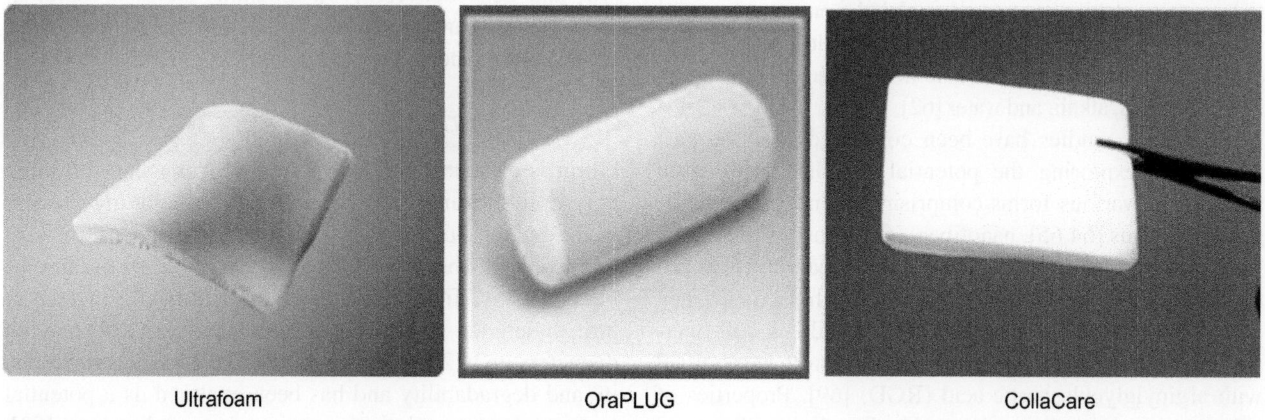

FIGURE 21.1 Photographs of some commercially available collagen scaffolds. (For color version of this figure, the reader is referred to the online version of this chapter.)

inadequate properties in comparison to native arteries in terms of thrombogenicity, mechanical properties, and cell interactions [46]. Due to the high resistance and insolubility of elastin to most of the treatment processes, decellularized elastin tissues have been proposed as an ideal material for various tissue engineering applications [47–49].

Another simple approach to make elastin-based biomaterials is the development of elastin-like polypeptides with pentapeptide repeating units. These polymers can easily be produced by *Escherichia coli* and are useful for various medical applications [50,51]. Elastin solubility can be improved by partial hydrolysis and can be used for elastic cartilage repair [52], bone regeneration [47], neural tissue engineering [53], and dermal regeneration [54]. A recent study demonstrated the combination of elastin/gelatin gels with excellent smooth cell adhesion and identical Young's modulus of native artery [55].

Synergistic effects of natural/synthetic polymer combinations were achieved by blending PLGA with gelatin and elastin to yield hydrolysis-resistant bioactive scaffolds, which required no further cross-linking [39,40]. Fibrous scaffolds with maximum surface area, controlled porosity, and mechanical strength can be achieved by electrospinning elastin/collagen to form multilayered nanofibrous scaffolds [56]. Elastin-like polypeptides are known to be excellent in tissue engineering, not only due to their immunogenicity but also due to a controllable amino acid sequence and molecular weight at the genetic or synthetic level [57]. Electrospun elastin fibers are found to be very effective in mimicking the internal mammary artery [58] and also for uniaxially directing smooth muscle cell in arterial media [59].

21.2.1.4 Silk

Natural silk contains two groups of proteins, fibroin and sericin [60] produced *by certain insect larvae to form cocoons*. Due to their specific molecular structure, silk fibers have significant durability, biocompatibility, degradability, and strength compared to other biomaterials [61]. Fibrous silk was shown to have more stability compared to globular proteins due to its considerable number of hydrogen bonds, hydrophobicity, and crystalline structure. It is insoluble in the majority of solvents like dilute acid, alkali, and water [62].

Numerous studies have been conducted over the past few years, exploring the potential of silk-fibroin-based scaffolds in various forms comprising membranes, hydrogels [63], films [64,65], nanofibers [66], sponges [67], mats [68], etc. Biomedical applications of silk can be enhanced through surface modifications with growth factors. Chen *et al.* reported an enhanced cell density and cell-cell interactions on silk when silk matrices were surface-modified with arginylglycylaspartic acid (RGD) [69]. Properties of silk films can be further enhanced by blending with natural and synthetic polymers [70,71].

Effects of silk fibroin in bone tissue engineering were clearly depicted by Fini *et al.* using rabbit model. In this, silk fibroins were successfully implanted to heal the critical-size cancellous bone defect [72]. In a separate study, *Bombyx mori* silkworm fibers having wire-rope design were used for anterior cruciate ligament repair, which supported the attachment and differentiation of progenitor cells of human bone marrow to ligament lineages [73]. Kim *et al.* state that the bioactivity of silk fibroin could be improved by making it into a composite scaffold. After seeding with human bone marrow stem cells, incorporation of calcium phosphate enhanced the osteoconductivity, apatite/bone morphogenetic protein (BMP2) content, and alkaline phosphatase activity [66]. Silk films can be modified to obtain slow degradability, biocompatibility, and transparency suitable for corneal stromal tissue architecture. In a recent study [74], silk films with 2 mm thickness were fabricated to emulate the collagen lamellae dimensions and surface-patterned to enhance translamellar diffusion of nutrients. Cellular interactions were achieved by creating 0.5-5 mm pores in the scaffold. *In vitro* studies with human and rabbit corneal fibroblasts revealed that silk films are potential candidates for corneal tissue engineering for both two-dimensional and 3D scaffold architectures.

21.2.1.5 Albumin

Albumin is the most abundant protein in human blood plasma. It is a water-soluble protein having the function to maintain the blood pH and carry fatty acid in the bloodstream. Albumin can be degraded by almost all the tissue types present in the body and belongs to a family of globular proteins with low-content tryptophan and methionine. Due to the presence of active functional groups and degradability, albumins can be molded into different forms such as nanofibers, nanospheres, and microparticles [75] and are effective for biomedical applications [76,77]. A well-known, FDA-approved albumin-based adhesive currently on the market is "BioGlue VR" by CryoLife. Apart from cardiovascular [78] and gene delivery [79], albumins are also useful in successful production of coating materials, payload delivery vehicles [80], and sutures [81].

21.2.1.6 Fibrin

Fibrin is a natural polymer similar to collagen with three polypeptide chains involved in the blood-clotting process in the human body. It has a molecular weight of 360 kDa consisting of fibrinopeptides A, B, and D. Enzyme thrombin can cleave fibrinopeptide to linear fibril [82], which in turn undergoes parallel arrangement to form fibers with diameters up to 200 nm. Fibrin has excellent biocompatibility and degradability and has been explored as a potential material for several tissue engineering applications [83]. The most common use of fibrin is fibrin glue as a sealant for hemostasis and surgical procedures. “Evicil VR” is an FDA-approved surgical fibrin sealant available on the market. Fibrin can also be used as a cell carrier [84], drug delivery vehicle [85], and bioactive molecule transporter. The cross-linking ability of fibrin can be used to tailor the material properties suitable for desired applications [86].

The most promising application of fibrin glue is in the healing of severe and chronic wounds [87,88]. A combination of keratinocytes with fibrin produced an efficient chronic wound healing product on the market called “Bioseeds.” The defects found during bone development and traumatic injury could be repaired using fibrin alone or a combination of fibrin with other factors. Yoo *et al.* reported bone healing in the proximal tibial physis of New Zealand white rabbits using autologous fibrin beads [89]. When fibrin combines with ceramic particles to get a porous mineralized morphology, it provides favorable osteogenic properties for bone healing by providing enhanced mechanical and morphological cues. As a result, such scaffolds could be able to provide angiogenesis, cell attachment, and proliferation compared to the ceramics alone [90]. Due to numerous advantages, fibrin is extensively used to take part in 3D scaffold development in tissue engineering for skin, cartilage, bone, liver, cardiac, ocular, ligament, tendons, and nervous system.

21.2.1.7 Keratin

Keratin belongs to the large family of structural proteins that makes the outer layer of human skin, component of hair, and nails. These tough and insoluble unmineralized tissues, arranged in the form of intermediate filament bundles, make the protective tissues of vertebrates. They are found in the cornified layer of epidermis and their properties rely on their supermolecular aggregation. The biomaterial application of keratin was initiated decades ago by using hair and nail as the source of production. Earlier, in 1910, Heinemann introduced the application of keratin in wound healing and drug delivery [91]. Three decades ago, more detailed applications of keratin were reported in cardiovascular applications [92]. Later, keratin-based research in turn moved to wound healing, drug delivery, and other tissue engineering-related fields [93,94]. Keratin-based biomaterials contain the potential cellular recognition sites, which can potentially mimic ECM. A recent study confirms the presence of similar recognition sites on the surface of keratin, making keratin biomaterials excellent for cell adhesion and proliferation [95,96].

21.2.2 Polysaccharides as Biomaterials

Polysaccharides are long-chain carbohydrate molecules composed of monosaccharides held together by ether linkages called glycoside bond. Their molecular formula is $C_x(H_2O)_y$ where x can be between 200 and 2500. The crystallinity and solubility of polysaccharides are distinct based on the nature of their monosaccharide building blocks. They are excellent biological polymers with structure- or storage-related functions in living organisms. Glycogen, a specific form of sugar molecule in the body system possesses an energy storage function in living animals. They are gaining more attention in research due to their unique biological functions ranging from immune recognition to cell signaling. Their tunable biodegradation and the feasibility to combine with synthetic biomaterials make them appropriate for numerous biomedical applications.

21.2.2.1 Chitosan

Chitosan is a derivative of chitin, a polysaccharide that can be found in the hard exoskeletons of shellfish-like shrimp and crab. It is a linear polysaccharide containing β-(1-4)-linked D-glucosamine and *N*-acetyl-D-glucosamine (Figure 21.2). Chitin is one of the most abundant and inexpensive polysaccharides found in nature, from which chitosan is derived by deacetylation [97]. It is a biodegradable cationic amino polysaccharide that can degrade into different products such as CS and glycosylated collagen [98]. Due to the ease of processability, controllable mechanical properties, availability of reactive functional groups, and minimal foreign body rejections, chitosan is desirable for several biomedical applications [99–102]. Major drawbacks of chitosan are its low mechanical strength and insolubility in water, which results in the brittle nature of the material. This could be balanced by blending chitosan with other natural and synthetic polymers with desired properties. One of the major features of chitosan is its structural similarity to glycosaminoglycans, which helps to improve cell adhesion compared to other synthetic polymers. Moreover, the positive charges on chitosan help to bind negatively charged growth factors and cell membranes to a greater extent.

Due to high antibacterial activity and minimal foreign body rejection, chitosan is well known for wound dressing applications [103,104]. It can combine with other natural or synthetic polymers to make bandages with suitable properties [105]. Various chitosan-based bandages like ExcelArrest® XT and HemCon VR are available on the market for wound dressing applications [106]. The ease of processability of chitosan to blend with other degradable polymers resulted in the formation of porous 3D poly(lactide-*co*-glycolide) acid (PLGA) composite scaffolds. These porous scaffolds were found to

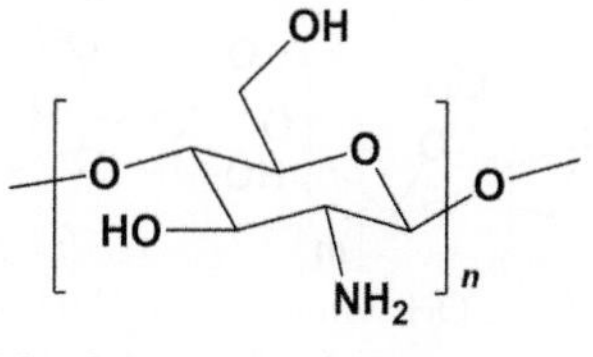

FIGURE 21.2 Molecular structure of chitosan.

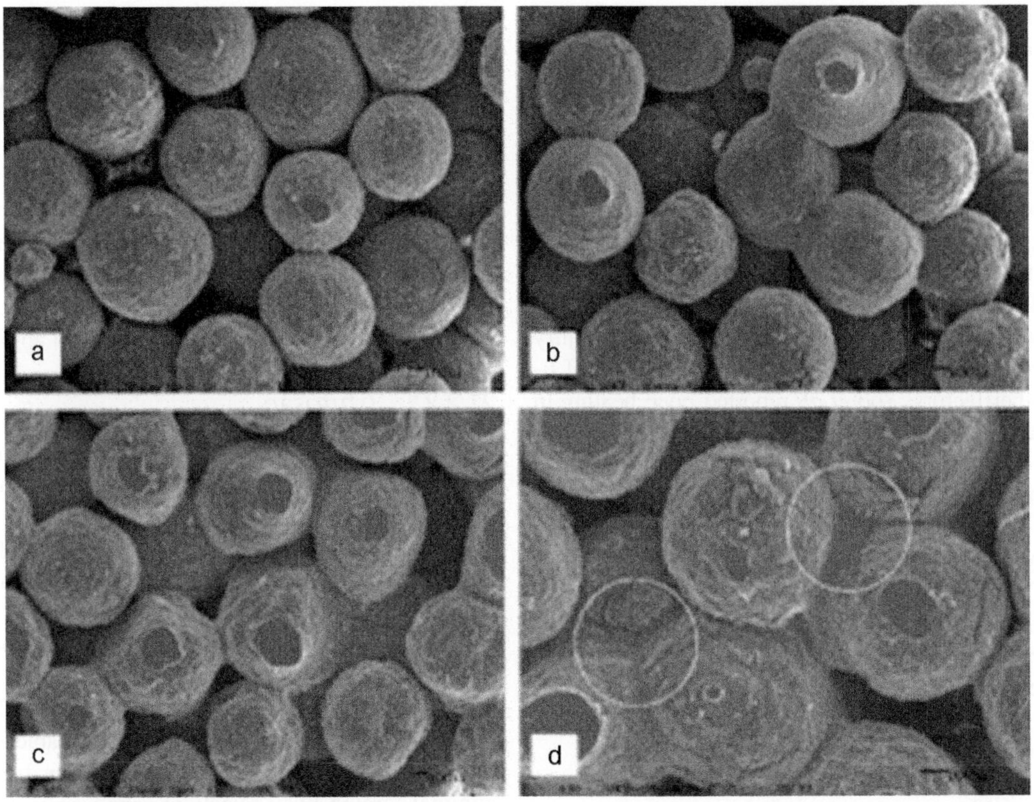

FIGURE 21.3 Scanning electron micrographs showing the morphology of the chitosan/PLAGA (1:4) scaffolds sintered for 4 h at (a) 100 °C, (b) 110 °C, (c) 115 °C, and (d) 115 °C. Greater fusion between adjacent microspheres was obtained by increasing sintering temperatures (a-c). Circles in (d) demonstrating pore closure among microspheres at the sintering temperature of 115 °C. Reprinted from Jiang *et al.* [107] with permission from Elsevier.

support the proliferation of osteoblast-like cells. Compared with PLGA scaffolds, chitosan/PLGA scaffolds show enhanced differentiation and maturation of osteoblast-like cells *in vitro* (Figure 21.3) [107]. By various processing methodologies, chitosan has been used in numerous regenerative medicine applications in the form of sponges [108], hydrogels [109], fibers [110], and membranes [111].

21.2.2.2 Alginic Acid

Alginate, a linear polysaccharide composed of b-D-mannuronic acid and a-L-guluronic acid, is derived from marine brown seaweed or algae (Figure 21.4). The structure and patterning of the polymer strongly depend on the source. Alginate has very good processability and moldability due to its inherent hydrophilicity and water solubility. As a relatively larger polymer with molecular weight up to 500 kDa and presence of carboxylate side groups, alginate is highly reactive to divalent cations to form gels. Its biocompatibility, degradability, and low immunogenicity enable its wide range of biomedical applications. As a biodegradable polymer, its structural similarity to glycosaminoglycans offers active sites for being connected to other molecules and also for increased cellular attachment [112].

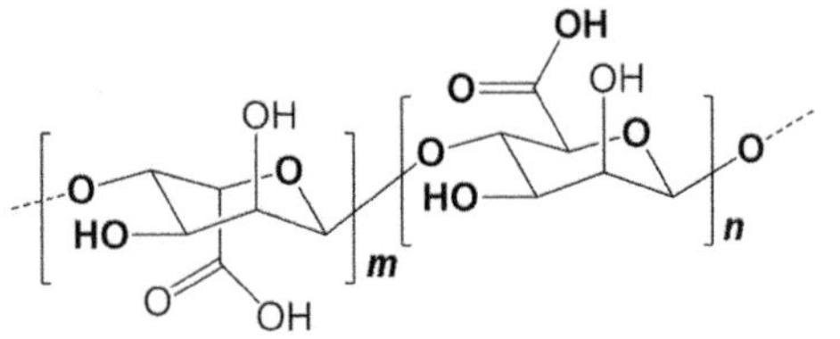

FIGURE 21.4 Molecular structure of alginic acid.

Alginates are capable of making hydrogels when combined with other divalent cationic compounds. During gelation, alginate can encapsulate cells and molecules with minimal adverse effect [113] to form hydrogels, which mimic natural ECM of tissues and enable widespread applications in tissue engineering. Alginates were thoroughly studied as drug delivery vehicles by chemical immobilization of drugs with reactive carboxylic groups in the polymer backbone. Such immobilizations could help to incorporate growth factors and protein to increase the efficiency of drug delivery [114]. In recent research, more attention has been given for the combined effect of alginate with other natural [115] and synthetic polymers [116–118] to be useful for various applications like regeneration of bone [119], for adipose-derived stem cell treatment [120], and cartilage regeneration [121].

Cross-linking reactions can develop between carboxylate groups in alginate and acetaldehydes to make pH-sensitive contracting gels [122], which can resolve some current limitations existing with alginate gels. Even though alginates are used for various biomedical applications, the major drawback with alginate is its incapability to degrade in enzymatic environments and poor cell adhesion. Studies are underway to eliminate the current limitations existing with alginates to popularize them as biomaterials.

21.2.2.3 Hyaluronic Acid

Hyaluronan, or hyaluronic acid, is another naturally occurring polysaccharide of alternating *N*-acetyl-D-glucosamine and D-glucuronic acid (Figure 21.5). It was first isolated in 1934 by Meyer *et al.* It is an anionic polymer with nonsulfated glycosaminoglycan linkages, which are widely distributed throughout neural and connective tissues. It is one of the major contents of ECM and helps for cell proliferation and migration. Hyaluronan is the core molecule for attachment of CS during aggrecan formation, one of the cartilage-specific core proteoglycans. This water-soluble polymer needs additional components for cross-linking when binding to a scaffold. Hyaluronan is well known for cartilage regeneration due to the ability to differentiate mesenchymal stem cells into chondrocytes. It is found in synovial fluid at a concentration of 0.1-1% (v/v). The presence of functional binding groups makes it easily cross-linkable with other amino acid polymers.

Hyaluronic acids are good free radical scavengers. Scaffold preparation using hyaluronic acid homopolymer is difficult due to its viscous nature. To overcome this, hyaluronan was cross-linked with ethyl esters to form hydrogels, commercially known by the name "HYAFF." HYAFF can be processed to different types of devices like tubes, membranes, gauzes, sponges, and nonwoven fabrics. HA has been used in knee joint repair as injectable material to treat osteoarthritis [123]. An early study with HYAFF11 showed excellent results in the regeneration of cartilage [124]. They are also used as postoperative care scaffolds to heal tissues after cataract surgery [125]. Using the fibronectin attachment ability of hyaluronan, suitable wound healing patches with provision for wound site cell migration can be achieved [126]. Blending of hyaluronan with monomers like methacrylic anhydride can produce injectable scaffolds, which can reduce pain caused from osteoarthritis. Being biocompatible and degradable, hyaluronan contributes a significant role in regenerative medicine to provide suitable scaffolds for neural [10], bone [127], cartilage [128], and corneal [129] tissue engineering applications.

21.2.2.4 Cellulose

Cellulose is the most abundant polysaccharide on Earth, with several thousand D-glucose units in a polymer chain (Figure 21.6). It is the main structural component of the cell wall of green plants. Cotton contains almost 90% of cellulose, whereas wood and dried hemp contain 50% and 45%, respectively. The first cellulose-based thermoplastic polymer was manufactured in 1870 and the first chemical synthesis was done by Kobayashi and Shoda in 1992 [130]. Solubility of cellulose in water depends on its chain length and it is degradable by enzymatic reaction [131]. Cellulose is easy to machine to form various shapes such as textiles, microsphere, sponges, and membranes.

Biocompatibility studies of different cellulose derivatives were well demonstrated by various research groups [132–135]. The mechanical stability of cellulose can be varied by combining with various polymers suitable for a range of different tissue types [136]. Another important area for cellulose in tissue engineering is the regeneration of cartilage for osteochondral defects by integration of cellulose by nearby tissues [137]. Mostly, cellulose acts as a weakly bioactive material for hard tissue regeneration like bone. In such cases, incorporation of calcium phosphate into the polymer matrix was found to be much more effective to enhance the osteoconductivity and tissue compatibility [138,139]. Cellulose fabricated in different forms can be used for various applications like development of surgical tools [140], dialysis membranes [141], and biosensors [142]. A combination of oxidized regenerated cellulose with isobutyl 2-cyanoacrylate and fibrin glue resulted in the development of excellent hemostatic agent for bleeding during vascular prosthesis. Another cellulose derivative called methyl cellulose is an excellent biocompatible material for treating traumatic brain injury. Mathew *et al.* reported that a viscous solution of methyl cellulose with a concentration up to 8% is compatible with rat astrocytes.

21.2.2.5 Chondroitin Sulfate

CSs are sulfated glycosaminoglycans with alternating chain of *N*-acetylgalactosamine and glucuronic acid. CS has a

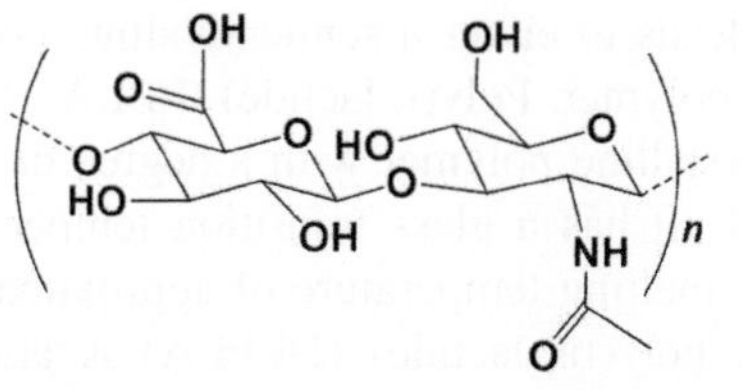

FIGURE 21.5 Molecular structure of hyaluronic acid.

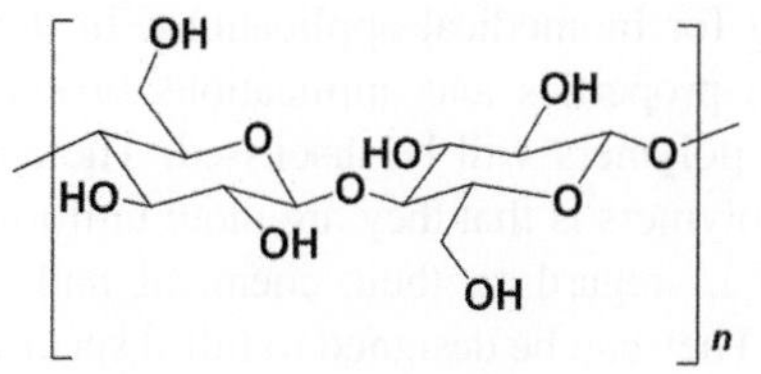

FIGURE 21.6 Molecular structure of cellulose.

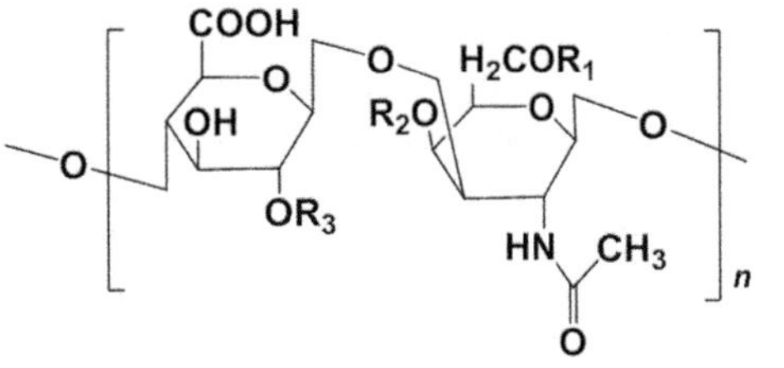

FIGURE 21.7 Molecular structure of chondroitin sulfate.

similar structure to hyaluronic acid (Figure 21.7). Based on the site of sulfation, CS is classified into four components as chondroitin-4-sulfate, chondroitin-6-sulfate, chondroitin-2,6-sulfate, and chondroitin-4,6-sulfate. CS has both structure and regulatory functions, which depend on the properties of proteoglycans of which it is a part. It is a main component of native ECM and plays a key role in maintaining the structural integrity of tissues. Being a major part of aggrecan, CS is very much important in the functioning of cartilage. Due to the electrostatic repulsive forces generating from sulfate groups in CS, cartilage has high compression resistance compared to other tissue types. Osteoarthritis is caused by the gradual loss of CS from the cartilage. CS is involved in cell signaling and recognition of ECM to cell surfaces [143] and also has anti-inflammatory properties to give more stimulation for chondrogenesis [144]. The importance of CS in tissue engineering was clearly depicted by Kosir *et al.* in 2009 [145] by showing the presence of cell-synthesized glycosaminoglycans in ECM. Due to the biocompatibility and degradability of CS scaffolds, a large number of studies were conducted to evaluate the wound healing capability of CS scaffolds [146–148]. Composite scaffolds of CS with a wide range of polymers were developed to tailor the application to different extents. When CS was combined with biodegradable polymers like chitosan [149], PLGA [150], and PCL [151], scaffolds with different morphologies and properties were obtained. In a different study, a tri-copolymer system, CS-HA-gelatin, was used to reproduce ECM for the cartilage. When seeded with chondrocytes for 5 weeks, they maintained their phenotypes and produced type II collagen [152].

21.3 SYNTHETIC POLYMERS IN TISSUE ENGINEERING AND REGENERATIVE MEDICINE

There is an increasing use of synthetic biomaterials in tissue engineering for biomedical applications. In the following section, the properties and applications of synthetic biodegradable polymers will be discussed. The advantage of synthetic polymers is that they are more uniform and more predictable in regard to their chemical and mechanical properties. They can be designed to fulfill specific purposes and are free from immunogenicity [153]. The behaviors of synthetic polymers are very much affected by their properties. The selection of different monomers, initiators, and reaction conditions, as well as the presence of additives, plays a role in determining polymer properties such as crystallinity, melting and glass transition temperatures, molecular weight, and side groups. These properties, in return, will affect polymer behaviors [154].

21.3.1 Poly(α-Esters)

21.3.1.1 Poly(Glycolic Acid)

PGA is a linear polymer of glycolic acid. Glycolic acid is produced during normal body metabolism and is known as hydroxyacetic acid [155]. PGA is a bulk degrading polymer with low solubility in water. The *in vivo* degradation of PGA is as follows: degradation and loss of material strength occurs around 1-2 months, with complete degradation of the total mass by 6 months [156]. PGA has a high crystallinity (45-55%), with a melting point of 220-225 °C and a glass transition temperature of 35-40 °C. Its high crystallinity is the main factor leading to its low solubility in organic solvents [153,157]. However, the polymer has relatively high strength due to its high crystallinity and orientation. It has a modulus around 12.8 GPa [156].

PGA has been fabricated into a variety of forms for sutures. The first commercially available biodegradable synthetic suture approved by the FDA in the United States was DEXON. Implants can also be made from PGA using a self-reinforcing technique and is used in the treatment of fractures and osteotomies [155]. PGA, PLA, and PDS are considered the three major polymers for bioabsorbable implants. The commercial product Bioflex® has been evaluated for use as a bone internal fixation device [156,153]. Other studies have explored the use of PGA combined with other polymers to improve overall material strength. Various methods, such as extrusion, injection, compression molding, particulate leaching, and solvent casting, are some of the techniques used to develop polyglycolide-based structures for biomedical applications [158].

21.3.1.2 Polylactic Acid

Lactide is the cyclic dimer of lactic acid, which exists as two optical active isomers: D and L. L-Lactide is the naturally occurring isomer, while DL-lactide is the synthetic blend of D-lactide and L-lactide. The polymerization of these monomers leads to either a semicrystalline polymer or an amorphous polymer. Poly(L-lactide) (PLLA), for example, is a semicrystalline polymer with a degree of crystallinity around 37%. It has a glass transition temperature of 60-65 °C and a melting temperature of approximately 175 °C. Conversely, poly(DL-lactide) (DLPLA) is an amorphous polymer with random distribution of both isomeric forms of lactic acid. Accordingly, DLPLA has lower tensile strength and faster degradation time and thus is used more in drug delivery systems [153,159–162].

PLLA is a relatively slow-degrading polymer compared to DLPLA with high tensile strength, low extension, and a high modulus (~4.8 GPa) and is thus considered an ideal biomaterial for load-bearing applications. It can be used as a high-strength fiber and is FDA-approved, making it a stronger alternative to DEXON® [153]. Additionally, PLLA has been considered as a scaffold material for developing ligament replacements such as Dacron [163,164]. Also commercially available PLLA-based products are orthopedic fixation devices including Phantom Soft Thread Soft Tissue Fixation Screws®, Phantom Suture Anchors® (DePuy), Full Thread Bio Interference Screws® (Arthrex), BioScrews®, Orthopaedics Phusiline® and Sysorb interference screws, and BIOFIX® and PL-FIX® pins. Other applications include scaffolds, biocomposite materials, and prostheses [159,160,162,165].

DLPLA has a lower strength (around 1.9 GPa); thus, it can be used for drug release vehicles or low-load-bearing tissue engineering scaffolds due to its amorphous properties and faster degradation rate [153]. Both PLLA and DLPLA nanofibers can be used for nerve grafts as they supported the neurite outgrowth and NSCs differentiated in the direction of fiber alignment [166,167].

21.3.1.3 Poly(Lactide-co-Glycolide)

It is always possible to copolymerize two monomers in order to get a new polymer composition with desired properties. For example, PLGA is copolymerized from polyglycolide with either L-lactide or D-lactide (Figure 21.8). The copolymer has been used in medical devices and drug applications.

The newest suture application available on the market is Vicryl Rapide®, a modified version of the multifilament suture Vicryls®, which is a copolymer containing 90% glycolic acid (GA) and 10% L-lactic acid (LA). The modification in Vicryl Rapide® allows for an increase in the rate of degradation of the suture. Another suture called Panacryl® used a higher LA/GA ratio for a decrease in the rate of degradation. In addition, VicrylMesh® mesh based on the copolymer was also applied in tissue-engineered skin graft named Dermagraft® [153]. As mentioned earlier, other applications for polymers include tissue engineering. Various 3D scaffolds were made with micro- and nanostructures based on PLGA as PLGA is a promising material for cell adhesion and proliferation. Kumbar *et al.* studied PLGA fiber matrices composed of scaffolds of varying fiber diameters for skin regeneration and discovered that human skin fibroblasts acquired a well-spread morphology and showed significant progressive growth on fiber matrices in the 350-1100 nm diameter range. Figure 21.9 is the SEM image of electrospun nonwoven PLAGA fiber matrices [166,167].

Other studies looked into the use of composite microspheres of PLGA/HA in bone regeneration; study results showed that the controlled release system of these microspheres functioned to improve osteoblast proliferation and also enabled upregulation of the key osteogenic enzyme alkaline phosphatase (ALP) [168].

21.3.1.4 Bacterial Polyesters

21.3.1.4.1 Polyhydroxybutyrate

PHB is a poly(hydroxyalkanoate) (PHA), a biodegradable plastic produced by microorganisms first discovered by Lemoigne in 1925. Later, it was found that several other bacterial strains could also produce PHB. Chemical synthesis has also been found by Shelton *et al.* [153,169]. Among all, polyhydroxybutyrate (PHB) is the most common polymer of OHB, a linear polyester of D(−)-3-hydroxybutyric acid [170]. In general, PHB is a semicrystalline surface-eroding polymer which undergoes hydrolytic cleavage of the ester bonds (Figure 21.10). It has a melting temperature in the range of 160-180 °C [169].

The homopolymer PHB is tough and brittle, has a relatively high melting point, and crystallizes rapidly. It makes entrapment of drugs technically difficult. The related copolymers of PHB and 3-hydroxyvalerate (3-HV) form PHBHV and are less tough or brittle and have similar semicrystalline properties as PHB but have lower melting temperatures, depending on HV content. PHB and PHBHV are both soluble in a wide range of solvents, making them processable in different shapes, spheres, and fibers. They are all potential candidates for controlled drug release vehicles. Piezoelectricity of PHBHV is very attractive for orthopedic applications since electrical stimulation is known to promote bone healing. It has also been investigated as a material for developing bone pins and plates [171]. For suture applications, FDA has

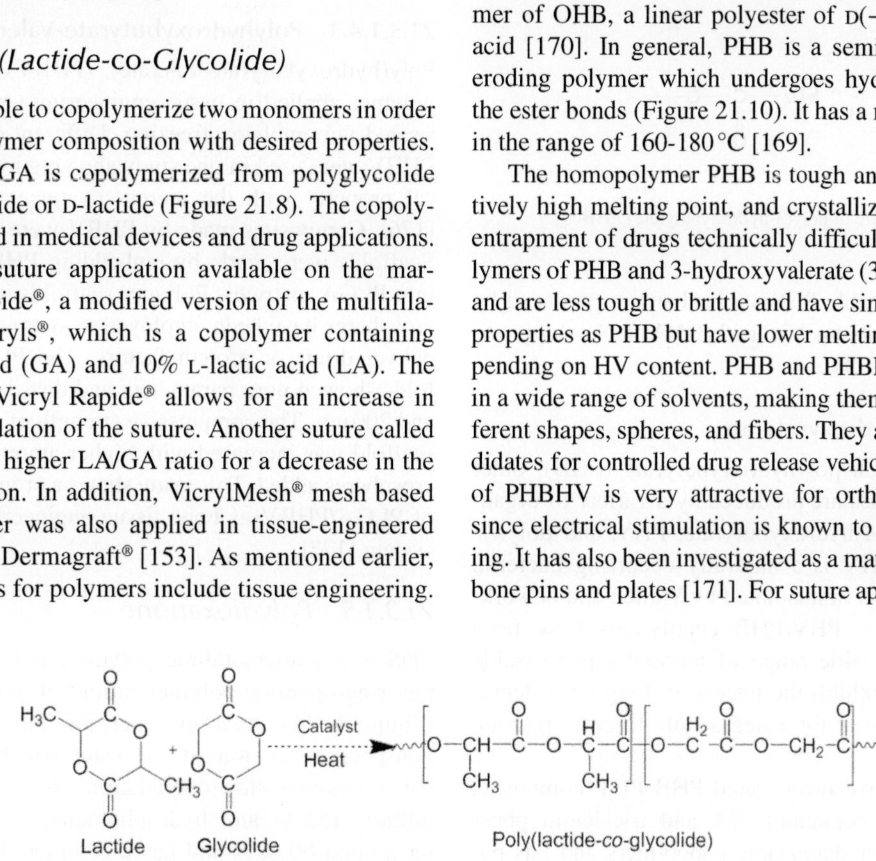

FIGURE 21.8 Synthesis of poly(lactide-*co*-glycolide) (PLG).

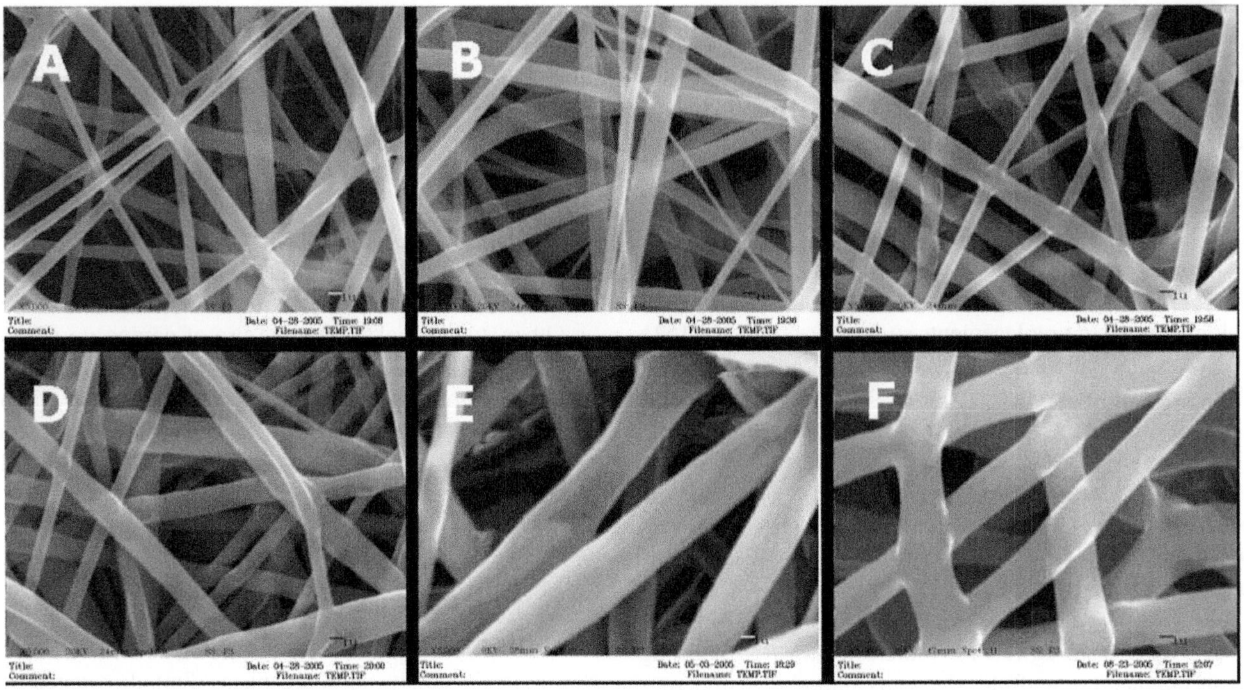

FIGURE 21.9 SEM of electrospun nonwoven PLAGA fiber matrices at a magnification of 5000× using constant electrospinning parameters of 20G needle, voltage gradient 1 kV/cm, and flow rate of 2 mL/h. The concentration and average fiber diameters of different samples are (a) Matrix-2: 0.225 g/mL, 200-300 nm; (b) Matrix-3: 0.24 g/mL, 250-467 nm; (c) Matrix-4: 0.27 g/mL, 500-900 nm; (d) Matrix-5: 0.3 g/mL, 600-1200 nm; (e) Matrix-6: 0.35 g/mL, 2500-3000 nm; and (f) Matrix-7: 0.42 g/mL, 3250-6000 nm. Kumbar *et al.* [166,167] with permission from Wiley Interscience.

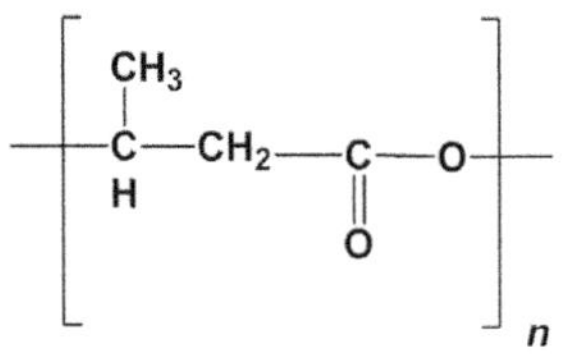

FIGURE 21.10 Structure of poly(3-hydroxybutyrate) PHB.

classified the absorbable PHB surgical suture produced by recombinant deoxyribonucleic acid (DNA) technology into class II (special controls).

21.3.1.4.2 Polyhydroxyvalerate

Besides the common polyhydroxybutyrate (PHB), other polymers of this class are produced by a variety of organisms such as poly-4-hydroxybutyrate, PHV, and polyhydroxyhexanoate. PHV is a naturally occurring bacterial polyester, which was first isolated by Wallen and coworkers [172,173]. Also, PHV/PHB copolymers have been studied to make a wide range of thermally processable polyesters, which exhibit the necessary long-term degradation profile required for a degradable fracture fixation device [174].

Other studies have investigated PHB-PHV composite, bioactive ceramics containing HA and tricalcium phosphate. This yields biodegradable copolymers and has potential for medical applications [175].

21.3.1.4.3 Polyhydroxybutyrate-Valerate

Poly(hydroxybutyrate-valerate) (PHBV) is a PHA-type polymer applied in tissue engineering scaffolds. It is fabricated via emulsion freezing. Different concentrations of PHBV were used in one study showing increased mechanical property with the increased concentration of PHBV [176]. Composites made by PHBV were also investigated. Scaffolds were made by embedding PHBV microspheres into PLGA matrices. Polymer scaffolds in the form of microspheres have been employed to support cell growth and deliver drugs or growth factors. The PLGA/PHBV scaffolds showed pore parameters and distribution between 50 and 200 μm. The compressive strength of the PLGA/PHBV scaffold was increased with higher amounts of PHBV microspheres added. This study shows a promising application of PLGA/PHBV as bone tissue engineering drug delivery system [177].

21.3.1.5 Polydioxanone

PDS is a semicrystalline polyester polymer prepared by the ring-opening polymerization of *p*-dioxanone [161] (Figure 21.11). As a polyester, the polymer goes through nonspecific scission of the ester backbone [153]. PDS has a relatively slow degradation rate due to its high crystallinity (55%) and hydrophobicity. It will degrade after around 60 days and has a complete loss of mass over 9-12 months. Polydioxanone inside the body will break

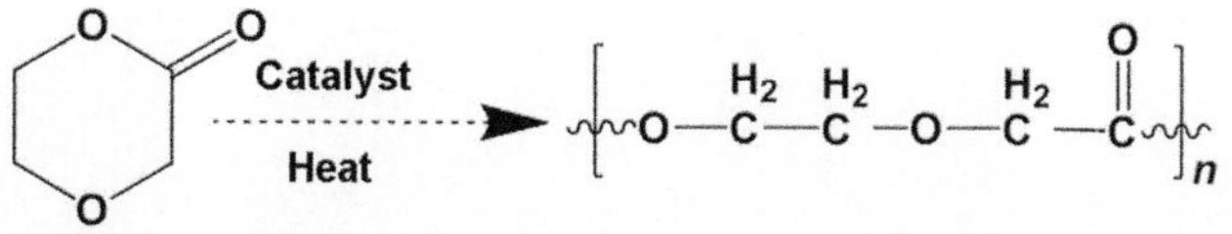

FIGURE 21.11 Synthesis of poly(dioxanone).

down into glyoxylate, be excreted in the urine or converted into glycine. It has no known acute or toxic effects upon implantation [156,157].

Polydioxanone was the first polymer used commercially as monofilament for biomedical applications under the trade name PDS®. As a monofilament, it has a smaller risk of infection with use and causes less friction when penetrating tissues [153]. Also, the material has been used by companies like Johnson & Johnson Orthopedics as an absorbable pin for fracture fixation composed of poly(dioxanone) [178].

21.3.1.6 Polycaprolactone

PCL is a semicrystalline polyester. Similar to poly(dioxanone), it can be made from ring-opening polymerization of "e-caprolactone." The homopolymer is easy to process. It has a melting point of 59-64 °C and a glass-transition temperature of −60 °C and can be blended with a wide range of polymers to form miscible solutions. The homopolymer is increasingly used in research labs for various applications due to its slow degradation rate, good permeability, and nontoxicity. Compared to other polymers, it has a relatively long degradation time of up to 2 or 3 years [153,161]. As for its copolymers, e-caprolactone has been used with glycolide to form a monofilament suture (MONOCRYLs) for commercial use. Further applications such as a drug/vaccine carrier and a long-term contraceptive with zero-order drug release are discussed in various papers [179].

Also, in recent years, PCL has been widely studied as materials for tissue engineering applications. Yoshimoto prepared electrospun PCL as a potential candidate scaffold for bone tissue engineering [180]. PCL blended with PLLA, PLGA/HA, and gelatin electrospun sheets was studied and characterized [36,181,182]. Advanced 3D prototyped blend composites made for hard-tissue engineering were investigated based on PCL/organic-inorganic hybrid fillers [183].

21.3.2 Polyurethanes

PU is a polymer composed of a chain of organic units joined by carbamate (urethane) links. It is generally a class of polymers derived from condensation of polyisocyanates and polyalcohols [184]. They are categorized as either thermoplastics or thermosets. PUs were first produced and investigated by Dr. Otto Bayer in 1937 [185]. It is a polymer in which the repeating unit contains a urethane

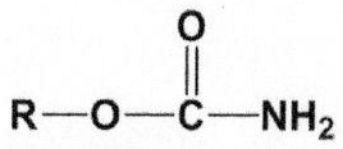

FIGURE 21.12 Urethane linkage.

moiety. Urethanes are derivatives of carbamic acids, which exist only in the form of their esters. The basic unit of urethane linkage is shown in Figure 21.12. Variations in the R group and substitutions of the amide hydrogen produce multiple urethanes. Although all PUs contain repeating urethane groups, other moieties such as urea, ester, ether, and aromatic may be included. The addition of these functional groups may result in fewer urethane moieties in the polymer than functional groups.

PUs represent a large group of polymers in the medical field due to their diversity in structure and properties [185]. They are investigated for use in long-term implants such as cardiac pacemakers, vascular grafts, and tissue replacements due to their excellent biocompatibility, high resistance to macromolecular oxidation [184], hydrolysis, and calcification properties. PUs also have superior mechanical properties such as increased tensile strength [153]. For tissue engineering applications, 3D scaffolds of superimposed square-meshed PU grids were prepared by using a rapid prototyping technique and tested *in vivo*. The scaffolds were chemically modified in order to enhance specific binding interactions between materials and biological environments. The cellular analysis showed that the PU scaffolds were not cytotoxic and significant alterations in their morphology and apoptotic process were not observed. Seeding fibroblasts on the scaffolds showed good adhesion and no change in cell morphology, as shown on Figure 21.13 [186].

21.3.3 Polyphosphazenes

Polyphosphazenes are inorganic-organic hybrid polymers with a backbone of alternating phosphorus and nitrogen atoms containing two organic side groups attached to each phosphorus atom. They have the advantage of synthetic flexibility and controllable mechanical properties [187]. Figure 21.14 shows the general structure of polyphosphazenes where R represents a variety of organic or organometallic side groups.

The polymer was first developed during the late 1960s by Allcock *et al.* and there are around 500 different types of polyphosphazenes synthesized so far [188]. The same group also developed biodegradable polyphosphazene

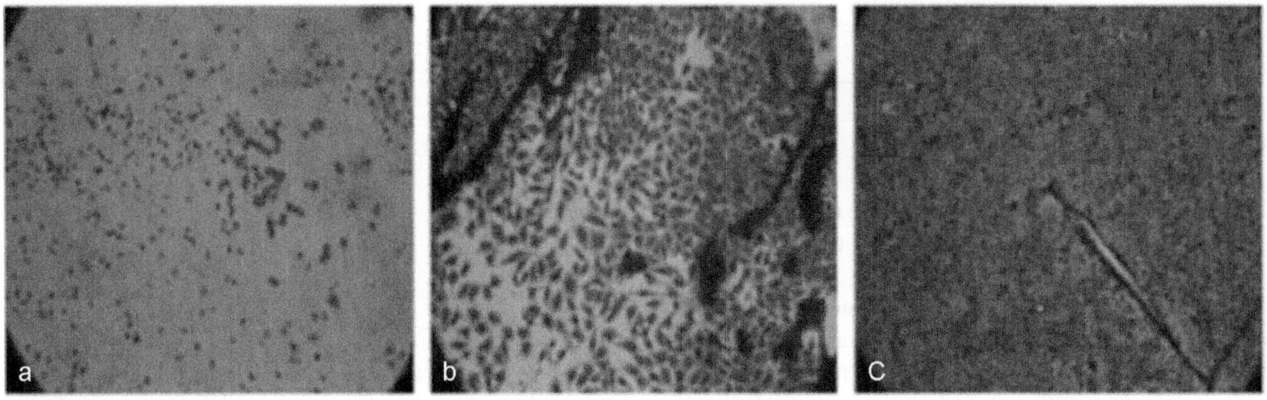

FIGURE 21.13 NIH-3T3 mouse fibroblasts adhered on PU films at 4, 24, and 48 h after cell seeding. With permission from Elsevier [186].

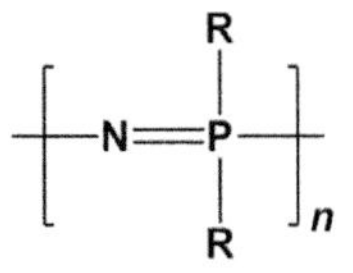

FIGURE 21.14 General structure of polyphosphazene.

materials. Their structural diversity and unique properties made them ideal materials for various applications such as scaffolds for tissue engineering, microencapsulating agents, biodegradable materials, biocompatible coatings, and carriers for gene delivery [189–191]. Dr. Laurencin's group has extensively studied the biocompatibility of polyphosphazene both *in vivo* and *in vitro* for various tissue engineering applications [187]. These included self-setting bone cements in which composites were made via an interesting reaction between polyphosphazene side groups and calcium phosphate ceramics [192]. Polyphosphazene/nano-HA composite microsphere scaffolds showed good osteoblast cell adhesion, proliferation, and ALP expression and are potential suitors for bone tissue engineering applications [193] (Figure 21.15).

Polyphosphazenes can also be blended with other polymers such as polyphosphazene-PLAGA; the functional group can regulate cell behavior and tissue regeneration. Additionally, the fact that polyphosphazene can be dissolved in a wide range of solvents and it has tunable degradation kinetics with nontoxic degradation products has made it suitable for drug delivery applications [192].

21.3.4 Polyanhydrides

Polyanhydrides are one of the most considered and investigated biodegradable surface-eroding polymers. It is synthesized by the dehydration of diacid molecules by melt polycondensation. In structure, the polymer has anhydride bonds that connect repeat units of the polymer backbone chain. *In vivo*, polyanhydrides degrade into nontoxic biocompatible diacid monomers that can be metabolized and eliminated from the body. Degradation times can be altered from days to years depending on the choices of hydrophobicity of monomers. Kumar *et al.* list the advantages of polyanhydrides as being relatively cheap with a simple, one-step synthesis procedure without further purification.

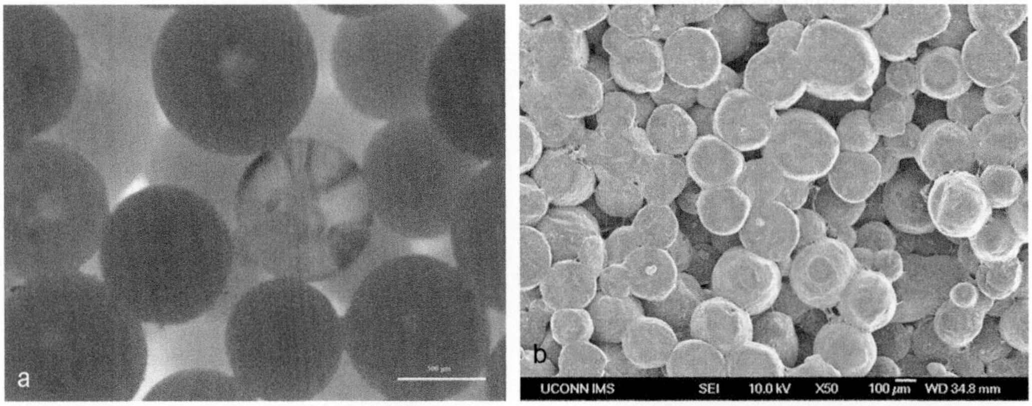

FIGURE 21.15 Polyphosphazene-based 3D matrices for tissue engineering. (a) Sintered polyphosphazene scaffolds. (b) SEM image of polyphosphazene microsphere scaffolds. (For color version of this figure, the reader is referred to the online version of this chapter.)

In addition, polyanhydrides have a predicable release rate as they are surface-eroding polymers [194].

So far, polyanhydrides have only been approved by the FDA as a drug delivery system commercially produced by Guilford. Chemotherapeutic agent BCNU (carmustine) loaded in poly(CPP-SA) 20:80 wafer (Gliadel®) was used for treating brain tumors [161,194,195]. More applications in controlled release include eye disorder, local anesthetics, and anticoagulants [194]. Further biocompatibility analyses need to be conducted for the use of polyanhydrides in other tissue engineering applications.

21.3.5 Poly(propylene Fumarate)

Resection of primary and metastatic tumors, bone loss after skeletal trauma, total joint arthroplasty with bone deficiency, spinal arthrodesis, and trabecular voids may all demand regeneration of bone tissue [196]. The ideal polymer material properties for bone tissue include biocompatibility, mechanical properties, osteoinductivity, sterilizability, and handling characteristics. PPF belongs to polyester branch and can be cross-linked through all the unsaturated bonds in the backbone as shown in Figure 21.16. It is an injectable liquid and will become solid during cross-linking. It can serve as an ideal filler for bone defects, making it an ideal candidate for various orthopedic applications [182,196,197].

Apart from bone fillers, other applications include ocular drugs and estrogenic tissue engineering. In the latter case, PPF is usually mixed with ceramics such as hydroxyapatite or alumoxane to create mechanically stronger, more bioactive scaffolds [197–201]. Recent applications used PPF to fill irregular-shaped bone defects such as ear ossicles or mandibular defects with different degrees of cross-linking depending on the required mechanical strength [197,202,203].

21.3.6 Poly(ethylene Glycol)

Hydrogels are a new, fast developing group of polymers, having wide applications in many fields such as pharmacy, medicine, and agriculture. The release systems are usually considered "smart" as they can respond to the stimulation of physical or chemical/biological reactions. Among them, PEG is one of the most widely used hydrogels in the biomedical field as it can provide a highly swollen 3D environment that mimics soft tissues and allows transport of nutrients or cellular waste in and out of the elastic networks [204,205].

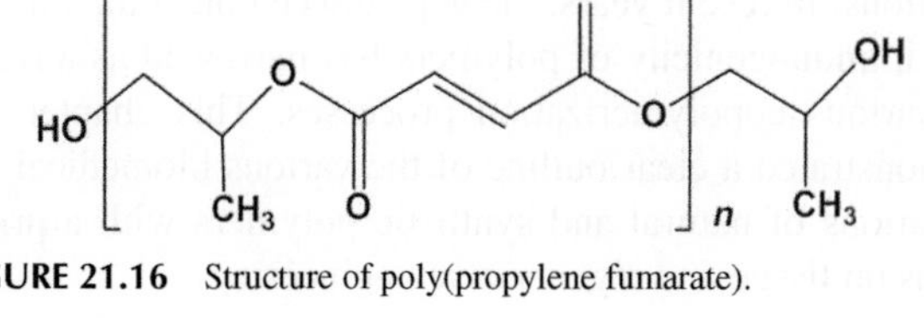

FIGURE 21.16 Structure of poly(propylene fumarate).

PEG is a water-soluble polymer composed of repeating ethylene oxide units flanked by alcohols. It is a hydrophilic polymer due to its polar and uncharged character. In addition, as the surface is flexible and neutrally charged, it prevents protein binding, which gives PEG excellent properties such as low protein adsorption, low toxicity, and nonimmunogenicity. Modifications with molecular weight, initiator, and geometries to change the polymer's physical and chemical properties have lead to a wide variety of biomaterial applications [206]. For drug delivery applications, hydrogels like PEG and PEG-based derivatives such as polyethylene glycol methacrylate and polyethylene glycol diacrylate proved to be ideal as they showed high biocompatibility, lack of toxic influence on surrounding tissue, and solubility in water [207].

As for tissue engineering applications, PEG hydrogels are usually modified or copolymerized to make them suitable for biomedical purposes. In one study, adhesive Arg-Gly-Asp (RGD) peptide sequences were introduced to facilitate the adhesion, spreading, and consequently the cytoskeletal organization of rat calvarial osteoblasts. The hydrogels served as encapsulation matrices for osteoblasts to assess their applicability in promoting bone tissue engineering [208]. Tessmar *et al.* also reported several PEG-derived hydrogel copolymers like photo-cross-linkable PEG acrylate hydrogels, PEG/PLA with acrylic end groups, and artificial ECM networks. For artificial ECM networks, further improvements of hydrophilic PEG-based hydrogels can be achieved by incorporation of more sophisticated cross-links that mimic the natural components of the tissue's ECM, like specific peptide sequences cleaved by proteinases. These hydrogels can be additionally equipped with peptide sequences containing integrin receptor substrates, which support the cellular adhesion onto these hydrophilic materials. The addition of modified growth factors or tissue morphogenes further allows the controlled release of these factors regulated by enzymatically catalyzed cleavage from the polymer backbone [209,210].

21.3.7 Poly(Ortho Esters)

Poly(ortho esters) were first developed by the ALZA corporation (Alzamer) in 1970 in order to seek new synthetic polymer for drug delivery applications. These polymers degrade by surface erosion and degradation rates may be controlled by incorporation of acidic or basic excipients. The polymer is hydrophobic enough such that its erosion in aqueous environments is very slow. The unique features of poly(ortho esters), in addition to their surface erosion mechanism, is the rate of degradation for these polymers, pH sensitivity, and glass transition temperatures, which can be controlled by using diols with varying levels of chain flexibility [153,161]. There are several generations of POE developed. Among the four generations of POE, the third (POE III) and fourth (POE IV) are promising viscous and injectable materials. Fourth-generation POE is distinguishable by its highly reproducible and controlled synthesis, a higher hydrophobicity, and excellent biocompatibility. Poly(ortho esters) are currently under development for a variety of applications, such as ocular delivery, periodontal disease treatment, and applications in veterinary medicine [211].

21.3.8 Polyphosphoesters

Polyphosphoesters (PPEs) are a branch of polymers with repeating phosphoester bonds in the backbone. The polymers are amorphous to semicrystalline and are structurally versatile, biocompatible, and biodegradable through hydrolysis and enzymatic digestion under physiological conditions. PPEs are appealing for biomedical applications because of their potential biocompatibility due to their similarity to nucleic acids, which can be further degraded inside the human body. Several synthesis methods of PPEs have been discussed. Zhao *et al.* mentioned structure synthesized by extending oligomeric lactide prepolymers with ethylphosphate groups. Microspheres were made for delivering anticancer therapeutics. Zhao *et al.* performed conjugation of charged groups to the side chain of the phosphate, which is one of the few biodegradable cationic polymers in the field for nonviral gene delivery. Capable of delivering exogenous genes to a cell nucleus or providing an extracellular sustained release of DNA, these cationic polyphosphoesters also serve as a valuable model to understand the important characteristics that render a polymer an effective gene carrier [212].

The potential applications of PPE are expanding with respect to the accessibility of different macromolecular structures, the capability of self-assembly to form nanostructures, the variety of functionality, and the corresponding diversity of properties on controlled synthetic methods of PPEs. However, further *in vivo* studies are required, where osteoconductivity, for example, can be determined [213].

21.3.9 Poly(ester Amide)s

Poly(ester amide)s (PEAs) are biodegradable polymers that have both ester and amide groups on their chemical structure. Continuous efforts have been made in the development of PEAs since the preparation of the first derivatives in the 1970s. PEAs can be prepared from different monomers following different synthetic methodologies such as ring-opening polymerization and condensation polymerization, which lead to polymers with random, blocky, or ordered microstructures. Hydrophilic/hydrophobic ratio and biodegradability of the polymer can be easily modified. There are numerous studies in the past decade focused on the functionalization of PEAs through incorporation of α-amino acids with hydroxyl, carboxyl, and amine pendant groups and also by incorporation of carbon-carbon double bonds in both the polymer main chain and the side groups [214].

PEAs are a branch of relatively new polymers. Research in the biomedical field is just starting to investigate possibilities in areas such as controlled drug release systems [215], hydrogels [216], tissue engineering [217,218], and other uses like adhesives and smart materials, together with the main families of functionalized PEAs that have been developed to date [214].

For tissue engineering applications, amino-functionalized PEA best supports endothelial cell adhesion, growth, and monolayer formation. Mechanical testing showed linearly elastic behavior at small strains (<10%), while on larger strains, the elastic moduli of these materials are strongly dependent on the charge formulation. PEA is shown as a promising candidate for engineering applications, especially for the use of vascular graft [219].

New biodegradable elastomeric PEAs have also been synthesized for tissue engineering applications. The conventional cross-linked aliphatic polyesters have disadvantages such as high cross-link densities resulting in stiffness, limited chemical moieties for chemical modification, and fast degradation. The new PEA elastomers are able to overcome these issues. For example, poly(1,3-diamino-2-hydroxypropane-*co*-polyol sebacate)s formed cross-linked networks through the hydroxyl group and polyol (glycerol and/or D,L-threitol) and displayed tensile Young's modulus on the order of 1 MPa and reversible elongations up to 92%. These polymers exhibited *in vitro* and *in vivo* biocompatibility and relatively long-projected degradation half-lives of up to 20 months *in vivo* [217].

21.4 CONCLUSIONS

A wide variety of natural and synthetic biodegradable polymers are currently available on the market for regenerative/tissue engineering applications. Most of them are either collagen-based or polyester-based materials. Use of advanced processing techniques and numerous synthetic organic routes still allows the modification of existing polymers to alter their properties suitable for tissue/organ repair and other biomedical applications. Appropriate biocompatibility and degradability could be achieved by combining different polymers by physical blending or chemical modifications. In recent years, the gap between biocompatibility and immunogenicity of polymers has narrowed as a result of various copolymerization processes. This chapter has demonstrated a clear outline of the various biomedical applications of natural and synthetic polymers with a prime focus on tissue and regenerative engineering.

ACKNOWLEDGMENTS

The authors gratefully acknowledge funding from the Raymond and Beverly Sackler Center for Biomedical, Biological, Physical and Engineering Sciences. Authors also acknowledge the funding from National Science Foundation (IIP-1311907 and EFRI-1332329) and Early Career Translational Research Award in Biomedical Engineering from Wallace H. Coulter Foundation.

REFERENCES

[1] R. Langer, J.P. Vacanti, Tissue engineering. Science 260 (5110) (1993) 920, doi:10.1126/science.8493529.

[2] C.T. Laurencin, A.M. Ambrosio, M.D. Borden, J.A. Cooper Jr., Tissue engineering: orthopedic applications. Annu. Rev. Biomed. Eng. 1 (1999) 19–46, doi:10.1146/annurev.bioeng.1.1.19.

[3] C.T. Laurencin, S.G. Kumbar, M. Deng, R. James, Nano-structured scaffolds for regenerative engineering, in: Honorary Series in Translational Research in Biomaterials, 2010 AICHE Annual Meeting, Salt Lake City, Utah, USA, 2010.

[4] L.S. Nair, C.T. Laurencin, Biodegradable polymers as biomaterials. Polym. Biomed. Appl. 32 (8-9) (2007) 762–798, doi:10.1016/j.progpolymsci.2007.05.017.

[5] D.W. Hutmacher, Scaffolds in tissue engineering bone and cartilage, Biomaterials 21 (24) (2000) 2529–2543.

[6] A. Aravamudhan, D.M. Ramos, J. Nip, M.D. Harmon, R. James, M. Deng, S.G. Kumbar, et al., Cellulose and collagen derived micro-nano structured scaffolds for bone tissue engineering, J. Biomed. Nanotechnol. 9 (4) (2013) 719–731.

[7] C.H. Lee, A. Singla, Y. Lee, Biomedical applications of collagen. Int. J. Pharm. 221 (1-2) (2001) 1–22, doi:10.1016/S0378-5173(01)00691-3.

[8] C.J. Doillon, C. DeBlois, M.-F. Côté, N. Fournier, Bioactive collagen sponge as connective tissue substitute. Mater. Sci. Eng. C 2 (1-2) (1994) 43–49, doi:10.1016/0928-4931(94)90028-0.

[9] K.S. Rho, L. Jeong, G. Lee, B.-M. Seo, Y.J. Park, S.-D. Hong, B.-M. Min, et al., Electrospinning of collagen nanofibers: effects on the behavior of normal human keratinocytes and early-stage wound healing. Biomaterials 27 (8) (2006) 1452–1461, doi:10.1016/j.biomaterials.2005.08.004.

[10] S.M. Willerth, S.E. Sakiyama-Elbert, Approaches to neural tissue engineering using scaffolds for drug delivery. Matrices Scaffolds Drug Deliv. Tissue Eng. 59 (4-5) (2007) 325–338, doi:10.1016/j.addr.2007.03.014.

[11] A.G. Schauss, D.J. Merkel, S.M. Glaza, S.R. Sorenson, Acute and subchronic oral toxicity studies in rats of a hydrolyzed chicken sternal cartilage preparation. Food Chem. Toxicol. 45 (2) (2007) 315–321, doi:10.1016/j.fct.2006.08.011.

[12] S.J. Roberts, L. Geris, G. Kerckhofs, E. Desmet, J. Schrooten, F.P. Luyten, The combined bone forming capacity of human periosteal derived cells and calcium phosphates. Biomaterials 32 (19) (2011) 4393–4405, doi:10.1016/j. biomaterials .2011.02.047.

[13] N. van Gastel, S. Torrekens, S.J. Roberts, K. Moermans, J. Schrooten, P. Carmeliet, G. Carmeliet, et al., Engineering vascularized bone: osteogenic and proangiogenic potential of murine periosteal cells. Stem Cells 30 (11) (2012) 2460–2471, doi:10.1002/stem.1210.

[14] B. Swarnalatha, S.L. Nair, K.T. Shalumon, L.C. Milbauer, R. Jayakumar, B. Paul-Prasanth, S.V. Nair, et al., Poly (lactic acid)–chitosan–collagen composite nanofibers as substrates for blood outgrowth endothelial cells. Int. J. Biol. Macromol. 58 (2013) 220–224, doi:10.1016/j.ijbiomac.2013.03.060.

[15] M. Ngiam, S. Liao, A.J. Patil, Z. Cheng, C.K. Chan, S. Ramakrishna, The fabrication of nano-hydroxyapatite on PLGA and PLGA/collagen nanofibrous composite scaffolds and their effects in osteoblastic behavior for bone tissue engineering. Bone 45 (1) (2009) 4–16, doi:10.1016/j.bone.2009.03.674.

[16] B. Guo, Y. Sun, A. Finne-Wistrand, K. Mustafa, A.-C. Albertsson, Electroactive porous tubular scaffolds with degradability and non-cytotoxicity for neural tissue regeneration. Acta Biomater. 8 (1) (2012) 144–153, doi:10.1016/j.actbio.2011.09.027.

[17] C. Huang, R. Chen, Q. Ke, Y. Morsi, K. Zhang, X. Mo, Electrospun collagen–chitosan–TPU nanofibrous scaffolds for tissue engineered tubular grafts. Colloids Surf. B Biointerfaces 82 (2) (2011) 307–315, doi:10.1016/j.colsurfb.2010.09.002.

[18] W. Shen, X. Chen, J. Chen, Z. Yin, B.C. Heng, W. Chen, H.-W. Ouyang, The effect of incorporation of exogenous stromal cell-derived factor-1 alpha within a knitted silk-collagen sponge scaffold on tendon regeneration. Biomaterials 31 (28) (2010) 7239–7249, doi:10.1016/j.biomaterials.2010.05.040.

[19] Y. Sumita, M.J. Honda, T. Ohara, S. Tsuchiya, H. Sagara, H. Kagami, M. Ueda, Performance of collagen sponge as a 3-D scaffold for tooth-tissue engineering. Biomaterials 27 (17) (2006) 3238–3248, doi:10.1016/j.biomaterials.2006.01.055.

[20] G. Tomoaia, O. Soritau, M. Tomoaia-Cotisel, L.B. Pop, A. Pop, A. Mocanu, et al., Scaffolds made of nanostructured phosphates, collagen and chitosan for cell culture. in: Special Issue: 5th International Granulation Workshop Granulation Across the Length Scale 2011, vol. 238(0), 2013, pp. 99–107, doi:10.1016/j.powtec.2012.05.023.

[21] K. Yamauchi, T. Goda, N. Takeuchi, H. Einaga, T. Tanabe, Preparation of collagen/calcium phosphate multilayer sheet using enzymatic mineralization. Biomaterials 25 (24) (2004) 5481–5489, doi:10.1016/j.biomaterials.2003.12.057.

[22] J.L. Drury, D.J. Mooney, Hydrogels for tissue engineering: scaffold design variables and applications, Biomaterials 24 (24) (2003) 4337–4351.

[23] S.A. Sell, M.J. McClure, K. Garg, P.S. Wolfe, G.L. Bowlin, Electrospinning of collagen/biopolymers for regenerative medicine and cardiovascular tissue engineering. Adv. Drug Deliv. Rev. 61 (12) (2009) 1007–1019, doi:10.1016/j.addr.2009.07.012.

[24] Y. Liu, L. Ma, C. Gao, Facile fabrication of the glutaraldehyde cross-linked collagen/chitosan porous scaffold for skin tissue engineering. Mater. Sci. Eng. C 32 (8) (2012) 2361–2366, doi:10.1016/j.msec.2012.07.008.

[25] M.-C. Ronzière, S. Roche, J. Gouttenoire, O. Démarteau, D. Herbage, A.-M. Freyria, Ascorbate modulation of bovine chondrocyte growth, matrix protein gene expression and synthesis in three-dimensional collagen sponges. Biomaterials 24 (5) (2003) 851–861, doi:10.1016/S0142-9612(02)00418-0.

[26] S.J. Kew, J.H. Gwynne, D. Enea, M. Abu-Rub, A. Pandit, D. Zeugolis, R.E. Cameron, et al., Regeneration and repair of tendon and ligament tissue using collagen fibre biomaterials. Acta Biomater. 7 (9) (2011) 3237–3247, doi:10.1016/j.actbio.2011.06.002.

[27] C.T. Laurencin, J.W. Freeman, Ligament tissue engineering: an evolutionary materials science approach. Biomaterials 26 (36) (2005) 7530–7536, doi:10.1016/j.biomaterials.2005.05.073.

[28] G.A. Di Lullo, S.M. Sweeney, J. Körkkö, L. Ala-Kokko, J.D. San Antonio, Mapping the ligand-binding sites and disease-associated

mutations on the most abundant protein in the human, type I collagen, J. Biol. Chem. 277 (6) (2002) 4223–4231.

[29] W. Xia, W. Liu, L. Cui, Y. Liu, W. Zhong, D. Liu, Y. Cao, et al., Tissue engineering of cartilage with the use of chitosan-gelatin complex scaffolds, J. Biomed. Mater. Res. B Appl. Biomater. 71 (2) (2004) 373–380.

[30] F.A. Johnston-Banks, Gelatine Food Gels, Springer, Heidelberg, 1990, pp. 233–289.

[31] X. Liu, L.A. Smith, J. Hu, P.X. Ma, Biomimetic nanofibrous gelatin/apatite composite scaffolds for bone tissue engineering. Biomaterials 30 (12) (2009) 2252–2258, doi:10.1016/j.biomaterials.2008.12.068.

[32] S. Kanokpanont, S. Damrongsakkul, J. Ratanavaraporn, P. Aramwit, An innovative bi-layered wound dressing made of silk and gelatin for accelerated wound healing. Int. J. Pharm. 436 (1-2) (2012) 141–153, doi:10.1016/j.ijpharm.2012.06.046.

[33] E.J. Chong, T.T. Phan, I.J. Lim, Y.Z. Zhang, B.H. Bay, S. Ramakrishna, C.T. Lim, Evaluation of electrospun PCL/gelatin nanofibrous scaffold for wound healing and layered dermal reconstitution. Acta Biomater. 2nd TMS Symp. Biol. Mater. Sci. 3 (3) (2007) 321–330, doi:10.1016/j.actbio.2007.01.002.

[34] H.-W. Kang, Y. Tabata, Y. Ikada, Fabrication of porous gelatin scaffolds for tissue engineering. Biomaterials 20 (14) (1999) 1339–1344, doi:10.1016/S0142-9612(99)00036-8.

[35] F. Zhao, Y. Yin, W.W. Lu, J.C. Leong, W. Zhang, J. Zhang, K. Yao, et al., Preparation and histological evaluation of biomimetic three-dimensional hydroxyapatite/chitosan-gelatin network composite scaffolds, Biomaterials 23 (15) (2002) 3227–3234.

[36] L. Ghasemi-Mobarakeh, M.P. Prabhakaran, M. Morshed, M.-H. Nasr-Esfahani, S. Ramakrishna, Electrospun poly (ε-caprolactone)/gelatin nanofibrous scaffolds for nerve tissue engineering, Biomaterials 29 (34) (2008) 4532–4539.

[37] S.R. Caliari, M.A. Ramirez, B.A.C. Harley, The development of collagen-GAG scaffold-membrane composites for tendon tissue engineering, Biomaterials 32 (34) (2011) 8990–8998.

[38] A. Jaklenec, E. Wan, M.E. Murray, E. Mathiowitz, Novel scaffolds fabricated from protein-loaded microspheres for tissue engineering, Biomaterials 29 (2) (2008) 185–192.

[39] M. Li, M.J. Mondrinos, X. Chen, M.R. Gandhi, F.K. Ko, P.I. Lelkes, Co-electrospun poly (lactide-co-glycolide), gelatin, and elastin blends for tissue engineering scaffolds, J. Biomed. Mater. Res. A 79 (4) (2006) 963–973.

[40] X. Li, L. Jin, G. Balian, C.T. Laurencin, D. Greg Anderson, Demineralized bone matrix gelatin as scaffold for osteochondral tissue engineering, Biomaterials 27 (11) (2006) 2426–2433.

[41] D. Puppi, F. Chiellini, A.M. Piras, E. Chiellini, Polymeric materials for bone and cartilage repair, Prog. Polym. Sci. 35 (4) (2010) 403–440.

[42] R.E. Neuman, M.A. Logan, The determination of collagen and elastin in tissues, J. Biol. Chem. 186 (2) (1950) 549–556.

[43] S.M. Mithieux, J.E.J. Rasko, A.S. Weiss, Synthetic elastin hydrogels derived from massive elastic assemblies of self-organized human protein monomers, Biomaterials 25 (20) (2004) 4921–4927.

[44] E.D. Boland, J.A. Matthews, K.J. Pawlowski, D.G. Simpson, G.E. Wnek, G.L. Bowlin, Electrospinning collagen and elastin: preliminary vascular tissue engineering, Front. Biosci. 9 (1422) (2004) C1432.

[45] M.J. Smith, M.J. McClure, S.A. Sell, C.P. Barnes, B.H. Walpoth, D.G. Simpson, G.L. Bowlin, Suture-reinforced electrospun polydioxanone–elastin small-diameter tubes for use in vascular tissue engineering: a feasibility study, Acta Biomater. 4 (1) (2008) 58–66.

[46] J. Chlupáč, E. Filova, L. Bačáková, Blood vessel replacement: 50 years of development and tissue engineering paradigms in vascular surgery, Physiol. Res. 58 (2009) S119.

[47] W.F. Daamen, J.H. Veerkamp, J.C.M. Van Hest, T.H. Van Kuppevelt, Elastin as a biomaterial for tissue engineering, Biomaterials 28 (30) (2007) 4378–4398.

[48] T.H. Petersen, E.A. Calle, L. Zhao, E.J. Lee, L. Gui, M.B. Raredon, C. Breuer, et al., Tissue-engineered lungs for in vivo implantation, Science 329 (5991) (2010) 538–541.

[49] A.P. Price, K.A. England, A.M. Matson, B.R. Blazar, A. Panoskaltsis-Mortari, Development of a decellularized lung bioreactor system for bioengineering the lung: the matrix reloaded, Tissue Eng. A 16 (8) (2010) 2581–2591.

[50] S.B. Adams, M.F. Shamji, D.L. Nettles, P. Hwang, L.A. Setton, Sustained release of antibiotics from injectable and thermally responsive polypeptide depots, J. Biomed. Mater. Res. B Appl. Biomater. 90 (1) (2009) 67–74.

[51] I. Massodi, E. Thomas, D. Raucher, Application of thermally responsive elastin-like polypeptide fused to a lactoferrin-derived peptide for treatment of pancreatic cancer, Molecules 14 (6) (2009) 1999–2015.

[52] H. Betre, L.A. Setton, D.E. Meyer, A. Chilkoti, Characterization of a genetically engineered elastin-like polypeptide for cartilaginous tissue repair, Biomacromolecules 3 (5) (2002) 910–916.

[53] J.F. Mano, G.A. Silva, H.S. Azevedo, P.B. Malafaya, R.A. Sousa, S.S. Silva, A.P. Marques, et al., Natural origin biodegradable systems in tissue engineering and regenerative medicine: present status and some moving trends, J. R. Soc. Interface 4 (17) (2007) 999–1030.

[54] W. Haslik, D.B. Lumenta, L.P. Kamolz, M. Frey, The use of a collagen–elastin matrix as dermal regeneration template for the treatment of full-thickness skin defects, Adv. Wound Care 1 (2010) 438–444.

[55] D. Lamprou, P. Zhdan, F. Labeed, C. Lekakou, Gelatine and gelatine/elastin nanocomposites for vascular grafts: processing and characterization, J. Biomater. Appl. 26 (2) (2011) 209–226.

[56] L. Buttafoco, N.G. Kolkman, P. Engbers-Buijtenhuijs, A.A. Poot, P.J. Dijkstra, I. Vermes, J. Feijen, Electrospinning of collagen and elastin for tissue engineering applications, Biomaterials 27 (5) (2006) 724–734. http://dx.doi.org/10.1016/j.biomaterials.2005.06.024.

[57] D.L. Nettles, A. Chilkoti, L.A. Setton, Applications of elastin-like polypeptides in tissue engineering, Adv. Drug Deliv. Rev. 62 (15) (2010) 1479–1485. http://dx.doi.org/10.1016/j.addr.2010.04.002.

[58] S.G. Wise, M.J. Byrom, A. Waterhouse, P.G. Bannon, M.K. Ng, A.S. Weiss, A multilayered synthetic human elastin/polycaprolactone hybrid vascular graft with tailored mechanical properties, Acta Biomater. 7 (1) (2011) 295–303.

[59] L. Nivison-Smith, J. Rnjak, A.S. Weiss, Synthetic human elastin microfibers: stable cross-linked tropoelastin and cell interactive constructs for tissue engineering applications, Acta Biomater. 6 (2) (2010) 354–359.

[60] C. Acharya, S.K. Ghosh, S.C. Kundu, Silk fibroin film from non-mulberry tropical tasar silkworms: a novel substrate for in vitro fibroblast culture, Acta Biomater. 5 (1) (2009) 429–437. http://dx.doi.org/10.1016/j.actbio.2008.07.003.

[61] J. Huang, C. Wong Po Foo, D.L. Kaplan, Biosynthesis and applications of silk-like and collagen-like proteins, J. Macromol. Sci. C Polym. Rev. 47 (1) (2007) 29–62.

[62] G.H. Altman, F. Diaz, C. Jakuba, T. Calabro, R.L. Horan, J. Chen, D.L. Kaplan, et al., Silk-based biomaterials, Biomaterials 24 (3) (2003) 401–416.

[63] H. Ayub, M. Arai, K. Hirabayashi, Mechanism of the gelation of fibroin solution, Biosci. Biotechnol. Biochem. 57 (11) (1993) 1910–1912.

[64] V. Karageorgiou, L. Meinel, S. Hofmann, A. Malhotra, V. Volloch, D. Kaplan, Bone morphogenetic protein-2 decorated silk fibroin films induce osteogenic differentiation of human bone marrow stromal cells, J. Biomed. Mater. Res. A 71 (3) (2004) 528–537.

[65] X. Wang, H.J. Kim, P. Xu, A. Matsumoto, D.L. Kaplan, Biomaterial coatings by stepwise deposition of silk fibroin, Langmuir 21 (24) (2005) 11335–11341.

[66] K.-H. Kim, L. Jeong, H.-N. Park, S.-Y. Shin, W.-H. Park, S.-C. Lee, Y.-M. Lee, et al., Biological efficacy of silk fibroin nanofiber membranes for guided bone regeneration, J. Biotechnol. 120 (3) (2005) 327–339.

[67] L. Meinel, S. Hofmann, V. Karageorgiou, C. Kirker-Head, J. McCool, G. Gronowicz, D.L. Kaplan, et al., The inflammatory responses to silk films in vitro and in vivo, Biomaterials 26 (2) (2005) 147–155.

[68] J. Ayutsede, M. Gandhi, S. Sukigara, M. Micklus, H.-E. Chen, F. Ko, Regeneration of *Bombyx mori* silk by electrospinning. Part 3: characterization of electrospun nonwoven mat, Polymer 46 (5) (2005) 1625–1634.

[69] J. Chen, G.H. Altman, V. Karageorgiou, R. Horan, A. Collette, V. Volloch, D.L. Kaplan, et al., Human bone marrow stromal cell and ligament fibroblast responses on RGD-modified silk fibers, J. Biomed. Mater. Res. A 67 (2) (2003) 559–570.

[70] X. Chen, W. Li, W. Zhong, Y. Lu, T. Yu, pH sensitivity and ion sensitivity of hydrogels based on complex-forming chitosan/silk fibroin interpenetrating polymer network, J. Appl. Polym. Sci. 65 (11) (1997) 2257–2262.

[71] G. Freddi, M. Tsukada, S. Beretta, Structure and physical properties of silk fibroin/polyacrylamide blend films, J. Appl. Polym. Sci. 71 (10) (1999) 1563–1571.

[72] M. Fini, A. Motta, P. Torricelli, G. Giavaresi, N. Nicoli Aldini, M. Tschon, C. Migliaresi, et al., The healing of confined critical size cancellous defects in the presence of silk fibroin hydrogel, Biomaterials 26 (17) (2005) 3527.

[73] G.H. Altman, R.L. Horan, H.H. Lu, J. Moreau, I. Martin, J.C. Richmond, D.L. Kaplan, Silk matrix for tissue engineered anterior cruciate ligaments, Biomaterials 23 (20) (2002) 4131–4141.

[74] B.D. Lawrence, J.K. Marchant, M.A. Pindrus, F.G. Omenetto, D.L. Kaplan, Silk film biomaterials for cornea tissue engineering, Biomaterials 30 (7) (2009) 1299–1308.

[75] Y. Dror, T. Ziv, V. Makarov, H. Wolf, A. Admon, E. Zussman, Nanofibers made of globular proteins, Biomacromolecules 9 (10) (2008) 2749–2754.

[76] T. Peters Jr., All About Albumin: Biochemistry, Genetics, and Medical Applications, in: Academic Press, San Diego, 1995.

[77] B.H. Prinsen, M.G. de Sain-van der Velden, Albumin turnover: experimental approach and its application in health and renal diseases, Clin. Chim. Acta 347 (1) (2004) 1–14.

[78] M. Uchida, A. Ito, K.S. Furukawa, K. Nakamura, Y. Onimura, A. Oyane, T. Tateishi, et al., Reduced platelet adhesion to titanium metal coated with apatite, albumin–apatite composite or laminin–apatite composite, Biomaterials 26 (34) (2005) 6924–6931.

[79] V.T.G. Chuang, U. Kragh-Hansen, M. Otagiri, Pharmaceutical strategies utilizing recombinant human serum albumin, Pharm. Res. 19 (5) (2002) 569–577.

[80] J.-M. Li, W. Chen, H. Wang, C. Jin, W.-Y. Lu, H.-M. Hou, et al., Preparation of albumin nanospheres loaded with gemcitabine and their cytotoxicity against BXPC-3 cells in vitro, Acta Pharmacol. Sin. 30 (9) (2009) 1337–1343.

[81] F. De Somer, J. Delanghe, P. Somers, M. Debrouwere, G. Van Nooten, Mechanical and chemical characteristics of an autologous glue, J. Biomed. Mater. Res. A 86 (4) (2008) 1106–1112.

[82] S.S.A. An, I. Suh, Binding of thrombin activatable fibrinolysis inhibitor (TAFI) to plasminogen may play a role in the fibrinolytic pathway, Bull. Korean Chem. Soc. 29 (11) (2008) 2209.

[83] F.M. Shaikh, A. Callanan, E.G. Kavanagh, P.E. Burke, P.A. Grace, T.M. McGloughlin, Fibrin: a natural biodegradable scaffold in vascular tissue engineering, Cells Tissues Organs 188 (4) (2008) 333–346.

[84] T. Hou, J. Xu, Q. Li, J. Feng, L. Zen, In vitro evaluation of a fibrin gel antibiotic delivery system containing mesenchymal stem cells and vancomycin alginate beads for treating bone infections and facilitating bone formation, Tissue Eng. A 14 (7) (2008) 1173–1182.

[85] U. Schillinger, G. Wexel, C. Hacker, M. Kullmer, C. Koch, M. Gerg, D. Hensler, et al., A fibrin glue composition as carrier for nucleic acid vectors, Pharm. Res. 25 (12) (2008) 2946–2962.

[86] M. Mana, M. Cole, S. Cox, B. Tawil, Human U937 monocyte behavior and protein expression on various formulations of three-dimensional fibrin clots, Wound Repair Regen. 14 (1) (2006) 72–80.

[87] S. Johnsen, T. Ermuth, E. Tanczos, H. Bannasch, R. Horch, I. Zschocke, M. Augustin, et al., Treatment of therapy-refractive ulcera cruris of various origins with autologous keratinocytes in fibrin sealant, VASA. Z. Gefasskrankheiten. J. Vasc. Dis. 34 (1) (2005) 25.

[88] J. Kopp, M.G. Jeschke, A.D. Bach, U. Kneser, R.E. Horch, Applied tissue engineering in the closure of severe burns and chronic wounds using cultured human autologous keratinocytes in a natural fibrin matrix, Cell Tissue Bank. 5 (2) (2004) 89–96.

[89] W.J. Yoo, I.H. Choi, C.Y. Chung, T.-J. Cho, I.-O. Kim, C.J. Kim, Implantation of perichondrium-derived chondrocytes in physeal defects of rabbit tibiae, Acta Orthop. 76 (5) (2005) 628–636.

[90] D. Le Nihouannen, L.L. Guehennec, T. Rouillon, P. Pilet, M. Bilban, P. Layrolle, G. Daculsi, Micro-architecture of calcium phosphate granules and fibrin glue composites for bone tissue engineering, Biomaterials 27 (13) (2006) 2716–2722.

[91] A. Heinemann, USA Patent No, 1910.

[92] Y. Noishiki, H. Ito, T. Miyamoto, H. Inagaki, Application of denatured wool keratin derivatives to an antithrombogenic biomaterial: vascular graft coated with a heparinized keratin derivative, Kobunshi Ronbunshu 39 (4) (1982) 221–227.

[93] R.L. Mauck, M.A. Soltz, C.C. Wang, D.D. Wong, P.-H.G. Chao, W.B. Valhmu, G.A. Ateshian, et al., Functional tissue engineering of articular cartilage through dynamic loading of chondrocyte-seeded agarose gels, Trans. Am. Soc. Mech. Eng. J. Biomech. Eng. 122 (3) (2000) 252–260.

[94] S.F. Timmons, C.R. Blanchard,R.A. Smith, Keratin-based tissue engineering scaffold: Google Patents, 2002.

[95] A. Tachibana, Y. Furuta, H. Takeshima, T. Tanabe, K. Yamauchi, Fabrication of wool keratin sponge scaffolds for long-term cell cultivation, J. Biotechnol. 93 (2) (2002) 165–170.

[96] A. Tachibana, S. Kaneko, T. Tanabe, K. Yamauchi, Rapid fabrication of keratin–hydroxyapatite hybrid sponges toward osteoblast cultivation and differentiation, Biomaterials 26 (3) (2005) 297–302.

[97] Q. Hesheng, Deacetylation of Chitin, J. Donghua Univ. Nat. Sci. 2 (1998) 030.

[98] S.M. Lim, D.K. Song, S.H. Oh, D.S. Lee-Yoon, E.H. Bae, J.H. Lee, In vitro and in vivo degradation behavior of acetylated chitosan porous beads, J. Biomater. Sci. Polym. Ed. 19 (4) (2008) 453–466.

[99] J.-K. Francis Suh, H.W. Matthew, Application of chitosan-based polysaccharide biomaterials in cartilage tissue engineering: a review, Biomaterials 21 (24) (2000) 2589–2598.

[100] S. Kumbar, A. Kulkarni, T. Aminabhavi, Crosslinked chitosan microspheres for encapsulation of diclofenac sodium: effect of crosslinking agent, J. Microencapsul. 19 (2) (2002) 173–180.

[101] S.G. Kumbar, C.T. Laurencin, Natural polymer-based orthopedic fixation screw for bone repair and regeneration: EP Patent 2,538,861, 2013.

[102] S.V. Madihally, H.W. Matthew, Porous chitosan scaffolds for tissue engineering, Biomaterials 20 (12) (1999) 1133–1142.

[103] S.-Y. Ong, J. Wu, S.M. Moochhala, M.-H. Tan, J. Lu, Development of a chitosan-based wound dressing with improved hemostatic and antimicrobial properties, Biomaterials 29 (32) (2008) 4323–4332.

[104] W. Paul, C.P. Sharma, Chitosan and alginate wound dressings: a short review, Trends Biomater. Artif Organs 18 (1) (2004) 18–23.

[105] G. Kratz, C. Arnander, J. Swedenborg, M. Back, C. Falk, I. Gouda, O. Larm, Heparin-chitosan complexes stimulate wound healing in human skin, Scand. J. Plast. Reconstr. Surg. Hand Surg. 31 (2) (1997) 119–123.

[106] B.G. Amsden, A. Sukarto, D.K. Knight, S.N. Shapka, Methacrylated glycol chitosan as a photopolymerizable biomaterial, Biomacromolecules 8 (12) (2007) 3758–3766.

[107] T. Jiang, W.I. Abdel-Fattah, C.T. Laurencin, In vitro evaluation of chitosan/poly(lactic acid-glycolic acid) sintered microsphere scaffolds for bone tissue engineering, Biomaterials 27 (28) (2006) 4894–4903. http://dx.doi.org/10.1016/j.biomaterials.2006.05.025.

[108] S.M. Oliveira, I.F. Amaral, M.A. Barbosa, C.C. Teixeira, Engineering endochondral bone: in vitro studies, Tissue Eng. A 15 (3) (2008) 625–634.

[109] L.A. Pfister, M. Papaloïzos, H.P. Merkle, B. Gander, Hydrogel nerve conduits produced from alginate/chitosan complexes, J. Biomed. Mater. Res. A 80 (4) (2007) 932–937.

[110] R. Jayakumar, M. Prabaharan, S. Nair, H. Tamura, Novel chitin and chitosan nanofibers in biomedical applications, Biotechnol. Adv. 28 (1) (2010) 142–150.

[111] K.-Y. Chen, W.-J. Liao, S.-M. Kuo, F.-J. Tsai, Y.-S. Chen, C.-Y. Huang, C.-H. Yao, Asymmetric chitosan membrane containing collagen I nanospheres for skin tissue engineering, Biomacromolecules 10 (6) (2009) 1642–1649.

[112] J.A. Rowley, G. Madlambayan, D.J. Mooney, Alginate hydrogels as synthetic extracellular matrix materials, Biomaterials 20 (1) (1999) 45–53.

[113] G. Klöck, A. Pfeffermann, C. Ryser, P. Gröhn, B. Kuttler, H.-J. Hahn, U. Zimmermann, Biocompatibility of mannuronic acid-rich alginates, Biomaterials 18 (10) (1997) 707–713.

[114] A.D. Augst, H.J. Kong, D.J. Mooney, Alginate hydrogels as biomaterials, Macromol. Biosci. 6 (8) (2006) 623–633.

[115] L. Zhao, M. Tang, M.D. Weir, M.S. Detamore, H.H. Xu, Osteogenic media and rhBMP-2-induced differentiation of umbilical cord mesenchymal stem cells encapsulated in alginate microbeads and integrated in an injectable calcium phosphate-chitosan fibrous scaffold, Tissue Eng. A 17 (7-8) (2010) 969–979.

[116] I. Colinet, V. Dulong, G. Mocanu, L. Picton, D. Le Cerf, New amphiphilic and pH-sensitive hydrogel for controlled release of a model poorly water-soluble drug, Eur. J. Pharm. Biopharm. 73 (3) (2009) 345–350.

[117] D.-H. Kim, D.C. Martin, Sustained release of dexamethasone from hydrophilic matrices using PLGA nanoparticles for neural drug delivery, Biomaterials 27 (15) (2006) 3031–3037.

[118] D. Shastri, S. Prajapati, L. Patel, Design and development of thermoreversible ophthalmic in situ hydrogel of moxifloxacin HCl, Curr. Drug Deliv. 7 (3) (2010) 238–243.

[119] H.H. Xu, M.D. Weir, L. Sun, Calcium and phosphate ion releasing composite: effect of pH on release and mechanical properties, Dent. Mater. 25 (4) (2009) 535–542.

[120] H.R. Moyer, R.C. Kinney, K.A. Singh, J.K. Williams, Z. Schwartz, B.D. Boyan, Alginate microencapsulation technology for the percutaneous delivery of adipose-derived stem cells, Ann. Plast. Surg. 65 (5) (2010) 497–503.

[121] E. Marsich, M. Borgogna, I. Donati, P. Mozetic, B.L. Strand, S.G. Salvador, S. Paoletti, et al., Alginate/lactose-modified chitosan hydrogels: a bioactive biomaterial for chondrocyte encapsulation, J. Biomed. Mater. Res. A 84 (2) (2008) 364–376.

[122] A.W. Chan, R.A. Whitney, R.J. Neufeld, Semisynthesis of a controlled stimuli-responsive alginate hydrogel, Biomacromolecules 10 (3) (2009) 609–616.

[123] O. Namiki, H. Toyoshima, N. Morisaki, Therapeutic effect of intra-articular injection of high molecular weight hyaluronic acid on osteoarthritis of the knee, Int. J. Clin. Pharmacol. Therapy Toxicol. 20 (11) (1982) 501.

[124] J. Aigner, J. Tegeler, P. Hutzler, D. Campoccia, A. Pavesio, C. Hammer, A. Naumann, et al., Cartilage tissue engineering with novel nonwoven structured biomaterial based on hyaluronic acid benzyl ester, J. Biomed. Mater. Res. 42 (2) (1998) 172–181.

[125] A. De Andres-Santos, A. Velasco-Martín, E. Hernández-Velasco, J. Martín-Gil, F. Martín-Gil, Thermal behaviour of aqueous solutions of sodium hyaluronate from different commercial sources, Thermochim. Acta 242 (1994) 153–160.

[126] X.Z. Shu, K. Ghosh, Y. Liu, F.S. Palumbo, Y. Luo, R.A. Clark, G.D. Prestwich, Attachment and spreading of fibroblasts on an RGD peptide-modified injectable hyaluronan hydrogel, J. Biomed. Mater. Res. A 68 (2) (2004) 365–375.

[127] L.A. Solchaga, J.E. Dennis, V.M. Goldberg, A.I. Caplan, Hyaluronic acid-based polymers as cell carriers for tissue-engineered repair of bone and cartilage, J. Orthop. Res. 17 (2) (1999) 205–213.

[128] J.L. Ifkovits, J.A. Burdick, Review: photopolymerizable and degradable biomaterials for tissue engineering applications, Tissue Eng. 13 (10) (2007) 2369–2385.

[129] G. Kogan, L. Šoltés, R. Stern, P. Gemeiner, Hyaluronic acid: a natural biopolymer with a broad range of biomedical and industrial applications, Biotechnol. Lett. 29 (1) (2007) 17–25.

[130] S. Kobayashi, K. Kashiwa, J. Shimada, T. Kawasaki, S. i Shoda, Enzymatic polymerization: the first in vitro synthesis of cellulose via nonbiosynthetic path catalyzed by cellulase, in: Paper Presented at the Makromolekulare Chemie. Macromolecular Symposia, 1992.

[131] M. Märtson, J. Viljanto, T. Hurme, P. Laippala, P. Saukko, Is cellulose sponge degradable or stable as implantation material? An in vivo subcutaneous study in the rat, Biomaterials 20 (21) (1999) 1989–1995.

[132] E. Entcheva, H. Bien, L. Yin, C.-Y. Chung, M. Farrell, Y. Kostov, Functional cardiac cell constructs on cellulose-based scaffolding, Biomaterials 25 (26) (2004) 5753–5762.

[133] G. Helenius, H. Bäckdahl, A. Bodin, U. Nannmark, P. Gatenholm, B. Risberg, In vivo biocompatibility of bacterial cellulose, J. Biomed. Mater. Res. A 76 (2) (2006) 431–438.

[134] M. Märtson, J. Viljanto, T. Hurme, P. Saukko, Biocompatibility of cellulose sponge with bone, Eur. Surg. Res. 30 (6) (1998) 426–432.

[135] F.A. Müller, L. Müller, I. Hofmann, P. Greil, M.M. Wenzel, R. Staudenmaier, Cellulose-based scaffold materials for cartilage tissue engineering, Biomaterials 27 (21) (2006) 3955–3963.

[136] J. Poustis, C. Baquey, D. Chauveaux, Mechanical properties of cellulose in orthopaedic devices and related environments, Clin. Mater. 16 (2) (1994) 119–124.
[137] C. Vinatier, O. Gauthier, A. Fatimi, C. Merceron, M. Masson, A. Moreau, J. Guicheux, et al., An injectable cellulose-based hydrogel for the transfer of autologous nasal chondrocytes in articular cartilage defects, Biotechnol. Bioeng. 102 (4) (2009) 1259–1267.
[138] P.L. Granja, M.A. Barbosa, L. Pouységu, B. De Jéso, F. Rouais, C. Baquey, Cellulose phosphates as biomaterials. Mineralization of chemically modified regenerated cellulose hydrogels, J. Mater. Sci. 36 (9) (2001) 2163–2172.
[139] Y. Wan, L. Hong, S. Jia, Y. Huang, Y. Zhu, Y. Wang, H. Jiang, Synthesis and characterization of hydroxyapatite–bacterial cellulose nanocomposites, Compos. Sci. Technol. 66 (11) (2006) 1825–1832.
[140] G.P. Andrews, S.P. Gorman, D.S. Jones, Rheological characterisation of primary and binary interactive bioadhesive gels composed of cellulose derivatives designed as ophthalmic viscosurgical devices, Biomaterials 26 (5) (2005) 571–580.
[141] M.V. Risbud, R.R. Bhonde, Suitability of cellulose molecular dialysis membrane for bioartificial pancreas: in vitro biocompatibility studies, J. Biomed. Mater. Res. 54 (3) (2001) 436–444.
[142] R. Vaidya, E. Wilkins, Effect of interference on amperometric glucose biosensors with cellulose acetate membranes, Electroanalysis 6 (8) (1994) 677–682.
[143] N.S. Hwang, S. Varghese, H.J. Lee, P. Theprungsirikul, A. Canver, B. Sharma, J. Elisseeff, Response of zonal chondrocytes to extracellular matrix-hydrogels, FEBS Lett. 581 (22) (2007) 4172–4178.
[144] P. Chan, J. Caron, G. Rosa, M. Orth, Glucosamine and chondroitin sulfate regulate gene expression and synthesis of nitric oxide and prostaglandin E (2) in articular cartilage explants, Osteoarthr. Cartil./OARS Osteoarthr. Res. Soc. 13 (5) (2005) 387.
[145] M.A. Kosir, C.C. Quinn, W. Wang, G. Tromp, Matrix glycosaminoglycans in the growth phase of fibroblasts: more of the story in wound healing, J. Surg. Res. 92 (1) (2000) 45–52.
[146] S.T. Boyce, M.C. Glafkides, T.J. Foreman, J.F. Hansbrough, Reduced wound contraction after grafting of full-thickness burns with a collagen and chondroitin-6-sulfate (GAG) dermal skin substitute and coverage with biobrane, J. Burn Care Res. 9 (4) (1988) 364–370.
[147] S. Dawlee, A. Sugandhi, B. Balakrishnan, D. Labarre, A. Jayakrishnan, Oxidized chondroitin sulfate-cross-linked gelatin matrixes: a new class of hydrogels, Biomacromolecules 6 (4) (2005) 2040–2048.
[148] M.E. Gilbert, K.R. Kirker, S.D. Gray, P.D. Ward, J.G. Szakacs, G.D. Prestwich, R.R. Orlandi, Chondroitin sulfate hydrogel and wound healing in rabbit maxillary sinus mucosa, Laryngoscope 114 (8) (2004) 1406–1409.
[149] Y.-L. Chen, H.-P. Lee, H.-Y. Chan, L.-Y. Sung, H.-C. Chen, Y.-C. Hu, Composite chondroitin-6-sulfate/dermatan sulfate/chitosan scaffolds for cartilage tissue engineering, Biomaterials 28 (14) (2007) 2294–2305.
[150] C.-T. Lee, C.-P. Huang, Y.-D. Lee, Biomimetic porous scaffolds made from poly (L-lactide)-g-chondroitin sulfate blend with poly (L-lactide) for cartilage tissue engineering, Biomacromolecules 7 (7) (2006) 2200–2209.
[151] K.-Y. Chang, L.-W. Cheng, G.-H. Ho, Y.-P. Huang, Y.-D. Lee, Fabrication and characterization of poly (γ-glutamic acid)-graft-chondroitin sulfate/polycaprolactone porous scaffolds for cartilage tissue engineering, Acta Biomater. 5 (6) (2009) 1937–1947.
[152] C.-H. Chang, H.-C. Liu, C.-C. Lin, C.-H. Chou, F.-H. Lin, Gelatin–chondroitin–hyaluronan tri-copolymer scaffold for cartilage tissue engineering, Biomaterials 24 (26) (2003) 4853–4858. http://dx.doi.org/10.1016/S0142-9612(03)00383-1.
[153] L.S. Nair, C.T. Laurencin, Biodegradable polymers as biomaterials. Prog. Polym. Sci. 32 (8-9) (2007) 762–798, doi:10.1016/j.progpolymsci.2007.05.017.
[154] G. Odian, Principles of Polymerization, fourth ed., Wiley-Interscience, New Jersey, 2004.
[155] N. Ashammakhi, P. Rokkanen, Absorbable polyglycolide devices in trauma and bone surgery, Biomaterials 18 (1) (1997) 3–9.
[156] P.B. Maurus, C.C. Kaeding, Bioabsorbable implant material review. Oper. Tech. Sports Med. 12 (3) (2004) 158–160, doi:10.1053/j.otsm.2004.07.015.
[157] C.A. Finch, Biomedical polymers: designed-to-degrade systems, in: S.W. Shalaby (Ed.), Polymer International, vol. 37, Hanser Publishers, Munich, Vienna, New York, 1994, p. 263.
[158] A.S. Koelling, N.J. Ballintyn, A. Salehi, D.J. Darden, M.E. Taylor, J. Varnavas, D.R. Melton, In vitro real-time aging and characterization of poly (L/D-lactic acid), in: Paper presented at the Biomedical Engineering Conference, 1997, Proceedings of the 1997 Sixteenth Southern, 4–6 April 1997, 1997.
[159] N. Ignjatovic, D. Uskokovic, Synthesis and application of hydroxyapatite/polylactide composite biomaterial, Appl. Surf. Sci. 238 (1) (2004) 314–319.
[160] K. Kesenci, L. Fambri, C. Migliaresi, E. Piskin, Preparation and properties of poly (L-lactide)/hydroxyapatite composites, J. Biomater. Sci. Polym. Ed. 11 (6) (2000) 617–632.
[161] J.C. Middleton, A.J. Tipton, Synthetic biodegradable polymers as orthopedic devices, Biomaterials 21 (23) (2000) 2335–2346.
[162] R. Zhang, P.X. Ma, Biomimetic polymer/apatite composite scaffolds for mineralized tissue engineering, Macromol. Biosci. 4 (2) (2004) 100–111.
[163] J.A. Cooper, H.H. Lu, F.K. Ko, J.W. Freeman, C.T. Laurencin, Fiber-based tissue-engineered scaffold for ligament replacement: design considerations and in vitro evaluation, Biomaterials 26 (13) (2005) 1523–1532.
[164] H.H. Lu, J.A. Cooper Jr., S. Manuel, J.W. Freeman, M.A. Attawia, F.K. Ko, C.T. Laurencin, Anterior cruciate ligament regeneration using braided biodegradable scaffolds: in vitro optimization studies, Biomaterials 26 (23) (2005) 4805–4816.
[165] R. Mehta, V. Kumar, H. Bhunia, S.N. Upadhyay, Synthesis of poly(lactic acid): a review. J. Macromol. Sci. C Polym. Rev. 45 (4) (2005) 325–349, doi:10.1080/15321790500304148.
[166] S. Kumbar, R. James, S. Nukavarapu, C. Laurencin, Electrospun nanofiber scaffolds: engineering soft tissues, Biomed. Mater. 3 (2008) 034002.
[167] S.G. Kumbar, S.P. Nukavarapu, R. James, L.S. Nair, C.T. Laurencin, Electrospun poly(lactic acid-co-glycolic acid) scaffolds for skin tissue engineering, Biomaterials 29 (30) (2008) 4100–4107. http://dx.doi.org/10.1016/j.biomaterials.2008.06.028.
[168] X. Shi, Y. Wang, L. Ren, Y. Gong, D.-A. Wang, Enhancing alendronate release from a novel PLGA/hydroxyapatite microspheric system for bone repairing applications. Pharm. Res. 26 (2) (2009) 422–430, doi:10.1007/s11095-008-9759-0.
[169] J.R. Shelton, J.B. Lando, D.E. Agostini, Synthesis and characterization of poly(β-hydroxybutyrate). J. Polym. Sci. B Polym. Lett. 9 (3) (1971) 173–178, doi:10.1002/pol.1971.110090303.
[170] E.A. Dawes, Polyhydroxybutyrate: an intriguing biopolymer, Biosci. Rep. 8 (6) (1988) 537–547.

[171] C.W. Pouton, S. Akhtar, Biosynthetic polyhydroxyalkanoates and their potential in drug delivery, Adv. Drug Deliv. Rev. 18 (2) (1996) 133–162. http://dx.doi.org/10.1016/0169-409X(95)00092-L.

[172] L.L.W.a.N, Davis, Biopolymers of activated sludge, Environ. Sci. Technol, 6 (2) (1972) 161–164.

[173] A.L. Pundsack, T.L. Bluhm, Technique for determining unit-cell constants of polyhydroxyvalerate using electron diffraction, J. Mater. Sci. 16 (2) (1981) 545–547.

[174] J. Knowles, G. Hastings, In vitro degradation of a polyhydroxybutyrate/polyhydroxyvalerate copolymer, J. Mater. Sci. Mater. Med. 3 (5) (1992) 352–358.

[175] L.J. Chen, M. Wang, Production and evaluation of biodegradable composites based on PHB–PHV copolymer, Biomaterials 23 (13) (2002) 2631–2639. http://dx.doi.org/10.1016/S0142-9612(01)00394-5.

[176] N. Sultana, M. Wang, Fabrication of HA/PHBV composite scaffolds through the emulsion freezing/freeze-drying process and characterisation of the scaffolds, J. Mater. Sci. Mater. Med. 19 (7) (2008) 2555–2561.

[177] W. Huang, X. Shi, L. Ren, C. Du, Y. Wang, PHBV microspheres—PLGA matrix composite scaffold for bone tissue engineering. Biomaterials 31 (15) (2010) 4278–4285, doi:10.1016/j.biomaterials.2010.01.059.

[178] W.S. Pietrzak, M.L. Verstynen, D.R. Sarver, Bioabsorbable fixation devices: status for the craniomaxillofacial surgeon, J. Craniofac. Surg. 8 (2) (1997) 92–96.

[179] L.S. Nair, C.T. Laurencin, Polymers as biomaterials for tissue engineering and controlled drug delivery, in: Tissue Engineering I, Springer, Berlin, Heidelberg, 2006, pp. 47–90.

[180] H. Yoshimoto, Y.M. Shin, H. Terai, J.P. Vacanti, A biodegradable nanofiber scaffold by electrospinning and its potential for bone tissue engineering, Biomaterials 24 (12) (2003) 2077–2082. http://dx.doi.org/10.1016/S0142-9612(02)00635-X.

[181] N.T. Hiep, B.-T. Lee, Electro-spinning of PLGA/PCL blends for tissue engineering and their biocompatibility, J. Mater. Sci. Mater. Med. 21 (6) (2010) 1969–1978.

[182] H. Kweon, M.K. Yoo, I.K. Park, T.H. Kim, H.C. Lee, H.-S. Lee, C.-S. Cho, et al., A novel degradable polycaprolactone networks for tissue engineering, Biomaterials 24 (5) (2003) 801–808.

[183] R. De Santis, A. Gloria, T. Russo, U. D'Amora, V. D'Antò, F. Bollino, L. Ambrosio, et al., Advanced composites for hard-tissue engineering based on PCL/organic–inorganic hybrid fillers: from the design of 2D substrates to 3D rapid prototyped scaffolds. Polymer Compos. 34 (2013) 1413–1417, doi:10.1002/pc.22446.

[184] G.T. Howard, Biodegradation of polyurethane: a review, Int. Biodeterior. Biodegrad. 49 (4) (2002) 245–252. http://dx.doi.org/10.1016/S0964-8305(02)00051-3.

[185] R.J. Zdrahala, I.J. Zdrahala, Biomedical applications of polyurethanes: a review of past promises, present realities, and a vibrant future, J. Biomater. Appl. 14 (1) (1999) 67–90.

[186] G. Ciardelli, A. Rechichi, S. Sartori, M. D'Acunto, A. Caporale, E. Peggion, P. Giusti, et al., Bioactive polyurethanes in clinical applications. Polym. Adv. Technol. 17 (9-10) (2006) 786–789, doi:10.1002/pat.781.

[187] S.G. Kumbar, S. Bhattacharyya, S.P. Nukavarapu, Y.M. Khan, L.S. Nair, C.T. Laurencin, In vitro and in vivo characterization of biodegradable poly (organophosphazenes) for biomedical applications, J. Inorg. Organomet. Polym. Mater. 16 (4) (2006) 365–385.

[188] H.R. Allcock, H. Allcock, Chemistry and Applications of Polyphosphazenes, Wiley-Interscience, Hoboken, 2003.

[189] H.R. Allcock, Expanding options in polyphosphazene biomedical research, in: Polyphosphazenes for Biomedical Applications, John Wiley & Sons, Inc, New Jersey, 2008, pp. 15–43.

[190] A.K. Andrianov, Polyphosphazene vaccine delivery vehicles: state of development and perspectives, in: Polyphosphazenes for Biomedical Applications, John Wiley & Sons, Inc, New Jersey, 2008, pp. 45–63.

[191] K. Johansen, J. Hinkula, C. Istrate, E. Johansson, D. Poncet, L. Svensson, Polyphosphazenes as adjuvants for inactivated and subunit rotavirus vaccines in adult and infant mice, in: Polyphosphazenes for Biomedical Applications, John Wiley & Sons, Inc, New Jersey, 2008, pp. 85–99.

[192] M. Deng, S.G. Kumbar, Y. Wan, U.S. Toti, H.R. Allcock, C.T. Laurencin, Polyphosphazene polymers for tissue engineering: an analysis of material synthesis, characterization and applications, Soft Matter 6 (14) (2010) 3119–3132.

[193] S.P. Nukavarapu, S.G. Kumbar, J.L. Brown, N.R. Krogman, A.L. Weikel, M.D. Hindenlang, C.T. Laurencin, et al., Polyphosphazene/nano-hydroxyapatite composite microsphere scaffolds for bone tissue engineering. Biomacromolecules 9 (7) (2008) 1818–1825, doi:10.1021/bm800031t.

[194] N. Kumar, R.S. Langer, A.J. Domb, Polyanhydrides: an overview, Adv. Drug Deliv. Rev. 54 (7) (2002) 889–910. http://dx.doi.org/10.1016/S0169-409X(02)00050-9.

[195] J. Kohn, R. Langer, Bioresorbable and bioerodible materials, Biomaterials Science: An Introduction to Materials in Medicine, Academic Press, San Diego, 1996, pp. 64–72.

[196] S. Wang, L. Lu, M.J. Yaszemski, Bone-tissue-engineering material poly(propylene fumarate): correlation between molecular weight, chain dimensions, and physical properties. Biomacromolecules 7 (6) (2006) 1976–1982, doi:10.1021/bm060096a.

[197] B.D. Ulery, L.S. Nair, C.T. Laurencin, Biomedical applications of biodegradable polymers. J. Polymer Sci. B Polym. Phys. 49 (12) (2011) 832–864, doi:10.1002/polb.22259.

[198] M.C. Hacker, A. Haesslein, H. Ueda, W.J. Foster, C.A. Garcia, D.M. Ammon, A.G. Mikos, et al., Biodegradable fumarate-based drug-delivery systems for ophthalmic applications. J. Biomed. Mater. Res. A 88 (4) (2009) 976–989, doi:10.1002/jbm.a.31942.

[199] K.W. Lee, S. Wang, M.J. Yaszemski, L. Lu, Physical properties and cellular responses to crosslinkable poly(propylene fumarate)/hydroxyapatite nanocomposites. Biomaterials 29 (19) (2008) 2839–2848, doi:10.1016/j.biomaterials.2008.03.030.

[200] A.S. Mistry, S.H. Cheng, T. Yeh, E. Christenson, J.A. Jansen, A.G. Mikos, Fabrication and in vitro degradation of porous fumarate-based polymer/alumoxane nanocomposite scaffolds for bone tissue engineering. J. Biomed. Mater. Res. A 89 (1) (2009) 68–79, doi:10.1002/jbm.a.32010.

[201] A.S. Mistry, Q.P. Pham, C. Schouten, T. Yeh, E.M. Christenson, A.G. Mikos, J.A. Jansen, In vivo bone biocompatibility and degradation of porous fumarate-based polymer/alumoxane nanocomposites for bone tissue engineering. J. Biomed. Mater. Res. A 92 (2) (2010) 451–462, doi:10.1002/jbm.a.32371.

[202] S. Danti, D. D'Alessandro, A. Pietrabissa, M. Petrini, S. Berrettini, Development of tissue-engineered substitutes of the ear ossicles: PORP-shaped poly(propylene fumarate)-based scaffolds cultured with human mesenchymal stromal cells. J. Biomed. Mater. Res. A 92 (4) (2010) 1343–1356, doi:10.1002/jbm.a.32447.

[203] C. Nguyen, S. Young, J.D. Kretlow, A.G. Mikos, M. Wong, Surface characteristics of biomaterials used for space maintenance in a mandibular defect: a pilot animal study. J. Oral Maxillofac. Surg. 69 (1) (2011) 11–18, doi:10.1016/j.joms.2010.02.026.

[204] S. Kizilel, T.I. Ergenc, Recent advances in the modeling of PEG hydrogel membranes for biomedical applications, Biomedical Engineering, Trends in Materials Science, InTech, New York, 2011.

[205] K.S. Soppimath, T.M. Aminabhavi, A.M. Dave, S.G. Kumbar, W.E. Rudzinski, Stimulus-responsive "Smart" hydrogels as novel drug delivery systems, Drug Dev. Ind. Pharm. 28 (8) (2002) 957.

[206] J.M. Harris, Poly (Ethylene Glycol) Chemistry: Biotechnical and Biomedical Applications, Plenum Publishing Corporation, New York, 1992.

[207] H. Tian, Z. Tang, X. Zhuang, X. Chen, X. Jing, Biodegradable synthetic polymers: preparation, functionalization and biomedical application. Prog. Polym. Sci. 37 (2) (2012) 237–280, doi:10.1016/j.progpolymsci.2011.06.004.

[208] J.A. Burdick, K.S. Anseth, Photoencapsulation of osteoblasts in injectable RGD-modified PEG hydrogels for bone tissue engineering, Biomaterials 23 (22) (2002) 4315–4323. http://dx.doi.org/10.1016/S0142-9612(02)00176-X.

[209] M.P. Lutolf, J.A. Hubbell, Synthetic biomaterials as instructive extracellular microenvironments for morphogenesis in tissue engineering. Nat. Biotechnol. 23 (1) (2005) 47–55, doi:10.1038/nbt1055.

[210] J.K. Tessmar, A.M. Göpferich, Customized PEG-derived copolymers for tissue-engineering applications. Macromol. Biosci. 7 (1) (2007) 23–39, doi:10.1002/mabi.200600096.

[211] S. Einmahl, S. Capancioni, K. Schwach-Abdellaoui, M. Moeller, F. Behar-Cohen, R. Gurny, Therapeutic applications of viscous and injectable poly(ortho esters), Adv. Drug Deliv. Rev. 53 (1) (2001) 45–73. http://dx.doi.org/10.1016/S0169-409X(01)00220-4.

[212] Z. Zhao, J. Wang, H.-Q. Mao, K.W. Leong, Polyphosphoesters in drug and gene delivery, Adv. Drug Deliv. Rev. 55 (4) (2003) 483–499. http://dx.doi.org/10.1016/S0169-409X(03)00040-1.

[213] Y.C. Wang, Y.Y. Yuan, J.Z. Du, X.Z. Yang, J. Wang, Recent progress in polyphosphoesters: from controlled synthesis to biomedical applications. Macromol. Biosci. 9 (12) (2009) 1154–1164, doi:10.1002/mabi.200900253.

[214] A. Rodriguez-Galan, L. Franco, J. Puiggali, Degradable poly(ester amide)s for biomedical applications, Polymers 3 (1) (2010) 65–99.

[215] G. John, S. Tsuda, M. Morita, Synthesis and modification of new biodegradable copolymers: serine/glycolic acid based copolymers, J. Polymer Sci. A Polymer Chem. 35 (10) (1997) 1901–1907.

[216] X. Pang, C.-C. Chu, Synthesis, characterization and biodegradation of poly (ester amide)s based hydrogels, Polymer 51 (18) (2010) 4200–4210.

[217] C.J. Bettinger, J.P. Bruggeman, J.T. Borenstein, R.S. Langer, Amino alcohol-based degradable poly(ester amide) elastomers, Biomaterials 29 (15) (2008) 2315–2325. http://dx.doi.org/10.1016/j.biomaterials.2008.01.029.

[218] P. Karimi, A.S. Rizkalla, K. Mequanint, Versatile biodegradable poly (ester amide)s derived from α-amino acids for vascular tissue engineering, Materials 3 (4) (2010) 2346–2368.

[219] J.A. Horwitz, K.M. Shum, J.C. Bodle, M. Deng, C.C. Chu, C.A. Reinhart-King, Biological performance of biodegradable amino acid-based poly (ester amide)s: endothelial cell adhesion and inflammation in vitro, J. Biomed. Mater. Res. A 95 (2) (2010) 371–380.

Chapter 22

Polymeric Biomaterials for Medical Diagnostics in the Central Nervous System

Yuan Yin, Dina Rassias, Anjana Jain

Biomedical Engineering Department, Worcester Polytechnic Institute, Worcester, Massachusetts, USA

Chapter Outline

22.1 INTRODUCTION

The major components for diagnostics entail utilization of sensing and imaging systems. Polymeric biomaterials can aid in enhancing image quality, increasing sensitivity and specificity, preventing degradation, and decreasing toxicity [1–4]. Polymeric biomaterials are used for several types of nanocarrier systems including solid nanoparticles, open-core particles, dendrimer-based nanoparticles, lipid-based systems, metallic crystalline lattices, and polymer shells, all having the capacity to be utilized in conjunction with diagnostic modalities. These polymeric biomaterials can be used for image enhancement, biosensing, and reduction of negative bioresponse instigated by the introduction of foreign agents. In this chapter, a comprehensive review of the current state of research in polymeric biomaterials for use to enhance current medical diagnostic systems within the CNS will be presented.

Despite the many technological advances made with diagnostic systems, several challenges still exist with diagnostic delivery, resolution, and targeting necessary to obtain vital information for implementing appropriate treatment. Many of the difficulties stem from physiology of the central nervous system (CNS), which is the body's main processing unit composed of the brain and spinal cord. It relays information to and from all parts of the body through a complex network of neurons and supports cells that make up the peripheral nervous system (PNS). Both the brain and spinal cord are protected by skeletal structures, meningeal layer, and the cerebrospinal fluid. In the vasculature that surrounds the CNS, tight junctions between capillary endothelial cells form a semipermeable membrane referred to as the blood-brain barrier (BBB) and the blood-spinal cord barrier [5]. These barriers prevent certain components of the blood or potentially harmful foreign molecules from entering the CNS while allowing passage of few essential molecules through active transport [6,7]. The barrier strictly regulates the passage of blood cells into nerve tissue of the CNS; particularly, the transendothelial migration of immune cells across the barriers is highly regulated for the development of an immune-privileged environment in the CNS [8,9]. While this highly evolved system is beneficial in preventing the entry of pathogens and macromolecules into the CNS, it also restricts the access of diagnostic agents and limits the reach of various standard diagnostic systems.

Along with a better understanding of biological processes involved in CNS disorders, many advances have been made in material science in combination with polymeric biomaterials for improving diagnostic modalities. Nano-sized carriers are made up of a variety of synthetic and natural polymers, which have been extensively studied for their multifunctionality and for their ability to extravasate to regions that are deep in the CNS. With the ability to

Natural and Synthetic Biomedical Polymers

incorporate multiple agents for diagnostics and therapeutics, they can be targeted and modified for specific pathologies, greatly improving localized delivery and reducing peripheral side effects [10]. Polymer choice and delivery agent synthesis vary widely depending on pathological applications within the CNS, but it is also critical to take into consideration the chemical composition of polymeric biomaterials and their interaction with biological systems.

22.2 CURRENT STANDARDS FOR MEDICAL DIAGNOSTICS IN THE CNS

Researchers have utilized and developed a variety of diagnostic standards for use with diseases and injuries within the CNS. Imaging and monitoring systems like magnetic resonance imaging (MRI), computed tomography (CT), positron emission tomography (PET), single-photon emission CT (SPECT), cerebral microdialysis, and implantable biosensors have all dramatically enhanced the understanding of CNS disorders, specifically in tailoring patient-specific therapeutic management [11–18]. These types of diagnostic modalities have led to numerous advancements in understanding CNS disorders including traumatic brain injury (TBI), spinal cord injury (SCI), CNS cancers, Alzheimer's disease (AD), Parkinson's diseases (PDs), and numerous other CNS injuries and diseases. However, these techniques still remain limited in resolution, degree of invasiveness, toxicity, and lack of targeting ability [19,20]. For example, MRI offers excellent image contrast of soft tissue with excellent resolution, but without the use of contrast-enhancing agents, extravasation into the brain parenchyma across the BBB remains difficult and free chelating agents subject the delicate cellular population of the CNS to an unnecessary high level of cytotoxicity [21–23]. CT offers excellent diagnostic capabilities for large-scale injuries including that of hematomas and skull fractures but remains limited in identifying characteristics associated with other CNS disorders like inflammation and tracking secondary injury [24,25]. Both PET and SPECT have instead of valuable applications for CNS disorders; with the ability to track blood flow, metabolic activity, and oxygen diffusion, these imaging modalities have offered an alternative method for diagnostics, but with the use of radioactive isotopes and gamma irradiation, cytotoxicity remains a critical issue of importance especially due to the delicate nature of the CNS [26–31]. Implantable monitoring devices have been at the forefront of recent research; with the ability to manufacture biosensors at the nanoscale, these systems offer real-time data monitoring and critical information often unattainable by simple imaging diagnostics [32]. However, the placement for these biosensors and microcerebral dialysis catheters to ensure accurate recordings makes these devices highly invasive and leads to more disadvantages than advantages. Implantable systems are also subjected to the foreign body response, otherwise known as glial scarring, and can lead to improper or distorted signal collection [32]. As a result, there is an increased need for more effective and safer alternatives to aid standard imaging and monitoring systems.

By utilizing polymeric biomaterials in conjunction with current generation diagnostic systems, researchers aim to reduce many of the drawbacks and limitations apparent in these systems. For example, the incorporation of contrast agents into polymeric complexes can further reduce toxicity by requiring less chelating ions, to achieve the same imaging enhancement characteristics of free contrast agent. Polymeric coatings on electrodes and implants extend the usable life of the device and prevent immune response at the implantation site, further reducing damage caused by the invasive implantation of biosensors. Traditional and current diagnostic systems have offered significant insight in understanding various diseases, injuries, and disorders within the CNS but still contain drawbacks and limitations.

The limited regenerative capabilities of neurons within the CNS highlight the need for efficient diagnostic systems capable of identifying areas of injury or diseases for treatment before irreversible damage occurs. The fragile state of the CNS also highlights the need for polymeric-incorporated diagnostic systems that will not further perpetuate damage to the site of injury or disease. Also, time between diagnosis and treatment can have a critical impact on recovery. Recently, a field of study termed "theranostics" combines therapeutics with diagnostic tools, providing a more immediate and personalized treatment strategy for various diseases and injuries. Especially in the CNS, not only it is critical to acquire immediate and accurate diagnostic information, but it is also crucial to employ a therapeutic regiment that is capable of protecting or destroying a cellular population of the CNS. Through the development of polymeric biomaterials, researchers have been able to further advance traditional diagnostic systems with direct applications within the CNS.

22.3 THE CHALLENGE OF DIAGNOSTICS IN THE CNS

As stated previously, the CNS is an immune-privileged zone where influx and efflux of compounds are highly regulated. The BBB plays an important role in CNS diseases, leading to significant neurodegenerative characteristics, often usually accompanied by neuroinflammation or neoplastic growth [10,33]. The inability to directly access the CNS makes imaging and treatment very challenging. Diagnostic and therapeutic strategies are faced with the difficulty of transporting contrast enhancements, genetic material, or therapeutics across the BBB. Numerous neurological diseases can cause dysfunction of the BBB leading to increased permeability of substances in systemic circulation [9]. While this phenomenon partly contributes

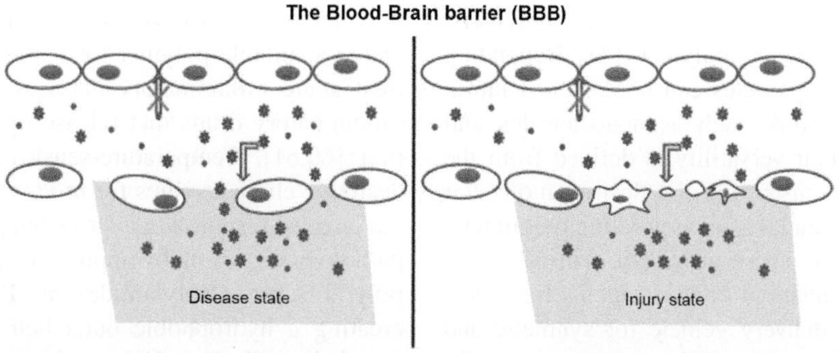

FIGURE 22.1 Breaches in the blood-brain barrier (BBB). In various CNS diseases, the BBB becomes leaky to macromolecules, and in CNS injuries, destruction of the endothelial lining creates permeability. By taking advantage of these characteristics, extravasation by nanoparticles, that is, liposomes, dendrimers, polymer-based diagnostics systems, can reach deep within the CNS. (For color version of this figure, the reader is referred to the online version of this chapter.) Figure adapted from Ref. [34].

to pathologies of CNS disorders, it can also be used as an advantage to develop nanoscale polymer systems for diagnostics and drug delivery modalities to take advantage of these physiological changes resulting from the disease or injury (Figure 22.1).

Extensive neuronal damage or neuronal loss occurs in neurodegenerative diseases such as PD or AD. Although these diseases differ, their pathology shares similarity with the involvement of oxidative stress, abnormal protein aggregation, and abnormal increase in neuronal apoptosis [35]. The accumulation of β-amyloid proteins is characteristic of AD, and α-synuclein aggregation and the presence of Lewy bodies are characteristics of PD [36,37]. Neurotransmitters, such as glutamate, mediate signal transduction in the CNS but in excess can result in neurotoxicity [38]. There is recent evidence that the integrity of the BBB is altered in AD and the passage of amyloid-beta peptides across the BBB accounts for this alteration [39]. Similarly, dysfunction of the BBB may be involved in the disease progression of PD [40]. Multiple sclerosis (MS) is another neurodegenerative disease consisting of characteristic chronic neural inflammation as a result of an autoimmune disorder where the body creates antibodies against myelin, the protective sheath around Schwann cells [41]. This chronic inflammation results from a host immune response to what the body considers an infection, and without infectious agents present, the body attacks this structural support component of the PNS resulting in demyelination of CNS neurons [42]. Similar to AD and PD, abnormalities of the BBB are evident and can lead to transendothelial migration of immune cells in brains affected by MS and the release of inflammatory cytokines and chemokines [43].

Neoplasms of the CNS, the majority being brain tumors, include benign brain tumors and malignant brain tumors ranging from low grade to invasive, metastatic high grade [44]. Benign intracranial tumors of the CNS, due to their potential adverse side effects, must follow the same course of treatment as malignant tumors [45]. Regardless of the level of severity, surgical removal or total resection remains a challenge to achieve, because it is difficult to distinguish the borders between normal and malignant tissue of the brain and precision is of great importance in cortical regions [46]. Also, delivery of chemotherapies, genetic material, and contrast agents for MRI across the BBB can be a challenge as in all other neurological disorders but is surmountable when the integrity of the BBB is disrupted [10]. Neoplasms of the brain have been shown to cause dysfunction of the BBB leading to increased permeability or gaps in the endothelial lining referred to as a leaky vasculature.

Injuries like TBI and SCI are a few of the leading causes of paralysis, morbidity, and mortality in the United States. Current technology is capable of diagnosing large-scale trauma including bone fractures, hematoma, and significant tissue damage, but not secondary damage; symptoms may not manifest until weeks, months, or even years after the original insult. With the misregulated homeostasis and extensive scarring occurring around the injury sites, these abnormalities further perpetuate cellular loss and prevent the minor regeneration the CNS is capable of. Through the use of polymer-based agents, it is possible to target specific cell populations and further identify areas of damage, which normally cannot be determined by conventional diagnostic systems.

22.4 POLYMERIC NANOPARTICLES

Polymeric nanoparticles have been gaining interest for the diagnosis and treatment of neurological disorders. With average size of 100 nm or smaller, the polymeric biomaterials can be used for optimized delivery across a compromised BBB, and they have large surface-to-volume ratio providing greater surface area for improved absorption [10]. Nanoparticles can be made from natural polymers derived from plants (i.e., agarose, alginate, and chitosan) or animals (i.e., collagen, fibrin, and albumin) and synthetic polymers, including, but not limited to, polyesters, polyamides, phosphorous-based polymers, silicone, acrylic polymers, and numerous combinations of copolymers [47]. Depending on their application, nanoparticles can be designed into a continuous polymer network such as nanocapsules and polymer shells [48]. Their versatility is derived from the ability to tailor the chemistry for the desired molecular composition with the potential of incorporating biomimetic or bioresponsive features. These polymeric particles offer the ability to develop enhanced contrast agents for image enhancement and as a delivery vehicle for synthetic and biological agents including DNA, RNA, and proteins [49].

Polyethylene glycol (PEG) is one of the first market-approved polymers used for protein conjugation and plays a pivotal role in diagnostics [50]. PEG has been used extensively as hydrophilic building blocks of copolymers for a wider range of particles like chelates, radioisotopes, gene therapy, proteins, and drugs being delivered [51]. PEG-based polymers fall into a class of nanoparticles consisting of both hydrophilic and hydrophobic components, creating an amphiphilic block copolymer. When exposed to aqueous solutions like plasma, these types of amphiphilic copolymers exhibit high stability with the ability to orient hydrophobic and hydrophilic components based on environment [52]. Poly(L-lysine) (PLL) and polyethyleneimine are a few other examples of polymers used for nanoparticle delivery of siRNA [53]. PLL chain polymers have been synthesized to chelating ions like diethylenetriaminepentaacetic acid (DTPA) for modification of metal atom-loaded diagnostic monoclonal antibodies to increase image enhancement at shorter time intervals [54]. Block copolymers of methoxy poly(ethylene glycol) and iodine-substituted PLL offered excellent blood pool CT imaging enhancement by allowing stable circulation of highly loaded nanoparticles [54]. Polymer choices can also affect *in vivo* responses, both immune and cellular, compatibility, and toxicity. One special feature of copolymers is the ability to tailor degradation properties and tune systems depending on application and need. Polylactic acid (PLA) and poly(lactic-*co*-glycolic acid) (PLGA) are some of the most widely used polymers for nanocarrier drug delivery due to their biocompatibility and tunable degradation properties [55,56]. Conjugation of PEG to polymeric nanoparticles or coatings to the surface of polymeric nanoparticles can prolong circulation time and reduce mononuclear phagocytic uptake [57,58].

Nanoparticles made up of smart polymers that are responsive to physical stimuli such as pH, temperature, and light are also highly beneficial for use in diagnostics. This type of feedback from a polymer is a very attractive feature for diagnostic strategies. One such example is poly(*N*,*N*-dimethylaminoethyl methacrylate (DMAEMA)/2-hydroxyethyl methacrylate (HEMA)) nanoparticles that can be actively triggered to release their payload when exposed to changes in pH [59]. Poly(1,4-phenyleneacetone dimethylene ketal) is another pH-sensitive, biodegradable polymer that hydrolyzes rapidly in response to acidic pH and is a suitable polymer for nanocarriers that are used to treat acute inflammatory diseases by incorporating anti-inflammatory drugs that release upon hydrolytic degradation [60,61]. Temperature-sensitive polymers also have desirable characteristics for *in vivo* diagnostic applications that involve response to internal temperature changes due to pathology or externally applied temperature. Incorporating poly(N isopropylacrylamide) into PLA nanoparticles, thus creating a hydrophobic outershell and a hydrophilic inner shell, will allow leakage from the outer shell when it is heated above phase transition. Therefore, becoming an advantageous dual-mode MRI-tracked anti-cancer drug carrier [62]. Nanoparticles can be targeted to a specific cell population by conjugating recognition molecules that have specificity for cell surface receptors that may be overexpressed on that cell type. Targeting using aptamers, antibodies, ligand, or recognition peptides helps to provide more efficient localized therapeutic delivery [63]. For example, targeted delivery of contrast agent and chemotherapies using nanoparticle systems is a trend in cancer therapeutics that results in higher doses to tumor cells and reduced cytotoxicity to healthy cells [64].

Specially designed nanoparticles containing ferric compounds like iron oxides and gold nanoparticles can be tracked utilizing MRI and CT, and through the addition of a focused magnetic field, these ferric nanoparticles can be used in heat ablation therapy [65]. Although iron oxide has excellent contrast-enhancing features, their toxicity has led to focus on encapsulation in nanoparticles synthesized with biocompatible and biodegradable polymers such as PLA or PLGA. Another highly beneficial capability of polymer-based diagnostics systems is the ability to track cellular movement and migration. One large area of study in the CNS is tracking regeneration of implanted stem cells. Utilizing magnetically coupled contrast agents allows for on-demand long-term tracking of cell transplant therapies through MRI [66]. Superparamagnetic iron oxides (SPIO)-conjugated nanoparticles known as magnetodendrimers are capable of labeling neural stem cells implanted into rodent models with no adverse effects of survival, growth, or differentiation characteristics [67].

Recent progress in metallic nanoparticles, in particular gold nanoparticles, has also led to enhanced optical properties including light scattering, luminescence imaging, and surface-enhanced spectroscopy [68]. Nanoparticles can become light responsive by conjugating a photoreactive moiety to a pH- or temperature-responsive polymeric material allowing for tracking and a secondary capacity to create active release capabilities [69]. Technologies involving lasers and optics along with biomaterials that can temporarily alter shape and configuration in response to stimuli, such as different wavelengths of UV light, can have very practical biological applications. Azobenzene-containing polymers

TABLE 22.1 Polymeric Nanocarriers for CNS Diagnostics

Polymeric Nanocarrier	CNS Application	References
Polysorbate 80 (PS80)-coated poly(*n*-butyl cyanoacrylate) nanoparticles (PBCA-NP)	Transient disruption of the BBB	[82]
Long-circulating PEGylated polycyanoacrylate (PEG-PHDCA) nanoparticles	Drug carrier for brain delivery and has been shown to modify the permeability of the BBB	[83]
Fluorescent polystyrene nanospheres	Monitor BBB permeability following cerebral ischemia using implanted microdialysis probe	[84]
Gold nanoparticles (AuNPs)	Spectroscopic advantages of visualization of AuNPs when bound to diseased cells/tissues	[85]
Magnetofluorescent particles	Bimodal imaging of target system	[86]
Fluorescent silica	Can be loaded with fluorescent dye for identification of target cells/tissues	[87]
Semiconductor quantum dots	Stable high-intensity fluorescent emission characteristics for imaging	[88]
PLGA and poly(e-caprolactone) (PCL)	Attenuate secondary neuronal death and scarring in acute TBI	[89]
Poly(L-lactic acid) (PLA) nanoparticles with 2-methacryloyloxyethyl phosphorylcholine (MPC), *n*-butyl methacrylate, and *p*-nitrophenyl ester-bearing methacrylate	Enzymes can be immobilized on the nanoparticle surface for use with a microdialysis system with electrode for a diagnostic system capable of detecting a target molecule *in situ*	[90]

have been shown to contract in response to light [70]. Thermoresponsive polyacrylamides can be synthesized with photosensitive side groups, that is, azobenzene, salicylideneaniline, or fulgimide, to formulate a copolymer that is both light- and thermal-responsive [71]. By altering ferritin, a well-known iron-storage protein, researchers have been able to create H-chain ferritin cages containing 24 subunits that are capable of targeting tumor tissue [72]. With catalyzing peroxidase substrates incorporated into the core, once bound to the oncological tissue, color change can be utilized as a diagnostic standard with this natural polymer system. One other type of polymeric system for diagnostics growing in research includes protein cage-like viral nanoparticles. Cowpea mosaic virus assembles much like the protein ferritin, and with the use of RT-PCR and fluorescence imaging, biodistribution of viral nanoparticles can be tracked through these fluorescently conjugated viral particles [73].

Aside from nanoparticles, biomaterial polymers are key components of biocompatible hydrogels and scaffolds for CNS implantation. Utilizing these types of polymers, it is possible to provide mechanical and chemical cues for reconstruction and regeneration. Some of the most common molecules used as chemoattractants include extracellular matrix proteins such as collagen and laminin, growth factors including neurotrophins like brain-derived neurotrophic factor, and other bioactive molecules like Rho GTPases [74–76]. These polymeric hydrogel systems have also allowed researchers to design multiplexed detection systems for a variety of biological agents. Porous hydrogels synthesized from polyacrylamide have been used as a metabolite-sensing system. Utilizing transcriptional regulator protein HucR, which readily dissociates in proportion to urate concentration, researchers have developed an *in vivo* platform to access of metabolites and other biological agents at the local implant site [77]. Hydrogel systems have also been used to take benchtop systems like gene, protein, and aptamer detection *in vivo*. By conjugating active particles capable of sensing and binding to various agents, these hydrogel-embedded polymers with fluorescence can be used as diagnostic systems replacing systems like HPLC, Western blot, and PCR and further advancing newer developments like microfluidic devices [78–81]. A summary list of polymeric biomaterial-based systems and their application for CNS diagnostics is presented in Table 22.1.

22.5 LIPID-BASED NANOCARRIER DIAGNOSTIC SYSTEMS

Lipid-based nanoparticles come in a variety of shapes and sizes, offering the ability to encapsulate various agents, functionalize a variety of ligands for cellular targeting, and bypass the immune system in systemic circulation until they have reached their target site [91–95]. With improvements in natural and synthetic lipid synthesis, researchers are able to create liposomes, micelles, and even lipid microtubes by utilizing different combinations of lipids, allowing for the encapsulation of various

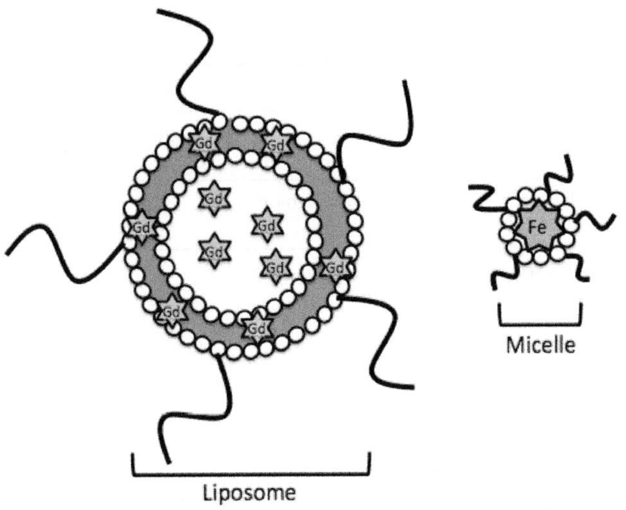

FIGURE 22.2 Schematic of paramagnetic and superparamagnetic lipid-based contrast agents for MRI. Encapsulated chelating agents allow for either T1 or T2 image enhancement and PEGylation allows for increased circulation times and functionalization for targeting capabilities. (For color version of this figure, the reader is referred to the online version of this chapter.)

agents including contrast agents to improve MRI imaging. By chemically altering the polar head group and fatty acid chains, it is possible to tune lipid-based carriers for specific applications. Traditionally, paramagnetic and superparamagnetic chelates have been used to increase T1 and T2 relaxivity times during MRI, respectively (Figure 22.2). T1-enhancing contrast agents are known as "hot spot" enhancers, displaying as areas of high intensity when localized like that of polymeric systems with gadolinium chelates. T2-enhancing contrast agents create "cold spots" in areas of aggregation like that of super iron oxide micelles [96]. FDA-approved contrast agents like gadolinium contrast agent Magnevist and super iron oxide contrast agent Lumirem have been used for many years, but other commercially available contrast agents have been removed from the market and further research halted due to safety concerns due to the toxic nature of the chelates [97]. Additionally, the lack of a targeting moeity on these contrast-enhancement chelates makes it extremely difficult to utilize contrast agents to pinpoint affected areas and they are often only used to increase resolution of vascularization or soft tissue near the injection site.

Liposomes and micelles have been well studied as contrast agents for use with MRI. Gadolinium chelates have been incorporated into the bilayer membrane and encapsulated within the aqueous core of the liposome as a T1-enhancing agent [98–100]. SPIOs have been encapsulated within micelles as a T2-enhancing agent [101,102]. Research into lipid-based delivery systems has led to numerous commercially available lipid chelate agents like 1,2-dimyristoyl-*sn*-glycero-3-phosphoethanolamine-*N*-DTPA (DSPE-Gd-DTPA) from Avanti Polar Lipids. Substantial research has indicated that by incorporating chelating agents encapsulated into lipid vesicles, systemic toxicity was significantly reduced compared to that of free contrast agent and also allows for a variety of targeting strategies to be employed [96,103,104].

For imaging applications in the CNS, the BBB remains one of the main hurdles for diagnostic aid delivery. Signal enhancement in MRI applications generally requires a relativity enhancement of, at minimum, $0.5\,s^{-1}$ to induce an observable change in imaging intensity [105]. This exemplifies the notion that for a successful diagnostic system to be employed in the CNS, delivery systems must be able to cross the BBB, accumulate at the optimum concentration at the site of interest, and preferably act on the tissue or site of interest, allowing for a targeted modality. Initial researcher developing targeted chelating agents opted to conjugate antibody ligands directly to Gd-DTPA contrast agents, but due to the low accumulation at target sites and the low sensitivity of clinical MRI systems, minimal image enhancement was achieved when targeting tumors and other diseases. Research into targeted delivery contrast agents has expanded into incorporating multiple chelates into single-carrier systems to increase concentration and therefore image enhancement characteristics. One example of this research utilized biotinylated anti-HER2 antibodies targeted for HER2 overexpressing breast cancer [106,107]. Utilizing gadolinium chelates functionalized to avidin, multiple chelates are able to bind with the anti-HER2 antibody and then subsequently bind to oncological cells within the tumor [107].

Lipid-based nanocarriers are capable of incorporating multiple chelates in one carrier, making it possible to achieve observable signal enhancement with a lower concentration of lipid carriers than with just free contrast agent [108]. In CNS cancers, the leaky vasculature allows for extravasation through the BBB at the tumor site of the 50-200 nm diameters of lipid carriers [109,110]. Additional antibody ligands like OX26, which targets the transferrin receptor, can also be functionalized to the surface of the lipid carriers to aid in receptor-mediated endocytosis across the BBB [111]. Lipid nanocarriers are capable of a wide plethora of modifications, and with the incorporation of PEG, long circulating nanocarriers are less susceptible to the mononuclear phagocyte system or otherwise known as the reticuloendothelial system [18,25]. These stealth nanocarrier diagnostic agents can also evade significant uptake into nontarget organs like the spleen, liver, and lungs [25]. Applications for MRI imaging contrast agents not only are limited to oncological activity, by varying the targeting moiety to neurological disorders like Alzheimer's, Parkinson's, and multiple sclerosis, but also can be readily diagnosed utilizing these lipid-based nanocarriers.

22.6 DENDRIMERS

Dendrimers are highly structured three-dimensional polymer compounds having branches that emanate from an inner core and vary in size with the number of branching layers. Made of repeating monomer units arranged in a spheroid structure, dendrimers offer a unique set of attributes for use in medical diagnostics. Dendrimers are capable of both encapsulating chemical compounds and containing numerous terminal branch sites for functionalization. By acting as polymeric cages, dendrimers have been widely used as transfection agents for gene therapy, delivery agents for therapeutics, and contrast agents, being highly tunable in size, surface chemistry, and encapsulation space.

A few different approaches are used in dendrimer synthesis, including divergent or convergent growth methods [112]. In the divergent growth method, each branch emanates from the inner core through a sequence of iterative covalent reactions [113]. The convergent growth method works in the opposite direction, from outer to inner, in which branching layers form through noncovalent self-assembly between (dendrimers) surface units and monomers resulting in a number of outer layers equal to the branching multiplicity [112,114,115]. The generation number (G_N) of a dendrimer begins with the core as the first generation, denoted G_0, and increases with each branching layer having double the number of surface groups with each generation [112]. The appeal of utilizing dendritic systems as a polymeric biomaterial to enhance diagnostics is primarily due to the multivalency of the system. The numerous binding sites on the outer shell and the ability to encapsulate compounds within the dendrimer itself provide several options for loading the system.

There are several different types of dendrimers used in diagnostics, including polypropylenimine (PPI) dendrimers, polylysine dendrimers, phosphorus dendrimers, and polyamidoamine (PAMAM) dendrimers. PAMAM dendrimers are one of the most widely used dendrimers that carry a positive surface charge and allow for surface modifications due to the availability of amine terminal groups. Some have been modified to have a neutral charge on the surface and a cationic charge on the inner part of the dendrimer [116]. The surface charge, along with their branched shape, and ability to enter the cell make them a desirable vehicle for delivery of negatively charged biological materials such as DNA or RNA for the modulation of gene expression [117,118]. Metal ions can also be incorporated during the aqueous synthesis phase of the dendrimers, and when metallic ions are incorporated, dendrimeric nanoparticles are created, which can be used in conjunction with a variety of *in vitro* and *in vivo* diagnostic systems. A widely studied use of PAMAM dendrimers is with the incorporation of gadolinium to be used as an MRI contrast agent. It was discovered that after synthesizing multiple generations of PAMAM dendrimers with Gd(III) DTPA, the sixth generation yielded a sixfold increase in the longitudinal relaxivity [119]. Generation-5 PAMAM dendrimers have also been used as dual-mode diagnostic agents, where gold

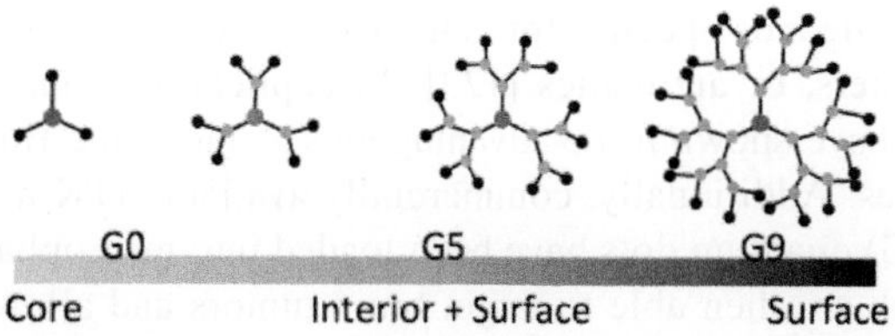

FIGURE 22.3 Schematic of sequential dendrimer generations. With increasing generations, availability of surface groups increase; however, high-generation dendrimers act more like solid particles incapable of encapsulation. Middle-generation dendrimers allow for both core encapsulation and surface modification.

and gadolinium have been incorporated into the dendrimeric system for image enhancement in MRI and CT [120]. PAMAM dendrimers with graphene cores functionalized with gold particles have been used as DNA hybridization assays that are capable of assessing for complementary double-stranded DNA, unhybridized DNA, and single-nucleotide polymorphisms [121]. Another notable dendrimer system includes the use of Gd(III)DTPA-terminated PPI dendrimers for blood pool angiography to detect vascular permeability through MRI [122] (Figure 22.3).

22.7 QUANTUM DOTS

Quantum dots, only a few nanometers in size, are fluorescent crystals that act as semiconductors between bulk materials and materials with electric properties [123]. They exhibit size-dependent absorption and emission in which excitation shifts to higher energy as size decreases [124]. Among the many applications of quantum dots, they are ideal for use in diagnostic imaging and immunofluorescence as they are resistant to bleaching and only a small number are needed to generate a strong signal [123]. The optical properties of quantum dots are directly affected by size, with the larger crystal lattice-like structures producing energy signatures in the red (low energy) spectrum. With the incorporation of thicker shells, like that of cadmium selenide (CdSe) quantum dots, it is possible to tune emission characteristics and stability for use as a diagnostic agent. Quantum dots are coated with different polymers depending on their intended application. ABC triblock polymer-coated CdSe-ZnS quantum dots consisting of a poly(butyl acrylate) segment, a poly(ethyl acrylate) segment, a polymethacrylic acid segment, and hydrocarbon side chains have been used successfully for an *in vivo* tumor-targeting platform for use with whole-body macroillumination system [125]. Quantum dots have also been used significantly as replacements for traditional dyes for *in vitro* imaging. These are 20-100 times brighter than traditional fluorophore, and with high extinction coefficient, these are far less susceptible to photobleaching [123]. They can also be functionalized to improve specificity for the cell population targeted using ligands specific for cell surface receptors, charged aptamers, or antibodies [123]. In applications, quantum dots have shown to be advantageous as inorganic fluorophores. Additionally, commercially available ITK amino (PEG) quantum dots have been loaded into macrophages, which are then able to target brain tumors and allow for subsequent *in vivo* imaging [126].

22.8 MICROBUBBLES

Gas-filled microbubbles with a thin shell, either lipid- or polymer-based, are often used for contrast enhancement in ultrasound to detect the trace of blood flow. These specialized nanoparticles made of polymers like PLGA or PEG and lipid-based constructs have the special ability of encapsulating bioactive gasses. They are on the micrometer-scale size, usually between 0.5 and 4 μm, which is small enough not to block tiny capillaries yet too large to pass through the endothelial lining of blood vessels allowing them to remain in circulation of the vasculature [127]. Microbubbles containing acoustic properties have the ability to reflect ultrasound waves from one substance to another, so when exposed to ultrasound frequency, an echo is generated through the reflected sound wave oscillations allowing for focal disruption of the particles and image enhancement capabilities [128]. This, along with the use of a contrast agent, can give a clear distinction between the blood and surrounding tissue, making it a very effective tool for imaging diagnostics in the vasculature [129]. Microbubbles can be made out of a variety of polymers with the ability to tune physical characteristics to generate image enhancement or allow for a variety of applications. Lipid- and polymer-coated microbubbles increase stability, allowing for systemic delivery while still allowing for release of bioactive agents when ultrasound frequencies are applied. What makes them attractive is that they provide real-time, noninvasive, radioactive-free imaging; they eliminate the need for a contrast agent; and they are cleared from the bloodstream rapidly when gas bubbles diffuse into the surrounding bloodstream [127,130]. Air-filled and polymer-shelled poly(vinyl alcohol) (PVA) microbubbles have been widely used as ultrasound contrast agents due to their unique polymer-based construction. Commercially available Definity® microbubble contrast agent composed of a lipid shell with encapsulated octafluoropropane (C_3F_8) gas has been shown to breach the BBB through the application of focused ultrasound (FUS) [131]. With the application of FUS, these and other types of microbubbles contract and expand, allowing for extravasation across the BBB to allow for image enhancement through ultrasound and payload delivery [132,133]. Another application of FUS is for targeted ablation of oncological tissue. Utilizing dual-mode MRI and ultrasound image enhancement-capable microbubbles, researchers are able to track microbubble delivery through MRI and utilize FUS to ablate tissue and delivery-encapsulated reagents. Gadolinium chelates that have been bound to DOTA within the lipid shells of microbubbles retain stability while allowing for MRI and FUS applications [134]. Ultrasound, being noninvasive and free of radiation, has promising potential for use with delivery of drugs or genes to treat CNS disorders, such as brain metastases or AD [135]. Microbubbles can rupture when exposed to higher ultrasound frequency, making them attractive for localized drug delivery [136]. They can also improve imaging capabilities to better locate vascular damage in neurodegenerative diseases or vascularization in tumors [130].

22.9 BIOSENSORS

One of the largest challenges of monitoring patients with CNS disorders is the inability to predict the devastating decline of neuronal degradation, which can begin at any time and progress unnoticeably. Patients need to be monitored continuously so that if complications were to arise, the appropriate actions may be taken immediately to minimize neuronal loss and possibly reverse it. For patients with severe CNS injuries, confined to the intensive care unit, this type of monitoring can be achieved with invasive cerebral microdialysis [137]. By delivering on-demand data of key analyte levels, linked with the prognosis of CNS injury patients, it is possible to treat accordingly and provide a detailed prognosis for the each patient.

The implantable biosensor is a recent nanotechnology-based platform that allows for minimally invasive monitoring of metabolite levels, pressure, gas exchange, electrical signaling, and more. Utilizing photoelectrical, piezoelectric, electrochemical, and other detection methods, biosensors are able to convert distinct biological elements into measurable and quantifiable data [138–141]. The development of the enzymatic biosensor, driven by diabetes research and initially designed for measuring glucose levels within circulation, has had a significant impact in neural monitoring for CNS injury patients [139]. Utilizing an oxidation-reduction reaction with a specific enzyme corresponding to the analyte, the biosensor is capable of collecting electrical signals proportional to the concentration of the specified analyte [138]. This type of enzymatic-based reaction and electrochemical detection enables continuous, on-demand monitoring of numerous factors like metabolites, oxygenation, pressure, pH, and temperature, which have been implicated in CNS injuries and diseases [142]. With further developments in the quantification of biological analytes, elements like receptors, nucleic acids, cells, and antibodies often found to be misregulated can be utilized as a neural monitoring modality.

Implantable biosensors have further gained value for use in the CNS by advancements in the field of wireless communication and further miniaturization. This has allowed for a wide array of available conformations unique for any implant site and application. Wireless communications also allow for a monitoring system that is not confined to a clinical setting and may allow individuals with only mild to moderate CNS injuries to resume daily activities while still offering continuous monitoring. A novel biosensor based on a lab-on-a-tube spiral design compacts multiple sensors, capable of measuring pH, oxygen, pressure, and other elements [143]. Utilizing concepts like these, it is possible to create multimodal monitoring biosensors for numerous important elements implicated in CNS injury. In addition, miniaturization of the device allows for implantation into areas near the injury site, while the high sensitivity allows sensors also to be implanted systemically, without eliciting a severe immunologic response.

Although advancements in material science and miniaturization have significantly reduced implant rejection, the susceptibility of biosensors to the normal immune response is still one of the most significant hurdles for recording accurate data. Anti-inflammatory coatings designed to reduce foreign body response while allowing for the normal function of the sensor have also seen significant advancements in the recent years [144,145]. Utilizing hydrogel coatings of appropriate thickness and permeability can effectively prevent biofouling: the aggregation of protein and immune cells that leads to a full foreign body response [146,147]. Coatings made up of hydrophilic polymers like PEG can prevent protein absorption, while other coatings consisting of agents like heparin can prevent leukocyte aggregation [144]. Coatings of lipid-based films onto implantable biosensors have also been tested to increase recording stability and reduce implantation site inflammation and biofouling. Poly(*N*,*N*-DMAEMA/2-HEMA) has also been used in biomimetic hydrogels for improved biocompatibility of microdisk electrode arrays for *in vivo* electrochemical biosensing that respond to changes in pH for the continuous remote monitoring of lactate and glucose [148]. Within the CNS, the normal inflammatory response leads to glial scarring at the implant site. This not only reduces sensor sensitivity but also impacts the healthy population of cells at the local site and creates further neuronal loss. The glial scar also prevents neuronal regrowth by physically and chemically preventing axons from rebridging the affected site (Figure 22.4).

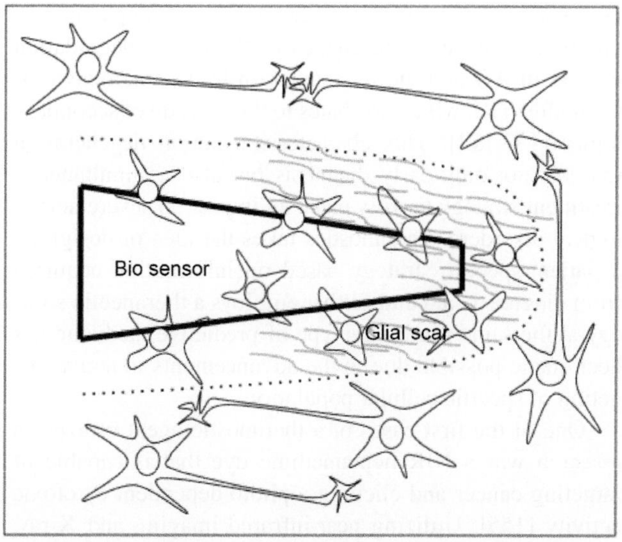

FIGURE 22.4 Schematic of glial scarring around implantable biosensor. Over time, reactive astrocytes for a physical fibrotic capsule around the implant, separating it from the tissue of interest, reducing sensor effectiveness and further perpetuating damage at the implant site.

22.10 TOXICITY

A major issue in many diagnostic systems that utilize various contrast-enhancing agents to increase resolution is local toxicity to an already injured area and systemic toxicity due to a lack of specificity. Toxicokinetics describes the rate that an agent will enter the body and, based on a variety of factors including adsorption, metabolism, and excretion, what happens to the agents once within body. Toxicity is an important factor to consider when designing polymeric materials and agents for use within the CNS, as they may cause acute and prolonged toxic effects to the fragile cells of the CNS. Though toxicity information on numerous biopolymer nanoparticles remains limited, several of the delivery systems listed throughout this chapter have had considerable systemic toxicity studies completed. Lipid-based systems are generally considered nontoxic because of their composition from natural lipid components found in most cell membranes [149]. Consequently, the act of bypassing the BBB itself may also contribute to the cytotoxicity of the agent. Poly(*n*-butyl cyanoacrylate) dextran polymers coated with polysorbate 80 (PBCA nanoparticles) have been well studied to extravasate through the BBB, but it is this exact action that also attributed to the significant cytotoxicity of this particle [150]. Dendrimer toxicity is largely dependent on generation and charge characteristics. PAMAM dendrimers at low generations exhibit lower toxicity at all dosage ranges [151]. Quantum dots have a wide variety of toxicity profiles due to the various surface coatings used by researchers. Generally, conditions including charge, size, dosage, biocompatibility, and stability all contribute to QDs' effects of cytotoxicity [152]. With almost all biomaterials used for medical diagnostics in the CNS, PEGylation has been employed to increase half-life, reduce toxicity, and allow for active targeting of a specific target cell population.

22.11 THERANOSTICS

An extremely disadvantageous trait of neurological disorders is the irreversible damage that occurs during the development and progression of CNS diseases and injuries.

For example, it is well known that amyloid-beta plaque formation significantly contributes to the neural degradation present in AD and the misregulation in the microenvironment after TBI, which attributes to the cascade of secondary injury [153,154]. This characteristic neural degeneration requires not only early diagnosis but also a simultaneous treatment strategy to truly make an impact on severe neurological disorders. Theranostics takes the idea of designing a patient-specific strategy based on information acquired from diagnostic aid and further employs a therapeutic strategy at the same time. This type of predictive medicine has been made possible due to the advancements in active targeting of specific cellular populations.

One of the first cases of a theranostic agent utilized in research was a NIR heptamethine dye that is capable of targeting cancer and eliciting a photo-dependent cytotoxic activity [155]. Utilizing near-infrared imaging and X-ray, researchers were able to image the tumor site and, using a LED laser for light irradiation, generated a cytotoxic profile for the dye conjugant. Since this development, photodynamic therapies have been utilized extensively within the CNS. A PDT agent can generate singlet oxygen to induce cytotoxicity, but because it can also emit fluorescence, it allows for simultaneous therapy and imaging [156]. Metallic nanoparticles used for the delivery of phthalocyanine to brain tumors have been used to study pharmacokinetics in real time [157].

Theranostics has been studied extensively for use in oncological applications. Metallic nanoparticles, capable of acting as contrast agent enhancers in MRI and CT applications, can function as stand-alone dual diagnostic and therapeutic systems. For example, iron and gold nanoparticles can be imaged and used in thermotherapy or heat ablation therapy and even hypothermia treatment [158,159]. By conjugating a ligand-specific antibody or other targeting moiety to the surface of the nanoparticle, it is possible to image a specific cell population and target that cell population for destruction. Anticancer drug doxorubicin-loaded liposomes can also be used to treat and, through fluorescent imaging, identify uptake and distribution [160]. There have also been several published applications of MRI contrast agents combined with chemotherapeutics for a dual imaging and treatment modality. One example is the use of polymer-coated SPIOs used for MRI that were also designed to deliver the anticancer therapeutic eprubicin [161,162]. Microbubbles utilized in ultrasound enhancement have also been targeted to the overexpressed epidermal growth factor receptor on cancer cells to deliver siRNA upon destruction of the bubbles by ultrasound [163,164]. This allows for the imaging of the tumor site and conformation of a successfully systemic delivery of a gene therapy. Microbubbles coated with albumin and poly(allylamine hydrochloride) allow a cationic charge for binding with oligonucleotides and gene delivery in addition with use with ultrasound diagnostic systems [165,166]. Microbubbles consisting of PVA can also be used to deliver biologically active gasses like nitrous oxide.

Theranostics not only is useful in cancer treatment but also can be utilized in eliciting neural protection in a variety of neurodegenerative diseases. The goal of neural protection is to prevent the further loss of neurons after injury or disease and aid in the regeneration of the CNS. Free radical species are known to cause mitochondrial damage in neurons and numerous polymeric compounds have been identified that are capable of eliminating ROS within the brain [167,168]. Cerium-, yttrium-, and fullerene-modified nanoparticles have particular excellent ROS elimination characteristics [169]. CdSe-coated quantum dots exhibit high cytotoxicity through the excitation of UV light and offer an excellent phototherapeutic agent in targeting tumors.

By encapsulating the drug within the different nanoparticle systems, these delivery systems are capable of more efficient biodistribution and reduce the effects of the cancer therapies on healthy cells and tissues. In most cases, researchers utilize a polymeric biomaterial along with their delivery system to incorporate a contrast agent and covalently bind a therapeutic agent to treat and diagnose simultaneously.

22.12 CONCLUSION AND FUTURE OPPORTUNITIES

The increased occurrence of neurological diseases, injuries, and disorders has led to significant research and progress in medical diagnostics tailored for the CNS. Current standards of diagnostics are excellent at identifying large-scale damage and quantifiable cognitive defects but remain lacking in the ability to identify the progression up to these dramatic symptoms. Targeted polymeric biomaterials capable of enhancing current imaging and diagnostic modalities have given researchers the ability to diagnose and treat CNS disorders in the early stages and the ability to treat patients before irreversible damage occurs. Enhanced diagnostics will lead to the new generation of tailored and predictive medicine, which offers the best opportunity to reduce morbidity and mortality, allowing individuals to regain their quality of life after severe CNS injury or disease.

REFERENCES

[1] B.B. Youan, Impact of nanoscience and nanotechnology on controlled drug delivery, Nanomedicine (Lond.) 3 (4) (2008) 401–406.

[2] R. Sinha, et al., Nanotechnology in cancer therapeutics: bioconjugated nanoparticles for drug delivery, Mol. Cancer Ther. 5 (8) (2006) 1909–1917.

[3] S.K. Sahoo, V. Labhasetwar, Nanotech approaches to drug delivery and imaging, Drug Discov. Today 8 (24) (2003) 1112–1120.

[4] O.C. Farokhzad, R. Langer, Impact of nanotechnology on drug delivery, ACS Nano 3 (1) (2009) 16–20.

[5] S. Baratchi, et al., Promises of nanotechnology for drug delivery to brain in neurodegenerative diseases, Curr. Nanosci. 5 (1) (2009) 15–25.

[6] N.J. Abbott, L. Rönnbäck, E. Hansson, Astrocyte–endothelial interactions at the blood–brain barrier, Nat. Rev. Neurosci. 7 (1) (2006) 41–53.
[7] M.A. Petty, E.H. Lo, Junctional complexes of the blood–brain barrier: permeability changes in neuroinflammation, Prog. Neurobiol. 68 (5) (2002) 311–323.
[8] J.S. Pachter, H.E. de Vries, Z. Fabry, The blood-brain barrier and its role in immune privilege in the central nervous system, J. Neuropathol. Exp. Neurol. 62 (6) (2003) 593–604.
[9] N. Weiss, et al., The blood-brain barrier in brain homeostasis and neurological diseases, Biochim. Biophys Acta 1788 (4) (2009) 842–857.
[10] J.R. Kanwar, B. Sriramoju, R.K. Kanwar, Neurological disorders and therapeutics targeted to surmount the blood–brain barrier, Int. J. Nanomed. 7 (2012) 3259.
[11] R.D. Stevens, R. Sutter, Prognosis in severe brain injury, Crit. Care Med. 41 (4) (2013) 1104–1123.
[12] C.H. Sotak, The role of diffusion tensor imaging in the evaluation of ischemic brain injury—a review, NMR Biomed. 15 (7–8) (2002) 561–569.
[13] M. Mascalchi, et al., Diffusion-weighted MR of the brain: methodology and clinical application, Radiol. Med. 109 (3) (2005) 155–197.
[14] D.E. Henry, A.E. Chiodo, W. Yang, Central nervous system reorganization in a variety of chronic pain states: a review, PM R 3 (12) (2011) 1116–1125.
[15] D. Goldberg-Zimring, et al., Diffusion tensor magnetic resonance imaging in multiple sclerosis, J. Neuroimaging 15 (4 Suppl.) (2005) 68S–81S.
[16] B.C. Chu, K. Miyasaka, The clinical application of diffusion weighted magnetic resonance imaging to acute cerebrovascular disorders, No To Shinkei 50 (9) (1998) 787–795.
[17] T. Abraham, J. Feng, Evolution of brain imaging instrumentation, Semin. Nucl. Med. 41 (3) (2011) 202–219.
[18] M.A. Horsfield, D.K. Jones, Applications of diffusion-weighted and diffusion tensor MRI to white matter diseases—a review, NMR Biomed. 15 (7–8) (2002) 570–577.
[19] A. Nunes, K.T. Al-Jamal, K. Kostarelos, Therapeutics, imaging and toxicity of nanomaterials in the central nervous system, J. Control Release 161 (2) (2012) 290–306.
[20] J.L. Gilmore, et al., Novel nanomaterials for clinical neuroscience, J. Neuroimmune Pharmacol. 3 (2) (2008) 83–94.
[21] F. Hu, Y.S. Zhao, Inorganic nanoparticle-based T1 and T1/T2 magnetic resonance contrast probes, Nanoscale 4 (20) (2012) 6235–6243.
[22] I.P. Grudzinski, Safety of medicinal nanoproducts: new areas of toxicological research, Rocz. Panstw. Zakl. Hig. 62 (3) (2011) 239–246.
[23] M.A. Cywinska, I.P. Grudzinski, Modern toxicology of magnetic nanomaterials, Rocz. Panstw. Zakl. Hig. 63 (3) (2012) 247–256.
[24] B.M. Rabin, et al., Radiation-induced changes in the central nervous system and head and neck, Radiographics 16 (5) (1996) 1055–1072.
[25] L.S. Constine, et al., Adverse effects of brain irradiation correlated with MR and CT imaging, Int. J. Radiat. Oncol. Biol. Phys. 15 (2) (1988) 319–330.
[26] M. Ono, Development of positron-emission tomography/single-photon emission computed tomography imaging probes for in vivo detection of beta-amyloid plaques in Alzheimer's brains, Chem. Pharm. Bull. (Tokyo) 57 (10) (2009) 1029–1039.
[27] B. Maziere, J. Delforge, Pharmacokinetic considerations in the PET and SPECT evaluation of CNS receptors, Q. J. Nucl. Med. 41 (2) (1997) 119–126.
[28] C. la Fougere, et al., PET and SPECT in epilepsy: a critical review, Epilepsy Behav. 15 (1) (2009) 50–55.
[29] M. Ichise, J.H. Meyer, Y. Yonekura, An introduction to PET and SPECT neuroreceptor quantification models, J. Nucl. Med. 42 (5) (2001) 755–763.
[30] J.J. Frost, Receptor imaging by positron emission tomography and single-photon emission computed tomography, Invest. Radiol. 27 (Suppl. 2) (1992) S54–S58.
[31] M. Benadiba, et al., New molecular targets for PET and SPECT imaging in neurodegenerative diseases, Rev. Bras. Psiquiatr. 34 (Suppl. 2) (2012) S125–S136.
[32] M. Perry, Q. Li, R.T. Kennedy, Review of recent advances in analytical techniques for the determination of neurotransmitters, Anal. Chim. Acta 653 (1) (2009) 1–22.
[33] M. Tansey, et al., Neuroinflammation in Parkinson's disease: is there sufficient evidence for mechanism-based interventional therapy? Front. Biosci. 13 (2008) 709.
[34] V.P. Torchilin, Targeted pharmaceutical nanocarriers for cancer therapy and imaging, AAPS J. 9 (2) (2007) E128–E147.
[35] J. Emerit, M. Edeas, F. Bricaire, Neurodegenerative diseases and oxidative stress, Biomed. Pharmacother. 58 (1) (2004) 39–46.
[36] G. Merlini, V. Bellotti, Molecular mechanisms of amyloidosis, New England J. Med. 349 (6) (2003) 583–596.
[37] V.N. Uversky, J. Li, A.L. Fink, Evidence for a partially folded intermediate in α-synuclein fibril formation, J. Biol. Chem. 276 (14) (2001) 10737–10744.
[38] A. Vaarmann, et al., Dopamine protects neurons against glutamate-induced excitotoxicity, Cell Death Dis. 4 (1) (2013) e455.
[39] H.S. Sharma, et al., The blood-brain barrier in Alzheimer's disease: novel therapeutic targets and nanodrug delivery, Int. Rev. Neurobiol. 102 (2012) 47–90.
[40] B.S. Desai, et al., Blood-brain barrier pathology in Alzheimer's and Parkinson's disease: implications for drug therapy, Cell Transplant. 16 (3) (2007) 285–299.
[41] B.D. Trapp, K.-A. Nave, Multiple sclerosis: an immune or neurodegenerative disorder? Annu. Rev. Neurosci. 31 (2008) 247–269.
[42] F. Zipp, O. Aktas, The brain as a target of inflammation: common pathways link inflammatory and neurodegenerative diseases, Trends Neurosci. 29 (9) (2006) 518–527.
[43] A. Minagar, J.S. Alexander, Blood-brain barrier disruption in multiple sclerosis, Mult. Scler. 9 (6) (2003) 540–549.
[44] T.S. Surawicz, et al., Descriptive epidemiology of primary brain and CNS tumors: results from the central brain tumor registry of the United States, 1990-1994, Neuro Oncol. 1 (1) (1999) 14–25.
[45] J.G. Gurney, et al., The contribution of nonmalignant tumors to CNS tumor incidence rates among children in the United States, Cancer Causes Control 10 (2) (1999) 101–105.
[46] G.C. Feigl, et al., Resection of malignant brain tumors in eloquent cortical areas: a new multimodal approach combining 5-aminolevulinic acid and intraoperative monitoring: clinical article, J. Neurosurg. 113 (2) (2010) 352–357.
[47] S. Kulkarni Vishakha, et al., Natural polymers—a comprehensive review, Int. J. Res. Pharmac. Biomed. Sci. 3 (4) (2012) 1597–1613.
[48] M. Hamidi, A. Azadi, P. Rafiei, Hydrogel nanoparticles in drug delivery, Adv. Drug Deliv. Rev. 60 (15) (2008) 1638–1649.
[49] R. Duncan, The dawning era of polymer therapeutics, Nat. Rev. Drug Discov. 2 (5) (2003) 347–360.
[50] F. Fuertges, A. Abuchowski, The clinical efficacy of poly (ethylene glycol)-modified proteins, J. Control. Release 11 (1) (1990) 139–148.
[51] G.S. Kwon, Diblock copolymer nanoparticles for drug delivery, Crit. Rev. Ther. Drug Carrier Syst. 15 (5) (1998) 481–512.

[52] M. Jones, J. Leroux, Polymeric micelles—a new generation of colloidal drug carriers, Eur. J. Pharm. Biopharm. 48 (2) (1999) 101–111.
[53] D.J. Gary, N. Puri, Y.-Y. Won, Polymer-based siRNA delivery: perspectives on the fundamental and phenomenological distinctions from polymer-based DNA delivery, J. Control. Release 121 (1) (2007) 64–73.
[54] V.P. Torchilin, Polymeric contrast agents for medical imaging, Curr. Pharm. Biotechnol. 1 (2) (2000) 183–215.
[55] M. Vert, The complexity of PLAGA-based drug delivery systems, in: Proceedings of the International Conference on Advances in Controlled Delivery, 1996.
[56] R.A. Jain, The manufacturing techniques of various drug loaded biodegradable poly (lactide-*co*-glycolide) (PLGA) devices, Biomaterials 21 (23) (2000) 2475–2490.
[57] G. Storm, et al., Surface modification of nanoparticles to oppose uptake by the mononuclear phagocyte system, Adv. Drug Deliv. Rev. 17 (1) (1995) 31–48.
[58] D. Bazile, et al., Stealth Me. PEG-PLA nanoparticles avoid uptake by the mononuclear phagocytes system, J. Pharm. Sci. 84 (4) (1995) 493–498.
[59] J.-O. You, D.T. Auguste, Feedback-regulated paclitaxel delivery based on poly (*N*, *N*-dimethylaminoethyl methacrylate-*co*-2-hydroxyethyl methacrylate) nanoparticles, Biomaterials 29 (12) (2008) 1950–1957.
[60] S.C. Yang, et al., Polyketal copolymers: a new acid-sensitive delivery vehicle for treating acute inflammatory diseases, Bioconjug. Chem. 19 (6) (2008) 1164–1169.
[61] M.J. Heffernan, N. Murthy, Polyketal nanoparticles: a new pH-sensitive biodegradable drug delivery vehicle, Bioconjug. Chem. 16 (6) (2005) 1340–1342.
[62] C.-L. Lo, K.-M. Lin, G.-H. Hsiue, Preparation and characterization of intelligent core-shell nanoparticles based on poly (d, l-lactide)-g-poly (*N*-isopropyl acrylamide-co-methacrylic acid), J. Control. Release 104 (3) (2005) 477–488.
[63] M. Das, C. Mohanty, S.K. Sahoo, Ligand-based targeted therapy for cancer tissue, Expert Opin. Drug Deliv. 6 (2009) 285–304.
[64] J.M. Chan, et al., Polymeric nanoparticles for drug delivery, in: Cancer Nanotechnology, Humana Press, NY, 2010, pp. 163–175.
[65] J.S. Weinstein, et al., Superparamagnetic iron oxide nanoparticles: diagnostic magnetic resonance imaging and potential therapeutic applications in neurooncology and central nervous system inflammatory pathologies, a review, J. Cereb. Blood Flow Metab. 30 (1) (2010) 15–35.
[66] H. Nejadnik, R. Castillo, H.E. Daldrup-Link, Magnetic resonance imaging and tracking of stem cells, Methods Mol. Biol. 1052 (2013) 1–10.
[67] E. Sykova, P. Jendelova, In vivo tracking of stem cells in brain and spinal cord injury, Prog. Brain Res. 161 (2007) 367–383.
[68] X. Huang, et al., Gold nanoparticles: interesting optical properties and recent applications in cancer diagnostics and therapy, Nanomedicine 2 (5) (2007) 681–693.
[69] J.-O. You, et al., Bioresponsive matrices in drug delivery, J. Biol. Eng. 4 (1) (2010) 1–12.
[70] X. Chen, et al., Photo-controlled molecular recognition of α-cyclodextrin with azobenzene containing polydiacetylene vesicles, Chem. Commun. 11 (2009) 1356–1358.
[71] F.D. Jochum, P. Theato, Temperature-and light-responsive polyacrylamides prepared by a double polymer analogous reaction of activated ester polymers, Macromolecules 42 (16) (2009) 5941–5945.
[72] K. Fan, L. Gao, X. Yan, Human ferritin for tumor detection and therapy, Wiley Interdiscip. Rev. Nanomed. Nanobiotechnol. 5 (4) (2013) 287–298.
[73] M. Manchester, P. Singh, Virus-based nanoparticles (VNPs): platform technologies for diagnostic imaging, Adv. Drug Deliv. Rev. 58 (14) (2006) 1505–1522.
[74] A. Jain, et al., Sustained delivery of activated Rho GTPases and BDNF promotes axon growth in CSPG-rich regions following spinal cord injury, PLoS One 6 (1) (2011) e16135.
[75] A. Jain, et al., In situ gelling hydrogels for conformal repair of spinal cord defects, and local delivery of BDNF after spinal cord injury, Biomaterials 27 (3) (2006) 497–504.
[76] A. Jain, S.M. Brady-Kalnay, R.V. Bellamkonda, Modulation of Rho GTPase activity alleviates chondroitin sulfate proteoglycan-dependent inhibition of neurite extension, J. Neurosci. Res. 77 (2) (2004) 299–307.
[77] X. Xiong, et al., Responsive DNA-based hydrogels and their applications, Macromol. Rapid Commun. 34 (2013) 1271–1283.
[78] K.S. Soppimath, et al., Stimulus-responsive "smart" hydrogels as novel drug delivery systems, Drug Dev. Ind. Pharm. 28 (8) (2002) 957–974.
[79] M. Prabaharan, J.F. Mano, Stimuli-responsive hydrogels based on polysaccharides incorporated with thermo-responsive polymers as novel biomaterials, Macromol. Biosci. 6 (12) (2006) 991–1008.
[80] X.J. Ju, et al., Biodegradable 'intelligent' materials in response to chemical stimuli for biomedical applications, Expert Opin. Ther. Pat. 19 (5) (2009) 683–696.
[81] J. Hu, G. Zhang, S. Liu, Enzyme-responsive polymeric assemblies, nanoparticles and hydrogels, Chem. Soc. Rev. 41 (18) (2012) 5933–5949.
[82] R. Rempe, et al., Transport of Poly (*n*-butylcyano-acrylate) nanoparticles across the blood–brain barrier *in vitro* and their influence on barrier integrity, Biochem. Biophys. Res. Commun. 406 (1) (2011) 64–69.
[83] P. Calvo, et al., Long-circulating PEGylated polycyanoacrylate nanoparticles as new drug carrier for brain delivery, Pharma. Res. 18 (8) (2001) 1157–1166.
[84] C.-S. Yang, et al., Nanoparticle-based in vivo investigation on blood-brain barrier permeability following ischemia and reperfusion, Anal. Chem. 76 (15) (2004) 4465–4471.
[85] C.D. Medley, et al., Gold nanoparticle-based colorimetric assay for the direct detection of cancerous cells, Anal. Chem. 80 (4) (2008) 1067–1072.
[86] F. Corsi, et al., Towards ideal magnetofluorescent nanoparticles for bimodal detection of breast-cancer cells, Small 5 (22) (2009) 2555–2564.
[87] S. Santra, Fluorescent silica nanoparticles for cancer imaging, Methods Mol. Biol. 624 (2010) 151–162.
[88] A.M. Smith, et al., Multicolor quantum dots for molecular diagnostics of cancer, Expert. Rev. Mol. Diagn. 6 (2) (2006) 231–244.
[89] D.Y. Wong, et al., Poly (∊-caprolactone) and poly (L-lactic-co-glycolic acid) degradable polymer sponges attenuate astrocyte response and lesion growth in acute traumatic brain injury, Tissue Eng. 13 (10) (2007) 2515–2523.
[90] T. Konno, J. Watanabe, K. Ishihara, Conjugation of enzymes on polymer nanoparticles covered with phosphorylcholine groups, Biomacromolecules 5 (2) (2004) 342–347.
[91] G. Stoll, M. Bendszus, Imaging of inflammation in the peripheral and central nervous system by magnetic resonance imaging, Neuroscience 158 (3) (2009) 1151–1160.

[92] H. Sarin, On the future development of optimally-sized lipid-insoluble systemic therapies for CNS solid tumors and other neuropathologies, Recent Pat. CNS Drug Discov. 5 (3) (2010) 239–252.
[93] E. Peira, et al., In vitro and in vivo study of solid lipid nanoparticles loaded with superparamagnetic iron oxide, J. Drug Target 11 (1) (2003) 19–24.
[94] M.R. Gasco, L. Priano, G.P. Zara, Chapter 10—solid lipid nanoparticles and microemulsions for drug delivery. The CNS, Prog. Brain Res. 180 (2009) 181–192.
[95] L. Biddlestone-Thorpe, et al., Nanomaterial-mediated CNS delivery of diagnostic and therapeutic agents, Adv. Drug Deliv. Rev. 64 (7) (2012) 605–613.
[96] A.M. Morawski, G.A. Lanza, S.A. Wickline, Targeted contrast agents for magnetic resonance imaging and ultrasound, Curr. Opin. Biotechnol. 16 (1) (2005) 89–92.
[97] P. Caravan, et al., Gadolinium(III) chelates as MRI contrast agents: structure, dynamics, and applications, Chem. Rev. 99 (9) (1999) 2293–2352.
[98] V.P. Torchilin, Recent advances with liposomes as pharmaceutical carriers, Nat. Rev. Drug Discov. 4 (2) (2005) 145–160.
[99] M.L. Matteucci, D.E. Thrall, The role of liposomes in drug delivery and diagnostic imaging: a review, Vet. Radiol. Ultrasound 41 (2) (2000) 100–107.
[100] K.B. Ghaghada, et al., New dual mode gadolinium nanoparticle contrast agent for magnetic resonance imaging, PLoS One 4 (10) (2009) e7628.
[101] M.M. Lin, et al., Development of superparamagnetic iron oxide nanoparticles (SPIONS) for translation to clinical applications, IEEE Trans. Nanobiosci. 7 (4) (2008) 298–305.
[102] S. Laurent, et al., Magnetic iron oxide nanoparticles: synthesis, stabilization, vectorization, physicochemical characterizations, and biological applications, Chem. Rev. 108 (6) (2008) 2064–2110.
[103] M. Oostendorp, et al., Pharmacokinetics of contrast agents targeted to the tumor vasculature in molecular magnetic resonance imaging, Contrast Media Mol. Imaging 5 (1) (2010) 9–17.
[104] D. Artemov, Molecular magnetic resonance imaging with targeted contrast agents, J. Cell. Biochem. 90 (3) (2003) 518–524.
[105] S. Ogawa, et al., Intrinsic signal changes accompanying sensory stimulation: functional brain mapping with magnetic resonance imaging, Proc. Natl. Acad. Sci. U.S.A. 89 (13) (1992) 5951–5955.
[106] M.A. Pysz, S.S. Gambhir, J.K. Willmann, Molecular imaging: current status and emerging strategies, Clin. Radiol. 65 (7) (2010) 500–516.
[107] W.J. Mulder, et al., Lipid-based nanoparticles for contrast-enhanced MRI and molecular imaging, NMR Biomed. 19 (1) (2006) 142–164.
[108] M.S. Martina, et al., Generation of superparamagnetic liposomes revealed as highly efficient MRI contrast agents for in vivo imaging, J. Am. Chem. Soc. 127 (30) (2005) 10676–10685.
[109] M. Law, et al., Comparison of cerebral blood volume and vascular permeability from dynamic susceptibility contrast-enhanced perfusion MR imaging with glioma grade, AJNR Am. J. Neuroradiol. 25 (5) (2004) 746–755.
[110] G. Johnson, et al., Measuring blood volume and vascular transfer constant from dynamic, T(2)*-weighted contrast-enhanced MRI, Magn. Reson. Med. 51 (5) (2004) 961–968.
[111] K. Ulbrich, et al., Transferrin- and transferrin-receptor-antibody-modified nanoparticles enable drug delivery across the blood-brain barrier (BBB), Eur. J. Pharm. Biopharm. 71 (2) (2009) 251–256.
[112] R. Esfand, D.A. Tomalia, Poly (amidoamine)(PAMAM) dendrimers: from biomimicry to drug delivery and biomedical applications, Drug Discov. Today 6 (8) (2001) 427–436.
[113] M. Fischer, F. Vögtle, Dendrimers: from design to application—a progress report, Angew. Chem. Int. Ed. 38 (7) (1999) 884–905.
[114] B.K. Nanjwade, et al., Dendrimers: emerging polymers for drug-delivery systems, Eur. J. Pharm. Sci. 38 (3) (2009) 185–196.
[115] U. Boas, J.B. Christensen, P.M. Heegaard, Dendrimers: design, synthesis and chemical properties, J. Mater. Chem. 16 (38) (2006) 3785–3798.
[116] M.L. Patil, et al., Surface-modified and internally cationic polyamidoamine dendrimers for efficient siRNA delivery, Bioconjug. Chem. 19 (7) (2008) 1396–1403.
[117] A. Bielinska, et al., Regulation of in vitro gene expression using antisense oligonucleotides or antisense expression plasmids transfected using starburst PAMAM dendrimers, Nucleic Acids Res. 24 (11) (1996) 2176–2182.
[118] J.D. Eichman, et al., The use of PAMAM dendrimers in the efficient transfer of genetic material into cells, Pharm. Sci. Technol. Today 3 (7) (2000) 232–245.
[119] L.H. Bryant Jr., et al., Synthesis and relaxometry of high-generation (G = 5, 7, 9, and 10) PAMAM dendrimer-DOTA-gadolinium chelates, J. Magn. Reson. Imaging 9 (2) (1999) 348–352.
[120] S. Wen, et al., Multifunctional dendrimer-entrapped gold nanoparticles for dual mode CT/MR imaging applications, Biomaterials 34 (5) (2013) 1570–1580.
[121] K. Jayakumar, et al., Gold nano particle decorated graphene core first generation PAMAM dendrimer for label free electrochemical DNA hybridization sensing, Biosens. Bioelectron. 31 (1) (2012) 406–412.
[122] S. Langereis, et al., Evaluation of Gd(III)DTPA-terminated poly(propylene imine) dendrimers as contrast agents for MR imaging, NMR Biomed. 19 (1) (2006) 133–141.
[123] X. Michalet, et al., Quantum dots for live cells, in vivo imaging, and diagnostics, Science 307 (5709) (2005) 538–544.
[124] W.K. Leutwyler, S.L. Bürgi, H. Burgl, Semiconductor clusters, nanocrystals, and quantum dots, Science 271 (1996) 933.
[125] X. Gao, et al., In vivo cancer targeting and imaging with semiconductor quantum dots, Nat. Biotechnol. 22 (8) (2004) 969–976.
[126] H. Jackson, et al., Quantum dots are phagocytized by macrophages and colocalize with experimental gliomas, Neurosurgery 60 (3) (2007) 524–529, discussion 529–30.
[127] A.L. Klibanov, Preparation of targeted microbubbles: ultrasound contrast agents for molecular imaging, Medical Biol. Eng. Comput. 47 (8) (2009) 875–882.
[128] M.L. Calvisi, et al., Shape stability and violent collapse of microbubbles in acoustic traveling waves, Phys. Fluids 19 (2007) 047101.
[129] B. Schrope, V. Newhouse, V. Uhlendorf, Simulated capillary blood flow measurement using a nonlinear ultrasonic contrast agent, Ultrason. Imaging 14 (2) (1992) 134–158.
[130] G.E. Weller, et al., Ultrasonic imaging of tumor angiogenesis using contrast microbubbles targeted via the tumor-binding peptide arginine-arginine-leucine, Cancer Res. 65 (2) (2005) 533–539.
[131] M.A. O'Reilly, Y. Huang, K. Hynynen, The impact of standing wave effects on transcranial focused ultrasound disruption of the blood-brain barrier in a rat model, Phys. Med. Biol. 55 (18) (2010) 5251–5267.
[132] K. Hynynen, et al., Noninvasive MR imaging–guided focal opening of the blood-brain barrier in rabbits1, Radiology 220 (3) (2001) 640–646.

[133] S. Meairs, A. Alonso, Ultrasound, microbubbles and the blood–brain barrier, Prog. Biophys. Molecular Biol. 93 (1) (2007) 354–362.
[134] J.A. Feshitan, et al., Theranostic Gd(III)-lipid microbubbles for MRI-guided focused ultrasound surgery, Biomaterials 33 (1) (2012) 247–255.
[135] E.C. Unger, et al., Therapeutic applications of lipid-coated microbubbles, Adv. Drug Deliv. Rev. 56 (9) (2004) 1291–1314.
[136] D.M. Skyba, et al., Direct in vivo visualization of intravascular destruction of microbubbles by ultrasound and its local effects on tissue, Circulation 98 (4) (1998) 290–293.
[137] L. Hillered, P.M. Vespa, D.A. Hovda, Translational neurochemical research in acute human brain injury: the current status and potential future for cerebral microdialysis, J. Neurotrauma 22 (1) (2005) 3–41.
[138] X. Zhang, Q. Guo, D. Cui, Recent advances in nanotechnology applied to biosensors, Sensors (Basel) 9 (2) (2009) 1033–1053.
[139] S. Vaddiraju, et al., Emerging synergy between nanotechnology and implantable biosensors: a review, Biosens. Bioelectron. 25 (7) (2010) 1553–1565.
[140] C. Nicolini, T. Bezerra, E. Pechkova, Protein nanotechnology for the new design and development of biocrystals and biosensors, Nanomedicine (Lond.) 7 (8) (2012) 1117–1120.
[141] C. Jianrong, et al., Nanotechnology and biosensors, Biotechnol. Adv. 22 (7) (2004) 505–518.
[142] N. Dale, et al., Listening to the brain: microelectrode biosensors for neurochemicals, Trends Biotechnol. 23 (8) (2005) 420–428.
[143] J. Williams, Cutting edge: a novel lab-on-a-tube for multimodality neuromonitoring of patients with traumatic brain injury (TBI), Lab Chip 9 (14) (2009) 1987.
[144] N. Wisniewski, M. Reichert, Methods for reducing biosensor membrane biofouling, Colloids Surf. B Biointerfaces 18 (3–4) (2000) 197–219.
[145] N. Wisniewski, F. Moussy, W.M. Reichert, Characterization of implantable biosensor membrane biofouling, Fresenius J. Anal. Chem. 366 (6–7) (2000) 611–621.
[146] Q. Yu, et al., Anti-fouling bioactive surfaces, Acta Biomater. 7 (4) (2011) 1550–1557.
[147] S. Brahim, D. Narinesingh, A. Guiseppi-Elie, Bio-smart hydrogels: co-joined molecular recognition and signal transduction in biosensor fabrication and drug delivery, Biosens. Bioelectron. 17 (11–12) (2002) 973–981.
[148] F. Meng, Z. Zhong, J. Feijen, Stimuli-responsive polymersomes for programmed drug delivery, Biomacromolecules 10 (2) (2009) 197–209.
[149] R.H. Muller, K. Mader, S. Gohla, Solid lipid nanoparticles (SLN) for controlled drug delivery—a review of the state of the art, Eur. J. Pharm. Biopharm. 50 (1) (2000) 161–177.
[150] J. Kreuter, et al., Direct evidence that polysorbate-80-coated poly(butylcyanoacrylate) nanoparticles deliver drugs to the CNS via specific mechanisms requiring prior binding of drug to the nanoparticles, Pharm. Res. 20 (3) (2003) 409–416.
[151] K. Jain, et al., Dendrimer toxicity: let's meet the challenge, Int. J. Pharm. 394 (1–2) (2010) 122–142.
[152] R. Hardman, A toxicologic review of quantum dots: toxicity depends on physicochemical and environmental factors, Environ. Health Perspect. 114 (2) (2006) 165–172.
[153] D.M. Walsh, et al., Amyloid-beta oligomers: their production, toxicity and therapeutic inhibition, Biochem. Soc. Trans. 30 (4) (2002) 552–557.
[154] A.I. Faden, Neuroprotection and traumatic brain injury: the search continues, Arch. Neurol. 58 (10) (2001) 1553–1555.
[155] X. Tan, et al., A NIR heptamethine dye with intrinsic cancer targeting, imaging and photosensitizing properties, Biomaterials 33 (7) (2012) 2230–2239.
[156] J.P. Celli, et al., Imaging and photodynamic therapy: mechanisms, monitoring, and optimization, Chem. Rev. 110 (5) (2010) 2795–2838.
[157] D. Bechet, et al., Nanoparticles as vehicles for delivery of photodynamic therapy agents, Trends Biotechnol. 26 (11) (2008) 612–621.
[158] T. Kobayashi, Cancer hyperthermia using magnetic nanoparticles, Biotechnol. J. 6 (11) (2011) 1342–1347.
[159] I. Hilger, et al., Thermal ablation of tumors using magnetic nanoparticles: an in vivo feasibility study, Invest. Radiol. 37 (10) (2002) 580–586.
[160] A.N. Lukyanov, et al., Tumor-targeted liposomes: doxorubicin-loaded long-circulating liposomes modified with anti-cancer antibody, J Control Release 100 (1) (2004) 135–144.
[161] H.L. Liu, et al., Magnetic resonance monitoring of focused ultrasound/magnetic nanoparticle targeting delivery of therapeutic agents to the brain, Proc. Natl. Acad. Sci. U.S.A. 107 (34) (2010) 15205–15210.
[162] C.H. Fan, et al., SPIO-conjugated, doxorubicin-loaded microbubbles for concurrent MRI and focused-ultrasound enhanced brain-tumor drug delivery, Biomaterials 34 (14) (2013) 3706–3715.
[163] R. Deckers, C.T. Moonen, Ultrasound triggered, image guided, local drug delivery, J. Control Release 148 (1) (2010) 25–33.
[164] M.R. Bohmer, et al., Ultrasound triggered image-guided drug delivery, Eur. J. Radiol. 70 (2) (2009) 242–253.
[165] P.G. Sanches, H. Grull, O.C. Steinbach, See, reach, treat: ultrasound-triggered image-guided drug delivery, Ther. Deliv. 2 (7) (2011) 919–934.
[166] C. Lorenzato, et al., MRI contrast variation of thermosensitive magnetoliposomes triggered by focused ultrasound: a tool for image-guided local drug delivery, Contrast Media Mol. Imaging 8 (2) (2013) 185–192.
[167] B. Radwanska-Wala, E. Buszman, D. Druzba, Reactive oxygen species in pathogenesis of central nervous system diseases, Wiad. Lek. 61 (1–3) (2008) 67–73.
[168] M. Gutowicz, The influence of reactive oxygen species on the central nervous system, Postepy Hig. Med. Dosw. (Online) 65 (2011) 104–113.
[169] D. Schubert, et al., Cerium and yttrium oxide nanoparticles are neuroprotective, Biochem. Biophys. Res. Commun. 342 (1) (2006) 86–91.

Chapter 23

Polymeric Biomaterials in Nanomedicine

Brittany L. Banik, Justin L. Brown

Department of Bioengineering, The Pennsylvania State University, University Park, Pennsylvania, USA

Chapter Outline

23.1 INTRODUCTION

Nanoscale is defined in literature as a level of measurement in the 1-100nm range [1–3]. The nanotechnology revolution is projected to be a $1 trillion market by 2015 [4]; the surge in expected expenditure and funding indicates the significance of investigations related to the nanoscale dimension. Nanoscale research has gained recognition based on the observations finding critically different responses at the cellular level and novel structural and functional advantages with the use of nanosized materials. The notable changes in surface properties and energetics due to the increase in surface area-to-volume ratio, surface roughness, surface defects, and change in electron distributions [2], among other properties, make nanomaterials an interesting area of consideration across the field of biomedical engineering.

23.1.1 Nanomedicine

At the most basic definition, nanomedicine is simply the application of nanotechnology to medicine [5,6]. A more conclusive definition described by El-Ansary and Al-Daihan explains nanomedicine as a field with applications to monitor, repair, construct, and control human biological systems through utilization of devices and structures at the nanodimension [7]. Novel therapeutic and diagnostic modalities engineered in the nanometer range have physicochemical properties [6] that promote host responses at the protein and cellular level [2], which are different from bulk materials of the same composition and provide nanomodalities with notable advantages over conventional systems—enhanced and controlled delivery, extended drug bioactivity [8], improved circulation times, enhanced dissolution and degradation rates [9], and high-throughput and performance. Additionally, at the nanometer level, high surface-to-volume ratio, surface tailorability, and multifunctionality options create encouraging prospects for these materials in the biomedical field [10]. Ramsden suggests a distinction in the use of nanotechnology in medicine as "direct" or "auxiliary" [11]. The paradigm of "auxiliary" nanomedicine mainly includes drug discovery, manufacture, and delivery and the indirect application to medical diagnostics; conversely, "direct" nanomedicine encompasses tissue engineering applications, such as scaffolds and tissues for implantation and lab-on-chip technology for clinical fluid analyses [11].

23.1.2 Natural and Synthetic Polymers

Polymeric materials have been substantially used in the nanomedicine field for pharmaceutical and biotechnology products [12]. Polymers offer a number of advantages over other material options like metals and ceramics, specifically the biodegradability of some polymers, flexible synthesis methods to modify properties for mechanical and physiological needs, and the ability to be fabricated into various morphologies (i.e., fibers, films, particles, and gels) [13]. The molecular weight, charge, branching, material constituents, and other structural factors influence polymeric nanomaterial functions including degradation, mechanical parameters, stability, cellular internalization, and release rates [5]. However, there are challenges in the use of polymeric biomaterials, which include considerations regarding sterilization methods, the potential for material swelling, debris from wear, leached or resorbed products, and durability and reliability for applications with higher mechanical loading. Particularly for polymer-drug applications, water penetration is a concern with the potential of chemical reactions, causing deactivation of the drug [14]. To combat some of these challenges, the addition of particulates or shortened and unidirectional fibers, depending on the application, may be used as reinforcement to improve mechanical integrity, durability, and reliability of polymeric nanostructures.

Polymeric biomaterials used in nanomedicine are generally classified as either natural or synthetic polymers. Natural polymers are raw materials that naturally occur in the biological environment; examples include chitosan, albumin, heparin, silk, and collagen [6,8]. Typically, these materials are biocompatible and enzymatically biodegradable [15], meaning that the material is readily recognized, accepted, and mechanically processed by the body. This suggests that natural polymers are prime candidates for nanomedicine; however, disadvantages involving immunogenicity, variability in purity across groups, complex structure, strength inadequacies, and difficulty in controlling material degradability [15] hinder complete acceptability in biomedical engineering applications. Interest in synthetic polymers, such as poly(lactic acid), poly(glycolic acid), copolymer poly(lactide-*co*-glycolide), polyesters, polyurethanes, and poly(ethylene glycol) (PEG), lies in the controllable chemical, structural, and mechanical properties with minimal variations between the batches of synthesized nanomaterials. Disadvantages of synthetic polymers include less predictability in material biocompatibility along with the additional possibility of toxic and inflammatory biological responses.

23.2 POLYMERIC NANOMEDICINE CONSIDERATIONS

Nanomedicine envelops a wide variety of structures—nanoparticles, nanofibers, nanoporous scaffolds, nanocages, nanoposts, nanodots, nanotubes, and nanowires, among other newly defined and developing configurations. Integration of polymeric biomaterials into biomedical applications requires investigations to understand the physicochemical properties and how they relate to biological interactions and functions [1]. Of particular interest in polymeric nanomedicine applications are polymers that resist nonspecific adsorption of proteins to reduce the negative biological responses [16] and structures that introduce beneficial components due to size and dimension. This section focuses on nanoparticles and nanofibers, which are gaining immense ground and acceptability for nanomedicine applications.

23.2.1 Nanoparticles

Polymeric nanoparticles have gained considerable attention in nanomedicine due to the potential for surface modification, pharmacokinetic control, suitability for targeted delivery of therapeutics [17], mechanical properties [5], and design flexibility. More specifically, the size, surface morphology, chemistry, and charge, porosity, and drug diffusivity and encapsulation efficiency are properties that push polymeric nanoparticles to the forefront for nanomedicine applications [18]. The polymers used for polymeric nanoparticles, such as poly(lactic-*co*-glycolic acid) (PLGA) and poly(caprolactone) (PCL) [5], often also have the favorable properties of biodegradability through clearance of the body's homeostatic metabolic pathways [12] and biocompatibility.

Nanoparticles navigate to the biological substrate through passive targeting, active targeting, or endocytosis. In passive targeting, the accumulation of nanoparticles to a site is based on the inherent biophysicochemical properties of the nanoparticles (i.e., size, morphology, and charge), blood hemodynamic forces, and diffusive mechanisms [12,19]; no active ligand is added for specificity or directionality of the particle. In passive targeting, nanoparticles are not taken up via vesicles but rather by van der Waals forces, electrostatic charges, steric interactions, or interfacial tension effects and are achieved based on the pathophysiological characteristics of the tissue [13,19]. An example of passive targeting is the internalization of nanocarriers by cancer cells, which is based on extravasation of nanoparticles through the "leaky" endothelium of tumor tissue; certain nanosized particles can diffuse and penetrate the endothelium openings [20]. Active targeting is an internalization method that uses receptors or surface modifications with affinity moieties such as ligands, antigen-antibody combinations, or aptamers [19] with specificity to enter diseased tissues or cells [12].

Another method of nanoparticle transport into the cell is engulfment or endocytosis by a phagocytic member, such as macrophage, through actin polymerization [21] or by various cell types, including endothelial cells, pulmonary epithelium, red blood cells, and nerve cells [13]. Following endocytosis, the vesicles housing the nanoparticles are characterized as late endosomes and then lysosomes; a distinct change in proton concentration occurs changing from physiological conditions of pH 7.4 to an acidic environment with a pH around 5.0 [22]. Additionally, differences in pH occurring peripheral to tumors are significantly different from the surrounding healthy tissue, as the pH in tumor tissues has been characterized to be ~6.8; this distinction has been utilized for targeting and drug release cancer therapy studies [23].

Nanoparticle fabrication includes methods such as solvent evaporation, spontaneous emulsification, solvent diffusion, salting out/emulsification diffusion, and polymerization techniques [24]. The synthesis method can greatly impact the particle's physical, chemical, and biological properties and the dispersion and stability of the particles, which are important considerations in biomedical applications. The nanoparticle size and shape is of great importance because the internalization, circulation, distribution, and targeting aspects of the system can be affected by these characteristics [5].

The nanoparticle surface chemistry can be tailored to be more effective for systemic circulation and targeted delivery through surface modification or functionalization. Surface changes may be necessary to increase stability, dispersion, specificity, and steric effects of the particles [10]. PEG has been included in many research studies as an outer polymeric layer to increase passive retention by suppressing opsonization [4] via steric repulsion forces [19], which in turn reduces rapid clearance and unwanted recognition by the mononuclear phagocyte system [1,17,25]. Nanoparticles with a hydrophilic, nontoxic, nonimmunogenic PEG layer have been coined as "stealth nanoparticles" [12] since the PEG coating allows for the particles to escape being distinguished as foreign particulates; additionally, PEG-coated particles show better shelf stability and capability for controlled therapeutic release [26].

Nanoparticles have the capability to advance nanomedicine research, but there are still challenges to overcome—inconsistency in synthesis, nanotoxicity effects, and degradation by-products. Continuous efforts are being made to produce polymeric nanoparticles that are more efficient and effective, tissue or disease specific, and nontoxic [6]. The ideal nanoparticle would ultimately be a "responsive particle"—able to adapt under environmental conditions to carry out a task by sensing the target's presence, binding to the target upon recognition, and fulfilling the applicable response to the present situation [11]. For example, if a concentration is low in the surrounding system, the nanoparticle could detect the deficiency and release the agent into the area to acquire more appropriate levels.

23.2.2 Nanofibers

Nanofibers are of specific interest in the field of nanomedicine due to their similarity to the extracellular matrix (ECM). Like nanoparticles, the higher surface-to-volume ratio and the tunability of the polymers used in synthesis are beneficial. The architecture can be tailored in regard to porosity, fiber diameter, mechanical properties, structure or fiber arrangement, and functionalization, among other variables.

A technique that has gained great popularity for nanofiber synthesis is electrospinning, which can be used to fabricate filtration membranes, catalytic nanofibers, and tissue engineering scaffolds [27]. Electrospinning is a well-documented process involving the application of voltage to draw a jet of polymer solution from a syringe source toward a collector. While traveling the distance to the target, the solvent evaporates and polymeric fibers collect on the target. Various polymer and solvent systems have been tested along with different collecting mechanisms to create nanofiber arrangements. The numerous technical parameters, such as polymer molecular weight, distance from target, flow rate, and applied voltage, not only can aid in fine-tuning the fiber fabrication but also can create issues in obtaining consistent fibers. Synthetic polymers are most often used in electrospinning schemes because electrospinning of natural polymers does not offer the versatility or processability of synthetic polymers and specific solvents must be identified so as not to affect the natural material integrity [27]. Nanofiber diameters in the range of 3 nm to 5 μm spun with either natural or synthetic fibers have been produced with electrospinning [27].

23.3 POLYMERIC NANOMEDICINE APPLICATIONS

23.3.1 Drug Delivery

In drug delivery, nanoparticles show great promise by altering the bioavailability, pharmacokinetic, and pharmacodynamic properties of drug molecules to improve therapeutic delivery; however, clinical translation has been slow with the lack of ideal and established solutions for precise targeting, cell internalization, and controlled drug solubility and release [24]. Polymeric nanomedicine can address these challenges.

23.3.1.1 Multifunctional Nanoparticles

The focus of nanoparticles in drug delivery hinges on the concept of creating single particles with the ability to perform multiple tasks—targeting/detection, imaging to monitor progression, and therapeutic/treatment functionalities [28]. Multifunctional nanoparticles are the next generation of nanomedicine that will be used to customize therapy for patient-to-patient treatment and eliminate the need for several drug dosages and the potential for deleterious effects of surrounding healthy cells and tissues. The general design rationale is based on the use of a polymer, linker, drug or protein, targeting group, and imaging agent. Various synthesis methods are underway with the hope of developing a logical scheme for the components to work in concert. Ideally, degradation mechanisms will be appropriate for the required drug or protein release, targeting agents will be successful for enhanced and effective imaging supplements, and overall manufacturing will be minimal in cost, time, complexity, and risk.

Flash NanoPrecipitation (FNP) is a novel technique that has been advantageous for investigations into the development of multifunctional nanoparticles. Johnson and Prud'homme first introduced FNP as a technology to create nanoparticles of a controllable size distribution and high drug loading rate based on a polymer-protected hydrophobic core for hydrophobic drug encapsulation [29]. Akbulut *et al.* demonstrated the use of FNP to prepare multifunctional particles with surface functionalization that allow for simultaneous drug delivery and fluorescent biological imaging at desired wavelengths. The use of PEG-PCL block copolymers to coencapsulate the hydrophobic fluorophore and therapeutic via FNP exhibited improved biocompatibility in comparison to approaches using semiconductor quantum dots for fluorescent nanoparticle synthesis while offering design flexibility of imaging, targeting, and drug loading in applications of biomedical imaging and fluid flow [30]. Success in flexibility of the FNP process was demonstrated upon inclusion of dual fluorescent molecules with combinations based on Nile red, coumarin, pyrene, perylene, anthracene, and porphyrin and with a fluorophore and nonfluorescent hydrophobic compound such as vitamin E. This study suggests the ability for multifunctionality in drug delivery and fluorescence targeting.

Multifunctionality, termed chemoradiation, is the combination of chemotherapy and radiotherapy. Wang *et al.* suggest the term ChemoRad nanoparticles to define this platform [31]. The difficulty of developing a chemoradiation methodology was to include both components without affecting nanoparticle physicochemical properties—size, morphology, surface charge, stability, and drug release profile—and circulation and distribution behaviors. The ChemoRad nanoparticle platform was based on a hydrophobic and biodegradable poly(D,L-lactic-*co*-glycolic acid) (PLGA) core (for chemotherapeutic docetaxel drug encapsulation), lecithin lipid monolayer on the surface (for docetaxel drug retention), hydrophilic PEG shell (for stability and circulation), and DMPE-DTPA lipid-chelator layer (to capture radioisotope yttrium-90) with prostate cancer as the model disease. The ChemoRad system was tested via active targeting with the conjugation of A10 RNA aptamer to target a prostate-specific antigen. Results indicated successful delivery of chemoradiation with high therapeutic efficiency as demonstrated by effective docetaxel delivery with a controlled release profile and efficient chelation of yttrium-90. Werner *et al.* capitalized on the use of ChemoRad nanoparticles for use in the treatment of ovarian cancer peritoneal metastasis; the group added a folate-targeting ligand to encourage uptake of nanoparticles into tumor cells [32]. Folate is a common targeting ligand applied to cancer nanomedicine because the folate receptor has been identified to be overexpressed in tumor cells [32,33].

The multidrug resistance effect is a challenging factor in cancer therapy as drug-resistant cells of tumors recycle the input of chemotherapeutic drugs via pumps identified as P-glycoprotein and multidrug resistance-associated protein; polymeric carriers, either for encapsulation or conjugation of the drug, can help reduce the outflux of drug for better cancer therapeutic performance [22,34]. Khdair *et al.* introduced another type of multifunctional nanoparticle system based on the natural polymer sodium alginate to counteract the multidrug resistance effect by targeting with two drugs in one particle—chemotherapeutic drug doxorubicin and methylene blue associated with photodynamic therapy [35]. Photodynamic therapy for cancer therapy uses light with the combination of certain chemicals to treat cancerous tumors through three main mechanisms upon exposure to light: inducing cell death through the generation of cytotoxic reactive oxygen species (ROS), damaging tumor vasculature, or activating an immune response against tumor cells by inhibiting the multidrug resistance pumps [35,36]. Single-dose studies of the combination chemotherapy and photodynamic therapy were performed in a mouse adenocarcinoma tumor model and showed significant improvement for drug accumulation and ROS production to impede tumor cell proliferation. Future studies are necessary to better understand the antitumor responses due to the combination therapy and to test multiple dosages in another animal model.

23.3.1.2 "Smart" Nanoparticles

The creation of "smart" nanoparticles is an emerging trend in nanomedicine. In a recent review on polymeric nanoparticles by Brewer *et al.*, "smart" nanoparticle research was divided into two categories: "site targeting" through the use of molecules such as ligands, antibodies, and aptamers and "site triggering" by chemical or physical environmental changes [22]. Great strides toward the fabrication of stimuli-sensitive nanomaterials are being made as a way to create combination therapeutics specific for drug delivery systems [9]. Nanocarriers are designed to respond to physiological changes (i.e., pH, ionic strength, solvent, and temperature) and external stimuli (i.e., electrical, magnetic, or optical) to aid in more efficient and effective drug delivery [9,37]. Polymer properties such as molecular weight and side-chain derivatives can be adjusted to increase the sensitivity to stimuli. Ester bonds are sensitive to changes in pH, and degradation of these bonds accelerates in acidic conditions [9]; therefore, functional groups containing ester bonds would be of interest and consideration for "smart" pH-responsive nanoparticles.

Current studies investigating various combinations of nanosystem properties are beginning to surface, specifically to overcome challenges in cancer therapy. Ding *et al.* present intelligent multifunctional multiblock polyurethane nanocarriers with controllable release capabilities of paclitaxel based on the acidic tumor environment [38]. The multiblock polyurethane design is based on soft segments of poly(ε-caprolactone) with pH-responsive hydrazone bonds to house lipophilic components and for drug release and hard segments of L-lysine ethyl ester diisocyanate and 1,3-propanedial for molecular weight control and biocompatibility. The core-shell-corona architecture was formed by self-assembly into nanomicelles; the components are shown in Figure 23.1: the hydrophobic biodegradable core with pH sensitivity for drug release and a shell-corona to improve cell internalization and circulation characterized by gemini quaternary ammonium cationic groups, tripeptide, and hydrazone linkage methoxyl-PEG. The stimuli for these "smart" nanoparticles are the acidic environment and lower pH level present in tumors; the drug is released due to micelle degradation from the cleavage of hydrazone bonds and partial detachment of the PEG shell. *In vivo* experimentation revealed higher antitumor activity and reduced tumor volume after 20 days for the micelle nanocarriers compared to free paclitaxel, which suggests the improvement in targeting specificity, biodistribution, and drug protection capable from the polyurethane formulations.

In a research study conducted by van Vlerken *et al.*, pH responsiveness was utilized for targeting breast cancer with the use of a polymer-blend nanoparticle system of 30% by weight pH-responsive polymer poly(beta-amino ester) and 70% by weight poly(D,L lactide-*co*-glycolide) to administer paclitaxel and C_6-ceramide, an apoptotic signaling molecule, respectively [39]. The polymer blend allowed for a delay in the release of C_6-ceramide, which based on preliminary results improved therapeutic efficacy in this specific combination therapy scheme for multidrug-resistant breast and ovarian cancer. Expected conclusions were generated for the accumulation of particles based on the enhanced permeability and retention (EPR) effect; however, the aspects of multidrug resistance pumps appeared to quickly remove the paclitaxel from the tumor site, diminishing potential for reliable multidrug-resistant conclusions [39].

23.3.1.3 Coatings

Along with PEG, as previously discussed for its stealth properties, *N*-(2-hydroxypropyl) methacrylamide (HPMA) is another hydrophilic linear, nonimmunogenic polymer used in polymer-drug conjugates. Drug attachment or conjugation with targeting ligands is possible due to the functionalizable side chains on HPMA; the biodegradability of HPMA is an additional ideal property for drug release [3].

The promising outlook of nanoparticle cancer therapy research stems initially from the ability for polymeric drug carriers to avert recognition by the reticuloendothelial system and, furthermore, by exploiting the EPR effect and using tumor-specific targeting methods [10]. The EPR effect is based on the hyperpermeable vasculature combined with lack of a lymphatic drainage system present in tumor tissue. Under these conditions, nanoparticles below 400 nm in diameter diffuse through the tumor vasculature and accumulate in the cancerous tissue [9]. PEG surface coatings on nanoparticles are often used for cancer therapy to extend the circulation time in order for nanoparticles to deposit in cancerous tumors via the EPR effect. The unfavorable property of nonsite-specific targeting commonly seen in anticancer chemotherapeutic agents can also be suppressed through appropriate surface coatings for active targeting mechanisms [3].

In the area of using polymer-drug conjugates, specifically for delivery of anticancer drugs, attention is given to designing a system that ensures a stable covalent linkage between drug and polymer, endocytotic uptake into tumor cells, improved drug load and retention efficiency, and increased specificity in targeting [20]. Polymeric coatings are becoming the primary option to improve biocompatibility, targeting, and tethering imaging agents in nanomedicine

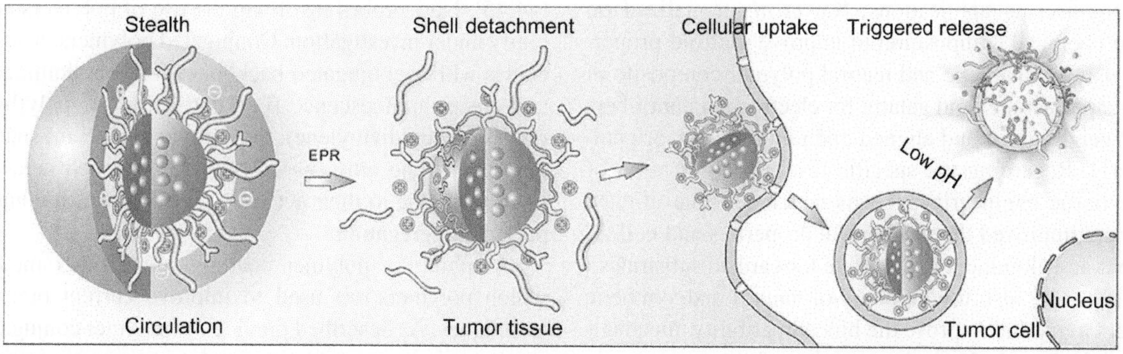

FIGURE 23.1 Mechanism of targeted drug delivery and cell internalization into tumor cells using multiblock multifunctional polyurethane nanoparticles. (For color version of this figure, the reader is referred to the online version of this chapter.) Reprinted with permission from Ref. [38]. Copyright 2013. American Chemical Society.

applications. Sahoo *et al.* fully utilized a smart polymer shell around a magnetic nanoparticle core to attach folic acid as the active targeting moiety for cancer cell and rhodamine B isothiocyanate as a fluorescent conjugation, creating a targeting and imaging multifunctionality [23]. The nanoparticle system was additionally enhanced as a "smart" system based on the polymeric coating because the dual-responsive poly(*N*-isopropylacrylamide-*block*-poly(acrylic acid)) copolymer was able to deliver the doxorubicin drug via thermal and pH-responsive stimuli. With a thermosensitive polymeric material, the principle was based on the idea that through application of a magnetic field, which would cause the magnetic nanoparticles to generate heat, a conformational change in the polymer could open pores for drug release; these materials were classified as magnetothermally responsive materials. It was noted that drug elution rate increased preferentially at the increase in temperature to 37 °C and decrease in pH to 5.0.

23.3.2 Regenerative Medicine

Regenerative medicine, similar to tissue engineering, involves biomaterial considerations with the intent to repair, replace, restore, or regenerate damaged or diseased tissue.

23.3.2.1 Scaffolding

Scaffold design is a niche in regenerative medicine that involves creating a foundation for cell adherence that directs proliferation in an appropriate configuration and differentiation scheme [14]. Nanoscale fibers have shown considerable success in the reparation and regeneration of soft tissues through tissue scaffolding in the skin, blood vessels, nerves, tendons, and cartilage applications [40]. Common design criteria include biocompatibility, porosity for cell growth and nutrient and waste flow, natural ECM architecture, biodegradability at a rate consistent with new tissue growth, and mechanical support [27].

Random or aligned nanofiber arrangements have been suggested to play a role in cell positioning and function in tissue engineering applications. Kai *et al.* capitalized on scaffold chemical composition to improve scaffold properties by using a synthetic and natural polymer composite of poly(ε-caprolactone) and gelatin for electrospun nanofibers (50:50 weight ratio) and aligned and random fiber orientations [41]. Regarding the specific research application of cardiac tissue engineering, it was concluded that aligned nanofibers improved the mechanical properties and cell attachment and alignment appropriate for cardiac substrates.

Zhang *et al.* also used a blend of natural and synthetic polymers to try and improve the biocompatibility, mechanical, chemical, and physical properties of tissue engineering scaffolds [42]. In this study, the natural polymer, silk fibroin, was selected for biocompatibility, permeability, and biodegradability characteristics and synthetic polymer poly(L-lactic acid-*co*-ε-caprolactone) for controllable degradation rate and mechanical properties was included in the scaffold fabrication. Fibers were electrospun in a random design at varying weight ratios.

23.3.2.2 Therapeutic Delivery

Drug delivery systems using nanofibers can provide growth factor and drug delivery directly to the site to encourage reparation and regeneration and prevent infection [27]. The surface of nanofibers can be modified for drug delivery, as shown in Figure 23.2, by plasma treatment or wet chemical method, surface graft polymerization, or coelectrospinning [43]. Drug release profiles, such as burst effect or controlled, are not well known in correlation to drug incorporation into the nanofibers; factors of polymer degradation, nanofiber mesh porosity, and surface coating can all play a role in drug release [27]. Additional research investigations include effects of fiber diameter, uniformity, and cell interaction.

23.3.3 Imaging

With the advent of nanotechnology, biomedical optical imaging has seen significant improvements in spatial resolution through the use of contrast agents and fluorescent nanoparticle probes for more sensitive and site-specific detection. Polymers are also utilized as a coating of metallic nanoparticles, such as superparamagnetic iron oxide particles, to increase uptake and reduce compatibility issues. To produce high-resolution images with a targeted component, it was suggested that imaging agents have characteristics of water dispersibility, photostability, low cytotoxicity, and ability for bioconjugation [44]. Bioimaging allows for molecular-level investigation into disease processes, causes, and effects.

Currently, proteins or organic dyes, low-molecular-weight imaging compounds, and quantum dots are of interest for bioimaging applications; however, each option demonstrates the specific drawbacks of limited photostability, poor sensitivity and specificity, and toxicity, respectively [44,45]. New probes involving the use of polymers are currently under investigation. Conjugated polymers, macromolecules with π-conjugated backbones for photoluminescence and electroluminescence (i.e., derivatives of polyfluorene, poly(fluorenyldivinylene), poly(*p*-phenylenevinylene), and poly(phenylene ethylene)) [44], are one such area that is interesting due to their activation mechanism of conjugated polymer aggregation.

Amphiphilic polymer coatings are another method in which polymers are used to improve current bioimaging techniques. As described previously, polymer conjugation or coating can increase blood circulation, provide attachment sites for probe molecules, reduce toxicity and immunogenicity, prevent aggregation, and stabilize otherwise unstable

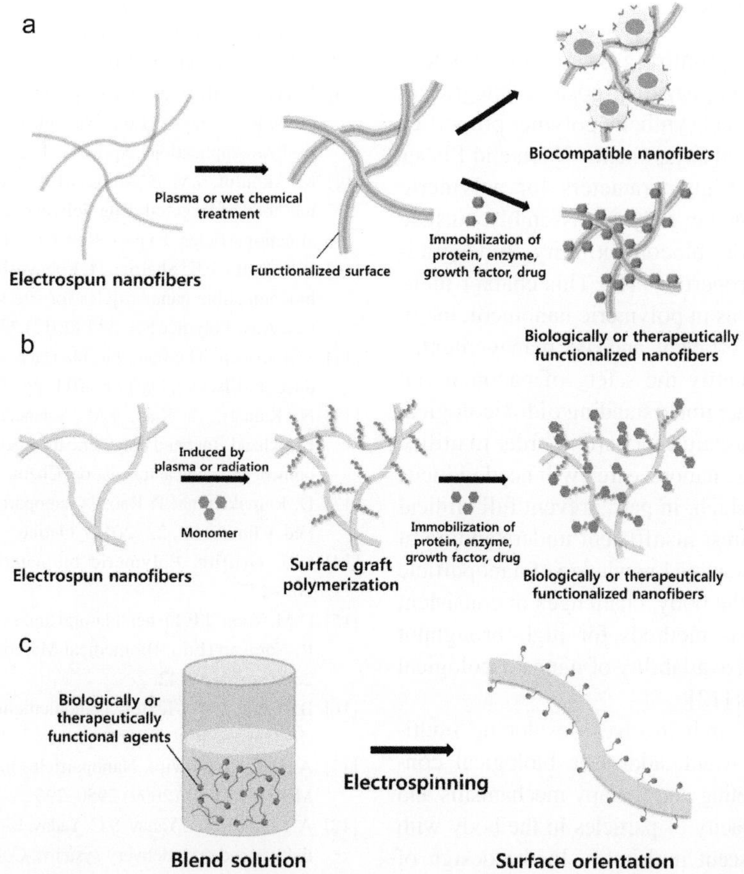

FIGURE 23.2 Surface modification techniques of electrospun nanofibers. (a) Plasma treatment or wet chemical method. (b) Surface graft polymerization. (c) Coelectrospinning. (For color version of this figure, the reader is referred to the online version of this chapter.) Reprinted with permission from Ref. [43]. Copyright 2009. Elsevier.

targeting nanoparticles [45]. The advantages of using polymers indicate possible improvements in both signal-to-noise ratio reduction and spatial resolution for bioimaging applications.

23.4 NANOTOXICITY AND POLYMERIC CHALLENGES

The exciting prospects that stem from using nanoscale features to enhance cell permeability, augment therapeutic delivery, reduce dosage amounts and frequency, and improve the efficacy of agents used in nanomedicine are clouded by the potential of nanotoxic effects, immunogenicity, stability issues, and cost-effective and scalable manufacturing processes [9,46]. At the nanoscale, the physicochemical properties can change dramatically in relation to size (increase in ratio of surface area to volume), surface structure (changes in surface charge and additional coatings), and bioavailability and pharmacological actions (enhanced solubility and various shapes) [4]. The changes due to nanomaterial exposure may be the source for negative toxic effects, collectively known as nanotoxicity, and unintended opsonization increased susceptibility for engulfment by phagocytosis. Nanotoxicity studies encompass a wide range of areas with consideration given to the physiological, physicochemical, and molecular aspects—biodistribution, molecular determinants [4], induced toxic biological responses, and biokinetics [7,47]. Major concerns relate to the redox activity of nanoparticles and their ability to cross cell membranes into organelles such as the mitochondria [48]. Overproduction of ROS has been demonstrated to correlate with the cytotoxic function. As the dimensions of nanostructures begin to approach the size of proteins, chemical messenger delivery and protein organization may also be affected.

Protein accumulation to the surface of nanoparticles can hinder delivery, targeting, and degradation and drug release rates. Additionally, monomer constituents for biodegradable polymeric nanoparticles may raise concern for cell death through apoptosis or necrosis depending on how the chemical environment is changed upon degradation, such as in lowering the pH level in surrounding tissue. Along with the complexity of multifunctional and "smart" nanoparticles comes the question of cost of manufacturing, reproducibility, and consistency of the nanoparticle synthesis [9].

23.5 CONCLUSIONS

Polymeric biomaterials are promising avenues in nanomedicine with natural polymers providing positive integration with biological substrates and synthetic polymer properties available for fine-tuning and adjustments. Yoon and Fisher summarize the critical design parameters for polymeric nanomedicine as the following: polymer assembly, surface properties, macrostructure, biocompatibility, biodegradability, and mechanical properties [15]. This chapter identified the research directions in polymeric nanomedicine in relation to quality, safety, efficacy, and risk management.

There is a need to identify the safety of nanomaterial use in order to create a better understanding of toxicological concerns and move nanoresearch forward in order to utilize it to its full potential. Future nanoresearch will need to focus on the following issues, which, in part, prevent full clinical translation of nanostructures: insufficient understanding at the nanobiointerface, inadequate knowledge of nanoparticle fate after introduction to the body, challenges in consistent and reproducible synthesis methods for high-throughput development, and reduced availability of nanotoxicological screening of nanoparticles [12].

The future of nanoresearch involves developing multifunctional nanoparticles, which addresses biological concerns with the use of targeting and therapy mechanisms and improves the imaging capacity of particles in the body with contrast agents or fluorescent molecules. In the design of nanomedicine, physicochemical (nanoparticle properties), biopharmaceutical (biobarriers), and pharmacological (biokinetics, clearance, and biodistribution) issues should be of great consideration [28]. Findings from the nanofield will fuel other research areas including insights into complex regulatory and signaling networks [49] and biological processes such as cell migration, metastasis, and immune function.

REFERENCES

[1] A. Albanese, P.S. Tang, W.C.W. Chan, The effect of nanoparticle size, shape, and surface chemistry on biological systems, Annu. Rev. Biomed. Eng. 14 (2012) 1–16.

[2] H. Liu, T.J. Webster, Nanomedicine for implants: a review of studies and necessary experimental tools, Biomaterials 28 (2007) 354–369.

[3] L. Zhang, F.X. Gu, J.M. Chan, A.Z. Wang, R.S. Langer, O.C. Farokhzad, Nanoparticles in medicine: therapeutic applications and developments, Clin. Pharmacol. Ther. 83 (2008) 761–769.

[4] S. Arora, J.M. Rajwade, K.M. Paknikar, Nanotoxicology and in vitro studies: the need of the hour, Toxicol. Appl. Pharmacol. 258 (2012) 151–165.

[5] N. Doshi, S. Mitragotri, Designer biomaterials for nanomedicine, Adv. Funct. Mater. 19 (2009) 3843–3854.

[6] A.-I. Moreno-Vega, T. Gómez-Quintero, R.-E. Nuñez-Anita, L.-S. Acosta-Torres, V. Castaño, Polymeric and ceramic nanoparticles in biomedical applications. J. Nanotechnol. 2012 (2012) 1–10, doi:10.1155/2012/936041.

[7] A. El-Ansary, S. Al-Daihan, On the toxicity of therapeutically used nanoparticles: an overview. J. Toxicol. 2009 (2009) 1–9, doi:10.1155/2009/754810.

[8] N. Hasirci, Micro and nano systems in biomedicine and drug delivery, in: M.R. Mozafari (Ed.), Nanomaterials and Nanosystems for Biomedical Applications, Springer, The Netherlands, 2007, pp. 1–26.

[9] M. Akbulut, S.M. D'Addio, M.E. Gindy, R.K. Prud'homme, Novel methods of targeted drug delivery: the potential of multifunctional nanoparticles, Expert. Rev. Clin. Pharmacol. 2 (2009) 265–282.

[10] R.K. Dutta, P.K. Sharma, H. Kobayashi, A.C. Pandey, Functionalized biocompatible nanoparticles for site-specific imaging and therapeutics. Adv. Polymer Sci. 247 (2012) 233–276, doi:10.1007/12.

[11] J. Ramsden, The nano/bio interface. in: Nanotechnology: An Introduction, Elsevier, Oxford, 2011, pp. 53–71.

[12] N. Kamaly, Z. Xiao, P.M. Valencia, A.F. Radovic-Moreno, O.C. Farokhzad, Targeted polymeric therapeutic nanoparticles: design, development and clinical translation, Chem. Soc. Rev. 41 (2012) 2971–3010.

[13] D. Ramakrishna, P. Rao, Is nanoparticle toxicity a concern? J. Int. Fed. Clin. Chem. 22 (2011). Online.

[14] L.G. Griffith, Polymeric biomaterials, Acta Mater. 48 (2000) 263–277.

[15] D.M. Yoon, J.P. Fisher, Natural and synthetic polymeric scaffolds. in: R. Narayan (Ed.), Biomedical Materials, Springer US, Boston, MA, 2009, pp. 415–442.

[16] B. Thierry, M. Textor, Nanomedicine in focus: opportunities and challenges ahead, Biointerphases 7 (2012) 1–4.

[17] A.H. Faraji, P. Wipf, Nanoparticles in cellular drug delivery, Bioorg. Med. Chem. 17 (2009) 2950–2962.

[18] A. Kumari, S.K. Yadav, S.C. Yadav, Biodegradable polymeric nanoparticles based drug delivery systems, Colloids Surf. B 75 (2010) 1–18.

[19] R. Dinarvand, N. Sepehri, S. Manoochehri, H. Rouhani, F. Atyabi, Polylactide-co-glycolide nanoparticles for controlled delivery of anticancer agents, Int. J. Nanomedicine 6 (2011) 877–895.

[20] J. Khandare, M. Calderón, N.M. Dagia, R. Haag, Multifunctional dendritic polymers in nanomedicine: opportunities and challenges, Chem. Soc. Rev. 41 (2012) 2824–2848.

[21] J. Champion, S. Mitragotri, Role of target geometry in phagocytosis, Proc. Natl. Acad. Sci. U.S.A. 103 (2006) 4930–4934.

[22] E. Brewer, J. Coleman, A. Lowman, Emerging technologies of polymeric nanoparticles in cancer drug delivery. J. Nanomater. 2011 (2011) 1–10, doi:10.1155/2011/408675.

[23] B. Sahoo, K.S.P. Devi, R. Banerjee, T.K. Maiti, P. Pramanik, D. Dhara, Thermal and pH responsive polymer-tethered multifunctional magnetic nanoparticles for targeted delivery of anticancer drug, ACS Appl. Mater. Interfaces 5 (2013) 3884–3893.

[24] B.V.N. Nagavarma, H.K.S. Yadav, A. Ayaz, L.S. Vasudha, H.G. Shivakumar, Different techniques for preparation of polymeric nanoparticles: a review, Asian J. Pharm. Clin. Res. 5 (2012) 16–23.

[25] N. Sanvicens, M.P. Marco, Multifunctional nanoparticles—properties and prospects for their use in human medicine, Trends Biotechnol. 26 (2008) 425–433.

[26] Y. Li, Y. Pei, X. Zhang, Z. Gu, Z. Zhou, W. Yuan, et al., PEGylated PLGA nanoparticles as protein carriers: synthesis, preparation and biodistribution in rats, J. Control. Release 71 (2001) 203–211.

[27] Q.P. Pham, U. Sharma, A.G. Mikos, Electrospinning of polymeric nanofibers for tissue engineering applications: a review, Tissue Eng. 12 (2006) 1197–1211.

[28] M. Wang, M. Thanou, Targeting nanoparticles to cancer, Pharmacol. Res. 62 (2010) 90–99.

[29] B.K. Johnson, R.K. Prud'homme, Flash NanoPrecipitation of organic actives and block copolymers using a confined impinging jets mixer, Aust. J. Chem. 56 (2003) 1021.
[30] M. Akbulut, P. Ginart, M.E. Gindy, C. Theriault, K.H. Chin, W. Soboyejo, et al., Generic method of preparing multifunctional fluorescent nanoparticles using Flash NanoPrecipitation, Adv. Funct. Mater. 19 (2009) 718–725.
[31] A. Wang, K. Yuet, L. Zhang, M. Huynh-Le, A. Radovic-Moreno, P. Kantoff, et al., ChemoRad nanoparticles: a novel multifunctional nanoparticle platform for targeted delivery of concurrent chemoradiation, Nanomedicine 5 (2010) 361–368.
[32] M.E. Werner, S. Karve, R. Sukumar, N.D. Cummings, J.A. Copp, R.C. Chen, et al., Folate-targeted nanoparticle delivery of chemo- and radiotherapeutics for the treatment of ovarian cancer peritoneal metastasis, Biomaterials 32 (2011) 8548–8554.
[33] L. Zhang, C.-H. Hu, S.-X. Cheng, R.-X. Zhuo, Hyperbranched amphiphilic polymer with folate mediated targeting property, Colloids Surf. B 79 (2010) 427–433.
[34] A. Persidis, Cancer multidrug resistance, Nat. Biotechnol. 17 (1999) 94–95.
[35] A. Khdair, D. Chen, Y. Patil, L. Ma, Q.P. Dou, M.P.V. Shekhar, et al., Nanoparticle-mediated combination chemotherapy and photodynamic therapy overcomes tumor drug resistance, J. Control. Release 141 (2010) 137–144.
[36] D. Dolmans, D. Fukumura, R. Jain, Photodynamic therapy for cancer, Nat. Rev. Cancer 3 (2003) 380–387.
[37] T.M. Ruhland, P.M. Reichstein, A.P. Majewski, A. Walther, A.H.E. Müller, Superparamagnetic and fluorescent thermo-responsive core-shell-corona hybrid nanogels with a protective silica shell, J. Colloid Interface Sci. 374 (2012) 45–53.
[38] M. Ding, N. Song, X. He, J. Li, L. Zhou, H. Tan, et al., Toward the next-generation nanomedicines: design of multifunctional multiblock polyurethanes for effective cancer treatment, ACS Nano 7 (2013) 1918–1928.
[39] L.E. van Vlerken, Z. Duan, S.R. Little, M.V. Seiden, M.M. Amiji, Biodistribution and pharmacokinetic analysis of paclitaxel and ceramide administered in multifunctional polymer-blend nanoparticles in drug resistant breast cancer model. Mol. Pharm. 5 (2008) 516–526.
[40] R. James, U.S. Toti, C.T. Laurencin, S.G. Kumbar, Electrospun nanofibrous scaffolds for engineering soft connective tissues, in: S.J. Hurst (Ed.), Biomedical Nanotechnology: Methods and Protocols, vol. 726, Humana Press, Totowa, NJ, 2011, pp. 243–258.
[41] D. Kai, M.P. Prabhakaran, G. Jin, S. Ramakrishna, Guided orientation of cardiomyocytes on electrospun aligned nanofibers for cardiac tissue engineering, J. Biomed. Mater. Res. B 98B (2011) 379–386.
[42] K. Zhang, H. Wang, C. Huang, Y. Su, X. Mo, Y. Ikada, Fabrication of silk fibroin blended P(LLA-CL) nanofibrous scaffolds for tissue engineering, J. Biomed. Mater. Res. A 93 (2010) 984–993.
[43] H.S. Yoo, T.G. Kim, T.G. Park, Surface-functionalized electrospun nanofibers for tissue engineering and drug delivery, Adv. Drug Deliv. Rev. 61 (2009) 1033–1042.
[44] K. Li, B. Liu, Polymer encapsulated conjugated polymer nanoparticles for fluorescence bioimaging, J. Mater. Chem. 22 (2012) 1257–1264.
[45] J.-H. Kim, K. Park, H.Y. Nam, S. Lee, K. Kim, I.C. Kwon, Polymers for bioimaging, Prog. Polym. Sci. 32 (2007) 1031–1053.
[46] J.E. Gagner, S. Shrivastava, X. Qian, J.S. Dordick, R.W. Siegel, Engineering nanomaterials for biomedical applications requires understanding the nano-bio interface: a perspective, J. Phys. Chem. Lett. 3 (2012) 3149–3158.
[47] B. Semete, L. Booysen, Y. Lemmer, L. Kalombo, L. Katata, J. Verschoor, et al., In vivo evaluation of the biodistribution and safety of PLGA nanoparticles as drug delivery systems, Nanomedicine 6 (2010) 662–671.
[48] S.M. Hussain, K.L. Hess, J.M. Gearhart, K.T. Geiss, J.J. Schlager, In vitro toxicity of nanoparticles in BRL 3A rat liver cells, Toxicol. In Vitro 19 (2005) 975–983.
[49] S.M. Moghimi, A.C. Hunter, J.C. Murray, Nanomedicine: current status and future prospects, FASEB J. 19 (2005) 311–330.

Index

Note: Page numbers followed by *f* indicate figures and *t* indicate tables.

U

V

W

X

Y

Printed and bound by CPI Group (UK) Ltd, Croydon, CR0 4YY
08/05/2025
01864928-0002